2013 NATIONAL HOME IMPROVEMENT ESTIMATOR

Edited by Ray F. Hicks

Includes inside the back cover:

- An estimating CD with all the costs in this book, plus,
- An estimating program that makes it easy to use these costs,
- An interactive video guide to the National Estimator program,
- A program that converts your estimates into invoices,
- A program that exports your estimates to QuickBooks.

Quarterly price updates on the Web are free and automatic all during 2013. You'll be prompted when it's time to collect the next update. A connection to the Web is required.

Download all of Craftsman's most popular costbooks for one low price with the Craftsman Site License. http://CraftsmanSiteLicense.com

Craftsman Book Company
6058 Corte del Cedro / P.O. Box 6500 / Carlsbad, CA 92018

Acknowledgments

Portions of chapters 8, 9 and 15 first appeared in the book *Renovating & Restyling Older Homes* by Lawrence Dworin. Mr. Dworin served as a resource in the development of this manuscript and contributed many valuable insights from his years of experience in the industry. An order form for *Renovating & Restyling Older Homes,* and his other construction reference *Profits in Buying & Renovating Homes* appears on the final pages of this manual.

Contributing Editor: *Walter James O'Brien*
Cover design by: *Bill Grote*
Photos: *Shutterstock*
Graphics by: *Devona Quindoy and Christal Stimpson*
Production Manager: *Christine Bruneau*
Layout by: *Christine Bruneau*
Software production: *Timothy Campbell*

ISBN 978-1-57218-281-3
Published November 2012 for the year 2013.

Contents

Pricing Home Improvement Jobs

1

Estimating home improvement costs requires specialized skills. You can't price home improvement work the same way you'd price new construction. The proportion of labor expense is greater. There's far more risk because there are far more variables and unknowns. An example will make this clear.

Hanging doors in new construction is a 1-2-3 affair. You know ahead of time exactly what's required. You built the wall, installed the frame and know each door will fit right the first time. There won't be any surprises. An experienced estimator can forecast the cost of hanging a door in a new home with a high degree of certainty.

Now let's look at the same task on a home improvement project. First, remember that you may be hanging only one door. There's no chance to improve the production rate on second or later doors. Installing this door is probably different from the last door you installed on a similar project and will be different from your next door installation.

Work starts with removing the old door. That's probably not too hard — once you break through six coats of accumulated paint on the hinges. Let's assume you work carefully and don't damage the casing or trim. The next step is to install a blind Dutchman where the original hinges were attached. If the frame's badly chewed up from years of neglect and abuse, you'll have to remove and replace the entire frame. Don't forget that the building has settled over the years. The frame has probably twisted out of plumb. More time will be needed to shim and level the frame.

Finally the door is installed and swings smoothly. But the job still isn't finished. You have to paint the door to match and will certainly have to haul away the old door and the debris.

Every home improvement cost estimate comes with dozens of chances to make an expensive mistake. Surprises are inevitable and nearly all surprises will add to the cost, not reduce the cost. Estimating new construction is a snap by comparison.

All home improvement has similar problems:

- ❖ No chance for economies of scale (mass production),
- ❖ Difficulty in removing just enough of the old,
- ❖ Work doesn't follow the normal (from the ground, up) sequence of construction,
- ❖ Difficulty adding new materials to deteriorated or nonstandard existing materials,
- ❖ The need to match designs, colors and textures,
- ❖ Covering up for another contractor's mistakes,
- ❖ Struggling to get access to the place where the work is to be done,
- ❖ Protecting adjacent surfaces and pathway to the work area,
- ❖ Initial uncertainty about how much work is needed.

If you're an experienced home improvement contractor, you understand this already. Whether experienced or not, you know full well that accurate pricing is crucial to survival in the home improvement business. Unfortunately, most home improvement specialists have far less experience in pricing their work than in getting the work done. No wonder so many home improvement companies sink into obscurity.

But it doesn't have to be that way.

The chapters that follow explain how to price each type of home improvement work. Emphasis will be on avoiding risk — pricing pitfalls that can turn any home improvement contract into a financial nightmare.

No Price Fits All Jobs

There's no single way to arrive at the correct price for home improvement work. Neither is there a single correct price for most home improvement projects. But there are both good and bad ways to estimate prices and there are good and bad prices for any proposed job. Your task as a home improvement estimator is to produce consistently good estimates on most jobs. If you're already doing that, congratulations. You don't need this book. Return it where you bought it and get a refund. But if you're not, information in the following chapters will make or save you many times what you paid for the book.

It's accepted wisdom among home improvement specialists that salespeople need authority to quote prices when closing a sale. When trying to wrap up a deal, there's no substitute for having current cost information at your fingertips. That's especially true on larger jobs where you've prepared detailed plans and a written estimate. The owner will have questions and suggestions that change the job specs. If you want to close the sale then and there, you'd better know how much to add or subtract for each change the owner wants.

Most home improvement companies authorize salespeople to quote prices from an approved list in a price book. That simplifies the salesperson's job, eliminates most major errors, and saves the owner of the company from approving every item in every estimate. In our opinion, quoting from a price book is the only way to build sales volume in the home improvement business. That means every home improvement contractor needs a good price book.

Of course, the best price book for your company would be based on your actual cost experience — work done by your crews on your jobs with materials from your dealers and installed by your subcontractors. Since every contractor uses different crews, subcontractors and suppliers, every price book should be different. And, of course, prices in the company price book should be revised regularly to reflect current labor and material costs.

Having admitted that there's no substitute for developing your own price book, we'll suggest that you not bother. Most home improvement contractors don't have the time or patience to maintain current installed prices for thousands of repair and remodeling items. Even if you did, spending hundreds of hours a year keeping a price book current would be a waste of time. Prices for home improvement work are usually negotiated on site. There's little value in keeping a book of exact costs if you have to cut a special deal to close each sale.

Instead, we suggest that you let this manual serve as your company price book. Using the prices in this book (and on the disk in the back of this manual) will eliminate most of the common estimating mistakes. If your labor costs are higher or if your crews aren't as skilled as most tradesmen, you may have to increase the selling prices listed here to make a reasonable profit. And, of course, sometimes you're going to have a job with costs that exceed what any reasonable estimate could have predicted. This manual isn't a substitute for the exercise of good judgment. That's always your job.

Cost Plus Markup Equals Selling Price

The figures in this manual show both costs to the contractor and a recommended selling price. The difference between your cost and your selling price is your markup. Markup varies widely, far more than either material cost or labor cost. You're the final authority on markup. You decide what markup fits best — based on market conditions, your client, and the profit you feel is reasonable. Once you decide on the "right" markup, it's easy to plug that percentage into the bid, assuming you use National Estimator.

Every construction contractor needs to distinguish between "markup" and "profit." Markup is what you add to estimated labor and material costs (usually called "hard costs") to find the selling price. Profit is what you have left when all bills have been paid. The two are very different. Profit is just the frosting on the cake.

Markup is also different from "margin." Figure 1-1 shows percentages of markup for various levels of hard costs. For example, if your markup on a $20,000 job is 50 percent, the selling price will be $30,000. The margin on that job is 33 percent, $10,000 divided by the selling price of $30,000. (Note the bottom row in Figure 1-1.) Margin is what's left after hard costs are recovered, and includes profit.

Markup vs. Margin				
Hard Cost	**Price at 50% Markup**	**Price at 70% Markup**	**Price at 100% Markup**	**Price at 150% Markup**
$50	$75	$85	$100	$125
$100	$150	$170	$200	$250
$200	$300	$340	$400	$500
$300	$450	$510	$600	$750
$400	$600	$680	$800	$1,000
$500	$750	$850	$1,000	$1,250
$1,000	$1,500	$1,700	$2,000	$2,500
$2,000	$3,000	$3,400	$4,000	$5,000
$3,000	$4,500	$5,100	$6,000	$7,500
$4,000	$6,000	$6,800	$8,000	$10,000
$5,000	$7,500	$8,500	$10,000	$12,500
$6,000	$9,000	$10,200	$12,000	$15,000
$7,000	$10,500	$11,900	$14,000	$17,500
$8,000	$12,000	$13,600	$16,000	$20,000
$9,000	$13,500	$15,300	$18,000	$22,500
$10,000	$15,000	$17,000	$20,000	$25,000
$11,000	$16,500	$18,700	$22,000	$27,500
$12,000	$18,000	$20,600	$24,000	$30,000
$13,000	$19,500	$22,100	$26,000	$32,500
$14,000	$21,000	$23,800	$28,000	$35,000
$15,000	$22,500	$25,500	$30,000	$37,500
$16,000	$24,000	$27,200	$32,000	$40,000
$17,000	$25,500	$28,900	$34,000	$42,500
$18,000	$27,000	$30,600	$36,000	$45,000
$19,000	$28,500	$32,300	$38,000	$47,500
$20,000	$30,000	$34,000	$40,000	$50,000
$30,000	$45,000	$51,000	$60,000	$75,000
$40,000	$60,000	$68,000	$80,000	$100,000
$50,000	$75,000	$85,000	$100,000	$125,000
Hard Cost	**33% Margin**	**41% Margin**	**50% Margin**	**60% Margin**

Figure 1-1

Markup (based on cost) vs. Margin (based on selling price)

Many successful home improvement contractors find they can stay in business if hard costs are 59 percent of selling price. Hard costs include material expense, subcontract costs and labor (including taxes and insurance). The other 41 percent of selling price ("margin") compensates the sales staff, covers overhead, supervision and contingency, and should yield a modest profit. From Figure 1-1, you can see that markup on hard costs has to be 70 percent to yield a margin of 41 percent.

To put this formula to work in your home improvement company, add 70 percent to your hard costs to find the selling price. Of course, some jobs need more markup and others can get by on less. A smaller job with more risk done for a demanding client may require greater markup. A larger job done mostly by subcontractors can usually carry a smaller markup.

Home improvement contractors have all the overhead of any business: office rent, telephone, owner's salary, office salaries, legal and accounting expense, insurance, auto and truck expense, and more. But unlike other contractors, home improvement specialists routinely deal with high risk from both the unknown and the unknowable — at least until work actually begins. Since most of what you didn't or couldn't anticipate will inflate your costs, you're assuming significant risk. That's why markups for home improvement work have to be higher than for new construction.

Naturally, competition dictates markup. If you're not getting enough work at 70 percent markup, maybe 70 percent is too much for your area. But remember that you shouldn't have to bid remodeling work on a level playing field. New construction usually goes to the lowest responsible bidder. A creative salesman who follows the recommendations in this book has an advantage over lowball bidders who rely on price alone to sell their services.

If you have trouble using National Estimator, we'll be glad to help, and we don't charge you for it.

Free telephone assistance is available from 8 a.m. until 5 p.m. Pacific time Monday through Friday (except holidays). Call 760-438-7828, Ext. 2.

This Book Works Two Ways

The disk in the back of this book has all the information that appears in the printed book, but with one advantage. The National Estimator program makes it easy to copy and paste these costs into an estimate (or bid) and then add whatever markup you select. **Quarterly price updates on the Web are free** and automatic all during 2013. You'll be prompted when it's time to collect the next update. A connection to the Web is required. To get started with the National Estimator disk, insert it into any computer that uses Microsoft Windows. Then follow instructions on the screen.

Before installation is complete, you'll be invited to watch ShowMe, an interactive video guide to the National Estimator program. ShowMe is designed to speed your mastery of National Estimator. You can start ShowMe at any time while using National Estimator. Click "**Help**" on the National Estimator menu bar. Then click "**Launch The Tutorial**." The disk has to be in your computer when ShowMe is running.

Estimates That Work Two Ways

This manual is designed for use by both the owner of a home improvement business and by company sales staff. Prices shown in this manual include both hard costs (labor and material) and a recommended selling price (usually based on 70 percent markup). Sales staff will use the selling price.

Figure 1-2 shows the last page of an estimate for the Stillwel room addition, including 70 percent markup. The company estimator created this estimate from a job survey prepared by a sales representative. Figure 1-2 is for company use and was created in the National Estimator program.

National Estimator [STILLWEL.EST]

Qty	Craft@Hours	Unit	Material	Labor	Equipment	Total
Roll & brush 1 coat of water base undercoat on interior flush doors						
Medium .3 MH/Door, 11.5 Doors/Gal						
3.00	2P@.9000	Ea	5.07	21.36	0.00	26.43
Roll & brush 1st finish coat of water base enamel on interior flush doors						
Medium .25 MH/Door, 12.5 Doors/Gal						
3.00	2P@.7500	Ea	5.39	17.80	0.00	23.18
Roll & brush 1 coat water base undercoat on vanity cabinet						
Medium 93 SF/MH, 250 SF/Gal						
0.90	2P@.9720	CSF	6.99	23.06	0.00	30.05
Roll & brush 1st finish coat of water base enamel on vanity cabinet						
Medium 103 SF/MH, 288 SF/Gal						
0.90	2P@.8740	CSF	7.02	20.70	0.00	27.72
Roll 1st coat of water base stain on rough sawn or resawn wood siding						
Medium 225 SF/MH, 213 SF/Gal						
17.00	2P@7.548	CSF	157.08	178.50	0.00	335.58
Roll 2nd coat of water base stain on rough sawn or resawn wood siding						
Medium 275 SF/MH, 273 SF/Gal						
17.00	2P@6.188	CSF	122.63	146.73	0.00	269.36
**Subtotal: Painting						
	40.7		1,014.54	973.46	0.00	1,988.00
Total Manhours, Material, Labor, and Equipment:						
	456.0		18,641.82	11,675.42	540.29	30,857.53
Total Only (Subcontract) Costs:						4,028.33
				Subtotal:		34,885.85
				70.00% Markup:		24,420.10
				Estimate Total:		59,305.95

Figure 1-2
Estimate with a 70 percent markup shown

Figure 1-3 shows the last page of a proposal for the same job. This is what the customer sees. It was created by the Job Cost Wizard program. Both National Estimator and Job Cost Wizard are on the disk in the back of this book. Notice that the total is the same in both documents, $59,305.95. Figure 1-2, the estimate, shows 70 percent markup on hard costs. Figure 1-3, the proposal, doesn't show the markup at all. Instead, markup has been distributed proportionately throughout each cost item in the job. Job Cost Wizard does this distribution at the click of a button.

Craftsman Construction
6058 Corte del Cedro
Box 6500
Fairfield, GA 30456
856-3806

Proposal

Date	Estimate #
8/3/13	494
Customer	**Job**
Bill Stillwel	Room Addition

Description	Qty	Rate	Amount
Roll & brush 1st finish coat of water base enamel on vanity cabinet. Medium 103 SF/MH, 288 SF/Gal			
Material, per CSF	0.90	13.26	11.94
Labor, per CSF	0.90	39.09	35.18
Roll 1st coat of water base stain on rough sawn or resawn wood siding. Medium 225 SF/MH, 213 SF/Gal			
Material, per CSF	17	15.71	267.04
Labor, per CSF	17	17.85	303.45
Roll 2nd coat of water base stain on rough sawn or resawn wood siding. Medium 275 SF/MH, 273 SF/Gal			
Material, per CSF	17	12.26	208.47
Labor, per CSF	17	14.67	249.44
*Painting subtotal			3,379.60
————————			
*Project Subtotal			59,305.96
*Project Total			59,305.96

Figure 1-3
Proposal for the same job not showing the markup

Once an estimate is finished in the National Estimator program and saved to computer disk, press Ctrl on your keyboard and tap the letter J to convert the estimate into a proposal in Job Cost Wizard. You can't make changes in the Job Cost Wizard screen. But it's easy to toggle back to National Estimator program (press Alt and tap the Tab key), make a change, and then press Ctrl-J once again.

Job Cost Wizard offers dozens of choices on what you show and don't show in written proposals. Your bids can be long (full description for everything in the estimate) or short (only a summary of each category). You can show or hide labor and material cost detail. You can show or hide markup and profit.

Once work begins, you'll want to monitor job expenses to be sure actual costs remain consistent with estimated costs. If you use QuickBooks to pay bills and figure payroll, let QuickBooks do the comparisons for you. Job Cost Wizard exports the proposal to QuickBooks, where you can prepare progress invoices and track expenses against estimates.

The National Estimator program lets you change anything in the costbook or even add your own estimated costs. Make the *National Home Improvement Estimator* your collection point for all estimating and pricing information. Anything you add to the costbook shows up in red and can be migrated to later editions of this manual when available. The *National Home Improvement Estimator* starts out as your most useful estimating reference. What you add to the costbook will make it even more valuable.

Job Survey (Scope of Work)

Of course, neither a good price book nor a computer estimating program will solve all of your estimating problems. Computers seldom make mistakes in addition or multiplication. But nothing prevents you from making expensive estimating mistakes on your own. By far the most common estimating mistake will be omitting something essential to the job.

Get in the habit of completing an exhaustive job survey before beginning any estimate for home improvement work. Some of your estimates on a job may be too high. Other estimates on a job may be too low. With any luck, your over-estimates will roughly balance with your under-estimates, leaving the job total about where it should be. But the estimated price for anything omitted from a job survey is always zero. That's a 100 percent miss. It's hard to balance a complete miss with anything. A few of those can create a financial disaster.

Material Costs in This Book

Material costs for each item are listed in the column headed Material. These are neither retail nor wholesale prices. They are estimates of what most contractors who buy in moderate volume will pay suppliers as of mid-2013. Discounts may be available for purchases in larger volumes.

Add Delivery Expense to the material cost for other than local delivery of reasonably large quantities. Cost of delivery varies with the distance from source of supply, method of transportation, and quantity to be delivered. But most material dealers absorb the delivery cost on local delivery (5 to 15 miles) of larger quantities to good customers. Add the expense of job site delivery when it's a significant part of the material cost.

Add Sales Tax when sales tax will be charged to the contractor buying the materials. In some states, contractors have to collect sales tax based on the contract price. No matter what your state (or county) requires, National Estimator can handle the task. Click **Edit** on the National Estimator menu bar. Then click **Sales Tax**.

Waste and Coverage Loss is included in the installed material cost. The cost of many materials per unit after installation is greater than the purchase price for the same unit because of waste, shrinkage or coverage loss during installation. For example, about 120 square feet of nominal 1" x 4" square edge boards will be needed to cover 100 square feet of floor or wall. There's no coverage loss with plywood sheathing, but waste due to cutting and fitting will average about 6 percent.

Costs in the Material column of this book assume normal waste and coverage loss. Small and irregular jobs may require a greater waste allowance. Materials priced without installation (with no labor cost) don't include an allowance for waste and coverage except as noted.

Labor Costs for installing the material or doing the work described are listed in the column headed Labor. The labor cost per unit is the labor cost per hour multiplied by the manhours per unit shown after the @ sign in the Craft@Hrs column. Labor cost includes the basic wage, the employer's contribution to welfare, pension, vacation and apprentice funds, and all tax and insurance charges based on wages.

Supervision Expense to the general contractor isn't included in the labor cost. The cost of supervision and non-productive labor varies widely from job to job. Calculate the cost of supervision and non-productive labor and add this to the estimate.

Payroll Taxes and Insurance are included in the labor cost. See the following section for more on the "contractor's burden."

Manhours per Unit and the Craft performing the work are listed in the Craft@Hrs column. To find the units of work done per worker in an 8-hour day, divide 8 by the manhours per unit. To find the units done by a crew in an 8-hour day, multiply the units per worker per 8-hour day by the number of crew members.

Manhours include all productive labor normally associated with installing the materials described.

This will usually include tasks such as:

- Unloading and storing construction materials, tools and equipment on site.
- Moving tools and equipment from a storage area or truck on site at the beginning of the day.
- Returning tools and equipment to a storage area or truck on site at the end of the day.
- Normal time lost for work breaks.
- Planning and discussing the work to be performed.
- Normal handling, measuring, cutting and fitting.
- Keeping a record of the time spent and work done.
- Regular cleanup of construction debris.
- Infrequent correction or repairs required because of faulty installation.

Adjust the Labor Cost to the job you're figuring when your actual hourly labor cost is known or can be estimated. The labor costs listed in Figure 1-5 will apply within a few percent on many jobs. But labor costs may be much higher or much lower on the job you're estimating.

If the hourly wage rates listed in Figure 1-5 aren't accurate, divide your known or estimated cost per hour by the listed cost per hour. The result is your adjustment for any figure in the Labor column for that craft.

Adjust for Unusual Labor Productivity. Costs in the Labor column are for normal conditions: experienced craftsmen working on reasonably well-planned and managed home improvement projects with fair to good productivity. Labor estimates assume that materials are standard grade, appropriate tools are on hand, work done by other crafts is adequate, layout and installation are relatively uncomplicated, and working conditions don't slow progress.

Working conditions at the jobsite have a major effect on labor cost. Estimating experience and careful analysis can help you predict the effect of most changes in working conditions. Obviously, no single adjustment will apply on all jobs. But the adjustments that follow should help you produce more accurate labor estimates. More than one condition may apply on a job.

Jobsite conditions affect labor costs

- Add 10% to 15% when working temperatures are below 40 degrees or above 95 degrees.
- Add 30% to 35% when temperatures are below 20 degrees. Materials and tools are hard to handle. Bulky clothing restricts freedom of movement.
- Add 15% to 25% for work on a ladder or a scaffold, in a 36" crawl space, in a congested area, or remote from the material storage point.
- Add 50% for work in an 18" to 36" crawl space. Allow extra time for cleaning out the area before work begins.
- Add 200% if portions of the crawl space are less than 18". Allow extra time for passing tools and materials. Few contractors bid on work like this.
- Deduct 10% when the work is in an open area with excellent access and good light.
- Add 1% for each 10 feet that materials must be lifted above ground level.
- Add 5% to 50% for tradesmen with below-average skills. Deduct 5% to 25% for highly-motivated, highly-skilled tradesmen.
- Deduct 10% to 20% when an identical task is repeated many times for several days at the same site.
- Add 20% to 50% on small jobs where fitting and matching of materials is required, adjacent surfaces have to be protected and the job site is occupied during construction.
- Add 25% to 50% for work done following a major flood, fire, earthquake, hurricane or tornado while skilled tradesmen are not readily available. Material costs may also be higher after a major disaster.
- Add 10% to 35% for demanding specs, rigid inspections, unreliable suppliers, a difficult owner or an inexperienced architect.

Use an Area Modification Factor from pages 18 through 26 if your material, hourly labor or equipment costs are unknown and can't be estimated. Here's how: Use the labor and material costs in this manual without modification. Then add or deduct the percentages shown for labor, material and equipment.

Equipment Costs for small tools and expendable supplies (such as saws and tape) are usually considered overhead expense and are covered by your markup. Equipment costs for larger equipment (such as a compressor or backhoe) should be based on the rental rate for the period needed. Equipment rental costs are included in sections where heavy equipment may be needed.

Labor and Material Costs Change. These costs were compiled in the fall of 2012 and projected to mid-2013 by adding a small percentage. This projection will be accurate for some materials and inaccurate for others. No one can predict material price changes accurately.

How Accurate Are These Figures? They're as accurate as possible considering that the estimators who wrote this book don't know your subcontractors or material suppliers, haven't seen the plans or specifications, don't know what building code applies or where the job is, had to project material costs at least six months into the future, and had no record of how much work the crew that will be assigned to the job can handle.

You wouldn't bid a job under those conditions. And we don't claim that all construction is done at these prices.

Estimating Is an Art, not a science. On many jobs the range between high and low bid will be 20 percent or more. There's room for legitimate disagreement on what the correct costs are, even when complete plans and specifications are available, the date and site are established, and labor and material costs are identical for all bidders.

No cost fits all jobs. Good estimates are custom-made for a particular project and a single contractor through judgment, analysis and experience.

This book isn't a substitute for judgment, analysis and sound estimating practice. It's an *aid* in developing an *informed opinion* of cost. If you're using this book as your sole cost authority for contract bids, you're reading more into these pages than the editors intend.

Use These Figures to compile preliminary estimates, when a "snap" bid is needed to close the deal, and when no actual costs are available. This book will reduce the chance of error or omission on bid estimates, speed "ball park" estimates, and be a good guide when there's no time to get a quote.

Where Do We Get These Figures? From the same sources all professional estimators use: material suppliers, material price services, analysis of plans, specifications, estimates and completed project costs, and both published and unpublished cost studies compiled by both private and government agencies. In addition, we conduct nationwide mail and phone surveys and have the use of several major national estimating databases.

We'll Answer Your Questions about any part of this book and explain how to apply these costs. Free telephone assistance is available from 8 a.m. until 5 p.m. Pacific time Monday through Friday except holidays. Phone 760-438-7828, Ext. 2. We don't accept collect calls and can't estimate the job for you. But if you need clarification on something in this manual, we can help.

Labor Costs and Crews

Throughout this manual you'll see a column headed *Craft@Hrs*. Letters and numbers in this column show our estimates of:

- ❖ Who will do the work (the craft code)
- ❖ An @ symbol which means at
- ❖ How long the work will take (manhours).

For example, in Chapter 4, page 70, you'll find estimates for installing BC plywood wall sheathing by the square foot. The Craft@Hrs column opposite 1/2" plywood wall sheathing shows:

B1@.016

That means we estimate the installation rate for crew B1 at 0.016 manhours per square foot. That's the same as 16 manhours per 1,000 square feet.

Figure 1-4 is a table that defines each of the craft codes used in this book. Notice that crew B1 is composed of two craftsmen: one laborer and one carpenter. To install 1,000 square feet of 1/2" BC wall sheathing at 0.016 manhours per square foot, that crew would need 16 manhours (one 8-hour day for a crew of two).

Notice also in the table that the cost per manhour for crew B1 is listed as $32.53. That's the average for a laborer (listed at $29.05 per hour) and a carpenter (listed at $36.01 per hour): $29.05 plus $36.01 is $65.06. Divide by 2 to get $32.53, the average cost per manhour for crew B1.

In the table, the cost per manhour is the sum of hourly costs of all crew members divided by the number of crew members. That's the average cost per manhour.

Costs in the Labor column in this book are the product of the installation time (in manhours) multiplied by the cost per manhour. For example, in Chapter 4, the labor cost listed for 1/2" BC wall sheathing is $0.52 per square foot. That's the installation time (0.016 manhours per square foot) multiplied by $32.53, the average cost per manhour for crew B1.

Figure 1-5 shows hourly labor costs components — base wage, typical fringe benefits, payroll taxes, insurance and the total hourly cost.

The labor costs shown in Column 6 were used to compute the manhour costs for crews and the figures in the Labor column.

Craft Codes

Craft Code	Cost Per Manhour	Crew Composition
B1	$32.53	1 laborer, 1 carpenter
B2	$33.69	1 laborer, 2 carpenters
B3	$31.37	2 laborers, 1 carpenter
B4	$34.84	1 laborer, 1 operating engineer, 1 reinforcing iron worker
B5	$34.88	1 laborer, 1 carpenter, 1 cement mason, 1 operating engineer, 1 reinforcing iron worker
B6	$31.46	1 laborer, 1 cement mason
B7	$29.39	1 laborer, 1 truck driver
B8	$34.90	1 laborer, 1 operating engineer
B9	$31.95	1 bricklayer, 1 bricklayer's helper
BB	$36.80	1 bricklayer
BC	$36.01	1 carpenter
BE	$39.05	1 electrician
BF	$32.73	1 floor layer
BG	$33.97	1 glazier
BH	$27.09	1 bricklayer's helper
BL	$29.05	1 laborer
BR	$34.90	1 lather
BS	$31.36	1 marblesetter
CF	$33.86	1 cement mason
CT	$33.77	1 mosaic & terrazzo worker
D1	$34.97	1 drywall installer, 1 drywall taper
DI	$34.64	1 drywall installer
DT	$35.50	1 drywall taper
HC	$27.98	1 plasterer helper
OE	$40.74	1 operating engineer
P1	$35.16	1 laborer, 1 plumber
PM	$41.27	1 plumber
PP	$32.91	1 painter, 1 laborer
PR	$36.41	1 plasterer
PT	$36.77	1 painter
R1	$34.74	1 roofer, 1 laborer
RI	$34.72	1 reinforcing iron worker
RR	$40.43	1 roofer
SW	$40.43	1 sheet metal worker
T1	$31.35	1 tile layer, 1 laborer
TL	$33.64	1 tile layer
TR	$29.73	1 truck driver

Figure 1-4
Craft codes used in this manual

Hourly Labor Cost						
	1	2	3	4	5	6
Craft	Base wage per hour	Taxable fringe benefits (5.15% of base wage)	Insurance and employer taxes (%)	Insurance and employer taxes ($)	Non-Taxable fringe benefits (4.55% of base wage)	Total hourly cost used in this book
Bricklayer	$26.94	$1.39	25.54%	$7.24	$1.23	$36.80
Bricklayer's Helper	19.84	1.02	25.54	5.33	0.90	27.09
Building Laborer	20.12	1.04	32.93	6.97	0.92	29.05
Carpenter	25.15	1.30	31.83	8.42	1.14	36.01
Cement Mason	25.22	1.30	23.35	6.19	1.15	33.81
Drywall Installer	25.72	1.32	23.77	6.43	1.17	34.64
Drywall Taper	26.21	1.35	23.77	6.55	1.19	35.30
Electrician	29.86	1.54	20.04	6.29	1.36	39.05
Floor Layer	24.25	1.25	24.02	6.13	1.10	32.73
Glazier	24.79	1.28	25.98	6.77	1.13	33.97
Lather	26.38	1.36	21.48	5.96	1.20	34.90
Marble Setter	23.69	1.22	21.56	5.37	1.08	31.36
Millwright	25.58	1.32	21.44	5.77	1.16	33.83
Mosiac & Terrazzo Worker	25.52	1.31	21.56	5.78	1.16	33.77
Operating Engineer	29.86	1.54	25.42	7.98	1.36	40.74
Painter	27.02	1.39	25.08	7.13	1.23	36.77
Plasterer	26.02	1.34	28.78	7.87	1.18	36.41
Plasterer Helper	19.99	1.03	28.78	6.05	0.91	27.98
Plumber	30.47	1.57	24.47	7.84	1.39	41.27
Reinforcing Ironworker	24.80	1.28	28.81	7.51	1.13	34.72
Roofer	25.86	1.33	44.34	12.06	1.18	40.43
Sheet Metal Worker	29.45	1.52	26.21	8.12	1.34	40.43
Sprinkler Fitter	29.93	1.54	25.28	7.96	1.36	40.79
Tile Layer	25.41	1.31	21.56	5.76	1.16	33.64
Truck Driver	21.63	1.11	26.42	6.01	0.98	29.73

Figure 1-5

Components of hourly labor costs

It's important that you understand what's included in the figures in each of the six columns in the table. Here's an explanation:

Column 1, the base wage per hour, is the craftsman's hourly wage. These figures are representative of what many contractors will be paying craftsmen working on home improvement jobs in 2013.

Column 2, taxable fringe benefits, includes vacation pay, sick leave and other taxable benefits. These fringe benefits average 5.15 percent of the base wage for many construction contractors. This benefit is in addition to the base wage.

Column 3, insurance and employer taxes in percent, shows the insurance and tax rate for construction trades. The cost of insurance in this column includes workers' compensation and contractor's casualty and liability coverage. Insurance rates vary widely from state to state and depend on a contractor's loss experience. Note that taxes and insurance increase the hourly labor cost by 30 to 35 percent for most trades. There is no legal way to avoid these costs.

Column 4, insurance and employer taxes in dollars, shows the hourly cost of taxes and insurance for each construction trade. Insurance and taxes are paid on the costs in both columns 1 and 2.

Column 5, non-taxable fringe benefits, includes employer paid non-taxable benefits such as medical coverage and tax-deferred pension and profit sharing plans. These fringe benefits average 4.55 percent of the base wage for many construction contractors. The employer pays no taxes or insurance on these benefits.

Column 6, the total hourly cost in dollars, is the sum of columns 1, 2, 4 and 5.

These hourly labor costs will apply within a few percent on many jobs. But wage rates may be much higher or lower in the area where you do business. We recommend using your actual labor cost rather than national averages. That's easy with the National Estimator program. When copying and pasting any cost to your estimate, adjust the assumed hourly labor cost to your actual cost. You need do this only once for each trade. And you can make this adjustment at any time. Any change you make is applied to that trade throughout the estimate.

Abbreviations

4WD	four-wheel drive	**EPA**	Environmental Protection Agency	**NEMA**	National Electrical Manufacturer's Assoc.
ABS	acrylonitrile butadiene styrene	**F**	Fahrenheit	**OC**	spacing from center to center
AC	alternating current	**GFCI**	ground fault circuit interrupter	**OD**	outside diameter
ACQ	Alkaline Copper Quat	**GRS**	galvanized rigid steel	**OSB**	oriented strand board
ADA	Americans with Disabilities Act	**H**	height	**OSHA**	Occupational Safety & Health Admin.
ANSI	American Nat. Standards Institute	**HOM**	higher order mode	**oz.**	ounce(s)
APP	atactic polypropylene	**HP**	horsepower	**PSI**	pounds per square inch
ASTM	Amer. Society for Testing Materials	**HVAC**	heating, ventilating & air cond.	**PVC**	polyvinyl chloride
AWG	American Wire Gauge	**ID**	inside diameter	**R**	thermal resistance
BF	board foot	**IMC**	intermediate metal conduit	**S4S**	surfaced 4 sides
Btr	better	**KD**	kiln dried	**SBS**	styrene butadiene styrene
Btu(s)	British thermal unit(s)	**KO**	knockout	**SF**	square foot
C	Celsius/Centigrade	**kW**	kilowatt(s)	**SFCA**	(per) square foot of contact area
CF	cubic foot	**L**	length	**Sq**	100 square feet
CFM	cubic feet per minute	**Lb(s)**	pound(s)	**Std**	standard
CLF	100 linear feet	**LF**	linear foot	**SY**	square yard
CPM	cycles per minute	**LS**	lump sum	**T**	thick
CSF	100 square feet	**MBF**	1,000 board feet	**T&G**	tongue & groove edge
CY	cubic yard	**MDF**	medium density fiberboard	**UL**	Underwriter's Laboratory
d	penny	**mm**	millimeter(s)	**USDA**	United States Dept. of Agriculture
D	depth	**Mo**	month	**UV**	ultraviolet
DWV	drain, waste & vent	**MPH**	miles per hour	**W**	width
Ea	each	**MSF**	1,000 square feet	**X**	by or times
EMT	electric metallic tube	**NEC**	National Electrical Code		

Figure 1-6

Abbreviations used in this manual

Area Modification Factors

Construction costs are higher in some areas than in other areas. Add or deduct the percentages shown on the following pages to adapt the costs in this book to your jobsite. Adjust your cost estimate by the appropriate percentages in this table to find the estimated cost for the site selected. Where 0% is shown, it means no modification is required.

Modification factors are listed by state and province. Areas within each state are listed by the first three digits of the postal zip code. For convenience, one representative city is identified in each three-digit zip or range of zips. Percentages are based on the average of all data points in the table. Factors listed for each state and province are the average of all data in that state or province. Figures for the zips are the average of all information in that area. And figures in the Total column are the weighted average of factors for Labor, Material and Equipment.

The National Estimator program will apply an area modification factor for any five-digit zip you select. Click Utilities. Click Options. Then select the Area Modification Factors tab.

These percentages are composites of many costs and will not necessarily be accurate when estimating the cost of any particular part of a building. But when used to modify costs for an entire structure, they should improve the accuracy of your estimates.

Area Modification Factors

Location	Zip	Mat.	Lab.	Equip.	Total Wtd. Avg.
Alabama Average		**-1**	**-11**	**0**	**-6%**
Anniston	362	-3	-17	-1	-9%
Auburn	368	0	-14	0	-6%
Bellamy	369	-2	-6	-1	-4%
Birmingham	350-352	-3	10	-1	3%
Dothan	363	-1	-8	0	-4%
Evergreen	364	-1	-26	0	-12%
Gadsden	359	-4	-21	-1	-12%
Huntsville	358	1	-9	0	-4%
Jasper	355	-1	-21	0	-10%
Mobile	365-366	-1	-4	0	-2%
Montgomery	360-361	-1	-8	0	-4%
Scottsboro	357	0	-15	0	-7%
Selma	367	-1	-10	0	-5%
Sheffield	356	0	-2	0	-1%
Tuscaloosa	354	1	-10	0	-4%
Alaska Average		**14**	**27**	**4**	**20%**
Anchorage	995	16	36	5	25%
Fairbanks	997	16	32	5	23%
Juneau	998	17	25	6	21%
Ketchikan	999	3	14	1	8%
King Salmon	996	16	30	5	22%

Location	Zip	Mat.	Lab.	Equip.	Total Wtd. Avg.
Arizona Average		**1**	**-12**	**0**	**-5%**
Chambers	865	1	-12	0	-5%
Douglas	855	0	-15	0	-7%
Flagstaff	860	2	-21	1	-9%
Kingman	864	1	-12	0	-5%
Mesa	852	1	0	0	1%
Phoenix	850	1	1	0	1%
Prescott	863	3	-21	1	-8%
Show Low	859	1	-22	0	-10%
Tucson	856-857	0	-15	0	-7%
Yuma	853	0	0	0	0%
Arkansas Average		**-2**	**-13**	**0**	**-7%**
Batesville	725	0	-23	0	-11%
Camden	717	-4	-8	-1	-6%
Fayetteville	727	1	-9	0	-4%
Fort Smith	729	-1	-12	0	-6%
Harrison	726	0	-31	0	-14%
Hope	718	-3	7	-1	2%
Hot Springs	719	-2	-29	-1	-14%
Jonesboro	724	-1	-17	0	-8%
Little Rock	720-722	-1	-6	0	-3%
Pine Bluff	716	-4	-4	-1	-4%

Area Modification Factors

Location	Zip	Mat.	Lab.	Equip.	Total Wtd. Avg.
Russellville	728	0	-10	0	-5%
West Memphis	723	-3	-9	-1	-6%
California Average		**2**	**14**	**1**	**7%**
Alhambra	917-918	3	14	1	8%
Bakersfield	932-933	0	4	0	2%
El Centro	922	1	0	0	1%
Eureka	955	1	-4	0	-1%
Fresno	936-938	0	-3	0	-1%
Herlong	961	2	-3	1	0%
Inglewood	902-905	3	15	1	8%
Irvine	926-927	3	21	1	11%
Lompoc	934	3	10	1	6%
Long Beach	907-908	3	16	1	9%
Los Angeles	900-901	3	13	1	8%
Marysville	959	1	0	0	1%
Modesto	953	1	-3	0	-1%
Mojave	935	0	11	0	5%
Novato	949	3	27	1	14%
Oakland	945-947	3	35	1	18%
Orange	928	3	19	1	10%
Oxnard	930	3	7	1	5%
Pasadena	910-912	3	15	1	8%
Rancho Cordova	956-957	2	12	1	7%
Redding	960	1	-7	0	-3%
Richmond	948	2	38	1	19%
Riverside	925	1	7	0	4%
Sacramento	958	1	12	0	6%
Salinas	939	3	1	1	2%
San Bernardino	923-924	0	5	0	2%
San Diego	919-921	3	15	1	8%
San Francisco	941	3	58	1	28%
San Jose	901-951	3	35	1	18%
San Mateo	943-944	4	34	1	18%
Santa Barbara	931	3	11	1	7%
Santa Rosa	954	3	10	1	6%
Stockton	952	1	2	0	1%
Sunnyvale	940	3	38	1	19%
Van Nuys	913-916	3	13	1	8%
Whittier	906	3	13	1	8%
Colorado Average		**1**	**0**	**1**	**1%**
Aurora	800-801	2	6	1	4%
Boulder	803-804	3	-1	1	1%
Colorado Springs	808-809	2	-6	1	-2%
Denver	802	2	11	1	6%
Durango	813	1	-5	0	-2%
Fort Morgan	807	1	2	0	1%
Glenwood Springs	816	2	7	1	4%
Grand Junction	814-815	1	-4	0	-1%
Greeley	806	2	5	1	3%
Longmont	805	3	2	1	3%
Pagosa Springs	811	0	-10	0	-5%
Pueblo	810	-1	8	0	3%
Salida	812	1	-11	0	-5%

Location	Zip	Mat.	Lab.	Equip.	Total Wtd. Avg.
Connecticut Average		**1**	**21**	**0**	**10%**
Bridgeport	66	1	21	0	10%
Bristol	60	1	25	0	12%
Fairfield	64	2	24	1	12%
Hartford	61	0	22	0	10%
New Haven	65	1	21	0	10%
Norwich	63	0	14	0	6%
Stamford	068-069	4	28	1	15%
Waterbury	67	1	19	0	9%
West Hartford	62	1	15	0	7%
Delaware Average		**1**	**6**	**0**	**3%**
Dover	199	1	-8	0	-3%
Newark	197	2	14	1	8%
Wilmington	198	0	13	0	6%
District of Columbia Average		**3**	**25**	**1**	**13%**
Washington	200-205	3	25	1	13%
Florida Average		**-1**	**-13**	**0**	**-6%**
Altamonte Springs	327	-1	-12	0	-6%
Bradenton	342	0	-16	0	-7%
Brooksville	346	-1	-16	0	-8%
Daytona Beach	321	-2	-21	-1	-11%
Fort Lauderdale	333	2	-3	1	0%
Fort Myers	339	0	-20	0	-9%
Fort Pierce	349	-2	-26	-1	-13%
Gainesville	326	-1	-20	0	-10%
Jacksonville	322	-1	-5	0	-3%
Lakeland	338	-3	-15	-1	-8%
Melbourne	329	-2	-19	-1	-10%
Miami	330-332	2	-5	1	-1%
Naples	341	3	-12	1	-4%
Ocala	344	-2	-24	-1	-12%
Orlando	328	0	-2	0	-1%
Panama City	324	-2	-5	-1	-3%
Pensacola	325	0	-15	0	-7%
Saint Augustine	320	-1	-11	0	-6%
Saint Cloud	347	0	-9	0	-4%
St Petersburg	337	-1	-12	0	-6%
Tallahassee	323	0	-17	0	-8%
Tampa	335-336	-1	-2	0	-1%
West Palm Beach	334	1	-7	0	-3%
Georgia Average		**-1**	**-10**	**0**	**-5%**
Albany	317	-2	-19	-1	-10%
Athens	306	0	-16	0	-7%
Atlanta	303	3	23	1	12%
Augusta	308-309	-2	-10	-1	-6%
Buford	305	0	-8	0	-4%
Calhoun	307	-1	-21	0	-10%
Columbus	318-319	-1	-10	0	-5%
Dublin/Fort Valley	310	-3	-15	-1	-8%
Hinesville	313	-2	-10	-1	-6%
Kings Bay	315	-2	-17	-1	-9%

Area Modification Factors

Location	Zip	Mat.	Lab.	Equip.	Total Wtd. Avg.
Macon	312	-2	-6	-1	-4%
Marietta	300-302	1	7	0	4%
Savannah	314	-2	-11	-1	-6%
Statesboro	304	-2	-22	-1	-11%
Valdosta	316	-1	-8	0	-4%
Hawaii Average		**17**	**32**	**6**	**24%**
Aliamanu	968	17	35	6	25%
Ewa	967	17	31	6	23%
Halawa Heights	967	17	31	6	23%
Hilo	967	17	31	6	23%
Honolulu	968	17	35	6	25%
Kailua	968	17	35	6	25%
Lualualei	967	17	31	6	23%
Mililani Town	967	17	31	6	23%
Pearl City	967	17	31	6	23%
Wahiawa	967	17	31	6	23%
Waianae	967	17	31	6	23%
Wailuku (Maui)	967	17	31	6	23%
Idaho Average		**0**	**-16**	**0**	**-8%**
Boise	837	1	-5	0	-2%
Coeur d'Alene	838	0	-21	0	-10%
Idaho Falls	834	-1	-16	0	-8%
Lewiston	835	0	-22	0	-10%
Meridian	836	0	-15	0	-7%
Pocatello	832	-1	-14	0	-7%
Sun Valley	833	-1	-21	0	-10%
Illinois Average		**-1**	**12**	**0**	**5%**
Arlington Heights	600	1	31	0	15%
Aurora	605	2	30	1	15%
Belleville	622	-2	12	-1	4%
Bloomington	617	1	4	0	2%
Carbondale	629	-3	-10	-1	-6%
Carol Stream	601	2	30	1	15%
Centralia	628	-3	-6	-1	-4%
Champaign	618	-1	5	0	2%
Chicago	606-608	2	35	1	17%
Decatur	623	-2	-12	-1	-7%
Galesburg	614	-2	-9	-1	-5%
Granite City	620	-3	15	-1	5%
Green River	612	-2	10	-1	4%
Joliet	604	0	33	0	15%
Kankakee	609	-2	2	-1	0%
Lawrenceville	624	-4	-5	-1	-4%
Oak Park	603	3	38	1	19%
Peoria	615-616	-1	13	0	5%
Peru	613	0	15	0	7%
Quincy	602	3	34	1	17%
Rockford	610-611	-2	12	-1	4%
Springfield	625-627	-2	0	-1	-1%
Urbana	619	-3	-8	-1	-5%

Location	Zip	Mat.	Lab.	Equip.	Total Wtd. Avg.
Indiana Average		**-2**	**0**	**-1**	**-1%**
Aurora	470	-1	-4	0	-2%
Bloomington	474	1	-6	0	-2%
Columbus	472	0	-9	0	-4%
Elkhart	465	-2	-4	-1	-3%
Evansville	476-477	-2	25	-1	10%
Fort Wayne	467-468	-3	-4	-1	-3%
Gary	463-464	-4	43	-1	18%
Indianapolis	460-462	-1	12	0	5%
Jasper	475	-1	-14	0	-7%
Jeffersonville	471	-1	-6	0	-3%
Kokomo	469	-2	-15	-1	-8%
Lafayette	479	-1	-9	0	-5%
Muncie	473	-4	-13	-1	-8%
South Bend	466	-4	-1	-1	-3%
Terre Haute	478	-4	0	-1	-2%
Iowa Average		**-2**	**-7**	**-1**	**-4%**
Burlington	526	0	-6	0	-3%
Carroll	514	-3	-18	-1	-10%
Cedar Falls	506	-1	-6	0	-3%
Cedar Rapids	522-524	0	3	0	1%
Cherokee	510	-3	-5	-1	-4%
Council Bluffs	515	-2	-11	-1	-6%
Creston	508	-3	-11	-1	-7%
Davenport	527-528	-1	2	0	0%
Decorah	521	-2	-7	-1	-4%
Des Moines	500-503	-2	10	-1	4%
Dubuque	520	-1	1	0	0%
Fort Dodge	505	-2	-7	-1	-4%
Mason City	504	0	-7	0	-3%
Ottumwa	525	0	-13	0	-6%
Sheldon	512	0	-17	0	-8%
Shenandoah	516	-3	-27	-1	-14%
Sioux City	511	-2	1	-1	-1%
Spencer	513	-1	-14	0	-7%
Waterloo	507	-4	-6	-1	-5%
Kansas Average		**-2**	**-12**	**-1**	**-7%**
Colby	677	-1	-22	0	-11%
Concordia	669	-1	-26	0	-12%
Dodge City	678	-2	-8	-1	-5%
Emporia	668	-3	-23	-1	-12%
Fort Scott	667	-2	-12	-1	-7%
Hays	676	-2	-28	-1	-14%
Hutchinson	675	-3	-12	-1	-7%
Independence	673	-3	0	-1	-2%
Kansas City	660-662	0	11	0	5%
Liberal	679	-2	-19	-1	-10%
Salina	674	-3	-11	-1	-7%
Topeka	664-666	-3	-4	-1	-3%
Wichita	670-672	-2	-8	-1	-5%

Area Modification Factors

Location	Zip	Mat.	Lab.	Equip.	Total Wtd. Avg.
Kentucky Average		**-1**	**-8**	**0**	**-4%**
Ashland	411-412	-3	-2	-1	-3%
Bowling Green	421	0	-12	0	-6%
Campton	413-414	-1	-16	0	-8%
Covington	410	-1	4	0	1%
Elizabethtown	427	-1	-16	0	-8%
Frankfort	406	1	-8	0	-3%
Hazard	417-418	-1	-9	0	-5%
Hopkinsville	422	-2	-11	-1	-6%
Lexington	403-405	1	2	0	1%
London	407-409	-1	-8	0	-4%
Louisville	400-402	-1	-1	0	-1%
Owensboro	423	-2	-3	-1	-2%
Paducah	420	-2	-6	-1	-4%
Pikeville	415-416	-3	-5	-1	-4%
Somerset	425-426	0	-25	0	-12%
White Plains	424	-2	-6	-1	-4%
Louisiana Average		**-1**	**2**	**-1**	**0%**
Alexandria	713-714	-3	-10	-1	-6%
Baton Rouge	707-708	0	22	0	10%
Houma	703	-2	16	-1	6%
Lafayette	705	0	5	0	2%
Lake Charles	706	-3	4	-1	0%
Mandeville	704	0	-1	0	0%
Minden	710	-2	-7	-1	-4%
Monroe	712	-2	-13	-1	-7%
New Orleans	700-701	0	11	0	5%
Shreveport	711	-2	-3	-1	-2%
Maine Average		**-1**	**-12**	**0**	**-6%**
Auburn	42	-1	-10	0	-5%
Augusta	43	-1	-12	0	-6%
Bangor	44	-2	-14	-1	-8%
Bath	45	1	-13	0	-5%
Brunswick	39-40	1	-5	0	-2%
Camden	48	-1	-19	0	-9%
Cutler	46	-1	-21	0	-10%
Dexter	49	-2	-10	-1	-6%
Northern Area	47	-2	-20	-1	-10%
Portland	41	2	0	1	1%
Maryland Average		**1**	**6**	**0**	**3%**
Annapolis	214	3	10	1	6%
Baltimore	210-212	-1	18	0	8%
Bethesda	208-209	3	26	1	14%
Church Hill	216	2	-10	1	-4%
Cumberland	215	-4	-11	-1	-7%
Elkton	219	2	4	1	3%
Frederick	217	1	5	0	3%
Laurel	206-207	2	18	1	9%
Salisbury	218	1	-9	0	-4%

Location	Zip	Mat.	Lab.	Equip.	Total Wtd. Avg.
Massachusetts Average		**2**	**25**	**1**	**12%**
Ayer	015-016	1	16	0	8%
Bedford	017	3	36	1	18%
Boston	021-022	3	75	1	36%
Brockton	023-024	2	44	1	21%
Cape Cod	026	2	3	1	2%
Chicopee	010	1	14	0	7%
Dedham	019	3	33	1	17%
Fitchburg	014	2	27	1	13%
Hingham	020	3	41	1	20%
Lawrence	018	2	32	1	16%
Nantucket	025	3	18	1	10%
New Bedford	027	2	15	1	8%
Northfield	013	1	0	0	1%
Pittsfield	012	1	2	0	1%
Springfield	011	-1	17	0	7%
Michigan Average		**-2**	**5**	**-1**	**1%**
Battle Creek	490-491	-3	4	-1	0%
Detroit	481-482	0	23	0	11%
Flint	484-485	-3	-3	-1	-3%
Grand Rapids	493-495	-1	5	0	2%
Grayling	497	1	-15	0	-6%
Jackson	492	-3	0	-1	-2%
Lansing	488-489	0	3	0	1%
Marquette	498-499	-1	-2	0	-1%
Pontiac	483	-3	25	-1	10%
Royal Oak	480	-2	22	-1	9%
Saginaw	486-487	-2	-5	-1	-3%
Traverse City	496	-1	2	0	0%
Minnesota Average		**0**	**1**	**0**	**0%**
Bemidji	566	-1	-8	0	-4%
Brainerd	564	0	-4	0	-2%
Duluth	556-558	-2	2	-1	0%
Fergus Falls	565	-1	-13	0	-7%
Magnolia	561	0	-15	0	-7%
Mankato	560	0	-7	0	-3%
Minneapolis	553-555	1	23	0	11%
Rochester	559	0	-5	0	-2%
St Cloud	563	-1	15	0	6%
St Paul	550-551	1	23	0	11%
Thief River Falls	567	0	-1	0	0%
Willmar	562	-1	-3	0	-2%
Mississippi Average		**-2**	**-15**	**-1**	**-8%**
Clarksdale	386	-3	-19	-1	-10%
Columbus	397	-1	-5	0	-3%
Greenville	387	-4	-25	-1	-14%
Greenwood	389	-3	-18	-1	-10%
Gulfport	395	-2	-6	-1	-4%
Jackson	390-392	-3	-8	-1	-5%
Laurel	394	-3	-10	-1	-6%

Area Modification Factors

Location	Zip	Mat.	Lab.	Equip.	Total Wtd. Avg.
McComb	396	-2	-18	-1	-9%
Meridian	393	-2	-16	-1	-8%
Tupelo	388	-1	-20	0	-10%
Missouri Average		**-1**	**-8**	**-1**	**-4%**
Cape Girardeau	637	-2	-8	-1	-5%
Caruthersville	638	-1	-22	0	-11%
Chillicothe	646	-2	-17	-1	-9%
Columbia	652	1	-10	0	-4%
East Lynne	647	-1	-11	0	-6%
Farmington	636	-3	-16	-1	-9%
Hannibal	634	1	-10	0	-4%
Independence	640	-2	13	-1	5%
Jefferson City	650-651	1	-10	0	-4%
Joplin	648	-2	-15	-1	-8%
Kansas City	641	-2	16	-1	6%
Kirksville	635	0	-21	0	-10%
Knob Noster	653	0	-10	0	-5%
Lebanon	654-655	-2	-21	-1	-11%
Poplar Bluff	639	-1	-12	0	-6%
Saint Charles	633	1	1	0	1%
Saint Joseph	644-645	-3	6	-1	1%
Springfield	656-658	-2	-16	-1	-8%
St Louis	630-631	-2	14	-1	5%
Montana Average		**0**	**-6**	**0**	**-3%**
Billings	590-591	0	3	0	1%
Butte	597	1	-10	0	-4%
Fairview	592	-1	1	0	0%
Great Falls	594	-1	-5	0	-3%
Havre	595	-1	-14	0	-7%
Helena	596	0	-5	0	-2%
Kalispell	599	1	-12	0	-5%
Miles City	593	-1	-3	0	-2%
Missoula	598	1	-12	0	-5%
Nebraska Average		**-1**	**-19**	**0**	**-9%**
Alliance	693	-1	-23	0	-11%
Columbus	686	0	-14	0	-6%
Grand Island	688	0	-18	0	-8%
Hastings	689	0	-21	0	-10%
Lincoln	683-685	0	-10	0	-5%
McCook	690	0	-27	0	-12%
Norfolk	687	-3	-18	-1	-10%
North Platte	691	0	-21	0	-10%
Omaha	680-681	-1	-3	0	-2%
Valentine	692	-2	-33	-1	-16%
Nevada Average		**2**	**11**	**1**	**6%**
Carson City	897	2	-9	1	-3%
Elko	898	1	65	0	30%
Ely	893	2	-7	1	-2%
Fallon	894	2	6	1	4%
Las Vegas	889-891	2	8	1	5%
Reno	895	2	1	1	2%

Location	Zip	Mat.	Lab.	Equip.	Total Wtd. Avg.
New Hampshire Average		**1**	**0**	**0**	**0%**
Charlestown	036	1	-9	0	-4%
Concord	034	1	-4	0	-1%
Dover	038	0	7	0	3%
Lebanon	037	2	-7	1	-2%
Littleton	035	-1	-6	0	-3%
Manchester	032-033	0	5	0	2%
New Boston	030-031	1	11	1	6%
New Jersey Average		**1**	**27**	**0**	**13%**
Atlantic City	080-084	-2	22	-1	9%
Brick	087	2	10	1	6%
Dover	078	1	26	0	12%
Edison	088-089	1	32	0	15%
Hackensack	076	3	26	1	14%
Monmouth	077	3	31	1	16%
Newark	071-073	2	29	1	14%
Passaic	070	2	30	1	15%
Paterson	074-075	2	23	1	12%
Princeton	085	-2	33	-1	14%
Summit	079	3	37	1	19%
Trenton	086	-3	29	-1	12%
New Mexico Average		**0**	**-20**	**0**	**-9%**
Alamogordo	883	-1	-21	0	-10%
Albuquerque	870-871	2	-8	1	-3%
Clovis	881	-2	-26	-1	-13%
Farmington	874	2	-2	1	0%
Fort Sumner	882	-3	-1	-1	-2%
Gallup	873	1	-18	0	-8%
Holman	877	2	-22	1	-9%
Las Cruces	880	-2	-23	-1	-12%
Santa Fe	875	3	-23	1	-9%
Socorro	878	1	-34	0	-15%
Truth or onsequences	879	-2	-38	-1	-19%
Tucumcari	884	-1	-20	0	-10%
New York Average		**0**	**19**	**0**	**9%**
Albany	120-123	0	22	0	10%
Amityville	117	2	25	1	13%
Batavia	140	-3	5	-1	1%
Binghamton	137-139	-3	6	-1	1%
Bronx	104	2	32	1	16%
Brooklyn	112	3	22	1	12%
Buffalo	142	-4	6	-1	1%
Elmira	149	-4	4	-1	0%
Flushing	113	3	43	1	21%
Garden City	115	3	37	1	19%
Hicksville	118	3	33	1	17%
Ithaca	148	-4	-2	-1	-3%
Jamaica	114	3	40	1	20%
Jamestown	147	-4	-5	-1	-4%
Kingston	124	0	-7	0	-3%
Long Island	111	3	72	1	35%

Area Modification Factors

Location	Zip	Mat.	Lab.	Equip.	Total Wtd. Avg.
Montauk	119	1	22	0	11%
New York (Manhattan)	100-102	3	74	1	36%
New York City	100-102	3	74	1	36%
Newcomb	128	-1	1	0	0%
Niagara Falls	143	-4	-10	-1	-7%
Plattsburgh	129	1	-2	0	0%
Poughkeepsie	125-126	0	5	0	2%
Queens	110	4	44	1	22%
Rochester	144-146	-3	7	-1	2%
Rockaway	116	3	30	1	15%
Rome	133-134	-3	-1	-1	-2%
Staten Island	103	3	22	1	12%
Stewart	127	-1	-9	0	-5%
Syracuse	130-132	-3	8	-1	2%
Tonawanda	141	-4	2	-1	-1%
Utica	135	-4	-3	-1	-3%
Watertown	136	-2	26	-1	11%
West Point	109	1	22	0	11%
White Plains	105-108	3	34	1	17%
North Carolina Average		**1**	**-10**	**0**	**-4%**
Asheville	287-289	1	-19	0	-8%
Charlotte	280-282	1	4	0	2%
Durham	277	2	-1	1	1%
Elizabeth City	279	1	-15	0	-6%
Fayetteville	283	-1	-17	0	-8%
Goldsboro	275	1	-4	0	-1%
Greensboro	274	1	-8	0	-3%
Hickory	286	-1	-17	0	-8%
Kinston	285	-1	-21	0	-10%
Raleigh	276	3	-1	1	1%
Rocky Mount	278	-1	-15	0	-7%
Wilmington	284	1	-4	0	-1%
Winston-Salem	270-273	0	-10	0	-5%
North Dakota Average		**-1**	**6**	**0**	**2%**
Bismarck	585	0	4	0	2%
Dickinson	586	-1	21	0	9%
Fargo	580-581	0	4	0	2%
Grand Forks	582	0	1	0	0%
Jamestown	584	-1	-5	0	-3%
Minot	587	-1	9	0	4%
Nekoma	583	-1	-18	0	-9%
Williston	588	-1	32	0	14%
Ohio Average		**-2**	**1**	**-1**	**-1%**
Akron	442-443	-2	2	-1	0%
Canton	446-447	-2	-4	-1	-3%
Chillicothe	456	-2	-8	-1	-5%
Cincinnati	450-452	-1	8	0	3%
Cleveland	440-441	-3	11	-1	3%
Columbus	432	0	9	0	4%
Dayton	453-455	-3	-2	-1	-3%

Location	Zip	Mat.	Lab.	Equip.	Total Wtd. Avg.
Lima	458	-3	-8	-1	-5%
Marietta	457	-2	-6	-1	-4%
Marion	433	-3	-9	-1	-6%
Newark	430-431	-1	7	0	3%
Sandusky	448-449	-1	-2	0	-1%
Steubenville	439	-3	-2	-1	-3%
Toledo	434-436	-1	12	0	5%
Warren	444	-4	-4	-1	-4%
Youngstown	445	-5	3	-2	-1%
Zanesville	437-438	-2	5	-1	1%
Oklahoma Average		**-3**	**-12**	**-1**	**-7%**
Adams	739	-2	-20	-1	-10%
Ardmore	734	-3	-11	-1	-7%
Clinton	736	-3	-11	-1	-7%
Durant	747	-4	-19	-1	-11%
Enid	737	-4	-13	-1	-8%
Lawton	735	-3	-15	-1	-8%
McAlester	745	-4	-19	-1	-11%
Muskogee	744	-2	-23	-1	-12%
Norman	730	-2	-4	-1	-3%
Oklahoma City	731	-2	-2	-1	-2%
Ponca City	746	-3	-5	-1	-4%
Poteau	749	-2	-19	-1	-10%
Pryor	743	-2	-22	-1	-11%
Shawnee	748	-4	-18	-1	-10%
Tulsa	740-741	-1	-3	0	-2%
Woodward	738	-4	8	-1	2%
Oregon Average		**1**	**-9**	**1**	**-3%**
Adrian	979	-1	-17	0	-8%
Bend	977	1	-15	0	-6%
Eugene	974	2	-5	1	-1%
Grants Pass	975	2	-14	1	-5%
Klamath Falls	976	2	-21	1	-9%
Pendleton	978	0	-7	0	-3%
Portland	970-972	2	19	1	10%
Salem	973	2	-8	1	-3%
Pennsylvania Average		**-3**	**2**	**-1**	**-1%**
Allentown	181	-2	11	-1	4%
Altoona	166	-3	-11	-1	-7%
Beaver Springs	178	-3	-6	-1	-4%
Bethlehem	180	-1	14	0	6%
Bradford	167	-4	-9	-1	-6%
Butler	160	-4	5	-1	0%
Chambersburg	172	-1	-13	0	-7%
Clearfield	168	2	-3	1	0%
DuBois	158	-2	-18	-1	-9%
East Stroudsburg	183	0	-7	0	-3%
Erie	164-165	-3	-9	-1	-6%
Genesee	169	-4	0	-1	-2%
Greensburg	156	-4	1	-1	-2%
Harrisburg	170-171	-2	6	-1	2%

Area Modification Factors

Location	Zip	Mat.	Lab.	Equip.	Total Wtd. Avg.
Hazleton	182	-3	-3	-1	-3%
Johnstown	159	-4	-15	-1	-9%
Kittanning	162	-4	-11	-1	-7%
Lancaster	175-176	-2	2	-1	0%
Meadville	163	-4	-21	-1	-12%
Montrose	188	-3	-3	-1	-3%
New Castle	161	-4	-3	-1	-3%
Philadelphia	190-191	-3	34	-1	14%
Pittsburgh	152	-4	22	-1	8%
Pottsville	179	-4	-12	-1	-8%
Punxsutawney	157	-4	-6	-1	-5%
Reading	195-196	-4	10	-1	2%
Scranton	184-185	-2	-2	-1	-2%
Somerset	155	-4	-12	-1	-8%
Southeastern	193	0	19	0	9%
Uniontown	154	-4	-7	-1	-5%
Valley Forge	194	-3	35	-1	15%
Warminster	189	-1	26	0	11%
Warrendale	150-151	-4	21	-1	8%
Washington	153	-3	23	-1	9%
Wilkes Barre	186-187	-3	1	-1	-1%
Williamsport	177	-3	5	-1	1%
York	173-174	-3	4	-1	0%
Rhode Island Average		**1**	**12**	**0**	**6%**
Bristol	28	1	11	0	6%
Coventry	028	1	11	0	6%
Cranston	029	1	14	0	7%
Davisville	028	1	11	0	6%
Narragansett	028	1	11	0	6%
Newport	028	1	11	0	6%
Providence	029	1	14	0	7%
Warwick	028	1	11	0	6%
South Carolina Average		**0**	**-10**	**0**	**-5%**
Aiken	298	0	9	0	4%
Beaufort	299	-1	-16	0	-8%
Charleston	294	-1	-5	0	-3%
Columbia	290-292	0	-9	0	-4%
Greenville	296	0	-9	0	-4%
Myrtle Beach	295	0	-20	0	-9%
Rock Hill	297	0	-18	0	-8%
Spartanburg	293	-1	-14	0	-7%
South Dakota Average		**-1**	**-13**	**0**	**-6%**
Aberdeen	574	-1	-11	0	-6%
Mitchell	573	0	-15	0	-7%
Mobridge	576	-1	-22	0	-11%
Pierre	575	-1	-21	0	-10%
Rapid City	577	-1	-11	0	-6%

Location	Zip	Mat.	Lab.	Equip.	Total Wtd. Avg.
Sioux Falls	570-571	0	-2	0	-1%
Watertown	572	-1	-11	0	-6%
Tennessee Average		**-1**	**-3**	**0**	**-2%**
Chattanooga	374	-1	5	0	2%
Clarksville	370	1	4	0	2%
Cleveland	373	-1	-1	0	-1%
Columbia	384	0	-15	0	-7%
Cookeville	385	0	-16	0	-7%
Jackson	383	-1	-3	0	-2%
Kingsport	376	0	-8	0	-4%
Knoxville	377-379	-1	-2	0	-1%
McKenzie	382	-2	-17	-1	-9%
Memphis	380-381	-2	8	-1	3%
Nashville	371-372	1	10	0	5%
Texas Average		**-2**	**-2**	**-1**	**-2%**
Abilene	795-796	-4	-3	-1	-3%
Amarillo	790-791	-2	-5	-1	-3%
Arlington	760	-1	1	0	0%
Austin	786-787	1	4	0	2%
Bay City	774	-2	53	-1	23%
Beaumont	776-777	-3	19	-1	7%
Brownwood	768	-3	-18	-1	-10%
Bryan	778	0	-11	0	-5%
Childress	792	-3	-31	-1	-16%
Corpus Christi	783-784	-3	6	-1	1%
Dallas	751-753	-1	10	0	4%
Del Rio	788	-3	-15	-1	-8%
El Paso	798-799	-3	-22	-1	-12%
Fort Worth	761-762	-2	1	-1	-1%
Galveston	775	-3	21	-1	8%
Giddings	789	0	-9	0	-4%
Greenville	754	-3	8	-1	2%
Houston	770-772	-1	23	0	10%
Huntsville	773	-2	25	-1	10%
Longview	756	-3	-4	-1	-3%
Lubbock	793-794	-3	-15	-1	-8%
Lufkin	759	-3	-4	-1	-3%
McAllen	785	-3	-27	-1	-14%
Midland	797	-3	14	-1	5%
Palestine	758	-2	-10	-1	-6%
Plano	750	-1	11	0	5%
San Angelo	769	-3	-14	-1	-8%
San Antonio	780-782	-3	-1	-1	-2%
Texarkana	755	-3	-16	-1	-9%
Tyler	757	-1	-14	0	-7%
Victoria	779	-3	-4	-1	-3%
Waco	765-767	-3	-10	-1	-6%
Wichita Falls	763	-3	-16	-1	-9%
Woodson	764	-3	-16	-1	-9%

Area Modification Factors

Location	Zip	Mat.	Lab.	Equip.	Total Wtd. Avg.
Utah Average		**1**	**-8**	**1**	**-3%**
Clearfield	840	2	-2	1	0%
Green River	845	1	-2	0	0%
Ogden	843-844	0	-16	0	-7%
Provo	846-847	2	-22	1	-9%
Salt Lake City	841	2	1	1	2%
Vermont Average		**1**	**-9**	**0**	**-4%**
Albany	058	1	-12	0	-5%
Battleboro	053	1	-8	0	-3%
Beecher Falls	059	1	-13	0	-5%
Bennington	052	-1	-14	0	-7%
Burlington	054	2	4	1	3%
Montpelier	056	2	-9	1	-3%
Rutland	057	-1	-15	0	-7%
Springfield	051	-1	-9	0	-5%
White River Junction	050	1	-9	0	-4%
Virginia Average		**0**	**-9**	**0**	**-5%**
Abingdon	242	-2	-9	-1	-5%
Alexandria	220-223	3	24	1	13%
Charlottesville	229	1	-13	0	-5%
Chesapeake	233	-1	-1	0	-1%
Culpeper	227	2	-12	1	-4%
Farmville	239	-2	-24	-1	-12%
Fredericksburg	224-225	1	-12	0	-5%
Galax	243	-2	-20	-1	-10%
Harrisonburg	228	1	-16	0	-7%
Lynchburg	245	-2	-18	-1	-9%
Norfolk	235-237	0	-1	0	0%
Petersburg	238	-2	-10	-1	-6%
Radford	241	-1	-22	0	-11%
Reston	201	3	18	1	10%
Richmond	232	-1	3	0	1%
Roanoke	240	-1	-22	-1	-11%
Staunton	244	0	-18	0	-8%
Tazewell	246	-3	-13	-1	-8%
Virginia Beach	234	0	-8	0	-4%
Williamsburg	230-231	0	-9	0	-4%
Winchester	226	0	-16	0	-7%
Washington Average		**1**	**-1**	**0**	**0%**
Clarkston	994	0	-19	0	-9%
Everett	982	2	3	1	2%
Olympia	985	2	-7	1	-2%
Pasco	993	0	9	0	4%
Seattle	980-981	3	24	1	13%
Spokane	990-992	0	-2	0	-1%
Tacoma	983-984	1	4	0	2%
Vancouver	986	2	1	1	2%

Location	Zip	Mat.	Lab.	Equip.	Total Wtd. Avg.
Wenatchee	988	1	-11	0	-5%
Yakima	989	0	-9	0	-4%
West Virginia Average		**-2**	**-9**	**-1**	**-5%**
Beckley	258-259	1	-10	0	-4%
Bluefield	247-248	-1	-2	0	-1%
Charleston	250-253	1	11	0	6%
Clarksburg	263-264	-3	-5	-1	-4%
Fairmont	266	0	-23	0	-11%
Huntington	255-257	-1	0	0	-1%
Lewisburg	249	-2	-30	-1	-15%
Martinsburg	254	-1	-16	0	-8%
Morgantown	265	-3	0	-1	-2%
New Martinsville	262	-3	-13	-1	-8%
Parkersburg	261	-3	2	-1	-1%
Romney	267	-4	-11	-1	-7%
Sugar Grove	268	-3	-23	-1	-12%
Wheeling	260	-3	-3	-1	-3%
Wisconsin Average		**-1**	**0**	**0**	**0%**
Amery	540	0	-1	0	0%
Beloit	535	0	9	0	4%
Clam Lake	545	-1	-16	0	-8%
Eau Claire	547	-1	0	0	-1%
Green Bay	541-543	0	2	0	1%
La Crosse	546	-2	-2	-1	-2%
Ladysmith	548	-2	-9	-1	-5%
Madison	537	2	15	1	8%
Milwaukee	530-534	0	15	0	7%
Oshkosh	549	-1	3	0	1%
Portage	539	0	3	0	1%
Prairie du Chien	538	-2	-8	-1	-5%
Wausau	544	-1	-5	0	-3%
Wyoming Average		**0**	**-5**	**0**	**-2%**
Casper	826	-2	7	-1	2%
Cheyenne/Laramie	820	1	-12	0	-5%
Gillette	827	-1	6	0	2%
Powell	824	0	-13	0	-6%
Rawlins	823	0	1	0	0%
Riverton	825	-1	-9	0	-5%
Rock Springs	829-831	0	6	0	3%
Sheridan	828	0	-8	0	-4%
Wheatland	822	0	-21	0	-10%
UNITED STATES TERRITORIES					
Guam		53	-21	-5	18%
Puerto Rico		2	-47	-5	-21%
VIRGIN ISLANDS (U.S.)					
St. Croix		18	-15	-4	2%
St. John		52	-15	-4	20%
St. Thomas		23	-15	-4	5%

Canadian Area Modification Factors

To find the cost in Canada in Canadian dollars, increase the costs in this book by the appropriate percentage listed below. These figures convert costs to Canadian dollars based on $1.00 Canadian to $1.00 U.S.

Location	Mat.	Lab.	Equip.	Total Wtd. Avg.
Alberta Average	**23**	**8**	**10**	**16%**
Calgary	23	10	10	17%
Edmonton	22	11	10	17%
Fort McMurray	25	2	10	14%
British Columbia Average	**23**	**-6**	**10**	**10%**
Fraser Valley	23	-7	10	9%
Okanagan	24	-9	10	9%
Vancouver	23	-2	10	11%
Manitoba Average	**17**	**-13**	**8**	**3%**
North Manitoba	17	-13	8	3%
South Manitoba	17	-13	8	3%
Selkirk	17	-13	8	3%
Winnipeg	17	-13	8	3%
New Brunswick Average	**13**	**-38**	**6**	**-11%**
Moncton	13	-38	6	-11%

Location	Mat.	Lab.	Equip.	Total Wtd. Avg.
Nova Scotia Average	**13**	**-27**	**6**	**-5%**
Amherst	13	-27	6	-6%
Nova Scotia	14	-26	7	-5%
Sydney	14	-28	7	-6%
Newfoundland/Labrador Average	**16**	**-32**	**7**	**-6%**
Ontario Average	**19**	**-2**	**8**	**9%**
London	19	-1	8	10%
Thunder Bay	21	-5	9	9%
Toronto	18	0	8	10%
Quebec Average	**16**	**-15**	**7**	**2%**
Montreal	16	-15	7	2%
Quebec City	16	-15	7	2%
Saskatchewan Average	**20**	**-9**	**9**	**6%**
La Ronge	21	-12	9	6%
Prince Albert	19	-11	8	5%
Saskatoon	19	-5	8	8%

Demolition

2

Nearly all home improvement projects require some breaking out and removing of existing materials and disposal of debris. Most will require protection of adjacent surfaces, taking safety precautions (such as setting up barricades), closing off doorways or windows, and normal cleanup. The demolition figures in this chapter include these tasks and will apply on most home improvement jobs that don't involve unusual conditions or complications. Costs will be higher when access to the work is limited, when your crew doesn't have complete control of the construction site, or when debris must be moved longer distances or around obstacles. All figures assume that debris is piled on site or in a bin. Add the cost of hauling to the nearest disposal site and tippage charges, if required. No salvage value is assumed except as noted.

Many of the following chapters include estimated costs to remove the old and replace with new materials of a similar description. For example, you'll find costs for removing shingles and replacing shingles in the roofing chapter. That's appropriate because they're part of the same task and usually performed by the same contractor. The estimates in this chapter are for demolition only and don't consider the cost of replacing what has been removed.

Of all construction tasks, demolition is probably the most difficult to estimate with any certainty. You're never sure what's in a wall to be demolished until the job begins. That's why it's best to exclude from your demolition bid what isn't in your estimate. Use the checklist to the right to help identify potential problems before they develop.

Demolition Checklist

- ❒ Can you anticipate any problem getting a demolition permit for this job?
- ❒ Is there a noise ordinance which will limit hours of operation or selection of equipment?
- ❒ Can the debris box be located close to the work?
- ❒ Can a trencher get to the job site?
- ❒ Is the site accessible to wheeled equipment?
- ❒ Will the debris have to be hand-carried to the roll-off box or gondola?
- ❒ Is there a direct route from the demolition site to the debris box?
- ❒ Will you need to re-route any plumbing, electrical, phone or gas lines?
- ❒ Will you need to remove and replace lawn sprinklers?
- ❒ Will you need to cut back or remove any bushes or trees?
- ❒ Will you have to remove and re-install a fence or gate?
- ❒ Does a septic tank or drain field extend into the site of construction?
- ❒ Can (must) materials generated from your project be recycled or salvaged?

	Day	Week	Month
Equipment Rental for Demolition.			
Air compressors, wheel-mounted			
16 CFM, shop type, electric	65.50	158.00	472.00
30 CFM, shop type, electric	72.10	216.00	650.00
80 CFM, shop type, electric	84.50	257.00	767.00
100 CFM, gasoline unit	105.00	315.00	886.00
125 CFM, gasoline unit	131.00	394.00	1,180.00
Paving breakers (no bits included) hand-held, pneumatic			
To 40 lb	30.00	85.00	230.00
41 - 55 lb	35.00	85.00	250.00
56 - 70 lb	35.00	105.00	280.00
71 - 90 lb	35.00	105.00	300.00
Paving breakers jackhammer bits			
Moil points, 15" to 18"	6.00	14.00	30.00
Chisels, 3"	7.00	17.00	44.00
Clay spades, 5-1/2"	10.00	23.00	62.00
Asphalt cutters, 5"	8.00	24.00	51.00
Pneumatic chippers, medium weight, 10 lb	35.00	110.00	290.00
Air hose rental, 50 LF section			
5/8" air hose	6.00	15.00	40.00
3/4" air hose	7.00	20.00	51.00
1" air hose	10.00	30.00	65.00
1-1/2" air hose	20.00	60.00	150.00
Dump truck rental rate plus mileage			
3 CY	225.00	700.00	1,640.00
5 CY	290.00	840.00	2,670.00
Hammer rental			
Electric brute breaker	60.00	235.00	725.00
Gas breaker	65.00	255.00	760.00
Demolition hammer, electric	53.00	210.00	585.00
Roto hammer, 7/8", electric	55.00	220.00	630.00
Roto hammer, 1-1/2", electric	53.00	189.00	525.00
Stump grinder, 9 HP	115.00	450.00	1,290.00
Brush chipper, trailer-mounted 40 HP	200.00	550.00	1,500.00
Chain saw, 18", gasoline	59.00	230.00	663.00
Chop saw, 14", electric	39.00	155.00	449.00
Backhoe/loader, wheel-mounted, diesel or gasoline			
1/2 CY bucket capacity, 55 HP	185.00	650.00	1,800.00
1 CY bucket capacity, 65 HP	320.00	880.00	2,550.00
1-1/4 CY bucket capacity, 75 HP	275.00	715.00	2,100.00
1-1/2 CY bucket capacity, 100 HP	270.00	830.00	2,420.00
Wheel loader, front-end load and dump, diesel			
3/4 CY bucket, 4WD, articulated	215.00	625.00	1,825.00
1 CY bucket, 4WD, articulated	285.00	830.00	2,420.00
2 CY bucket, 4WD, articulated	355.00	1,060.00	3,230.00

	Craft@Hrs	Unit	Material	Labor	Total	Sell
Sitework Demolition						
Debris disposal and removal. Tippage charges for solid waste disposal at the dump vary from $30 to $120 per ton. For planning purposes, estimate waste disposal at $75 per ton, plus the hauling cost. Call the dump or trash disposal company for actual charges. Typical costs are shown below.						
Dumpster, 3 CY trash bin, emptied weekly	—	Mo	—	—	330.00	—
Dumpster, 40 CY solid waste bin (lumber, drywall, roofing)						
Hauling cost, per load	—	Ea	—	—	230.00	—
Add to per load charge, per ton	—	Ton	—	—	55.00	—
Low-boy, 14 CY solid waste container (asphalt, dirt, masonry, concrete)						
Hauling cost, per load (7 CY maximum load)	—	Ea	—	—	230.00	—
Add to per load charge, per ton	—	Ton	—	—	55.00	—
Recycler fees. Recycling construction waste materials can substantially reduce disposal costs. Recycling charges vary from $95 to $120 per load, depending on the type of material and the size of the load. Add the cost of hauling to a recycling facility.						
Greenwaste	—	Ton	—	—	30.00	—
Asphalt, per load (7 CY)	—	Ea	—	—	100.00	—
Concrete, masonry or rock, per load (7 CY)	—	Ea	—	—	100.00	—
Dirt, per load (7 CY)	—	Ea	—	—	100.00	—
Mixed loads, per load (7 CY)	—	Ea	—	—	100.00	—
Bush and tree removal. Includes cutting into manageable pieces with power hand tools and dumping debris in a trash bin on site. Add the cost of power equipment rental, if needed.						
Shrubs and bushes, including stump removal.						
4' high, per each	BL@.750	Ea	—	21.79	21.79	36.60
Tree removal. Costs will vary widely depending on the condition, size, location and accessibility of the tree. Includes cutting into manageable pieces with power hand tools and dumping debris in a trash bin or chipper on site. Does not include stump removal or grinding. Add the cost of power equipment rental, bucket truck, and other specialized equipment if needed. Large shade tree removal may require a crew of 5 or more. Use $2,000 as a minimum job charge for trees with a 36" diameter trunk and larger.						
8" to 12" diameter trunk	B8@2.50	Ea	—	87.30	87.30	148.00
13" to 18" diameter trunk	B8@3.50	Ea	—	122.00	122.00	207.00
19" to 24" diameter trunk	B8@5.50	Ea	—	192.00	192.00	326.00
25" to 36" diameter trunk	B8@7.00	Ea	—	244.00	244.00	415.00
Stump grinding, using a 9 HP wheel-mounted stump grinder						
6" to 10" diameter stump	B8@.800	Ea	—	27.90	27.90	47.40
11" to 14" diameter stump	B8@1.04	Ea	—	36.30	36.30	61.70
15" to 18" diameter stump	B8@1.30	Ea	—	45.40	45.40	77.20
19" to 24" diameter stump	B8@1.50	Ea	—	52.40	52.40	89.10
Fencing demolition. Remove chain link fence and cemented posts for disposal. These figures assume fencing is removed by cutting ties at posts and rails and rolling the fabric. These rolls can be heavy. A larger crew will be needed on larger jobs.						
To 6' high	BL@.100	LF	—	2.91	2.91	4.95
Remove chain link fence and cemented posts for salvage,						
To 6' high	BL@.120	LF	—	3.49	3.49	5.93
Remove board fence and cemented posts for disposal,						
To 6' high	BL@.100	LF	—	2.91	2.91	4.95

Building Demolition

Single-story room demolition. Includes breaking out a concrete or masonry foundation using pneumatic tools, breaking materials into manageable pieces with hand tools and handling debris to a trash bin on site. Includes the cost of erecting a temporary vapor barrier to seal the wall opening in the existing structure. Add the cost of wall patching on the remaining structure. Add the cost of equipment rental.

	Craft@Hrs	Unit	Material	Labor	Total	Sell
Demolish a wood-frame addition built on a concrete slab, including grade beams. Use these figures to estimate the cost of demolishing an attached garage.						
Per square foot of floor area demolished	BL@.100	SF	—	2.91	2.91	4.95
Demolish a wood-frame addition built on a conventional foundation, including wood deck, stairs to grade, and the concrete or concrete block foundation.						
Per square foot of floor area demolished	BL@.120	SF	—	3.49	3.49	5.93

Detached garage demolition. Includes breaking materials into manageable pieces with hand tools and handling debris to a trash bin on site. No slab or foundation demolition included. No salvage of materials assumed.

	Craft@Hrs	Unit	Material	Labor	Total	Sell
Frame garage with wood or aluminum siding						
One-car garage, to 10' wide by 22' deep	BL@16.5	Ea	—	479.00	479.00	814.00
Two-car garage, to 20' wide by 28' deep	BL@23.0	Ea	—	668.00	668.00	1,140.00
Frame garage with stucco siding						
One-car garage to 12' wide by 22' deep	BL@23.5	Ea	—	683.00	683.00	1,160.00
One-car to 16' wide by 22' deep	BL@28.5	Ea	—	828.00	828.00	1,410.00
Two-car garage to 22' wide by 28' deep	BL@42.5	Ea	—	1,230.00	1,230.00	2,090.00
Brick or block garage						
One-car garage to 12' wide by 22' deep	BL@23.0	Ea	—	668.00	668.00	1,140.00
One-car garage to 16' wide by 22' deep	BL@28.0	Ea	—	813.00	813.00	1,380.00
Two-car garage to 20' wide by 22' deep	BL@33.0	Ea	—	959.00	959.00	1,630.00

Gutting a building. Interior finish stripped back to the structural walls. Building structure to remain. No allowance for salvage value. These costs include loading and hauling up to 6 miles but no dump fees. Costs are per square foot of floor area based on 8' ceiling height. Add the cost of equipment rental. Figures in parentheses show the approximate "loose" volume of materials after demolition.

	Craft@Hrs	Unit	Material	Labor	Total	Sell
Residential building (125 SF per CY)	BL@.100	SF	—	2.91	2.91	4.95

Building Component Demolition. Itemized costs for demolition of building components when building is being remodeled, repaired or rehabilitated and not completely demolished. Costs include protecting adjacent areas and normal clean-up. Costs are to break out the items listed and pile debris on site or in a bin.

Brick wall demolition. Cost per square foot removed measured on one face. Removed using a pneumatic breaker. Add the cost of equipment rental. Figures in parentheses show the volume before and after demolition and weight of the materials after demolition.

	Craft@Hrs	Unit	Material	Labor	Total	Sell
4" thick walls						
(60 SF per CY and 36 lbs. per SF)	BL@.061	SF	—	1.77	1.77	3.01
8" thick walls						
(30 SF per CY and 72 lbs. per SF)	BL@.110	SF	—	3.20	3.20	5.44

	Craft@Hrs	Unit	Material	Labor	Total	Sell
Brick sidewalk demolition. Cost per square foot removed. Add the cost of Bobcat rental. Larger jobs will require a crew of 2 or more. Figures in parentheses show the volume before and after demolition and weight of materials after demolition.						
2-1/2" thick brick on sand base, no mortar bed, removed using hand tools						
(100 SF per CY and 28 lbs. per SF)	BL@.010	SF	—	.29	.29	.49
Brick pavers up to 4-1/2" thick with mortar bed, removed using a pneumatic breaker						
(55 SF per CY and 50 lbs. per SF)	BL@.050	SF	—	1.45	1.45	2.47
Concrete masonry wall demolition. Cost per square foot of wall removed measured on one side. Removed using a pneumatic breaker. Add the cost of equipment rental. Figures in parentheses show the volume before and after demolition and weight of materials after demolition.						
4" thick walls						
(60 SF per CY and 19 lbs. per SF)	BL@.066	SF	—	1.92	1.92	3.26
6" thick walls						
(40 SF per CY and 28 lbs. per SF)	BL@.075	SF	—	2.18	2.18	3.71
8" thick walls						
(30 SF per CY and 34 lbs. per SF)	BL@.098	SF	—	2.85	2.85	4.85
12" thick walls						
(20 SF per CY and 46 lbs. per SF)	BL@.140	SF	—	4.07	4.07	6.92
Reinforced or grouted walls add	—	%	—	—	50.0	—
Concrete foundation demolition (footings). Steel reinforced, removed using a pneumatic breaker. Add the cost of equipment rental. Figures in parentheses show the volume before and after demolition.						
Cost per CY (.75 CY per CY)	BL@3.96	CY	—	115.00	115.00	196.00
Cost per LF with width and depth as shown						
6" W x 12" D (35 LF per CY)	BL@.075	LF	—	2.18	2.18	3.71
8" W x 12" D (30 LF per CY)	BL@.098	LF	—	2.85	2.85	4.85
8" W x 16" D (20 LF per CY)	BL@.133	LF	—	3.86	3.86	6.56
8" W x 18" D (18 LF per CY)	BL@.147	LF	—	4.27	4.27	7.26
10" W x 12" D (21 LF per CY)	BL@.121	LF	—	3.52	3.52	5.98
10" W x 16" D (16 LF per CY)	BL@.165	LF	—	4.79	4.79	8.14
10" W x 18" D (14 LF per CY)	BL@.185	LF	—	5.37	5.37	9.13
12" W x 12" D (20 LF per CY)	BL@.147	LF	—	4.27	4.27	7.26
12" W x 16" D (13 LF per CY)	BL@.196	LF	—	5.69	5.69	9.67
12" W x 20" D (11 LF per CY)	BL@.245	LF	—	7.12	7.12	12.10
12" W x 24" D (9 LF per CY)	BL@.294	LF	—	8.54	8.54	14.50
Concrete sidewalk demolition. To 4" thick, cost per SF removed. Figures in parentheses show the volume before and after demolition.						
Non-reinforced, removed by hand with sledge						
(60 SF per CY)	BL@.050	SF	—	1.45	1.45	2.47
Reinforced with wire mesh. Mesh cut into manageable pieces, then removed using pneumatic breaker. Add the cost of equipment rental.						
(55 SF per CY)	BL@.060	SF	—	1.74	1.74	2.96

	Craft@Hrs	Unit	Material	Labor	Total	Sell
Porch and deck demolition. Includes breaking materials into manageable pieces with hand tools and handling debris to a trash bin on site.						
Demolish an enclosed wood-frame porch. Includes breaking out the concrete or masonry pier foundation. Per square foot of floor.						
One story	BL@.250	SF	—	7.26	7.26	12.30
Two story	BL@.225	SF	—	6.54	6.54	11.10
Three story	BL@.196	SF	—	5.69	5.69	9.67
Demolish a screened porch built on a concrete slab. Includes demolition of one concrete or wood frame step to grade. Per square foot of floor area.						
Porch and one step (no slab demolition)	BL@.100	SF	—	2.91	2.91	4.95
Demolish a wood deck with railing or kneewall. Includes demolition of the wood deck and up to 7 steps to grade.						
Deck to 150 SF	BL@18.5	Ea	—	537.00	537.00	913.00
Add per SF for deck over 150 SF	BL@.100	SF	—	2.91	2.91	4.95
Concrete slab demolition. Break up and remove concrete slab using a compressor and pneumatic hammer. Includes loosening concrete with a pick and handling debris to a trash bin on site. Costs for demolition of reinforced concrete slab include burning off rebars with an acetylene torch. Add the rental cost of a compressor, pneumatic hammer and a torch, as required. Figures in parentheses show the volume before and after demolition.						
4" non-reinforced concrete (60 SF per CY)	BL@.057	SF	—	1.66	1.66	2.82
4" reinforced concrete (60 SF per CY)	BL@.071	SF	—	2.06	2.06	3.50
6" non-reinforced concrete (45 SF per CY)	BL@.083	SF	—	2.41	2.41	4.10
6" reinforced concrete (45 SF per CY)	BL@.103	SF	—	2.99	2.99	5.08
8" non-reinforced concrete (30 SF per CY)	BL@.109	SF	—	3.17	3.17	5.39
8" reinforced concrete (30 SF per CY)	BL@.122	SF	—	3.54	3.54	6.02
Asphalt pavement demolition. Asphaltic concrete (bituminous) 3" thick, removed using a pneumatic spade. Add the cost of equipment rental and dump fees. Figures in parentheses show the volume before and after demolition and weight of debris.						
Break up and shovel on to a truck (4 SY per CY and 660 lbs. per SY)	BL@.140	SY	—	4.07	4.07	6.92
Concrete wall demolition. Steel reinforced, removed using a compressor and pneumatic breaker. Add the cost of equipment rental. Figures in parentheses show the volume before and after demolition.						
Cost per CY (1.33 CY per CY)	BL@4.76	CY	—	138.00	138.00	235.00
Cost per SF with thickness as shown						
3" wall thickness (80 SF per CY)	BL@.045	SF	—	1.31	1.31	2.23
4" wall thickness (55 SF per CY)	BL@.058	SF	—	1.68	1.68	2.86
5" wall thickness (45 SF per CY)	BL@.075	SF	—	2.18	2.18	3.71
6" wall thickness (40 SF per CY)	BL@.090	SF	—	2.61	2.61	4.44
8" wall thickness (30 SF per CY)	BL@.120	SF	—	3.49	3.49	5.93
10" wall thickness (25 SF per CY)	BL@.147	SF	—	4.27	4.27	7.26
12" wall thickness (20 SF per CY)	BL@.176	SF	—	5.11	5.11	8.69

	Craft@Hrs	Unit	Material	Labor	Total	Sell
Wood Framing Demolition. Typical costs for demolition of wood frame structural components using hand tools. Normal center-to-center spacing (12" thru 24" OC) is assumed. Includes breaking materials into manageable pieces with hand tools and handling debris to a trash bin on site. Costs include removal of Romex cable as necessary, protecting adjacent areas, and normal clean-up. Costs do not include temporary shoring to support structural elements above or adjacent to the work area. (Add for demolition of flooring, roofing, sheathing, etc.) Figures in parentheses give the approximate "loose" volume and weight of the materials after being demolished.						
Ceiling joist demolition. Per in-place SF of ceiling area removed.						
2" x 4" (720 SF per CY and 1.18 lbs. per SF)	BL@.009	SF	—	.26	.26	.44
2" x 6" (410 SF per CY and 1.76 lbs. per SF)	BL@.013	SF	—	.38	.38	.65
2" x 8" (290 SF per CY and 2.34 lbs. per SF)	BL@.018	SF	—	.52	.52	.88
2" x 10" (220 SF per CY and 2.94 lbs. per SF)	BL@.023	SF	—	.67	.67	1.14
2" x 12" (190 SF per CY and 3.52 lbs. per SF)	BL@.027	SF	—	.78	.78	1.33
Floor joist demolition. Per in-place SF of floor area removed.						
2" x 6" (290 SF per CY and 2.04 lbs. per SF)	BL@.016	SF	—	.46	.46	.78
2" x 8" (190 SF per CY and 2.72 lbs. per SF)	BL@.021	SF	—	.61	.61	1.04
2" x 10" (160 SF per CY and 3.42 lbs. per SF)	BL@.026	SF	—	.76	.76	1.29
2" x 12" (120 SF per CY and 4.10 lbs. per SF)	BL@.031	SF	—	.90	.90	1.53
Rafter demolition. Per in-place SF of actual roof area removed.						
2" x 4" (610 SF per CY and 1.42 lbs. per SF)	BL@.011	SF	—	.32	.32	.54
2" x 6" (360 SF per CY and 2.04 lbs. per SF)	BL@.016	SF	—	.46	.46	.78
2" x 8" (270 SF per CY and 2.68 lbs. per SF)	BL@.020	SF	—	.58	.58	.99
2" x 10" (210 SF per CY and 3.36 lbs. per SF)	BL@.026	SF	—	.76	.76	1.29
2" x 12" (180 SF per CY and 3.94 lbs. per SF)	BL@.030	SF	—	.87	.87	1.48
Stud wall demolition. Interior or exterior, includes allowance for plates and blocking, per in-place SF of wall area removed, measured on one face.						
2" x 3" (430 SF per CY and 1.92 lbs. per SF)	BL@.013	SF	—	.38	.38	.65
2" x 4" (310 SF per CY and 2.58 lbs. per SF)	BL@.017	SF	—	.49	.49	.83
2" x 6" (190 SF per CY and 3.74 lbs. per SF)	BL@.025	SF	—	.73	.73	1.24
Stud wall demolition, salvage condition. Remove wall cover (nailed hardboard or fiberboard) from partition wall and salvage the stud framing. Per linear foot of stud wall.						
2" x 4" x 8', 16" on center	BL@.156	LF	—	4.53	4.53	7.70
Roof cover demolition. Includes breaking materials into manageable pieces with hand tools and handling debris to a trash bin on site. Cost per Sq or "square". (1 square = 100 square feet.) Figures in parentheses show the approximate "loose" volume and weight of materials after being demolished.						
Built-up, 5 ply (2.5 Sq per CY and 250 lbs per Sq)	BL@1.50	Sq	—	43.60	43.60	74.10
Shingles, asphalt, single layer (2.5 Sq per CY and 240 lbs per Sq)	BL@1.33	Sq	—	38.60	38.60	65.60
Shingles, asphalt, double layer (1.2 Sq per CY and 480 lbs per Sq)	BL@2.00	Sq	—	58.10	58.10	98.80
Shingles, slate. Weight ranges from 600 lbs. to 1,200 lbs. per Sq (1 Sq per CY and 900 lbs per Sq)	BL@1.79	Sq	—	52.00	52.00	88.40

	Craft@Hrs	Unit	Material	Labor	Total	Sell
Shingles, wood (1.7 Sq per CY and 400 lbs per Sq)	BL@2.02	Sq	—	58.70	58.70	99.80
Clay or concrete tile (.70 Sq per CY)	BL@1.65	Sq	—	47.90	47.90	81.40
Remove gravel stop or flashing	BL@.070	LF	—	2.03	2.03	3.45

Interior and exterior finish demolition. Includes breaking materials into manageable pieces with hand tools and handling debris to a trash bin on site. Figures in parentheses show the approximate "loose" volume and weight of the materials after being demolished. Demolition for disposal. No salvage of materials included.

	Craft@Hrs	Unit	Material	Labor	Total	Sell
Plywood sheathing, up to 1" thick, cost per in-place SF (200 SF per CY and 2 lbs. per SF)	BL@.017	SF	—	.49	.49	.83
Wood board sheathing (250 SF per CY and 2 lbs. per SF)	BL@.030	SF	—	.87	.87	1.48
Metal siding, using hand tools (200 SF per CY and 2 lbs. per SF)	BL@.027	SF	—	.78	.78	1.33
Stucco on walls or soffits, cost per in-place SF, removed using hand tools (150 SF per CY and 8 lbs. per SF)	BL@.036	SF	—	1.05	1.05	1.79
Wallboard 1/2" thick, cost per in-place SF, removed using hand tools. Add 50% to labor costs for 1" wallboard.						
Gypsum, walls or ceilings (250 SF per CY and 2.3 lbs. per SF)	BL@.010	SF	—	.29	.29	.49
Including strip furring	BL@.015	SF	—	.44	.44	.75
Plywood or insulation board (200 SF per CY and 2 lbs. per SF)	BL@.018	SF	—	.52	.52	.88
Insulation, fiberglass batts or rolls, cost per square foot removed (500 SF per CY and .3 lb per SF)	BL@.003	SF	—	.09	.09	.15
Plaster on walls, cost per square foot removed by hand (150 SF per CY and 8 lbs. per SF)	BL@.015	SF	—	.44	.44	.75
Remove baseboard molding						
Single member base, no salvage	BL@.022	LF	—	.64	.64	1.09
Base and base shoe, no salvage	BL@.038	LF	—	1.10	1.10	1.87
Remove and salvage single member base	BL@.025	LF	—	.73	.73	1.24
Remove and salvage base and base shoe	BL@.044	LF	—	1.28	1.28	2.18

Ceiling demolition. Knock down with hand tools at heights to 9' and handle debris to a trash bin on site. Building structure to remain. No allowance for salvage value. Costs shown are per square foot of ceiling. Figures in parentheses show the approximate "loose" volume after demolition.

	Craft@Hrs	Unit	Material	Labor	Total	Sell
Plaster ceiling (typically 175 to 200 SF per CY)						
Including lath and furring	BL@.025	SF	—	.73	.73	1.24
Including suspended grid	BL@.020	SF	—	.58	.58	.99
Acoustic tile ceiling (typically 200 to 250 SF per CY)						
Including suspended grid	BL@.010	SF	—	.29	.29	.49
Including grid in salvage condition	BL@.019	SF	—	.55	.55	.94
Tile glued or stapled to ceiling	BL@.015	SF	—	.44	.44	.75
Tile on strip furring, including furring	BL@.025	SF	—	.73	.73	1.24
Drywall ceiling (typically 250 to 300 SF per CY)						
Nailed or attached with screws to joists	BL@.010	SF	—	.29	.29	.49

Floor cover demolition. Includes breaking materials into manageable pieces with hand tools and handling debris to a trash bin on site. Cost per square yard of floor removed. (1 square yard = 9 square feet.) Figures in parentheses show the approximate "loose" volume and weight of the materials after demolition.

	Craft@Hrs	Unit	Material	Labor	Total	Sell
Ceramic tile						
(25 SY per CY and 34 lbs. per SY)	BL@.263	SY	—	7.64	7.64	13.00
Hardwood, nailed						
(25 SY per CY and 18 lbs. per SY)	BL@.290	SY	—	8.42	8.42	14.30
Hardwood, glued						
(25 SY per CY and 18 lbs. per SY)	BL@.503	SY	—	14.60	14.60	24.80
Linoleum, sheet						
(30 SY per CY and 3 lbs. per SY)	BL@.056	SY	—	1.63	1.63	2.77
Resilient tile						
(30 SY per CY and 3 lbs. per SY)	BL@.300	SY	—	8.72	8.72	14.80
Terrazzo						
(25 SY per CY and 34 lbs. per SY)	BL@.286	SY	—	8.31	8.31	14.10
Carpet on tack strip						
(40 SY per CY and 5 lbs. per SY)						
BL@.028	SY	—	.81	.81	1.38	
Glue-down carpet with rubber backing						
(35 SY per CY and 1.7 lbs. per SY)	BL@.160	SY	—	4.65	4.65	7.91
Add for each LF of room perimeter	BL@.008	LF	—	.23	.23	.39
Carpet pad						
(35 SY per CY and 1.7 lbs. per SY)	BL@.014	SY	—	.41	.41	.70

Removal of interior items. Includes breaking materials into manageable pieces with hand tools and handling debris to a trash bin on site. Items removed in salvageable condition do not include an allowance for salvage value. Figures in parentheses give the approximate "loose" volume of the materials after demolition. Removed in non-salvageable condition (demolished) except as noted.

	Craft@Hrs	Unit	Material	Labor	Total	Sell
Windows, typical cost to remove window, frame and hardware, cost per SF of window						
Metal (36 SF per CY)	BL@.058	SF	—	1.68	1.68	2.86
Wood (36 SF per CY)	BL@.063	SF	—	1.83	1.83	3.11
Wood stairs, cost per in-place SF or LF						
Risers (25 SF per CY)	BL@.116	SF	—	3.37	3.37	5.73
Landings (50 SF per CY)	BL@.021	SF	—	.61	.61	1.04
Handrails (100 LF per CY)	BL@.044	LF	—	1.28	1.28	2.18
Posts (200 LF per CY)	BL@.075	LF	—	2.18	2.18	3.71
Hollow metal door and frame in a masonry wall. (2 doors per CY)						
Single door to 4' x 7'	BL@1.00	Ea	—	29.10	29.10	49.50
Two doors, per opening to 8' x 7'	BL@1.50	Ea	—	43.60	43.60	74.10
Wood door and frame in a wood frame wall (2 doors per CY)						
Single door to 4' x 7'	BL@.500	Ea	—	14.50	14.50	24.70
Two doors, per opening to 8' x 7'	BL@.750	Ea	—	21.80	21.80	37.10
Metal door and frame in masonry wall, salvage condition						
Hollow metal door to 4' x 7'	BL@2.00	Ea	—	58.10	58.10	98.80
Wood door to 4' x 7'	BL@1.00	Ea	—	29.10	29.10	49.50

Foundations and Slabs

3

Many older homes have foundations that have cracked or settled. Check the foundation wall for deterioration that could allow water to enter the basement. Check both foundation walls and piers for settling. Windows or door frames out of square or loosened interior wall finish suggest that the foundation has settled. The next chapter explains how to correct minor settling by jacking and re-leveling beams and floor joists. Individual piers can be replaced. But if the pier has stopped settling, jack the supported girder or joist and add a block to the top of the pier.

Most concrete foundation walls develop minor hairline cracks that have no effect on the structure. However, open cracks may indicate a failure of the foundation that's getting progressively worse. To find out if a crack is active or dormant, scratch a line at the end of the crack and wedge a nail tightly into the crack. If the crack grows beyond the scratch mark or if the nail can be removed easily several months later, the crack is probably active.

If a crack is dormant, it can be repaired by routing and sealing. Enlarge the crack with a concrete saw or by chipping with hand tools. The crack should be routed $1^1/4$" or more in width and about the same depth. Rinse the joint clean and let it dry. Then apply a joint sealer such as an epoxy-cement compound in accord with the manufacturer's instructions.

Active cracks require an elastic sealant. Again, follow the manufacturer's instructions. Good-quality concrete sealant will remain pliable for many years. The minimum routing depth and width for these sealants is $^3/4$". The elastic material will deform but maintain a tight seal as the crack moves. You could also apply a strip sealant over the crack. But these protrude from the surface and make a poor choice if the wall is visible from the building exterior.

Repair loose mortar by brushing thoroughly to remove dust and loose particles. Before applying new mortar, dampen the clean surface so that it won't absorb water during repair. You can buy premixed mortar with the consistency of putty. Apply mortar over the cavity as if you were filling a void with painter's caulk. For a good bond, force mortar into the crack. Then smooth the surface with a trowel. Cover with a vapor barrier for a few days to keep the mortar from drying too fast.

Damp or leaky basement walls are usually caused by clogged drain tile, clogged or broken downspouts, cracks in walls, or by water that puddles against the foundation. Look for downspouts that empty against the foundation wall or surface drainage channeling by the foundation. For a dry basement, keep water away from the foundation by proper grading.

Foundation and Slab Pre-Construction Checklist

- ❒ Is the site accessible to wheeled equipment?
- ❒ Is the site accessible to a concrete truck?
- ❒ Will you need to re-route plumbing, electrical, gas or communication lines?
- ❒ Will you need to remove and replace lawn sprinklers?
- ❒ Will you need to cut back or remove bushes or trees?
- ❒ Will you have to remove and re-install a fence or gate?
- ❒ Does a septic tank or drain field extend into the site of construction?
- ❒ Can excess soil be spread on the premises?
- ❒ Will dirt need to be imported or exported for site balance?
- ❒ Will heavy equipment damage the roots of trees?
- ❒ Is a permit needed to move heavy equipment across city sidewalks?
- ❒ Do city sidewalks have to be planked?
- ❒ Will swales or culverts pose problems for equipment?
- ❒ Will a septic system, catch basins or wells have to be relocated?
- ❒ Will barricades and warning lights be needed?

A high water table is a more serious problem. A basement will never be completely dry if the water table extends above the basement floor. Heavy foundation waterproofing or footing drains may help, but they're unlikely to do more than minimize the problem.

	Day	Week	Month
Equipment Rental for Foundations and Slabs.			
Buggies, push type, 7 CF	60.00	220.00	610.00
Buggies, walking type, 12 CF	75.00	260.00	750.00
Buggies, riding type, 14 CF	85.00	300.00	810.00
Buggies, riding type, 18 CF	115.00	350.00	725.00
Vibrator, electric, 3 HP, flexible shaft	40.00	130.00	335.00
Vibrator, gasoline, 6 HP, flexible shaft	64.00	188.00	562.00
Troweling machine, 36", 4 paddle	50.00	180.00	470.00
Cement mixer, 6 CF	55.00	200.00	590.00
Towable mixer (10+ CF)	70.00	155.00	640.00
Power trowel 36", gas, with standard handle	50.00	180.00	470.00
Bull float	15.00	45.00	110.00
Concrete saws, gas powered, excluding blade cost			
10 HP, push type	65.00	215.00	595.00
20 HP, self propelled	90.00	325.00	910.00
37 HP, self propelled	150.00	515.00	1,480.00
Concrete conveyor, belt type, portable, gas powered, all sizes	150.00	525.00	1,470.00
Vibrating screed, 16', single beam	65.00	220.00	640.00
Column clamps, to 48", per set	—	—	6.60
Skid steer loaders			
Bobcat 643, 1,000-lb. capacity	155.00	540.00	1,450.00
Bobcat 753, 1,350-lb. capacity	180.00	580.00	1,640.00
Bobcat 763, 1,750-lb. capacity	195.00	640.00	1,830.00
Bobcat 863, 1,900-lb. capacity	220.00	680.00	2,000.00
Skid steer attachments			
Auger	90.00	280.00	785.00
Hydraulic breaker	170.00	530.00	1,580.00
Backhoe	100.00	340.00	1,130.00
Sweeper	120.00	370.00	1,090.00
Grapple bucket	55.00	166.00	490.00
Trenchers, inclined chain boom type, pneumatic tired			
15 HP, 12" wide, 48" max. depth, walking	190.00	630.00	1,800.00
20 HP, 12" wide, 60" max. depth, riding	230.00	775.00	2,200.00
Wheel loaders, front-end load and dump, diesel			
3/4 CY bucket, 4WD, articulated	215.00	625.00	1,825.00
1 CY bucket, 4WD, articulated	285.00	830.00	2,420.00
2 CY bucket, 4WD, articulated	355.00	1,060.00	3,230.00
Vibro plate, 300 lb, 24" plate width, gas	82.00	265.00	775.00
Vibro plate, 600 lb, 32" plate width, gas	150.00	490.00	1,440.00
Rammer, 60 CPM, 200 lb, gas powered	70.00	220.00	610.00

	Craft@Hrs	Unit	Material	Labor	Total	Sell
Concrete pumping. Using a trailer-mounted gasoline pump.						
3/8" aggregate mix (pea gravel), using hose to 200'						
Cost per cubic yard (pumping only)	—	CY	—	—	13.67	—
Add for hose over 200', per LF	—	LF	—	—	1.75	—
3/4" aggregate mix, using hose to 200'						
Cost per cubic yard (pumping only)	—	CY	—	—	14.00	—
Add for hose over 200', per LF	—	LF	—	—	1.75	—

80lb bag = .6 cu ft.

Concrete Flatwork

Most concrete work on home improvement jobs will be slabs such as sidewalks, driveways and floors. All concrete flatwork requires fine grading, edge forms, concrete pouring, and finishing.

4" thick concrete slab. Typical cost for a 100-square-foot slab, based on costs in this chapter. Use the figures that follow to adjust for other conditions: grading, base, forms, concrete thickness, concrete mix characteristics and finishing

	Craft@Hrs	Unit	Material	Labor	Total	Sell
Fine grading and shaping, 100 SF	BL@.800	CSF	—	23.20	23.20	39.40
Aggregate 4" base (1.23 CY per CSF)	BL@.600	CSF	39.00	17.40	56.40	95.90
Lay out and set edge forms, 50 LF, 1 use	B2@2.50	50LF	28.50	84.20	112.70	192.00
.006" polyethylene vapor barrier	B5@.118	CSF	8.00	4.12	12.12	20.60
Place W2.9 x W2.9 x 6" x 6" mesh, 100 SF	RI@.400	CSF	62.00	13.90	75.90	129.00
Place ready-mix concrete, from chute 1.25 CY	B5@.540	CSF	152.00	18.80	170.80	290.00
Steel trowel finish	B6@1.00	CSF	—	31.50	31.50	53.60
Acrylic concrete cure and seal	B6@.300	CSF	18.00	9.44	27.44	46.60
Strip edge forms, 50 LF, no re-use	BL@.600	50LF	—	17.40	17.40	29.60
Total job cost for 4" thick 100 SF slab	—@6.85	CSF	307.50	333.00	640.50	1,090.00
Cost per SF for 4" thick 100 SF job	—@.600	SF	3.08	29.10	32.18	54.70

Excavation and Backfill by Hand. Using hand tools.

	Craft@Hrs	Unit	Material	Labor	Total	Sell
General excavation, using a pick and shovel (loosening and one throw)						
Light soil	BL@1.10	CY	—	32.00	32.00	54.40
Average soil	BL@1.70	CY	—	49.40	49.40	84.00
Heavy soil or loose rock	BL@2.25	CY	—	65.40	65.40	111.00
Backfilling (one shovel throw from stockpile)						
Sand	BL@.367	CY	—	10.70	10.70	18.20
Average soil	BL@.467	CY	—	13.60	13.60	23.10
Rock or clay	BL@.625	CY	—	18.20	18.20	30.90
Add for compaction, average soil or sand	BL@.400	CY	—	11.60	11.60	19.70
Fine grading	BL@.008	SF	—	.23	.23	.39
Footings, average soil, using a pick and shovel						
6" deep x 12" wide (1.85 CY per CLF)	BL@.034	LF	—	.99	.99	1.68
8" deep x 12" wide (2.47 CY per CLF)	BL@.050	LF	—	1.45	1.45	2.47
8" deep x 16" wide (3.29 CY per CLF)	BL@.055	LF	—	1.60	1.60	2.72
8" deep x 18" wide (3.70 CY per CLF)	BL@.060	LF	—	1.74	1.74	2.96
10" deep x 12" wide (3.09 CY per CLF)	BL@.050	LF	—	1.45	1.45	2.47
10" deep x 16" wide (4.12 CY per CLF)	BL@.067	LF	—	1.95	1.95	3.32
10" deep x 18" wide (4.63 CY per CLF)	BL@.075	LF	—	2.18	2.18	3.71
12" deep x 12" wide (3.70 CY per CLF)	BL@.060	LF	—	1.74	1.74	2.96
12" deep x 16" wide (4.94 CY per CLF)	BL@.081	LF	—	2.35	2.35	4.00
12" deep x 20" wide (6.17 CY per CLF)	BL@.100	LF	—	2.91	2.91	4.95
12" deep x 24" wide (7.41 CY per CLF)	BL@.125	LF	—	3.63	3.63	6.17
16" deep x 16" wide (6.59 CY per CLF)	BL@.110	LF	—	3.20	3.20	5.44
Loading trucks (shoveling, one throw)						
Average soil	BL@1.45	CY	—	42.10	42.10	71.60
Rock or clay	BL@2.68	CY	—	77.90	77.90	132.00
Pits to 5'. Pits over 5' deep require special consideration: shoring, liners and a way to lift the soil.						
Light soil	BL@1.34	CY	—	38.90	38.90	66.10
Average soil	BL@2.00	CY	—	58.10	58.10	98.80
Heavy soil	BL@2.75	CY	—	79.90	79.90	136.00
Shaping trench bottom for pipe						
To 10" pipe	BL@.024	LF	—	.70	.70	1.19
12" to 20" pipe	BL@.076	LF	—	2.21	2.21	3.76

	Craft@Hrs	Unit	Material	Labor	Total	Sell
Shaping embankment slopes						
Up to 1-in-4 slope	BL@.060	SY	—	1.74	1.74	2.96
Over 1-in-4 slope	BL@.075	SY	—	2.18	2.18	3.71
Add for top crown or toe	—	%	—	50.0	—	—
Add for swales	—	%	—	90.0	—	—
Spreading material piled on site						
Average soil	BL@.367	CY	—	10.70	10.70	18.20
Stone or clay	BL@.468	CY	—	13.60	13.60	23.10
Strip and pile top soil	BL@.024	SF	—	.70	.70	1.19
Tamping, hand tamp only	BL@.612	CY	—	17.80	17.80	30.30
Trenches to 5', soil piled beside trench						
Light soil	BL@1.13	CY	—	32.80	32.80	55.80
Average soil	BL@1.84	CY	—	53.50	53.50	91.00
Heavy soil or loose rock	BL@2.86	CY	—	83.10	83.10	141.00
Add for depth over 5' to 9'	BL@.250	CY	—	7.26	7.26	12.30

Trenching and backfill with power equipment. Using a 55 HP wheel loader with integral backhoe to excavate utility line trenches and continuous footings where soil is piled adjacent to the trench. Linear feet (LF) and cubic yards (CY) per hour shown are based on a 2-man crew. Reduce productivity by 10 to 25 percent when soil is loaded in trucks. Shoring, dewatering and unusual conditions will increase these costs. Add the cost of equipment rental.

	Craft@Hrs	Unit	Material	Labor	Total	Sell
12" wide bucket, for 12" wide trench. Depths 3' to 5'						
Light soil (60 LF per hour)	B8@.033	LF	—	1.15	1.15	1.96
Medium soil (55 LF per hour)	B8@.036	LF	—	1.26	1.26	2.14
Heavy or wet soil (35 LF per hour)	B8@.057	LF	—	1.99	1.99	3.38
18" wide bucket, for 18" wide trench. Depths 3' to 5'						
Light soil (55 LF per hour)	B8@.036	LF	—	1.26	1.26	2.14
Medium soil (50 LF per hour)	B8@.040	LF	—	1.40	1.40	2.38
Heavy or wet soil (30 LF per hour)	B8@.067	LF	—	2.34	2.34	3.98
24" wide bucket, for 24" wide trench. Depths 3' to 5'						
Light soil (50 LF per hour)	B8@.040	LF	—	1.40	1.40	2.38
Medium soil (45 LF per hour)	B8@.044	LF	—	1.54	1.54	2.62
Heavy or wet soil (25 LF per hour)	B8@.080	LF	—	2.79	2.79	4.74

Soil compaction. Add the cost of equipment rental.

	Craft@Hrs	Unit	Material	Labor	Total	Sell
Compaction of soil in trenches in 8" layers.						
Pneumatic tampers						
(40 CY per hour)	BL@.050	CY	—	1.45	1.45	2.47
Vibrating rammers, gasoline powered "Jumping Jack"						
(20 CY per hour)	BL@.100	CY	—	2.91	2.91	4.95

Calculating slab area. Patio slabs and driveways are often irregular in shape. For example, a driveway may be wider at the garage end and narrower at the street. The easy way to figure the area is to measure the width at both ends and then divide by two. Then multiply the result by the length of the driveway. For example, assume:

The length of the driveway is 30 feet
The driveway width at the apron end is 22 feet
The driveway width at the street tapers to 8 feet
Add width at the two ends: 20' + 8' = 28'
Then divide by two: 28 feet divided by 2 is 14 feet
Then multiply by the average width by the length: 14 feet times 30 feet is 420 square feet.

A cubic yard of concrete will cover the following area, assuming no waste:
At 2" thick, 1 cubic yard covers 162 square feet
At 3" thick, 1 cubic yard covers 108 square feet
At 4" thick, 1 cubic yard covers 81 square feet
At 5" thick, 1 cubic yard covers 64.8 square feet
At 6" thick, 1 cubic yard covers 54 square feet
At 7" thick, 1 cubic yard covers 46.2 square feet
At 8" thick, 1 cubic yard covers 40.5 square feet.

If the job requires 420 square feet of 4" concrete, divide 420 by 81 to determine that 5.18 cubic yards are needed. It's good estimating practice to allow 5 percent for waste and over-excavation (removing more soil than needed to maintain slab or footing depth). Add 5 percent to find the order quantity of 5.44 cubic yards.

Garage slabs. Slabs are poured monolithic and should be pitched from rear to the front opening so snow melt drains to the street. Garage slab floor drains are poor practice in colder climates, as melted snow freezes in the drain and tends to crack the slab. Keep the slab surface 4" above grade to provide good drainage. When grade beams are required by code, excavate the additional depth around the perimeter of the slab. No additional forming is needed. The beam is poured as an integral part of the garage floor. Wire mesh or rod reinforcement is good practice, even if not required by code. If the driveway includes a change in grade, be sure the incline, including any swale, won't be too steep for most cars. Be sure the grade at the street or alley matches the existing grade.

Consider the drainage pattern before quoting concrete work. Water draining around the perimeter of a slab can cause damage that's expensive to repair.

Concrete forms. Most concrete form lumber can be used more than once, either several times for concrete forming or once for forming and then on some other part of the job when forming is done. The figures that follow include options for multiple use of form lumber. These options assume that forms can be removed, partially disassembled and cleaned. Then 75 percent of the form material will be reused. Costs include assembly, oiling, setting, stripping and cleaning. Adjust material costs in this section to reflect your actual cost of lumber. Here's how: Divide your actual lumber cost per MBF by the assumed lumber cost. Then multiply the cost in the material column by this adjustment factor.

	Craft@Hrs	Unit	Material	Labor	Total	Sell

Driveway and walkway edge forms. Material costs include stakes, nails, form oil and 5 percent waste. Per LF of edge form. When forms are required on both sides of the concrete, include the length of each side plus the end widths.

	Craft@Hrs	Unit	Material	Labor	Total	Sell
Using 2" x 4" lumber, .7 BF per LF	—	MBF	579.00	—	579.00	—
Using nails, ties and form oil, per SF	—	LF	.22	—	.22	—
1 use	B2@.050	LF	.57	1.68	2.25	3.83
3 use	B2@.050	LF	.34	1.68	2.02	3.43
5 use	B2@.050	LF	.29	1.68	1.97	3.35
2" x 6" edge form						
Using 2" x 6" lumber, 1.05 BF per LF	—	MBF	594.00	—	594.00	—
1 use	B2@.050	LF	.78	1.68	2.46	4.18
3 use	B2@.050	LF	.44	1.68	2.12	3.60
5 use	B2@.050	LF	.37	1.68	2.05	3.49
2" x 8" edge form						
Using 2" x 8" lumber, 1.4 BF per LF	—	MBF	571.00	—	571.00	—
1 use	B2@.055	LF	1.01	1.85	2.86	4.86
3 use	B2@.055	LF	.54	1.85	2.39	4.06
5 use	B2@.055	LF	.44	1.85	2.29	3.89
2" x 10" edge form						
Using 2" x 10" lumber, 1.75 BF per LF	—	MBF	587.00	—	587.00	—
1 use	B2@.055	LF	1.19	1.85	3.04	5.17
3 use	B2@.055	LF	.62	1.85	2.47	4.20
5 use	B2@.055	LF	.50	1.85	2.35	4.00
2" x 12" edge form						
Using 2" x 12" lumber, 2.1 BF per LF	—	MBF	664.00	—	664.00	—
1 use	B2@.055	LF	1.55	1.85	3.40	5.78
3 use	B2@.055	LF	.79	1.85	2.64	4.49
5 use	B2@.055	LF	.62	1.85	2.47	4.20

Plywood forming. For foundation walls, building walls and retaining walls, using 3/4" plyform with 10 percent waste, and 2" bracing with 20 percent waste. All material costs include nails, ties, clamps, form oil and stripping. Costs shown are per square foot of contact area (SFCA). When forms are required on both sides of the concrete, include the contact surface area for each side.

	Craft@Hrs	Unit	Material	Labor	Total	Sell
Using 3/4" plyform, per MSF	—	MSF	1,900.00	—	1,900.00	—
Using 2" bracing, per MBF	—	MBF	586.00	—	586.00	—
Using nails, ties, clamps & form oil, per SF	—	SFCA	.22	—	.22	—
Walls up to 4' high (1.10 SF of plywood and .42 BF of bracing per SF of form area)						
1 use	B2@.051	SFCA	2.56	1.72	4.28	7.28
3 use	B2@.051	SFCA	1.27	1.72	2.99	5.08
5 use	B2@.051	SFCA	.99	1.72	2.71	4.61
Walls 4' to 8' high (1.10 SF of plywood and .60 BF of bracing per SF of form area)						
1 use	B2@.060	SFCA	2.66	2.02	4.68	7.96
3 use	B2@.060	SFCA	1.32	2.02	3.34	5.68
5 use	B2@.060	SFCA	1.03	2.02	3.05	5.19
Walls 8' to 12' high (1.10 SF of plywood and .90 BF of bracing per SF of form area)						
1 use	B2@.095	SFCA	2.84	3.20	6.04	10.30
3 use	B2@.095	SFCA	1.40	3.20	4.60	7.82
5 use	B2@.095	SFCA	1.08	3.20	4.28	7.28

	Craft@Hrs	Unit	Material	Labor	Total	Sell
Form stripping. Labor to remove forms and bracing, clean, and stack on job site.						
Board forms at wall footings, grade beams and column footings. Per SF of contact area						
1" thick lumber	BL@.010	SF	—	.29	.29	.49
2" thick lumber	BL@.012	SF	—	.35	.35	.60
Removing keyways, 2" x 4" and 2" x 6"	BL@.006	LF	—	.17	.17	.29
Slab edge forms. Per LF of edge form						
2" x 4" to 2" x 6"	BL@.012	LF	—	.35	.35	.60
2" x 8" to 2" x 12"	BL@.013	LF	—	.38	.38	.65
Walls, plywood forms. Per SF of contact area stripped						
To 4' high	BL@.017	SFCA	—	.49	.49	.83
Over 4' to 8' high	BL@.018	SFCA	—	.52	.52	.88
Over 8' to 12' high	BL@.020	SFCA	—	.58	.58	.99
Over 12' to 16' high	BL@.027	SFCA	—	.78	.78	1.33
Over 16' high	BL@.040	SFCA	—	1.16	1.16	1.97
Board forming and stripping. For wall footings, grade beams, column footings, site curbs and steps. Includes 5 percent waste. Per SF of contact area. When forms are required on both sides of the concrete, include the contact surface for each side.						
2" thick forms and bracing, using 2.85 BF of form lumber per SF. Includes nails, ties, and form oil						
Using 2" lumber, per MBF	—	MBF	586.00	—	586.00	—
Using nails, ties and form oil, per SF	—	SFCA	.22	—	.22	—
1 use	B2@.115	SFCA	1.95	3.87	5.82	9.89
3 use	B2@.115	SFCA	.95	3.87	4.82	8.19
5 use	B2@.115	SFCA	.75	3.87	4.62	7.85
Add for keyway beveled on two edges, one use. No stripping included						
2" x 4"	B2@.027	LF	.40	.91	1.31	2.23
2" x 6"	B2@.027	LF	.62	.91	1.53	2.60
Concrete Reinforcing Steel. Steel reinforcing bars (rebar), ASTM A615 Grade 60. Material costs are for deformed steel reinforcing rebars, including 10 percent lap allowance, cutting and bending. Add for epoxy or galvanized coating of rebars, chairs and splicing, if required. Costs per pound (Lb) and per linear foot (LF) including tie wire and tying.						
Reinforcing steel placed and tied in footings and grade beams						
1/4" diameter, #2 rebar	RI@.015	Lb	1.06	.52	1.58	2.69
1/4" diameter, #2 rebar (.17 lb per LF)	RI@.003	LF	.18	.10	.28	.48
3/8" diameter, #3 rebar	RI@.011	Lb	.67	.38	1.05	1.79
3/8" diameter, #3 rebar (.38 lb per LF)	RI@.004	LF	.26	.14	.40	.68
1/2" diameter, #4 rebar	RI@.010	Lb	.65	.35	1.00	1.70
1/2" diameter, #4 rebar (.67 lb per LF)	RI@.007	LF	.44	.24	.68	1.16
5/8" diameter, #5 rebar	RI@.009	Lb	.58	.31	.89	1.51
5/8" diameter, #5 rebar (1.04 lb per LF)	RI@.009	LF	.60	.31	.91	1.55
3/4" diameter, #6 rebar	RI@.008	Lb	.56	.28	.84	1.43
3/4" diameter, #6 rebar (1.50 lb per LF)	RI@.012	LF	.84	.42	1.26	2.14
7/8" diameter, #7 rebar	RI@.008	Lb	.67	.28	.95	1.62
7/8" diameter, #7 rebar (2.04 lb per LF)	RI@.016	LF	1.37	.56	1.93	3.28
1" diameter, #8 rebar	RI@.008	Lb	.58	.28	.86	1.46
1" diameter, #8 rebar (2.67 lb per LF)	RI@.021	LF	1.56	.73	2.29	3.89
Reinforcing steel placed and tied in structural slabs						
1/4" diameter, #2 rebar	RI@.014	Lb	1.06	.49	1.55	2.64
1/4" diameter, #2 rebar (.17 lb per LF)	RI@.002	LF	.18	.07	.25	.43
3/8" diameter, #3 rebar	RI@.010	Lb	.67	.35	1.02	1.73
3/8" diameter, #3 rebar (.38 lb per LF)	RI@.004	LF	.26	.14	.40	.68

	Craft@Hrs	Unit	Material	Labor	Total	Sell
1/2" diameter, #4 rebar	RI@.009	Lb	.65	.31	.96	1.63
1/2" diameter, #4 rebar (.67 lb per LF)	RI@.006	LF	.44	.21	.65	1.11
5/8" diameter, #5 rebar	RI@.008	Lb	.58	.28	.86	1.46
5/8" diameter, #5 rebar (1.04 lb per LF)	RI@.008	LF	.60	.28	.88	1.50
3/4" diameter, #6 rebar	RI@.007	Lb	.56	.24	.80	1.36
3/4" diameter, #6 rebar (1.50 lb per LF)	RI@.011	LF	.84	.38	1.22	2.07
7/8" diameter, #7 rebar	RI@.007	Lb	.67	.24	.91	1.55
7/8" diameter, #7 rebar (2.04 lb per LF)	RI@.014	LF	1.37	.49	1.86	3.16
1" diameter, #8 rebar	RI@.007	Lb	.58	.24	.82	1.39
1" diameter, #8 rebar (2.67 lb per LF)	RI@.019	LF	1.56	.66	2.22	3.77
Reinforcing steel placed and tied in walls						
1/4" diameter, #2 rebar	RI@.017	Lb	1.06	.59	1.65	2.81
1/4" diameter, #2 rebar (.17 lb per LF)	RI@.003	LF	.18	.10	.28	.48
3/8" diameter, #3 rebar	RI@.012	Lb	.67	.42	1.09	1.85
3/8" diameter, #3 rebar (.38 lb per LF)	RI@.005	LF	.26	.17	.43	.73
1/2" diameter, #4 rebar	RI@.011	Lb	.65	.38	1.03	1.75
1/2" diameter, #4 rebar (.67 lb per LF)	RI@.007	LF	.44	.24	.68	1.16
5/8" diameter, #5 rebar	RI@.010	Lb	.58	.35	.93	1.58
5/8" diameter, #5 rebar (1.04 lb per LF)	RI@.010	LF	.60	.35	.95	1.62
3/4" diameter, #6 rebar	RI@.009	Lb	.56	.31	.87	1.48
3/4" diameter, #6 rebar (1.50 lb per LF)	RI@.014	LF	.84	.49	1.33	2.26
7/8" diameter, #7 rebar	RI@.009	Lb	.67	.31	.98	1.67
7/8" diameter, #7 rebar (2.04 lb per LF)	RI@.018	LF	1.37	.62	1.99	3.38
1" diameter, #8 rebar	RI@.009	Lb	.58	.31	.89	1.51
1" diameter, #8 rebar (2.67 lb per LF)	RI@.024	LF	1.56	.83	2.39	4.06
Add for reinforcing steel coating. Coating applied prior to shipment to job site. Labor for placing coated rebar is the same as that shown for plain rebar above. Cost per pound of reinforcing steel.						
Epoxy coating						
Any size rebar, add	—	Lb	.30	—	.30	—
Galvanized coating						
Any #2 rebar, #3 rebar or #4 rebar, add	—	Lb	.40	—	.40	—
Any #5 rebar or #6 rebar, add	—	Lb	.38	—	.38	—
Any #7 rebar or #8 rebar, add	—	Lb	.36	—	.36	—
Welded wire mesh. Electric weld, including 15 percent waste and overlap.						
2" x 2" W.9 x W.9 (#12 x #12), slabs	RI@.004	SF	1.02	.14	1.16	1.97
2" x 2" W.9 x W.9 (#12 x #12), beams and columns	RI@.020	SF	.99	.69	1.68	2.86
4" x 4" W1.4 x W1.4 (#10 x #10), slabs	RI@.003	SF	.47	.10	.57	.97
4" x 4" W2.0 x W2.0 (#8 x #8), slabs	RI@.004	SF	.55	.14	.69	1.17
4" x 4" W2.9 x W2.9 (#6 x #6), slabs	RI@.005	SF	.62	.17	.79	1.34
4" x 4" W4.0 x W4.0 (#4 x #4), slabs	RI@.006	SF	.82	.21	1.03	1.75
6" x 6" W1.4 x W1.4 (#10 x #10), slabs	RI@.003	SF	.21	.10	.31	.53
6" x 6" W2.0 x W2.0 (#8 x #8), slabs	RI@.004	SF	.36	.14	.50	.85
6" x 6" W2.9 x W2.9 (#6 x #6), slabs	RI@.004	SF	.42	.14	.56	.95
6" x 6" W4.0 x W4.0 (#4 x #4), slabs	RI@.005	SF	.58	.17	.75	1.28
Add for lengthwise cut, LF of cut	RI@.002	LF	—	.07	.07	.12

	Craft@Hrs	Unit	Material	Labor	Total	Sell
Concrete expansion joints. Fiber or asphaltic-felt sided. Per linear foot.						
1/4" wall thickness						
3" deep	B1@.030	LF	.18	.98	1.16	1.97
3-1/2" deep	B1@.030	LF	.22	.98	1.20	2.04
4" deep	B1@.030	LF	.43	.98	1.41	2.40
6" deep	B1@.030	LF	.59	.98	1.57	2.67
3/8" wall thickness						
3" deep	B1@.030	LF	.23	.98	1.21	2.06
3-1/2" deep	B1@.030	LF	.30	.98	1.28	2.18
4" deep	B1@.030	LF	.57	.98	1.55	2.64
6" deep	B1@.030	LF	.79	.98	1.77	3.01
1/2" wall thickness						
3" deep	B1@.030	LF	.30	.98	1.28	2.18
3-1/2" deep	B1@.030	LF	.37	.98	1.35	2.30
4" deep	B1@.030	LF	.72	.98	1.70	2.89
6" deep	B1@.030	LF	.99	.98	1.97	3.35

Concrete. Ready-mix delivered by truck. Typical prices for most cities. Includes delivery up to 20 miles for 10 CY or more, 3" to 4" slump. Material cost only, no placing or pumping included. See forming and finishing costs on the following pages.

	Craft@Hrs	Unit	Material	Labor	Total	Sell
Footing and foundation concrete, 1-1/2" aggregate						
2,000 PSI, 4.8 sack mix	—	CY	95.50	—	95.50	—
2,500 PSI, 5.2 sack mix	—	CY	98.00	—	98.00	—
3,000 PSI, 5.7 sack mix	—	CY	99.70	—	99.70	—
3,500 PSI, 6.3 sack mix	—	CY	103.00	—	103.00	—
4,000 PSI, 6.9 sack mix	—	CY	106.00	—	106.00	—
Slab, sidewalk, and driveway concrete, 1" aggregate						
2,000 PSI, 5.0 sack mix	—	CY	97.10	—	97.10	—
2,500 PSI, 5.5 sack mix	—	CY	99.70	—	99.70	—
3,000 PSI, 6.0 sack mix	—	CY	101.00	—	101.00	—
3,500 PSI, 6.6 sack mix	—	CY	103.00	—	103.00	—
4,000 PSI, 7.1 sack mix	—	CY	105.00	—	105.00	—
Pea gravel pump mix / grout mix, 3/8" aggregate						
2,000 PSI, 6.0 sack mix	—	CY	105.00	—	105.00	—
2,500 PSI, 6.5 sack mix	—	CY	106.00	—	106.00	—
3,000 PSI, 7.2 sack mix	—	CY	110.00	—	110.00	—
3,500 PSI, 7.9 sack mix	—	CY	113.00	—	113.00	—
5,000 PSI, 8.5 sack mix	—	CY	120.00	—	120.00	—
Extra delivery costs for ready-mix concrete						
Add for delivery over 20 miles	—	Mile	8.00	—	8.00	—
Add for standby charge in excess of 5 minutes per CY delivered per minute of extra time	—	Ea	1.50	—	1.50	—
Add for less than 10 CY per load delivered	—	CY	25.00	—	25.00	—
Extra costs for non-standard aggregates						
Add for lightweight aggregate, typical	—	CY	44.50	—	44.50	—
Add for lightweight aggregate, pump mix	—	CY	46.50	—	46.50	—
Add for granite aggregate, typical	—	CY	6.98	—	6.98	—

	Craft@Hrs	Unit	Material	Labor	Total	Sell
Extra costs for non-standard mix additives						
High early strength 5 sack mix	—	CY	12.60	—	12.60	—
High early strength 6 sack mix	—	CY	16.70	—	16.70	—
Add for white cement (architectural)	—	CY	57.60	—	57.60	—
Add for 1% calcium chloride	—	CY	1.62	—	1.62	—
Add for chemical compensated shrinkage	—	CY	21.20	—	21.20	—

Add for colored concrete, typical prices. The ready-mix supplier charges for cleanup of the ready-mix truck used for delivery of colored concrete. The usual practice is to provide one truck per day for delivery of all colored concrete for a particular job. Add the cost below to the cost per cubic yard for the design mix required. Also add the cost for truck cleanup.

	Craft@Hrs	Unit	Material	Labor	Total	Sell
Adobe	—	CY	20.20	—	20.20	—
Black	—	CY	31.90	—	31.90	—
Blended red	—	CY	18.10	—	18.10	—
Brown	—	CY	23.40	—	23.40	—
Green	—	CY	36.10	—	36.10	—
Yellow	—	CY	24.50	—	24.50	—
Colored concrete, truck cleanup, per day	—	LS	78.80	—	78.80	—

Concrete for walls. Cast-in-place concrete walls for buildings or retaining walls. Material costs for concrete placed direct from chute are based on a 2,000 PSI, 5.0 sack mix, with 1" aggregate, including 5 percent waste. Pump mix cost includes an additional $10.00 per CY for the pump. Labor costs are for placing only. Add the cost of excavation, formwork, steel reinforcing, finishes and curing. Square foot costs are based on SF of wall measured on one face only. Costs don't include engineering, design or foundations.

	Craft@Hrs	Unit	Material	Labor	Total	Sell
Using concrete (before 5% waste allowance) at	—	CY	97.10	—	97.10	—
4" thick walls (1.23 CY per CSF)						
To 4' high, direct from chute	B1@.013	SF	1.25	.42	1.67	2.84
4' to 8' high, pumped	B3@.015	SF	1.44	.47	1.91	3.25
8' to 12' high, pumped	B3@.017	SF	1.44	.53	1.97	3.35
12' to 16' high, pumped	B3@.018	SF	1.44	.56	2.00	3.40
16' high, pumped	B3@.020	SF	1.44	.63	2.07	3.52
6" thick walls (1.85 CY per CSF)						
To 4' high, direct from chute	B1@.020	SF	1.88	.65	2.53	4.30
4' to 8' high, pumped	B3@.022	SF	2.16	.69	2.85	4.85
8' to 12' high, pumped	B3@.025	SF	2.16	.78	2.94	5.00
12' to 16' high, pumped	B3@.027	SF	2.16	.85	3.01	5.12
16' high, pumped	B3@.030	SF	2.16	.94	3.10	5.27
8" thick walls (2.47 CY per CSF)						
To 4' high, direct from chute	B1@.026	SF	2.52	.85	3.37	5.73
4' to 8' high, pumped	B3@.030	SF	2.89	.94	3.83	6.51
8' to 12' high, pumped	B3@.033	SF	2.89	1.04	3.93	6.68
12' to 16' high, pumped	B3@.036	SF	2.89	1.13	4.02	6.83
16' high, pumped	B3@.040	SF	2.89	1.25	4.14	7.04
10" thick walls (3.09 CY per CSF)						
To 4' high, direct from chute	B1@.032	SF	3.15	1.04	4.19	7.12
4' to 8' high, pumped	B3@.037	SF	3.62	1.16	4.78	8.13
8' to 12' high, pumped	B3@.041	SF	3.62	1.29	4.91	8.35
12' to 16' high, pumped	B3@.046	SF	3.62	1.44	5.06	8.60
16' high, pumped	B3@.050	SF	3.62	1.57	5.19	8.82
12" thick walls (3.70 CY per CSF)						
To 4' high, placed direct from chute	B1@.040	SF	3.77	1.30	5.07	8.62
4' to 8' high, pumped	B3@.045	SF	4.33	1.41	5.74	9.76
8' to 12' high, pumped	B3@.050	SF	4.33	1.57	5.90	10.00
12' to 16' high, pumped	B3@.055	SF	4.33	1.73	6.06	10.30
16' high, pumped	B3@.060	SF	4.33	1.88	6.21	10.60

Concrete Footings, Grade Beams and Stem Walls. Use the figures below for preliminary estimates. Concrete costs are based on a 2,000 PSI, 5.0 sack mix with 1" aggregate placed directly from the chute of a ready-mix truck. Figures in parentheses show the cubic yards of concrete per linear foot of foundation including 5 percent waste. Costs shown include concrete, 60 pounds of reinforcing per CY of concrete, and typical excavation using a 3/4 CY backhoe with excess backfill spread on site.

Concrete footings and grade beams. Cast directly against the earth, including excavation, steel reinforcing and concrete but no forming or finishing. These costs assume the top of the foundation will be at finished grade. For scheduling purposes estimate that a crew of 3 can lay out, excavate, place and tie the reinforcing steel and place 13 CY of concrete in an 8-hour day. Add the cost of equipment rental. Use $900.00 as a minimum job charge.

	Craft@Hrs	Unit	Material	Labor	Total	Sell
Using 2,000 PSI concrete, per CY	—	CY	97.10	—	97.10	—
Using reinforcing bars, per pound	—	Lb	.65	—	.65	—
Typical cost per CY	B4@1.80	CY	136.00	62.70	198.70	338.00
12" W x 6" D (0.019CY per LF)	B4@.034	LF	2.60	1.18	3.78	6.43
12" W x 8" D (0.025CY per LF)	B4@.045	LF	3.41	1.57	4.98	8.47
12" W x 10" D (0.032 CY per LF)	B4@.058	LF	4.36	2.02	6.38	10.80
12" W x 12" D (0.039 CY per LF)	B4@.070	LF	5.32	2.44	7.76	13.20
12" W x 18" D (0.058 CY per LF)	B4@.104	LF	7.91	3.62	11.53	19.60
16" W x 8" D (0.035 CY per LF)	B4@.063	LF	4.78	2.19	6.97	11.80
16" W x 10" D (0.043 CY per LF)	B4@.077	LF	5.87	2.68	8.55	14.50
16" W x 12" D (0.052 CY per LF)	B4@.094	LF	7.09	3.27	10.36	17.60
18" W x 8" D (0.037 CY per LF)	B4@.067	LF	5.05	2.33	7.38	12.50
18" W x 10" D (0.049 CY per LF)	B4@.088	LF	6.69	3.07	9.76	16.60
18" W x 24" D (0.117 CY per LF)	B4@.216	LF	16.00	7.53	23.53	40.00
20" W x 12" D (0.065 CY per LF)	B4@.117	LF	8.87	4.08	12.95	22.00
24" W x 12" D (0.078 CY per LF)	B4@.140	LF	10.60	4.88	15.48	26.30
24" W x 24" D (0.156 CY per LF)	B4@.281	LF	21.20	9.79	30.99	52.70
24" W x 30" D (0.195 CY per LF)	B4@.351	LF	26.60	12.20	38.80	66.00
24" W x 36" D (0.234 CY per LF)	B4@.421	LF	31.90	14.70	46.60	79.20
30" W x 36" D (0.292 CY per LF)	B4@.526	LF	39.80	18.30	58.10	98.80
30" W x 42" D (0.341 CY per LF)	B4@.614	LF	46.50	21.40	67.90	115.00
36" W x 48" D (0.467 CY per LF)	B4@.841	LF	63.60	29.30	92.90	158.00

To estimate the cost of footing or grade beam sizes not shown, multiply the width in inches by the depth in inches and divide the result by 3700. This is the CY of concrete per LF of footing including 5 percent waste. Multiply by the "Typical cost per CY" in the prior table to find the cost per LF.

Continuous concrete footing with foundation stem wall. These figures assume the foundation stem wall projects 24" above the finished grade and extends into the soil 18" to the top of the footing. Costs shown include typical excavation using a 3/4 CY backhoe with excess backfill spread on site, forming both sides of the foundation wall and the footing, based on three uses of the forms and 2 #4 rebar. Use $1,200.00 as a minimum cost for this type work. Add the cost of equipment rental.

	Craft@Hrs	Unit	Material	Labor	Total	Sell
Typical cost per CY	B5@7.16	CY	153.00	250.00	403.00	685.00
Concrete footing and stem wall for single-story structure.						
Typical single-story structure, footing 12" W x 8" D, wall 6" T x 42" D (.10 CY per LF)	B5@.716	LF	15.30	25.00	40.30	68.50
Concrete footing and stem wall for two-story structure.						
Typical two-story structure, footing 18" W x 10" D, wall 8" T x 42" D (.14 CY per LF)	B5@1.00	LF	21.30	34.90	56.20	95.50
Concrete footing and stem wall for three-story structure.						
Typical three-story structure, footing 24" W x 12" D, wall 10" T x 42" D (.19 CY per LF)	B5@1.36	LF	29.00	47.40	76.40	130.00

	Craft@Hrs	Unit	Material	Labor	Total	Sell
Concrete slabs, walks and driveways. Typical costs for reinforced concrete slabs-on-grade including fine grading, slab base, forms, vapor barrier, wire mesh, 3,000 PSI concrete, finishing and curing. For thickened edge slabs, add the area of the thickened edge. Use 500 square feet as the minimum job size.						
2" thick	B5@.067	SF	1.81	2.34	4.15	7.06
3" thick	B5@.068	SF	2.12	2.37	4.49	7.63
4" thick	B5@.069	SF	2.43	2.41	4.84	8.23
5" thick	B5@.070	SF	2.74	2.44	5.18	8.81
6" thick	B5@.071	SF	3.05	2.48	5.53	9.40
Slab Base. Aggregate base for slabs. No waste included. Labor costs are for spreading aggregate from piles only. Add for fine grading, using hand tools.						
Crushed stone base 1.4 tons equal one cubic yard						
Using crushed stone, per CY	—	CY	31.60	—	31.60	—
1" base (.309 CY per CSF)	BL@.001	SF	.10	.03	.13	.22
2" base (.617 CY per CSF)	BL@.003	SF	.20	.09	.29	.49
3" base (.926 CY per CSF)	BL@.004	SF	.29	.12	.41	.70
4" base (1.23 CY per CSF)	BL@.006	SF	.39	.17	.56	.95
5" base (1.54 CY per CSF)	BL@.007	SF	.49	.20	.69	1.17
6" base (1.85 CY per CSF)	BL@.008	SF	.59	.23	.82	1.39
Sand fill base 1.40 tons equal one cubic yard						
Using sand, per CY	—	CY	14.70	—	14.70	—
1" fill (.309 CY per CSF)	BL@.001	SF	.05	.03	.08	.14
2" fill (.617 CY per CSF)	BL@.002	SF	.09	.06	.15	.26
3" fill (.926 CY per CSF)	BL@.003	SF	.14	.09	.23	.39
4" fill (1.23 CY per CSF)	BL@.004	SF	.18	.12	.30	.51
5" fill (1.54 CY per CSF)	BL@.006	SF	.23	.17	.40	.68
6" fill (1.85 CY per CSF)	BL@.007	SF	.27	.20	.47	.80
Add for fine grading for slab on grade	BL@.008	SF	—	.23	.23	.39
Concrete finishing.						
Slab finishes						
Broom finish	B6@.012	SF	—	.38	.38	.65
Float finish	B6@.010	SF	—	.31	.31	.53
Trowel finishing						
Steel, machine work	B6@.015	SF	—	.47	.47	.80
Steel, hand work	B6@.018	SF	—	.57	.57	.97
Finish treads and risers						
No abrasives, no plastering, per LF of tread	B6@.042	LF	—	1.32	1.32	2.24
With abrasives, plastered, per LF of tread	B6@.063	LF	.60	1.98	2.58	4.39
Scoring concrete surface, hand work	B6@.005	LF	—	.16	.16	.27
Sweep, scrub and wash down	B6@.006	SF	.02	.19	.21	.36
Liquid curing and sealing compound, spray-on, acrylic concrete cure and seal,						
400 SF per gallon	B6@.003	SF	.18	.09	.27	.46
Exposed aggregate (washed, including finishing),						
no disposal of slurry	B6@.017	SF	.30	.53	.83	1.41
Non-metallic color and concrete hardener, troweled on, 2 applications, red, gray or black,						
60 pounds per 100 SF	B6@.021	SF	.44	.66	1.10	1.87
100 pounds per 100 SF	B6@.025	SF	.72	.79	1.51	2.57

	Craft@Hrs	Unit	Material	Labor	Total	Sell
Concrete wall finish.						
Cut back ties and patch	B6@.011	SF	.13	.35	.48	.82
Remove fins	B6@.008	LF	.05	.25	.30	.51
Grind smooth	B6@.021	SF	.10	.66	.76	1.29
Sack, burlap grout rub	B6@.013	SF	.06	.41	.47	.80
Wire brush, green	B6@.015	SF	.05	.47	.52	.88
Wash with acid and rinse	B6@.004	SF	.20	.13	.33	.56
Break fins, patch voids, carborundum rub	B6@.035	SF	.13	1.10	1.23	2.09
Break fins, patch voids, burlap grout rub	B6@.026	SF	.07	.82	.89	1.51
Specialty finishes						
Monolithic natural aggregate topping						
3/16"	B6@.020	SF	.26	.63	.89	1.51
1/2"	B6@.022	SF	.30	.69	.99	1.68
Integral colors						
Primary colors, based on 25 lb sack	—	Lb	5.10	—	5.10	—
Blends, based on 25 lb sack	—	Lb	2.55	—	2.55	—
Colored release agent	—	Lb	2.00	—	2.00	—
Dry shake colored hardener	—	Lb	.45	—	.45	—
Stamped finish (embossed concrete) of joints, if required						
Diamond, square, octagonal patterns	B6@.047	SF	—	1.48	1.48	2.52
Spanish paver pattern	B6@.053	SF	—	1.67	1.67	2.84
Add for grouting	B6@.023	SF	.30	.72	1.02	1.73

Concrete slab sawing, Subcontract. Typical rates quoted by concrete sawing subcontractors. Using a gasoline powered saw. Costs per linear foot for cured concrete assuming a level surface with good access and saw cut lines laid out and pre-marked by others. Costs include local travel time. Minimum cost will be $250. Electric powered slab sawing will be approximately 40 percent higher for the same depth.

Depth	Unit	Under 200'	200' to 1,000'	Over 1,000'
1" deep	LF	.70	.61	.52
1-1/2" deep	LF	1.03	.96	.80
2" deep	LF	1.38	1.20	1.03
2-1/2" deep	LF	1.73	1.55	1.29
3" deep	LF	2.07	1.80	1.55
3-1/2" deep	LF	2.41	2.16	1.80
4" deep	LF	2.74	2.41	2.07
5" deep	LF	3.44	3.02	2.58
6" deep	LF	4.14	3.62	3.09
7" deep	LF	5.00	4.29	3.71
8" deep	LF	5.67	4.98	4.29
9" deep	LF	6.46	5.67	4.91
10" deep	LF	7.32	6.46	5.60
11" deep	LF	8.17	7.22	6.27
12" deep	LF	9.05	8.00	6.97

	Craft@Hrs	Unit	Material	Labor	Total	Sell
Asphalt paving. Small jobs and repairs. Using hand tools, except as noted. Add the cost of equipment rental (pneumatic hammer, compressor, front end loader, roller), hauling and dump fees, as required.						
Break up and remove asphalt pavement up to 3" thick using a pneumatic hammer, loosen and load bituminous debris on truck with shovel						
Per square yard of pavement	B8@.147	SY	—	5.13	5.13	8.72
Trim area, break up pavement with a pneumatic hammer, load debris on truck with a front end loader, remove existing base to depth of 3"						
Per square yard of pavement	B8@.470	SY	—	16.40	16.40	27.90
Add for job setup	B8@.053	SY	—	1.85	1.85	3.15
Spread hot mix 3" thick with shovel, rake smooth and hand tamp						
Per square yard of pavement	BL@.266	SY	29.90	7.73	37.63	64.00
Add for job setup	BL@.026	SY	—	.76	.76	1.29
Shovel, rake and hand tamp 3" thick base material, sweep and tack coat, shovel, rake and hand tamp 3" bituminous pavement						
Per square yard of pavement	BL@.316	SY	37.30	9.18	46.48	79.00
Add for extra 1" of thickness	BL@.024	SY	9.99	.70	10.69	18.20
Add for job setup	BL@.039	SY	—	1.13	1.13	1.92
Shovel, rake and machine roll 3" thick base material, sweep and tack coat area, shovel, rake and machine roll 3" bituminous mix						
Per square yard of pavement	B8@.407	SY	37.30	14.20	51.50	87.60
Add for each 1" of thickness	B8@.022	SY	9.99	.77	10.76	18.30
Add for job setup	B8@.104	SY	—	3.63	3.63	6.17

Rough Carpentry

4

Before bidding a home improvement job, spend a few minutes checking the integrity of the framing. While wood framing doesn't just wear out (some wood-frame structures in Japan have been in daily use for over a thousand years), there are destructive organisms that attack wood. Framing that has serious decay, mold, or termite infestation may be deteriorated past the point of economical repair, and makes the home more a candidate for demolition than renovation. Financing for home repair may not be available from conventional lenders when a home has serious structural defects.

Fortunately, the organisms that attack wood can be controlled with wood preservative and by keeping wood dry. With a little care and attention, nearly any wood-frame structure will reach functional obsolescence long before the wood deteriorates.

Floor Beams

Figure 4-1 shows the principal components of a framed floor. If the house has a basement, at least one wood or steel floor beam will support joists under the first floor.

Wood posts that support floor beams should rest on pedestals. Posts embedded in a concrete slab or foundation tend to absorb moisture and eventually decay. See Figure 4-2. Examine the base of wood support posts for decay, even if not in contact with concrete. Steel posts are normally supported on metal plates. Any wood in contact with concrete should be pressure-treated with a preservative such as ACQ (Alkaline Copper Quat), a product made from recycled copper waste. Use lumber treated with 0.25 pounds of ACQ per cubic foot for applications not in direct contact with the ground. For lumber in direct contact with the ground, use 0.40 ACQ-treated lumber. Use 0.60 ACQ-treated lumber for wood foundation systems.

Decayed beams and sills can be replaced, but it's slow, difficult work, especially when done in a crawlspace. However difficult the repairs, they may be the only alternative to demolishing the home. Using temporary needle beams, most homes can be lifted enough to remove weight from the floor beam to make repairs. Insert a needle beam (6" x 6" or greater) through the foundation wall. Then lift the joists off the decayed beam with bottle jacks. All lifting in basements and crawlspaces should be done very slowly to minimize stress damage to the structure. A good rule is to lift no more than 1/4" per day.

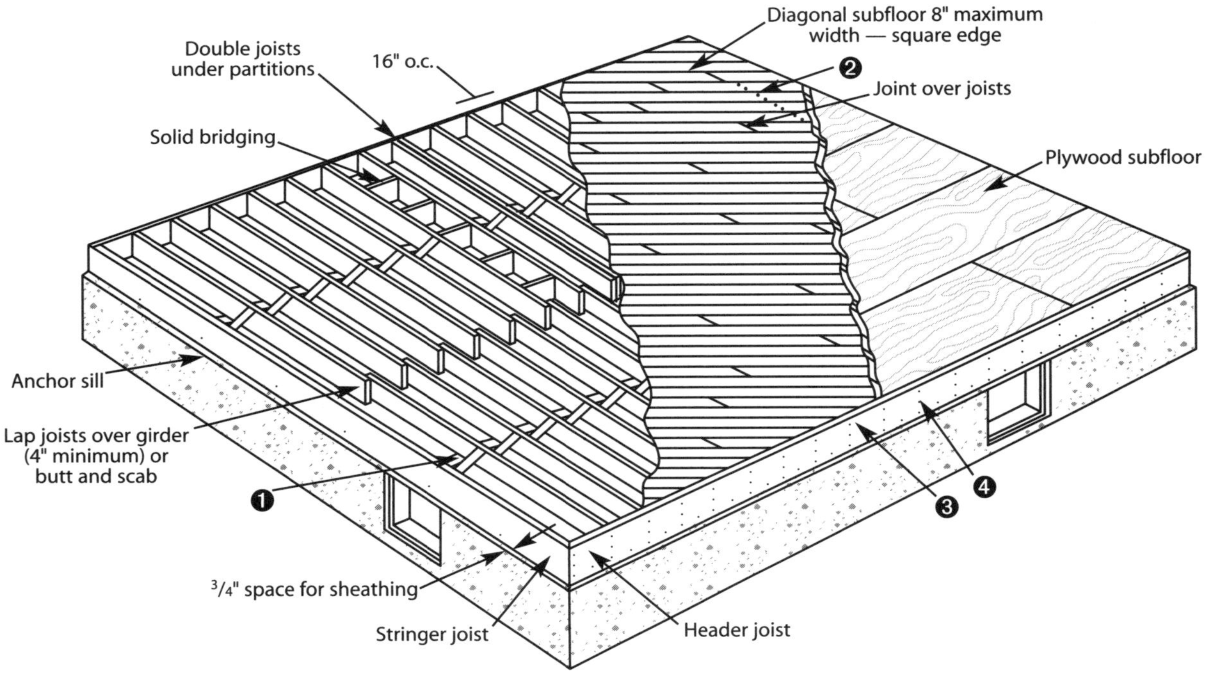

Figure 4-1

*Floor framing: **(1)** nailing bridging to joists; **(2)** nailing board subfloor to joists; **(3)** nailing header to joists; **(4)** toenailing header to sill*

When the weight is off the beam, cut out the decayed member and replace it. Then lower the needle beam slowly, again not more than 1/4" per day. This work is easier if you set additional support piers or posts and replace the beam in sections shorter than the original member.

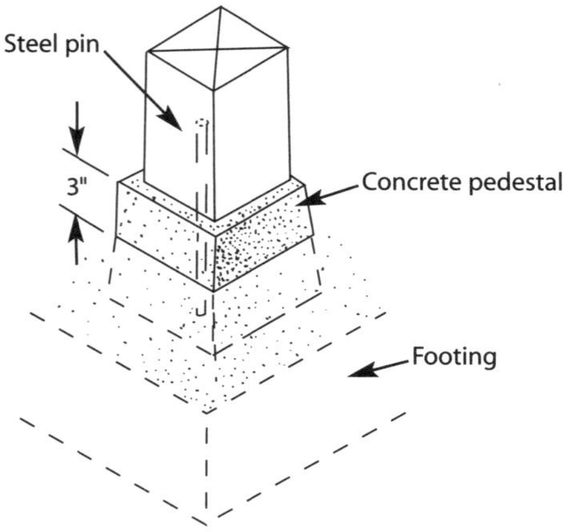

Figure 4-2

Basement post on pedestal above the floor

Sagging Floors

It's seldom practical to replace decayed floor joists, especially longer joists set with adhesive to the floor sheathing. It's much easier and just as effective to spike a new joist into position beside the old joist. First, remove the joist blocking to make room for the new joist. Then lift the section of decayed joists just enough to level the floor. A heavy 6" x 6" crossarm on top of a hydraulic jack will support a 4' to 6' width of a single-story house. Where you need additional support, add more jacks and crossarms. Raise the jacks slowly, over several days, and only just enough to allow you to insert a new pressure-treated joist beside the old joist. Excessive jacking will pull the building frame out of square. Insert the new joist, supporting it with joist hangers, a ledger, or by spiking it firmly to the old joist. Lower the jack. Then replace the joist bridges.

If only the joist end is decayed, brush wood preservative on the affected area. Then jack the existing joist up to the correct level and nail a short length of new material to the side of the existing joist. See Figure 4-3.

Some sag is normal in permanently-loaded wood beams. It's not a structural problem unless the walls above are obviously distorted. Deflection that doesn't exceed 3/8" in 10' is generally acceptable for structural purposes.

Sagging floor joists aren't serious unless the foundation system has settled unevenly, causing excessive deflection in the floor. Look for deflection caused by poor support of heavy partition walls or by joist sections cut away by a plumbing, HVAC or electrical crew. If the floor appears to be reasonably level, joist sag is probably acceptable. Otherwise, jack the floor back to a level position and double the joists or add a floor beam supported at each end by concrete piers.

You can reinforce a sagging floor beam with jack posts. See Figures 4-4 and 4-5. Jack posts used to stiffen a springy floor can set directly on concrete. Where a jack post has to carry a heavy load, install a steel plate to distribute the load over a larger area of floor slab. Jack posts aren't made for heavy jacking. If heavy jacking is required, use cribbing and an automotive bottle jack to lift the load. Then slip the jack post into place.

If the floor bounces under foot traffic, check to be sure joist bridges are installed between joists every eight feet. If you see adequate bridging, consider adding extra joists alongside every second joist in the affected area. Be sure to place new joists with the crown up. It's more work, but you could also add a supplemental beam at mid-span under the joists. Support the beam on both ends by installing concrete piers or posts. Sagging or bouncy floors may also indicate one or more broken floor joists. Inspect the area around the furnace — furnace heat may have caused excessive drying and cracking of floor joists.

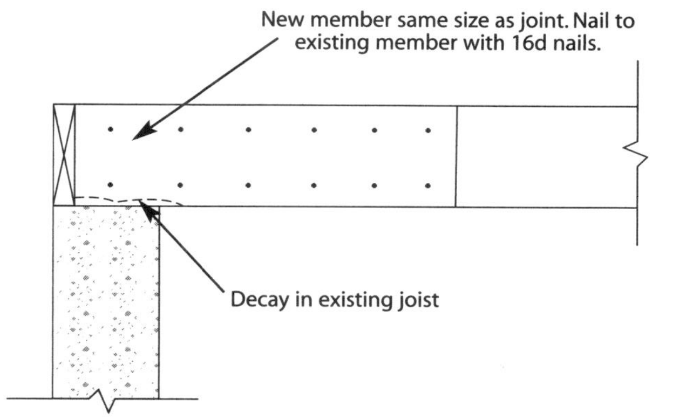

Figure 4-3

Repair of joist with decay in end contacting the foundation

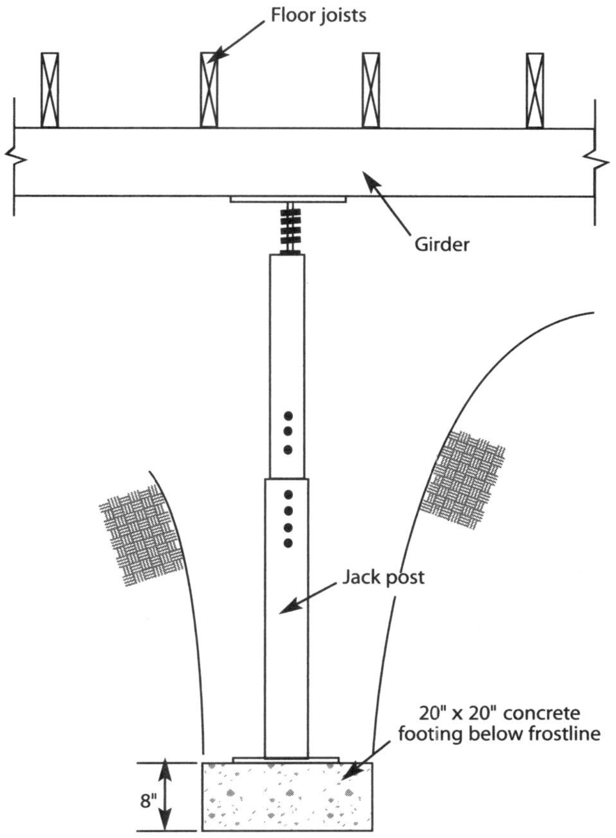

Figure 4-4

Jack post supporting a sagging girder in a crawl space house

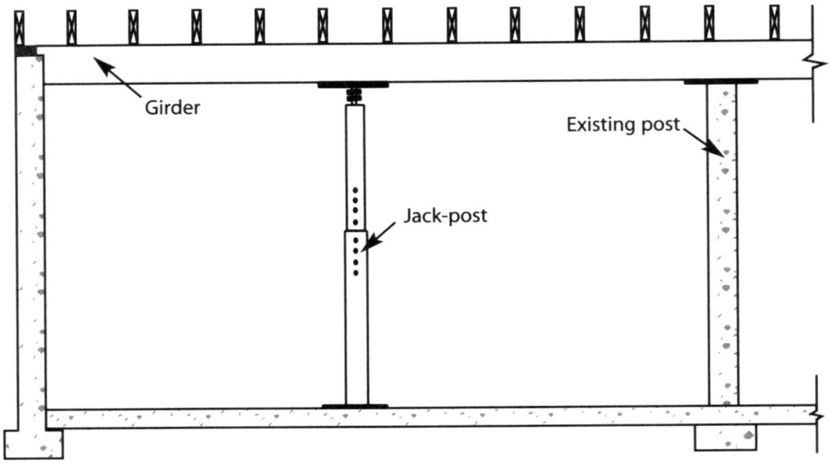

Figure 4-5

Jack post used to level a sagging girder in a basement

Squeaky Wood Floors

Floor squeaks are usually caused by two edges rubbing together, such as the tongue of one flooring strip rubbing against the groove in the adjacent strip. Applying a small amount of mineral oil to the joint may solve the problem.

A sagging floor joist can pull away from the subfloor, leaving a small gap between the two. If you suspect this is the cause of squeaks, open a joint and squeeze construction mastic into the cavity. An alternate remedy is to drive small wedges into the space between the joists and subfloor. See Figure 4-6. Drive wedges only far enough for a snug fit. This repair only works for a small area. Too many wedges will make the finish floor look bumpy.

Floor joists that are too small for the span will squeak as the floor deflects under load. The best solution is to add a girder to shorten the joist span.

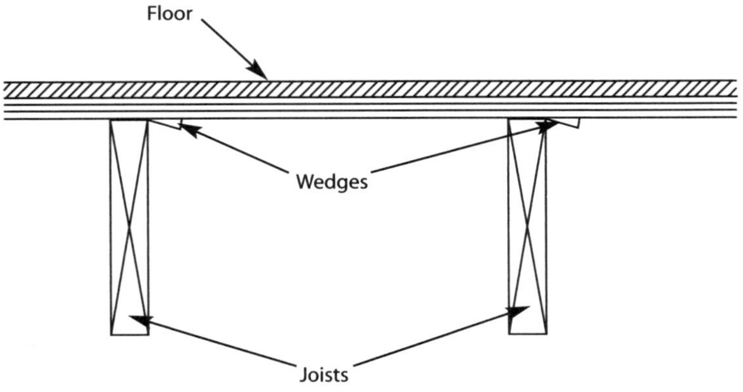

Figure 4-6

Wedges driven between joists and subfloor to stop squeaks

Strip flooring installed parallel to the joists usually deflects more under load than flooring installed perpendicular. If this is the cause of squeaks, solid blocking nailed between joists and fitted snugly against the subfloor (Figure 4-7) will limit deflection and help silence the floor.

Another common cause of squeaking is insufficient nails. If you decide to face nail the floor where it squeaks, countersink the nails and then fill each nail hole with wood putty. A better choice is to drill a small pilot hole through the tongue edge of a floor strip. Then nail through the pilot hole into a joist,

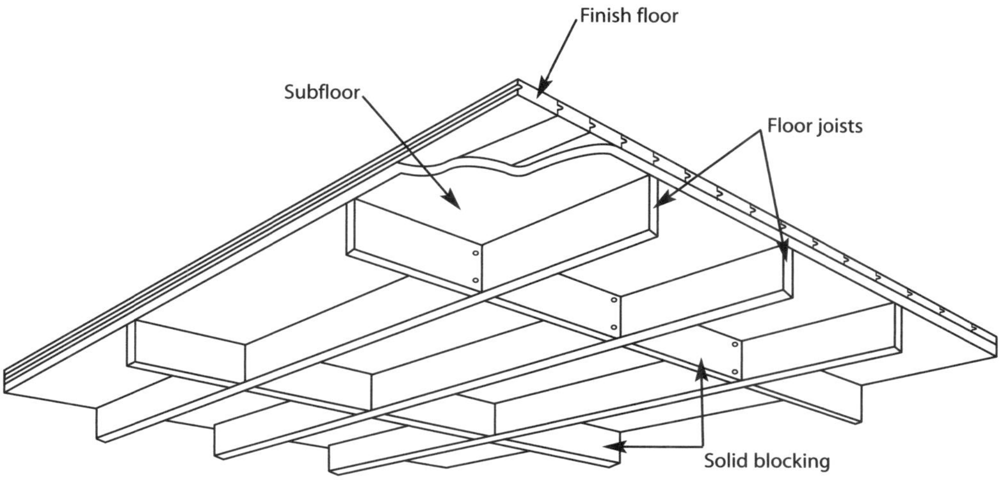

Figure 4-7

Solid blocking between floor joists where finish floor is laid parallel to joists

if possible. Even better, work from under the floor using screws driven through the subfloor into the finish floor. This method will also bring warped flooring flush with the subfloor.

Wall Framing

Stud walls made of 2" x 4" lumber set 16" on center (Figure 4-8) and stiffened by fireblocks seldom need to be repaired. The studs are stiff enough to hold the roof up for many years. But individual studs may have to be replaced if they've been weakened, either by cuts made for plumbing or electrical lines, or by insect damage.

Attic conversions or adding a second story to a home generally require that the studs be doubled — another stud added between the original studs set 16" on center. Just installing ceramic tile in a second-story bathroom or adding a bathtub on a second floor can increase the load carried by studs below by several thousand pounds. When installing new ceramic tile, you should also add solid blocking to the floor joists directly under the bathroom. A ceramic tile floor can't tolerate any movement. Any flexing of the floor will cause tile popping or cracking. Consider doubling studs in the first floor support wall when adding fixtures to the room above.

Occasionally you'll see a remodeled home where a partition wall, supported only by floor joists, is carrying a load from the second story. This is common when an attic has been converted to a storage room or an extra bedroom. The additional weight will tend to deform the main floor downward. Think of partition walls as curtains that simply divide up space. They aren't intended to support roof loads or any weight from the floor above. Doubling the support studs won't help if those

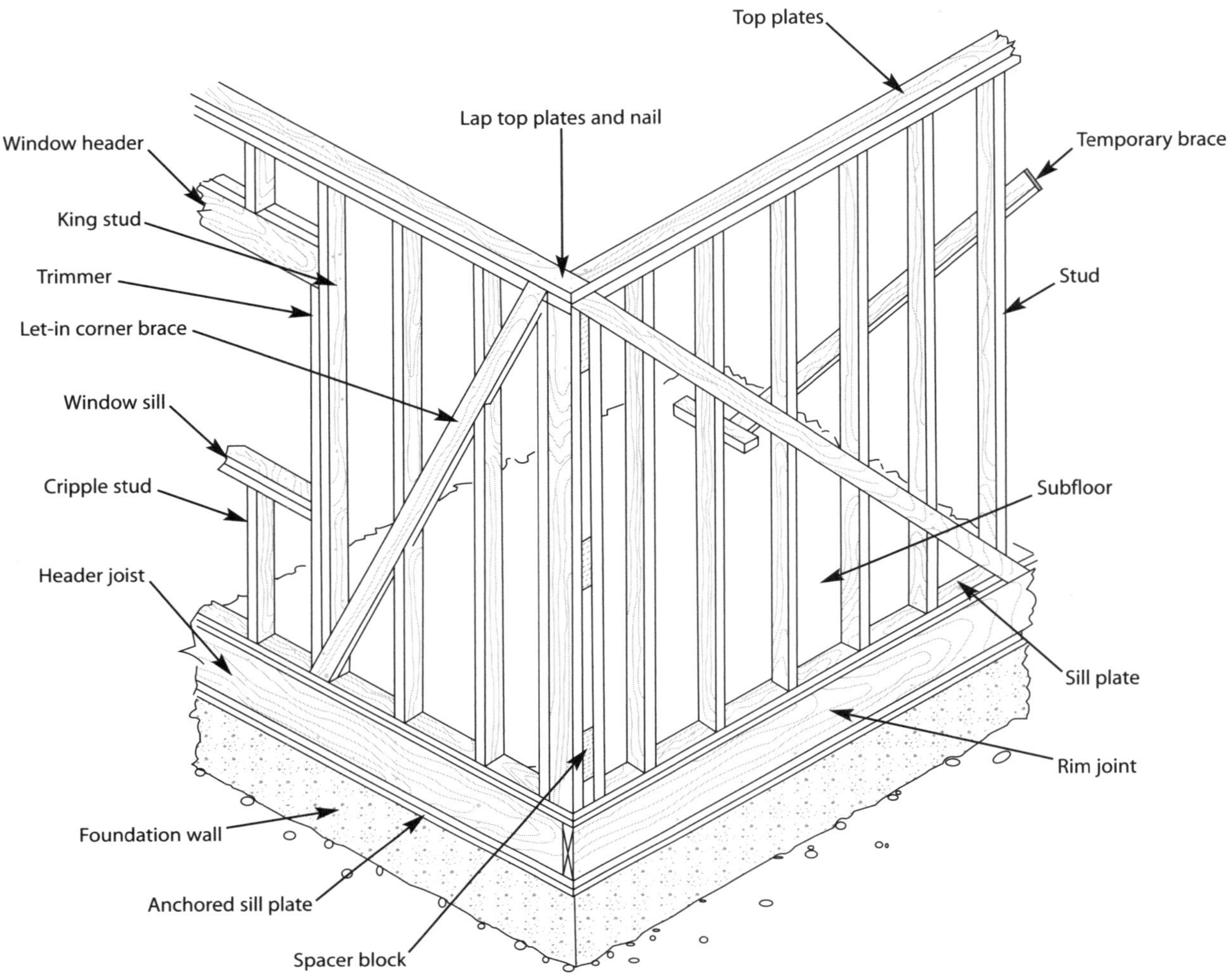

Figure 4-8
Wall framing used with platform construction

studs rest on a plate supported only by floor joists. You have to convert the partition wall to a load-bearing wall by adding a support beam and posts in the crawlspace or basement below.

Defects in the Roof

Roof leaks are usually caused by a failure of the roof cover, and can lead to rot damage to both the sheathing and the rafters. It would be poor practice to replace roof cover without making repairs to a sagging or decayed roof deck. Unlike problems with floor joists and floor beams, nearly all roof framing problems can be repaired economically. See Figure 4-9. Examine the roof for sagging at the ridge,

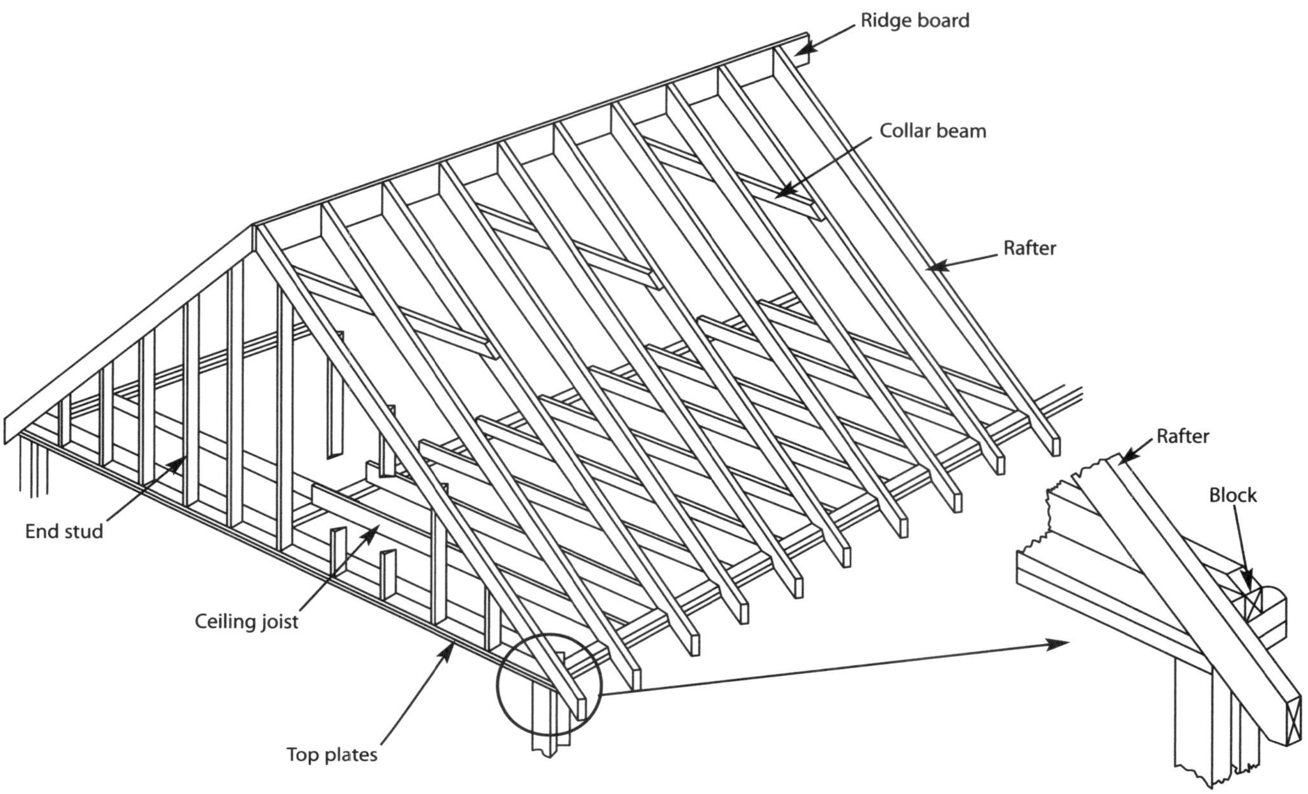

Figure 4-9
Ceiling and roof framing: Overall gable roof framing

rafters and in the sheathing. A ridge board can sag due to improper support, inadequate ties at the plate level, or even from sagging rafters. Rafters sag if the lumber wasn't well seasoned or if the lumber didn't have enough strength for the span. Sheathing sag is the result of rafter spacing that's too wide or sheathing material that's too thin. Occasionally, plywood sheathing will de-laminate and lose load-carrying capacity.

You can level a sagging ridge board or roof truss by jacking up points between supports and then installing braces to hold the member in place. When jacking any roof component, be sure to set the jack where the load will be transferred directly down through the structure to the foundation. Otherwise, pressure from the jack will deform the ceiling joists downward, damaging the ceiling below. If there's no bearing wall available, lay a beam in position where it can transfer loads to remote bearing points. When a ridge or truss chord is jacked into level position, cut a 2 x 4 just long enough to fit between the ceiling joist and ridge or chord. Then nail the prop at both ends. See Figure 4-10. For a short ridge board, one prop may be enough. Add more props as needed. If a rafter is sagging, jack the rafter into alignment. Then nail a new rafter into position at the side of the sagging rafter. That won't remove the sag from the old rafter, but it will transfer the load to a new, straight rafter.

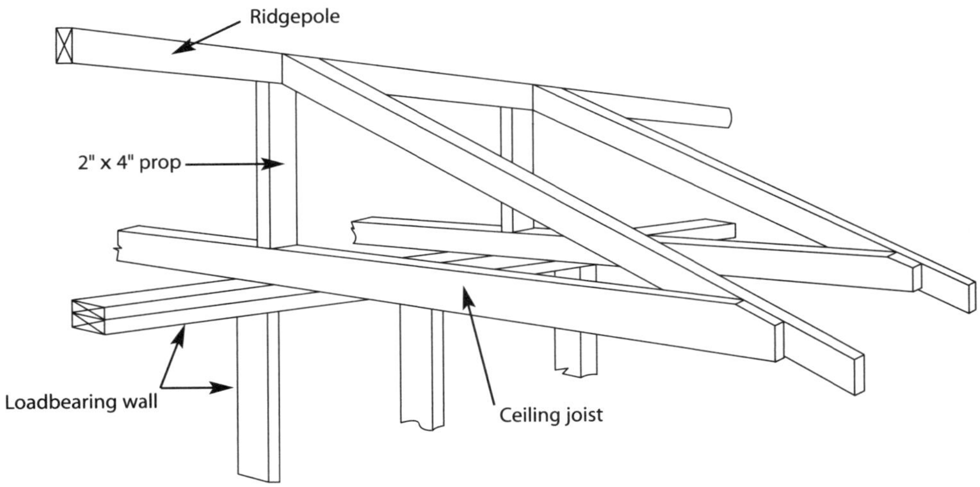

Figure 4-10

Prop to hold sagging ridgepole in level position

Sheathing

A wavy roof surface is evidence that the sheathing has sagged. The only practical repair is to remove and replace the sheathing. However, you may be able to nail new sheathing right over the old. This saves both the labor of removing the old sheathing and a big cleanup job. Of course, you must remove old sheathing that shows signs of decay or mold.

New sheathing that's nailed over existing sheathing must be secured with longer nails than normally used on roof sheathing, because the nails still need to penetrate the framing 1¼" to 1½". You should nail the sheathing edges at 6" spacing and nail to the intermediate framing members at 12" spacing. Install sheathing with the long edge at right angles to the rafters. For a built-up roof, join panel edges with plywood clips where edges don't meet over a rafter. You can use ⅜" thick sheathing for 16" rafter spacing, which is the minimum thickness required. However, ½" sheathing plywood is better, and is required for heavy roofing materials such as clay or concrete tile. You'll find that Oriented Strand Board (OSB) sheathing costs less and works just as well as plywood sheathing.

Strip sheathing helps promote air circulation, which is an important consideration in any climate. Slate and tile roofs are commonly installed on furring strips rather than plywood sheathing. However, strip sheathing is not appropriate for asphalt shingle applications.

When replacing the roof cover, roof sheathing or roof supports, check the roof flashing for rust and physical damage. Inadequate or damaged flashing could result in leaks. Replace any flashing that shows signs of age. Then renew the caulking around chimneys and parapet walls.

Adding a Roof Overhang

Many older homes have little or no roof overhang. Without an adequate overhang, rain and melting snow trickle down the wall finish, creating moisture problems in siding and trim, and at window and door openings, as well as increasing the moisture content at the foundation.

Extending the roof overhang will both improve the appearance of the house and reduce the maintenance required on siding and exterior trim. If you're adding new sheathing, consider extending the sheathing an inch or two beyond the edge of the existing roof. This is both an inexpensive and comparatively easy improvement.

	Craft@Hrs	Unit	Material	Labor	Total	Sell

Framing Repairs

Pricing for floor framing repairs. Place temporary 6" x 6" x 10' needle beam under joists for floor jacking, using 4" x 4" pre-cut cribbing for end supports. Includes removing the needle beam when lifting is complete. All jacking materials to be salvaged. No concrete work included.

	Craft@Hrs	Unit	Material	Labor	Total	Sell
Beam set in a basement or first floor	B1@1.72	Ea	—	56.00	56.00	95.20
Beam set in a crawl space	B1@3.44	Ea	—	112.00	112.00	190.00

Remove a floor beam. Includes jacking and cutting out the old beam. Add the cost of repairs to sheathing, siding and interior finish. Assumes all jacking materials are salvaged. Add the cost of the beam from the section that follows. Add needle beam placement costs from the lines above. Lifting and lowering should be done slowly. A good rule is to lift no more than 1/4" per day.

	Craft@Hrs	Unit	Material	Labor	Total	Sell
Per 8' section, basement or first floor	B1@6.87	Ea	—	223.00	223.00	379.00
Per 16' section, basement or first floor	B1@9.34	Ea	—	304.00	304.00	517.00
Per 8' section, work in a crawl space	B1@10.3	Ea	—	335.00	335.00	570.00
Per 16' section, work in a crawl space	B1@14.0	Ea	—	455.00	455.00	774.00
Needle beam, purchase						
6" x 6" x 10', green No. 1 Douglas fir	—	Ea	35.90	—	35.90	—
Cribbing, 4" x 4" x 8' No. 2 fir						
Per crib, purchase	—	Ea	7.36	—	7.36	—
Bottle (hydraulic) jack. Two-piece handle with internal overload valve, 9" to 18" lift						
6 ton capacity, purchase	—	Ea	28.50	—	28.50	—
12 ton capacity, purchase	—	Ea	46.40	—	46.40	—
20 ton capacity, purchase	—	Ea	65.00	—	65.00	—

Beams and girders. Std & Btr. First floor work. See beam removal costs in the previous section. Crawl space work assumes a minimum 24" high reasonably accessible area. Work done in crawl spaces less than 24" high may cost up to 100% more. Figures in parentheses show board feet per linear foot of beam or girder, including 7% waste.

	Craft@Hrs	Unit	Material	Labor	Total	Sell
4" x 6", per MBF	B1@16.0	MBF	708.00	520.00	1,228.00	2,090.00
4" x 8", 10", 12", per MBF	B1@16.0	MBF	882.00	520.00	1,402.00	2,380.00
6" x 6", per MBF	B1@16.0	MBF	1,220.00	520.00	1,740.00	2,960.00
6" x 8", 10", 12", 8" x 8", per MBF	B1@16.0	MBF	1,220.00	520.00	1,740.00	2,960.00
4" x 6" (2.15 BF per LF)	B1@.034	LF	1.52	1.11	2.63	4.47
4" x 8" (2.85 BF per LF)	B1@.045	LF	2.52	1.46	3.98	6.77
4" x 10" (3.58 BF per LF)	B1@.057	LF	3.16	1.85	5.01	8.52
4" x 12" (4.28 BF per LF)	B1@.067	LF	3.78	2.18	5.96	10.10
6" x 6" (3.21 BF per LF)	B1@.051	LF	3.92	1.66	5.58	9.49
6" x 8" (4.28 BF per LF)	B1@.067	LF	5.22	2.18	7.40	12.60
6" x 10" (5.35 BF per LF)	B1@.083	LF	6.53	2.70	9.23	15.70
6" x 12" (6.42 BF per LF)	B1@.098	LF	7.84	3.19	11.03	18.80
8" x 8" (5.71 BF per LF)	B1@.088	LF	6.97	2.86	9.83	16.70
Add if replacing a beam, basement or first floor	B1@.047	LF	—	1.53	1.53	2.60
Add if replacing a beam in a crawl space	B1@.095	LF	—	3.09	3.09	5.25

Shim up a floor beam. Jack a floor beam into position with a crossarm mounted on a hydraulic jack. Insert a pressure-treated shim above the pier or column and remove the jack. Per jacking point. Work done on sloping or unstable ground will cost more. Assumes all jacking materials are salvaged and includes cost of removing the jack. Lifting and lowering should not exceed 1/4" per day.

	Craft@Hrs	Unit	Material	Labor	Total	Sell
Jacking done in a basement	B1@.700	Ea	2.00	22.80	24.80	42.20
Jacking done in a crawl space	B1@1.25	Ea	2.00	40.70	42.70	72.60

	Craft@Hrs	Unit	Material	Labor	Total	Sell
Adjustable steel jack posts. Temporary or permanent support for beams during home improvement work. Two height-adjustment pins seated in separate pockets. Top and bottom plates fit steel or wood beams. 8" turning bar. 5" screw height adjustment. 14,300 to 15,500 pound capacity. Set on an existing slab or pier.						
6' 6" to 6' 10", 15,500 lb capacity	B1@.250	Ea	69.40	8.13	77.53	132.00
6' 9" to 7', 15,500 lb capacity	B1@.250	Ea	72.40	8.13	80.53	137.00
7' to 7' 4", 15,500 lb capacity	B1@.250	Ea	74.50	8.13	82.63	140.00
7' 3" to 7' 7", 15,500 lb capacity	B1@.250	Ea	77.50	8.13	85.63	146.00
7' 6" to 7' 10", 14,300 lb capacity	B1@.250	Ea	81.60	8.13	89.73	153.00
7' 9" to 8' 1", 14,300 lb capacity	B1@.250	Ea	91.80	8.13	99.93	170.00
8' to 8' 4", 14,300 lb capacity	B1@.250	Ea	100.00	8.13	108.13	184.00
Remove basement post and replace concrete pier. Add the cost of placing a temporary jack post from the previous section and setting the post itself from the following section.						
Remove an existing 4" x 4" x 7' wood basement post						
Per post	B1@.250	Ea	—	8.13	8.13	13.80
Break out an existing isolated concrete pier						
Using hand tools, per cubic foot of concrete	B1@.435	CF	—	14.20	14.20	24.10
Hand excavate for a concrete pad or pier, average soil						
Per cubic foot of soil	B1@.100	CF	—	3.25	3.25	5.53
Set and strip square concrete form for pier						
Per square foot of contact area, 1 use	B1@.090	SF	2.10	2.93	5.03	8.55
Mix, pour and finish concrete for an isolated pier						
Per cubic foot of concrete	B1@.400	CF	8.45	13.00	21.45	36.50
Place post anchor in concrete pier						
Per anchor	B1@.132	Ea	6.16	4.29	10.45	17.80
Wood support posts. S4S, green. Set in basement or first floor. Material costs include 10% waste. For posts incorporated in wall framing, see stud wall costs.						
4" x 4", 4" x 6" posts, per BF	—	BF	.69	—	.69	—
4" x 8" to 4" x 12" posts, per BF	—	BF	.88	—	.88	—
6" x 6" to 8" x 12" posts, per BF	—	BF	1.22	—	1.22	—
4" x 4" per LF	B1@.110	LF	1.01	3.58	4.59	7.80
4" x 6" per LF	B1@.120	LF	1.53	3.90	5.43	9.23
4" x 8" per LF	B1@.140	LF	2.59	4.55	7.14	12.10
4" x 10" per LF	B1@.143	LF	3.24	4.65	7.89	13.40
4" x 12" per LF	B1@.145	LF	3.88	4.72	8.60	14.60
6" x 6" per LF	B1@.145	LF	4.03	4.72	8.75	14.90
6" x 8" per LF	B1@.145	LF	4.44	4.72	9.16	15.60
6" x 10" per LF	B1@.147	LF	6.71	4.78	11.49	19.50
6" x 12" per LF	B1@.150	LF	8.06	4.88	12.94	22.00
8" x 8" per LF	B1@.150	LF	7.16	4.88	12.04	20.50
8" x 10" per LF	B1@.155	LF	8.95	5.04	13.99	23.80
Remove a sill plate. Includes jacking and cutting out the old plate. Add the cost of repairs to sheathing, siding and interior finish. All jacking materials salvaged. Add the cost of placing a temporary needle beam from the previous section. Add the cost of setting the sill plate from the following section.						
Per 8' section, work in a basement	B1@6.87	Ea	—	223.00	223.00	379.00
Per 16' section, work in a basement	B1@9.34	Ea	—	304.00	304.00	517.00
Per 8' section, work in a crawl space	B1@10.3	Ea	—	335.00	335.00	570.00
Per 16' section, work in a crawl space	B1@14.0	Ea	—	455.00	455.00	774.00

Sill plate. Mud sill at foundation, #2 pressure treated lumber, drilled and installed with foundation bolts at 48" OC, no bolts, nuts or washers included. Figures in parentheses show board feet per LF of foundation, including 5% waste.

	Craft@Hrs	Unit	Material	Labor	Total	Sell
Sill plates, per MBF	B1@22.9	MBF	890.00	745.00	1,635.00	2,780.00
2" x 3" (.53 BF per LF)	B1@.020	LF	.38	.65	1.03	1.75
2" x 4" (.70 BF per LF)	B1@.023	LF	.63	.75	1.38	2.35
2" x 6" (1.05 BF per LF)	B1@.024	LF	.83	.78	1.61	2.74
2" x 8" (1.40 BF per LF)	B1@.031	LF	1.38	1.01	2.39	4.06
Add for installation in a basement	B1@.095	LF	—	3.09	3.09	5.25
Add for installation in a crawl space	B1@.158	LF	—	5.14	5.14	8.74

Framing Room Additions

Floor joists — room addition. Per SF of area covered. Figures in parentheses indicate board feet per square foot of floor including box or band joist, typical double joists (under partition walls), and 6% waste. No beams, blocking or bridging included. Deduct for openings over 25 SF. Costs shown are based on a job with 1,000 SF of area covered. For scheduling purposes, estimate that a crew of two can complete 750 SF of area per 8-hour day for 12" center-to-center framing; 925 SF for 16" OC; 1,100 SF for 20" OC; or 1,250 SF for 24" OC. See the following section for costs of repairing floor joists in an existing building. For more complete coverage, see *National Framing and Finish Carpentry Estimator* at http://CraftsmanSiteLicense.com.

	Craft@Hrs	Unit	Material	Labor	Total	Sell
2" x 6" Std & Btr						
2" x 6" floor joists, per MBF	—	MBF	594.00	—	594.00	—
12" centers (1.28 BF per SF)	B1@.021	SF	.76	.68	1.44	2.45
16" centers (1.02 BF per SF)	B1@.017	SF	.61	.55	1.16	1.97
20" centers (.88 BF per SF)	B1@.014	SF	.52	.46	.98	1.67
24" centers (.73 BF per SF)	B1@.012	SF	.43	.39	.82	1.39
2" x 8" Std & Btr						
2" x 8" floor joists, per MBF	—	MBF	571.00	—	571.00	—
12" centers (1.71 BF per SF)	B1@.023	SF	.98	.75	1.73	2.94
16" centers (1.36 BF per SF)	B1@.018	SF	.78	.59	1.37	2.33
20" centers (1.17 BF per SF)	B1@.015	SF	.67	.49	1.16	1.97
24" centers (1.03 BF per SF)	B1@.013	SF	.59	.42	1.01	1.72
2" x 10" Std & Btr						
2" x 10" floor joists, per MBF	—	MBF	587.00	—	587.00	—
12" centers (2.14 BF per SF)	B1@.025	SF	1.26	.81	2.07	3.52
16" centers (1.71 BF per SF)	B1@.020	SF	1.00	.65	1.65	2.81
20" centers (1.48 BF per SF)	B1@.016	SF	.87	.52	1.39	2.36
24" centers (1.30 BF per SF)	B1@.014	SF	.76	.46	1.22	2.07
2" x 12" Std & Btr						
2" x 12" floor joists, per MBF	—	MBF	664.00	—	664.00	—
12" centers (2.56 BF per SF)	B1@.026	SF	1.70	.85	2.55	4.34
16" centers (2.05 BF per SF)	B1@.021	SF	1.36	.68	2.04	3.47
20" centers (1.77 BF per SF)	B1@.017	SF	1.18	.55	1.73	2.94
24" centers (1.56 BF per SF)	B1@.015	SF	1.04	.49	1.53	2.60

Add floor joists under an existing building. Per linear foot of new joist set in joist hangers or "doubled" by spiking to an existing joist.

Jack two or three adjacent joists into position with a crossarm mounted on a hydraulic jack. Per jacking point. Work done on sloping or unstable ground will cost more. Includes the cost of removing the jack.

	Craft@Hrs	Unit	Material	Labor	Total	Sell
Jacking done in a basement	B1@.700	Ea	—	22.80	22.80	38.80
Jacking done in a crawl space	B1@1.25	Ea	—	40.70	40.70	69.20
Set 2" x 6" floor joist						
Work done in a basement	B1@.042	LF	.71	1.37	2.08	3.54
Work done in a crawl space	B1@.070	LF	.71	2.28	2.99	5.08

	Craft@Hrs	Unit	Material	Labor	Total	Sell
Set 2" x 8" floor joist						
Work done in a basement	B1@.046	LF	.93	1.50	2.43	4.13
Work done in a crawl space	B1@.077	LF	.93	2.50	3.43	5.83
Set 2" x 10" floor joist						
Work done in a basement	B1@.050	LF	1.25	1.63	2.88	4.90
Work done in a crawl space	B1@.083	LF	1.25	2.70	3.95	6.72
Set 2" x 12" floor joist						
Work done in a basement	B1@.052	LF	1.79	1.69	3.48	5.92
Work done in a crawl space	B1@.087	LF	1.79	2.83	4.62	7.85

Floor joist wood, TJI truss type. Initial use, 50 PSF floor load design. Costs shown are per square foot (SF) of floor area, based on joists at 16" OC, for a job with 1,000 SF of floor area. Figure 1.22 SF of floor area for each LF of joist. Add the cost of beams, supports and blocking. For scheduling purposes, estimate that a crew of two can install 900 to 950 SF of joists in an 8-hour day.

9-1/2" TJI/15	B1@.017	SF	1.99	.55	2.54	4.32
14" TJI/35	B1@.018	SF	3.15	.59	3.74	6.36

Bridging or blocking. Installed between 2" x 6" thru 2" x 12" joists. Costs shown are per each set of cross bridges or per each block for solid bridging and include normal waste. Spacing between the bridging or blocking, sometimes called a "bay", depends on job requirements. Labor costs assume bridging is cut to size on site.

1" x 4" cross						
Joist bridging, per MBF	—	MBF	1,330.00	—	1,330.00	—
Joists on 12" centers	B1@.034	Ea	.66	1.11	1.77	3.01
Joists on 16" centers	B1@.034	Ea	.91	1.11	2.02	3.43
Joists on 20" centers	B1@.034	Ea	1.15	1.11	2.26	3.84
Joists on 24" centers	B1@.034	Ea	1.40	1.11	2.51	4.27
Add for work in a basement or first floor	B1@.068	Ea	—	2.21	2.21	3.76
Add for work in a crawl space	B1@.115	Ea	—	3.74	3.74	6.36
2" x 6" solid, Std & Btr						
2" x 6" blocking, per MBF	—	MBF	594.00	—	594.00	—
Joists on 12" centers	B1@.042	Ea	.65	1.37	2.02	3.43
Joists on 16" centers	B1@.042	Ea	.87	1.37	2.24	3.81
Joists on 20" centers	B1@.042	Ea	1.09	1.37	2.46	4.18
Joists on 24" centers	B1@.042	Ea	1.31	1.37	2.68	4.56
Add for work in a basement or first floor	B1@.084	Ea	—	2.73	2.73	4.64
Add for work in a crawl space	B1@.140	Ea	—	4.55	4.55	7.74
2" x 8" solid, Std & Btr						
2" x 8" blocking, per MBF	—	MBF	571.00	—	571.00	—
Joists on 12" centers	B1@.042	Ea	.84	1.37	2.21	3.76
Joists on 16" centers	B1@.042	Ea	1.12	1.37	2.49	4.23
Joists on 20" centers	B1@.042	Ea	1.40	1.37	2.77	4.71
Joists on 24" centers	B1@.042	Ea	1.68	1.37	3.05	5.19
Add for work in a basement or first floor	B1@.084	Ea	—	2.73	2.73	4.64
Add for work in a crawl space	B1@.140	Ea	—	4.55	4.55	7.74
2" x 10" solid, Std & Btr						
2" x 10" blocking, per MBF	—	MBF	587.00	—	587.00	—
Joists on 12" centers	B1@.057	Ea	1.08	1.85	2.93	4.98
Joists on 16" centers	B1@.057	Ea	1.43	1.85	3.28	5.58
Joists on 20" centers	B1@.057	Ea	1.79	1.85	3.64	6.19
Joists on 24" centers	B1@.057	Ea	2.15	1.85	4.00	6.80
Add for work in a basement or first floor	B1@.114	Ea	—	3.71	3.71	6.31
Add for work in a crawl space	B1@.190	Ea	—	6.18	6.18	10.50

	Craft@Hrs	Unit	Material	Labor	Total	Sell
2" x 12" solid, Std & Btr						
2" x 12" blocking, per MBF	—	MBF	664.00	—	664.00	—
Joists on 12" centers	B1@.057	Ea	1.46	1.85	3.31	5.63
Joists on 16" centers	B1@.057	Ea	1.95	1.85	3.80	6.46
Joists on 20" centers	B1@.057	Ea	2.43	1.85	4.28	7.28
Joists on 24" centers	B1@.057	Ea	2.92	1.85	4.77	8.11
Add for work in a basement or first floor	B1@.114	Ea	—	3.71	3.71	6.31
Add for work in a crawl space	B1@.190	Ea	—	6.18	6.18	10.50
Steel, no nail type, cross						
Wood joists on 12" centers	B1@.020	Ea	3.19	.65	3.84	6.53
Wood joists on 16" centers	B1@.020	Ea	3.32	.65	3.97	6.75
Wood joists on 20" centers	B1@.020	Ea	3.63	.65	4.28	7.28
Wood joists on 24" centers	B1@.020	Ea	4.15	.65	4.80	8.16

Subflooring.

Plywood sheathing, CD standard exterior grade. Material costs shown include fasteners and 5% for waste.

	Craft@Hrs	Unit	Material	Labor	Total	Sell
3/8"	B1@.011	SF	.47	.36	.83	1.41
1/2"	B1@.012	SF	.62	.39	1.01	1.72
5/8"	B1@.012	SF	.71	.39	1.10	1.87
3/4"	B1@.013	SF	.83	.42	1.25	2.13
Remove plywood floor sheathing	B1@.020	SF	—	.65	.65	1.11

OSB sheathing, Material costs shown include fasteners and 5% for waste.

	Craft@Hrs	Unit	Material	Labor	Total	Sell
3/8"	B1@.011	SF	.39	.36	.75	1.28
1/2"	B1@.012	SF	.47	.39	.86	1.46
5/8"	B1@.012	SF	.48	.39	.87	1.48
3/4"	B1@.013	SF	.57	.42	.99	1.68
7/8"	B1@.014	SF	1.09	.46	1.55	2.64
1"	B1@.014	SF	1.58	.46	2.04	3.47
Remove OSB floor sheathing	B1@.020	SF	—	.65	.65	1.11

Board sheathing, laid diagonal, 1" x 6" #3 & Btr, 1.18 board feet per square foot. Costs shown include nails, 12% shrinkage, and 5% waste.

	Craft@Hrs	Unit	Material	Labor	Total	Sell
Board sheathing, per MBF	—	MBF	2,010.00	—	2,010.00	—
1" x 6", joists 16" on center	B1@.027	SF	2.61	.88	3.49	5.93
1" x 6", joists 24" on center	B1@.030	SF	2.61	.98	3.59	6.10
1" x 8", joists 16" on center	B1@.016	SF	2.61	.52	3.13	5.32
1" x 8", joists 24" on center	B1@.017	SF	2.61	.55	3.16	5.37
Remove board floor sheathing	B1@.030	SF	—	.98	.98	1.67

Wall studs — room addition. Per square foot of wall area. Do not subtract for openings less than 16' wide. Figures in parentheses indicate typical board feet per SF of wall area, measured on one side, and include normal waste. Add for each corner and partition from below. Costs include studding and nails. Add for plates from the sections above and below. Add for door and window opening framing, backing, let-in bracing, fire blocking, and sheathing for shear walls from the sections that follow. Labor includes layout, plumb and align.

	Craft@Hrs	Unit	Material	Labor	Total	Sell
2" x 3", Std & Btr						
2" x 3" studs, 16" centers, per MBF	—	MBF	584.00	—	584.00	—
12" centers (.55 BF per SF)	B1@.024	SF	.32	.78	1.10	1.87
16" centers (.41 BF per SF)	B1@.018	SF	.24	.59	.83	1.41
20" centers (.33 BF per SF)	B1@.015	SF	.19	.49	.68	1.16
24" centers (.28 BF per SF)	B1@.012	SF	.16	.39	.55	.94
Add for each corner or partition	B1@.083	Ea	7.58	2.70	10.28	17.50
2" x 4", Std & Btr						
2" x 4" studs, 16" centers, per MBF	—	MBF	579.00	—	579.00	—
12" centers (.73 BF per SF)	B1@.031	SF	.42	1.01	1.43	2.43
16" centers (.54 BF per SF)	B1@.023	SF	.31	.75	1.06	1.80

	Craft@Hrs	Unit	Material	Labor	Total	Sell
20" centers (.45 BF per SF)	B1@.020	SF	.26	.65	.91	1.55
24" centers (.37 BF per SF)	B1@.016	SF	.21	.52	.73	1.24
Add for each corner or partition	B1@.083	Ea	9.82	2.70	12.52	21.30
2" x 6", Std & Btr						
2" x 6" studs, 24" centers, per MBF	—	MBF	594.00	—	594.00	—
12" centers (1.10 BF per SF)	B1@.044	SF	.65	1.43	2.08	3.54
16" centers (.83 BF per SF)	B1@.033	SF	.49	1.07	1.56	2.65
20" centers (.67 BF per SF)	B1@.027	SF	.40	.88	1.28	2.18
24" centers (.55 BF per SF)	B1@.022	SF	.33	.72	1.05	1.79
Add for each corner or partition	B1@.083	Ea	14.20	2.70	16.90	28.70
4" x 4", Std & Btr						
Installed in wall framing, with 10% waste	B1@.064	LF	1.02	2.08	3.10	5.27

Stud wall repair. Costs per stud added or repaired. Costs will be higher when plumbing or electrical lines run through the stud wall affected.

	Craft@Hrs	Unit	Material	Labor	Total	Sell
Straighten stud by cutting saw kerf. Drive shim or screw into the kerf						
Reinforce with a plywood scab	BC@.300	Ea	—	10.80	10.80	18.40
Remove and replace a badly warped stud	BC@.300	Ea	—	10.80	10.80	18.40

Add stud to an existing stud wall. Remove blocking, insert an additional stud in an existing wall, replace blocking.

	Craft@Hrs	Unit	Material	Labor	Total	Sell
2" x 3" x 8' stud	B1@.300	Ea	2.00	9.76	11.76	20.00
2" x 4" x 8' stud	B1@.346	Ea	3.00	11.30	14.30	24.30
2" x 6" x 8' stud	B1@.392	Ea	5.00	12.80	17.80	30.30

Plates. (Wall plates), Std & Btr, untreated. For pressure-treated plates see also Sill Plates in this chapter. Figures in parentheses indicate board feet per LF. Costs shown include 10% for waste and nails.

	Craft@Hrs	Unit	Material	Labor	Total	Sell
2" x 3" Wall plates, per MBF	—	MBF	584.00	—	584.00	—
2" x 3" (.55 BF per LF)	B1@.010	LF	.32	.33	.65	1.11
2" x 4" Wall plates, per MBF	—	MBF	504.00	—	504.00	—
2" x 4" (.73 BF per LF)	B1@.012	LF	.42	.39	.81	1.38
2" x 6" Wall plates, per MBF	—	MBF	517.00	—	517.00	—
2" x 6" (1.10 BF per LF)	B1@.018	LF	.65	.59	1.24	2.11

Wall Bracing. See also sheathing for plywood bracing and shear panels.

	Craft@Hrs	Unit	Material	Labor	Total	Sell
Let-in wall bracing, using Std & Btr lumber						
1" x 4"	B1@.021	LF	.60	.68	1.28	2.18
1" x 6"	B1@.027	LF	.87	.88	1.75	2.98
2" x 4"	B1@.035	LF	.39	1.14	1.53	2.60
Steel strap bracing, 1-1/4" wide						
9'6" or 11'6" lengths	B1@.010	LF	.53	.33	.86	1.46
Steel V-bracing, 3/4" x 3/4"						
9'6" or 11'6" lengths	B1@.010	LF	.83	.33	1.16	1.97
Temporary wood frame wall bracing, assumes salvage at 50% and 3 uses						
1" x 4" Std & Btr	B1@.006	LF	.15	.20	.35	.60
1" x 6" Std & Btr	B1@.010	LF	.22	.33	.55	.94
2" x 4" utility	B1@.012	LF	.10	.39	.49	.83
2" x 6" utility	B1@.018	LF	.15	.59	.74	1.26

	Craft@Hrs	Unit	Material	Labor	Total	Sell

Fireblocks. Installed in wood frame walls, per LF of wall to be blocked. Figures in parentheses indicate board feet of fire blocking per linear foot of wall, including 10% cutting waste. See also Bridging and Backing and Nailers.

	Craft@Hrs	Unit	Material	Labor	Total	Sell
2" x 3" blocking, Std & Btr						
2" x 3" fireblocks, 16" centers, per MBF	—	MBF	584.00	—	584.00	—
12" OC members (.48 BF per LF)	B1@.026	LF	.28	.85	1.13	1.92
16" OC members (.50 BF per LF)	B1@.025	LF	.29	.81	1.10	1.87
20" OC members (.51 BF per LF)	B1@.022	LF	.30	.72	1.02	1.73
24" OC members (.52 BF per LF)	B1@.020	LF	.30	.65	.95	1.62
Add for removing and replacing, per block	B1@.026	LF	—	.85	.85	1.45
2" x 4" blocking, Std & Btr						
2" x 4" fireblocks, 16" centers, per MBF	—	MBF	579.00	—	579.00	—
12" OC members (.64 BF per LF)	B1@.030	LF	.37	.98	1.35	2.30
16" OC members (.67 BF per LF)	B1@.025	LF	.39	.81	1.20	2.04
20" OC members (.68 BF per LF)	B1@.022	LF	.39	.72	1.11	1.89
24" OC members (.69 BF per LF)	B1@.020	LF	.40	.65	1.05	1.79
Add to remove and replace, per block	B1@.030	LF	—	.98	.98	1.67
2" x 6" blocking, Std & Btr						
2" x 6" fireblocks, 24" centers, per MBF	—	MBF	594.00	—	594.00	—
12" OC members (.96 BF per LF)	B1@.031	LF	.57	1.01	1.58	2.69
16" OC members (1.00 BF per LF)	B1@.026	LF	.59	.85	1.44	2.45
20" OC members (1.02 BF per LF)	B1@.022	LF	.61	.72	1.33	2.26
24" OC members (1.03 BF per LF)	B1@.020	LF	.61	.65	1.26	2.14
Add to remove and replace, per block	B1@.034	LF	—	1.11	1.11	1.89

Ceiling beams — room addition. Std & Btr. Installed over wall openings and around floor, ceiling and roof openings or where a flush beam is called out in the plans, including 10% waste. Don't use these cost estimates for door or window headers in framed walls. See the door or window opening framing assemblies costs that follow.

	Craft@Hrs	Unit	Material	Labor	Total	Sell
2" x 6" beams, per MBF	B1@24.5	MBF	594.00	797.00	1,391.00	2,360.00
2" x 8" beams, per MBF	B1@25.2	MBF	571.00	820.00	1,391.00	2,360.00
2" x 10" beams, per MBF	B1@25.1	MBF	587.00	817.00	1,404.00	2,390.00
2" x 12" beams, per MBF	B1@25.9	MBF	664.00	843.00	1,507.00	2,560.00
4" x 6" beams, per MBF	B1@25.9	MBF	708.00	843.00	1,551.00	2,640.00
4" x 8", 10", 12", 14" beams, per MBF	B1@25.5	MBF	882.00	830.00	1,712.00	2,910.00
6" x 6", 8", 10", 12", 14" beams, per MBF	B1@19.1	MBF	1,220.00	621.00	1,841.00	3,130.00
2" x 6" (1.10 BF per LF)	B1@.028	LF	.65	.91	1.56	2.65
2" x 8" (1.47 BF per LF)	B1@.037	LF	.84	1.20	2.04	3.47
2" x 10" (1.83 BF per LF)	B1@.046	LF	1.07	1.50	2.57	4.37
2" x 12" (2.20 BF per LF)	B1@.057	LF	1.46	1.85	3.31	5.63
4" x 6" (2.20 BF per LF)	B1@.057	LF	1.56	1.85	3.41	5.80
4" x 8" (2.93 BF per LF)	B1@.073	LF	2.59	2.37	4.96	8.43
4" x 10" (3.67 BF per LF)	B1@.094	LF	3.24	3.06	6.30	10.70
4" x 12" (4.40 BF per LF)	B1@.112	LF	3.88	3.64	7.52	12.80
4" x 14" (5.13 BF per LF)	B1@.115	LF	4.53	3.74	8.27	14.10
6" x 6" (3.30 BF per LF)	B1@.060	LF	4.03	1.95	5.98	10.20
6" x 8" (4.40 BF per LF)	B1@.080	LF	5.37	2.60	7.97	13.50
6" x 10" (5.50 BF per LF)	B1@.105	LF	6.71	3.42	10.13	17.20
6" x 12" (6.60 BF per LF)	B1@.115	LF	8.06	3.74	11.80	20.10
6" x 14" (7.70 BF per LF)	B1@.120	LF	9.40	3.90	13.30	22.60

	Craft@Hrs	Unit	Material	Labor	Total	Sell
Window opening framing. In 2" x 4" stud wall, based on 8' wall height. Costs shown are per window opening framed into a new stud wall and include a header of appropriate size, sub-sill plate (double sub-sill if opening is 8' wide or wider), double vertical studs each side of openings less than 8' wide (triple vertical studs each side of openings 8' wide or wider), top and bottom cripples, blocking, nails, and normal waste. Figures in parentheses indicate header size.						
To 2' wide (4" x 4" header)	B1@1.00	Ea	15.80	32.50	48.30	82.10
Over 2' to 3' wide (4" x 4" header)	B1@1.17	Ea	19.60	38.10	57.70	98.10
Over 3' to 4' wide (4" x 6" header)	B1@1.45	Ea	24.10	47.20	71.30	121.00
Over 4' to 5' wide (4" x 6" header)	B1@1.73	Ea	26.90	56.30	83.20	141.00
Over 5' to 6' wide (4" x 8" header)	B1@2.01	Ea	35.00	65.40	100.40	171.00
Over 6' to 7' wide (4" x 8" header)	B1@2.29	Ea	44.30	74.50	118.80	202.00
Over 7' to 8' wide (4" x 10" header)	B1@2.57	Ea	43.70	83.60	127.30	216.00
Over 8' to 10' wide (4" x 12" header)	B1@2.57	Ea	67.90	83.60	151.50	258.00
Over 10' to 12' wide (4" x 14" header)	B1@2.85	Ea	88.80	92.70	181.50	309.00
Add per foot of height for walls over 8' high	—	Ea	2.31	—	2.31	—
In 2" x 6" stud wall, based on 8' wall height. Costs shown are per window opening framed into a new stud wall and include a header of appropriate size, sub-sill plate (double sub-sill if opening is 8' wide or wider), double vertical studs each side of openings less than 8' wide (triple vertical studs each side of openings 8' wide or wider), top and bottom cripples, blocking, nails, and normal waste. Figures in parentheses indicate header size.						
To 2' wide (4" x 4" header)	B1@1.17	Ea	20.40	38.10	58.50	99.50
Over 2' to 3' wide (6" x 4" header)	B1@1.45	Ea	25.90	47.20	73.10	124.00
Over 3' to 4' wide (6" x 6" header)	B1@1.73	Ea	38.30	56.30	94.60	161.00
Over 4' to 5' wide (6" x 6" header)	B1@2.01	Ea	45.50	65.40	110.90	189.00
Over 5' to 6' wide (6" x 8" header)	B1@2.29	Ea	55.70	74.50	130.20	221.00
Over 6' to 8' wide (6" x 8" header)	B1@2.85	Ea	86.80	92.70	179.50	305.00
Over 8' to 10' wide (6" x 12" header)	B1@2.85	Ea	118.00	92.70	210.70	358.00
Over 8' to 12' wide (6" x 14" header)	B1@3.13	Ea	151.00	102.00	253.00	430.00
Add per foot of height for walls over 8' high	—	Ea	3.56	—	3.56	—
Door opening framing. In 2" x 4" stud wall, based on 8' wall height. Costs shown are per each door opening framed into a new stud wall and include a header of appropriate size, double vertical studs each side of the opening less than 8' wide (triple vertical studs each side of openings 8' wide or wider), cripples, blocking, nails, and normal waste. Width shown is size of finished opening. Figures in parentheses indicate header size.						
To 3' wide (4" x 4" header)	B1@.830	Ea	17.20	27.00	44.20	75.10
Over 3' to 4' wide (4" x 6" header)	B1@1.11	Ea	20.90	36.10	57.00	96.90
Over 4' to 5' wide (4" x 6" header)	B1@1.39	Ea	22.90	45.20	68.10	116.00
Over 5' to 6' wide (4" x 8" header)	B1@1.66	Ea	30.30	54.00	84.30	143.00
Over 6' to 8' wide (4" x 10" header)	B1@1.94	Ea	47.60	63.10	110.70	188.00
Over 8' to 10' wide (4" x 12" header)	B1@1.94	Ea	60.20	63.10	123.30	210.00
Over 10' to 12' wide (4" x 14" header)	B1@2.22	Ea	74.90	72.20	147.10	250.00
Add per foot of height for walls over 8' high	—	Ea	2.31	—	2.31	—
In 2" x 6" stud wall, based on 8' wall height. Costs shown are per each door opening framed into a new stud wall and include a header of appropriate size, double vertical studs each side of the opening less than 8' wide (triple vertical studs each side of openings 8' wide or wider), cripples, blocking, nails, and normal waste. Width shown is size of finished opening. Figures in parentheses indicate header size.						
To 3' wide (6" x 4" header)	B1@1.11	Ea	27.80	36.10	63.90	109.00
Over 3' to 4' wide (4" x 6" header)	B1@1.39	Ea	35.10	45.20	80.30	137.00
Over 4' to 5' wide (4" x 6" header)	B1@1.66	Ea	39.40	54.00	93.40	159.00
Over 5' to 6' wide (4" x 8" header)	B1@1.94	Ea	50.90	63.10	114.00	194.00
Over 6' to 8' wide (4" x 10" header)	B1@2.22	Ea	80.50	72.20	152.70	260.00
Over 8' to 10' wide (4" x 12" header)	B1@2.22	Ea	107.00	72.20	179.20	305.00
Over 10' to 12' wide (4" x 14" header)	B1@2.50	Ea	137.00	81.30	218.30	371.00
Add per foot of height for walls over 8' high	—	Ea	3.56	—	3.56	—

	Craft@Hrs	Unit	Material	Labor	Total	Sell
OSB sheathing. Oriented strand board. Compressed wood strands bonded by phenolic resin. By nominal thickness.						
3/8" x 4' x 8'	B1@.014	SF	.41	.46	.87	1.48
7/16" x 4' x 8'	B1@.015	SF	.38	.49	.87	1.48
15/32" x 4' x 8'	B1@.016	SF	.43	.52	.95	1.62
1/2" x 4' x 8'	B1@.016	SF	.45	.52	.97	1.65
5/8" x 4' x 8', square edge	B1@.018	SF	.54	.59	1.13	1.92
5/8" x 4' x 8', tongue and groove	B1@.018	SF	.86	.59	1.45	2.47
3/4" x 4' x 8', square edge	B1@.018	SF	.70	.59	1.29	2.19
3/4" x 4' x 8', tongue and groove	B1@.018	SF	.67	.59	1.26	2.14
Remove OSB sheathing	B1@.017	SF	—	.55	.55	.94
Wall sheathing. Per square foot of wall surface, including 5% waste. See also OSB sheathing.						
BC plywood wall sheathing, plugged and touch sanded, interior grade						
1/4"	B1@.013	SF	.74	.42	1.16	1.97
3/8"	B1@.015	SF	.75	.49	1.24	2.11
1/2"	B1@.016	SF	.88	.52	1.40	2.38
5/8"	B1@.018	SF	1.01	.59	1.60	2.72
3/4"	B1@.020	SF	1.17	.65	1.82	3.09
Remove plywood wall sheathing	B1@.017	SF	—	.55	.55	.94
Board wall sheathing 1" x 6" or 1" x 8" utility T&G						
1" x 6" or 8" utility T&G, per MBF	—	MBF	1,440.00	—	1,440.00	—
(1.13 BF per SF)	B1@.020	SF	1.63	.65	2.28	3.88
Add for diagonal patterns	B1@.006	SF	—	.20	.20	.34
Remove board sheathing	B1@.030	SF	—	.98	.98	1.67
Patch wall sheathing. Studs 16" on center, remove and replace panels in several places on wall surface.						
1" x 8" x 12' long, per board	B1@.248	Ea	10.20	8.07	18.27	31.10
1" x 8", per square foot	B1@.031	SF	1.28	1.01	2.29	3.89
1/2", 4-ply plywood, per 4' x 8' sheet	B1@.960	Ea	20.30	31.20	51.50	87.60
1/2", 4-ply plywood, per square foot	B1@.030	SF	.63	.98	1.61	2.74

Changes to Walls and Ceilings

	Craft@Hrs	Unit	Material	Labor	Total	Sell
Remove and replace interior partition wall. Use 16 linear feet as the minimum job size. No interior floor finish included.						
2" x 4" stud walls with 1/2" gypsum drywall both sides, ready for painting						
Cost per square foot of wall	B1@.105	SF	.93	3.42	4.35	7.40
Cost per running foot, 8' high wall	B1@.842	LF	7.45	27.40	34.85	59.20
2" x 4" stud walls with 5/8" gypsum fire-rated drywall both sides, ready for painting						
Cost per square foot of wall	B1@.108	SF	1.05	3.51	4.56	7.75
Cost per running foot, 8' high wall	B1@.864	LF	8.41	28.10	36.51	62.10
2" x 6" stud walls with 1/2" gypsum drywall both sides, ready for painting						
Cost per square foot of wall	B1@.112	SF	1.20	3.64	4.84	8.23
Cost per running foot, 8' high wall	B1@.896	LF	9.58	29.10	38.68	65.80
2" x 6" stud walls with 5/8" gypsum fire-rated drywall both sides, ready for painting						
Cost per square foot of wall	B1@.116	SF	1.32	3.77	5.09	8.65
Cost per running foot, 8' high wall	B1@.928	LF	10.50	30.20	40.70	69.20

	Craft@Hrs	Unit	Material	Labor	Total	Sell
Remove and replace exterior wall. Add temporary joist support costs from below. Use 16 linear feet as the minimum job size. No interior floor finish included.						
2" x 4" stud walls with drywall interior, wood siding exterior, 1/2" gypsum drywall inside face ready for painting, over 3-1/2" R-11 insulation with 5/8" thick rough sawn T-1-11 exterior grade plywood siding on the outside face						
Using 5/8" rough sawn T-1-11 siding at	—	MSF	1,190.00	—	1,190.00	—
Cost per square foot of wall	B1@.108	SF	2.30	3.51	5.80	9.31
Cost per running foot, 8' high wall	B1@.864	LF	18.40	28.10	46.50	79.10
2" x 6" stud walls with drywall interior, wood siding exterior, same construction as above, except with 6-1/4" R-19 insulation.						
Cost per square foot of wall	B1@.117	SF	2.64	3.81	6.45	11.00
Cost per running foot, 8' high wall	B1@.936	LF	21.10	30.40	51.50	87.60
2" x 4" stud walls with drywall interior, 1/2" gypsum drywall inside face ready for painting, over 3-1/2" R-11 insulation with .042" vinyl lap siding on the outside face.						
Using .042" vinyl lap siding at	—	MSF	700.00	—	700.00	—
Cost per square foot of wall	B1@.112	SF	1.97	3.64	5.61	9.54
Cost per running foot, 8' high wall	B1@.896	LF	15.80	29.10	44.90	76.30
2" x 4" stud walls with drywall interior, 1/2" gypsum drywall on inside face ready for painting, over 3-1/2" R-11 insulation with 1" x 6" southern yellow pine drop siding, D Grade, 1.19 BF per SF at 5-1/4" exposure on the outside face.						
Using D grade yellow pine drop siding at	—	MSF	2,610.00	—	2,610.00	—
Cost per square foot of wall	B1@.114	SF	3.72	3.71	7.43	12.60
Cost per running foot, 8' high wall	B1@.912	LF	29.80	29.70	59.50	101.00
2" x 6" stud walls with drywall interior, 1" x 6" drop siding exterior, same construction as above, except with 6-1/4" R-19 insulation						
Cost per square foot of wall	B1@.123	SF	4.06	4.00	8.06	13.70
Cost per running foot, 8' high wall	B1@.984	LF	32.50	32.00	64.50	110.00
2" x 4" stud walls with drywall interior, stucco exterior, 1/2" gypsum drywall on inside face ready for painting, over 3-1/2" R-11 insulation and a three-coat exterior plaster (stucco) finish with integral color on the outside face						
Cost per square foot of wall	B1@.090	SF	3.39	2.93	6.32	10.70
Cost per running foot, 8' high wall	B1@.720	LF	27.10	23.40	50.50	85.90
2" x 6" stud walls with drywall interior, stucco exterior, same construction as above, except with 6-1/4" R-19 insulation						
Cost per square foot of wall	B1@.099	SF	3.73	3.22	6.95	11.80
Cost per running foot, 8' high wall	B1@.792	LF	29.80	25.80	55.60	94.50
Add for other types of gypsum board						
1/2" or 5/8" moisture-resistant greenboard						
Cost per SF, greenboard per side, add	—	SF	.21	—	.21	—
1/2" or 5/8" moisture-resistant greenboard						
Cost per running foot per side 8' high	—	LF	1.62	—	1.62	—
5/8" thick fire-rated Type X gypsum drywall						
Cost per SF, per side, add	—	SF	.06	—	.06	—
5/8" thick fire-rated Type X gypsum drywall						
Cost per running foot per side 8' high	—	LF	.49	—	.49	—
Support ceiling joists during replacement of a stud wall, using one 4" x 4" post, 2" x 8" pads and wedges, 4" x 4" strong back. All materials salvaged. Allow one post per each 5' of wall.						
Per support post set and removed	B1@.567	Ea	—	18.40	18.40	31.30

	Craft@Hrs	Unit	Material	Labor	Total	Sell

Breakthrough wall opening. Includes breaking an opening in an existing interior or exterior wall, cutting out the existing wall frame and cover, installing a header where studs or brick are removed, and trimming and casing the opening. Add the cost of floor and wall finish to match the existing surfaces on both sides of the new opening. In multi-story construction or in wide openings, a steel lintel may be required to support the framing above the opening. Add the cost of engineering, the lintel, and one hour of carpenter time for each foot of lintel length.

	Craft@Hrs	Unit	Material	Labor	Total	Sell
Door or cased opening to 3' wide x 7' high. Includes the door frame. Add the cost of the door and jamb, as required.						
Frame wall	B1@5.25	Ea	—	171.00	171.00	291.00
Add for each foot of width over 3'	B1@.700	LF	—	22.80	22.80	38.80
Brick veneer over frame wall	B1@6.25	Ea	—	203.00	203.00	345.00
Add for each foot of width over 3'	B1@.750	LF	—	24.40	24.40	41.50
Block or brick wall	B1@6.75	Ea	—	220.00	220.00	374.00
Add for each foot of width over 3'	B1@.850	LF	—	27.70	27.70	47.10
Wall breakthrough for window or wall air conditioner, to 3' wide x 4' high. Includes the window frame and cripple stud fill below the new opening. Add the cost of the window, as required.						
Frame wall	B1@6.50	Ea	—	211.00	211.00	359.00
Brick veneer over frame wall	B1@10.5	Ea	—	342.00	342.00	581.00
Block or brick	B1@11.8	Ea	—	384.00	384.00	653.00

Furring. Per SF of surface area to be covered, including typical wedging. Figures in parentheses show coverage, including 7% waste, typical job.

	Craft@Hrs	Unit	Material	Labor	Total	Sell
Over masonry, utility grade						
1" furring, per MBF	—	MBF	1,330.00	—	1,330.00	—
12" OC, 1" x 2" (.24 BF per SF)	B1@.025	SF	.32	.81	1.13	1.92
16" OC, 1" x 2" (.20 BF per SF)	B1@.020	SF	.27	.65	.92	1.56
20" OC, 1" x 2" (.17 BF per SF)	B1@.018	SF	.23	.59	.82	1.39
24" OC, 1" x 2" (.15 BF per SF)	B1@.016	SF	.20	.52	.72	1.22
12" OC, 1" x 3" (.36 BF per SF)	B1@.025	SF	.48	.81	1.29	2.19
16" OC, 1" x 3" (.29 BF per SF)	B1@.020	SF	.39	.65	1.04	1.77
20" OC, 1" x 3" (.25 BF per SF)	B1@.018	SF	.33	.59	.92	1.56
24" OC, 1" x 3" (.22 BF per SF)	B1@.016	SF	.29	.52	.81	1.38
Over wood frame, utility grade						
1" furring, per MBF	—	MBF	1,330.00	—	1,330.00	—
12" OC, 1" x 2" (.24 BF per SF)	B1@.016	SF	.32	.52	.84	1.43
16" OC, 1" x 2" (.20 BF per SF)	B1@.013	SF	.27	.42	.69	1.17
20" OC, 1" x 2" (.17 BF per SF)	B1@.011	SF	.23	.36	.59	1.00
24" OC, 1" x 2" (.15 BF per SF)	B1@.010	SF	.20	.33	.53	.90
12" OC, 1" x 3" (.36 BF per SF)	B1@.016	SF	.48	.52	1.00	1.70
16" OC, 1" x 3" (.29 BF per SF)	B1@.013	SF	.39	.42	.81	1.38
20" OC, 1" x 3" (.25 BF per SF)	B1@.011	SF	.33	.36	.69	1.17
24" OC, 1" x 3" (.22 BF per SF)	B1@.010	SF	.29	.33	.62	1.05
Over wood subfloor						
1" furring, per MBF	—	MBF	1,330.00	—	1,330.00	—
12" OC, 1" x 2" utility (.24 BF per SF)	B1@.033	SF	.32	1.07	1.39	2.36
16" OC, 1" x 2" utility (.20 BF per SF)	B1@.028	SF	.27	.91	1.18	2.01
20" OC, 1" x 2" utility (.17 BF per SF)	B1@.024	SF	.23	.78	1.01	1.72
24" OC, 1" x 2" utility (.15 BF per SF)	B1@.021	SF	.20	.68	.88	1.50
2" furring, per MBF	—	MBF	579.00	—	579.00	—
12" OC, 2" x 2" Std & Btr (.48 BF per SF)	B1@.033	SF	.28	1.07	1.35	2.30
16" OC, 2" x 2" Std & Btr (.39 BF per SF)	B1@.028	SF	.23	.91	1.14	1.94
20" OC, 2" x 2" Std & Btr (.34 BF per SF)	B1@.024	SF	.20	.78	.98	1.67
24" OC, 2" x 2" Std & Btr (.30 BF per SF)	B1@.021	SF	.17	.68	.85	1.45

	Craft@Hrs	Unit	Material	Labor	Total	Sell
Remove 1" furring. All nails pulled or driven flush. Per linear foot of 1" furring						
Furring installed on masonry	B1@.030	LF	—	.98	.98	1.67
Furring installed on wood frame	B1@.020	LF	—	.65	.65	1.11
Furring installed on subfloor	B1@.035	LF	—	1.14	1.14	1.94
Sleepers. Std & Btr pressure treated lumber, laid on concrete, including 5% waste.						
2" x 6" sleepers, per MBF	—	MBF	789.00	—	789.00	—
2" x 6" sleepers, per LF	B1@.017	LF	.83	.55	1.38	2.35
Add for taper cuts on sleepers, per cut	B1@.050	Ea	—	1.63	1.63	2.77
Backing and nailers. Std & Btr, for wall finishes, "floating" backing for drywall ceilings, trim, Z-bar, appliances and fixtures, etc. Figures in parentheses show board feet per LF, including 10% waste.						
1" x 4" (.37 BF per LF)	B1@.011	LF	.49	.36	.85	1.45
1" x 6" (.55 BF per LF)	B1@.017	LF	.79	.55	1.34	2.28
1" x 8" (.73 BF per LF)	B1@.022	LF	1.05	.72	1.77	3.01
2" x 4" (.73 BF per LF)	B1@.023	LF	.42	.75	1.17	1.99
2" x 6" (1.10 BF per LF)	B1@.034	LF	.65	1.11	1.76	2.99
2" x 8" (1.47 BF per LF)	B1@.045	LF	.84	1.46	2.30	3.91
2" x 10" (1.83 BF per LF)	B1@.057	LF	1.07	1.85	2.92	4.96
Ledger strips. Std & Btr, nailed to faces of studs, beams, girders, joists, etc. See also Ribbons (Ribbands) in this section for let-in type. Figures in parentheses indicate board feet per LF, including 10% waste.						
1" x 2", 3" or 4" ledger, per MBF	—	MBF	1,330.00	—	1,330.00	—
1" x 2" (.18 BF per LF)	B1@.010	LF	.24	.33	.57	.97
1" x 3" (.28 BF per LF)	B1@.010	LF	.37	.33	.70	1.19
1" x 4" (.37 BF per LF)	B1@.010	LF	.49	.33	.82	1.39
2" x 2", 3" or 4" ledger, per MBF	—	MBF	579.00	—	579.00	—
2" x 2" (.37 BF per LF)	B1@.010	LF	.21	.33	.54	.92
2" x 3" (.55 BF per LF)	B1@.010	LF	.32	.33	.65	1.11
2" x 4" (.73 BF per LF)	B1@.010	LF	.42	.33	.75	1.28
Ribbons (Ribbands). Let-in to wall framing. Figures in parentheses indicate board feet per LF, including 10% waste.						
1" x 3", 4" or 6" ribbon, per MBF	—	MBF	1,330.00	—	1,330.00	—
1" x 3", Std & Btr (.28 BF per LF)	B1@.020	LF	.37	.65	1.02	1.73
1" x 4", Std & Btr (.37 BF per LF)	B1@.020	LF	.49	.65	1.14	1.94
1" x 6", Std & Btr (.55 BF per LF)	B1@.030	LF	.73	.98	1.71	2.91
2" x 3", 4" ribbon, per MBF	—	MBF	579.00	—	579.00	—
2" x 3", Std & Btr (.55 BF per LF)	B1@.041	LF	.32	1.33	1.65	2.81
2" x 4", Std & Btr (.73 BF per LF)	B1@.041	LF	.42	1.33	1.75	2.98
2" x 6" ribbon, per MBF	—	MBF	594.00	—	594.00	—
2" x 6", Std & Btr (1.10 BF per LF)	B1@.045	LF	.65	1.46	2.11	3.59
2" x 8" ribbon, per MBF	—	MBF	571.00	—	571.00	—
2" x 8", Std & Btr (1.47 BF per LF)	B1@.045	LF	.84	1.46	2.30	3.91
Ceiling joists — room addition. Per SF of area covered. Figures in parentheses indicate board feet per square foot of ceiling, including end joists, header joists, and 5% waste. No beams, bridging, blocking, or ledger strips included. Deduct for openings over 25 SF. For scheduling purposes, estimate that a crew of two can complete 650 SF of area per 8-hour day for 12" center-to-center framing; 800 SF for 16" OC; 950 SF for 20" OC; or 1,100 SF for 24" OC.						
2" x 4", Std & Btr grade						
2" x 4" ceiling joists, 16" centers, per MBF	—	MBF	579.00	—	579.00	—
12" centers (.78 BF per SF)	B1@.026	SF	.45	.85	1.30	2.21
16" centers (.59 BF per SF)	B1@.020	SF	.34	.65	.99	1.68
20" centers (.48 BF per SF)	B1@.016	SF	.28	.52	.80	1.36
24" centers (.42 BF per SF)	B1@.014	SF	.24	.46	.70	1.19

	Craft@Hrs	Unit	Material	Labor	Total	Sell
2" x 6", Std & Btr grade						
2" x 6" ceiling joists, 16" centers, per MBF	—	MBF	594.00	—	594.00	—
12" centers (1.15 BF per SF)	B1@.026	SF	.68	.85	1.53	2.60
16" centers (.88 BF per SF)	B1@.020	SF	.52	.65	1.17	1.99
20" centers (.72 BF per SF)	B1@.016	SF	.43	.52	.95	1.62
24" centers (.63 BF per SF)	B1@.014	SF	.37	.46	.83	1.41
2" x 8", Std & Btr grade						
2" x 8" ceiling joists, 16" centers, per MBF	—	MBF	571.00	—	571.00	—
12" centers (1.53 BF per SF)	B1@.028	SF	.87	.91	1.78	3.03
16" centers (1.17 BF per SF)	B1@.022	SF	.67	.72	1.39	2.36
20" centers (.96 BF per SF)	B1@.018	SF	.55	.59	1.14	1.94
24" centers (.84 BF per SF)	B1@.016	SF	.48	.52	1.00	1.70
2" x 10", Std & Btr grade						
2" x 10" ceiling joists, 16" centers, per MBF	—	MBF	587.00	—	587.00	—
12" centers (1.94 BF per SF)	B1@.030	SF	1.14	.98	2.12	3.60
16" centers (1.47 BF per SF)	B1@.023	SF	.86	.75	1.61	2.74
20" centers (1.21 BF per SF)	B1@.019	SF	.71	.62	1.33	2.26
24" centers (1.04 BF per SF)	B1@.016	SF	.61	.52	1.13	1.92
2" x 12", Std & Btr grade						
2" x 12" ceiling joists, 16" centers, per MBF	—	MBF	664.00	—	664.00	—
12" centers (2.30 BF per SF)	B1@.033	SF	1.53	1.07	2.60	4.42
16" centers (1.76 BF per SF)	B1@.025	SF	1.17	.81	1.98	3.37
20" centers (1.44 BF per SF)	B1@.020	SF	.86	.65	1.51	2.57
24" centers (1.26 BF per SF)	B1@.018	SF	.84	.59	1.43	2.43

Adding ceiling joists. Per linear foot of new joist set in joist hangers or "doubled" by spiking to an existing joist. Work done from the floor below. Add the cost of removing ceiling cover to expose the joists.

Jack two or three adjacent joists into position with a crossarm mounted on a hydraulic jack. Includes the cost of removing the jack.

	Craft@Hrs	Unit	Material	Labor	Total	Sell
Per jacking point	B1@.700	Ea	—	22.80	22.80	38.80
Add 2" x 6" ceiling joist						
Per linear foot of joist	B1@.042	LF	.59	1.37	1.96	3.33
Add 2" x 8" ceiling joist						
Per linear foot of joist	B1@.046	LF	.76	1.50	2.26	3.84
Add 2" x 10" ceiling joist						
Per linear foot of joist	B1@.050	LF	.97	1.63	2.60	4.42
Add 2" x 12" ceiling joist						
Per linear foot of joist	B1@.052	LF	1.32	1.69	3.01	5.12

Changes to Roof Framing

Rafters. Flat, shed, or gable roofs, up to 5 in 12 slope (5/24 pitch), maximum 25' span. Figures in parentheses indicate board feet per square foot of actual roof surface area (not roof plan area), including rafters, ridge boards, collar beams and normal waste, but no blocking, bracing, purlins, curbs, or gable walls.

	Craft@Hrs	Unit	Material	Labor	Total	Sell
2" x 4", Std & Btr						
2" x 4" rafters, 16" centers, per MBF	—	MBF	579.00	—	579.00	—
12" centers (.89 BF per SF)	B1@.021	SF	.52	.68	1.20	2.04
16" centers (.71 BF per SF)	B1@.017	SF	.41	.55	.96	1.63
24" centers (.53 BF per SF)	B1@.013	SF	.31	.42	.73	1.24
Remove and replace 2" x 4" rafter, per LF	B1@.050	LF	.39	1.63	2.02	3.43
2" x 6", Std & Btr						
2" x 6" rafters, 16" centers, per MBF	—	MBF	594.00	—	594.00	—
12" centers (1.29 BF per SF)	B1@.029	SF	.77	.94	1.71	2.91
16" centers (1.02 BF per SF)	B1@.023	SF	.61	.75	1.36	2.31
24" centers (.75 BF per SF)	B1@.017	SF	.45	.55	1.00	1.70
Remove and replace 2" x 6" rafter, per LF	B1@.068	LF	.59	2.21	2.80	4.76
2" x 8", Std & Btr						
2" x 8" rafters, 16" centers, per MBF	—	MBF	571.00	—	571.00	—
12" centers (1.71 BF per SF)	B1@.036	SF	.98	1.17	2.15	3.66
16" centers (1.34 BF per SF)	B1@.028	SF	.77	.91	1.68	2.86
24" centers (1.12 BF per SF)	B1@.024	SF	.64	.78	1.42	2.41
Remove and replace 2" x 8" rafter, per LF	B1@.087	LF	.76	2.83	3.59	6.10
2" x 10", Std & Btr						
2" x 10" rafters, 16" centers, per MBF	—	MBF	587.00	—	587.00	—
12" centers (2.12 BF per SF)	B1@.039	SF	1.24	1.27	2.51	4.27
16" centers (1.97 BF per SF)	B1@.036	SF	1.16	1.17	2.33	3.96
24" centers (1.21 BF per SF)	B1@.022	SF	.71	.72	1.43	2.43
Remove and replace 2" x 10" rafter, per LF	B1@.098	LF	.98	3.19	4.17	7.09
2" x 12", Std & Btr						
2" x 12" rafters, 16" centers, per MBF	—	MBF	664.00	—	664.00	—
12" centers (2.52 BF per SF)	B1@.045	SF	1.67	1.46	3.13	5.32
16" centers (1.97 BF per SF)	B1@.035	SF	1.31	1.14	2.45	4.17
24" centers (1.43 BF per SF)	B1@.026	SF	.95	.85	1.80	3.06
Remove and replace 2" x 12" rafter, per LF	B1@.113	LF	1.32	3.68	5.00	8.50
Add for hip roof	B1@.007	SF	—	.23	.23	.39
Add for slope over 5 in 12	B1@.015	SF	—	.49	.49	.83

Roof frame jacking. Repair sag in ridge board by jacking and installing shoring supported by bearing walls.

	Craft@Hrs	Unit	Material	Labor	Total	Sell
Per support installed	B1@7.27	Ea	5.30	236.00	241.30	410.00

Trimmers and curbs. At stairwells, skylights, dormers, etc. Figures in parentheses show board feet per LF, Std & Btr grade, including 10% waste.

	Craft@Hrs	Unit	Material	Labor	Total	Sell
2" x 4" (.73 BF per LF)	B1@.018	LF	.42	.59	1.01	1.72
2" x 6" (1.10 BF per LF)	B1@.028	LF	.65	.91	1.56	2.65
2" x 8" (1.47 BF per LF)	B1@.038	LF	.84	1.24	2.08	3.54
2" x 10" (1.83 BF per LF)	B1@.047	LF	1.07	1.53	2.60	4.42
2" x 12" (2.20 BF per LF)	B1@.057	LF	1.46	1.85	3.31	5.63

Collar beams & collar ties. Std & Btr grade, including 10% waste.

	Craft@Hrs	Unit	Material	Labor	Total	Sell
Collar beams, 2" x 6"	B1@.013	LF	.65	.42	1.07	1.82
Collar ties, 1" x 6"	B1@.006	LF	.79	.20	.99	1.68

	Craft@Hrs	Unit	Material	Labor	Total	Sell
Purlins (Purling). Std & Btr, installed below roof rafters. Figures in parentheses indicate board feet per LF including 5% waste.						
2" x 4" (.70 BF per LF)	B1@.012	LF	.41	.39	.80	1.36
2" x 6" (1.05 BF per LF)	B1@.017	LF	.62	.55	1.17	1.99
2" x 8" (1.40 BF per LF)	B1@.023	LF	.80	.75	1.55	2.64
2" x 10" (1.75 BF per LF)	B1@.027	LF	1.03	.88	1.91	3.25
2" x 12" (2.10 BF per LF)	B1@.030	LF	1.39	.98	2.37	4.03
4" x 6" (2.10 BF per LF)	B1@.034	LF	1.49	1.11	2.60	4.42
4" x 8" (2.80 BF per LF)	B1@.045	LF	2.27	1.46	3.73	6.34
Dormer studs. Std & Btr, per square foot of wall area, including 10% waste.						
2" x 4", (.84 BF per SF)	B1@.033	SF	.49	1.07	1.56	2.65
Roof trusses. 24" OC, any slope from 3 in 12 to 12 in 12, total height not to exceed 12' high from bottom chord to highest point on truss. Prices for trusses over 12' high will be up to 100% higher. Square foot (SF) costs, where shown, are per square foot of roof area to be covered.						
Scissor truss						
2" x 4" top and bottom chords						
Up to 38' span	B1@.022	SF	2.36	.72	3.08	5.24
40' to 50' span	B1@.028	SF	2.95	.91	3.86	6.56
Fink truss W (conventional roof truss)						
2" x 4" top and bottom chords						
Up to 38' span	B1@.017	SF	2.05	.55	2.60	4.42
40' to 50' span	B1@.022	SF	2.47	.72	3.19	5.42
2" x 6" top and bottom chords						
Up to 38' span	B1@.020	SF	2.53	.65	3.18	5.41
40' to 50' span	B1@.026	SF	3.04	.85	3.89	6.61
Truss with gable fill at 16" OC						
28' span, 5 in 12 slope	B1@.958	Ea	170.00	31.20	201.20	342.00
32' span, 5 in 12 slope	B1@1.26	Ea	210.00	41.00	251.00	427.00
40' span, 5 in 12 slope	B1@1.73	Ea	295.00	56.30	351.30	597.00
Roof truss repairs.						
Repair splits in 2" x 6" diagonal or 2" x 6" vertical member by installing five prefabricated clamps						
per truss	B1@.492	Ea	—	16.00	16.00	27.20
Repair splits at various locations by installing splice plates						
per splice plate	B1@.104	Ea	—	3.38	3.38	5.75
Repair splits in bottom chord by installing two prefabricated clamps near joints, drill holes to arrest split						
per truss	B1@.280	Ea	—	9.11	9.11	15.50
Repair split ends in 2" x 6" bottom chord by installing splice plate and four 1/2" stitch bolts						
per truss	B1@.617	Ea	—	20.10	20.10	34.20
Repair vertical compression in 2" x 4" member by scabbing pieces to each side and bolting						
per scab repair	B1@1.00	Ea	—	32.50	32.50	55.30
Fascia board. Installed on roof eaves working at 8' height from ladders.						
Pry off nailed fascia and drop debris to the ground						
per linear foot of fascia	B1@.007	LF	—	.23	.23	.39
Install 1" x 6" fascia board by nailing at roof eaves						
per linear foot of fascia	B1@.044	LF	1.00	1.43	2.43	4.13
Remove and replace fascia board at roof eaves						
per linear foot of fascia	B1@.051	LF	1.00	1.66	2.66	4.52
Add for heights to 20', per linear foot	B1@.015	LF	—	.49	.49	.83

	Craft@Hrs	Unit	Material	Labor	Total	Sell
Roof sheathing.						
CDX plywood sheathing, rough, power nailed, including normal waste.						
1/2"	B1@.013	SF	.60	.42	1.02	1.73
1/2", 4-ply	B1@.013	SF	.65	.42	1.07	1.82
1/2", 5-ply	B1@.013	SF	.69	.42	1.11	1.89
5/8", 4-ply	B1@.013	SF	.74	.42	1.16	1.97
5/8", 5-ply	B1@.013	SF	.80	.42	1.22	2.07
3/4"	B1@.013	SF	.86	.42	1.28	2.18
Add for hip roof	B1@.007	SF	—	.23	.23	.39
Add for steep pitch or cut-up roof	B1@.015	SF	—	.49	.49	.83
Remove existing sheathing	B1@.017	SF	—	.55	.55	.94
OSB sheathing, rough, power nailed, including normal waste.						
1/2"	B1@.013	SF	.50	.42	.92	1.56
5/8"	B1@.013	SF	.51	.42	.93	1.58
3/4"	B1@.013	SF	.60	.42	1.02	1.73
7/8"	B1@.014	SF	1.14	.46	1.60	2.72
1"	B1@.015	SF	1.65	.49	2.14	3.64
Add for hip roof	B1@.007	SF	—	.23	.23	.39
Add for steep pitch or cut-up roof	B1@.015	SF	—	.49	.49	.83
Remove existing sheathing	B1@.017	SF	—	.55	.55	.94
Board sheathing, 1" x 6" or 1" x 8" utility T&G laid diagonal						
1" utility T&G lumber, per MBF	—	MBF	1,590.00	—	1,590.00	—
(1.13 BF per SF)	B1@.026	SF	1.97	.85	2.82	4.79
Add for hip roof	B1@.007	SF	—	.23	.23	.39
Add for steep pitch or cut-up roof	B1@.015	SF	—	.49	.49	.83
Remove board sheathing	B1@.030	SF	—	.98	.98	1.67
Remove deteriorated boards	B1@.034	SF	—	1.11	1.11	1.89
Remove boards in salvage condition	B1@.036	SF	—	1.17	1.17	1.99
Remove and patch roof sheathing, rafters 24" on center, 3 in 12 pitch.						
1" x 8" x 12' long, per board	B1@.400	Ea	12.20	13.00	25.20	42.80
1" x 8", per square foot	B1@.050	SF	1.52	1.63	3.15	5.36
1/2", 4-ply plywood, per 4' x 8' sheet	B1@.960	Ea	20.80	31.20	52.00	88.40
1/2", 4-ply plywood, per square foot	B1@.030	SF	.65	.98	1.63	2.77

Insulation and Moisture Control

5

Homes built before the early 1970s were designed in an era when gasoline was less than 30 cents a gallon, heating oil was cheap and natural gas supplies were plentiful. Few people worried about energy efficiency because there was energy to spare. Builders and homeowners didn't spend money on insulating houses because it was cheaper just to turn up the heat. But times have changed. Modern insulation standards are considerably higher.

Ceiling Insulation

Most homes have an accessible attic with exposed ceiling framing. That makes it easy to check the depth of insulation and apply more when needed. Before about 1970, the most common attic insulation, if there was any, was reflective foil (in warmer climates) or a few inches of rock wool (in colder climates).

Optimum insulation still varies with location. But homes in milder climates can benefit from 6" of fiberglass above the ceiling. In colder climates, you need 12". If the home is heated by electricity or a means more expensive than gas, propane or heating oil, extra insulation usually makes good economic sense.

> The effectiveness of insulation is measured in R-value, which is the resistance of the material to heat transfer. The higher the rating, the more insulating value. For example, 12" of fiberglass insulation usually carries an R-38 rating. Six inches of the same fiberglass will be rated at R-19.

Batt and roll blanket insulation is made to fit between joists and studs installed either 16" or 24" on center. When you see coverage figures for insulation, those numbers include framing area. For example a roll of insulation described as "covers 80 SF" will fill a wall area measuring 8' by 10'.

If a vapor barrier is already in place, or if you're installing a separate vapor barrier, use unfaced rolls or batts. Otherwise, install Kraft-faced insulation. However, there's one exception when adding more insulation in an attic space. Even if the existing ceiling insulation doesn't have a vapor barrier, install unfaced insulation.

Loose-fill insulation is just as effective as rolls or batts and it's easy to install in an attic. Simply pour insulation between the joists and screed it off to the right thickness (Figure 5-1).

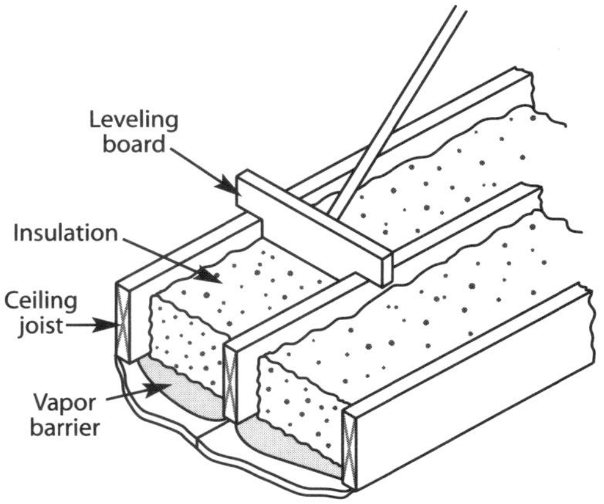

Figure 5-1

Installation of loose-fill ceiling insulation

Wall Insulation

Any time you remove interior wall finish, check the insulation. Walls in mild climates need R-13 insulation. That's 3$^1/_2$" of fiberglass, which fits perfectly in the cavity of a 2 x 4 stud wall. R-19 wall insulation is appropriate in severely cold climates. But that requires a 6" wall cavity and 2 x 6 studs. If you need more than R-19 from an existing 2 x 4 stud wall, 1" polyurethane sheathing can add R-6.

Adding insulation is easy when the wall cavity is open. When the cavity is closed, you can blow cellulose insulation through a 1"-diameter hole cut in siding and sheathing. Open a hole at the top of each stud space. Lower a plumb bob into the cavity to find where the fire block is installed. Then drill another hole in the same cavity below the fire block. Blow cellulose into the stud space both above and below the block. When the cavity is filled, patch entry holes with either plastic or wood plugs.

Floor Insulation

Houses with a furnace in the basement don't need insulation between the floor joists. But homes should have floor insulation equivalent to the wall insulation if the floors are located above an unheated crawlspace or any other unheated space. Fit insulation batts between the floor joists. Hold the batts in place with wood or wire strips cut slightly longer than the joist space so they spring into place (Figure 5-2). You could also staple wire netting under the floor joists to hold either batt or blanket insulation tight against the underside of the floor sheathing. A vapor barrier has to be on the top side of insulation placed between floor joists. That presents a problem when installing insulation above a crawlspace. Insulation manufacturers now offer roll insulation with inverse tabs so insulation can be stapled between floor joists with the poly face-up against the floor sheathing.

Board Insulation

You can insulate brick, block and concrete walls by applying insulation board to the interior surface. The thicker the board, the higher the R-value. Polystyrene insulated sheathing board has an R-value of 3.5 per inch of thickness. Polyurethane board carries a 6.25 rating per inch of thickness. Many types of insulated sheathing can be installed over the existing siding. No tear-off is necessary, though additional edge trim will be needed around windows and doors. If the owner wants 2" of foam insulation on a basement wall, consider attaching 2" x 2" furring strips to the wall at 16" on center. Then install 2" insulation board between the strips and hang drywall or other wall cover on the furring. Most insulation boards, such as polystyrene, must be covered with drywall. They're fragile, and won't hold paint or wallpaper. Also, most insulation boards aren't fire-resistant. Fire codes require that interior wall surfaces be covered with fire-resistant material, such as drywall.

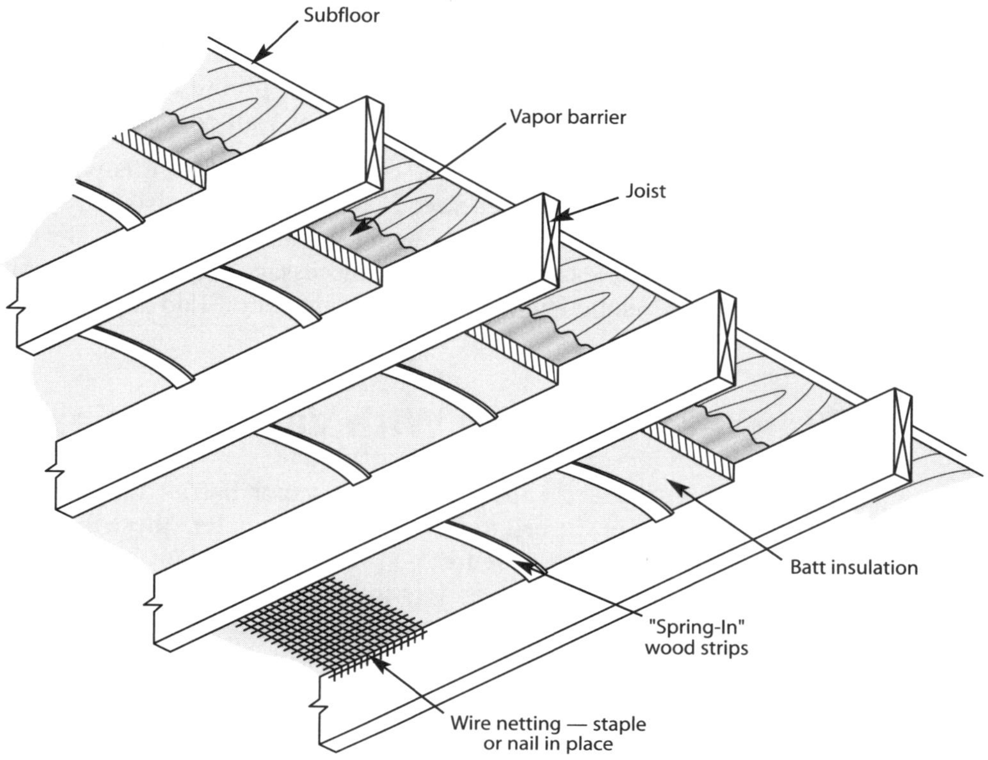

Figure 5-2
Installing insulating batts in floor

Vapor Barrier

Good insulation cuts heating costs and adds to comfort by making the temperature in the house more uniform. But simply adding insulation to an old home can create more problems than it solves. In poorly-insulated older homes, water vapor escapes to the exterior without interference. But in a tighter home with fewer air leaks and better insulation, water vapor tends to collect in walls. There, it cools and condenses into liquid. The result can be saturated insulation and siding. Running a humidifier, cooking, bathing, and simply breathing aggravate the problem. Liquid moisture in a wall reduces the effectiveness of insulation, promotes decay and supports the growth of mold. To control moisture problems, be sure there is a vapor barrier on the (winter) warm side of walls and ceilings.

Most homes built before the mid-1930s don't have a vapor barrier in either the walls or the ceilings. If attic insulation has been upgraded since construction, you'll probably find batts or blankets with a Kraft face that resist the passage of moisture. The Kraft face should face down, against the ceiling finish. But if the ceiling insulation is loose fill, there should be a separate vapor barrier of coated paper, aluminum foil, or plastic film below the insulation.

Older homes develop thousands of tiny air leaks around doors and windows, at the sill, around electrical outlets and where interior and exterior walls have cracked.

> Vapor movement through any barrier is measured in perms. The lower the perm rating, the more effective the barrier. Materials with good perm ratings include polyethylene at least 2 mils thick, asphalt-impregnated and surface-coated Kraft papers, and asphalt-laminated paper.

To be truly effective, the vapor barrier should be continuous over the inside of studs and joists. That's easy in new construction. In home improvement work, applying continuous vapor barrier is practical only when interior wall cover is being replaced.

If removing the ceiling or wall finish is part of the job, staple vapor barrier directly to the studs or joists. Lay vapor barrier on a subfloor directly under the new floor finish. Be sure to lap joints at least 2" and avoid unnecessary punctures. Nails or screws driven snug against the vapor barrier don't do any harm.

Blanket Insulation With Vapor Barrier

Most roll (or "blanket") insulation has a vapor barrier on one side that extends beyond the edges of the roll to form a stapling tab. Lap these tabs over the 2" thickness of studs and joists. (Figure 5-3). Tabs on adjacent blankets should overlap. Your drywall crew probably prefers to see these tabs stapled to the inside face of studs or joists. That makes it easier to hang drywall with a smooth, even surface. But stapling tabs inside the stud or joist cavity will also reduce the effectiveness of the vapor barrier. Moisture will seep through openings between the tabs and framing members. It's better to make the drywall finishers work a little harder and protect the studs and joists from decay, mold and termites.

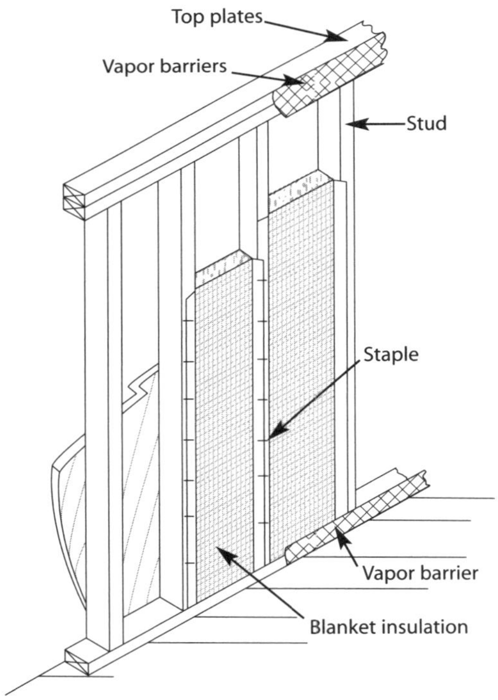

Figure 5-3

Installation of blanket insulation with vapor barriers on one side

Vapor-Resistant Coating

If walls and ceilings don't have a vapor barrier, the next best thing is a vapor-resistant interior coating such as aluminum primer. Two coats of aluminum primer covered with acrylic latex offer moderate vapor resistance, though not as much as a true membrane.

Crawlspace Ventilation

Damp soil in a crawlspace will transfer moisture to the floor framing and even to occupied rooms above. Good ventilation helps keep the crawlspace dry. Promote good cross-ventilation by installing a vent at each corner of the foundation. Taken together, foundation vent "free circulation area" should be $^{1}/_{150}{}^{th}$ of the crawlspace surface (10 square feet of vent for a 1,500-square-foot crawlspace). Free vent area excludes space occupied by the vent frame, screen wire and louvers. Most manufactured vents are identified by the overall dimensions and the net free area. If there's a partial basement, the crawlspace can be vented to the basement instead of the exterior.

If soil in the crawlspace is covered with a vapor barrier, foundation vent area can be as little as $^1/_{1,500}$th of the ground area (1 square foot for a 1,500-square-foot crawlspace). It's usually much cheaper to cover the soil in a crawlspace with membrane than to add more foundation vents. Ground cover should lap at least 2" at all joints (Figure 5-4). Lay brick or stones on top of the membrane to hold it down and prevent curling.

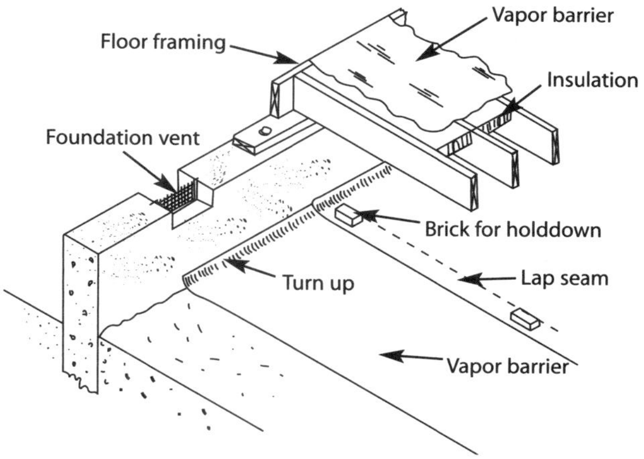

Figure 5-4

Vapor barrier for crawlspace (ground cover)

Attic and Roof Ventilation

Insulation laid between ceiling joists helps hold heat inside the home. But above that insulation, air should circulate freely. Excess moisture accumulation will damage insulation and other attic materials. Good air circulation is especially important in colder climates and when the roof cover is less permeable. With built-up roofing or asphalt shingles, the only way to get moisture out of an attic is with vents.

Inadequately-vented attics can reach temperatures above 140 degrees F on a sunny day. Some of the heat will work its way into the house, overburdening the air conditioning system. Properly-vented attics will stay cooler and can save up to 30 percent on cooling costs. In addition, a cooler attic will keep asphalt shingles from degrading prematurely in the heat.

Air circulation works best with soffit vents around the edge of the roof and outlet vents located high, near the ridge. Warm air rises in the attic, exits at the peak, and draws fresh air through soffit vents at the perimeter. Under those conditions, inlet vents should be $^1/_{900}$th of attic floor area. Outlet vents should have the same area. A home with only gable vents needs more free vent area. Air circulation through gable vents depends on wind blowing the right direction. Provide one square foot of free vent area for each 300' of attic area.

Hip roofs don't have gables and flat roofs don't have peaks. That reduces your venting options. The estimating section that follows includes power ventilators that can be installed on nearly any roof.

Flat roofs with no attic require some type of ventilation above the ceiling insulation. If this space is divided by joists, each joist space should be ventilated. A continuous soffit vent strip is the best choice. Drill through all headers that impede passage of air to the opposite eave.

A post and beam roof with a roof-plank ceiling has no attic and needs no ventilation. But the roof surface requires insulation board coated with a walkable sprayed-on insulator such as urethane.

	Craft@Hrs	Unit	Material	Labor	Total	Sell

Fiberglass Insulation

Kraft-faced fiberglass roll insulation. Stapled between studs or joists. Coverage includes framing area.

	Craft@Hrs	Unit	Material	Labor	Total	Sell
R-11, 15" x 70'6" x 3-1/2" thick	BC@.010	SF	.24	.36	.60	1.02
R-11, 23" x 70'6" x 3-1/2" thick	BC@.009	SF	.23	.32	.55	.94
R-13, 15" x 32' x 3-1/2" thick	BC@.010	SF	.24	.36	.60	1.02
R-19, 15" x 39' x 6-1/4" thick	BC@.010	SF	.33	.36	.69	1.17
R-19, 23" x 39' x 6-1/4" thick	BC@.009	SF	.47	.32	.79	1.34

Unfaced fiberglass roll insulation. Laid between ceiling joists. Coverage includes framing area.

	Craft@Hrs	Unit	Material	Labor	Total	Sell
R-6.7, 15" x 48" x 2" thick handy roll	BC@.005	SF	.95	.18	1.13	1.92
R-19, 15" x 39' x 6-1/4" thick	BC@.005	SF	.35	.18	.53	.90
R-19, 23" x 39' x 6-1/4" thick	BC@.005	SF	.35	.18	.53	.90
R-25, 23" x 17' x 8" thick	BC@.005	SF	.53	.18	.71	1.21
R-25, 15" x 18' x 8" thick	BC@.005	SF	.59	.18	.77	1.31
R-30, 15" x 25' x 9-1/2" thick	BC@.005	SF	.52	.18	.70	1.19
R-30, 23" x 25' x 9-1/2" thick	BC@.005	SF	.52	.18	.70	1.19
Add for rolls in floor joists above a crawl space	BC@.008	SF	.05	.29	.34	.58

Kraft-faced fiberglass batt insulation. Stapled between studs and joists. 16" widths and 24" widths are laid over ceiling joists.

	Craft@Hrs	Unit	Material	Labor	Total	Sell
R-11, 15" x 93" x 3" thick	BC@.012	SF	.27	.43	.70	1.19
R-13, 15" x 93" x 3-1/2" thick	BC@.012	SF	.35	.43	.78	1.33
R-13, 23" x 93" x 3-1/2" thick	BC@.013	SF	.35	.47	.82	1.39
R-19, 15" x 93" x 6-1/4" thick	BC@.009	SF	.66	.32	.98	1.67
R-19, 23" x 93" x 6-1/4" thick	BC@.013	SF	.66	.47	1.13	1.92
R-30, 16" x 48" x 10" thick	BC@.005	SF	.79	.18	.97	1.65
R-30, 24" x 48" x 10" thick	BC@.005	SF	.79	.18	.97	1.65
R-38, 16" x 48" x 12-1/2" thick	BC@.007	SF	.99	.25	1.24	2.11
R-38, 24" x 48" x 12-1/2" thick	BC@.005	SF	.99	.18	1.17	1.99

Unfaced fiberglass batt insulation. Fit between studs and joists. 16" and 24" widths are laid over ceiling joists. Coverage includes framing area.

	Craft@Hrs	Unit	Material	Labor	Total	Sell
R-11, 15" x 94" x 3-1/2" thick	BC@.007	SF	.29	.25	.54	.92
R-11, 16" x 96" x 3-1/2" thick	BC@.006	SF	.29	.22	.51	.87
R-13, 15" x 93" x 3-1/2" thick	BC@.007	SF	.35	.25	.60	1.02
R-19, 15" x 93" x 6-1/4" thick	BC@.007	SF	.44	.25	.69	1.17
R-19, 23" x 93" x 6-1/4" thick	BC@.006	SF	.47	.22	.69	1.17
R-21, 15" x 93" x 5-1/2" thick	BC@.007	SF	.60	.25	.85	1.45
R-21, 23" x 93" x 5-1/2" thick	BC@.006	SF	.61	.22	.83	1.41
R-30, 16" x 48" x 9-1/2" thick	BC@.005	SF	.82	.18	1.00	1.70
R-30, 16" x 48" x 10-1/4" thick	BC@.005	SF	.80	.18	.98	1.67
R-30, 24" x 48" x 9-1/2" thick	BC@.005	SF	.77	.18	.95	1.62
R-30, 24" x 48" x 10-1/4" thick	BC@.005	SF	.82	.18	1.00	1.70
R-38, 16" x 48" x 13" thick	BC@.007	SF	.95	.25	1.20	2.04
R-38, 24" x 48" x 13" thick	BC@.007	SF	.97	.25	1.22	2.07
Add for batts in floor joists above a crawl space	BC@.008	SF	.05	.29	.34	.58

	Craft@Hrs	Unit	Material	Labor	Total	Sell
Encapsulated roll fiberglass insulation. Stapled between studs and joists. Coverage includes framing area. Fibers are coated to reduce irritation.						
R-13, 15" x 37'6" x 3-1/2" thick	BC@.010	SF	.45	.36	.81	1.38
R-13, 15" x 32' x 3-1/2" thick	BC@.010	SF	.45	.36	.81	1.38
R-25, 15" x 22' x 8-1/4" thick	BC@.010	SF	.71	.36	1.07	1.82
R-25, 23" x 22' x 8-1/4" thick	BC@.009	SF	.71	.32	1.03	1.75
Encapsulated batt fiberglass insulation. Fit between studs and joists. Coverage includes framing area. Fibers are coated to reduce irritation. 16" widths and 24" widths are laid over ceiling joists.						
R-13, 15" x 93" x 3-1/2" thick	BC@.007	SF	.49	.25	.74	1.26
R-19, 15" x 93" x 6-1/2" thick	BC@.007	SF	.60	.25	.85	1.45
R-19, 23" x 8' x 6-1/2" thick	BC@.005	SF	.60	.18	.78	1.33
R-30, 16" x 48" x 10-1/4" thick	BC@.005	SF	.87	.18	1.05	1.79
R-30, 24" x 48" x 10-1/4" thick	BC@.005	SF	.91	.18	1.09	1.85
Blow-in cellulose insulation. One bag covers 40 square feet at 6" deep and provides an R-19 rating.						
40 square foot bag of stabilized cellulose	BC@.400	Ea	12.20	14.40	26.60	45.20
Blower and hose, daily rental	—	Ea	50.00	—	50.00	—
Hose connector	—	Ea	7.00	—	7.00	—
2" to 1" hose reducer and nozzle	—	Ea	39.00	—	39.00	—
1" wood plugs, bag of 25	BC@.500	Ea	8.50	18.00	26.50	45.10
1" plastic plugs, bag of 25	BC@.500	Ea	7.75	18.00	25.75	43.80

Panel Insulation

	Craft@Hrs	Unit	Material	Labor	Total	Sell
Insulation board, 4' x 8' panels. Owens Corning Foamular XAE rigid extruded polystyrene foam board. Film-faced, stapled in place, including taped joints and 5% waste. 4' x 8' or 9' panels, tongue and groove or square edge.						
1/2" thick (R-3.2), $11.98 per panel	BC@.010	SF	.37	.36	.73	1.24
3/4" thick (R-3.8), $15.25 per panel	BC@.011	SF	.48	.40	.88	1.50
1" thick (R-5.0), $18.83 per panel	BC@.011	SF	.59	.40	.99	1.68
1-1/2" thick (R-7.5), $22.95 per panel	BC@.015	SF	.72	.54	1.26	2.14
2" thick (R-10), $29.95 per panel	BC@.015	SF	.94	.54	1.48	2.52
Extruded polystyrene insulated sheathing. Dow Blue Board, water resistant. Blue Board has a density of 1.6 pounds per cubic foot. Gray Board has a density of 1.3 pounds per cubic foot. Include 5% waste.						
3/4" x 2' x 8', tongue & groove, R-3.8	BC@.011	SF	.45	.40	.85	1.45
1" x 2' x 8', R-5.0	BC@.011	SF	.66	.40	1.06	1.80
1" x 2' x 8', tongue & groove, R-5.0	BC@.011	SF	.86	.40	1.26	2.14
1" x 4' x 8', R-5.0	BC@.011	SF	.62	.40	1.02	1.73
1" x 4' x 8', tongue & groove, R-5.0	BC@.011	SF	.81	.40	1.21	2.06
1-1/2" x 2' x 8', R-7.5	BC@.015	SF	.96	.54	1.50	2.55
1-1/2" x 2' x 8', tongue & groove, R-7.5	BC@.015	SF	1.05	.54	1.59	2.70
1-1/2" 4' x 8', square edge, R-7.5	BC@.015	SF	1.04	.54	1.58	2.69
2" x 2' x 8', R-10	BC@.015	SF	1.24	.54	1.78	3.03
2" x 4' x 8', R-10	BC@.015	SF	1.28	.54	1.82	3.09
1" x 2' x 8' gray (1.3 Lbs. per CF)	BC@.015	SF	.34	.54	.88	1.50
1-1/2" x 2' x 8' gray board	BC@.015	SF	.82	.54	1.36	2.31
2" x 2' x 8' gray board	BC@.015	SF	1.05	.54	1.59	2.70
Rigid foam insulated sheathing. Fiberglass polymer faced two sides, Celotex Tuff-R, 4' x 8' panels, including 5% waste.						
1/2" (R-3.2)	BC@.010	SF	.32	.36	.68	1.16
3/4" (R-5.6)	BC@.011	SF	.46	.40	.86	1.46
1" (R-7.2)	BC@.011	SF	.63	.40	1.03	1.75
2" (R-14.4)	BC@.015	SF	1.03	.54	1.57	2.67

	Craft@Hrs	Unit	Material	Labor	Total	Sell
Fanfold extruded polystyrene insulation. 4' x 50' panels, including 5% waste.						
1/4" (R-1)	BC@.011	SF	.23	.40	.63	1.07
3/8" (R-1)	BC@.011	SF	.26	.40	.66	1.12
Add for nails, per 50-pound carton (large square heads)						
2-1/2", for 1-1/2" boards	—	LS	150.00	—	150.00	—
3", for 2" boards	—	LS	150.00	—	150.00	—
Foil-faced urethane sheathing. Including 5% waste. 4' x 8' panels. Including 5% waste.						
1" (R-7.2)	BC@.011	SF	.63	.40	1.03	1.75
Polystyrene foam underlay. Polyethylene covered, R-Gard. Including 5% waste.						
3/8" 4' x 24' fanfold	BC@.011	SF	.28	.40	.68	1.16
1/2" x 4' x 8'	BC@.011	SF	.30	.40	.70	1.19
1" x 4' x 8'	BC@.011	SF	.49	.40	.89	1.51
1-1/2" x 2' x 8'	BC@.015	SF	.80	.54	1.34	2.28
2" x 2' x 8'	BC@.015	SF	.92	.54	1.46	2.48
Perlite roof insulation. 24" x 48" board.						
3/4" thick (R-2.08)	RR@.007	SF	.36	.28	.64	1.09
1" thick (R-2.78)	RR@.007	SF	.47	.28	.75	1.28
1-1/2" thick (R-4.17)	RR@.007	SF	.66	.28	.94	1.60
2" thick (R-5.26)	RR@.007	SF	.80	.28	1.08	1.84
Cant strips for insulated roof board. One piece wood, tapered.						
2" strip	RR@.018	LF	.21	.73	.94	1.60
3" strip	RR@.020	LF	.20	.81	1.01	1.72
4" strip	RR@.018	LF	.19	.73	.92	1.56
Sill plate gasket. Fills gaps between the foundation and the plate. Also used for sealing around windows and doors.						
1/4" x 50', 3-1/2" wide	BC@.003	LF	.10	.11	.21	.36
1/4" x 50', 5-1/2" wide	BC@.003	LF	.14	.11	.25	.43
Perimeter foundation insulation. Expanded polystyrene with polyethylene skinned surface. Low moisture retention. Minimum compressive strength of 1,440 pounds per SF.						
3/4" x 4' x 8'	BC@.011	SF	.24	.40	.64	1.09
1-1/2" x 2' x 4'	BC@.015	SF	.55	.54	1.09	1.85
2" x 2' x 4'	BC@.015	SF	.76	.54	1.30	2.21
2" x 4' x 4'	BC@.015	SF	.60	.54	1.14	1.94
FBX 1240 industrial board insulation. Rockwool, semi-rigid, no waste included. 24" on-center framing members.						
2" (R-8)	BC@.004	SF	.54	.14	.68	1.16
3" (R-12)	BC@.004	SF	.79	.14	.93	1.58
3-1/2" (R-14)	BC@.004	SF	.92	.14	1.06	1.80
Sound control insulation. Fiberglass acoustically designed to absorb sound vibration. For use between interior walls, floors and ceilings. Quietzone™ batts. Coverage allows for framing.						
3-1/2", 16" on-center framing	BC@.006	SF	.32	.22	.54	.92
3-1/2", 24" on-center framing	BC@.006	SF	.32	.22	.54	.92
Sound insulation board. 4' x 8' panels installed on walls.						
1/2" Homasote	BC@.013	SF	.85	.47	1.32	2.24
1/2" sound deadening board	BC@.013	SF	.31	.47	.78	1.33
Add for installation on ceilings	BC@.004	SF	—	.14	.14	.24

	Craft@Hrs	Unit	Material	Labor	Total	Sell
Sound Attenuation Fire Batt Insulation (SAFB). Rockwool, semi-rigid, no waste included, pressed between framing members.						
16" on-center framing members						
2" (R-8)	BC@.004	SF	.60	.14	.74	1.26
3" (R-12)	BC@.004	SF	.85	.14	.99	1.68
4" (R-16)	BC@.005	SF	1.11	.18	1.29	2.19
24" on-center framing members						
2" (R-8)	BC@.003	SF	.57	.11	.68	1.16
3" (R-12)	BC@.003	SF	.85	.11	.96	1.63
4" (R-16)	BC@.004	SF	1.11	.14	1.25	2.13

Loose Insulation

	Craft@Hrs	Unit	Material	Labor	Total	Sell
Vermiculite insulation. Poured over ceilings.						
Vermiculite, 25 lb sack (2 CF)	—	Ea	21.40	—	21.40	—
At 3" depth (120 sacks per 1,000 SF)	BL@.007	SF	2.57	.20	2.77	4.71
At 4" depth (144 sacks per 1,000 SF)	BL@.007	SF	3.08	.20	3.28	5.58
Masonry fill insulation. Poured in concrete block cores.						
Using 2 CF bags of perlite, per bag	—	Ea	17.30	—	17.30	—
4" wall, 8.1 SF per CF	B9@.006	SF	1.07	.19	1.26	2.14
6" wall, 5.4 SF per CF	B9@.006	SF	1.60	.19	1.79	3.04
8" wall, 3.6 SF per CF	B9@.006	SF	2.40	.19	2.59	4.40
Natural fiber building insulation. EPA registered anti-microbial agent that offers protection from mold, mildew, fungi and pests. Made from all natural cotton fibers. Friction fit between framing 16" or 24" on center. Based on Ultra Touch.						
16" x 48", R-8	BC@.005	SF	1.15	.18	1.33	2.26
16" x 94", R-13	BC@.006	SF	.76	.22	.98	1.67
24" x 94", R-13	BC@.006	SF	.72	.22	.94	1.60

Vapor Barrier

	Craft@Hrs	Unit	Material	Labor	Total	Sell
Polyethylene film. Clear or black. Includes 5% for waste and 10% for laps.						
4 mil (.004" thick)						
50 LF rolls, 3' to 20' wide	BL@.003	SF	.06	.09	.15	.26
6 mil (.006" thick)						
50 LF rolls, 3' to 40' wide	BL@.003	SF	.08	.09	.17	.29
Vapor barrier paper. Including 12% for overlap and waste.						
15 lb. asphalt felt, 432 SF roll, 36" x 144'	BC@.003	SF	.06	.11	.17	.29
30 lb. asphalt felt, 216 SF roll, 36" x 72'	BC@.003	SF	.12	.11	.23	.39
SuperJumbo Tex black building paper, asphalt saturated, (162 SF roll)	BC@.003	SF	.08	.11	.19	.32
Red rosin sized sheathing paper (duplex sheathing) 36" wide, (500 SF roll)	BC@.003	SF	.03	.11	.14	.24
Double Kraft Aquabar, Class A, 36" wide, 30-50-30, (1,000 SF roll)	BC@.003	SF	.04	.11	.15	.26

	Craft@Hrs	Unit	Material	Labor	Total	Sell
Double Kraft Aquabar, class B, 36" wide, 30-30-30, (500 SF roll)	BC@.003	SF	.05	.11	.16	.27
Vaporstop 298, fiberglass reinforcing and asphaltic adhesive between 2 layers of Kraft, 32" x 405' roll (1,080 SF roll)	BC@.003	SF	.11	.11	.22	.37
Pyro-Kure 600, 2 layers of heavy Kraft with fire-retardant adhesive edge reinforced with fiberglass, 32" x 405' roll (1,080 SF roll)	BC@.003	SF	.18	.11	.29	.49
Foil Barrier 718, fiberglass-reinforced aluminum foil and adhesive between 2 layers of Kraft, 52" x 231' roll (1,000 SF roll)	BC@.004	SF	.32	.14	.46	.78
Seekure, fiberglass reinforcing strands and non-staining adhesive between 2 layers of Kraft, 48" x 300' roll, (1,200 SF roll)	BC@.003	SF	.15	.11	.26	.44
Moistop, fiberglass-reinforced Kraft between 2 layers of polyethylene, 8' x 250' roll (2,000 SF roll)	BC@.003	SF	.20	.11	.31	.53

Ice and water shield. Self-adhesive rubberized asphalt and poly. Vycor.

225 SF roll	BC@.006	SF	.61	.22	.83	1.41

Roof flashing paper. Seals around skylights, dormers, vents, valleys and eaves. Rubberized, fiberglass reinforced, self-adhesive. GAF StormGuard.

Roll covers 200 SF at $99.00	RR@.006	SF	.50	.24	.74	1.26

Weatherproof roof underlay. For use under shingles. Granular surface.

Roll covers 200 SF at $80.00	RR@.006	SF	.40	.24	.64	1.09

Tyvek™ House Wrap. Air infiltration barrier. High-density polyethylene fiber sheet.

3' x 165' rolls or 9' x 100' rolls	BC@.003	SF	.24	.11	.35	.60
House Wrap tape, 2" x 165'	BC@.005	LF	.13	.18	.31	.53

Vents and Louvers

Rafter Bay Vent Channel. Keeps attic insulation away from soffit vents to promote good air circulation. Extruded polystyrene. 48" x 1-1/4".

Rafters 16" on center, 14" wide, 15" net-free vent area	BC@.220	Ea	1.99	7.92	9.91	16.80
Rafters 24" on center, 22" wide, 26" net-free vent area	BC@.220	Ea	2.53	7.92	10.45	17.80

Replacement automatic foundation vent. Fits 8" x 16" concrete block opening. Aluminum mesh on front and molded plastic screen on back. Opens at 72 degrees F and closes at and 38 degrees F. 1" frame each side.

Automatic open and close	SW@.255	Ea	16.20	10.30	26.50	45.10

Foundation vent with manual damper. Sheet metal.

4" x 16"	SW@.255	Ea	8.85	10.30	19.15	32.60
6" x 16"	SW@.255	Ea	9.64	10.30	19.94	33.90
8" x 16"	SW@.255	Ea	11.20	10.30	21.50	36.60

Foundation access door. Replacement cost only.

24" x 24"	BC@.363	Ea	29.10	13.10	42.20	71.70
32" x 24"	BC@.363	Ea	33.60	13.10	46.70	79.40

	Craft@Hrs	Unit	Material	Labor	Total	Sell
Foundation screen vent.						
14" x 6"	BC@.255	Ea	3.34	9.18	12.52	21.30
16" x 4"	BC@.255	Ea	2.86	9.18	12.04	20.50
16" x 6"	BC@.255	Ea	3.18	9.18	12.36	21.00
16" x 8"	BC@.255	Ea	3.58	9.18	12.76	21.70

Under eave vent. Decomesh vent strips. 26 square inches free air per linear foot. 3/32" diameter perforations. 0.016-gauge aluminum.

4-1/2" x 8', white	BC@.040	LF	1.20	1.44	2.64	4.49
4-1/2" x 8', brown	BC@.040	LF	1.20	1.44	2.64	4.49
6" x 8', white	BC@.040	LF	1.69	1.44	3.13	5.32

Continuous louvered soffit vent. Reversible for flush or recessed mounting.

2-5/8" x 8', mill finish aluminum	BC@.360	Ea	3.83	13.00	16.83	28.60
2-5/8" x 8', white finish aluminum	BC@.360	Ea	5.08	13.00	18.08	30.70

Louvered soffit vent. Finned louvers with galvanized steel frame and wire mesh. Replacement cost only. For new installations, add the cost of cutting a hole in an existing soffit.

14" x 4"	BC@.220	Ea	3.18	7.92	11.10	18.90
16" x 7"	BC@.220	Ea	3.58	7.92	11.50	19.60
14" x 6"	BC@.220	Ea	3.81	7.92	11.73	19.90
14" x 8"	BC@.220	Ea	4.25	7.92	12.17	20.70
22" x 3"	BC@.220	Ea	3.58	7.92	11.50	19.60

Plastic gable louver vent. With double baffle and screen. Replacement cost only. For new installations, add the cost of cutting a hole in an existing gable.

Half round, 16" x 24", white	BC@.500	Ea	52.20	18.00	70.20	119.00
Half round, 22" x 34", paintable	BC@.500	Ea	69.10	18.00	87.10	148.00
Half round, 22" x 34", white	BC@.500	Ea	69.10	18.00	87.10	148.00
Octagon, 18", #01 white	BC@.500	Ea	51.80	18.00	69.80	119.00
Octagon, 22", paintable	BC@.500	Ea	60.80	18.00	78.80	134.00
Octagon, 22", white	BC@.500	Ea	60.80	18.00	78.80	134.00
Rectangular, 12" x 18", ivory	BC@.500	Ea	39.30	18.00	57.30	97.40
Rectangular, 12" x 18", paintable	BC@.500	Ea	39.30	18.00	57.30	97.40
Rectangular, 18" x 24", paintable	BC@.500	Ea	49.70	18.00	67.70	115.00
Rectangular, 12" x 18", white	BC@.500	Ea	39.30	18.00	57.30	97.40
Rectangular, 18" x 24", white	BC@.500	Ea	41.00	18.00	59.00	100.00
Replace-A-Vent, 12" x 18", white	BC@.500	Ea	51.60	18.00	69.60	118.00
Round, 22", paintable	BC@.500	Ea	60.80	18.00	78.80	134.00
Round, 22", white	BC@.500	Ea	59.20	18.00	77.20	131.00
Square, 12" x 12", white	BC@.500	Ea	30.30	18.00	48.30	82.10

	Craft@Hrs	Unit	Material	Labor	Total	Sell
Metal gable louver vent. Replacement cost only. For new installations, add the cost of cutting a hole in an existing gable.						
12" x 12", brown	BC@.500	Ea	16.90	18.00	34.90	59.30
12" x 12", galvanized	BC@.500	Ea	14.30	18.00	32.30	54.90
12" x 12", white	BC@.500	Ea	17.80	18.00	35.80	60.90
12" x 18", brown	BC@.500	Ea	18.10	18.00	36.10	61.40
12" x 18", galvanized	BC@.500	Ea	19.30	18.00	37.30	63.40
12" x 18", white	BC@.500	Ea	21.40	18.00	39.40	67.00
14" x 12", white	BC@.500	Ea	14.10	18.00	32.10	54.60
14" x 18", brown	BC@.500	Ea	25.10	18.00	43.10	73.30
14" x 18", galvanized	BC@.500	Ea	22.60	18.00	40.60	69.00
14" x 18", white	BC@.500	Ea	22.80	18.00	40.80	69.40
14" x 24", brown	BC@.500	Ea	26.10	18.00	44.10	75.00
14" x 24", galvanized	BC@.500	Ea	22.40	18.00	40.40	68.70
14" x 24", white	BC@.500	Ea	26.40	18.00	44.40	75.50
Redwood gable vent. Replacement cost only. For new installations, add the cost of cutting a hole in an existing gable.						
14" x 18" rectangle	BC@.500	Ea	87.27	18.00	105.27	179.00
14" x 24" rectangle	BC@.500	Ea	101.00	18.00	119.00	202.00
18" x 24" rectangle	BC@.500	Ea	101.00	18.00	119.00	202.00
24" octagon	BC@.500	Ea	128.00	18.00	146.00	248.00
24" full round	BC@.500	Ea	168.00	18.00	186.00	316.00
Cedar louver vent. Replacement cost only.						
Arch top, 16" x 24"	BC@.500	Ea	103.00	18.00	121.00	206.00
Octagon, 18"	BC@.500	Ea	113.00	18.00	131.00	223.00
Round, 18"	BC@.500	Ea	108.00	18.00	126.00	214.00
Rectangle, 16" x 24"	BC@.500	Ea	96.10	18.00	114.10	194.00
Dormer louver vent. With galvanized wire mesh to keep birds out. Replacement cost only.						
19" x 3", low profile	BC@.500	Ea	34.40	18.00	52.40	88.70
Galvanized under-eave vent.						
14" x 5"	BC@.400	Ea	4.04	14.40	18.44	31.30
14" x 6"	BC@.400	Ea	3.52	14.40	17.92	30.50
22" x 3"	BC@.400	Ea	4.04	14.40	18.44	31.30
22" x 6"	BC@.400	Ea	5.31	14.40	19.71	33.50
22" x 7"	BC@.400	Ea	5.31	14.40	19.71	33.50
Aluminum ridge vent.						
10' long, black	BC@.850	Ea	18.20	30.60	48.80	83.00
10' long, brown	BC@.850	Ea	17.20	30.60	47.80	81.30
10' long, white	BC@.850	Ea	17.20	30.60	47.80	81.30
Joint strap	BC@.010	Ea	1.82	.36	2.18	3.71
End connector plug	BC@.010	Ea	2.22	.36	2.58	4.39
4' long, hinged for steep roof	BC@.355	Ea	14.50	12.80	27.30	46.40
Ridge vent coil.						
10.5" wide, 20' coil	BC@1.00	Ea	56.10	36.00	92.10	157.00
10.5" wide, 50' coil	BC@2.50	Ea	92.90	90.00	182.90	311.00

	Craft@Hrs	Unit	Material	Labor	Total	Sell
Slant-back aluminum roof louver. 60-square-inch net-free venting area.						
18" x 20', black	BC@1.00	Ea	13.20	36.00	49.20	83.60
18" x 20', brown	BC@1.00	Ea	13.20	36.00	49.20	83.60
18" x 20', mill	BC@1.00	Ea	11.20	36.00	47.20	80.20
18" x 20', shingle match weathered wood	BC@1.00	Ea	13.20	36.00	49.20	83.60
Aluminum wall louver vent.						
12" x 12"	BC@.450	Ea	10.20	16.20	26.40	44.90
12" x 18"	BC@.450	Ea	12.00	16.20	28.20	47.90
14" x 24"	BC@.450	Ea	18.90	16.20	35.10	59.70
18" x 24"	BC@.450	Ea	22.00	16.20	38.20	64.90
Galvanized wall louver vent.						
12" x 12"	BC@.450	Ea	9.00	16.20	25.20	42.80
12" x 18"	BC@.450	Ea	11.20	16.20	27.40	46.60
14" x 24"	BC@.450	Ea	14.80	16.20	31.00	52.70
Plastic wall louver vent.						
8" x 8"	BC@.450	Ea	5.39	16.20	21.59	36.70
12" x 12"	BC@.450	Ea	5.99	16.20	22.19	37.70
12" x 18"	BC@.450	Ea	8.13	16.20	24.33	41.40
18" x 24"	BC@.450	Ea	13.90	16.20	30.10	51.20
Aluminum midget louver. For venting eaves and soffits.						
2", vent area 1-3/16"	BC@.440	Ea	2.02	15.80	17.82	30.30
3", vent area 1-3/4"	BC@.440	Ea	2.93	15.80	18.73	31.80
4", vent area 3-1/2"	BC@.440	Ea	3.48	15.80	19.28	32.80
Combination attic vent.						
12" x 12", brown	BC@.450	Ea	10.80	16.20	27.00	45.90
12" x 12", galvanized	BC@.450	Ea	8.82	16.20	25.02	42.50
12" x 12", white	BC@.450	Ea	10.80	16.20	27.00	45.90
12" x 18", brown	BC@.450	Ea	12.10	16.20	28.30	48.10
12" x 18", white	BC@.450	Ea	12.10	16.20	28.30	48.10
14" x 24", brown	BC@.450	Ea	13.50	16.20	29.70	50.50
14" x 24", galvanized	BC@.450	Ea	12.10	16.20	28.30	48.10
14" x 24", white	BC@.450	Ea	13.30	16.20	29.50	50.20

Ventilators

	Craft@Hrs	Unit	Material	Labor	Total	Sell
Roof turbine vent with base.						
12", galvanized	SW@1.00	Ea	35.70	40.40	76.10	129.00
12", weathered wood	SW@1.00	Ea	38.20	40.40	78.60	134.00
12", white aluminum	SW@1.00	Ea	44.90	40.40	85.30	145.00
Add for steep pitch base	SW@.440	Ea	12.20	17.80	30.00	51.00
Add for weather cap	—	Ea	10.80	—	10.80	—
Belt drive attic exhaust fan. 1/3 HP, with aluminum shutter and plenum boards.						
30" diameter	SW@3.00	Ea	304.00	121.00	425.00	723.00
36" diameter	SW@3.00	Ea	315.00	121.00	436.00	741.00
12-hour timer switch	BE@.950	Ea	30.30	37.10	67.40	115.00

	Craft@Hrs	Unit	Material	Labor	Total	Sell
Power attic gable vent.						
1,280 CFM	SW@1.65	Ea	53.60	66.70	120.30	205.00
1,540 CFM	SW@1.65	Ea	85.70	66.70	152.40	259.00
1,600 CFM	SW@1.65	Ea	111.00	66.70	177.70	302.00
Shutter for gable vent	SW@.450	Ea	43.50	18.20	61.70	105.00
Humidistat control	BE@.550	Ea	30.30	21.50	51.80	88.10
Thermostat control	BE@.550	Ea	26.50	21.50	48.00	81.60
Roof-mount power ventilator.						
1,130 CFM	SW@1.65	Ea	52.00	66.70	118.70	202.00
1,250 CFM	SW@1.65	Ea	94.90	66.70	161.60	275.00
1,600 CFM	SW@1.65	Ea	131.00	66.70	197.70	336.00

Storm flood damage repair and prevention. Assumes compliance with recommendations of The International Hurricane Protection Association and Institute for Inspection, Cleaning and Restoration.

Mold Retarder. For concrete and drywall surfaces. FEMA and IICRC certified. Seal coats are applied over a concrete slab or an external sidewall. Vapor retarder is installed between the soil and a building wall. Biocidal coating can be applied on any interior surface.

	Craft@Hrs	Unit	Material	Labor	Total	Sell
Survey wall and floor for patchable cracks	B6@.001	SF	—	.03	.03	.05
Dry affected areas and patch cracks	B6@.005	SF	.02	.16	.18	.31
Quikrete 1 clear epoxy coating	PP@.008	SF	.18	.26	.44	.75
VaporBlock ASTM E-1745 retarder sheets	PP@.005	SF	.05	.16	.21	.36
Join retarder sheets with IC PRO tape	B6@.005	SF	.01	.16	.17	.29
Brush on Recuma 20 biocidal coating	B6@.008	SF	.03	.25	.28	.48

Dry rot remediation. Fungicidal agent is ShellGard RTU or equal. Add the cost of insurance inspection, if required.

	Craft@Hrs	Unit	Material	Labor	Total	Sell
Expose and dry water-soaked structural lumber	B1@.010	SF	—	.33	.33	.56
Apply ShellGard RTU fungicide	PP@.008	SF	.50	.26	.76	1.29
Clean away dead mold	B1@.005	SF	—	.16	.16	.27

Basement HVAC restoration. Required repairs will usually include installation of a waterproof custom metal 18-inch-high coffer dam around the base of all floor-standing HVAC components. Required repairs may also include replacement of wet insulation, installation of a new boiler burner head and electric controls and installation of a larger flow-rate floor drain. Complies with Factory Mutual requirements.

	Craft@Hrs	Unit	Material	Labor	Total	Sell
Survey of HVAC damage	PM@2.00	Ea	—	82.50	82.50	140.00
Report on repairs required to meet insurance specs	PM@16.0	Ea	—	660.00	660.00	1,120.00
Waterproof coffer dam (typical)	PM@2.00	Ea	250.00	82.50	332.50	565.00
Conduct carrier-witnessed startup	PM@4.00	Ea	—	165.00	165.00	281.00

Brace plate installation. Secure the structure to the foundation by adding steel straps between the concrete foundation and the bottom plate. Typical costs for a 1,600 to 2,400 SF single-family residence. Add the cost of insurance inspection and compliance documentation, if required.

	Craft@Hrs	Unit	Material	Labor	Total	Sell
Locate joints in bottom plates	B1@4.00	Ea	—	130.00	130.00	221.00
Mark and expose joining points	B1@4.00	Ea	—	130.00	130.00	221.00
Attach brace plates with masonry screws	B1@8.00	Ea	225.00	260.00	485.00	825.00
Patch, tape, finish and prime holes in drywall	B1@8.00	Ea	50.00	260.00	310.00	527.00

Corrosion coatings. Required for copper and aluminum pipe exposed to salt water. Coating is aerosol-applied industrial grade ZeroRust phenolic alkyd paint. Add the cost of surface protection, insurance inspection and compliance documentation, if required. Consider pipe 4" or smaller to have one square foot of surface for each linear foot.

	Craft@Hrs	Unit	Material	Labor	Total	Sell
Inspect and clean copper or aluminum	PP@.005	LF	—	.16	.16	.27
Apply anti-corrosion coating	PP@.008	LF	.50	.26	.76	1.29

Siding and Trim

6

Most older homes have wood siding, often referred to as *clapboard*. The term is dated now. But the functional equivalent is still around as *bevel*, *bungalow* or *lap* siding. All consist of tapered boards installed horizontally. The top edge is thinner than the bottom edge so moisture falling down the wall is diverted away from the joint between boards, keeping the wall interior dry. At least that was the theory. Figure 6-1 shows bevel (or clapboard) siding installed over lumber sheathing.

Unlike vinyl and aluminum siding or stucco, board siding needs to be painted about every ten years. But heavily deteriorated board siding, usually as a result of prolonged exposure to moisture, may not be worth painting. Look for signs of water coming from either the inside or outside of the building. Exterior moisture damage is probably caused by lack of a roof overhang or by siding in contact with damp concrete or soil. Damage from moisture originating inside the home is probably caused by an inadequate vapor barrier.

> Check siding for decay where two boards are butted together end to end, at corners, and around window and door openings. Look for gaps between horizontal siding boards by sighting along the wall.

Repair or Replace?

Fill gaps and cracks in the siding with stainable latex wood filler, and re-nail warped boards. Simple repairs like these can go a long way towards extending the life of board siding. When board siding is seriously damaged or decayed, it's fairly easy to replace a few boards. Slip a pry bar into the bevel and carefully pry up to loosen the nails. When the nails are loose, slip the damaged board down and off the wall. Then cut a replacement strip, leaving a $^{1}/_{16}$" gap at each end. If your local lumberyard doesn't stock exactly the pattern you need, a few minutes work with a tabletop router will yield boards indistinguishable from what's being replaced. Repaint, and your repair will be invisible.

Aluminum siding and mineral siding don't decay like wood siding. That's an advantage. However, both aluminum and mineral siding can be damaged fairly easily by impact, such as from an errant baseball. Vinyl siding has nearly replaced aluminum siding in many areas. Finding an aluminum replacement section may be a problem — and matching mineral siding may be even more difficult. If only a small section of aluminum siding is damaged, cut it out and patch with sheet aluminum set in construction adhesive. Then repaint. GAF offers a fiber-cement lookalike product sold under the name *Purity* that you may be able to use to match

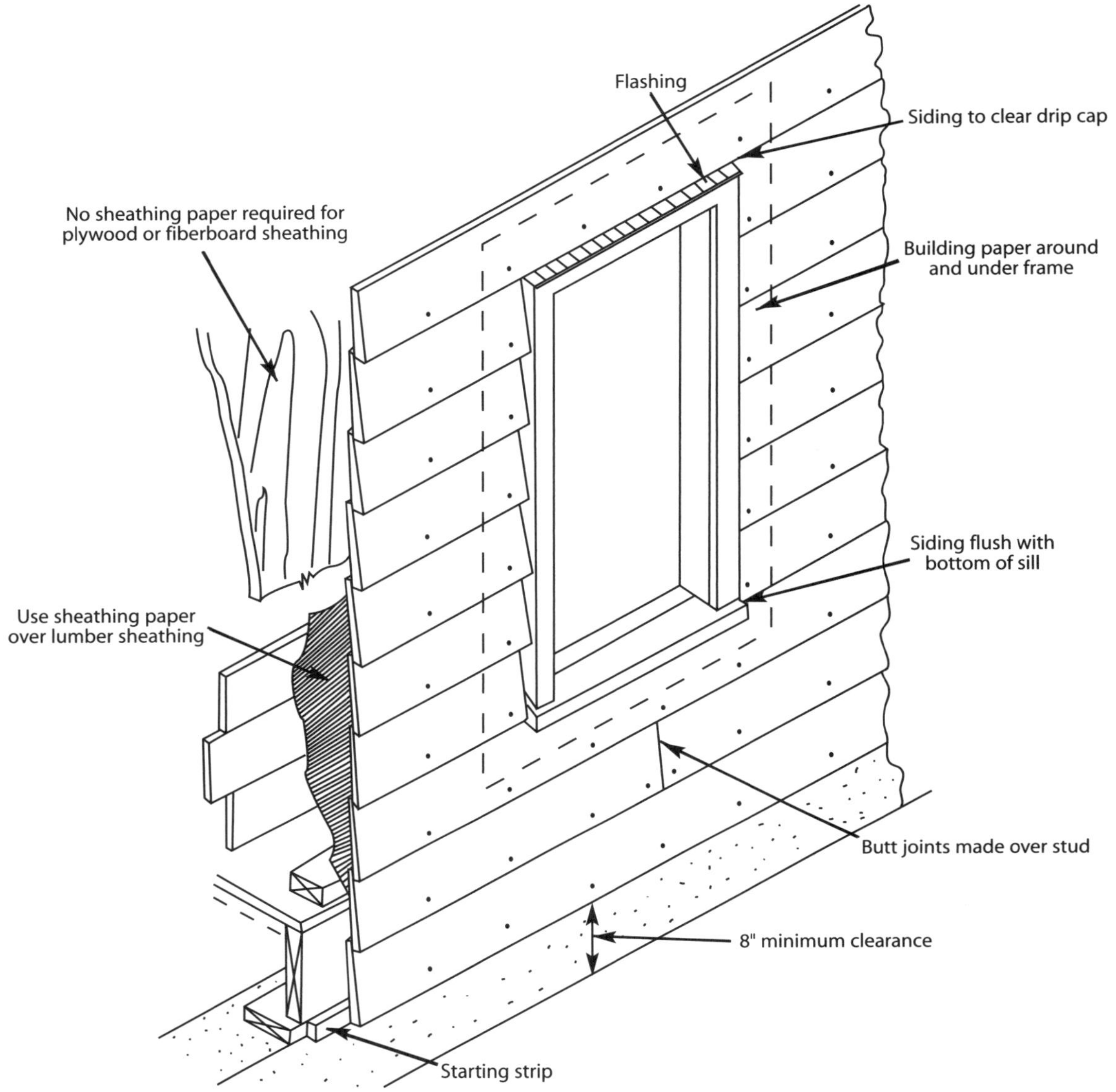

Figure 6-1

Application of bevel siding to coincide with window sill and drip cap

mineral siding. In a pinch, scavenge a few pieces from an inconspicuous location on the same building to make your repair. Then repair the inconspicuous area with a non-matching material. Both aluminum and mineral siding can be painted.

Incidentally, mineral siding was formerly sold as *asbestos-cement* siding. Some asbestos materials are known to pose a health risk and require special handling by qualified experts. But the asbestos in most cement-asbestos siding is non-friable. That means asbestos fibers aren't normally released into the air. If you're in doubt, have the siding checked for friable content. To find a testing laboratory, search for *asbestos testing* on the Internet. If the friable content is about 1 percent, removing the siding presents little or no danger. If friable asbestos content is high, removal is a task for the experts. Even better, leave asbestos siding undisturbed or cover it with another type of siding. Asbestos siding left alone presents no danger, though the owner may have to make disclosure to any prospective buyer.

New Siding Over Old?

If the building has only one layer of siding, and if that siding is sound (no decay problem), consider applying new siding over the old. This offers several advantages. First, you avoid the cost of demolition and disposal. Second, the extra layer of wall cover adds at least some insulation value. And third, installing siding over siding isn't much more effort than installing siding over sheathing.

In some cases, removing the old siding is the best choice. If wood siding has decayed to the point where it won't hold a nail, you could try using longer nails to attach it right into the sheathing, but it's probably a candidate for tear-off. Mineral siding fractures into pieces when nailed. It's easier to remove mineral siding than to try to nail through it. Also, if the home already has two layers of siding, it's probably best to do a tear-off.

Trim around windows and doors needs special attention when applying new siding over old. Extra trim is needed to compensate for the added wall thickness. Window sills aren't a problem. Generally, sills extend well beyond the existing siding in older homes. Look back at Figure 6-1. But if the window casing is nearly flush with existing siding, some type of extension molding will be needed. Build up the casing with an extra thickness of trim, as shown in Figure 6-2. You'll probably need a wider drip cap at the top of the window. If the drip cap is in good condition and can be removed in one piece, extend the cap by fitting a block where the cap was. Then re-nail the cap over the block. Compare Figure 6-3 A (before) and Figure 6-3 B (after). Building out trim on windows and doors is usually easier when vertical siding is applied over horizontal siding.

Rather than build up the depth of window casing, you could add a wider trim piece at the edge of existing trim. See Figure 6-4. A wider drip cap will also be required. Extend exterior door trim in the same way.

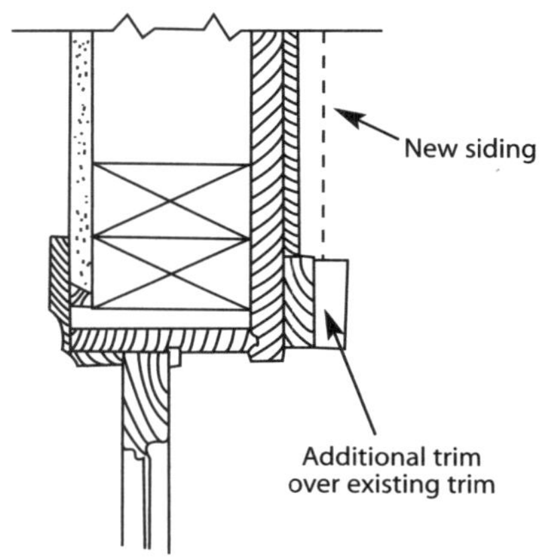

Figure 6-2

Top view of window casing extended by adding trim over existing trim

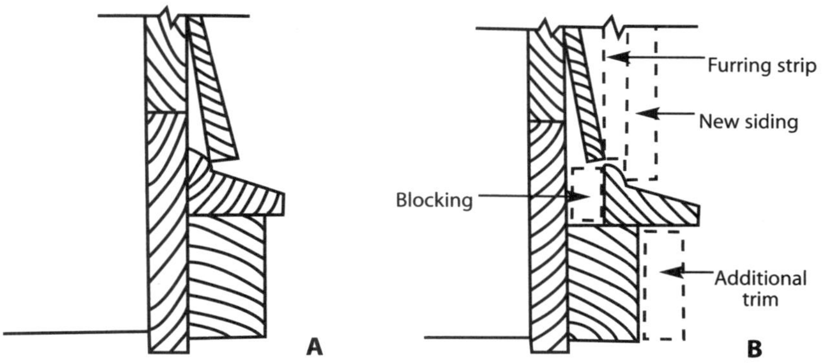

Figure 6-3

Change in drip cap with new siding: ***A****, existing drip cap and trim;* ***B****, drip cap blocked out to extend beyond new siding and added trim*

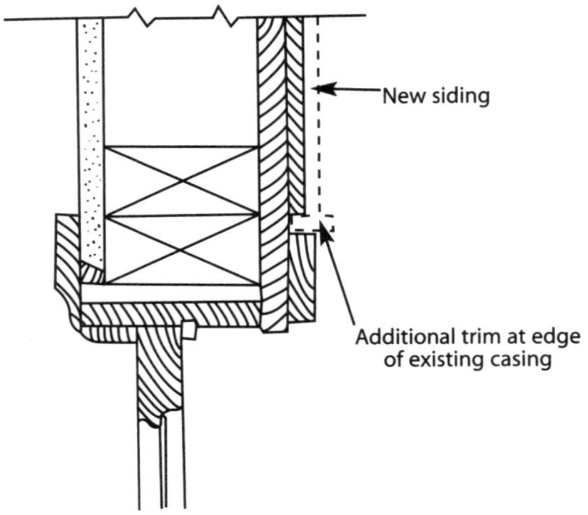

Figure 6-4

Top view of window casing extended for new siding by adding trim at the edge of existing casing

Vinyl Siding

Vinyl siding is by far the most popular siding in use today. It's relatively inexpensive, and comes in a wide variety of styles and colors. Vinyl siding is maintenance-free and resistant to most damage. It's also easy to install. With a little care, your first installation will look great. You can install vinyl siding directly over horizontal board siding if the wood surface is firm and flat. Or, even better, install fanfold polystyrene insulating board over the wood siding first. That provides extra insulation and assures a smooth, even surface over which to install the vinyl siding. Remove and replace any deteriorated boards. Re-nail loose boards. If the existing siding is irregular, install furring strips so the new siding is supported by a flush surface. Then re-caulk around doors and windows and renew the window flashing, if needed.

Follow the manufacturer's installation instructions. But note that all vinyl siding expands and contracts with heat and cold, as much as 1/2" over the length of a 12'6" panel. If nails or screws are driven tight against the nailing slot, the panel will buckle noticeably in warm weather. Use aluminum, galvanized steel or other corrosion-resistant nails or screws and leave 1/32" (the thickness of a dime) clearance between the head and the vinyl. Drive fasteners through the center of the nailing slot in the hem so horizontal siding can slide 1/2" left and right after installation. Drive a fastener every 16" on horizontal panels, every 12" on vertical panels and every 8" to 10" on accessories such as J-channel and corner posts.

Adjacent vinyl panels lock together to form a snug joint. Where panel ends meet, overlap the ends by about 1". Avoid end overlap near doors or windows where regular use is likely to put stress on the joint.

All vinyl siding manufacturers offer trim pieces designed for use with their siding materials. These include starter strip, drip cap, J-channel, molding, corner posts and soffits. Anything projecting from the wall, such as a doorbell, a faucet or a porch light, should be surrounded by a vinyl *mounting block*. The block forms a collar that surrounds the siding penetration. Setting a mounting block is much easier and neater than cutting a hole in a full-length vinyl siding panel. Mounting blocks also offer better protection from the elements.

> Vinyl siding can be installed over stucco. But first install 1" x 3" furring secured to the studs. Then install the siding, driving either ring-shank nails or screws into the furring. Nails or screws driven into the stucco alone will not secure the siding.

The trickiest part of any vinyl siding job is dealing with existing windows and doors. Most new windows are designed with vinyl siding in mind. They're contoured so that vinyl siding will trim out neatly around them. Older existing windows, however, are a serious problem. If you just trim around the window frame, you'll leave exposed wood, which will need to be painted. Most siding installers will cover the existing window frames with aluminum, which needs to be custom-bent on site using an aluminum brake. This is very time consuming. Also, the new siding will stand out further from the house than the existing siding, causing the windows to be slightly recessed. This alters the look of the house, which the homeowner may not like. In that case, you'll need to build out the window frames to restore the original depth.

If the homeowner is considering having new windows installed in his house, this would be an excellent time to do it. New windows will eliminate all these problems. You might want to offer the homeowner a package deal on new windows to go along with the new vinyl siding. It will save them money, and save you a lot of time and aggravation.

Horizontal Wood Siding

Horizontal wood siding patterns include bevel, Dolly Varden, drop, and channel siding. Common widths are 4", 6", 8", 10" and even 12". Smooth-finish wood siding can be stained or painted. Rough-sawn wood siding is usually stained.

Horizontal siding has to be applied over a smooth surface. If you decide to leave the existing siding in place, either cover it with OSB (oriented strand board panels) or nail furring strips over each stud. Nail siding at each stud with corrosion-resistant nails. Use 6d nails for siding less than $^{1}/_{2}$" thick and 8d nails for thicker siding. Try to avoid nailing through the brittle top edge of siding boards. Bevel siding joints should overlap no less than 1". Be sure the butt joints of horizontal siding boards fall over a stud. If possible, adjust the lap so butt edges coincide with the bottom of the sill and the top of the drip cap at window frames. Refer back to Figure 6-1. Finish interior corners by butting the siding against a corner board at least $1^{1}/_{8}$" square. See Figure 6-5. Exterior corners can be mitered, butted against corner boards at least $1^{1}/_{8}$" square and $1^{1}/_{2}$" wide (Figure 6-6), or covered with metal corners.

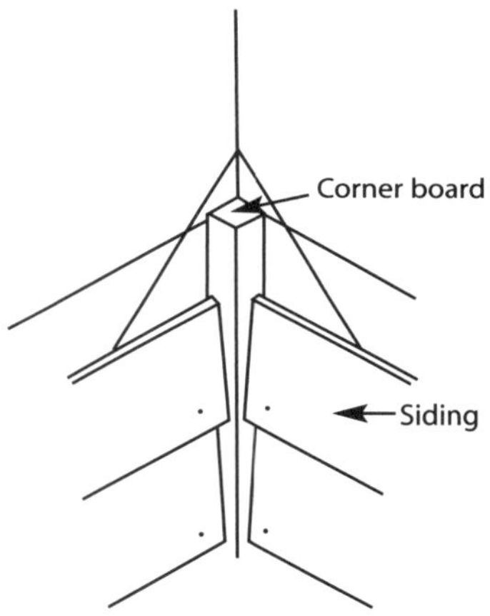

Figure 6-5

Corner board for application of horizontal siding at interior corner

Pre-finished hardboard plank siding is sold in widths from 6" to 16" and in lengths to 16'. Matching trim, corners and connectors are available. Nail hardboard plank siding at each vertical joint, the same as bevel siding.

Vertical Wood Siding

Vertical board siding applied over horizontal board siding makes a good combination. Probably the most popular vertical pattern is matched (tongue and groove) boards. Nail through the existing siding and into the sheathing. If you tear off the existing siding, you may need nailing strips. Sheathing that's $^{1}/_{2}$" or $^{3}/_{8}$" thick doesn't have enough nail-holding power. When the existing sheathing is thinner than $^{5}/_{8}$", apply 1" x 4" nailing strips horizontally across the wall, spaced 16" to 24" apart vertically. Then nail

> Wood siding can also be installed diagonally. Add 15 percent to the labor estimate and allow 5 percent more waste of materials.

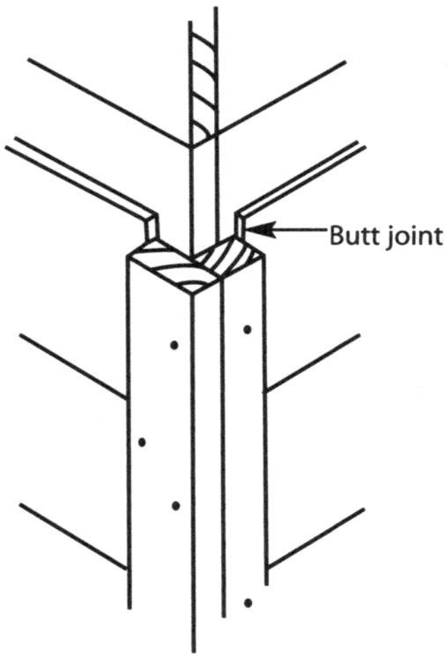

Figure 6-6

Corner boards for application of horizontal siding at exterior corner

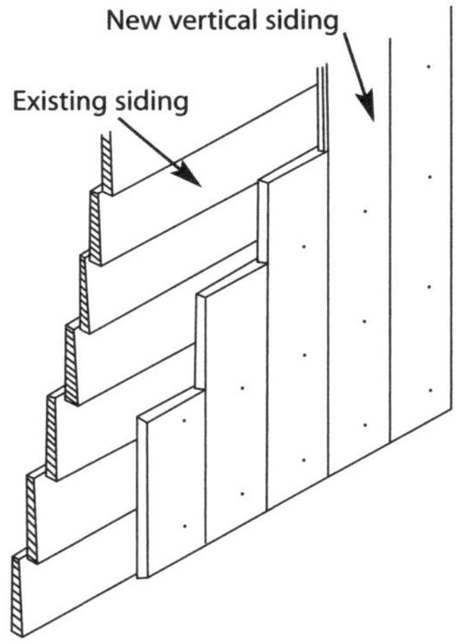

Figure 6-7

Application of vertical siding

the vertical siding to the strips. Blind-nail through the siding tongue at each strip with galvanized 7d finishing nails. When boards are nominal 6" or wider, also face-nail the middle of each board with an 8d galvanized nail. See Figure 6-7.

Board and batten (*board 'n bat*) is another popular choice for vertical siding. Again, apply horizontal nailing strips if sheathing is less than 5/8" thick. Nail each board at the top and bottom with a galvanized 8d nail. See Figure 6-8. Wider boards need two nails at the top and bottom, spaced 1" each side of the center line. Close spacing prevents splitting as the boards shrink. Cover the 1/2" gap between boards with the batten, secured with 12d finishing nails. Be careful to miss the underboard when nailing batts. Use only corrosion-resistant nails. Figure 6-8 shows two variations of board and batt siding.

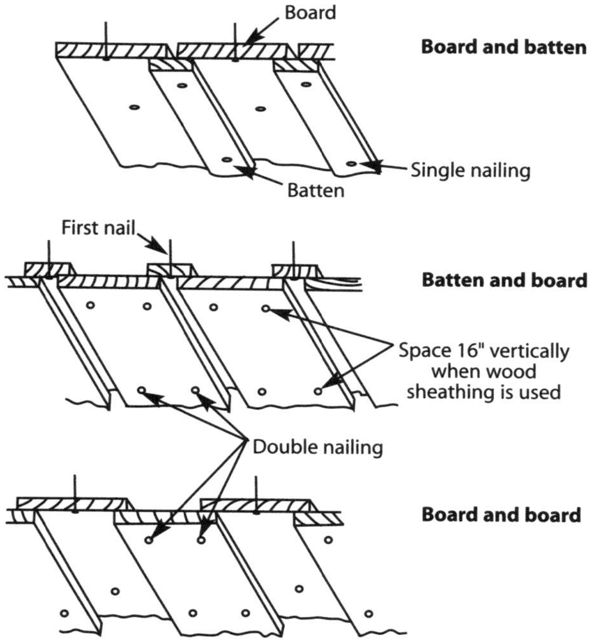

Figure 6-8

Application of vertical wood siding

Panel Siding

If existing siding is uneven, panel siding (plywood, hardboard or particleboard) is the best choice. Panel siding tends to smooth out any unevenness in the existing surface. And panel siding is probably the easiest to install. It can be applied over nearly any surface.

If you use plywood, it's best to select a plywood siding product that's specifically made for use as siding. Hardboard used for siding must be tempered. Most panel siding products are sold in both 4' x 8' and 4' x 9' sheets. Rough-textured plywood (such as Texture 1-11) soaks up water-repellent preservative stain better than smooth-finish plywood. Smooth plywood can be stained, but the finish isn't as durable as saw-texture plywood. Fiber-overlaid plywood (usually called MDO, for medium-density overlay) is particularly good if you plan to paint the surface. The resin-treated fiber overlay presents a very smooth surface that both bonds securely to paint and minimizes expansion and contraction due to moisture changes.

Panel siding with vertical grooves 4" or 8" on center adds visual appeal. Or, get a similar effect by nailing 1" x 2" vertical battens over each panel joint and stud.

Plywood Siding

In new construction, plywood siding can be installed directly over the studs without any sheathing. If there is no sheathing, the plywood should be at least $^3/_8$" thick for 16" stud spacing and $^1/_2$" thick for 24" stud spacing. Grooved plywood is normally $^5/_8$" thick with $^1/_4$" deep grooves.

Nail plywood siding every 7" to 8" around the perimeter of the panel and at each intermediate stud. Use galvanized or other rust-resistant nails. When installing panel siding over existing siding, use nails long enough to penetrate $^3/_4$" into the surface below.

Some plywood siding has shiplap joints. Treat the joint edges with a water-repellent preservative and then nail the siding on both sides of the joint. See Figure 6-9A. With shiplap joints, no caulking is needed. When hanging square-edge panels, you'll need to caulk the joints with sealant, as shown in Figure 6-9B. Again, nail the panels on both sides of the joint. If you install battens over panel joints and at intermediate studs, nail the battens with 8d galvanized nails spaced 12" apart. Use longer nails if you have to penetrate through existing siding to the sheathing below.

You'll pay a premium for pre-finished plywood siding, but the finish lasts longer and requires little or no maintenance. When installing pre-finished panel siding, use nails and connectors recommended by the manufacturer.

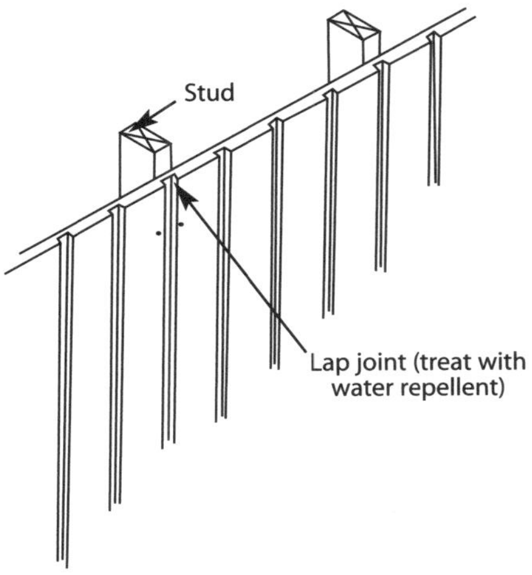

Figure 6-9A
Joint of plywood panel siding: shiplap joint

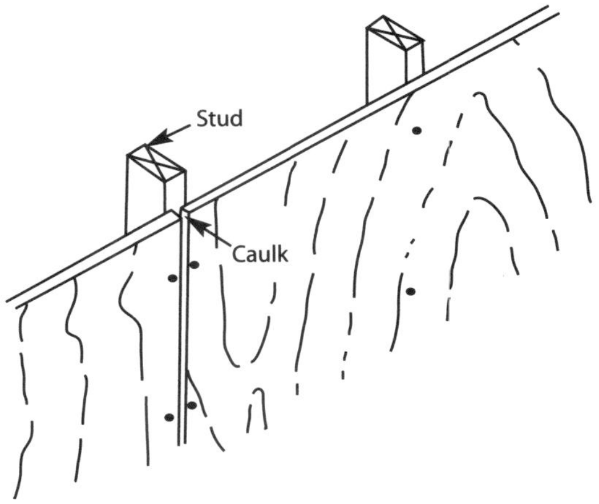

Figure 6-9B
Joint of plywood panel siding: square-edge joint

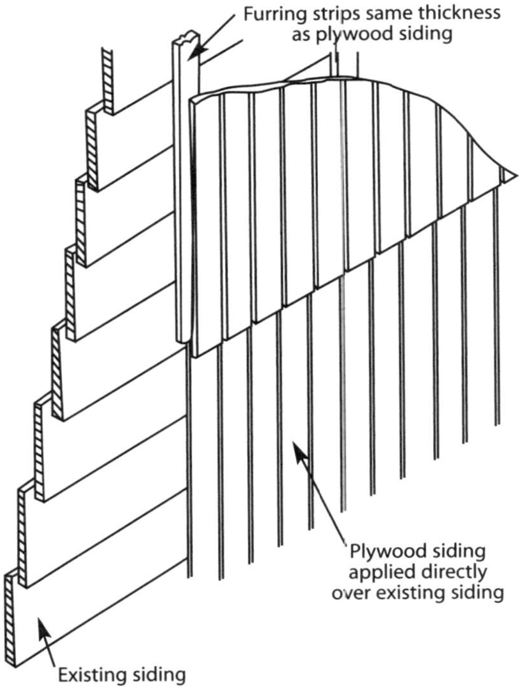

Figure 6-10

Application of plywood siding at gable end

Gable ends present a problem because panels aren't long enough to extend the full wall height. Horizontal butt joints in panels are an invitation to water damage, even when covered with trim. Be sure the joints are as snug as possible. Alternatively, you can lap siding panels over the top of panels below — although this may detract from the overall appearance. See Figure 6-10. Install furring strips on the gable that are the same thickness as the new siding below. Nail a furring strip over the existing siding or sheathing at each stud. Then install siding on the strips the way you'd apply siding directly to studs.

Hardboard Siding

Hardboard panel siding is available in lengths up to 16'. It's usually 1/4" thick, but may be thicker when grooved. Hardboard is usually factory primed. Apply a finish coat after installation. Hang hardboard siding the same way you install plywood siding.

Finish corners with corner boards, the same as for horizontal board siding. Use 1 1/8" x 1 1/8" corner boards at inside corners and 1 1/8" x 1 1/2" or 2 1/2" boards at outside corners. Apply caulking where siding butts against corner boards, window or door casings, and trim boards at gable ends.

Estimating Furring

Shingle Exposure	Linear Feet per 100 SF of Siding
4" exposure	300 feet
4 1/2" exposure	270 feet
5" exposure	240 feet
5 1/2" exposure	220 feet
6" exposure	200 feet
6 1/2" exposure	180 feet
7" exposure	171 feet
8" exposure	150 feet

Figure 6-11

Linear feet of furring per 100 square feet of shingle siding

Shingle and Shake Siding

It's not always easy to spot shingle siding that's reached the end of its useful life, especially if it's been painted. Sometimes rotted shingles appear fine visually, but will crumble if you touch them. Look closely for broken, warped, and upturned shingles.

It's common to find houses in which some areas of shingles are decayed, but the rest are fine. This is because exposure to the weather varies from one part of the house to the other. Areas that are more exposed, or are kept damp by overhanging trees, will tend to decay first. Removing damaged shingles can be tricky. Prying them up tends to damage the shingle directly above the bad one. You may find yourself replacing a vertical strip of shingles that extends substantially past the bad shingles.

Repairs made to individual shingles, or small sections of shingles, will be obvious for several years. Old shingles are bleached nearly gray by sun and weather. New shingles have the rich tone of freshly-cut cedar. It would be nicer to re-side the entire house so that all the shingles match, but budgets usually don't allow for this.

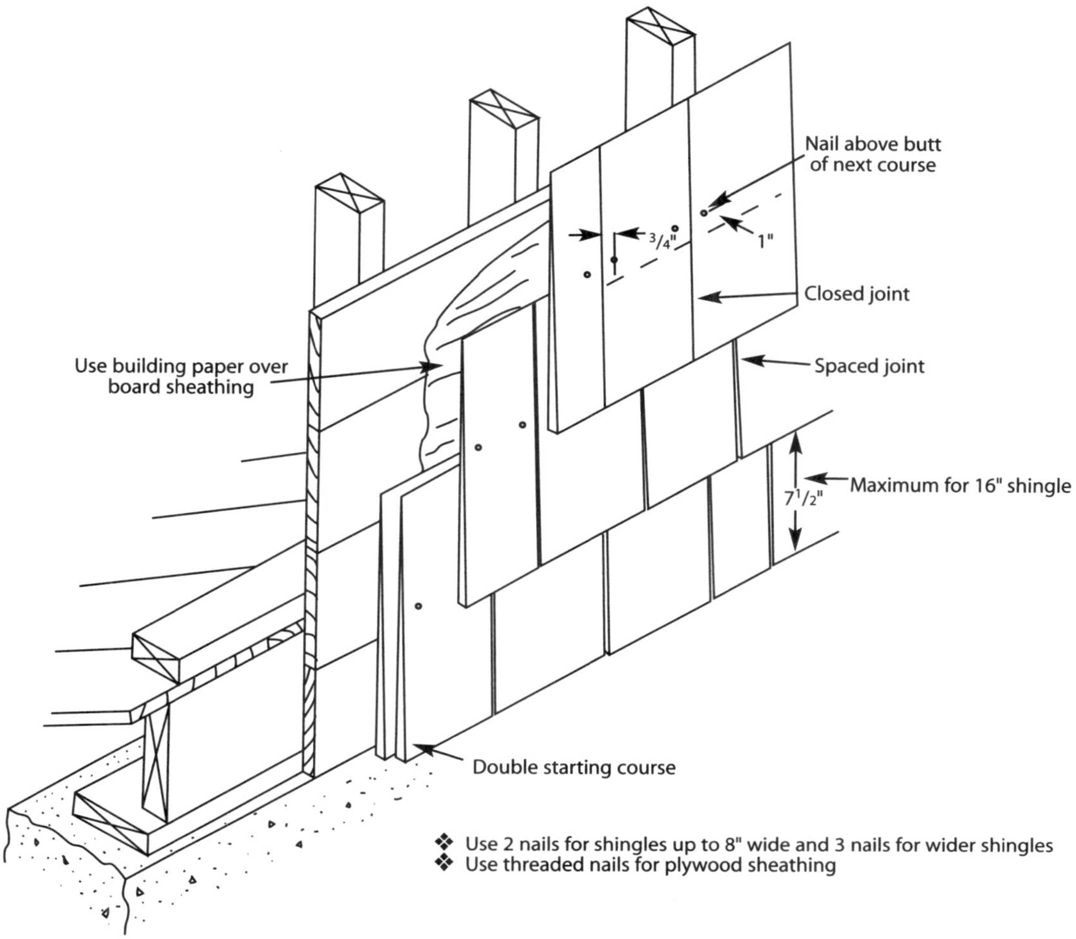

Figure 6-12
Single-course application of shingle siding

Siding shingles can be applied over either wood or plywood sheathing or over existing board siding. When installed over sheathing, apply 15-pound asphalt felt between the sheathing and the shingles. If the surface is irregular, apply horizontal furring strips of 1" x 3" lumber or 1" x 4" lumber. Spacing has to match the exposed shingle length. Use Figure 6-11 to estimate the furring required for various exposures per 100 square feet of shingle siding. Leave a gap of 1/8" to 1/4" between shingles to allow for expansion during wet weather.

Single-course shingles are applied like bevel siding. Each top course laps over the course below with all joints staggered. See Figure 6-12. Use second-grade shingles (rather than first or third grade) because only one-half or less of the butt portion is exposed.

Compare the double-course shingles in Figure 6-13. The first grade (No. 1) shingles cover the undercourse (third grade) shingles. The butt ends of top course shingles should project 1/4" to 1/2" beyond the butt ends of undercourse shingles.

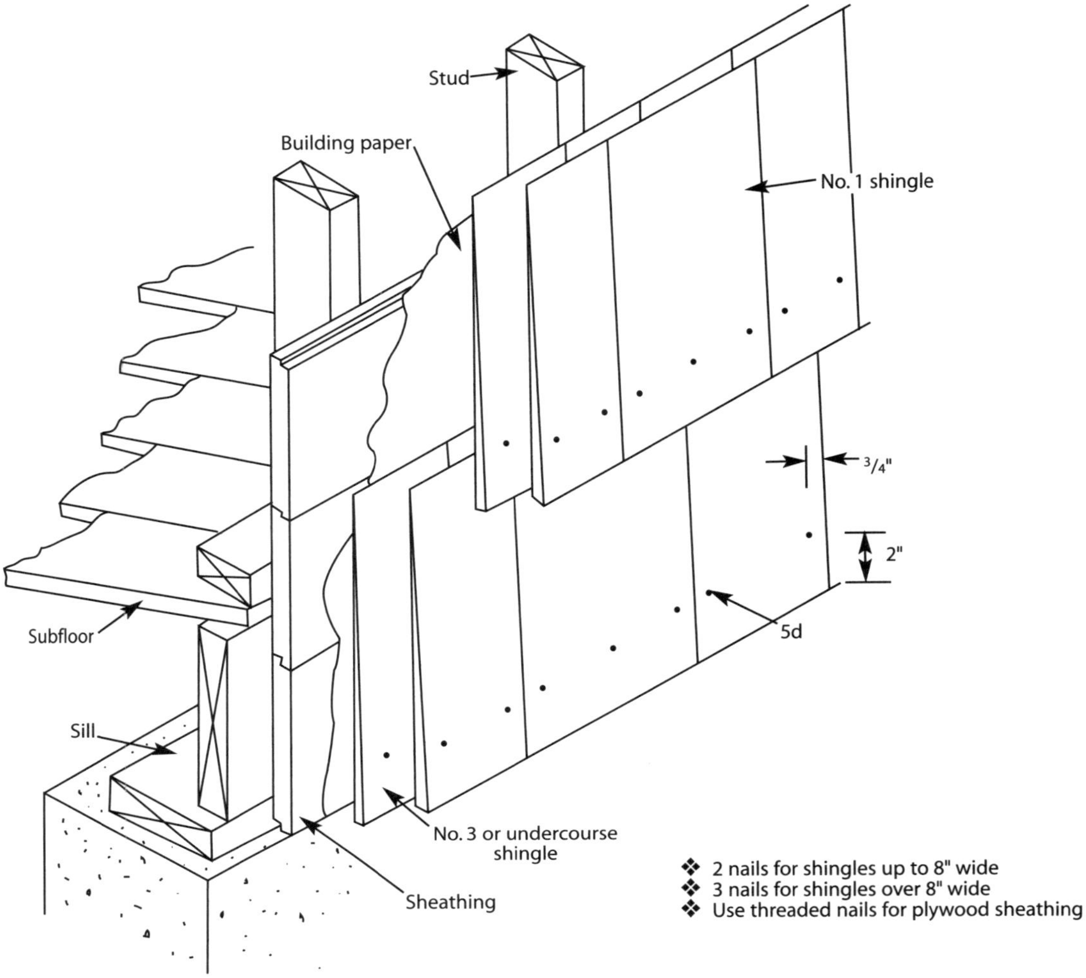

Figure 6-13

Double-course application of shingle siding

With double coursing, shingle exposure of the top course can be greater. Figure 6-14 shows recommended exposure distances for shingles.

Whether single-course or double-course, all joints should be staggered. Vertical joints of the upper shingle should be at least 1 1/2" from any joint in the lower shingle.

Use rust-resistant, anti-stain nails when installing shingles. Two nails are enough on shingles up to 8" wide. Use three nails on shingles over 8" wide. For single-coursing, 3d or 4d shingle nails are best. Nails with small flat heads are best for double-coursing because the nail heads are exposed. Use 5d nails for the top course and 3d or 4d nails on the lower course. When plywood sheathing is less than 3/4" thick, use threaded nails for increased holding power. Keep nails at least 3/4" from the shingle edge. When single-coursing, keep nails 1" above the horizontal butt line of the next higher course. See Figure 6-12. When double-coursing, nails should be 2" above the bottom butt.

Maximum Exposure for Shingles				
Material	**Length**	**Single Coursing**	**Double Coursing No. 1 grade**	**Double Coursing No. 2 grade**
Shingles	16"	$7^1/_2$"	12"	10"
	18"	$8^1/_2$"	14"	11"
	24"	$11^1/_2$"	16"	14"
Shakes (handsplit and resawn)	18"	$8^1/_2$"	14"	—
	24"	$11^1/_2$"	20"	—
	32"	15"	—	—

Figure 6-14

Exposure distances for wood shingles and shakes on sidewalls

Masonry Veneer

Brick or stone veneer siding won't wear out. But mortar can crumble and settling will cause cracks. In either case, apply new mortar to keep moisture out of the wall and to improve the overall appearance. Scratch out all loose mortar. Brush the joint to remove dust and loose particles. Then dampen the surface and apply mortar to fill the joint. Tamp mortar into the joint for a tight bond. Use pre-mixed joint compound and a jointing tool to keep the joint depths uniform. When repointing red brick, any mortar left on the face of individual brick will cause a permanent stain. Washing or brushing the brick after repointing tends to erode the joint you've just patched. Consider masking the face of brick when making joint repairs, especially when the brick is going to be left its natural color rather than painted. Let the mortar dry. Then pull off the masking tape. Finally, consider coating the entire surface with transparent waterproofing seal.

> Stained or discolored masonry can be cleaned with a water blaster or by brushing with diluted (22 percent) muriatic acid. Common household bleach will remove green mold from masonry.

Stucco Repair

Hairline cracks in stucco are normal as the stucco dries and ages. Simply recoat the surface with masonry paint. Wider cracks, especially cracks that run diagonally, are usually the result of the structure settling. To repair any crack too wide for paint, scribe out $^1/_4$" of stucco along the crack. A screwdriver makes a good scribing tool. Then fill the scribed line with stucco patch or caulk and recoat with masonry paint.

Occasionally, cracks will be the result of defects in the stucco itself. For example, a grid of horizontal and vertical cracks may indicate that the lath is working loose from the studs. A spider web of cracks indicates that the scratch or brown coat is too thin or dried too fast. Under those circumstances, there may be no alternative to removing and replacing the affected section.

When stucco is exposed to water for long periods, the color coat flakes off and turns to powder. Remove the source of the moisture. Then wire-brush the surface and apply a new color coat.

Estimating Procedure for Siding

When measuring quantities for siding jobs, it's good practice to ignore wall openings (such as windows and doors). Window and door openings result in only a modest saving of materials, but do increase the installation cost. Each wall corner will also increase the labor costs. The figures in this chapter assume a typical job consisting of six corners and 2,500 square feet of siding. Allow one additional installation hour for each corner over six. The accessories needed for most siding jobs are a substantial part of the costs: J-channel, starting strip, L-channel, undersill trim, corners, nails, and colored caulk to match the siding. These figures also assume that you'll be working with vinyl siding-ready windows, or that you'll simply be trimming around existing doors and windows. If it will be necessary to cover windows with aluminum trim bent on-site, add figures from the section Aluminum siding trim in this chapter.

On soffits, assume that 30 percent of the soffit will be vented. Trim on fascia and soffit jobs can include utility trim, starter, cove molding, divider for miter cuts, frieze molding, reversible frieze, overhanging drip edge, brick trim, matching nails and matching caulk.

The figures that follow assume that the work is done on the exterior of a one- or two-story dwelling. Work at higher levels requires scaffolding. Add the cost of scaffold or lifting equipment and about $^{1}/_{2}$ hour of extra time per 100 square feet of siding installed.

Masonry Fireplaces and Chimneys

Cracks in a masonry chimney will usually be the result of someone attaching an antenna or holiday decorations at the roof level. Cracks can also be the result of settlement of the fireplace foundation. A well-built chimney has a terra-cotta flue lining which keeps the chimney air-tight, even when the masonry exterior develops cracks. If the chimney does not have a fireproof lining, cracks can be a fire hazard. Chip out and replace any fractured brick. Seal all deep cracks and regrout (tuck point) the masonry joints. The combustion chamber (firebox) in a fireplace will be lined with firebrick. With age and frequent use, firebrick can spall and flake, leaving the common brick backing exposed to excessive heat. Chip out and reset any deteriorated firebrick. House framing should neither support nor be supported by the chimney. The fireplace and chimney should be supported on a separate foundation. If the fireplace has developed stress fractures due to settling of either the foundation or the house, demolition may be the only practical alternative.

Prefabricated metal fireplaces are a good choice for home improvement because they can be set on a wood-frame floor without a separate foundation and without any masonry.

Siding and Soffit Repairs

	Craft@Hrs	Unit	Material	Labor	Total	Sell
Repair of board siding. Using hand tools and working from a ladder.						
Cut out 3" x 3" or smaller damaged area with a chisel, fill with stainable latex wood filler, sand smooth, per area cut and filled	BC@.133	Ea	—	4.79	4.79	8.14
Stainable latex wood filler, 16 oz. tube	—	Ea	11.90	—	11.90	—
Re-nail split or warped siding board, per board repaired	BC@.067	Ea	2.00	2.41	4.41	7.50
Pry out 8' length of siding, cut replacement siding board to fit, nail new siding board in place, per board replaced	BC@.416	Ea	13.50	15.00	28.50	48.50
Remove and replace 8' vertical siding board or batt, per board or batt	BC@.350	Ea	13.50	12.60	26.10	44.40
Replace 1" x 2" x 8' siding corner board	BC@.400	Ea	1.75	14.40	16.15	27.50
Touch-up painting to match	PT@.333	Ea	4.50	12.20	16.70	28.40
Repair of vinyl or aluminum siding. Using hand tools and working from a ladder.						
Cut out damaged panel with a siding removal tool, loosen the panel above, remove damaged panel nails, insert replacement panel, per panel	BC@1.15	Ea	7.00	41.40	48.40	82.30
Siding removal tool	—	Ea	6.10	—	6.10	—
Scab patch vinyl or aluminum siding, trim away damaged material, cut patch from scrap, set patch in construction adhesive, per patch	BC@.300	Ea	.60	10.80	11.40	19.40
Remove and replace aluminum corner cap, pry off or trim off damaged cap, set new cap in panel adhesive, per corner cap	BC@.167	Ea	.99	6.01	7.00	11.90
Touch-up painting to match	PT@.333	Ea	4.50	12.20	16.70	28.40
Remove and replace a siding shake or shingle. Using hand tools and working from a ladder. Per section of 2-3 shingles removed and replaced.						
Single course. Chip out damaged shingle, hacksaw old nails, trim a new shingle and place 1/2" below correct alignment, nail at top of course, tap shingle into alignment	BC@.416	Ea	3.00	15.00	18.00	30.60
Double course. Chip out damaged shingle in two courses, pull old nails, cut undercourse and top course shingles to fit, place and nail both courses, per set of two shingles	BC@.500	Ea	6.00	18.00	24.00	40.80
Shingle removal tool 24"	—	Ea	36.00	—	36.00	—
Touch-up stain to match	PT@.333	Ea	4.50	12.20	16.70	28.40
Stucco repair. Using hand tools and working from a ladder.						
Fill crack by scribing out the crack to 1/4" depth, fill crack with elastomeric caulk, per linear foot filled	CF@.033	Ea	—	1.12	1.12	1.90
Elastomeric patch and caulk, 10 oz. tube	—	Ea	3.85	—	3.85	—
Patch hole in exterior stucco. Chip and wire brush loose stucco, moisten surface, apply stucco patch with a trowel, per 4" x 4" patch	CF@.250	Ea	.50	8.47	8.97	15.20
Premixed stucco patch, quart	—	Ea	7.43	—	7.43	—
Touch-up painting to match	PT@.333	Ea	4.50	12.20	16.70	28.40

	Craft@Hrs	Unit	Material	Labor	Total	Sell
Siding tear off. Remove the old siding, break debris into convenient size and pile on site. Using hand tools. Includes the cost of driving or pulling out nails to prepare the surface for installation of new siding. Add the cost of caulking and flashing, if needed. No salvage value assumed.						
Aluminum siding, horizontal	BL@.027	SF	—	.78	.78	1.33
Aluminum siding, vertical	BL@.031	SF	—	.90	.90	1.53
Asphalt siding, panel or roll	BL@.021	SF	—	.61	.61	1.04
Mineral siding	BL@.027	SF	—	.78	.78	1.33
Plywood siding	BL@.017	SF	—	.49	.49	.83
Plywood siding with battens	BL@.019	SF	—	.55	.55	.94
Stucco on metal lath	BL@.035	SF	—	1.02	1.02	1.73
Stucco on wood lath	BL@.040	SF	—	1.16	1.16	1.97
Vinyl siding, horizontal	BL@.021	SF	—	.61	.61	1.04
Vinyl siding, vertical	BL@.024	SF	—	.70	.70	1.19
Wood siding, horizontal	BL@.030	SF	—	.87	.87	1.48
Wood siding, vertical	BL@.033	SF	—	.96	.96	1.63
Wood shingles or shakes, single course	BL@.021	SF	—	.61	.61	1.04
Wood shingles or shakes, double course	BL@.026	SF	—	.76	.76	1.29
Add for work above 9' high	BL@.008	SF	—	.23	.23	.39
Remove 1" wood furring, per linear foot	BL@.020	LF	—	.58	.58	.99
Soffit tear off. Remove board, panel, vinyl or aluminum soffit and trim and pile debris on site. Using hand tools and working from a ladder at 8' to 10' height. Includes the cost of driving or pulling out nails to prepare the surface for installation of a new soffit. No salvage value assumed. Per linear foot of soffit.						
To 12" wide soffit	BL@.016	SF	—	.46	.46	.78
Over 12" wide soffit	BL@.020	SF	—	.58	.58	.99
Add for heights over 10' to 20'	BL@.008	SF	—	.23	.23	.39
Remove and replace for siding work. Remove in salvage condition using hand tools. Then replace when siding work is complete.						
Awning or hood over door or window	BC@1.75	Ea	—	63.00	63.00	107.00
Patio canopy, per square foot	BC@.088	SF	—	3.17	3.17	5.39
Fixed shutters, per pair	BC@1.00	Pair	—	36.00	36.00	61.20
Operable shutters, per pair	BC@1.50	Pair	—	54.00	54.00	91.80
Storm windows, each	BC@.625	Ea	—	22.50	22.50	38.30
Storm doors, each	BC@2.50	Ea	—	90.00	90.00	153.00
Gutters or downspouts, per linear foot	BC@.080	LF	—	2.88	2.88	4.90

Vinyl Siding

Vinyl lap siding. Solid vinyl double traditional lap profile. When calculating quantities, do not deduct for openings (such as doors and windows) less than 6' wide. Labor and material needed for accessories will be a large part of the total cost. Accessories include J-channel, starting strip, L-channel, undersill trim, corners, nails, and colored caulking to match the siding. Each corner adds about an hour of work to the job. These figures assume six walls are being covered. Add the cost of soffit, fascia and extra trim needed over doors and windows. These figures assume 5% waste of materials. Per 100 square feet (Square).

	Craft@Hrs	Unit	Material	Labor	Total	Sell
White/Colors, .042" x 8" exposure x 12'6"	B1@3.34	Sq	86.80	109.00	195.80	333.00
White/Colors, .042" x 9" exposure x 12'1"	B1@3.34	Sq	86.80	109.00	195.80	333.00
White/Colors, .042" x 10" exposure x 12'	B1@3.34	Sq	86.80	109.00	195.80	333.00
White/Colors, .040" x 8" exposure x 12'6"	B1@3.34	Sq	58.90	109.00	167.90	285.00
White/Colors, .040" x 9" exposure x 12'1"	B1@3.34	Sq	58.90	109.00	167.90	285.00
White/Colors, .040" x 10" exposure x 12'	B1@3.34	Sq	58.90	109.00	167.90	285.00
Durabuilt 440 Series Double 4" Traditional Lap	B1@3.34	Sq	120.00	109.00	229.00	389.00
Durabuilt 440 Series Double 5" Dutch Lap	B1@3.34	Sq	120.00	109.00	229.00	389.00
Durabuilt 410 Series Double 4" Traditional Lap	B1@3.34	Sq	97.30	109.00	206.30	351.00
Durabuilt 410 Series Double 5" Traditional Lap	B1@3.34	Sq	97.50	109.00	206.50	351.00
Starter strip, 5" wide	B1@.025	LF	.43	.81	1.24	2.11
J-channel trim at wall openings, 3/4"	B1@.033	LF	.31	1.07	1.38	2.35
Outside corner post, 3/4" x 3/4"	B1@.033	LF	2.09	1.07	3.16	5.37
Inside corner post, 3/4" x 3/4"	B1@.033	LF	1.48	1.07	2.55	4.34
Under sill trim, 1-1/2"	B1@.038	LF	.52	1.24	1.76	2.99
Casing and utility trim	B1@.038	LF	.52	1.24	1.76	2.99
Vinyl siding 1-3/4" nails, per 1 pound bag	—	Lb	16.60	—	16.60	—
Add for R-3 insulated vinyl siding	B1@.350	Sq	50.00	11.40	61.40	104.00

Siding fixture mounting block. For mounting accessories such as electrical outlets and hose bibs on wood, aluminum, vinyl, stucco, shake, or brick siding. Snap-on finishing ring. No caulking or J-channel required. 1-1/4" diameter fixture cutout.

	Craft@Hrs	Unit	Material	Labor	Total	Sell
Recessed, 6-1/4" x 4-3/4"	B1@.350	Ea	10.70	11.40	22.10	37.60
Surface mount, 6-3/4" x 6-3/4"	B1@.350	Ea	11.40	11.40	22.80	38.80
Jumbo, 12" x 8"	B1@.500	Ea	15.40	16.30	31.70	53.90

Thermoplastic resin siding. Nailite™ 9/10" thick panels installed over sheathing. Add for trim board, corners, starter strips, J-channels and mortar fill. Based on five nails per piece. Includes 10% waste.

	Craft@Hrs	Unit	Material	Labor	Total	Sell
Hand-split shake, 41-3/8" x 18-3/4"	B1@.043	SF	2.89	1.40	4.29	7.29
Hand-laid brick, 44-1/4" x 18-5/8"	B1@.043	SF	3.07	1.40	4.47	7.60
Hand-cut stone, 44-1/4" x 18-5/8"	B1@.043	SF	3.07	1.40	4.47	7.60
Rough sawn cedar, 59-1/4" x 15"	B1@.043	SF	3.20	1.40	4.60	7.82
Perfection plus cedar, 36-1/8" x 15"	B1@.043	SF	3.20	1.40	4.60	7.82
Ledge trim						
Brick ledge trim	B1@.033	LF	6.68	1.07	7.75	13.20
Stone ledge trim	B1@.033	LF	6.68	1.07	7.75	13.20
Corners						
Hand-split shake, 4" x 18"	B1@.030	Ea	12.70	.98	13.68	23.30
Rough sawn cedar, 3" x 26"	B1@.030	Ea	13.40	.98	14.38	24.40
Hand-laid brick, 4" x 18"	B1@.030	Ea	12.10	.98	13.08	22.20
Hand-cut stone, 4" x 18"	B1@.030	Ea	13.40	.98	14.38	24.40
Perfection plus cedar, 3" x 13"	B1@.030	Ea	11.10	.98	12.08	20.50
Starter strips						
12' universal	B1@.025	LF	.66	.81	1.47	2.50

	Craft@Hrs	Unit	Material	Labor	Total	Sell
J-channels						
Hand-split shake, 3/4" opening	B1@.025	LF	1.00	.81	1.81	3.08
Hand-split shake, 1-1/4" flexible opening	B1@.027	LF	2.00	.88	2.88	4.90
Rough sawn cedar, 3/4" opening	B1@.025	LF	.76	.81	1.57	2.67
Rough sawn cedar, 1-1/4" flexible opening	B1@.027	LF	2.00	.88	2.88	4.90
Hand-laid brick, 3/4" opening	B1@.025	LF	.76	.81	1.57	2.67
Hand-cut stone, 3/4" opening	B1@.025	LF	.76	.81	1.57	2.67
Perfection plus cedar, 3/4" opening	B1@.025	LF	.76	.81	1.57	2.67
Thermoplastic siding accessories						
Mortar fill, 12 tubes, 200 LF per carton	—	Ea	73.50	—	73.50	—
Touch up paint, 6 oz. aerosol can	—	Ea	6.17	—	6.17	—

Vinyl fascia. Installed on rafter tails with small tools, working from a ladder. Per linear foot, including 5% waste.

	Craft@Hrs	Unit	Material	Labor	Total	Sell
8" wide x 12' long panels	B1@.025	LF	1.38	.81	2.19	3.72

Vinyl soffit. Panel conceals the under side of roof overhang. Per linear foot including 5% waste.

	Craft@Hrs	Unit	Material	Labor	Total	Sell
Soffit, 12" x 12'	B1@.040	LF	1.32	1.30	2.62	4.45
J-channel support, 1/2" x 12'	B1@.033	LF	.53	1.07	1.60	2.72
F-channel support, 1/2" x 12'	B1@.033	LF	.73	1.07	1.80	3.06
H-bar double channel support, 1/2" x 12'	B1@.025	LF	.68	.81	1.49	2.53

Wood Board Siding

Pine bevel siding. 1/2" thick by 6" wide. Each linear foot of 6" wide bevel siding covers .38 square feet of wall. Do not deduct for wall openings less than 10 square feet. Add 5% for typical waste and end cutting.

	Craft@Hrs	Unit	Material	Labor	Total	Sell
8' lengths	B1@.038	SF	1.58	1.24	2.82	4.79
12' lengths	B1@.036	SF	1.58	1.17	2.75	4.68
16' lengths	B1@.033	SF	1.58	1.07	2.65	4.51

Western red cedar bevel siding. A Grade and better, 1/2" thick by 6" wide. Each linear foot of 6" wide bevel siding covers .38 square feet of wall. Do not deduct for wall openings less than 10 square feet. Add 5% for typical waste and end cutting.

	Craft@Hrs	Unit	Material	Labor	Total	Sell
8' lengths	B1@.038	SF	3.03	1.24	4.27	7.26
12' lengths	B1@.036	SF	3.05	1.17	4.22	7.17
16' lengths	B1@.033	SF	3.03	1.07	4.10	6.97

Clear cedar 1/2" bevel siding. Vertical grain, 1/2" thick by 6" wide, kiln dried. Each linear foot of 6" wide bevel siding covers .38 square feet of wall. Do not deduct for wall openings less than 10 square feet. Add 5% for typical waste and end cutting.

	Craft@Hrs	Unit	Material	Labor	Total	Sell
8' lengths	B1@.038	SF	3.03	1.24	4.27	7.26
12' lengths	B1@.036	SF	3.05	1.17	4.22	7.17
16' lengths	B1@.033	SF	3.03	1.07	5.44	9.25

Cedar 3/4" bevel siding. 3/4" thick by 8" wide. Each linear foot of 8" wide bevel siding covers .54 square feet of wall. Do not deduct for wall openings less than 10 square feet. Add 5% for typical waste and end cutting.

	Craft@Hrs	Unit	Material	Labor	Total	Sell
8' lengths	B1@.036	SF	1.61	1.16	2.77	4.71
12' lengths	B1@.034	SF	1.61	1.11	2.72	4.62
16' lengths	B1@.031	SF	1.60	1.01	2.61	4.44

Primed pine V rustic siding. 1" thick by 6" wide. Each linear foot of 6" wide channel siding covers .38 square feet of wall. Do not deduct for wall openings less than 10 square feet. Add 5% for typical waste and end cutting.

	Craft@Hrs	Unit	Material	Labor	Total	Sell
16' lengths	B1@.034	SF	4.44	1.11	5.55	9.44

	Craft@Hrs	Unit	Material	Labor	Total	Sell
Tongue and groove cedar siding. 1" thick by 6" wide. Kiln dried, Select Grade. Each linear foot of 6" wide tongue and groove siding covers .43 square feet of wall. Do not deduct for wall openings less than 10 square feet. Add 5% for typical waste and end cutting.						
8' lengths	B1@.036	SF	2.74	1.17	3.91	6.65
10' lengths	B1@.035	SF	2.74	1.14	3.88	6.60
12' lengths	B1@.034	SF	2.74	1.11	3.85	6.55
Redwood 5/8" shiplap siding. B Grade. 5/8" thick by 5-3/8" wide. Each linear foot of 5 3/8" wide shiplap siding covers .38 square feet of wall. Do not deduct for wall openings less than 10 square feet. Add 5% for typical waste and end cutting.						
8' lengths	B1@.038	SF	4.31	1.24	5.55	9.44
10' lengths	B1@.037	SF	4.30	1.20	5.50	9.35
12' lengths	B1@.036	SF	4.29	1.17	5.46	9.28
Southern yellow pine 6" pattern 105 siding. 1" thick by 6" wide. Select Grade except as noted. Each linear foot of 6" wide siding covers .38 square feet of wall. Do not deduct for wall openings less than 10 square feet. Add 5% for typical waste and end cutting.						
8' lengths	B1@.038	SF	3.62	1.24	4.86	8.26
10' lengths	B1@.037	SF	3.67	1.20	4.87	8.28
12' lengths, No. 2	B1@.036	SF	2.06	1.17	3.23	5.49
12' lengths	B1@.036	SF	3.64	1.17	4.81	8.18
Southern yellow pine 8" pattern 105 siding. 1" thick by 8" wide. Select Grade except as noted. Each linear foot of 8" wide siding covers .54 square feet of wall. Do not deduct for wall openings less than 10 square feet. Add 5% for typical waste and end cutting.						
8' lengths	B1@.036	SF	3.36	1.17	4.53	7.70
10' lengths	B1@.034	SF	3.24	1.11	4.35	7.40
12' lengths, No. 2	B1@.031	SF	1.66	1.01	2.67	4.54
12' lengths	B1@.031	SF	3.23	1.01	4.24	7.21
Southern yellow pine 6" pattern 116 siding. 1" thick by 6" wide. Select Grade except as noted. Each linear foot of 6" wide siding covers .38 square feet of wall. Do not deduct for wall openings less than 10 square feet. Add 5% for typical waste and end cutting.						
8' lengths	B1@.038	SF	3.55	1.24	4.79	8.14
10' lengths	B1@.037	SF	3.52	1.20	4.72	8.02
12' lengths	B1@.036	SF	3.57	1.17	4.74	8.06
Southern yellow pine 6" pattern 117 siding. 1" thick by 6" wide. Select Grade except as noted. Each linear foot of 6" wide siding covers .38 square feet of wall. Do not deduct for wall openings less than 10 square feet. Add 5% for typical waste and end cutting.						
8' lengths	B1@.038	SF	3.62	1.24	4.86	8.26
10' lengths	B1@.037	SF	3.72	1.20	4.92	8.36
12' lengths	B1@.036	SF	3.61	1.17	4.78	8.13

	Craft@Hrs	Unit	Material	Labor	Total	Sell
Southern yellow pine 8" pattern 131 siding. 1" thick by 8" wide. Select Grade except as noted. Each linear foot of 8" wide siding covers .54 square feet of wall. Do not deduct for wall openings less than 10 square feet. Add 5% for typical waste and end cutting.						
8' lengths	B1@.036	SF	3.44	1.17	4.61	7.84
10' lengths	B1@.034	SF	3.43	1.11	4.54	7.72
12' lengths	B1@.031	SF	3.42	1.01	4.43	7.53
Southern yellow pine 8" pattern 122 siding. 1" thick by 8" wide. V-joint. Select Grade except as noted. Each linear foot of 8" wide siding covers .54 square feet of wall. Do not deduct for wall openings less than 10 square feet. Add 5% for typical waste and end cutting.						
8' lengths	B1@.036	SF	3.09	1.17	4.26	7.24
10' lengths	B1@.034	SF	3.06	1.11	4.17	7.09
12' lengths	B1@.031	SF	3.04	1.01	4.05	6.89
Southern yellow pine 6" pattern 122 siding. 1" thick by 6" wide. Select Grade except as noted. Each linear foot of 6" wide siding covers .38 square feet of wall. Do not deduct for wall openings less than 10 square feet. Add 5% for typical waste and end cutting.						
8' lengths	B1@.038	SF	2.40	1.24	3.64	6.19
10' lengths	B1@.037	SF	2.43	1.20	3.63	6.17
12' lengths	B1@.036	SF	2.30	1.17	3.47	5.90
Southern yellow pine 8" shiplap siding. 1" thick by 8" wide. No. 2 Grade. Each linear foot of 8" wide siding covers .54 square feet of wall. Do not deduct for wall openings less than 10 square feet. Add 5% for typical waste and end cutting.						
8' lengths	B1@.036	SF	1.48	1.17	2.65	4.51
10' lengths	B1@.034	SF	1.49	1.11	2.60	4.42
12' lengths	B1@.031	SF	1.47	1.01	2.48	4.22
Rustic 8" pine siding. 1" thick by 8" wide. V-grooved. Each linear foot of 8" wide siding covers .54 square feet of wall. Do not deduct for wall openings less than 10 square feet. Add 5% for typical waste and end cutting.						
16' lengths	B1@.031	SF	3.52	1.01	4.53	7.70
20' lengths	B1@.031	SF	3.59	1.01	4.60	7.82
Log cabin siding. 1-1/2" thick. Each linear foot of 6" wide siding covers .41 square feet of wall. Each linear foot of 8" wide siding covers .55 square feet of wall. Do not deduct for wall openings less than 10 square feet. Add 5% for typical waste and end cutting.						
6" x 12', T&G latia	B1@.036	SF	2.79	1.17	3.96	6.73
8" x 12'	B1@.031	SF	2.73	1.01	3.74	6.36
8" x 16'	B1@.030	SF	2.74	.98	3.72	6.32
Log lap spruce siding. 1-1/2" thick. Each linear foot of 6" wide siding covers .41 square feet of wall. Each linear foot of 8" wide siding covers .55 square feet of wall. Each linear foot of 10" wide siding covers .72 square feet of wall. Do not deduct for wall openings less than 10 square feet. Add 5% for typical waste and end cutting.						
8" width	B1@.031	SF	4.49	1.01	5.50	9.35
10" width	B1@.030	SF	4.18	.98	5.16	8.77

	Craft@Hrs	Unit	Material	Labor	Total	Sell
Texture 1-11 plywood siding. Per square foot including 5% waste. Exterior Grade 4' x 8' panels.						
3/8" southern yellow pine, grooved 4" OC	B1@.022	SF	.87	.72	1.59	2.70
3/8" plain fir	B1@.022	SF	.77	.72	1.49	2.53
3/8" ACQ treated	B1@.022	SF	1.10	.72	1.82	3.09
5/8" borate treated	B1@.022	SF	1.50	.72	2.22	3.77
5/8" fir, grooved 8" OC	B1@.022	SF	1.34	.72	2.06	3.50
5/8" x 4' x 8' southern yellow pine, grooved 8" OC	B1@.022	SF	2.11	.72	2.83	4.81
5/8" ACQ treated southern, grooved 8" OC	B1@.022	SF	1.50	.72	2.22	3.77
5/8" borate treated, grooved 4" OC	B1@.022	SF	1.64	.72	2.36	4.01
Add for 4' x 9' panels	—	SF	.27	—	.27	—
Plywood panel siding. Per square foot including 5% waste. Exterior Grade 4' x 8' panels.						
3/8" plain southern yellow pine	B1@.022	SF	.90	.72	1.62	2.75
5/8" southern yellow pine, reverse board and batt 12" OC	B1@.022	SF	1.24	.72	1.96	3.33
5/8" ACQ treated southern yellow pine, grooved 12 OC	B1@.022	SF	1.44	.72	2.16	3.67
OSB panel siding. Oriented strand board ("Waferboard"). Compressed wood strands bonded by phenolic resin. Per square foot including 5% waste. Exterior Grade 4' x 8' panels.						
3/8" Smart Panel II	B1@.022	SF	.77	.72	1.49	2.53
7/16" textured and grooved 8" OC	B1@.022	SF	1.24	.72	1.96	3.33
5/8" borate treated, grooved 8" OC	B1@.022	SF	1.81	.72	2.53	4.30
Add for 4' x 9' panels	—	SF	.27	—	.27	—
OSB plank lap siding. Oriented strand board ("Waferboard"). Compressed wood strands bonded by phenolic resin. Per square foot including 8% waste. Textured.						
7/16" x 6" x 16'	B1@.043	SF	1.90	1.40	3.30	5.61
7/16" x 8" x 16'	B1@.040	SF	1.70	1.30	3.00	5.10

Hardboard Panel

	Craft@Hrs	Unit	Material	Labor	Total	Sell
Hardboard panel siding. Cost per square foot including 5% waste. 4' x 8' panels.						
7/16" plain panel	B1@.022	SF	.85	.72	1.57	2.67
7/16" grooved 8" OC	B1@.022	SF	.96	.72	1.68	2.86
7/16" Sturdi-Panel, grooved 8" OC	B1@.022	SF	.70	.72	1.42	2.41
7/16" textured board and batt 12" OC	B1@.022	SF	.74	.72	1.46	2.48
7/16" cedar, grooved 8" OC	B1@.022	SF	.65	.72	1.37	2.33
7/16" cedar, plain, textured	B1@.022	SF	.87	.72	1.59	2.70
15/32" Duratemp, grooved 8" OC 1/8" hardboard face, plywood back	BL@.022	SF	1.20	.64	1.84	3.13
Add for 4' x 9' panels	—	SF	.20	—	.20	—

	Craft@Hrs	Unit	Material	Labor	Total	Sell
Hardboard lap siding. Primed plank siding. With 8% waste. 6" x 16' covers 5.6 square feet, 8" x 16' covers 7.4 square feet, 16" x 16' covers 14.8 square feet.						
7/16" x 6" x 16', smooth lap	B1@.033	SF	1.56	1.07	2.63	4.47
7/16" x 6" x 16', textured cedar	B1@.033	SF	1.61	1.07	2.68	4.56
7/16" x 8" x 16', textured cedar	B1@.031	SF	1.61	1.01	2.62	4.45
7/16" x 8" x 16', smooth	B1@.031	SF	1.66	1.01	2.67	4.54
1/2" x 8" x 16', self-aligning	B1@.031	SF	2.20	1.01	3.21	5.46
1/2" x 8" x 16', old mill sure lock	B1@.031	SF	2.19	1.01	3.20	5.44
1/2" x 16" x 16', cottage beveled 5" OC	B1@.030	SF	2.01	.98	2.99	5.08
7/16" x 8", textured joint cover	—	Ea	.73	—	.73	—
4/4" x 4" x 16' reversible trim	—	LF	.76	—	.76	—
4/4" x 6" x 16' reversible trim	—	LF	.76	—	.76	—
4/5" x 4" x 16' reversible trim	—	LF	1.02	—	1.02	—
4/5" x 6" x 16' reversible trim	—	LF	1.29	—	1.29	—

Fiber Cement Siding

	Craft@Hrs	Unit	Material	Labor	Total	Sell
Fiber cement lap siding, Hardiplank®. 6-1/4" width requires 20 planks per 100 square feet of wall. 7-1/4" width requires 17 planks per 100 square feet of wall. 8-1/4" width requires 15 planks per 100 square feet of wall. 9-1/4" width requires 13 planks per 100 square feet of wall. 12" width requires 10 planks per 100 square feet of wall. Cost per square foot including 5% waste						
6-1/4" x 12', cedarmill	B1@.058	SF	1.23	1.89	3.12	5.30
7-1/4" x 12', cedarmill	B1@.052	SF	1.09	1.69	2.78	4.73
8-1/4" x 12', colonial roughsawn	B1@.048	SF	1.23	1.56	2.79	4.74
8-1/4" x 12', cedarmill	B1@.048	SF	.94	1.56	2.50	4.25
8-1/4" x 12', beaded cedarmill	B1@.048	SF	1.09	1.56	2.65	4.51
8-1/4" x 12', beaded	B1@.048	SF	1.22	1.56	2.78	4.73
8-1/4" x 12', smooth	B1@.048	SF	1.89	1.56	3.45	5.87
9-1/4" x 12', cedarmill	B1@.046	SF	1.79	1.50	3.29	5.59
12" x 12', smooth	B1@.042	SF	1.49	1.37	2.86	4.86
12" x 12', cedarmill	B1@.042	SF	1.50	1.37	2.87	4.88
Fiber cement panel siding, Hardipanel®. 5/16" thick. Cost per square foot including 5% waste.						
4' x 8', Sierra, grooved 8" OC	B1@.042	SF	.91	1.37	2.28	3.88
4' x 8', Stucco	B1@.042	SF	.93	1.37	2.30	3.91
4' x 8', cedarmill, grooved 8" OC	B1@.042	SF	.97	1.37	2.34	3.98
4' x 9', Sierra, grooved 8" OC	B1@.042	SF	.86	1.37	2.23	3.79
Trim for fiber cement siding, Hardipanel®. Per linear foot including 10% waste.						
5/4" x 4" x 10', smooth	B1@.035	LF	1.70	1.14	2.84	4.83
5/4" x 8" x 10', smooth	B1@.035	LF	3.22	1.14	4.36	7.41
5/4" x 12" x 12', smooth XLD	B1@.035	LF	3.32	1.14	4.46	7.58
5/4" x 6" x 12', smooth XLD	B1@.035	LF	1.81	1.14	2.95	5.02
Fiber cement soffit, Hardisoffit®. 1/4" thick. Per square foot including 10% waste. Vented.						
2' x 8', smooth	B1@.040	SF	1.46	1.30	2.76	4.69
4' x 8', cedarmill	B1@.040	SF	1.10	1.30	2.40	4.08
4' x 8', smooth	B1@.040	SF	1.17	1.30	2.47	4.20
12" x 12', smooth	B1@.044	SF	1.61	1.43	3.04	5.17
12" x 12', cedarmill	B1@.044	SF	1.70	1.43	3.13	5.32
16" x 12', cedarmill	B1@.044	SF	1.65	1.43	3.08	5.24

	Craft@Hrs	Unit	Material	Labor	Total	Sell
Fiber-cement siding shingle. 18 pieces cover 33.3 square feet at 11" exposure. Three cartons cover 100 square feet. Price per square foot.						
Purity shingle	B1@.030	SF	3.00	.98	3.98	6.77

Shingle Siding

	Craft@Hrs	Unit	Material	Labor	Total	Sell
Cedar sidewall shingle panels. Shakertown. Clear all-heart, KD. Applied over building paper to solid or spaced nailable sheathing. For use on walls or mansard roofs with minimum 20 in 12 pitch. Use rust resistant, 7d nails for penetration of studs at least 1/2".						
1 course shingle panels, 8' long, 7" exposure	B1@.033	SF	6.01	1.07	7.08	12.00
2 course shingle panels, 8' long, 7" exposure	B1@.031	SF	6.20	1.01	7.21	12.30
3 course shingle panels, 8' long, 7" exposure	B1@.028	SF	7.04	.91	7.95	13.50
Designer shingles, 7-1/2" exposure	B1@.040	SF	9.64	1.30	10.94	18.60
Striated sidewall shake, 18" with natural groove	B1@.030	SF	4.16	.98	5.14	8.74
1 or 2 course corner unit, 7" exposure	B1@.030	LF	9.65	.98	10.63	18.10
3 course corner unit, 7" exposure	B1@.030	LF	12.20	.98	13.18	22.40
Shingle siding. Cedar rebutted sidewall shingles, Western red cedar.						
No. 1 western red cedar rebutted and rejointed shingles. Carton covers 25 square feet at 7" exposure, $131 per carton	R1@.035	SF	5.24	1.22	6.46	11.00
Red cedar Perfects 5x green, 16" long, 7-1/2" exposure, 5 shingles are 2" thick (5/2), 3 bundles cover 100 square feet (one square) including waste. Costs include fasteners.						
#1 B1@.039	SF	3.91	1.27	5.18	8.81	
#2 red label	B1@.039	SF	3.15	1.27	4.42	7.51
#3 B1@.039	SF	1.81	1.27	3.08	5.24	
Panelized shingles, Shakertown, KD, regraded shingle on plywood backing, no waste.						
8' long x 7" wide single course						
Colonial 1	B1@.028	SF	4.35	.91	5.26	8.94
Cascade	B1@.028	SF	4.19	.91	5.10	8.67
8' long x 14" wide, single course						
Colonial 1	B1@.028	SF	4.35	.91	5.26	8.94
8' long x 14" wide, double course						
Colonial 2	B1@.028	SF	4.76	.91	5.67	9.64
Cascade Classic	B1@.028	SF	4.19	.91	5.10	8.67
Decorative shingles, 18" long x 5" wide, Shakertown, fancy cuts, 9 hand-shaped patterns.						
7-1/2" exposure	B1@.030	SF	5.09	.98	6.07	10.30
Deduct for 10" exposure	—	SF	−.55	—	−.55	—

Aluminum Siding and Soffit

	Craft@Hrs	Unit	Material	Labor	Total	Sell
Aluminum corrugated 4-V x 2-1/2", plain or embossed finish. Includes 15% waste.						
17 gauge, 26" x 6' to 24'	B1@.034	SF	2.57	1.11	3.68	6.26
19 gauge, 26" x 6' to 24'	B1@.034	SF	2.66	1.11	3.77	6.41
Rubber filler strip, 3/4" x 7/8" x 6'	—	LF	.36	—	.36	—
Flashing for corrugated aluminum siding, embossed.						
End wall, 10" x 52"	B1@.048	Ea	4.59	1.56	6.15	10.50
Side wall, 7-1/2" x 10'	B1@.015	LF	1.98	.49	2.47	4.20

	Craft@Hrs	Unit	Material	Labor	Total	Sell
Aluminum smooth 24 gauge, horizontal patterns, non-insulated.						
8" or double 4" widths, acrylic finish	B1@.028	SF	1.83	.91	2.74	4.66
12" widths, bonded vinyl finish	B1@.028	SF	1.85	.91	2.76	4.69
Add for foam backing	—	SF	.37	—	.37	—
Starter strip	B1@.030	LF	.44	.98	1.42	2.41
Inside and outside corners	B1@.033	LF	1.32	1.07	2.39	4.06
Casing and trim	B1@.033	LF	.44	1.07	1.51	2.57
Drip cap	B1@.044	LF	.45	1.43	1.88	3.20
Aluminum siding trim, fabrication and installation labor only.						
Cover frieze board, part of a siding job	B1@.055	LF	—	1.79	1.79	3.04
Cover frieze board only	B1@.061	LF	—	1.98	1.98	3.37
Cover gable trim	B1@.075	LF	—	2.44	2.44	4.15
Cover fascia board	B1@.061	LF	—	1.98	1.98	3.37
Cover porch ceiling, part of siding job	B1@.058	SF	—	1.89	1.89	3.21
Cover porch ceiling only	B1@.082	SF	—	2.67	2.67	4.54
Cover porch post and framework	B1@.073	SF	—	2.37	2.37	4.03
Cover window trim, masonry wall	B1@.610	Ea	—	19.80	19.80	33.70
Cover window trim, frame wall	B1@.720	Ea	—	23.40	23.40	39.80
Cover door trim	B1@1.00	Ea	—	32.50	32.50	55.30
Cover trim on one car garage, service door	B1@3.25	Ea	—	106.00	106.00	180.00
Cover trim on two car garage, service door	B1@5.00	Ea	—	163.00	163.00	277.00

Aluminum fascia. Installed on rafter tails with small tools, working from a ladder. Per linear foot, including 5% waste.

	Craft@Hrs	Unit	Material	Labor	Total	Sell
4" x 12'	B1@.025	LF	.93	.81	1.74	2.96
6" x 12'	B1@.025	LF	1.18	.81	1.99	3.38
8" x 12'	B1@.025	LF	1.48	.81	2.29	3.89
12" x 12'	B1@.025	LF	1.38	.81	2.19	3.72
Frieze molding, 1-1/2" x 12'	B1@.030	LF	.75	.98	1.73	2.94
Inside fascia, 12" corner	B1@.333	Ea	1.52	10.80	12.32	20.90
Outside fascia, 12" corner	B1@.333	Ea	1.52	10.80	12.32	20.90

Aluminum soffit. Installed on rafter tails with small tools, working from a ladder. Per linear foot, including 5% waste.

	Craft@Hrs	Unit	Material	Labor	Total	Sell
12" x 12' non-vented	B1@.031	LF	1.47	1.01	2.48	4.22
12" x 12' perforated	B1@.031	LF	1.47	1.01	2.48	4.22
12' miter divider	B1@.042	LF	1.13	1.37	2.50	4.25
J-channel support	B1@.033	LF	.53	1.07	1.60	2.72
Aluminum nails, 1-1/4", 1/4 pound, 250 nails	—	Ea	5.20	—	5.20	—

Stucco and Masonry Siding

Portland cement stucco. Including paperback lath, 3/8" scratch coat, 3/8" brown coat and 1/8" color coat. Per square yard covered.

	Craft@Hrs	Unit	Material	Labor	Total	Sell
Stucco on walls						
Natural gray, sand float finish	PR@.567	SY	7.74	20.60	28.34	48.20
Natural gray, trowel finish	PR@.640	SY	7.74	23.30	31.04	52.80
White cement, sand float finish	PR@.675	SY	8.72	24.60	33.32	56.60
White cement, trowel finish	PR@.735	SY	8.72	26.80	35.52	60.40
Stucco on soffits,						
Natural gray, sand float finish	PR@.678	SY	7.74	24.70	32.44	55.10
Natural gray, trowel finish	PR@.829	SY	7.74	30.20	37.94	64.50
White cement, sand float finish	PR@.878	SY	8.72	32.00	40.72	69.20
White cement, trowel finish	PR@1.17	SY	8.72	42.60	51.32	87.20

	Craft@Hrs	Unit	Material	Labor	Total	Sell

Cleaning and pointing masonry. These costs assume the masonry surface is in fair to good condition, with no unusual damage. Add the cost of protecting adjacent surfaces such as trim or the base of the wall, and the cost of scaffolding, if required. Labor required to presoak or saturate the area cleaned is included in the labor cost. Work more than 12' above floor level will increase costs.

	Craft@Hrs	Unit	Material	Labor	Total	Sell
Brushing (hand cleaning) masonry, includes the cost of detergent or chemical solution.						
Light cleanup (100 SF per manhour)	B9@.010	SF	.03	.32	.35	.60
Medium scrub (75 SF per manhour)	B9@.013	SF	.04	.42	.46	.78
Heavy (50 SF per manhour)	B9@.020	SF	.06	.64	.70	1.19
Water blasting masonry, using rented 400 to 700 PSI power washer with 3 to 8 gallon per minute flow rate, includes blaster rental at $50 per day.						
Smooth face (250 SF per manhour)	B9@.004	SF	—	.13	.13	.22
Rough face (200 SF per manhour)	B9@.005	SF	—	.16	.16	.27
Sandblasting masonry, using 150 PSI compressor (with accessories and sand) at $150.00 per day.						
Smooth face (50 SF per manhour)	B9@.020	SF	.31	.64	.95	1.62
Rough face (40 SF per manhour)	B9@.025	SF	.34	.80	1.14	1.94
Steam cleaning masonry, using rented steam cleaning rig (with accessories) at $52.50 per day.						
Smooth face (75 SF per manhour)	B9@.013	SF	.06	.42	.48	.82
Rough face (55 SF per manhour)	B9@.018	SF	.08	.58	.66	1.12
Add for masking adjacent surfaces	—	%	—	5.0	—	—
Add for difficult stain removal	—	%	50.0	50.0	—	—
Add for working from scaffold	—	%	—	20.0	—	—
Repointing brick, cut out joint, mask (blend-in), and regrout (tuck pointing).						
30 SF per manhour	B9@.033	SF	.08	1.05	1.13	1.92

Masonry cleaning. Costs to clean masonry surfaces using commercial cleaning agents, add for pressure washing and scaffolding equipment.

	Craft@Hrs	Unit	Material	Labor	Total	Sell
Typical cleaning of surfaces, granite, sandstone, terra cotta, brick	BL@.015	SF	.30	.44	.74	1.26
Cleaning surfaces of heavily carbonated limestone or cast stone	BL@.045	SF	.40	1.31	1.71	2.91
Typical wash with acid and rinse	BL@.004	SF	.46	.12	.58	.99

Masonry Fireplaces

Masonry fireplace demolition. Includes breaking a concrete block or brick fireplace and chimney into manageable pieces using a pneumatic chipper and piling on site. No fireplace foundation included. Add the cost of debris disposal.

	Craft@Hrs	Unit	Material	Labor	Total	Sell
Demolish interior fireplace and chimney, to 12' high	BL@8.00	Ea	—	232.00	232.00	394.00
Demolish fireplace and exterior chimney, to 12' high	BL@10.7	Ea	—	311.00	311.00	529.00
Demolish chimney over 12' high, per linear foot of height over 12'	BL@.700	Ea	—	20.30	20.30	34.50

Masonry fireplace repairs. Chip out using a pneumatic chipper.

	Craft@Hrs	Unit	Material	Labor	Total	Sell
Remove firebrick in a firebox wall	B9@.388	SF	—	12.40	12.40	21.10
Remove firebrick in a firebox floor	B9@.204	SF	—	6.52	6.52	11.10
Replace firebrick firebox wall or floor	B9@.186	SF	5.00	5.94	10.94	18.60
Remove and replace nonadjacent fireplace or chimney brick, per brick removed and replaced	B9@.360	Ea	1.00	11.50	12.50	21.30
Remove and replace brick in 4" fireplace or chimney wall, per square foot of wall	B9@.675	SF	4.00	21.60	25.60	43.50

	Craft@Hrs	Unit	Material	Labor	Total	Sell
Masonry fireplaces. Fire box lined with refractory firebrick, backed with common brick and enclosed in a common brick wall. Includes fireplace throat and smoke chamber. Overall height of the fireplace and throat is 5'. Chimney (flue) is made from common brick and a flue liner. Including lintels, damper and 12" thick reinforced concrete foundation.						
30" wide x 29" high x 16" deep fire box	B9@39.7	LS	651.00	1,270.00	1,921.00	3,270.00
36" wide x 29" high x 16" deep fire box	B9@43.5	LS	686.00	1,390.00	2,076.00	3,530.00
42" wide x 32" high x 16" deep fire box	B9@48.3	LS	759.00	1,540.00	2,299.00	3,910.00
48" wide x 32" high x 18" deep fire box	B9@52.8	LS	833.00	1,690.00	2,523.00	4,290.00
12" x 12" chimney to 12' high	B9@9.50	LS	385.00	304.00	689.00	1,170.00
Add for higher chimney, per LF	B9@1.30	LF	52.50	41.50	94.00	160.00
Add for 8" x 10" ash cleanout door	B9@.500	SF	20.90	16.00	36.90	62.70
Add for combustion air inlet kit	B9@.750	SF	31.40	24.00	55.40	94.20
Add for common brick veneer interior face	B9@.451	SF	4.81	14.40	19.21	32.70
Add for rubble or cobble stone veneer face	B9@.480	SF	13.10	15.30	28.40	48.30
Add for limestone veneer face or hearth	B9@.451	SF	24.90	14.40	39.30	66.80
Add for 1-1/4" marble face or hearth	B9@.451	SF	37.40	14.40	51.80	88.10
Add for modern design 11" x 77" pine mantel	B1@3.81	Ea	737.00	124.00	861.00	1,460.00
Add for gas valve, log lighter and key	PM@.750	Ea	84.80	31.00	115.80	197.00
Prefabricated fireplaces. Zero clearance factory-built fireplaces, including metal fireplace body, refractory interior and fuel grate. UL listed.						
Radiant heating, open front only, with fuel grate						
38" wide	B9@5.00	Ea	643.00	160.00	803.00	1,370.00
43" wide	B9@5.00	Ea	786.00	160.00	946.00	1,610.00
Radiant heating, open front and one end						
38" wide	B9@5.00	Ea	979.00	160.00	1,139.00	1,940.00
Radiant heating, three sides open, with brass and glass doors and refractory floor						
36" wide (long side)	B9@6.00	Ea	1,560.00	192.00	1,752.00	2,980.00
Radiant heating, air circulating, open front only, recessed screen panels						
38" screen size	B9@5.50	Ea	881.00	176.00	1,057.00	1,800.00
43" screen size	B9@5.50	Ea	1,020.00	176.00	1,196.00	2,030.00
Radiant heating simulated masonry fireplace						
45" wide	B9@5.00	Ea	1,490.00	160.00	1,650.00	2,810.00
Convection-type heat circulating prefabricated fireplace, open front						
38" screen size	B9@5.50	Ea	734.00	176.00	910.00	1,550.00
43" screen size	B9@5.50	Ea	797.00	176.00	973.00	1,650.00
Forced air heat circulating prefabricated fireplace, open front						
35" screen size	B9@5.50	Ea	647.00	176.00	823.00	1,400.00
45" screen size	B9@5.50	Ea	979.00	176.00	1,155.00	1,960.00
Radiant, see-through with brass and glass doors, designer model						
38" screen size	B9@6.50	Ea	1,560.00	208.00	1,768.00	3,010.00
Header plate,						
Connects venting system to gas fireplace	B9@.167	Ea	89.90	5.34	95.24	162.00
Fireplace flue (chimney), double wall, typical cost for straight vertical installation						
8" inside diameter	B9@.167	LF	27.10	5.34	32.44	55.10
10" inside diameter	B9@.167	LF	38.70	5.34	44.04	74.90
Add for flue spacers, for passing through ceiling						
1" or 2" clearance	B9@.167	Ea	16.10	5.34	21.44	36.40
Add for typical flue offset	B9@.250	Ea	101.00	7.99	108.99	185.00
Add for flashing and storm collar, one required per roof penetration						
Flat to 6/12 pitch roofs	B9@3.25	Ea	64.30	104.00	168.30	286.00
6/12 to 12/12 pitch roofs	B9@3.25	Ea	93.30	104.00	197.30	335.00
Add for spark arrestor top	B9@.500	Ea	131.00	16.00	147.00	250.00

Roofing and Flashing

7

Damage from a leaky roof is usually obvious. However, the precise source of the leak won't always be clear. The hole in the roof is seldom directly above the drip stains on the ceiling. Water can get through the roof surface and run along the top of the sheathing until it finds a panel edge. Then it may run along a rafter or the underside of the panel for a foot or so before dropping down to the attic, where it puddles on the vapor barrier. The puddle expands downhill, finds a seam, and then begins saturating the drywall or plaster below. The distance between the hole in the roof and the wet spot on the ceiling can be several feet or more.

So where do you start patching? If the attic crawlspace is readily accessible, start your search in the attic. Look for recent water stains. Once you have that information, you can go up on the roof and look for:

- *Missing shingles.* Leaks are nearly certain when roof components get blown away.
- *Physical damage.* Foot traffic can break seals and dislodge shingles. Was anyone working on the roof or in the attic before the last rain? Check for a tree branch that could be hitting the roof's surface during a high wind.
- *Deteriorated caulking.* Caulked joints harden with age and then fracture along the joint with movement.
- *Debris.* Anything resting on a roof can collect water and speed deterioration. Debris tends to accumulate on the high side of the roof next to penetrations such as a chimney or skylight.
- *Shiners.* Shiners are nails not covered by the shingle course above. If left exposed for many years, the nails can rust out, leaving a hole in the roof.
- *Broken mortar.* Any gap or opening in a chimney or parapet wall offers an opening to the attic.
- *Obstructed gutters.* Be sure rain runoff gear is working properly. Water accumulating in a gutter can back up over the top of the fascia, run along the soffit, and then down the wall interior.
- *Open roll roofing seams.* Even if you see bitumen or adhesive sticking out under a lap, the seam may be wicking moisture to the interior. Try running a putty knife or pocketknife blade under the lap. If the blade slides into the lap more than an inch, assume the seam needs to be resealed.

- *Roof blisters.* Blistering (air pockets trapped between the layers of roofing material) is a failure of the bond. Punctured blisters are an obvious source of leaks.
- *Splits in the membrane.* Walk the area with your feet close together, taking many small steps. Twist your feet as you walk. If a split is developing, you'll see the roof membrane separate between your feet.
- *Obstructed roof drains.* Drains are made to carry water one way. When backed up, the roof will leak.
- *Deteriorated expansion joints.* Seams on expansion joints may have to be replaced more often than the roof itself.
- *A leak when it isn't raining.* Condensate from roof-mounted coolers should be piped to a legal drain, not discharged on the roof.
- *Hundred year flood.* It's probably OK to ignore leaks first noticed after the heaviest rain of the century or when rain is driven by unusually heavy wind. Neither you nor the present owner may be around the next time that happens.

Flashing Leaks

If the roof surface appears sound, look for problems with the flashing. The term flashing is used to describe two types of roofing material. One is metal flashing, which is used at roof joints, such as in the valley of a hip roof where one surface meets another, or the metal counterflashing that extends down a parapet wall and over the roof surface below. The other type is non-metallic flashing. It's usually made from asphalt-coated materials, and adds an extra layer of protection under the roof cover, especially along the eave line. In cold climates, roll-roof flashing helps prevent damage from ice dams, as shown in Figure 7-1. Ice dams form when melting snow runs down the roof and re-freezes at the cornice. The ice forms a barrier, causing water to back up under the shingles. A layer of 45-pound smooth-surface roll roofing (flashing) won't prevent the dam from forming, but it will minimize the chance of water entering the attic and the wall below. Be sure this flashing extends 36" inside the warm wall. If two strips are needed, seal the joint and end joints with mastic. Many building departments require self-adhesive type roll roofing on eaves and gables.

Metal counterflashing is lapped over the top of bituminous base flashing to keep water above the roof surface. The most common metal counterflashing problems are:

- Corrosion and deterioration of the metal.
- Open joints between pieces of metal flashing.
- Separation of counterflashing from vertical surfaces.

- Air gaps between the counterflashing and base flashing. Counterflashing that ends too high above the base flashing.

Common problems with non-metallic base flashing include:

- Flashing that's too thin, such as 15-pound roofing felt used where 45-pound is needed.
- Flashing that's not wide enough, such as an 18" strip of felt used where 36" felt is needed.

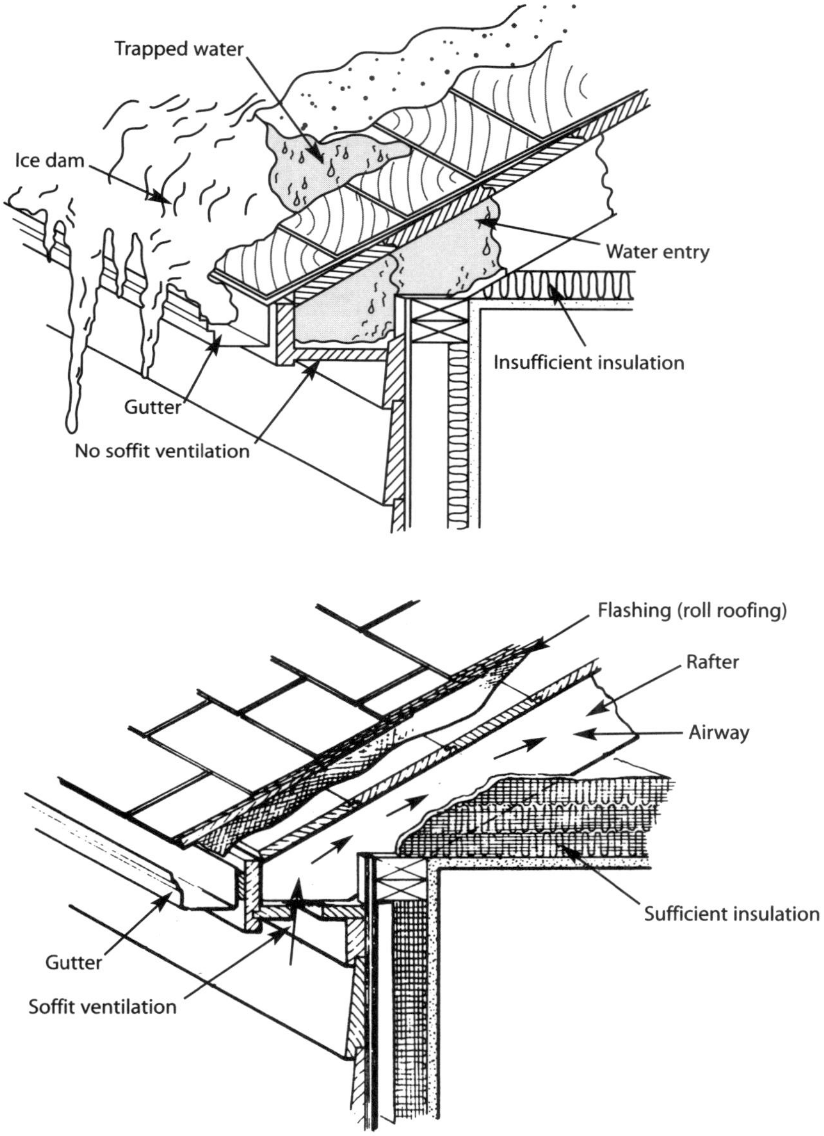

Figure 7-1

Flashing and ventilation to prevent damage from ice dams.

- Lack of a 45-degree cant strip where a roof joins the wall, leaving the flashing unsupported at the base of the wall.
- Poorly sealed seams and end laps.
- Poorly fastened base flashing at walls or curbs.
- Decay of the sheathing below.

Both metallic and non-metallic flashing can be used on the roof surface *(base)* with most types of roof cover. Built-up roofing is an exception. As the surface heats and cools during the day, metal flashing expands and contracts much more than roll roofing. The result will be ripples, tearing and, eventually, separation of the built-up roofing at the edges of the metal flashing. Use roll-roofing materials as flashing under a built-up roof.

All penetrations in the roof surface need flashing. Plumbing stacks and small vents should be flashed with a roof jack or flange placed directly on the last ply of roofing material and stripped in with felt and mastic or felt and bitumen. Prefabricated vinyl or plastic seals for plumbing stacks are effective, and are available in a wide variety of styles and sizes. Larger penetrations should be made through a roof-mounted curb built from 2 x 6 or larger framing lumber. Bituminous base flashing should run up the curb, and metal counterflashing should run down the curb from above.

> Attaching an antenna, flagpole, sign or bracing directly to the roof membrane is asking for trouble. Leaks are almost certain. The best repair is to remove the attachment. However, if it must be mounted on top of the roof, mount it using a raised curb-type support. Then install flashing to keep the roof watertight.

Leaks around a roof penetration are usually caused by:

- Open or broken seams in a metal curb due to expansion and contraction.
- Standing water behind a penetration.
- Sagging or separating base flashing.
- Missing or deteriorated counterflashing.
- Splitting or separation of felt stripping over the edge of metal flanges.
- Improper priming and stripping of metal surfaces.
- Fasteners working loose around the flashing.
- Movement between stack vents or pipes and flashing.

Can't Find the Leak? If All Else Fails . . .

If you're truly stumped about the source of a leak, or have been called back several times to make unsuccessful repairs of the same leak, try this: Take a garden hose up on the roof and direct a steady stream of water at every source likely to leak. When the leak is active again, trace the path of the leak from the building interior to the roof surface. Be sure to brief the owner before you test for leaks — and get the owner's help in reporting an active leak.

Roof Repairs

Damaged or missing shingles will need to be replaced. In other cases, once you've found the source of a leak, use trowel-grade wet patch plastic roofing cement to make repairs. First, wire brush or scrape away everything that's loose, including gravel, granules, membrane, dirt, dust and debris. Then trowel on plastic roofing cement so the patch extends at least 6" beyond the edge of the repair. Press the roofing cement firmly into the membrane. Then test the patch by laying the trowel flat on the cement surface. Plastic cement shouldn't lift off when you pull the trowel away; if it does, work the cement deeper into the membrane. Next, cut fiberglass-reinforcing fabric to fit the patch. Press the fabric into the first layer of plastic roofing cement. Fiberglass reinforces the plastic cement and extends the useful life of the repair. Over the embedded fabric, trowel on another layer of plastic roofing cement. Be sure to cover all the exposed fabric. Finally, dust the patch with mineral granules that match the existing surface. Granules make the patch more durable by adding sun and weather protection.

On a smooth-surface roof, such as built-up asphalt, EPDM or PVC thermoplastic, make emergency repairs with roof tape membrane. Start by cleaning the damaged surface with alcohol or a household cleaning solution such as window cleaner. Then wipe the surface with a clean cloth saturated with splice cleaner or white gas. (Use rubber gloves to protect your skin.) Apply roof tape to the damaged roofing material. Be sure to press the tape firmly over the entire surface. If you can't locate a source for roof tape, follow the procedure described here, but apply a smooth coat of butyl or polyurethane caulk.

Most flashing can be patched with plastic roofing cement, but on a new roof the useful life of the repair will usually be less than for the roof surface. Make more permanent flashing repairs by stripping off old flashing down to the roof surface. Then build up new flashing in layers of felt and adhesive. You can make flashing around scuppers and drains watertight with fresh caulk.

After high winds, on homes with a flat, gravel-covered roof, it's common to find roof gravel scattered, leaving some areas bare and unprotected. Before redistributing the gravel, inspect the membrane for open seams, punctures, and tears from flying debris. If you find a leak that's the result of a failed expansion joint or deteriorated caulk, remove and replace the defective material.

Life Expectancy of Roofing Materials

The life expectancy of roofing materials will vary considerably with the type and how well the roof is maintained. Common 3-tab asphalt shingles last about 20 years. Under favorable conditions, a wood shingle roof will last 30 years. A 3-ply built-up roof should last 15 years, and you can expect 20 years from a 4-ply or 5-ply built-up roof. Some architectural-grade laminated shingles are rated at 45 years. A well-maintained slate or tile roof can last 150 years.

Premature roof failure is usually caused by fungus, high winds, roof deck deterioration or failure of the flashing. In some climates, portions of a roof can develop fungus that discolors and eventually deteriorates the shingles. After a high wind (over 50 miles per hour), it's common to see shingles blown off. Replacing a shingle takes only minutes, but that's seldom the extent of the damage. High winds create a vacuum above portions of the roof, lifting and flexing the surface like the wings of an airplane. Such movement of the roof deck will cause seals to break and fasteners to work loose. It's this unseen damage you have to watch out for — it may not be apparent until the next rain.

When a roof is approaching the end of its useful life, you'll see obvious signs of weathering. Asphalt shingles lose granules, curl on the edges and get brittle. Wood shingles break, warp, curl and blow off in high winds. A good wood shingle roof forms a perfect mosaic. A worn wood shingle roof looks ragged. Excessively worn roofing loses the ability to shed water, making it a candidate for replacement — even if it hasn't yet begun to leak.

Choosing a New Roof Material

The roof is a major design element of most homes. If the roof cover has to be replaced, be sure to select material that fits the character of the building. You don't want to put Spanish tile on a Tudor house.

A paramount consideration in choosing a new roof covering is the weight of the material. You can't replace lighter roofing materials (such as wood shingles) with heavier roofing materials (such as concrete tile or slate) unless the owner is prepared to reinforce the roof to support the additional load. Select materials that are similar in weight to what you're replacing.

Wood shingles and composition shingles are still the most popular roofing materials for pitched roofs, although in an increasing number of areas in the United States, wood shingles untreated with fire-retardant may not be installed. Both of these can be applied directly over some types of roofing without doing a tear-off. But there are limits. First, ask your customer if the house has been previously reroofed. When in doubt, you'll need to get up on a ladder and peel back a corner of the existing shingles to see what's under them. Be sure you understand your local building department's requirements regarding layering and tear-offs.

Here are some other tips to help you decide whether or not you can avoid a tear-off. You can't install fiberglass strip shingles over hexagon, Dutch lap or T-lock shingles. However, you can install laminated fiberglass strip shingles over asphalt shingles without cutouts and over wood shingles, assuming the old surface is smooth or can be made smooth. T-lock asphalt shingles can't be installed over hexagonal, Dutch lap or wood shingles, or on a roof with 2 in 12 or 3 in 12 pitch. But T-locks can be installed over 3-tab shingles and shingles without cutouts. Slate, tile, wood shakes and cement fiber roofing must always be removed before reroofing.

Before signing a contract for reroofing over an old roof, consider:

- *Building codes.* Some building codes prohibit installing a new roof over old. Others allow either two or three layers of roofing. A few allow a one-layer tear-off — remove and replace just the top layer of a two- or three-layer roof. If in doubt, place a call to your local building department.
- *Bad decking.* Walk the deck. If you see sags between the rafters or trusses, or if the deck feels weak under foot, the sheathing may be deteriorating or damaged. You may find that only a small section of the deck needs to be replaced. This is often caused by a long-term leak. If possible, go into the attic space and inspect the deck from underneath. This may show you the extent of the problem. Do a tear-off and plan on repairing the deck. The deck under a wood shingle roof may use skip sheathing: 1" x 2" to 6" boards laid with a 1" to 3" gap between boards. Skip sheathing is fine for wood shakes and shingles, but asphalt shingles need a solid board or plywood deck.
- *Ice dams.* A full tear-off is best if there's a history of damage from ice dams.
- *Incompatible shingles.* Installing heavyweight architectural shingles over lightweight 3-tabs is nearly always successful. A lightweight shingle installed over a heavyweight shingle may leave obvious bumps and ridges.
- *Poor condition.* If the existing roof is in really bad shape, with curled tabs and crooked rows, a tear-off is in order.
- *Shorter life span.* Most roofing professionals agree that the life expectancy of a new roof is reduced by 10 to 20 percent when it's installed as an overlay on an old roof.

When installing new shingles over old wood or asphalt shingles, manufacturers may recommend removing a 6" wide strip of the old shingles along eaves and gables. Then nail down a nominal l" x 6" board where the shingles were cut away. If the old shingles were asphalt 3-tabs, use thinner boards. Remove the old covering from ridges and hips and place a strip of lumber over each valley to separate old metal flashing from new.

Wood Shingles

If you decide on wood shingles as the new roofing material, use Number 1 Grade shingles. Number 1 shingles are all heartwood, all edge grain and tapered. Western red cedar shingles and redwood shingles offer the best decay resistance and shrink very little as they age. Wood shingle widths vary. Narrower shingles are classified as either Number 2 or Number 3. Figure 7-2 shows common shingle sizes and recommended exposures.

Maximum Exposure for Wood Shingles			
Length	Thickness (green)	Slope* less than 4 in 12	Slope 4 in 12 or over
16"	5 butts in 2"	3 3/4"	5"
18"	5 butts in 2 1/4"	4 1/4"	5 1/2"
24"	4 butts in 2"	5 3/4"	7 1/2"

*Minimum slope for main roofs — 4 in 12.
Minimum slopes for porch roofs — 3 in 12.

Figure 7-2

Recommended exposure for wood shingles

Follow the standards illustrated in Figure 7-3 when you apply wood shingles:

1. Extend the shingles 1 1/2" to 2" beyond the eave line and about 3/4" beyond the rake *(gable)* edge.
2. Nail each shingle with two rust-resistant nails spaced about 3/4" from the edge and 1 1/2" above the butt line of the next course. Use 3d nails for 16" and 18" shingles and 4d nails for 24" shingles. If you're applying new shingles over old wood shingles, you'll need longer nails to penetrate through the old roofing and into the sheathing. Use ring-shank (threaded) nails on plywood roof sheathing that's less than 1/2" thick.
3. Allow a 1/8" to 1/4" expansion space between shingles. Stagger the joints in succeeding courses so the joint in one course isn't in line with either of the two courses above.
4. Shingle away from valleys, using the widest shingles in the valleys. The valley should be 4" wide at the top and increase in width at the rate of 1/8" per foot from the top to the bottom. Valley flashing with a standing seam (W valley) works best. Don't nail shingles through the metal. Valley flashing should be a minimum of 24" wide for roof slopes under 4 in 12; 18" wide for roof slopes of 4 in 12 to 7 in 12; and 12" wide for roof slopes of 7 in 12 and over.
5. Place metal edging along the gable end to direct water away from the end walls.

You can use the same instructions for applying wood shakes, except you'll need longer nails because shakes are thicker. Shakes are also longer and the exposure distances can be greater: 8" for 18" shakes, 10" for 24" shakes, and 13" for 32" shakes. For a rustic appearance, lay the butts unevenly.

Install an 18"-wide strip of 30-pound asphalt felt between each course to protect against wind-driven snow. The lower edge of the shake felt should be above the butt end by twice the exposure distance. If the exposure distance is less than one-third the total length, most shingle roofers don't bother with shake felt.

Asphalt-Fiberglass (Composition) Shingles

Asphalt shingles are made from a fiberglass mat saturated with asphalt. Fiberglass adds strength and durability, and asphalt adds moisture protection. But asphalt doesn't bind very well with fiberglass, so mineral fibers are added to the fiberglass before coating. The result is a "composition" shingle that is both durable and effective.

The most common type of asphalt shingle is the square-butt strip shingle. It measures 12" by 36", has three tabs, and is usually laid with 5" exposed to the weather. Top-quality laminated asphalt shingles weigh as much as 450 pounds per

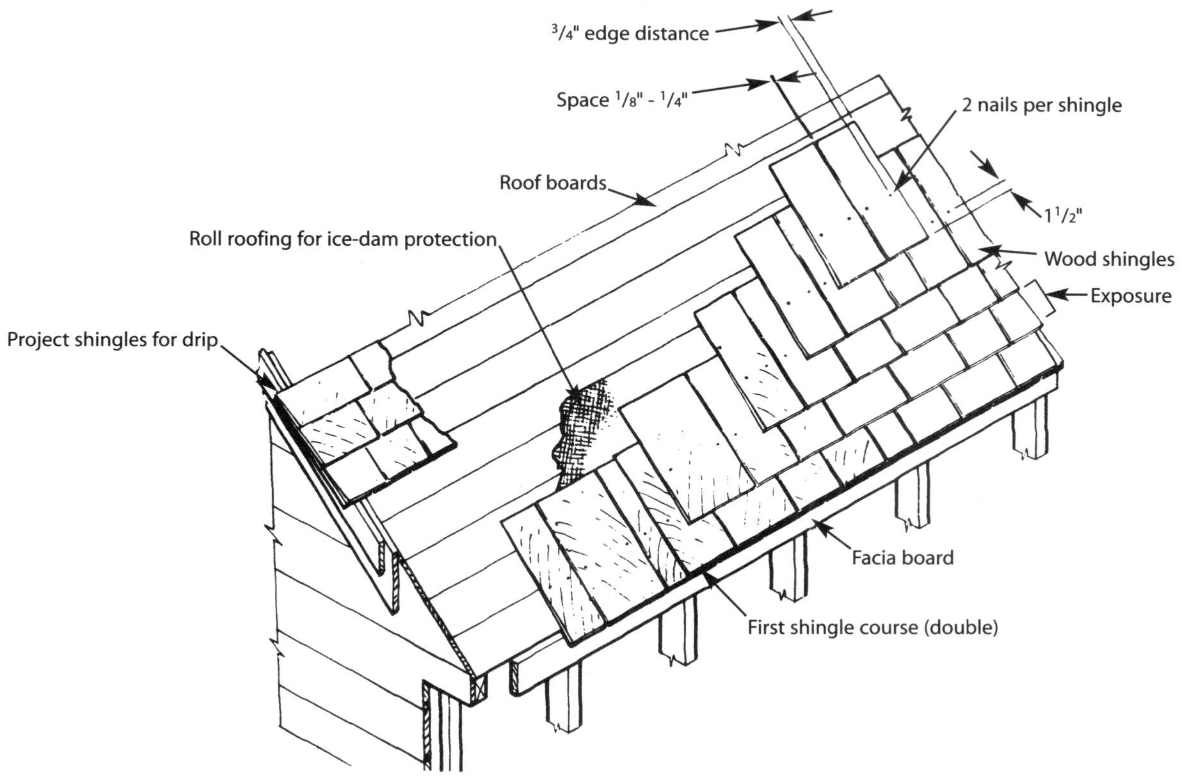

Figure 7-3
Application of wood-shingle roofing over boards

100 square feet (a square) and can be expected to last up to 45 years. Heavyweight laminated asphalt shingles are sold in four bundles to the square. Regular weight asphalt shingles are sold in three bundles to the square. Store the bundles flat so the strips don't curl when they're installed. Avoid storing them in the sun, which will activate the self-sealing tabs, causing the top shingles in each bundle to stick together.

Install 15-pound saturated felt before laying asphalt shingles on a roof with a 4 in 12 to 7 in 12 pitch. Start by laying the felt at the roof eave and lap each successive course by several inches. Stagger the felt courses to separate joints. Apply a double layer of 15-pound felt on a roof with less than a 4 in 12 pitch. Doubling the felt will take more time because it's harder to keep wrinkles out.

Begin application with metal edging along the eave line. Cut the exposure tabs off the first course of asphalt shingles and lay them along the roof edge. See Figure 7-4 A. Then apply another course on top of the first, extending the shingle exposure 1/2" over the metal edging at the rake to prevent water from backing up under the shingles (Figure 7-4 B). Apply the next course with the recommended exposure. Snap horizontal chalklines on the felt to help keep shingles in straight lines. Follow the manufacturer's recommendation on nailing. When a nail penetrates a crack or knothole, remove the nail, seal the hole, and drive another nail into sound wood. Any nail that doesn't hit sound wood will gradually work out and cause a hump in the shingle above. Use seal-tab or T-lock shingles in high wind areas.

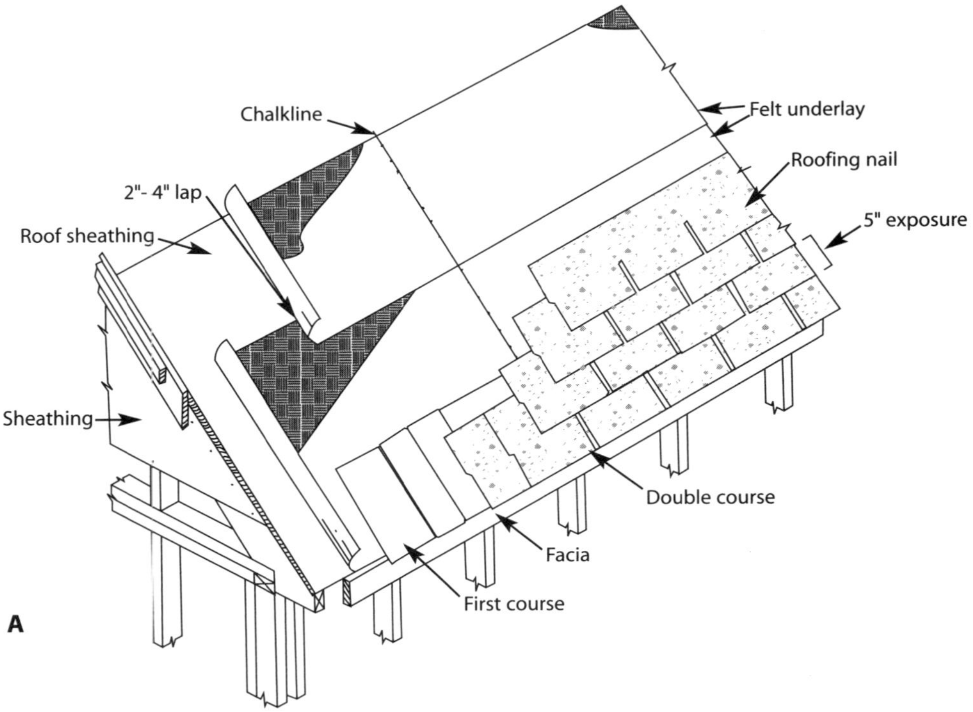

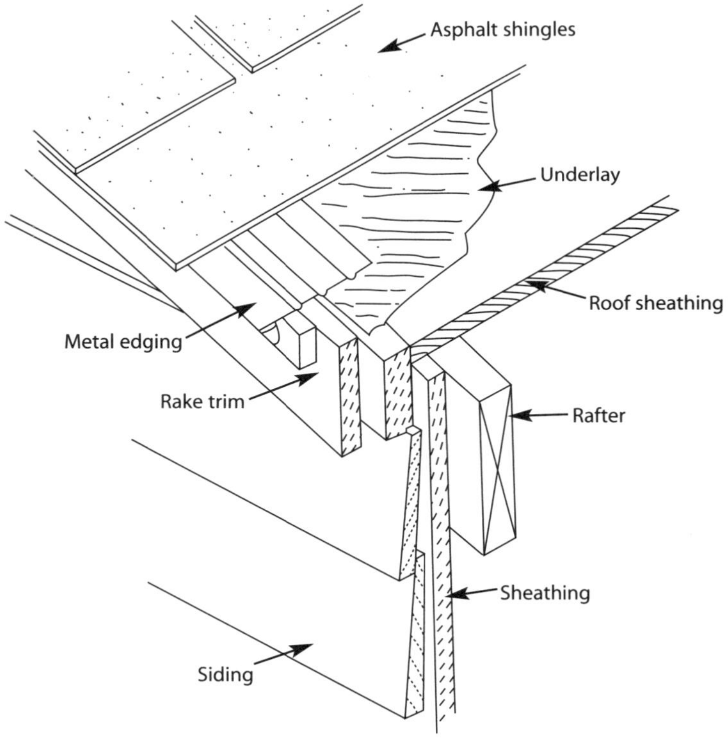

Figure 7-4

*Application of asphalt-shingle roofing over plywood: **A**, with strip shingles ; **B**, metal edging at gable end.*

Built-Up Roofing

A built-up roof that isn't worth repairing must be stripped off before new roofing is installed. Once the deck is clean, pound all exposed nails flush (or pull them and re-nail). Make repairs to the deck and flashing as needed.

The base sheet for a built-up roof can be either resin-coated or asphalt-coated organic felt or fiberglass. On a wood deck, nail the base sheet to the deck. Use a Type I or Type II venting base sheet on a concrete deck.

The felt layers are applied on top of the base sheet. Type I (15-pound) and Type II (30-pound) asphalt-saturated organic felts are the most common. Asphalt-impregnated fiberglass felts are Types III, IV and V. Type III fiberglass felt is lower quality than Type IV. Coal tar-saturated felts are popular in some areas. Polyester felts are stronger than fiberglass felts, and fiberglass felts are stronger than organic felts.

Lay each 100 square feet of felt in 25 to 30 pounds of hot asphalt, 20 to 25 pounds of coal tar, or 2 to 4 gallons of cold-applied roofing adhesive (lap cement). If the last layer is a mineral surface cap sheet, use the same quantity of asphalt, coal tar or adhesive. For a topping of gravel or slag, flood each 100 square feet of surface with 60 pounds of hot asphalt, 70 pounds of coal tar or 3 gallons of emulsion. Then install 400 to 500 pounds of gravel or 300 to 400 pounds of slag per 100 square feet of roof surface. If gravel or slag is set in emulsion, apply an aluminum reflective coating to protect against UV damage.

Modified Bitumen Roofing (APP and SBS)

In the 1970s, chemists in Europe discovered that asphalt took on rubber-like properties when combined with certain additives. These additives include APP (atactic polypropylene) and SBS (styrene butadiene styrene). Adding about 30 percent APP to asphalt creates a product that can be stretched up to 50 percent without breaking. SBS modified roofing was developed by French and German chemists. They found that asphalt took on the character of rubber when 10 to 15 percent SBS was added. SBS-modified asphalt will stretch up to six times its original length without breaking. Unlike APP, SBS-modified asphalt returns to its original size when released. This elasticity makes asphalt roofing more durable. The roof surface can shift or expand and contract without breaking.

APP roofing uses a polyester mat that matches the pliability of APP-modified asphalt better than a woven glass mat. You apply APP membranes using a torch. The back of the sheet has extra asphalt that bonds to the surface below when heated. That's convenient on smaller jobs with limited working area, because you need very little equipment.

You can use many mat materials with SBS, including fiberglass and polyester. Fiberglass SBS mats weigh from 1.0 to 2.5 pounds per 100 square feet. Polyester SBS mats weigh from 3.5 to 5.0 pounds per 100 square feet. You can apply SBS membranes with hot asphalt, a torch, or cold adhesive.

Modified bitumen membrane roofing comes in 100-square-foot rolls. Use one, two, or three plies to create a finished roof. Or combine modified bitumen membrane with fiberglass felt to form a hybrid roof system.

The minimum roof slope should be not less than 1/4" of rise per 12" of horizontal run. The surface can be finished with mineral granules or an aggregate, such as gravel or slag.

How to Measure Roof Area

Flat roofs are easy to measure. Just divide the area into rectangles, calculate the area of each, and add up the total. You can probably do that from the ground. Sloping roofs are a little more complex. The steeper the slope, the more the roof area differs from the horizontal (plane) area. To calculate the roof surface accurately, you have to multiply the rafter lengths by the width of each pitch. However, if the roof pitch is no more than 5 in 12, a simple rule of thumb will yield nearly the same result in a fraction of the time: Multiply the length and width of the building, including eaves and overhang. Divide by 100 to find the number of roofing "squares." Then add 10 percent for a gable roof, 15 percent for a hip roof and 20 percent for a roof with dormers.

	Craft@Hrs	Unit	Material	Labor	Total	Sell

Roof Repairs

Tear-off roofing. Includes removing the existing surface and throwing debris into a bin below a one-story or two-story building. Also includes pulling or driving all protruding nails but no repairs to the roof deck, flashing or rafters. Add the cost of hauling away debris and dump or recycling fees. Roofing tear-off will take longer when the pitch is steeper than 6 in 12.

	Craft@Hrs	Unit	Material	Labor	Total	Sell
Asphalt or fiberglass shingles, single layer	BL@1.00	Sq	—	29.10	29.10	49.50
Asphalt or fiberglass shingles, double layer	BL@1.45	Sq	—	42.10	42.10	71.60
Wood shingles or shakes	BL@1.50	Sq	—	43.60	43.60	74.10
Slate shingles	BL@1.79	Sq	—	52.00	52.00	88.40
Clay or concrete tile	BL@1.65	Sq	—	47.90	47.90	81.40
Remove gravel stop or flashing, per LF	BL@.070	LF	—	2.03	2.03	3.45
Remove chimney flashing, per chimney	BL@1.50	Ea	—	43.60	43.60	74.10
Roof insulation board	BL@.650	Sq	—	18.90	18.90	32.10
Gypsum plank	BL@3.16	Sq	—	91.80	91.80	156.00
Metal decking	BL@3.52	Sq	—	102.00	102.00	173.00
Remove vent flashing, per vent	BL@.400	Sq	—	11.60	11.60	19.70
Remove and reset skylight to 10 SF	BL@2.20	Ea	—	63.90	63.90	109.00

Tear-off built-up roofing. Includes removing the existing surface and throwing debris into a bin below a one-story or two-story building. Add the cost of hauling away debris and dump or recycling fees.

	Craft@Hrs	Unit	Material	Labor	Total	Sell
Roofing on wood deck	BL@1.08	Sq	—	31.40	31.40	53.40
Roofing and insulation on wood deck	BL@1.73	Sq	—	50.30	50.30	85.50
Roofing on concrete deck	BL@1.31	Sq	—	38.10	38.10	64.80
Roofing and insulation on concrete deck	BL@1.96	Sq	—	56.90	56.90	96.70

Repair roof deck — remove and replace. Includes removing the existing surface, throwing debris into a bin below a one-story or two-story building, pulling or driving all protruding nails but no repairs to flashing or rafters. Add the cost of hauling away debris and dump or recycling fees. Per square foot removed and replaced.

	Craft@Hrs	Unit	Material	Labor	Total	Sell
2" x 6" T&G boards	B1@.032	SF	1.83	1.04	2.87	4.88
2" x 10" T&G boards	B1@.029	SF	1.85	.94	2.79	4.74
1" x 6" square edge boards	B1@.029	SF	1.74	.94	2.68	4.56
1" x 8" square edge boards	B1@.027	SF	1.74	.88	2.62	4.45
1" x 12" square edge boards	B1@.026	SF	1.77	.85	2.62	4.45
Plywood 4' x 8' sheets or partial sheets	B1@.025	SF	.63	.81	1.44	2.45

Repair roof flashing — remove and replace.

	Craft@Hrs	Unit	Material	Labor	Total	Sell
Roof drain or vent	R1@.918	Ea	20.10	31.90	52.00	88.40
Flashing felt and counterflashing	R1@.160	LF	2.66	5.56	8.22	14.00
Pitch pocket, set of two	R1@.400	Set	23.00	13.90	36.90	62.70
1-ply membrane at headwall	R1@.053	LF	.44	1.84	2.28	3.88
2-ply membrane at headwall	R1@.076	LF	.85	2.64	3.49	5.93
1-ply felt and flashing at headwall	R1@.129	LF	.97	4.48	5.45	9.27
2-ply felt and flashing at headwall	R1@.152	LF	1.27	5.28	6.55	11.10

Repair roof waterproofing. Install membrane on smooth and clean vertical concrete or masonry surface, such as a parapet wall. Per square foot.

	Craft@Hrs	Unit	Material	Labor	Total	Sell
1-ply	R1@.025	SF	.91	.87	1.78	3.03
2-ply	R1@.041	SF	1.62	1.42	3.04	5.17
3-ply	R1@.056	SF	2.70	1.95	4.65	7.91
Vertical joint, 2-ply	R1@.046	LF	.55	1.60	2.15	3.66

Non-fibered roof and foundation coat. Moisture barrier on exterior concrete and masonry surfaces or heavy-duty roof primer.

	Craft@Hrs	Unit	Material	Labor	Total	Sell
1.5 gallons per square	R1@.006	SF	.18	.21	.39	.66

	Craft@Hrs	Unit	Material	Labor	Total	Sell
Remove and replace roof waterproofing. Chip away damaged membrane on a vertical surface, such as a parapet wall. Clean with wire brush. Replace with new waterproof membrane. Per square foot.						
1-ply	R1@.042	SF	.59	1.46	2.05	3.49
2-ply	R1@.058	SF	1.09	2.01	3.10	5.27
3-ply	R1@.074	SF	1.83	2.57	4.40	7.48
Waterproof roof vent. Per vent. Based on vents averaging 24" diameter.						
2-ply, using hot asphalt	R1@.797	Ea	2.27	27.70	29.97	50.90
4-ply, using hot asphalt	R1@1.59	Ea	3.05	55.20	58.25	99.00
2-ply, using plastic roofing cement	R1@1.28	Ea	3.82	44.50	48.32	82.10
4-ply, using plastic roofing cement	R1@2.54	Ea	3.99	88.20	92.19	157.00
Repair roof vent waterproofing. Chip away damaged membrane around a roof vent. Clean with a wire brush. Replace with new waterproof membrane. Per vent. Based on vents averaging 24" diameter.						
2-ply, using hot asphalt	R1@1.56	Ea	2.27	54.20	56.47	96.00
4-ply, using hot asphalt	R1@2.35	Ea	3.05	81.60	84.65	144.00
1-ply, using plastic roofing cement	R1@.764	Ea	2.27	26.50	28.77	48.90
2-ply, using plastic roofing cement	R1@2.04	Ea	3.82	70.90	74.72	127.00
4-ply, using plastic roofing cement	R1@3.31	Ea	5.35	115.00	120.35	205.00
Waterproof masonry wall. Waterproof masonry such as a parapet wall. Membrane set in asphaltic waterproofing. Per square foot of concrete or concrete block wall.						
1-ply, vertical wall	R1@.034	SF	.63	1.18	1.81	3.08
2-ply, vertical wall	R1@.041	SF	1.16	1.42	2.58	4.39
3-ply, vertical wall	R1@.056	SF	1.94	1.95	3.89	6.61
Remove and replace asphalt shingles.						
5 square foot patch (3 shingles)	R1@.171	Ea	4.82	5.94	10.76	18.30
12 square foot patch (10 shingles)	R1@.367	Ea	11.60	12.70	24.30	41.30
31 square foot patch (25 shingles)	R1@.804	Ea	29.90	27.90	57.80	98.30
Re-seal shingle with roofing cement, per shingle	R1@.033	Ea	.42	1.15	1.57	2.67

Composition Shingles

	Craft@Hrs	Unit	Material	Labor	Total	Sell
Architectural grade laminated shingles. Three bundles per square (100 square feet). Class A fire rating. 70 MPH wind resistance.						
Oakridge, 30 year	R1@1.83	Sq	111.00	63.60	174.60	297.00
Oakridge, 30 year, algae-resistant	R1@1.83	Sq	118.00	63.60	181.60	309.00
Oakridge, 50 year, algae-resistant	R1@1.83	Sq	133.00	63.60	196.60	334.00
Hip and ridge shingles with sealant. Per linear foot of hip or ridge.						
High ridge	R1@.028	LF	3.10	.97	4.07	6.92
High ridge, algae-resistant	R1@.028	LF	3.16	.97	4.13	7.02
High style	R1@.028	LF	3.43	.97	4.40	7.48
3-tab asphalt fiberglass shingles. Three bundles per square (100 square feet). Class A fire rating. 60 MPH wind resistance.						
Supreme, 25 year	R1@1.83	Sq	93.30	63.60	156.90	267.00
Supreme, 25 year, algae-resistant	R1@1.83	Sq	91.40	63.60	155.00	264.00
Classic, 20 year	R1@1.83	Sq	79.70	63.60	143.30	244.00
Classic, 20 year, algae-resistant	R1@1.83	Sq	70.90	63.60	134.50	229.00
Glaslock® interlocking shingles, Owens Corning. Three bundles per square (100 square feet). Class A fire rating. 60 MPH wind resistance.						
25 year rating	R1@1.83	Sq	84.80	63.60	148.40	252.00

	Craft@Hrs	Unit	Material	Labor	Total	Sell
Architectural grade laminated shingles, Timberline® Ultra®, GAF. Class A fire rating. 110 MPH wind resistance. Four bundles per square (100 square feet).						
Ultra, lifetime warranty white	R1@1.83	Sq	146.00	63.60	209.60	356.00
Select, 40 year	R1@1.83	Sq	111.00	63.60	174.60	297.00
Select, 40 year, algae-resistant	R1@1.83	Sq	161.00	63.60	224.60	382.00
Select, 30 year	R1@1.83	Sq	93.30	63.60	156.90	267.00
Select, 30 year, blend, fungus-resist.	R1@1.83	Sq	135.00	63.60	198.60	338.00
Hip and ridge shingles, GAF. Per linear foot of hip or ridge.						
High Ridge®	R1@.028	LF	3.04	.97	4.01	6.82
High Ridge® algae-resistant	R1@.028	LF	3.93	.97	4.90	8.33
High Style	R1@.028	LF	3.48	.97	4.45	7.57
3-tab fiberglass asphalt shingles, GAF. Class A fire rating. 60 MPH wind warranty. Three bundles cover 100 square feet at 5" exposure. Per square (100 square feet).						
Royal Sovereign®, 25 year	R1@1.83	Sq	91.50	63.60	155.10	264.00
Royal Sovereign®, 25 year, algae-resistant	R1@1.83	Sq	76.20	63.60	139.80	238.00
Sentinel®, 20 year	R1@1.83	Sq	64.30	63.60	127.90	217.00
Sentinel®, 20 year, fungus- and algae-resistant	R1@1.83	Sq	101.00	63.60	164.60	280.00
Shingle starter strip. Peel 'N Stick SBS membrane. Rubberized to seal around nails and resist wind lift. Per roll.						
7" x 33'	R1@.460	Ea	18.00	16.00	34.00	57.80
Asphalt roofing starter strip.						
9" x 36'	R1@.500	Ea	19.10	17.40	36.50	62.10

Built-up Roofing

	Craft@Hrs	Unit	Material	Labor	Total	Sell
Patch built-up roofing — concrete deck. Includes removing a section of the existing built-up roof on a concrete deck and replacing with built-up roofing. Per square (100 square feet) patched.						
3-ply roof and gravel	R1@2.94	Sq	110.00	102.00	212.00	360.00
4-ply roof and gravel	R1@3.11	Sq	128.00	108.00	236.00	401.00
5-ply roof and gravel	R1@3.28	Sq	179.00	114.00	293.00	498.00
Patch built-up roofing and insulation — concrete deck. Includes removing a section of the existing built-up roofing and insulation on a concrete deck and replacing with built-up roofing and 1" polystyrene insulation board. Per square (100 square feet) patched.						
3-ply roof, gravel and insulation	R1@3.18	Sq	178.00	110.00	288.00	490.00
4-ply roof, gravel and insulation	R1@3.35	Sq	243.00	116.00	359.00	610.00
5-ply roof, gravel and insulation	R1@3.52	Sq	262.00	122.00	384.00	653.00

	Craft@Hrs	Unit	Material	Labor	Total	Sell
Patch built-up roofing — wood deck. Includes removing a section of the existing built-up roof on a wood deck and replacing with built-up roofing. Per square (100 square feet) patched.						
2-ply roof and cap sheet	R1@2.66	Sq	93.70	92.40	186.10	316.00
3-ply roof and cap sheet	R1@2.84	Sq	110.00	98.70	208.70	355.00
4-ply roof and cap sheet	R1@3.01	Sq	128.00	105.00	233.00	396.00
Patch built-up roofing and insulation — wood deck. Includes removing a section of the existing built-up roofing and insulation on a wood deck and replacing with built-up roofing and 1" roof insulation board. Per square (100 square feet) patched.						
2-ply roof, cap sheet and insulation	R1@2.91	Sq	113.00	101.00	214.00	364.00
3-ply roof, cap sheet and insulation	R1@3.08	Sq	178.00	107.00	285.00	485.00
4-ply roof, cap sheet and insulation	R1@3.25	Sq	243.00	113.00	356.00	605.00
Built-up roofing. Installed over existing suitable substrate. Add the cost of equipment rental (the kettle) and consumable supplies (mops, brooms, gloves). Typical cost for equipment and consumable supplies is $15 per square applied.						
Type 1: 2-ply felt and 1-ply 90-lb. cap sheet, including 2 plies of hot mop asphalt.						
Base sheet, nailed down, per 100 SF	R1@.250	Sq	7.16	8.69	15.85	26.90
Hot mop asphalt, 30 lbs. per 100 SF	R1@.350	Sq	18.50	12.20	30.70	52.20
Asphalt felt, 15 lbs. per 100 SF	R1@.100	Sq	7.16	3.47	10.63	18.10
Hot mop asphalt, 30 lbs. per 100 SF	R1@.350	Sq	18.50	12.20	30.70	52.20
Mineral surface, 90 lbs. per 100 SF	R1@.200	Sq	42.40	6.95	49.35	83.90
Type 1, total per 100 SF	R1@1.25	Sq	93.72	43.40	137.12	233.00
Type 2: 3-ply asphalt, 1-ply 30 pound felt, 2 plies 15 pound felt, 3 coats hot mop asphalt and gravel.						
Asphalt felt, 30 lbs., nailed down	R1@.300	Sq	14.30	10.40	24.70	42.00
Hot mop asphalt, 30 lbs. per 100 SF	R1@.350	Sq	18.50	12.20	30.70	52.20
Asphalt felt, 15 lbs. per 100 SF	R1@.100	Sq	7.16	3.47	10.63	18.10
Hot mop asphalt, 30 lbs. per 100 SF	R1@.350	Sq	18.50	12.20	30.70	52.20
Asphalt felt, 15 lbs. per 100 SF	R1@.100	Sq	7.16	3.47	10.63	18.10
Hot mop asphalt, 60 lbs. per 100 SF	R1@.350	Sq	18.50	12.20	30.70	52.20
Gravel, 400 lbs. per 100 SF	R1@.600	Sq	25.50	20.80	46.30	78.70
Type 2, total per 100 SF	R1@2.15	Sq	109.62	74.70	184.32	313.00
Type 3: 4-ply asphalt, 4 plies 15 pound felt including 4 coats hot mop and gravel.						
Base sheet, nailed down, per 100 SF	R1@.250	Sq	7.16	8.69	15.85	26.90
Hot mop asphalt, 30 lbs. per 100 SF	R1@.350	Sq	18.50	12.20	30.70	52.20
Asphalt felt, 15 lbs. per 100 SF	R1@.100	Sq	7.16	3.47	10.63	18.10
Hot mop asphalt, 30 lbs. per 100 SF	R1@.350	Sq	18.50	12.20	30.70	52.20
Asphalt felt, 15 lbs. per 100 SF	R1@.100	Sq	7.16	3.47	10.63	18.10
Hot mop asphalt, 30 lbs. per 100 SF	R1@.350	Sq	18.50	12.20	30.70	52.20
Asphalt felt, 15 lbs. per 100 SF	R1@.100	Sq	7.16	3.47	10.63	18.10
Hot mop asphalt, 60 lbs. per 100 SF	R1@.350	Sq	18.50	12.20	30.70	52.20
Gravel, 400 lbs. per 100 SF	R1@.600	Sq	25.50	20.80	46.30	78.70
Type 3, total per 100 square feet	R1@2.55	Sq	128.14	88.60	216.74	368.00
Roofing asphalt. 100 pound keg.						
Type III, to 3 in 12 roof, 195-205 degree	—	Keg	56.10	—	56.10	—
Type III, at 30 pounds per 100 SF	R1@.350	Sq	18.70	12.20	30.90	52.50
Type IV, to 6 in 12 roof, 210-225 degree	—	Keg	52.00	—	52.00	—
Type IV, at 30 pounds per 100 SF	R1@.350	Sq	15.60	12.20	27.80	47.30

	Craft@Hrs	Unit	Material	Labor	Total	Sell
Mineral surface roll roofing. Roll covers one square (100 square feet).						
90 pound, fiberglass mat reinforced	R1@.200	Sq	34.00	6.95	40.95	69.60
GAF Mineral Guard	R1@.200	Sq	50.90	6.95	57.85	98.30
SBS-modified bitumen, Flintlastic	R1@.200	Sq	76.00	6.95	82.95	141.00
Add for mopping in asphalt	R1@.350	Sq	6.00	12.20	18.20	30.90
Add for adhesive application	R1@.350	Sq	10.00	12.20	22.20	37.70
Roofing membrane. Asphalt saturated. Per square. 15-pound roll covers 400 square feet. 30-pound roll covers 200 square feet. Nailed down.						
15-pound, Type I	R1@.140	Sq	4.00	4.86	8.86	15.10
15-pound, ASTM D226	R1@.140	Sq	6.50	4.86	11.36	19.30
30-pound, Type II	R1@.150	Sq	7.75	5.21	12.96	22.00
30-pound, ASTM D226	R1@.150	Sq	12.50	5.21	17.71	30.10
Add for mopping in asphalt	R1@.350	Sq	6.00	12.20	18.20	30.90
Add for adhesive application	R1@.350	Sq	10.00	12.20	22.20	37.70
Fiberglass Type IV roll roofing. 500-square-foot roll. ASTM D-2178. Fiberglass mat coated with asphalt and surfaced with a liquid parting agent that prevents rolls from sticking.						
Per 100 square feet	R1@.140	Sq	17.40	4.86	22.26	37.80
Add for mopping in asphalt	R1@.350	Sq	18..50	12.20	30.20	51.30
Add for adhesive application	R1@.350	Sq	10.00	12.20	22.20	37.70
Fiberglass base sheet. ASTM D-4601. 300-square-foot roll. Nailed down.						
Per 100 square feet	R1@.300	Sq	13.70	10.40	24.10	41.00
Modified bitumen adhesive. Rubberized. Use to adhere SBS modified bitumen membranes and rolled roofing. Blind nailing cement. Brush grade.						
2 gallons per 100 square feet	R1@.350	Sq	26.70	12.20	38.90	66.10
Cold process and lap cement. Bonds plies of built-up roofing and gravel.						
2 gallons per 100 square feet	R1@.350	Sq	16.90	12.20	29.10	49.50
Modified bitumen cold process adhesive. For placing shingles and cementing asphalt-coated roofing sheets together. Use for bonding felts, smooth and granule-surfaced roll roofing. Quick-setting, waterproof adhesive with strong bonding action.						
2 gallons per 100 square feet	R1@.350	Sq	21.50	12.20	33.70	57.30
Non-fibered asphalt emulsion. For metal and smooth-surface built-up roofs. Type III, Class I.						
3 gallons cover 100 square feet	R1@.350	Sq	25.90	12.20	38.10	64.80
Cold-Ap® built-up roofing cement, Henry. Bonds layers of conventional built-up roofing or SBS base and cap sheets. Replaces hot asphalt in built-up roofing. For embedding polyester fabric, gravel and granules. Not recommended for saturated felt. ASTM D3019 Type III.						
2 gallons per 100 square feet	R1@.350	Sq	28.40	12.20	40.60	69.00
PBA™ Permanent Bond Adhesive, Henry. Cold-process waterproofing roofing adhesive. ASTM D3019 Type III.						
2 gallons per 100 square feet	R1@.350	Sq	21.50	12.20	33.70	57.30

	Craft@Hrs	Unit	Material	Labor	Total	Sell
GAFGLAS torch and mop system built-up roofing, GAF. Low-slope roofing. Cap sheet made of asphalt-coated glass fiber mat. Cap sheet surfaced with mineral granules. Use as a surfacing ply in the application of hot-applied built-up roofs, and as a top ply in base flashing construction. Glass base resists moisture, and granule surface provides an ultraviolet protective surface. One 33.5' x 39.6" roll covers 1 square (100 square feet). One roll of felt covers 5 squares (500 square feet). Cost per 100 square feet.						
Cap sheet, white, 33.5' x 39.6" roll	RR@.550	Sq	37.50	22.20	59.70	101.00
Torch application, granular, black, 33.5' x 39.6" roll	RR@.350	Sq	81.10	14.20	95.30	162.00
Torch application, granular, white, 33.5' x 39.6" roll	RR@.350	Sq	81.60	14.20	95.80	163.00
Torch application, smooth, white, 33.5' x 39.6" roll	RR@.350	Sq	81.60	14.20	95.80	163.00
Mop SBS polyester mat, granular, black, 100 square foot roll	RR@.350	Sq	80.60	14.20	94.80	161.00
Mop SBS polyester mat, granular, white, 100 square foot roll	RR@.350	Sq	80.60	14.20	94.80	161.00
#75 glass mat base sheet, 300 square foot roll	RR@.500	Sq	46.00	20.20	66.20	113.00
3-ply SBS field adhesive, GAF. Low-slope roofing. GAF Torch and Mop System.						
5 gallons	—	Ea	33.00	—	33.00	—
GAFGLAS Type IV glass felt, GAF. Glass fiber mat coated with asphalt. Use as a ply felt or base sheet in built-up roof or as a flashing membrane. Class A, B, or C roofs. 39.4" x 161.8' roll covers 500 square feet.						
Per 100 square feet	R1@.350	Ea	9.20	12.20	21.40	36.40
Ruberoid Mop SBS granular white roll roofing. Non-woven polyester mat coated with SBS polymer-modified asphalt. For new roofing, reroofing, flashing, and built-up roofing. Roll size, 1 square (100 square feet).						
Granular, white, 0.160" thick	R1@.350	Sq	60.00	12.20	72.20	123.00
Building paper. Moisture barrier installed on a vertical or horizontal surface over sheathing. Per 100 square feet (square).						
1-ply, black, 40" x 97' roll	—	Sq	3.25	—	3.25	—
2-ply, 162 SF roll	—	Sq	7.00	—	7.00	—
2-ply Jumbo Tex, 162 SF roll	—	Sq	7.55	—	7.55	—
Rosin paper, 500 SF roll	—	Sq	3.25	—	3.25	—

Roof Patching

	Craft@Hrs	Unit	Material	Labor	Total	Sell
Self-adhesive roof flashing. Self-adhesive ice and water shield. For use as flashing around valleys, vents, chimneys and skylights. Adhesive allows for one-time repositioning but aggressive bond over time. Per 100 square feet (square).						
GAF StormGuard	R1@.500	Sq	44.40	17.40	61.80	105.00
GAF leak barrier	R1@.500	Sq	35.20	17.40	52.60	89.40
Owens Corning Weatherlock®	R1@.500	Sq	61.70	17.40	79.10	134.00
Tarco ice and water armor	R1@.500	Sq	70.00	17.40	87.40	149.00
Grace ice and water shield	R1@.500	Sq	72.10	17.40	89.50	152.00
Remove roof flashing. Chip off damaged waterproofing on a wood or concrete surface and clean with a wire brush.						
Per linear foot of joint	R1@.030	LF	—	1.04	1.04	1.77

	Craft@Hrs	Unit	Material	Labor	Total	Sell
Peel & Patch™ waterproofing. Aluminum foil, polymer film and rubberized asphalt with self-sticking back. For metal roof repair, gutter patch, valley patch, flashing and waterproofing.						
4" x 10'	—	Ea	8.50	—	8.50	—
6" x 10'	—	Ea	12.00	—	12.00	—
Leak Stopper™ rubberized roof patch, Gardner. Stops roof leaks on wet or dry surfaces. Fiber-reinforced. Trowel grade. One gallon covers 12 square feet at 1/8" thick.						
10.1 ounces	—	Ea	4.28	—	4.28	—
1 gallon	—	Ea	15.00	—	15.00	—
SEBS (styrene-ethylene-butylene-styrene) asphalt. Seals openings around chimneys, vents, pipes, air conditioning equipment, skylights, drains, flashings, and sheet metal. One gallon covers 12 square feet at 1/8" thick.						
All-weather, per gallon	—	Ea	11.30	—	11.30	—
Reflective, per gallon	—	Ea	15.80	—	15.80	—
Quick Roof, Cofair Products. Self-adhesive, reinforced, aluminum surface for flashing repairs on flat and metal roofs.						
3" x 16' roll	—	Ea	17.00	—	17.00	—
6" x 16' roll	—	Ea	26.00	—	26.00	—
Flashing cement. For repairing chimneys, shingle cracks, skylights, pipe penetrations and gutters						
1 gallon	—	Ea	14.90	—	14.90	—
3 gallons	—	Ea	39.00	—	39.00	—
5 gallons	—	Ea	52.50	—	52.50	—
Rubberized, 3 gallons	—	Ea	59.00	—	59.00	—
Patch roof or flashing. Apply roofing cement and reinforce with embedded fiberglass mat. Per linear foot of patch.						
One-ply cement and fiberglass	R1@.098	LF	1.50	3.40	4.90	8.33
Two-ply cement and fiberglass	R1@.121	LF	2.50	4.20	6.70	11.40
Rubber wet patch roof cement. For sealing leaks on wet or dry surfaces. One gallon covers 12 square feet at 1/8" thick.						
11 ounces	—	Ea	4.31	—	4.31	—
1 gallon	—	Ea	17.80	—	17.80	—
3-1/2 gallons	—	Ea	64.30	—	64.30	—
Wet Patch® roof cement, Henry. Seals around chimneys, vents, skylights, and flashings.						
11 ounce cartridge	—	Ea	3.98	—	3.98	—
1 gallon	—	Ea	17.10	—	17.10	—
3-1/2 gallons	—	Ea	53.00	—	53.00	—
5 gallons	—	Ea	65.00	—	65.00	—
Wet-Stick asphalt plastic roof cement, DeWitts. Adheres to wood, felt, plastics, glass, concrete, and all metals. One gallon covers 15 to 20 square feet when applied 1/8" thick.						
10-1/2 ounces	—	Ea	2.81	—	2.81	—
1 quart	—	Ea	5.36	—	5.36	—
1 gallon	—	Ea	13.50	—	13.50	—
5 gallons	—	Ea	39.80	—	39.80	—
Kool Patch™ white acrylic patching cement, Kool Seal. Stops leaks on damp or dry surfaces. Blends in on existing white roofs.						
11 ounce cartridge	—	Ea	3.55	—	3.55	—
1 gallon	—	Ea	17.90	—	17.90	—

	Craft@Hrs	Unit	Material	Labor	Total	Sell
Fiberglass reinforcing membrane. Fabric membrane for repairing cracks, holes, seams, and joints with reinforcing roof cements and coatings.						
Black, 6" wide, 50' long	—	Ea	11.50	—	11.50	—
Black, 6" wide, 150' long	—	Ea	21.10	—	21.10	—
Polyester stress-block fabric, Gardner. Asphalt-saturated. Repairs cracks, holes, seams, and joints with reinforcing roof cements and coatings.						
White, 4" wide, 25' long	—	Ea	11.50	—	11.50	—
Yellow resin coated glass fabric, Henry. Woven, open-mesh fiberglass. 20 x 10 thread count. ASTM D1668.						
4" wide, 150' long	—	Ea	19.80	—	19.80	—
6" wide, 25' long	—	Ea	11.70	—	11.70	—
Asphalt saturated cotton roof patch. Cotton fabric membrane for patching roof seams and holes.						
4" wide, 150' long	—	Ea	13.20	—	13.20	—
6" wide, 150' long	—	Ea	20.00	—	20.00	—
Elastotape, Henry. Non-woven polyester mat for use as mastic reinforcement. Lightweight, heat set, 100% spunbonded.						
4" wide, 150' long	—	Ea	16.90	—	16.90	—
Roofing granules.						
Colors, 60 pound pail	—	Ea	26.00	—	26.00	—
White, 100 pound bag	—	Ea	32.20	—	32.20	—
Siliconizer crack filler. Seals cracks and stops leaks in concrete, metal, asphalt, brick and polyurethane surfaces.						
10.1 ounce tube	—	Ea	2.79	—	2.79	—
1 quart	—	Ea	6.61	—	6.61	—
1 gallon	—	Ea	15.80	—	15.80	—
Elastocaulk®, Henry. White elastomeric acrylic roof patch. Apply by brush or trowel.						
11 ounce tube	—	Ea	4.03	—	4.03	—
1 gallon	—	Ea	30.50	—	30.50	—
Elastomastic, Henry. SEBS-modified sealant for roof areas subject to movement. Seals metal to metal joints. Also used to fill pitch pockets.						
11 ounce tube	—	Ea	7.95	—	7.95	—
1 gallon	—	Ea	38.40	—	38.40	—
Roof caulk, Karnak. Seals shingle tabs, flashing and joints at chimneys, skylights, pipe penetrations and gutters.						
Flashing cement, 10 ounces	—	Ea	2.50	—	2.50	—
Wet or dry roof caulk, 10 ounces	—	Ea	2.86	—	2.86	—
Rubberized wet/dry caulk, 10 ounces	—	Ea	4.40	—	4.40	—
PondPatch™, Henry. Fills roof ponds. Trowel on over or under roof membrane. For use on built-up, SBS or APP surfaces.						
6 Gallon pail covers 12 SF at 3/4" thick	R1@.085	SF	8.25	2.95	11.20	19.00
Roof repair tools.						
Roof mop, cotton	—	Ea	15.00	—	15.00	—
Roof cement trowel, 9"	—	Ea	3.00	—	3.00	—
Roof shingle remover	—	Ea	32.00	—	32.00	—
Poly roofing broom with scraper	—	Ea	36.00	—	36.00	—

	Craft@Hrs	Unit	Material	Labor	Total	Sell
Fibered roof coating. Moisture barrier for use over metal, composition and built-up roofs. Seals worn roof surfaces.						
On built-up roof, 50 SF per gallon	R1@.600	Sq	16.30	20.80	37.10	63.10
On metal roof, 80 SF per gallon	R1@.600	Sq	10.30	20.80	31.10	52.90
SolarFlex® white acrylic roof coating, Henry. Non-fibered elastomeric. Two gallons per 100 square feet using two coats.						
White, two coats	R1@1.00	Sq	31.40	34.70	66.10	112.00
Tan, two coats	R1@1.00	Sq	38.80	34.70	73.50	125.00
Fibered aluminum roof coating. For metal, smooth or mineral-surfaced built-up, composition, SBS and APP roofs.						
2 gallons cover 100 square feet	R1@.600	Sq	36.10	20.80	56.90	96.70
Snow Roof. White liquid roof coating for built-up, modified bitumen, composition, spray urethane foam, metal, asphalt, fiberglass, and polystyrene foam roofing.						
2 gallons cover 100 square feet	R1@.600	Sq	41.70	20.80	62.50	106.00
Ultra White elastomeric roof coating. White acrylic roof coating for built-up, modified bitumen, urethane foam, concrete, and previously coated roofs.						
60 square feet per gallon	R1@.600	Ea	31.60	20.80	52.40	89.10
Wet-Stick fibered roof coat. For coating old or new metal, built-up, composition or gravel roofs.						
2 gallons cover 100 square feet	R1@.600	Ea	12.50	20.80	33.30	56.60
Cant strip. 3" x 3" x 3-5/8" x 4' long.						
Per 4' piece	R1@.080	Ea	2.00	2.78	4.78	8.13
30-piece bundle, 120 linear feet	R1@2.40	Ea	36.40	83.40	119.80	204.00

Wood Shingle Roofing

	Craft@Hrs	Unit	Material	Labor	Total	Sell
Patch wood shingle roof. Remove wood shingles by chipping out with a chisel. Trim new shingles to fit. Nail down and tap shingle into alignment.						
3 shingle patch (1 square foot)	R1@.256	Ea	2.70	8.89	11.59	19.70
12 shingle patch (4 square feet)	R1@.551	Ea	11.00	19.10	30.10	51.20
30 shingle patch (10 square feet)	R1@1.20	Ea	27.00	41.70	68.70	117.00
Wood shakes and shingles. Red cedar. No pressure treating. Labor includes flashing and assumes a gable or hip roof to 5 in 12. Add 6% to the labor costs for a cut-up roof. Add 17% to the labor cost for a steeper roof or more complex hip roof. Add 25% to the labor cost when shakes or shingles are laid with staggered butts or in an irregular pattern such as thatch, serrated weave or ocean.						
Taper split No. 1 medium cedar shakes, 4 bundles cover 100 square feet (1 square) at 10" exposure,						
No. 1 medium shakes, $61.20 per bundle	R1@3.52	Sq	268.00	122.00	390.00	663.00
Add for pressure treated shakes	—	%	30.0	—	30.0	—
Deduct in the Pacific Northwest	—	%	-20.0	—	-20.0	—
Shake felt, 30 lb. (180 SF roll at $24.30)	R1@.500	Sq	14.10	17.40	31.50	53.60
Hip and ridge shakes, medium, 20 per bundle, covers 16.7 LF at 10" and 20 LF at 12" exposure,						
Per bundle	R1@1.00	Ea	49.90	34.70	84.60	144.00
Sawn shakes, sawn 1 side, Class C fire retardant,						
1/2" to 3/4" x 24" (4 bundle/sq at 10" exp.)	R1@3.52	Sq	364.00	122.00	486.00	826.00
3/4" to 5/4" x 24" (5 bundle/sq at 10" exp.)	R1@4.16	Sq	504.00	145.00	649.00	1,100.00

	Craft@Hrs	Unit	Material	Labor	Total	Sell
Cedar roofing shingles, No. 1 Grade.						
Four bundles of Perfection Grade 18" shingles cover 100 square feet (1 square) at 5-1/2" exposure. Five Perfections are 2-1/4" thick. Four bundles of 5x Perfect Grade shingles cover 100 square feet (1 square) at 5" exposure. Five Perfects are 2" thick.						
Perfections, $79.00 per bundle (Florida)	R1@3.52	Sq	316.00	122.00	438.00	745.00
Perfects, $60.50 per bundle	R1@3.52	Sq	242.00	122.00	364.00	619.00
No. 2 Perfections, $39.50 per bundle	R1@3.52	Sq	158.00	122.00	280.00	476.00
Add for pitch over 6 in 12	—	%	—	40.0	—	—
No. 2 cedar ridge shingles, medium, bundle covers 16 linear feet, Per bundle	R1@1.00	Ea	47.30	34.70	82.00	139.00
Fire-treated cedar shingles, 18", 5-1/2" exposure, 5 shingles are 2-1/4" thick (5/2-1/4),						
No. 1 (houses)	R1@3.52	Sq	429.00	122.00	551.00	937.00
No. 2, red label (houses or garages)	R1@2.00	Sq	358.00	69.50	427.50	727.00
No. 3 (garages)	R1@2.00	Sq	307.00	69.50	376.50	640.00
Eastern white cedar shingles, smooth butt edge, four bundles cover 100 square feet (1 square) at 5" exposure,						
White extras, $59.25 per bundle	R1@2.00	Sq	237.00	69.50	306.50	521.00
White clears, $47.25 per bundle	R1@2.00	Sq	189.00	69.50	258.50	439.00
White No. 2 clears, $27.54 per bundle	R1@2.00	Sq	110.00	69.50	179.50	305.00
Cedar shim shingles, tapered Western red cedar, Builder's 12-pack, per pack	—	Ea	4.56	—	4.56	—

Sheet Metal Roofing

Roofing sheets.

	Craft@Hrs	Unit	Material	Labor	Total	Sell
Metal sheet roofing, utility gauge.						
26" wide, 5-V crimp (includes 15% coverage loss), 6' to 12' lengths	R1@.027	SF	.67	.94	1.61	2.74
26" wide, corrugated (includes 15% coverage loss), 6' to 12' lengths	R1@.027	SF	.66	.94	1.60	2.72
27-1/2" wide, corrugated (includes 20% coverage loss), 6' to 12' lengths	R1@.027	SF	.68	.94	1.62	2.75
Ridge roll, plain, 10" wide	R1@.030	LF	2.24	1.04	3.28	5.58
Ridge cap, formed, plain, 12", 12" x 10' roll	R1@.061	LF	1.15	2.12	3.27	5.56
Sidewall flashing, plain, 3" x 4" x 10'	R1@.035	LF	.82	1.22	2.04	3.47
5-V crimp closure strips, 24" long	R1@.035	Ea	1.25	1.22	2.47	4.20
Endwall flashing, 2-1/2" corrugated, 10" x 28"	R1@.035	LF	1.66	1.22	2.88	4.90
Wood filler strip, corrugated, 7/8" x 7/8" x 6' (at $1.68 each)	R1@.035	LF	.28	1.22	1.50	2.55

	Craft@Hrs	Unit	Material	Labor	Total	Sell

Slate and Tile Roofing

Roofing slate. Local delivery included. Costs will be higher where slate is not mined. Add freight cost at 800 to 1,000 pounds per 100 square feet. Includes 20" long random width slate, 3/16" thick, 7-1/2" exposure. Meets Standard SS-S-451.

	Craft@Hrs	Unit	Material	Labor	Total	Sell
Semi-weathering green and gray	R1@11.3	Sq	471.00	393.00	864.00	1,470.00
Vermont black and gray black	R1@11.3	Sq	561.00	393.00	954.00	1,620.00
China black or gray	R1@11.3	Sq	636.00	393.00	1,029.00	1,750.00
Unfading and variegated purple	R1@11.3	Sq	636.00	393.00	1,029.00	1,750.00
Unfading mottled green and purple	R1@11.3	Sq	636.00	393.00	1,029.00	1,750.00
Unfading green	R1@11.3	Sq	521.00	393.00	914.00	1,550.00
Red slate	R1@13.6	Sq	1,630.00	472.00	2,102.00	3,570.00
Add for other specified widths and lengths	—	%	20.0	—	20.0	—

Fiber-cement slate roofing. Eternit™, Stonit™, or Thrutone™, various colors. Non-asbestos formulation. Laid over 1/2" sheathing. Includes 30-pound felt and one copper storm anchor per slate. Add for sheathing, flashing, hip, ridge and valley units.

	Craft@Hrs	Unit	Material	Labor	Total	Sell
23-5/8" x 11-7/8", English (420 lb, 30 year)						
2" head lap (113 pieces per Sq)	R1@5.50	Sq	441.00	191.00	632.00	1,070.00
3" head lap (119 pieces per Sq)	R1@5.75	Sq	463.00	200.00	663.00	1,130.00
4" head lap (124 pieces per Sq)	R1@6.00	Sq	480.00	208.00	688.00	1,170.00
Add for hip and ridge units, in mastic	R1@0.10	LF	9.01	3.47	12.48	21.20
15-3/4" x 10-5/8", Continental (420 lb, 30 year)						
2" head lap (197 pieces per Sq)	R1@6.50	Sq	455.00	226.00	681.00	1,160.00
3" head lap (214 pieces per Sq)	R1@6.75	Sq	496.00	234.00	730.00	1,240.00
4" head lap (231 pieces per Sq)	R1@7.00	Sq	537.00	243.00	780.00	1,330.00
Add for hip and ridge units, in mastic	R1@0.10	LF	8.27	3.47	11.74	20.00
Copper storm anchors, box of 1,000 ($40.00)	—	Ea	.04	—	.04	—
Stainless steel slate hooks	—	Ea	.40	—	.40	—
Add for extra felt under 5 in 12 pitch	R1@0.20	Sq	14.10	6.95	21.05	35.80

Roofing tile, clay. Material costs include felt and flashing. No freight or waste included. Normal waste is 5%. Roofing tiles are very heavy and may have to be transported long distances to the job site. Many roofing supply dealers don't stock the tile and quote prices F.O.B. at the factory. Cost of delivery to the job site can be a very substantial item.

	Craft@Hrs	Unit	Material	Labor	Total	Sell
Spanish tile, S-shaped, 88 pieces per square, 800 pounds per square at 11" centers and 15" exposure.						
Red clay tile	R1@3.46	Sq	283.00	120.00	403.00	685.00
Add for coloring	—	Sq	13.20	—	13.20	—
Red hip and ridge units	R1@.047	LF	1.00	1.63	2.63	4.47
Color hip and ridge units	R1@.047	LF	1.43	1.63	3.06	5.20
Red rake units	R1@.047	LF	1.91	1.63	3.54	6.02
Color rake units	R1@.047	LF	2.08	1.63	3.71	6.31
Red clay mission tile, 2-piece, 86 pans and 86 tops per square, 7-1/2" x 18" x 8-1/2" tiles at 11" centers and 15" exposure.						
Red clay tile	R1@5.84	Sq	330.00	203.00	533.00	906.00
Add for coloring (costs vary widely)	—	Sq	50.10	—	50.10	—
Red hip and ridge units	R1@.047	LF	1.05	1.63	2.68	4.56
Color hip and ridge units	R1@.047	LF	1.43	1.63	3.06	5.20
Red rake units	R1@.047	LF	1.91	1.63	3.54	6.02
Color rake units	R1@.047	LF	2.08	1.63	3.71	6.31

	Craft@Hrs	Unit	Material	Labor	Total	Sell
Concrete roof tile. Material includes felt, nails, and flashing. Approximately 90 pieces per square. Monier Lifetile.						
Shake, slurry color	R1@3.25	Sq	138.00	113.00	251.00	427.00
Slate, thru color	R1@3.25	Sq	167.00	113.00	280.00	476.00
Espana, slurry coated	R1@3.25	Sq	167.00	113.00	280.00	476.00
Monier 2000	R1@3.25	Sq	179.00	113.00	292.00	496.00
Vignette	R1@3.25	Sq	157.00	113.00	270.00	459.00
Collage	R1@3.25	Sq	157.00	113.00	270.00	459.00
Tapestry	R1@3.25	Sq	171.00	113.00	284.00	483.00
Split shake	R1@3.25	Sq	188.00	113.00	301.00	512.00
Trim tile						
Mansard V-ridge or rake, slurry coated	—	Ea	2.21	—	2.21	—
Mansard, ridge or rake, thru color	—	Ea	2.30	—	2.30	—
Hipstarters, slurry coated	—	Ea	18.80	—	18.80	—
Hipstarters, thru color	—	Ea	24.00	—	24.00	—
Accessories						
Eave closure or birdstop	R1@.030	LF	1.33	1.04	2.37	4.03
Hurricane or wind clips	R1@.030	Ea	.48	1.04	1.52	2.58
Underlayment or felt	R1@.050	Sq	17.30	1.74	19.04	32.40
Metal flashing and nails	R1@.306	Sq	10.00	10.60	20.60	35.00
Pre-formed plastic flashing						
Hip (13" long)	R1@.020	Ea	34.70	.69	35.39	60.20
Ridge (39" long)	R1@.030	Ea	35.50	1.04	36.54	62.10
Anti-ponding foam	R1@.030	LF	.31	1.04	1.35	2.30
Batten extenders	R1@.030	Ea	1.20	1.04	2.24	3.81
Roof loading						
Add to load tile and accessories on roof	R1@.822	Sq	—	28.60	28.60	48.60

Elastomeric Decking and Roofing

Roof walking deck. Includes cleaning of the existing wood, masonry or coated deck, filling fractures, masking adjacent surfaces, primer at .3 gallons per 100 SF, Deckgard elastomeric polyurethane texture coat at 1 gallon per 100 square feet, polyurethane top coat at 1 gallon per 100 square feet. Per 100 square feet (square). Includes the cost of spray equipment.

	Craft@Hrs	Unit	Material	Labor	Total	Sell
Based on 800 SF job	R1@5.00	Sq	78.00	174.00	252.00	428.00

Urethane spray foam roof. Applied over an existing low-slope urethane roof. Includes cleaning roof with a broom and air blast, caulking fractures with polyurethane, masking adjacent surfaces, emulsion primer at 2 gallons per 100 SF, 3.0-pound density urethane foam, 20 mil elastomeric acrylic top coat covered with 40 pounds of mineral granules per square. Per 100 square feet (square). Includes the cost of spray equipment.

	Craft@Hrs	Unit	Material	Labor	Total	Sell
Based on 2,500 SF job	R1@2.80	Sq	130.00	97.30	227.30	386.00

Fiberglass Roofing

Fiberglass panels. Nailed or screwed onto wood frame.

	Craft@Hrs	Unit	Material	Labor	Total	Sell
Corrugated, 8', 10', 12' panels, 2-1/2" corrugations, standard colors. Costs include 10% for waste,						
Polycarbonate, 26"-wide panels	BC@.012	SF	1.29	.43	1.72	2.92
Fiberglass, 26"-wide panels	BC@.012	SF	1.29	.43	1.72	2.92
PVC, 26"-wide panels	BC@.012	SF	.86	.43	1.29	2.19
Flat panels, clear, green or white,						
.06" flat sheets, 4' x 8', 10', 12'	BC@.012	SF	1.20	.43	1.63	2.77
.03" rolls, 24", 36", 48" x 50'	BC@.012	SF	1.28	.43	1.71	2.91
.037" rolls, 24", 36", 48" x 50'	BC@.012	SF	1.48	.43	1.91	3.25
Nails, ring shank with rubber washer, 1-3/4" (labor included with panels),						
100 per box (covers 130 SF)	—	SF	.08	—	.08	—

	Craft@Hrs	Unit	Material	Labor	Total	Sell
Self-tapping screws, 1-1/4", rust-proof (labor included with panels), 100 per box (covers 130 SF)	—	SF	.30	—	.30	—
Horizontal closure strips, corrugated (labor included with panels),						
Redwood, 2-1/2" x 1-1/2"	—	LF	.33	—	.33	—
Poly-foam, 1" x 1"	—	LF	.54	—	.54	—
Rubber, 1" x 1"	—	LF	.44	—	.44	—
Vertical crown molding,						
Redwood, 1-1/2" x 1" or poly-foam, 1" x 1"	—	LF	.33	—	.33	—
Rubber, 1" x 1"	—	LF	.83	—	.83	—

Roof Flashing

Aluminum roll flashing. Installs under shingles, along roof valley. Provides a waterproof valley at intersecting roof sections. 0.010" economy gauge. Mill finish.

	Craft@Hrs	Unit	Material	Labor	Total	Sell
6" wide, 10' long	RR@.200	Ea	6.41	8.09	14.50	24.70
6" wide, 50' long	RR@1.00	Ea	17.90	40.40	58.30	99.10
8" wide, 10' long	RR@.200	Ea	8.53	8.09	16.62	28.30
8" wide, 50' long	RR@1.00	Ea	23.40	40.40	63.80	108.00
10" wide, 10' long	RR@.200	Ea	7.91	8.09	16.00	27.20
10" wide, 50' long	RR@1.00	Ea	26.50	40.40	66.90	114.00
14" wide, 10' long	RR@.200	Ea	12.70	8.09	20.79	35.30
14" wide, 50' long	RR@1.00	Ea	35.60	40.40	76.00	129.00
20" wide, 10' long	RR@.200	Ea	13.20	8.09	21.29	36.20
20" wide, 50' long	RR@1.00	Ea	39.80	40.40	80.20	136.00

Aluminum roll valley flashing. Use for valley, hip and ridge flashing.

	Craft@Hrs	Unit	Material	Labor	Total	Sell
6" wide, 25' long	RR@.500	Ea	16.50	20.20	36.70	62.40
8" wide, 25' long	RR@.500	Ea	21.80	20.20	42.00	71.40
10" wide, 10' long	RR@.200	Ea	13.80	8.09	21.89	37.20
10" wide, 25' long	RR@.500	Ea	19.90	20.20	40.10	68.20
14" wide, 10' long	RR@.200	Ea	13.60	8.09	21.69	36.90
14" wide, 25' long	RR@.500	Ea	23.80	20.20	44.00	74.80
20" wide, 25' long	RR@.500	Ea	36.70	20.20	56.90	96.70

Aluminum window drip cap. Commercial gauge 0.0175". Metal edge cap over door and window frames.

	Craft@Hrs	Unit	Material	Labor	Total	Sell
1-1/4" x 10' long, white	BC@.400	Ea	4.01	14.40	18.41	31.30
1-5/8" x 10' long, white	BC@.350	Ea	5.07	12.60	17.67	30.00

Aluminum drip edge.

	Craft@Hrs	Unit	Material	Labor	Total	Sell
2-3/4" x 1-3/4" x 10', brown	RR@.200	Ea	4.98	8.09	13.07	22.20
2-3/4" x 1-3/4" x 10', white	RR@.200	Ea	5.09	8.09	13.18	22.40

Copper roll flashing.

	Craft@Hrs	Unit	Material	Labor	Total	Sell
8" x 20' roll	RR@1.00	Ea	40.30	40.40	80.70	137.00
12" x 20' roll	RR@1.25	Ea	59.10	50.50	109.60	186.00

	Craft@Hrs	Unit	Material	Labor	Total	Sell
Copper step flashing.						
5" x 7"	RR@.040	Ea	4.25	1.62	5.87	9.98
Copper formed flashing.						
1-1/2" x 2", roof edge	RR@.350	Ea	72.40	14.20	86.60	147.00
2" x 2", roof edge	RR@.350	Ea	79.60	14.20	93.80	159.00
4" x 6", roof-to-wall	RR@.250	Ea	193.00	10.10	203.10	345.00
5" x 7", shingle	—	Ea	7.86	—	7.86	—
8" x 12", shingle	—	Ea	22.70	—	22.70	—
Galvanized roll flashing. Ridge and valley flashing. 26 gauge.						
6" wide, 10' long	RR@.200	Ea	7.10	8.09	15.19	25.80
6" wide, 50' long	RR@1.00	Ea	25.50	40.40	65.90	112.00
8" wide, 50' long	RR@1.00	Ea	24.00	40.40	64.40	109.00
10" wide, 50' long	RR@1.00	Ea	36.40	40.40	76.80	131.00
14" wide, 10' long	RR@.200	Ea	14.90	8.09	22.99	39.10
14" wide, 50' long	RR@1.25	Ea	42.60	50.50	93.10	158.00
20" wide, 10' long	RR@.250	Ea	19.20	10.10	29.30	49.80
20" wide, 50' long	RR@1.40	Ea	70.10	56.60	126.70	215.00
Galvanized angle flashing. 10' lengths. 26 gauge.						
3" x 5", 90-degree	RR@.200	Ea	6.42	8.09	14.51	24.70
4" x 4"	RR@.200	Ea	10.20	8.09	18.29	31.10
4" x 4", hemmed	RR@.200	Ea	9.47	8.09	17.56	29.90
4" x 5"	RR@.200	Ea	10.50	8.09	18.59	31.60
4" x 5" x 1/2", turnback flashing	RR@.200	Ea	8.91	8.09	17.00	28.90
6" x 6"	RR@.200	Ea	16.30	8.09	24.39	41.50
7-1/2", apron	RR@.200	Ea	7.08	8.09	15.17	25.80
Galvanized drip edge. Used along roof edge to direct water away from the fascia or into rain gutter.						
1-1/4" x 1-1/4" x 10', white	RR@.200	Ea	5.52	8.09	13.61	23.10
2" x 2" x 10'	RR@.200	Ea	3.90	8.09	11.99	20.40
2-1/8" x 2" x 10', white	RR@.200	Ea	4.90	8.09	12.99	22.10
2-1/2" x 1" x 10', white	RR@.200	Ea	4.20	8.09	12.29	20.90
3" x 1" x 10', heavy duty	RR@.200	Ea	5.15	8.09	13.24	22.50
10' long, vented	RR@.200	Ea	19.90	8.09	27.99	47.60
Gravel stop. Raised 26-gauge metal bead nailed around the roof perimeter to keep gravel on the roof.						
2" x 4" x 1/2" x 10', bonderized	RR@.300	Ea	11.70	12.10	23.80	40.50
2" x 4" x 1/2" x 10', galvanized	RR@.330	Ea	5.80	13.30	19.10	32.50
3-1/2" x 3-1/2" x 10', galvanized	RR@.300	Ea	7.11	12.10	19.21	32.70
3" x 4" x 1/4" x 10', galvanized	RR@.330	Ea	9.62	13.30	22.92	39.00
4" x 4" x 1/2" x 10', bonderized	RR@.330	Ea	13.50	13.30	26.80	45.60

	Craft@Hrs	Unit	Material	Labor	Total	Sell
L-flashing. Galvanized 28 gauge, except as noted.						
1" x 1-1/2" x 10', galvanized	RR@.200	Ea	3.37	8.09	11.46	19.50
1" x 2" x 10', galvanized	RR@.200	Ea	4.26	8.09	12.35	21.00
1-1/2" x 1-1/2", galvanized	RR@.200	Ea	4.74	8.09	12.83	21.80
2" x 2" x 10', dark brown	RR@.240	Ea	6.97	9.70	16.67	28.30
2" x 2" x 10', galvanized	RR@.240	Ea	5.64	9.70	15.34	26.10
2" x 3" x 10', dark brown	RR@.240	Ea	8.54	9.70	18.24	31.00
2" x 3" x 10', galvanized	RR@.240	Ea	6.04	9.70	15.74	26.80
3" x 3" x 10', bonderized	RR@.250	Ea	11.00	10.10	21.10	35.90
3" x 3" x 10', galvanized	RR@.250	Ea	8.93	10.10	19.03	32.40
4" x 4" x 10', galvanized	RR@.250	Ea	12.20	10.10	22.30	37.90
4" x 6" x 10', white	RR@.270	Ea	22.50	10.90	33.40	56.80
Pipe flashing. Hard base.						
3" to 1"	RR@.046	Ea	4.30	1.86	6.16	10.50
3" to 4"	RR@.046	Ea	5.38	1.86	7.24	12.30
Lead flashing roll.						
8" x 30'	RR@.750	Ea	127.00	30.30	157.30	267.00
10" x 24'	RR@.672	Ea	127.00	27.20	154.20	262.00
12" x 20'	RR@.600	Ea	127.00	24.30	151.30	257.00
Galvanized roll valley flashing. Laid in the valley of a hip roof.						
7" x 10', galvanized	RR@.200	Ea	8.08	8.09	16.17	27.50
7" x 25', galvanized	RR@.500	Ea	17.50	20.20	37.70	64.10
7" x 50', galvanized	RR@1.00	Ea	35.70	40.40	76.10	129.00
8" x 10', brown	RR@.220	Ea	13.90	8.89	22.79	38.70
8" x 10', white	RR@.220	Ea	13.10	8.89	21.99	37.40
10" x 10', galvanized	RR@.250	Ea	12.40	10.10	22.50	38.30
10" x 25', galvanized	RR@.625	Ea	31.00	25.30	56.30	95.70
10" x 50', galvanized	RR@1.25	Ea	29.20	50.50	79.70	135.00
12" x 12', galvanized	RR@2.50	Ea	19.80	101.00	120.80	205.00
12" x 25', galvanized	RR@.500	Ea	40.70	20.20	60.90	104.00
14" x 10', galvanized	RR@.270	Ea	14.30	10.90	25.20	42.80
14" x 25', galvanized	RR@.675	Ea	43.50	27.30	70.80	120.00
20" x 10', galvanized	RR@.300	Ea	19.80	12.10	31.90	54.20
20" x 12', galvanized	RR@.500	Ea	23.70	20.20	43.90	74.60
20" x 25', green	RR@.625	Ea	76.00	25.30	101.30	172.00
20" x 25', galvanized	RR@.600	Ea	55.00	24.30	79.30	135.00
20" x 25', red	RR@.625	Ea	59.30	25.30	84.60	144.00
Galvanized roof apron flashing. Slips under shingles to extend drip beyond fascia board.						
10' long, brown	RR@.400	Ea	5.04	16.20	21.24	36.10
10' long, white	RR@.400	Ea	5.04	16.20	21.24	36.10
Galvanized roof edge flashing. Installs under shingles along roof edge to retard decay.						
1-1/2" x 1-1/2" x 10'	RR@.250	Ea	4.26	10.10	14.36	24.40
1-1/2" x 1-1/2" x 10', dark brown	RR@.250	Ea	4.67	10.10	14.77	25.10
2" x 2" x 10', white	RR@.250	Ea	5.07	10.10	15.17	25.80
2" x 3" x 10', bonderized	RR@.250	Ea	8.08	10.10	18.18	30.90
2" x 3" x 10'	RR@.250	Ea	6.03	10.10	16.13	27.40
2" x 8" x 10'	RR@.350	Ea	15.00	14.20	29.20	49.60
6" x 6" x 10'	RR@.350	Ea	14.80	14.20	29.00	49.30

	Craft@Hrs	Unit	Material	Labor	Total	Sell
Galvanized roof-to-wall flashing.						
2" x 3" x 10'	RR@.250	Ea	7.94	10.10	18.04	30.70
2" x 3" x 10'	RR@.250	Ea	6.39	10.10	16.49	28.00
2" x 4" x 10'	RR@.250	Ea	7.82	10.10	17.92	30.50
2" x 6" x 10', bonderized	RR@.250	Ea	15.70	10.10	25.80	43.90
2" x 6" x 10'	RR@.250	Ea	12.70	10.10	22.80	38.80
3" x 5" x 10', dark brown	RR@.250	Ea	12.00	10.10	22.10	37.60
3" x 5" x 10'	RR@.250	Ea	11.00	10.10	21.10	35.90
4" x 5" x 10'	RR@.250	Ea	15.90	10.10	26.00	44.20
4" x 6" x 10'	RR@.250	Ea	12.70	10.10	22.80	38.80

Galvanized step flashing. Sheet metal formed into two leafs meeting at a right angle. Placed so each step overlaps the step below. Used to waterproof the meeting point of two surfaces, such as joint between chimney and roof surface.

	Craft@Hrs	Unit	Material	Labor	Total	Sell
3" x 4" x 7"	RR@.150	Ea	.71	6.06	6.77	11.50
4" x 4" x 8"	RR@.150	Ea	.98	6.06	7.04	12.00
4" x 4" x 8"	RR@.150	Ea	.63	6.06	6.69	11.40
4" x 4" x 12"	RR@.150	Ea	1.17	6.06	7.23	12.30
4" x 6" x 14"	RR@.150	Ea	1.79	6.06	7.85	13.30

Tin repair shingles. Galvanized.

	Craft@Hrs	Unit	Material	Labor	Total	Sell
5" x 7"	—	Ea	.65	—	.65	—
8" x 12"	—	Ea	1.00	—	1.00	—
8" x 12", dark brown	—	Ea	1.41	—	1.41	—

Flashing shingles.

	Craft@Hrs	Unit	Material	Labor	Total	Sell
5" x 8", dark brown	—	Ea	.99	—	.99	—
5" x 8", galvanized	—	Ea	.63	—	.63	—
6" x 8", gray	—	Ea	.63	—	.63	—

W-valley flashing. 10' lengths. 28 gauge. Sheet metal formed into a flattened W-shape and laid into a roof valley. Channels water flow better than a smooth valley. Roof finish is applied over the valley from both sides.

	Craft@Hrs	Unit	Material	Labor	Total	Sell
18", brown	RR@.250	Ea	25.60	10.10	35.70	60.70
18", copper	RR@.250	Ea	171.00	10.10	181.10	308.00
18", galvanized	RR@.250	Ea	24.30	10.10	34.40	58.50
18", white	RR@.250	Ea	37.70	10.10	47.80	81.30
24", dark brown	RR@.250	Ea	32.90	10.10	43.00	73.10
24", galvanized	RR@.250	Ea	29.20	10.10	39.30	66.80

Counterflashing.

	Craft@Hrs	Unit	Material	Labor	Total	Sell
6" x 10', galvanized	RR@.500	Ea	8.15	20.20	28.35	48.20
6" x 10', grey	RR@.500	Ea	8.15	20.20	28.35	48.20

Galvanized dormer vent. Wire mesh screen keeps out birds and large insects. 28 gauge galvanized steel.

	Craft@Hrs	Unit	Material	Labor	Total	Sell
18" x 18", low profile	RR@.871	Ea	40.40	35.20	75.60	129.00
19" x 3", low profile	RR@.871	Ea	34.40	35.20	69.60	118.00

Ridge vent. For attic and rafter-bay ventilation.

	Craft@Hrs	Unit	Material	Labor	Total	Sell
1-1/4" x 11" x 10'	R1@.500	Ea	18.60	17.40	36.00	61.20

VentSure™ Ridge Vent, Owens Corning.

	Craft@Hrs	Unit	Material	Labor	Total	Sell
Rigid roll	R1@.028	LF	2.66	.97	3.50	5.95

	Craft@Hrs	Unit	Material	Labor	Total	Sell
Cobra® Rigid Vent II ridge vent, GAF. Polyester composite.						
4' long, .75" thick	R1@.255	Ea	11.90	8.86	20.76	35.30
Shingle-over ridge vent. Narrow opening keeps out insects and debris.						
4' vent	R1@.255	Ea	11.50	8.86	20.36	34.60
Vent end plug	R1@.025	Ea	5.19	.87	6.06	10.30
Galvanized roof jack.						
4" pipe, 4 in 12 pitch	RR@.311	Ea	19.00	12.60	31.60	53.70
6" pipe, 4 in 12 pitch	RR@.311	Ea	23.40	12.60	36.00	61.20
8" pipe, 4 in 12 pitch	RR@.311	Ea	36.30	12.60	48.90	83.10
Aluminum base roof jack.						
1" to 3" pipe	RR@.311	Ea	6.82	12.60	19.42	33.00
3" to 4" pipe	RR@.311	Ea	7.97	12.60	20.57	35.00
3-N-1 rubber gasket roof jack. Roof pitch to 12 in 12.						
1" to 3" pipe	RR@.311	Ea	6.34	12.60	18.94	32.20
3" to 4" pipe	RR@.311	Ea	7.35	12.60	19.95	33.90
All rubber roof jack.						
1-1/4" to 4" pipe	R1@.311	Ea	26.30	10.80	37.10	63.10
Roof jack lead flashing.						
1-1/2" pipe	RR@.271	Ea	34.30	11.00	45.30	77.00
2" pipe	RR@.271	Ea	37.00	11.00	48.00	81.60
4" pipe	RR@.271	Ea	44.00	11.00	55.00	93.50
Tile roof pipe flashing. For plumbing, electrical, and heating lines exiting a Spanish tile roof with pitch from 5 in 12 to 12 in 12. Aluminum base.						
1-1/2" pipe	RR@.311	Ea	9.05	12.60	21.65	36.80
2" pipe	RR@.311	Ea	8.84	12.60	21.44	36.40
3" pipe	RR@.311	Ea	11.10	12.60	23.70	40.30
4" pipe	RR@.311	Ea	12.80	12.60	25.40	43.20
Square roof vent. Duct outlet for kitchen or bathroom exhaust fan. 50-square-inch net free vent area.						
Aluminum	RR@.440	Ea	6.32	17.80	24.12	41.00
Plastic	RR@.440	Ea	8.86	17.80	26.66	45.30
Square slant back aluminum roof vent. 60-square-inch free vent area. 18" x 20" flashing.						
Aluminum	RR@.440	Ea	11.20	17.80	29.00	49.30
Plastic, low profile	RR@.440	Ea	9.81	17.80	27.61	46.90
T-top roof exhaust vent. For bath, dryer, kitchen or other exhaust ducts terminating through the roof.						
4", bonderized	RR@.440	Ea	11.70	17.80	29.50	50.20
4", galvanized	RR@.440	Ea	11.70	17.80	29.50	50.20
7", bonderized	RR@.440	Ea	15.50	17.80	33.30	56.60
7", galvanized	RR@.440	Ea	13.90	17.80	31.70	53.90
4", T-top sub-base flashing	RR@.100	Ea	8.90	4.04	12.94	22.00
7", T-top sub-base flashing	RR@.100	Ea	11.60	4.04	15.64	26.60

	Craft@Hrs	Unit	Material	Labor	Total	Sell
Chimney cap.						
9" x 9"	RR@.440	Ea	28.80	17.80	46.60	79.20
9" x 13"	RR@.440	Ea	32.10	17.80	49.90	84.80
8" x 8" to 13" x 13", adjustable	RR@.440	Ea	31.80	17.80	49.60	84.30
13" x 13"	RR@.440	Ea	33.30	17.80	51.10	86.90
13" x 18"	RR@.440	Ea	34.30	17.80	52.10	88.60
13" x 13", stainless steel	RR@.440	Ea	41.60	17.80	59.40	101.00
Roof turbine vent with base. Externally braced. Stainless steel ball bearings.						
12", galvanized base	RR@.694	Ea	36.40	28.10	64.50	110.00
12", wood base	RR@.694	Ea	40.60	28.10	68.70	117.00
12", combination aluminum base	RR@.694	Ea	49.80	28.10	77.90	132.00

Rain Handling Gear

	Craft@Hrs	Unit	Material	Labor	Total	Sell
Gutters and downspouts.						
Galvanized steel, 4" box type,						
Gutter, 5", 10' lengths, .015"	B1@.070	LF	.96	2.28	3.24	5.51
Inside or outside corner	B1@.091	Ea	6.86	2.96	9.82	16.70
Fascia support brackets	—	Ea	1.95	—	1.95	—
Spike and ferrule support, 7"	—	Ea	.80	—	.80	—
Add per 2" x 3" galvanized downspout, with drop outlet, 3 ells and straps,						
10' downspout	B1@.540	Ea	13.60	17.60	31.20	53.00
Aluminum rain gutter, 5" box type, primed,						
Gutter, 5", 10' lengths	B1@.070	LF	.90	2.28	3.18	5.41
Inside or outside corner	B1@.091	Ea	9.10	2.96	12.06	20.50
Fascia support brackets	—	Ea	1.91	—	1.91	—
Spike and ferrule support, 7"	—	Ea	.80	—	.80	—
Section joint kit	—	Ea	4.61	—	4.61	—
Add per 2" x 3" aluminum downspout, with drop outlet, strainer, 3 ells and straps,						
10' downspout	B1@.540	Ea	9.75	17.60	27.35	46.50
20' downspout, with connector	B1@.810	Ea	19.75	26.30	46.05	78.30
Vinyl rain gutter, white PVC, (Plastmo Vinyl), half round gutter, (10' lengths),						
Gutter, 4"	B1@.054	LF	.54	1.76	2.30	3.91
Add for fittings (connectors, hangers, etc)	—	LF	1.40	—	1.40	—
Add for inside or outside corner	B1@.091	Ea	6.80	2.96	9.76	16.60
Add for bonding kit (covers 150 LF)	—	Ea	6.50	—	6.50	—
Add for 5" gutter	—	%	15.0	—	15.0	—
Downspouts, 3" round	B1@.022	LF	1.03	.72	1.75	2.98
Add for 3 ells and 2 clamps, per downspout	—	Ea	34.70	—	34.70	—

Windows

8

Windows can be an important (and expensive) part of any home improvement project. Loose-fitting, single-glazed windows are a major source of heat loss and gain. Wood windows in older homes often look similar to the eyesore in Figure 8-1. If your first instinct is to replace the window, keep reading. That may not be the best choice. This chapter explains why.

Figure 8-1

Signs of excessive water damage are evident in the paint peeling off this window sill and sash and broken caulking around the window

About one quarter of the heating and cooling load in a well-insulated home is the result of heat transfer through windows. Old window frames and sashes tend to be very porous, admitting streams of outside air during both heating and cooling seasons; modern windows have superior weatherstripping and are much tighter. In fact, modern windows are so tight that a home can develop condensation problems and a musty odor after leaky, old windows are replaced. Modern windows have insulated frames and dual glazing with about 1/2" of dead air space between the panes. In better-quality insulated windows, this dead space is filled with argon gas, which transfers even less heat than dead air. Low-emissivity (Low-E) glass has two invisible layers of microscopically-thin silver sandwiched between anti-reflective layers of metal oxide. Low-E glass admits sunlight but blocks radiant heat from escaping during the winter. During the cooling season, Low-E glass reflects back to the exterior as much as 25 percent of unwanted solar heat. Low-E coatings also block more than 80 percent of the ultraviolet light that tends to fade fabrics. Glass edges in better-quality insulated windows are sealed with "warm edge" spacers that transfer even less heat around the glass perimeter.

Modern windows have clear advantages over windows installed in homes before about 1970. But these advantages come at a price, especially if windows have to be custom-made to fit the opening. Before recommending replacement, take a few minutes to examine the existing windows.

Check the tightness of fit. A window sash that rattles in the frame probably leaks cold air in the winter, hot air in the summer and water during heavy rain. Check the operation of the window. Old double-hung and single-hung wood windows seldom operate smoothly. You may find many that are painted shut. Years of re-painting has sealed and immobilized the sash in the frame. That may reduce leaking but it also eliminates venting as an option. Old casement and awning windows tend to warp at the top and bottom. Dark stains on the sash and sill are usually caused by condensation running down the glass. If you see condensation stains, check for softening of the molding and sill.

Brown or black discoloration near joints is a sign of decay. Untreated wood windows are particularly vulnerable to insect attack, either by termites, powder post beetles or carpenter ants. Termites usually leave sand-like pellets outside the wood they're destroying. Powder post beetles make the wood look as though it was hit by birdshot. Carpenter ants deposit neat piles of coarse sawdust where they're nesting. Fumigation by a professional exterminator is usually required to eliminate termites. Use spray insecticide to eliminate carpenter ants and powder post beetles.

Although aluminum windows don't decay, they're still candidates for replacement. Aluminum is highly conductive to heat and cold. Aluminum windows can get so cold in winter that frost will form inside of them, wasting heat while creating unpleasant drafts and moisture problems. Modern windows offer better insulating value against heat, cold, UV radiation and noise. They protect better against forced entry, and better resist both wind and rain. And if that's not enough, modern windows are easier to keep clean. Many tilt inward so both sides can be cleaned from the interior. Perhaps you're old enough to remember your mother or grandmother sitting on the window sills of the upstairs bedrooms, facing in, with her legs dangling inside the room, cleaning the outside of the windows. It wasn't a good idea then, and it's not a good idea now.

Windows and the Building Code

Most states and counties have energy codes that set energy-efficiency standards for both new construction and alteration work. To get a building permit for a significant home improvement project, you may have to submit energy calculations that demonstrate energy code compliance. State, city and county standards vary. But the U.S. Department of Energy offers free computer software to help you meet requirements of most model energy codes. The DOE website is http://www.energycodes.gov. Naturally, the size and orientation of windows affect energy use. Too many large windows in the wrong places increase both the heating and cooling load beyond what energy codes allow.

Building codes require that sleeping rooms below the fourth floor have at least one operable window or door approved for an emergency exit (egress). If your building department requires upgrading the home to modern egress standards, you'll have to install at least one egress window (or door) in each bedroom. An egress window measures at least 20" wide by at least 24" high and has a clear opening of at least 5.7 square feet. The sill can't be more than 44" above the floor.

Many building codes also require that certain windows have tempered glass. These include windows within 18" of the floor, windows on a stairway landing and windows used as a sidelight beside an entry door. Tempered glass fractures into pebble-like shards less likely to cause injury.

Storm Windows

In colder climates, storm windows *(storms)* reduce the heating load and prevent damage from condensation on window interiors. Storms also protect the windows from extreme weather. The damaged window in Figure 8-2 wasn't protected by a storm window. The sash joints are weak and need to be reinforced; the glazing putty has to be replaced; the swollen, rough wood has to be sanded smooth so the window will open and close without sticking; and the window needs paint. Most of this work could have been avoided if a storm window had been installed during winter months.

Figure 8-2
Weather-damaged window

When evaluating windows, note whether the house has removable storms and screens. Most homes built in cold climates between 1920 and 1980 have (or had) storm windows. Newer homes usually have insulated, double-pane glazing that offers weather protection similar to storm windows.

Galvanized self-storing storm-and-screen windows were introduced in the 1940s. Aluminum storms and screens became popular in the 1950s. Older aluminum storms will be oxidized and pitted and rubber glazing strips will be deteriorated. If you elect to replace windows, storms may not be needed. If you plan to repair windows, storms will be required in colder climates. It's usually better to replace, rather than repair, deteriorated storms, especially for standard size windows. Most storm windows have a 2" skirt that can be trimmed off to make minor adjustments in width and height. Storms can be custom fabricated for odd-size windows at moderate cost. For a little extra money, consider high-performance storms with Low-E glass.

Repair of Existing Wood Windows

Repair is easy when the only problem with a wood window is minor decay. A coat of water-repellent preservative on the exterior will stop further deterioration. This is a three-step process:

1. Use a heat gun or paint stripper to remove the existing paint.
2. Brush on preservative.
3. Repaint the window.

Paint can't be used over some types of wood preservative, so be sure to select a paintable preservative.

Next, consider aesthetics. This is especially important on older or historic homes. If you install modern windows, the house will have a new look. Is that new look going to be in harmony with the exterior trim? With the siding? With the roofing? With other windows in the house? If just a few windows need replacing, can you replace just those few without creating an eyesore? In such cases, repairing a window may be a much better choice than replacement, both in the money saved and in the preservation of a consistent style for the home.

Reproducing the leaded glass windows in Figure 8-3 would cost thousands of dollars today. Instead, repair fine-quality windows like these. Replacing the windows in Figure 8-4 with modern vinyl windows would be a mistake. The vinyl windows would stick out like sore thumbs.

Weigh the repair work needed against the cost of a replacement window. Nearly all old windows need new glazing putty. Cracked panes have to be replaced. Corners may need reinforcement. And you'll probably have to scrape off layers of paint so the window opens and closes properly. These are relatively minor repairs. More extensive decay may require that the window be rebuilt rather than repaired.

Figure 8-3

Fancy Victorian windows should be saved

Figure 8-4

These windows are historically accurate

Replacing Glazing Putty

Glazing putty lasts 40 to 50 years when protected by a storm window. Without protection from severe weather, glazing putty cracks in 5 to 10 years. Sections of putty fall out, leaving glass loose in the window frame. If a window pane is cracked and needs to be replaced, you'll have to remove the old putty to get the broken glass out.

When re-glazing, select putty that's smooth, creamy, and easy to work with. Avoid glazing putty that's hard to spread. You can apply glazing putty with a putty knife, but it's better to use a glazing tool that both applies new putty and removes old. Most of the old putty will come out easily. If some refuses to budge, use the glazing tool to cut under the old putty, removing a little sliver of wood, putty and all.

Allow about 15 minutes to remove and replace putty for one pane of glass, whether large or small. So figure four hours to renew putty for a pair of windows like the ones in Figure 8-5. If a pane is broken, take the sash out of the frame. Lay the window down flat on a newspaper. Then break the old glass out onto the newspaper. When you're done, just fold up the newspaper, broken glass and all, and throw it away. This keeps the jobsite free of broken glass. Measure carefully for replacement glass, including the 1/4" of glass area under the glazing putty.

Sash Repair

Joints tend to work loose and decay as windows age. Use wood glue and an angle brace to reinforce a weak corner. Figure 8-6 shows this type of repair on the weather-damaged window in Figure 8-2. When the window is painted, the braces won't be noticeable.

If a single- or double-hung wood window doesn't operate smoothly, try applying paste or paraffin wax to parts of the stop, jamb, and parting strip that come in contact during operation. If that doesn't solve the problem, abrasion marks on the sash will usually reveal where the sash is binding. Move nailed stops (Figure 8-7) away from the sash slightly. If stops are fastened with screws, remove the stop and sand or plane it lightly on the face in contact with the sash. But be careful not to plane away too much of the stop. Opening up too much space creates a path for air infiltration. If the sash is binding against the jamb, remove the sash and then plane the vertical edges slightly.

Figure 8-5
Multi-paned windows

If the sash rattles in the frame, glue a thin strip of wood veneer to one side of the sash. Keep veneer strips narrow enough so they're hidden by the stops.

The most common cause of window jamming is paint buildup. This is particularly true of the top sash, which is opened less than the bottom sash. Remove paint buildup with a heat gun. Start by removing the interior stop. Then lift out both top and bottom sash. Use a heat gun and a scraper to remove paint from the casing. Remove paint from the sash as well. Buildup may be jamming the sash against the stop. Be careful to keep the heat gun away from glass. Hot air from the gun can crack cold glass.

Fit the windows back into the frame and check operation. If the sash is still tight, a little sanding should solve the problem.

Once the window is working properly, it's OK to repaint the casing and sash. A single coat of paint won't do any harm. This entire job usually takes about two hours per window, including touch-up painting.

Figure 8-6
Steel angle brace used to reinforce sash

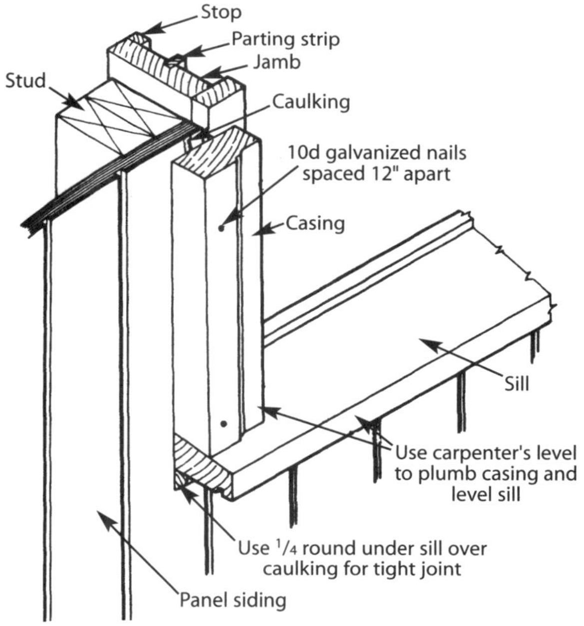

Figure 8-7
Installation of double-hung window frame

Counterweighted Windows

Quality wood windows built before about 1930 had sash balances inside the wall, usually two weights, one on each side of the window. See Figure 8-8. The weights and the window were perfectly balanced so the window would stay put when opened to any height. Most likely, at least one pulley and rope failed many years ago. Now, the only way to keep the window open is with a stick supporting the sash. Even worse, the balance mechanism may still be working on just one side. That immobilizes the sash at nearly any height.

If sash balances have failed on an otherwise sound window, consider installing a new spring balance or replacing the broken sash cord. Installing a spring balance only requires removing the parting strip and stops. Follow the spring balance manufacturer's installation instructions. Then replace the stops.

Replacing broken balance ropes is simple, but not easy. The weights are difficult to get at. Working from inside the house, remove the inside stop. Back out screws or carefully pry out the small nails anchoring the stop. Once loose, lift the sash and swivel it out of the window frame. With luck, you'll only need to remove the stop on one side of the window.

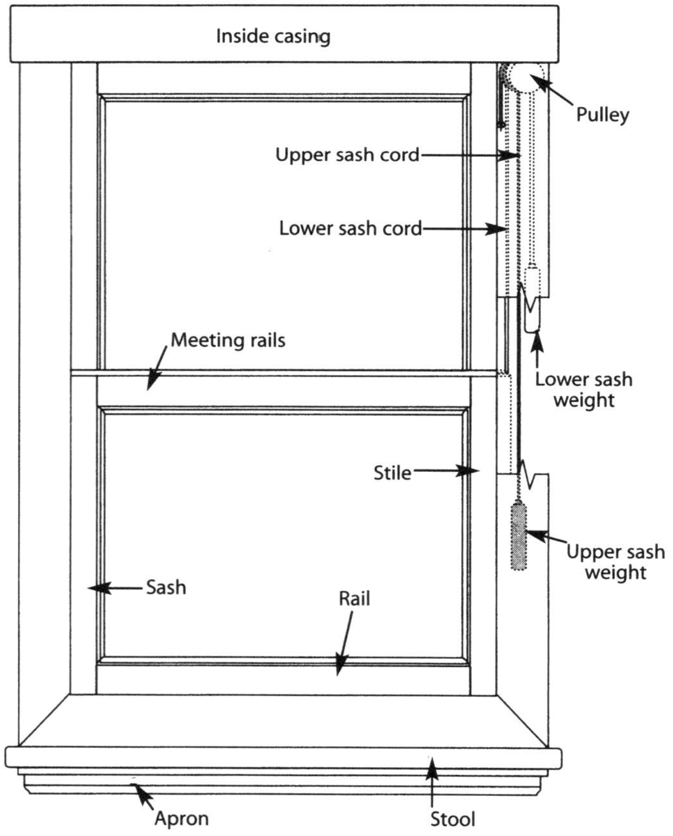

Figure 8-8
Weight assembly in counterweighted windows

If you see a little access cover under the inside stop on each side of the window, remove the cover to expose the channel where sash weights are supposed to operate. If you don't see an access cover, pry off the outer casing boards on each side of the window. Be careful not to crack the casing. Casing is usually attached with 4" trim-head nails. If these have rusted, they can be very hard to get loose. Don't put too much pressure on one spot. You'll either break the trim board or knock a hole in the wall. Work all around the board until you get it loose.

When the trim board is off, you won't see any weights. But have faith, the weights are in there. The window was installed before the furring strips and plaster went on. The outer casing boards were put on top of that. The weights are under the furring strips.

You'll need to cut through the plaster and furring strips at the bottom of the window frame. Don't worry, the hole won't show once you put the trim board back on. Cut carefully to avoid breaking any big chunks of plaster loose.

You may not see the weight at first. It's recessed a little below the bottom edge of the window frame. That's to give it someplace to go when the window is all the way up. If you stick your fingers into the hole, you'll feel the top of the sash weight. It's usually a rough cast-iron cylinder with a loop on top for attaching the sash cord.

Now you need to replace the sash cord. You can use cotton clothesline, because it looks just like the cord that was there originally. Feed an end over the pulley until it appears in the hole. Tie the end to the loop on top of the weight.

If you look at the outside edge of the sash frame on each side, you'll see keyhole-shaped cutouts where the cord is supposed to go. Cut new cord to the proper length and tie a knot in the end. The knot goes in the round-hole part of the keyhole-shaped cutout. The long thin part of the cutout acts as a guide for the rest of the rope. Once you fit the sash back in the frame, the cord will be held in place by pressure.

Before you put the window back together, check and see that it operates properly. If the repair has been done correctly, the window will slide smoothly and stay put when released in any position. If the window doesn't slide easily, do a little cleaning or sanding.

Replacing a set of four sash cords requires about two hours per window, assuming you don't break any trim boards or knock chunks out of the plaster. But those two hours will save several hundred dollars that a new custom-made insulated vinyl window would cost. Since counterweighted windows were made from high-quality materials like oak, a vinyl window might not look right in this location anyway. And a new window probably won't cut heating or cooling cost by more than a dollar or so a month. When in doubt, repair the existing window and buy new storms.

Adding or Moving a Window

Adding or moving a window will usually require both structural changes (studs and headers) and changes to the interior finish (wallboard, painting and trim). If the home improvement budget is modest, money should be spent elsewhere. But a strategically-placed window can add both livability and resale value to an older home. Figures 8-9 and 8-10 are before and after shots of a window re-sizing job. To take good advantage of a beautiful view, both openings were enlarged considerably. This job required about 12 hours per window. Chapter 3 includes estimates for cutting new wall openings for windows and doors.

Here are some rules to observe when adding, moving or re-sizing a window:

- Glass area should be equivalent to about 10 percent of floor area.
- Locate windows to promote good cross-ventilation.
- Group windows to eliminate undesirable contrasts in brightness.
- Window venting area should be about half of the window area.
- Provide screens for all sashes that open for ventilation.

Figure 8-9
Original windows

Figure 8-10
New larger windows take advantage of view

- Sill height can be lower in rooms used mostly for seating (living room).
- Sill heights can be higher in rooms used mostly for standing (kitchen).

Avoid small windows spaced evenly across a wall. Scattered windows cut up the wall space, making it unusable for larger pieces of furniture, such as a bookcase. It's better to cluster windows into one or two areas and leave other wall space undisturbed. See Figure 8-11. If you have a choice, put windows in south-facing walls, especially in colder climates. Winter sun is lower on the horizon, projecting sunlight deeper into the room through south-facing windows. Summer sun is higher in the sky and will be shaded even by a small overhang on the south side. In very warm climates, favor windows in north walls to avoid heat gain. Avoid west-facing windows to limit solar heating during the hottest part of the day. Even a large overhang is ineffective against a setting sun.

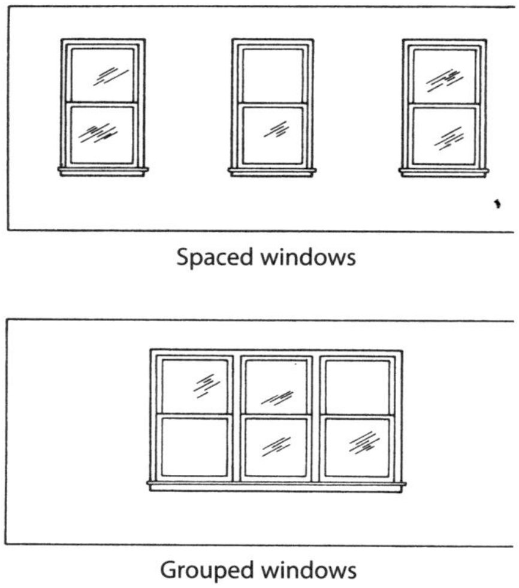

Figure 8-11
Window spacing

If you decide to close up a window opening, install vertical framing members spaced to match stud spacing below the window. Keeping the stud spacing consistent simplifies the nailing of both interior and exterior wall cover. Toenail new framing to the old window header and to the sill, using three 8d or 10d nails at each joint. Install sheathing of the same thickness as the existing sheathing. Then add insulation and apply a vapor barrier on the inside face of the framing.

Make sure the new vapor barrier laps over the vapor barrier in the remainder of the wall. Finally, apply interior and exterior wall covering to match the existing wall cover.

Replacement Windows

Figure 8-12 shows a double-casement window too decayed for repair. The replacement window (Figure 8-13) preserves the appearance of the old window but adds modern comfort and convenience.

Windows made for new construction have a nailing flange that laps over the exterior of the rough opening. For replacement purposes, that requires removal of the siding or stucco around the window perimeter and either a lot of patching or some very wide casing. Replacement windows fit into the existing jamb and have only a slightly wider casing to hide what's left of the old window. Be sure to specify the type of window you're replacing, either brick mold frame, wood frame or metal frame.

No special tools or equipment are needed to install a replacement window. But the window has to fit the opening exactly. If you're lucky, a stock size replacement window will fit. The tables that follow in this chapter list stock sizes for both replacement windows and new construction windows. If a stock size doesn't fit the opening, consider adding blocks and cripples so the opening fits a standard size replacement window. Of course, you'll need wider casing to cover the extra framing width at the head and jamb. If re-framing the window opening isn't practical, order a replacement window custom-made to size. For a price quote on the Web, go to http://www.thewindowsite.com.

Figure 8-12

This rotted window had to be replaced

Figure 8-13

New thermopane window retains the look of the original

	Craft@Hrs	Unit	Material	Labor	Total	Sell

Window Repairs

Close up window opening, wood frame wall. Remove and dispose of window and frame, install studs, sheathing, insulation, hang drywall interior (not taped or finished), install exterior wall finish and job clean-up. Near (not perfect) match of exterior wall finish. Per opening.

	Craft@Hrs	Unit	Material	Labor	Total	Sell
Stucco, wood, or vinyl siding exterior	B1@6.85	Ea	170.00	223.00	393.00	668.00
Brick veneer toothed into existing brick	B9@9.50	Ea	315.00	304.00	619.00	1,050.00

Close up window opening, masonry wall. Remove and dispose of window and frame, install brick or block backup, tooth face brick into the existing brick, install insulation, hang drywall interior (not taped or finished) and job clean-up. Near (not perfect) match of materials. Per opening.

	Craft@Hrs	Unit	Material	Labor	Total	Sell
Brick or brick and block	B9@12.0	Ea	310.00	383.00	693.00	1,180.00

Change window opening size, wood frame wall. Including cripples, shims and blocks as needed. Add the cost of patching the interior and exterior wall surfaces.

	Craft@Hrs	Unit	Material	Labor	Total	Sell
Opening to 20 square feet, per opening	B1@1.53	Ea	25.80	49.80	75.60	129.00
Opening over 20 square feet, per SF	B1@.077	SF	1.50	2.50	4.00	6.80

Change window opening size, masonry wall. Includes masonry patching and toothing of brick as needed.

	Craft@Hrs	Unit	Material	Labor	Total	Sell
Opening to 20 square feet, per opening	B6@1.95	Ea	23.30	61.30	84.60	144.00
Opening over 20 square feet, per SF	B6@.098	SF	3.02	3.08	6.10	10.40
Add for new cut stone sill, if needed	B6@.080	LF	5.65	2.52	8.17	13.90
Add for new steel lintel, if needed	B6@.250	Ea	9.43	7.87	17.30	29.40

Cut window opening and install new window. Cut window opening in an exterior frame wall, frame a new window opening, install new window and patch interior and exterior surfaces. Based on wall with 1" x 6" drop siding, diagonal 1" x 12" sheathing, 2" x 4" studs 18" on center, and drywall interior. Assumes a double 2" x 6" header for a 3' opening, double 2" x 8" header for a 5' opening, double 2" x 10" header for a 6' opening or a double 2" x 12" header for an 8' opening.

	Craft@Hrs	Unit	Material	Labor	Total	Sell
Window to 10 square feet	BC@9.50	Ea	30.00	342.00	372.00	632.00
Add per square foot for window over 10 SF	B1@.950	SF	2.70	30.90	33.60	57.10
Add per linear foot of 4" wide vinyl sill	BC@.050	LF	4.50	1.80	6.30	10.70

Remove and replace wood window. Pry off trim. Remove stops and sash. Loosen siding to expose jamb. Cut or pull nails securing window to wall framing. Lift out window. Level and install same-size jamb. Re-install trim. Add the cost of patching and finishing interior and exterior walls.

	Craft@Hrs	Unit	Material	Labor	Total	Sell
Window to 10 square feet	BC@3.12	Ea	—	112.00	112.00	190.00
Add per square foot for window over 10 SF	B1@.310	SF	—	10.10	10.10	17.20

Remove, repair and replace window sash. Wood double-hung window. Remove stops and sash. Plane and sand sash. Apply paraffin. Re-install sash and stops.

	Craft@Hrs	Unit	Material	Labor	Total	Sell
Per window sash	BC@.400	Ea	—	14.40	14.40	24.50

Break paint seal to free window sash. Using a hammer, wood block and putty knife blade. Add the cost of touch-up painting, if required.

	Craft@Hrs	Unit	Material	Labor	Total	Sell
Per sash	BC@.200	Ea	—	7.20	7.20	12.20

Remove and replace sash cord. Wood double-hung window. Remove window stop, sash and weight pocket covers. Remove weights and cords. Replace and attach new cords. Replace window sash and stop.

	Craft@Hrs	Unit	Material	Labor	Total	Sell
Per sash (2 cords)	BC@1.00	Ea	1.27	36.00	37.27	63.40

Remove exterior window trim. Working from a ladder.

	Craft@Hrs	Unit	Material	Labor	Total	Sell
12' trim length	BC@.220	Ea	—	7.92	7.92	13.50

	Craft@Hrs	Unit	Material	Labor	Total	Sell
Caulk window exterior. Caulk up to 12 linear feet of window perimeter with 5 ounces of paintable silicone sealant.						
Per window to 12 linear feet	BC@.104	Ea	6.09	3.75	9.84	16.70
Add for each additional LF	BC@.009	LF	.51	.32	.83	1.41
Remove and replace window caulk. Working from a ladder. Chip out old caulk around window perimeter and replace with 12 linear feet of paintable silicone sealant.						
Per window to 12 linear feet	BC@.178	Ea	6.09	6.41	12.50	21.30
Add for each additional LF	BC@.015	LF	.51	.54	1.05	1.79

Vinyl Windows

	Craft@Hrs	Unit	Material	Labor	Total	Sell
Custom-made vinyl replacement windows. Typical costs. Prices vary with window specifications and distance from the point of manufacture. Allow two weeks to a month for delivery. Insulating glass. Base cost per unit plus additional cost per square foot of opening size. Add the cost of Low-E glazing, grilles and screens below.						
Fixed (picture) vinyl window, base cost	BC@1.00	Ea	325.00	36.00	361.00	614.00
Add per square foot of opening	B1@.050	SF	8.17	1.63	9.80	16.70
Double-hung vinyl window, base cost	BC@1.00	Ea	351.00	36.00	387.00	658.00
Add per square foot of opening	B1@.050	SF	8.17	1.63	9.80	16.70
Two-lite sliding vinyl window, base cost	BC@1.00	Ea	374.00	36.00	410.00	697.00
Add per square foot of opening	B1@.050	SF	8.17	1.63	9.80	16.70
Three-lite sliding vinyl window, base cost	BC@1.00	Ea	405.00	36.00	441.00	750.00
Add per square foot of opening	B1@.050	SF	8.17	1.63	9.80	16.70
Add for Low-E glazing, per SF of opening	—	SF	2.43	—	2.43	—
Add for colonial grille, per SF of opening	—	SF	2.36	—	2.36	—
Add for screen, per screen	—	Ea	18.80	—	18.80	—
Vinyl double-hung stock replacement windows, 7/8" insulating Low-E glass. Fusion-welded white vinyl frame. By nominal (opening) size, width by height. Actual size is 1/2" less in both width and height. Tilt sash. With screen. Obscure glass and colonial grilles are available on some stock sizes. Labor includes removing the old stops and sash and setting new J-trim as needed.						
24" x 36"	BC@1.00	Ea	126.00	36.00	162.00	275.00
24" x 38"	BC@1.00	Ea	126.00	36.00	162.00	275.00
24" x 46"	BC@1.00	Ea	147.00	36.00	183.00	311.00
24" x 54"	BC@1.00	Ea	151.00	36.00	187.00	318.00
28" x 37"	BC@1.00	Ea	142.00	36.00	178.00	303.00
28" x 38"	BC@1.00	Ea	144.00	36.00	180.00	306.00
28" x 46"	BC@1.00	Ea	147.00	36.00	183.00	311.00
28" x 53"	BC@1.00	Ea	151.00	36.00	187.00	318.00
28" x 54"	B1@1.50	Ea	142.00	48.80	190.80	324.00
28" x 58"	B1@1.50	Ea	162.00	48.80	210.80	358.00
28" x 62"	B1@1.50	Ea	162.00	48.80	210.80	358.00
30" x 38"	B1@1.50	Ea	151.00	48.80	199.80	340.00
30" x 54"	B1@1.50	Ea	155.00	48.80	203.80	346.00
30" x 58"	B1@1.50	Ea	162.00	48.80	210.80	358.00
30" x 62"	B1@1.50	Ea	162.00	48.80	210.80	358.00
30" x 66"	B1@1.50	Ea	167.00	48.80	215.80	367.00
31" x 53"	B1@1.50	Ea	155.00	48.80	203.80	346.00
31" x 54"	B1@1.50	Ea	159.00	48.80	207.80	353.00
31" x 57"	B1@1.50	Ea	162.00	48.80	210.80	358.00
31" x 58"	B1@1.50	Ea	162.00	48.80	210.80	358.00
31" x 61"	B1@1.50	Ea	162.00	48.80	210.80	358.00
32" x 38"	B1@1.50	Ea	125.00	48.80	173.80	295.00

	Craft@Hrs	Unit	Material	Labor	Total	Sell
32" x 46"	B1@1.50	Ea	151.00	48.80	199.80	340.00
32" x 50"	B1@1.50	Ea	155.00	48.80	203.80	346.00
32" x 54"	B1@1.50	Ea	162.00	48.80	210.80	358.00
32" x 58"	B1@1.50	Ea	162.00	48.80	210.80	358.00
32" x 61"	B1@1.50	Ea	162.00	48.80	210.80	358.00
32" x 62"	B1@1.50	Ea	167.00	48.80	215.80	367.00
32" x 66"	B1@1.50	Ea	173.00	48.80	221.80	377.00
32" x 70"	B1@1.50	Ea	188.00	48.80	236.80	403.00
34" x 38"	B1@1.50	Ea	123.00	48.80	171.80	292.00
34" x 46"	B1@1.50	Ea	150.00	48.80	198.80	338.00
34" x 54"	B1@1.50	Ea	150.00	48.80	198.80	338.00
34" x 62"	B1@1.50	Ea	170.00	48.80	218.80	372.00
36" x 38"	B1@1.50	Ea	147.00	48.80	195.80	333.00
36" x 46"	B1@1.50	Ea	155.00	48.80	203.80	346.00
36" x 50"	B1@1.50	Ea	162.00	48.80	210.80	358.00
36" x 54"	B1@1.50	Ea	162.00	48.80	210.80	358.00
Add for colonial grille between the lites	—	%	15.0	—	—	—

Vinyl single-hung stock replacement windows, insulating glass. Tilt-in sash. Front flange frame for installation in masonry block openings. Lower sash opens. Upper sash is fixed. Includes half screen. By opening size, width by height. Grille pattern shows lites in upper and lower sash, such as 6/6. Labor includes removing the old stops and sash and setting new J-trim as needed.

	Craft@Hrs	Unit	Material	Labor	Total	Sell
35-7/8" x 24-7/8"	B1@1.00	Ea	92.80	32.50	125.30	213.00
35-7/8" x 37-1/4"	B1@1.00	Ea	112.00	32.50	144.50	246.00
35-7/8" x 49-1/2"	B1@1.00	Ea	123.00	32.50	155.50	264.00
35-7/8" x 61-7/8"	B1@1.50	Ea	134.00	48.80	182.80	311.00
52" x 37-1/4"	B1@1.50	Ea	130.00	48.80	178.80	304.00
52" x 49-1/2"	B1@1.50	Ea	146.00	48.80	194.80	331.00
35-7/8" x 37-1/4", 6/6 grille	B1@1.50	Ea	128.00	48.80	176.80	301.00
35-7/8" x 49-1/2", 6/6 grille	B1@1.50	Ea	139.00	48.80	187.80	319.00
35-7/8" x 61-7/8", 6/6 grille	B1@1.50	Ea	149.00	48.80	197.80	336.00
52" x 37-1/4", 8/8 grille	B1@1.50	Ea	145.00	48.80	193.80	329.00
52" x 49-1/2", 8/8 grille	B1@1.50	Ea	161.00	48.80	209.80	357.00

Vinyl casement bay windows, 7/8" insulating glass. Fixed center lite and two casement flankers. White vinyl. Factory assembled. By nominal (opening) size, width by height. Actual size is 1/2" less in both width and height. Includes two screens. Low-E glass is available on some stock sizes.

	Craft@Hrs	Unit	Material	Labor	Total	Sell
69" x 50"	B1@4.00	Ea	1,070.00	130.00	1,200.00	2,040.00
74" x 50"	B1@4.00	Ea	1,080.00	130.00	1,210.00	2,060.00
93" x 50"	B1@4.50	Ea	1,220.00	146.00	1,366.00	2,320.00
98" x 50"	B1@4.50	Ea	1,280.00	146.00	1,426.00	2,420.00

Vinyl garden windows, 7/8" insulating glass. Single front awning vent. White vinyl construction with wood frame. Tempered glass shelf. 5/4" vinyl laminated head and seat board. With hardware and screen. By nominal (opening) size, width by height. Replacement window. Labor includes removing the old stops and sash and setting new J-trim as needed.

	Craft@Hrs	Unit	Material	Labor	Total	Sell
36" x 36"	B1@2.75	Ea	801.00	89.50	890.50	1,510.00
36" x 36", Low-E glass	B1@2.75	Ea	591.00	89.50	680.50	1,160.00
36" x 48"	B1@2.75	Ea	606.00	89.50	695.50	1,180.00
36" x 48", Low-E glass	B1@2.75	Ea	730.00	89.50	819.50	1,390.00

	Craft@Hrs	Unit	Material	Labor	Total	Sell
Half-circle fixed vinyl windows. Installed in an opening that's already framed to the correct size. Dual glazed, white frame. Low-E insulating glass. Stock units for new construction. Dimensions are width by height.						
20" x 10"	B1@2.75	Ea	502.00	89.50	591.50	1,010.00
30" x 16"	B1@2.75	Ea	502.00	89.50	591.50	1,010.00
40" x 20"	B1@2.75	Ea	662.00	89.50	751.50	1,280.00
50" x 26"	B1@2.75	Ea	790.00	89.50	879.50	1,500.00
60" x 30"	B1@2.75	Ea	897.00	89.50	986.50	1,680.00
80" x 40"	B1@2.75	Ea	1,300.00	89.50	1,389.50	2,360.00
Vinyl jalousie windows. Installed in an opening that's already framed to the correct size. Positive locking handle for security. By opening size, width by height.						
24" x 35-1/2", clear	BC@1.00	Ea	261.00	36.00	297.00	505.00
24" x 35-1/2", obscure glass	BC@1.00	Ea	297.00	36.00	333.00	566.00
30" x 35-1/2", clear	BC@1.00	Ea	273.00	36.00	309.00	525.00
30" x 35-1/2", obscure glass	BC@1.00	Ea	318.00	36.00	354.00	602.00
36" x 35-1/2", clear	BC@1.00	Ea	286.00	36.00	322.00	547.00
36" x 35-1/2", obscure glass	BC@1.00	Ea	340.00	36.00	376.00	639.00
Hopper vinyl windows. White. Installed in an opening that's already framed to the correct size. Does not include finish or trim work. Low-E insulating glass. Argon filled. Frame depth 3-5/16". With screen. Stock units for new construction. By opening size, width by height.						
32" x 15"	B1@.500	Ea	67.00	16.30	83.30	142.00
32" x 17"	B1@.500	Ea	66.30	16.30	82.60	140.00
32" x 19"	B1@.500	Ea	66.30	16.30	82.60	140.00
32" x 23"	B1@.500	Ea	78.50	16.30	94.80	161.00

Vinyl-Clad Wood Windows

	Craft@Hrs	Unit	Material	Labor	Total	Sell
Vinyl-clad double-hung Low-E insulated glass wood windows. Prefinished pine interior. White vinyl exterior. Tilt sash. For new construction. By nominal (opening) size, width by height. Actual width is 1-5/8" more. Actual height is 3-1/4" more. Includes factory-applied extension jamb for 4-9/16" wall. Add the cost of screens and colonial grilles.						
24" x 36"	B1@.500	Ea	164.00	16.30	180.30	307.00
28" x 48"	B1@.500	Ea	206.00	16.30	222.30	378.00
28" x 36"	B1@.500	Ea	175.00	16.30	191.30	325.00
28" x 42"	B1@.500	Ea	190.00	16.30	206.30	351.00
32" x 48"	B1@.750	Ea	216.00	24.40	240.40	409.00
32" x 57"	B1@.750	Ea	239.00	24.40	263.40	448.00
32" x 36"	B1@.750	Ea	184.00	24.40	208.40	354.00
36" x 42"	B1@.750	Ea	210.00	24.40	234.40	398.00
36" x 48"	B1@.750	Ea	226.00	24.40	250.40	426.00
36" x 54"	B1@.750	Ea	242.00	24.40	266.40	453.00
36" x 57"	B1@.750	Ea	250.00	24.40	274.40	466.00
Deduct for non Low-E glazing, per SF of opening	—	SF	−1.70	—	−1.70	—

	Craft@Hrs	Unit	Material	Labor	Total	Sell
Screens for vinyl-clad wood windows. White. By window nominal size.						
2'0" x 3'0"	B1@.050	Ea	19.30	1.63	20.93	35.60
2'0" x 3'6"	B1@.050	Ea	21.80	1.63	23.43	39.80
2'4" x 3'0"	B1@.050	Ea	19.90	1.63	21.53	36.60
2'4" x 3'6"	B1@.050	Ea	21.90	1.63	23.53	40.00
2'4" x 4'0"	B1@.050	Ea	22.90	1.63	24.53	41.70
2'4" x 4'6"	B1@.050	Ea	28.10	1.63	29.73	50.50
2'4" x 4'9"	B1@.050	Ea	27.90	1.63	29.53	50.20
2'8" x 3'0"	B1@.050	Ea	21.90	1.63	23.53	40.00
2'8" x 3'6"	B1@.050	Ea	22.90	1.63	24.53	41.70
2'8" x 4'0"	B1@.050	Ea	24.90	1.63	26.53	45.10
2'8" x 4'6"	B1@.050	Ea	28.90	1.63	30.53	51.90
2'8" x 4'9"	B1@.050	Ea	30.20	1.63	31.83	54.10
3'0" x 3'0"	B1@.050	Ea	23.60	1.63	25.23	42.90
3'0" x 3'6"	B1@.050	Ea	23.90	1.63	25.53	43.40
3'0" x 4'0"	B1@.050	Ea	27.90	1.63	29.53	50.20
3'0" x 4'6"	B1@.050	Ea	29.90	1.63	31.53	53.60
3'0" x 4'9"	B1@.050	Ea	29.90	1.63	31.53	53.60
3'0" x 5'6"	B1@.050	Ea	27.90	1.63	29.53	50.20
Vinyl-clad wood awning windows, insulating glass. Prefinished pine interior. White vinyl exterior. For new construction. Add the cost of screens. By nominal (opening) size, width by height.						
2'1" x 2'0"	B1@.500	Ea	203.00	16.30	219.30	373.00
3'1" x 2'0"	B1@.500	Ea	248.00	16.30	264.30	449.00
4'1" x 2'0"	B1@.500	Ea	260.00	16.30	276.30	470.00
Vinyl-clad wood casement windows, insulating glass. Prefinished pine interior. White or tan vinyl exterior. Tilt sash. For new construction. By nominal (opening) size, width by height. Add the cost of screens and hardware. Hinge operation as viewed from exterior.						
2'1" x 3'0", right hinge	B1@.500	Ea	237.00	16.30	253.30	431.00
2'1" x 3'0", left hinge	B1@.500	Ea	237.00	16.30	253.30	431.00
2'1" x 3'6", right hinge	B1@.500	Ea	274.00	16.30	290.30	494.00
2'1" x 3'6", left hinge	B1@.500	Ea	274.00	16.30	290.30	494.00
2'1" x 4'0", right hinge	B1@.750	Ea	303.00	24.40	327.40	557.00
2'1" x 4'0", left hinge	B1@.750	Ea	303.00	24.40	327.40	557.00
2'1" x 5'0", right hinge	B1@.750	Ea	313.00	24.40	337.40	574.00
2'1" x 5'0", left hinge	B1@.750	Ea	313.00	24.40	337.40	574.00
4'1" x 3'0", left and right hinge	B1@.750	Ea	342.00	24.40	366.40	623.00
4'1" x 3'5", left and right hinge	B1@.750	Ea	475.00	24.40	499.40	849.00
4'1" x 4'0", left and right hinge	B1@.750	Ea	525.00	24.40	549.40	934.00
Add for casement hardware	—	Ea	12.00	—	12.00	—

Aluminum Windows

Aluminum single-hung vertical sliding white windows, insulating glass. White frame. Tilt-in removable sash. Dual sash weatherstripping. Includes screen. By nominal (opening) size, width by height. Nailing flange for new construction.

	Craft@Hrs	Unit	Material	Labor	Total	Sell
2' x 3'	B1@.500	Ea	50.90	16.30	67.20	114.00
2' x 5'	B1@1.00	Ea	74.80	32.50	107.30	182.00
2'8" x 5'	B1@1.00	Ea	83.10	32.50	115.60	197.00
3' x 3'	B1@1.00	Ea	65.60	32.50	98.10	167.00
3' x 4'	B1@1.00	Ea	77.00	32.50	109.50	186.00
3' x 5'	B1@1.25	Ea	88.30	40.70	129.00	219.00
3' x 6'	B1@1.25	Ea	99.00	40.70	139.70	237.00
4' x 3'	B1@1.00	Ea	85.80	32.50	118.30	201.00

	Craft@Hrs	Unit	Material	Labor	Total	Sell
4' x 4'	B1@1.50	Ea	105.00	48.80	153.80	261.00
5' x 3'	B1@1.50	Ea	101.00	48.80	149.80	255.00
5' x 4'	B3@2.00	Ea	120.00	62.70	182.70	311.00
6' x 4'	B3@2.00	Ea	141.00	62.70	203.70	346.00
8' x 4'	B3@2.00	Ea	183.00	62.70	245.70	418.00
Add for obscure glass in 2' windows	—	Ea	3.45	—	3.45	—

Aluminum single-hung vertical sliding bronze windows, insulating glass. Bronze frame. Better quality removable sash. Dual sash weatherstripping. Includes screen. By nominal (opening) size, width by height. Includes a window grid between the panes. Nailing flange for new construction.

	Craft@Hrs	Unit	Material	Labor	Total	Sell
2' x 3'	B1@.500	Ea	95.00	16.30	111.30	189.00
2' 8" x 3'	B1@.500	Ea	106.00	16.30	122.30	208.00
2' 8" x 4' 4"	B1@1.00	Ea	131.00	32.50	163.50	278.00
2' 8" x 5'	B1@1.00	Ea	145.00	32.50	177.50	302.00
3' x 3'	B1@1.00	Ea	115.00	32.50	147.50	251.00
3' x 4'	B1@1.00	Ea	131.00	32.50	163.50	278.00
3' x 4' 4"	B1@1.00	Ea	138.00	32.50	170.50	290.00
3' x 5'	B1@1.25	Ea	151.00	40.70	191.70	326.00
3' x 6'	B1@1.25	Ea	171.00	40.70	211.70	360.00

Aluminum single-hung vertical sliding windows, single glazed. Mill finish. Includes screen. By nominal (opening) size, width by height. Clear glass except as noted. Nailing flange for new construction.

	Craft@Hrs	Unit	Material	Labor	Total	Sell
2' x 1'	B1@.500	Ea	30.60	16.30	46.90	79.70
2' x 1', obscure glass	B1@.500	Ea	31.30	16.30	47.60	80.90
2' x 2'	B1@.500	Ea	24.60	16.30	40.90	69.50
2' x 2', obscure glass	B1@.500	Ea	42.10	16.30	58.40	99.30
2' x 3', obscure glass	B1@.500	Ea	59.90	16.30	76.20	130.00
3' x 2'	B1@.500	Ea	45.30	16.30	61.60	105.00
3' x 3'	B1@1.00	Ea	56.60	32.50	89.10	151.00
4' x 3'	B1@1.00	Ea	59.10	32.50	91.60	156.00
5' x 3'	B1@1.50	Ea	68.00	48.80	116.80	199.00
6' x 3'	B3@1.50	Ea	93.40	47.10	140.50	239.00

Aluminum horizontal sliding windows, single glazed. Right sash is operable when viewed from the exterior. With screen. 7/8" nailing fin setback for 3-coat stucco or wood siding. By nominal (opening) size, width by height. For new construction. Obscure glass is usually available in smaller (1' and 2') windows at no additional cost.

	Craft@Hrs	Unit	Material	Labor	Total	Sell
2' x 2', mill finish	B1@.500	Ea	70.40	16.30	86.70	147.00
2' x 2', bronze or white finish	B1@.500	Ea	66.30	16.30	82.60	140.00
3' x 1', mill finish	B1@.750	Ea	66.30	24.40	90.70	154.00
3' x 2', mill finish	B1@.750	Ea	77.50	24.40	101.90	173.00
3' x 3', mill finish	B1@1.00	Ea	85.70	32.50	118.20	201.00
3' x 3', bronze or white finish	B1@1.00	Ea	107.00	32.50	139.50	237.00
3' x 4', mill finish	B1@1.00	Ea	101.00	32.50	133.50	227.00
3' x 4', bronze or white finish	B1@1.00	Ea	111.00	32.50	143.50	244.00
4' x 3', mill finish	B1@1.00	Ea	100.00	32.50	132.50	225.00
4' x 3', bronze or white finish	B1@1.00	Ea	111.00	32.50	143.50	244.00
4' x 4', mill finish	B1@1.50	Ea	114.00	48.80	162.80	277.00
4' x 4', bronze or white finish	B1@1.50	Ea	132.00	48.80	180.80	307.00
5' x 3', mill finish	B3@1.50	Ea	138.00	47.10	185.10	315.00
5' x 4', mill finish	B3@1.50	Ea	138.00	47.10	185.10	315.00
5' x 4', bronze or white finish	B3@1.50	Ea	148.00	47.10	195.10	332.00
6' x 3', mill finish	B3@1.50	Ea	148.00	47.10	195.10	332.00
6' x 4', mill finish	B3@1.50	Ea	138.00	47.10	185.10	315.00
6' x 4', bronze or white finish	B3@1.50	Ea	172.00	47.10	219.10	372.00

	Craft@Hrs	Unit	Material	Labor	Total	Sell

Aluminum horizontal sliding windows, insulated glass. Right sash is operable when viewed from the exterior. With screen. Nailing fin for new construction. By nominal (opening) size, width by height. Obscure glass is usually available in smaller (1' and 2') windows at no additional cost. White or bronze finish.

	Craft@Hrs	Unit	Material	Labor	Total	Sell
2' x 2'	B1@.500	Ea	41.80	16.30	58.10	98.80
3' x 1'	B1@.750	Ea	41.80	24.40	66.20	113.00
3' x 2'	B1@.750	Ea	52.00	24.40	76.40	130.00
3' x 3'	B1@1.00	Ea	62.20	32.50	94.70	161.00
3' x 4'	B1@1.00	Ea	73.40	32.50	105.90	180.00
4' x 3'	B1@1.00	Ea	73.40	32.50	105.90	180.00
4' x 4'	B1@1.50	Ea	85.70	48.80	134.50	229.00
Add for colonial grid between the lites	—	%	15.0	—	—	—
Add for Low-E glass	—	%	15.0	—	—	—

Aluminum awning windows, insulated glass. With hardware and screen. Natural mill finish. For new construction. Nominal (opening) sizes, width by height.

	Craft@Hrs	Unit	Material	Labor	Total	Sell
19" x 26", two lites high	BC@.500	Ea	81.50	18.00	99.50	169.00
19" x 38-1/2", three lites high	BC@.750	Ea	98.50	27.00	125.50	213.00
19" x 50-1/2", four lites high	BC@.750	Ea	117.00	27.00	144.00	245.00
19" x 63", five lites high	BC@1.00	Ea	138.00	36.00	174.00	296.00
26" x 26", two lites high	BC@.750	Ea	101.00	27.00	128.00	218.00
26" x 38-1/2", three lites high	BC@.750	Ea	119.00	27.00	146.00	248.00
26" x 50-1/2", four lites high	BC@1.00	Ea	153.00	36.00	189.00	321.00
26" x 63", five lites high	BC@1.00	Ea	179.00	36.00	215.00	366.00
37" x 26", two lites high	BC@.750	Ea	117.00	27.00	144.00	245.00
37" x 38-1/2", three lites high	BC@1.00	Ea	135.00	36.00	171.00	291.00
37" x 50-1/2", four lites high	BC@1.00	Ea	156.00	36.00	192.00	326.00
37" x 63", five lites high	BC@1.00	Ea	211.00	36.00	247.00	420.00
53" x 26", two lites high	BC@1.00	Ea	139.00	36.00	175.00	298.00
53" x 38-1/2", three lites high	BC@1.00	Ea	159.00	36.00	195.00	332.00
53" x 50-1/2", four lites high	B3@1.50	Ea	206.00	47.10	253.10	430.00
53" x 63", five lites high	B3@1.50	Ea	242.00	47.10	289.10	491.00
Add for bronze finish	—	%	15.0	—	—	—
Add for enamel finish	—	%	10.0	—	—	—
Add for special order replacement window	—	%	100.0	—	—	—

Fixed aluminum windows (picture windows), insulating glass. Bronze or white enamel finish. For new construction. Wall opening (nominal) size, width by height.

	Craft@Hrs	Unit	Material	Labor	Total	Sell
24" x 20"	BC@.500	Ea	111.00	18.00	129.00	219.00
24" x 30"	BC@.500	Ea	129.00	18.00	147.00	250.00
24" x 40"	BC@.500	Ea	153.00	18.00	171.00	291.00
30" x 20"	BC@.750	Ea	137.00	27.00	164.00	279.00
30" x 30"	BC@.750	Ea	158.00	27.00	185.00	315.00
30" x 40"	BC@.750	Ea	186.00	27.00	213.00	362.00
30" x 50"	BC@1.00	Ea	135.00	36.00	171.00	291.00
30" x 60"	BC@1.00	Ea	247.00	36.00	283.00	481.00
36" x 20"	BC@.750	Ea	162.00	27.00	189.00	321.00
36" x 30"	BC@.750	Ea	192.00	27.00	219.00	372.00
36" x 40"	BC@1.00	Ea	236.00	36.00	272.00	462.00

	Craft@Hrs	Unit	Material	Labor	Total	Sell
36" x 50"	B1@1.50	Ea	207.00	48.80	255.80	435.00
36" x 60"	B1@1.50	Ea	246.00	48.80	294.80	501.00
40" x 20"	BC@.750	Ea	190.00	27.00	217.00	369.00
40" x 30"	BC@.750	Ea	237.00	27.00	264.00	449.00
40" x 40"	BC@1.00	Ea	280.00	36.00	316.00	537.00
48" x 20"	BC@.750	Ea	229.00	27.00	256.00	435.00
48" x 30"	BC@1.00	Ea	285.00	36.00	321.00	546.00
48" x 40"	BC@1.00	Ea	314.00	36.00	350.00	595.00
Add for storm sash, per square foot of glass	—	SF	8.22	—	8.22	—

Louvered aluminum (jalousie) windows. Single glazed, 4" clear glass louvers, including hardware. Jalousie windows are usually made to order and can be fabricated to any size.

	Craft@Hrs	Unit	Material	Labor	Total	Sell
18" x 24"	BC@1.00	Ea	61.30	36.00	97.30	165.00
18" x 36"	BC@1.00	Ea	88.60	36.00	124.60	212.00
18" x 48"	BC@1.00	Ea	110.00	36.00	146.00	248.00
18" x 60"	BC@1.00	Ea	143.00	36.00	179.00	304.00
24" x 24"	BC@1.00	Ea	69.00	36.00	105.00	179.00
24" x 36"	BC@1.00	Ea	100.00	36.00	136.00	231.00
24" x 48"	BC@1.00	Ea	125.00	36.00	161.00	274.00
24" x 60"	BC@1.50	Ea	167.00	54.00	221.00	376.00
30" x 24"	BC@1.00	Ea	82.50	36.00	118.50	201.00
30" x 36"	BC@1.00	Ea	118.00	36.00	154.00	262.00
30" x 48"	BC@1.50	Ea	149.00	54.00	203.00	345.00
30" x 60"	BC@1.50	Ea	192.00	54.00	246.00	418.00
36" x 24"	BC@1.00	Ea	89.50	36.00	125.50	213.00
36" x 36"	BC@1.50	Ea	129.00	54.00	183.00	311.00
36" x 48"	BC@1.50	Ea	167.00	54.00	221.00	376.00
36" x 60"	BC@1.50	Ea	215.00	54.00	269.00	457.00
40" x 24"	BC@1.00	Ea	104.00	36.00	140.00	238.00
40" x 36"	BC@1.50	Ea	147.00	54.00	201.00	342.00
40" x 48"	BC@1.50	Ea	186.00	54.00	240.00	408.00
40" x 60"	BC@1.50	Ea	237.00	54.00	291.00	495.00

Wood Windows

Wood double-hung windows. Treated western pine. Natural wood interior. 4-9/16" jamb width. 5/8" insulating glass. Weatherstripped. Stock units for new construction. Tilt-in, removable sash. With hardware. Nominal sizes, width by height.

	Craft@Hrs	Unit	Material	Labor	Total	Sell
2'0" x 3'2"	BC@.750	Ea	150.00	27.00	177.00	301.00
2'4" x 3'2"	BC@.750	Ea	186.00	27.00	213.00	362.00
2'4" x 4'6"	BC@.750	Ea	175.00	27.00	202.00	343.00
2'8" x 3'2"	BC@.750	Ea	170.00	27.00	197.00	335.00
2'8" x 3'10"	BC@.750	Ea	200.00	27.00	227.00	386.00
2'8" x 4'6"	BC@1.25	Ea	201.00	45.00	246.00	418.00
2'8" x 5'2"	BC@1.25	Ea	230.00	45.00	275.00	468.00
3'0" x 3'2"	BC@.750	Ea	174.00	27.00	201.00	342.00
3'0" x 4'6"	BC@1.25	Ea	230.00	45.00	275.00	468.00
3'0" x 5'2"	BC@1.25	Ea	208.00	45.00	253.00	430.00
2'8" x 4'6", twin	BC@1.25	Ea	417.00	45.00	462.00	785.00

Fixed octagon wood windows. Stock units for new construction. Installed in a pre-cut and framed opening.

	Craft@Hrs	Unit	Material	Labor	Total	Sell
24", insulated glass	BC@1.50	Ea	101.00	54.00	155.00	264.00
24", single glazed	BC@1.50	Ea	80.60	54.00	134.60	229.00
24", venting insulated glass	BC@1.50	Ea	142.00	54.00	196.00	333.00

Angle bay windows. Fixed center lite and two double-hung flankers. 30-degree angle. Pine with aluminum cladding, insulated glass, screens, white or bronze finish, 4-9/16" wall thickness. Stock units for new construction. Nominal (opening) sizes, width by height. Add the cost of interior trim and installing or framing the bay window roof. See the figures that follow.

	Craft@Hrs	Unit	Material	Labor	Total	Sell
6'4" x 4'8"	B1@4.00	Ea	1,580.00	130.00	1,710.00	2,910.00
7'2" x 4'8"	B1@4.00	Ea	1,590.00	130.00	1,720.00	2,920.00
7'10" x 4'8"	B1@4.50	Ea	1,720.00	146.00	1,866.00	3,170.00
8'6" x 4'8"	B1@4.50	Ea	1,980.00	146.00	2,126.00	3,610.00
8'10" x 4'8"	B1@4.50	Ea	2,010.00	146.00	2,156.00	3,670.00
9'2" x 4'8"	B1@5.00	Ea	1,940.00	163.00	2,103.00	3,580.00
9'10" x 4'8"	B1@5.00	Ea	2,040.00	163.00	2,203.00	3,750.00
Add for 45-degree angle windows	—	Ea	80.00	—	80.00	—
Add for roof framing kit	—	Ea	144.00	—	144.00	—

Bay window roof cover. Architectural copper roof system for standard angle bay windows. Includes hardware. Stock units for new construction.

	Craft@Hrs	Unit	Material	Labor	Total	Sell
Up to 48" width	SW@.500	Ea	670.00	20.20	690.20	1,170.00
49" to 60" width	SW@.500	Ea	762.00	20.20	782.20	1,330.00
61" to 72" width	SW@.600	Ea	854.00	24.30	878.30	1,490.00
73" to 84" width	SW@.600	Ea	950.00	24.30	974.30	1,660.00
85" to 96" width	SW@.600	Ea	1,040.00	24.30	1,064.30	1,810.00
97" to 108" width	SW@.700	Ea	1,130.00	28.30	1,158.30	1,970.00
109" to 120" width	SW@.700	Ea	1,230.00	28.30	1,258.30	2,140.00
121" to 132" width	SW@.800	Ea	1,400.00	32.30	1,432.30	2,430.00
133" to 144" width	SW@.800	Ea	1,500.00	32.30	1,532.30	2,600.00
145" to 156" width	SW@1.00	Ea	1,590.00	40.40	1,630.40	2,770.00
157" to 168" width	SW@1.00	Ea	1,680.00	40.40	1,720.40	2,920.00
Add for aged copper, brass or aluminum finish	—	%	20.0	—	—	—
Add for 24" to 26" projection	—	%	10.0	—	—	—
Add for 26" to 28" projection	—	%	15.0	—	—	—
Add for 28" to 30" projection	—	%	20.0	—	—	—
Add for brick soffit kit	—	%	45.0	—	—	—
Add for 90-degree bays	—	Ea	120.00	—	120.00	—

Casement bow windows. Insulated glass, assembled, with screens, weatherstripped, unfinished pine, 16" projection, 22" x 56" lites, 4-7/16" jambs. Add the cost of interior trim and installing or framing the bow window roof. Stock units for new construction. Generally a roof kit is required for projections of 12" or more. See the figures that follow.

	Craft@Hrs	Unit	Material	Labor	Total	Sell
8'1" wide x 4'8" (4 lites, 2 venting)	B1@4.75	Ea	1,570.00	155.00	1,725.00	2,930.00
8'1" wide x 5'4" (4 lites, 2 venting)	B1@4.75	Ea	1,690.00	155.00	1,845.00	3,140.00
8'1" wide x 6'0" (4 lites, 2 venting)	B1@4.75	Ea	1,800.00	155.00	1,955.00	3,320.00
10' wide x 4'8" (5 lites, 2 venting)	B1@4.75	Ea	1,910.00	155.00	2,065.00	3,510.00
10' wide x 5'4" (5 lites, 2 venting)	B1@4.75	Ea	2,060.00	155.00	2,215.00	3,770.00
10' wide x 6'0" (5 lites, 2 venting)	B1@4.75	Ea	2,150.00	155.00	2,305.00	3,920.00
Add for roof framing kit, 5 lites	—	Ea	121.00	—	121.00	—

Bow window wood roof kit. Framed wood roof kit for four panel bow window. Built with a 10 in 12 pitch, roof overhangs bay or bow 1" to 3". Includes hardware.

	Craft@Hrs	Unit	Material	Labor	Total	Sell
Up to 72" width	B1@.500	Ea	350.00	16.30	366.30	623.00
73" to 96" width	B1@.500	Ea	410.00	16.30	426.30	725.00
97" to 120" width	B1@.500	Ea	473.00	16.30	489.30	832.00
121" to 144" width	B1@.600	Ea	555.00	19.50	574.50	977.00
145" to 168" width	B1@.600	Ea	738.00	19.50	757.50	1,290.00
Add for 5 panel bow	—	Ea	54.00	—	54.00	—

	Craft@Hrs	Unit	Material	Labor	Total	Sell
Bow or bay window roof, framed on-site. 1/2" plywood sheathing, fiberglass batt insulation, fascia board, cove molding, roof flashing, 15-pound felt, 240-pound seal-tab asphalt shingles, and cleanup.						
Up to 120" width	B1@3.50	Ea	89.30	114.00	203.30	346.00
Add per linear foot of width over 10'	B1@.350	LF	8.93	11.40	20.33	34.60

Storm Windows

	Craft@Hrs	Unit	Material	Labor	Total	Sell
Basement storm windows. White aluminum frame with removable plexiglass window. Clips on over aluminum or wood basement window. Screen included. Nominal window sizes, width by height.						
32" x 14"	BC@.420	Ea	27.50	15.10	42.60	72.40
32" x 18"	BC@.420	Ea	29.60	15.10	44.70	76.00
32" x 22"	BC@.420	Ea	27.40	15.10	42.50	72.30
Aluminum storm windows. Removable glass inserts. Removable screen. Double-track. Includes weatherstripping. Adaptable to aluminum single-hung windows.						
24" x 39", mill finish	BC@.420	Ea	37.70	15.10	52.80	89.80
24" x 39", white finish	BC@.420	Ea	39.80	15.10	54.90	93.30
24" x 55", mill finish	BC@.420	Ea	55.10	15.10	70.20	119.00
24" x 55", white finish	BC@.420	Ea	50.00	15.10	65.10	111.00
28" x 39", mill finish	BC@.420	Ea	38.80	15.10	53.90	91.60
28" x 39", white finish	BC@.420	Ea	39.80	15.10	54.90	93.30
28" x 47", white finish	BC@.420	Ea	49.90	15.10	65.00	111.00
28" x 55", mill finish	BC@.420	Ea	47.90	15.10	63.00	107.00
28" x 55", white finish	BC@.420	Ea	50.00	15.10	65.10	111.00
32" x 39", mill finish	BC@.420	Ea	39.80	15.10	54.90	93.30
32" x 39", white finish	BC@.420	Ea	43.90	15.10	59.00	100.00
32" x 47", white finish	BC@.420	Ea	59.70	15.10	74.80	127.00
32" x 55", mill finish	BC@.420	Ea	50.00	15.10	65.10	111.00
32" x 55", white finish	BC@.420	Ea	53.00	15.10	68.10	116.00
32" x 63", mill finish	BC@.420	Ea	58.10	15.10	73.20	124.00
32" x 63", white finish	BC@.420	Ea	61.40	15.10	76.50	130.00
32" x 75", white finish	BC@.420	Ea	65.30	15.10	80.40	137.00
36" x 39", mill finish	BC@.420	Ea	43.90	15.10	59.00	100.00
36" x 39", white finish	BC@.420	Ea	45.90	15.10	61.00	104.00
36" x 55", mill finish	BC@.420	Ea	53.00	15.10	68.10	116.00
36" x 55", white finish	BC@.420	Ea	56.10	15.10	71.20	121.00
36" x 63", mill finish	BC@.420	Ea	61.40	15.10	76.50	130.00
36" x 63", white finish	BC@.420	Ea	60.20	15.10	75.30	128.00

Window Accessories

	Craft@Hrs	Unit	Material	Labor	Total	Sell
Window sills, EasySills. Solid vinyl. Set with latex or solvent base adhesive. Hard scratch-resistant surface. Ultraviolet protection to prevent yellowing. Gloss finish.						
3-1/2" x 3'	BC@.129	Ea	16.30	4.65	20.95	35.60
3-1/2" x 4'	BC@.172	Ea	20.40	6.19	26.59	45.20
3-1/2" x 5'	BC@.215	Ea	26.50	7.74	34.24	58.20
3-1/2" x 6'	BC@.258	Ea	29.60	9.29	38.89	66.10
4-1/4" x 3'	BC@.129	Ea	20.40	4.65	25.05	42.60
4-1/4" x 4'	BC@.172	Ea	25.60	6.19	31.79	54.00
4-1/4" x 5'	BC@.215	Ea	32.60	7.74	40.34	68.60
4-1/4" x 6'	BC@.258	Ea	37.70	9.29	46.99	79.90
5-1/2" x 3'	BC@.129	Ea	22.40	4.65	27.05	46.00
5-1/2" x 4'	BC@.172	Ea	29.60	6.19	35.79	60.80
5-1/2" x 5'	BC@.215	Ea	35.70	7.74	43.44	73.80
5-1/2" x 6'	BC@.258	Ea	42.80	9.29	52.09	88.60
6-1/4" x 3'	BC@.129	Ea	23.50	4.65	28.15	47.90

	Craft@Hrs	Unit	Material	Labor	Total	Sell
6-1/4" x 4'	BC@.172	Ea	32.40	6.19	38.59	65.60
6-1/4" x 5'	BC@.215	Ea	33.60	7.74	41.34	70.30
6-1/4" x 6'	BC@.258	Ea	47.40	9.29	56.69	96.40
Aluminum window mullions. White.						
38-1/4"	—	Ea	47.90	—	47.90	—
50-1/2"	—	Ea	58.10	—	58.10	—
62-7/8"	—	Ea	68.30	—	68.30	—
Replacement casement window operator handle. 9" arm. Casement opening direction when viewed from the exterior.						
Left, aluminum window	BC@.250	Ea	10.10	9.00	19.10	32.50
Right, aluminum window	BC@.250	Ea	10.10	9.00	19.10	32.50
Left, wood window	BC@.250	Ea	16.20	9.00	25.20	42.80
Right, wood window	BC@.250	Ea	16.20	9.00	25.20	42.80
Casement window lock handle. Universal casement window lock for aluminum or steel windows. Accepts a 1/8" security pin. Screws included.						
Aluminum	BC@.250	Ea	4.02	9.00	13.02	22.10
Bronze finish, 5/16" bore size	BC@.250	Ea	4.05	9.00	13.05	22.20
White	BC@.250	Ea	4.23	9.00	13.23	22.50
Awning window operator handle.						
Left	BC@.250	Ea	10.10	9.00	19.10	32.50
Right	BC@.250	Ea	10.10	9.00	19.10	32.50
Jalousie window operator handle.						
Crank handle, 5/16"	BC@.250	Ea	3.03	9.00	12.03	20.50
Crank handle, 3/8"	BC@.250	Ea	3.03	9.00	12.03	20.50
T-handle, 5/16"	BC@.250	Ea	3.03	9.00	12.03	20.50
T-handle, 3/8"	BC@.250	Ea	3.09	9.00	12.09	20.60
Wood window sash spring. Replaces broken sash cords and eliminates window rattle. Includes two sash springs and two nail fasteners.						
Per pair	BC@.750	Ea	4.56	27.00	31.56	53.70
Sash balance.						
4-1/2-pound	BC@.360	Ea	19.40	13.00	32.40	55.10
6-pound	BC@.360	Ea	20.10	13.00	33.10	56.30
8-pound	BC@.360	Ea	20.90	13.00	33.90	57.60
10-pound	BC@.360	Ea	22.90	13.00	35.90	61.00
12-pound	BC@.360	Ea	23.90	13.00	36.90	62.70
Sash lift pull.						
Oil rubbed bronze	BC@.060	Ea	10.40	2.16	12.56	21.40
Satin nickel	BC@.060	Ea	10.40	2.16	12.56	21.40
Double-hung sash lock.						
Bronze	BC@.350	Ea	9.53	12.60	22.13	37.60
Oil rubbed bronze	BC@.350	Ea	9.90	12.60	22.50	38.30
White	BC@.350	Ea	9.53	12.60	22.13	37.60
Satin nickel	BC@.350	Ea	9.70	12.60	22.30	37.90
Double-hung window bolt.						
Brass	BC@.350	Ea	2.70	12.60	15.30	26.00

	Craft@Hrs	Unit	Material	Labor	Total	Sell
Arch top louvered exterior shutters. Copolymer construction with molded-through color. Available in colors and paintable. Price per pair.						
15" x 36"	BC@.200	Ea	29.60	7.20	36.80	62.60
15" x 39"	BC@.200	Ea	32.60	7.20	39.80	67.70
15" x 43"	BC@.200	Ea	34.70	7.20	41.90	71.20
15" x 48"	BC@.200	Ea	34.70	7.20	41.90	71.20
15" x 52"	BC@.320	Ea	38.80	11.50	50.30	85.50
15" x 55"	BC@.320	Ea	42.80	11.50	54.30	92.30
15" x 60"	BC@.320	Ea	45.90	11.50	57.40	97.60
15" x 64"	BC@.320	Ea	44.90	11.50	56.40	95.90
15" x 67"	BC@.320	Ea	51.00	11.50	62.50	106.00
15" x 72"	BC@.320	Ea	47.90	11.50	59.40	101.00
15" x 75"	BC@.320	Ea	51.00	11.50	62.50	106.00
Decorative shutter hardware	—	Ea	8.30	—	8.30	—
Lock fasteners	BC@.125	Ea	5.30	4.50	9.80	16.70
Classic window guards. Welded steel 1/2" square tubes with spear points and ornamentation. Add screws and brackets below. Width by height.						
24" x 24"	BC@.500	Ea	17.30	18.00	35.30	60.00
24" x 36"	BC@.500	Ea	18.30	18.00	36.30	61.70
24" x 54"	BC@.500	Ea	28.60	18.00	46.60	79.20
30" x 54"	BC@.500	Ea	24.50	18.00	42.50	72.30
36" x 24"	BC@.500	Ea	21.40	18.00	39.40	67.00
36" x 36"	BC@.500	Ea	25.50	18.00	43.50	74.00
36" x 42"	BC@.750	Ea	25.50	27.00	52.50	89.30
36" x 48"	BC@.750	Ea	33.70	27.00	60.70	103.00
36" x 54"	BC@.750	Ea	31.60	27.00	58.60	99.60
42" x 42"	BC@.750	Ea	33.70	27.00	60.70	103.00
48" x 36"	BC@.750	Ea	33.70	27.00	60.70	103.00
48" x 48"	BC@.750	Ea	39.80	27.00	66.80	114.00
60" x 36"	BC@.750	Ea	42.80	27.00	69.80	119.00
60" x 48"	BC@.750	Ea	39.80	27.00	66.80	114.00
72" x 36"	BC@.750	Ea	39.80	27.00	66.80	114.00
1" one-way screws	—	Ea	3.59	—	3.59	—
3" one-way screws	—	Ea	4.53	—	4.53	—
4" one-way screws	—	Ea	4.53	—	4.53	—
Flush brackets	—	Ea	8.14	—	8.14	—
3" projection brackets	—	Ea	8.14	—	8.14	—
3" tee brackets	—	Ea	9.63	—	9.63	—
Connector pins	—	Ea	6.37	—	6.37	—
Aluminum window screen frame kits. For rescreening windows. White or bronze finish. Matching color corners snap into place. Add the cost of screen wire and spline. Labor includes removing the old screen, measuring, cutting and assembling the frame, measuring and cutting screen wire, pressing screen and spline into the frame channel, re-hanging the screen.						
5/16" frame, to 24" x 36" screen	BC@.750	Ea	10.20	27.00	37.20	63.20
5/16" frame, to 36" x 48" screen	BC@1.00	Ea	12.20	36.00	48.20	81.90
5/16" frame, to 48" x 60" screen	BC@1.25	Ea	12.60	45.00	57.60	97.90
7/16" frame, to 24" x 36" screen	BC@.750	Ea	10.10	27.00	37.10	63.10
7/16" frame, to 48" x 60" screen	BC@1.25	Ea	13.30	45.00	58.30	99.10

	Craft@Hrs	Unit	Material	Labor	Total	Sell
Window screens, aluminum. Aluminum mill finish frame. Fiberglass mesh. For double-hung windows. Labor includes removing the old screen and re-hanging the new screen.						
2'0" x 3'2"	BC@.350	Ea	14.30	12.60	26.90	45.70
2'4" x 3'2"	BC@.350	Ea	15.30	12.60	27.90	47.40
2'4" x 4'6"	BC@.350	Ea	17.30	12.60	29.90	50.80
2'8" x 3'1"	BC@.350	Ea	17.30	12.60	29.90	50.80
2'8" x 3'2"	BC@.350	Ea	16.30	12.60	28.90	49.10
2'8" x 4'6"	BC@.350	Ea	18.30	12.60	30.90	52.50
2'8" x 5'2"	BC@.350	Ea	19.40	12.60	32.00	54.40
2'8" x 6'2"	BC@.350	Ea	19.40	12.60	32.00	54.40
3'0" x 3'1"	BC@.350	Ea	18.50	12.60	31.10	52.90
3'0" x 3'2"	BC@.350	Ea	17.30	12.60	29.90	50.80
3'0" x 4'6"	BC@.350	Ea	19.40	12.60	32.00	54.40
3'0" x 5'2"	BC@.350	Ea	20.40	12.60	33.00	56.10
3'0" x 6'2"	BC@.350	Ea	19.40	12.60	32.00	54.40
Aluminum screen rolls. Replaces worn-out screening on windows, patio doors and screened-in porches. Bright or charcoal.						
36" x 7'	—	Ea	6.61	—	6.61	—
36" x 25'	—	Ea	20.40	—	20.40	—
48" x 7'	—	Ea	8.65	—	8.65	—
48" x 25'	—	Ea	26.50	—	26.50	—
Fiberglass screen rolls. Charcoal or gray.						
36" x 7'	—	Ea	8.14	—	8.14	—
36" x 25'	—	Ea	13.80	—	13.80	—
48" x 25'	—	Ea	17.30	—	17.30	—
60" x 25'	—	Ea	23.40	—	23.40	—
72" x 25'	—	Ea	28.50	—	28.50	—
84" x 25'	—	Ea	30.40	—	30.40	—
96" x 25'	—	Ea	32.60	—	32.60	—
Screen patch. Self-stick.						
3" x 3", aluminum	—	Ea	4.50	—	4.50	—
3" x 3", charcoal	—	Ea	4.50	—	4.50	—
Screen replacement splines. Black or gray.						
0.125" x 25'	—	Ea	4.06	—	4.06	—
0.125" x 100'	—	Ea	8.14	—	8.14	—
0.140" x 25'	—	Ea	4.06	—	4.06	—
0.140" x 100'	—	Ea	5.59	—	5.59	—
0.160" x 25'	—	Ea	4.06	—	4.06	—
0.160" x 100'	—	Ea	8.14	—	8.14	—
0.175" x 25'	—	Ea	4.06	—	4.06	—
0.175" x 100'	—	Ea	8.14	—	8.14	—
0.190" x 25'	—	Ea	4.06	—	4.06	—
0.190" x 100'	—	Ea	8.65	—	8.65	—
0.210" x 25'	—	Ea	4.65	—	4.65	—
0.210" x 100', flat	—	Ea	7.64	—	7.64	—
0.220" x 25'	—	Ea	4.06	—	4.06	—
0.220" x 100'	—	Ea	10.20	—	10.20	—
0.250" x 25'	—	Ea	3.56	—	3.56	—
Roller spline tool.						
Plastic handle	—	Ea	2.54	—	2.54	—

	Craft@Hrs	Unit	Material	Labor	Total	Sell
Tilt sash window grille sets. Removable window grille system fits double-hung wood windows.						
2'0" x 3'2", 4 x 4 grille	BC@.250	Ea	12.80	9.00	21.80	37.10
2'4" x 3'2", 6 x 6 grille	BC@.250	Ea	19.60	9.00	28.60	48.60
2'4" x 4'6", 6 x 6 grille	BC@.250	Ea	19.60	9.00	28.60	48.60
2'8" x 3'2", 6 x 6 grille	BC@.250	Ea	19.60	9.00	28.60	48.60
2'8" x 3'10", 6 x 6 grille	BC@.250	Ea	19.60	9.00	28.60	48.60
2'8" x 4'6", 6 x 6 grille	BC@.250	Ea	19.60	9.00	28.60	48.60
2'8" x 5'2", 6 x 6 grille	BC@.250	Ea	19.60	9.00	28.60	48.60
2'8" x 6'2", 9 x 9 grille	BC@.250	Ea	22.50	9.00	31.50	53.60
3'0" x 3'2", 6 x 6 grille	BC@.250	Ea	19.60	9.00	28.60	48.60
3'0" x 3'10", 6 x 6 grille	BC@.250	Ea	19.60	9.00	28.60	48.60
3'0" x 4'6", 6 x 6 grille	BC@.250	Ea	19.60	9.00	28.60	48.60
3'0" x 5'2", 6 x 6 grille	BC@.250	Ea	19.60	9.00	28.60	48.60

Window Glazing

	Craft@Hrs	Unit	Material	Labor	Total	Sell
Single-strength replacement glass.						
8" x 10", clear	BG@.246	Ea	1.65	8.36	10.01	17.00
9" x 12", clear	BG@.333	Ea	1.38	11.30	12.68	21.60
10" x 12", clear	BG@.370	Ea	1.49	12.60	14.09	24.00
12" x 14", clear	BG@.280	Ea	1.71	9.51	11.22	19.10
12" x 16", clear	BG@.322	Ea	2.78	10.90	13.68	23.30
12" x 18", clear	BG@.363	Ea	2.24	12.30	14.54	24.70
12" x 24", clear	BG@.296	Ea	4.77	10.10	14.87	25.30
12" x 30", clear	BG@.370	Ea	3.14	12.60	15.74	26.80
12" x 36", clear	BG@.444	Ea	4.98	15.10	20.08	34.10
12" x 36", obscure	BG@.444	Ea	6.12	15.10	21.22	36.10
14" x 20", clear	BG@.369	Ea	2.64	12.50	15.14	25.70
16" x 20", clear	BG@.329	Ea	4.01	11.20	15.21	25.90
16" x 28", clear	BG@.460	Ea	3.80	15.60	19.40	33.00
16" x 32", clear	BG@.525	Ea	4.39	17.80	22.19	37.70
16" x 36", clear	BG@.592	Ea	5.18	20.10	25.28	43.00
18" x 24", clear	BG@.444	Ea	3.24	15.10	18.34	31.20
18" x 36", clear	BG@.666	Ea	6.34	22.60	28.94	49.20
18" x 36", obscure	BG@.666	Ea	15.40	22.60	38.00	64.60
18" x 52", clear	BG@.962	Ea	9.36	32.70	42.06	71.50
20" x 24", clear	BG@.492	Ea	4.76	16.70	21.46	36.50
24" x 30", clear	BG@.555	Ea	7.52	18.90	26.42	44.90
24" x 36", clear	BG@.666	Ea	7.52	22.60	30.12	51.20
24" x 52", clear	BG@.961	Ea	12.20	32.60	44.80	76.20
28" x 36", clear	BG@.777	Ea	8.48	26.40	34.88	59.30
30" x 32", clear	BG@.739	Ea	7.97	25.10	33.07	56.20
30" x 36", clear	BG@.833	Ea	13.20	28.30	41.50	70.60
32" x 40", clear	BG@.985	Ea	10.80	33.50	44.30	75.30
36" x 48", clear	BG@1.30	Ea	17.20	44.20	61.40	104.00
Jalousie glass. Replacement horizontal panes or slats for a jalousie window.						
4" x 36-1/4", clear	BG@.200	Ea	5.49	6.79	12.28	20.90
4-1/2" x 36-1/4", clear	BG@.200	Ea	5.40	6.79	12.19	20.70
4" x 24-7/32", obscure	BG@.200	Ea	4.47	6.79	11.26	19.10
4" x 36-7/32", obscure	BG@.200	Ea	5.49	6.79	12.28	20.90
4-1/2" x 36-7/32", obscure	BG@.200	Ea	6.60	6.79	13.39	22.80

	Craft@Hrs	Unit	Material	Labor	Total	Sell
Remove and replace glazing putty. Using the existing glass pane. Heat old putty with a heat gun. Scrape out old putty. Refill the channel with putty.						
Pane to 4 square feet	BG@.250	Ea	—	8.49	8.49	14.40
Add per square foot over 4 SF	BG@.063	SF	—	2.14	2.14	3.64
Measure and cut glass to size. Single- or double-strength window glass.						
Per pane, to 4 square feet	BG@.167	Ea	—	5.67	5.67	9.64
Add for panes over 4 SF, per SF	BG@.042	SF	—	1.43	1.43	2.43
Remove and replace window glass set in putty. Remove sash. Hack out glass fragments. Heat old putty with a heat gun. Scrape out old putty. Pull glazing points. Set glass in putty bead. Reset glazing points. Fill channel with putty.						
Remove sash from frame and replace	BG@.400	Ea	—	13.60	13.60	23.10
Hack out glass fragments and putty	BG@.250	Ea	—	8.49	8.49	14.40
Set glass and putty, pane to 4 square feet	BG@.250	Ea	1.50	8.49	9.99	17.00
Set glass and putty, per SF over 4 SF	BG@.063	SF	.50	2.14	2.64	4.49
Remove window glass set in vinyl stops. Remove sash. Hack out glass fragments. No glass included. Per pane, width plus height. For example, the width plus height of a window pane measuring 24" x 36" is 60".						
To 40" width plus height	BG@.580	Ea	—	19.70	19.70	33.50
41" to 50" width plus height	BG@.640	Ea	—	21.70	21.70	36.90
51" to 60" width plus height	BG@.710	Ea	—	24.10	24.10	41.00
61" to 70" width plus height	BG@.900	Ea	—	30.60	30.60	52.00
81" to 90" width plus height	BG@1.00	Ea	—	34.00	34.00	57.80
Over 91" width plus height	BG@1.10	Ea	—	37.40	37.40	63.60
Set window glass in vinyl stops. No glass included. Add the cost of new vinyl bead, if required. Per pane, width plus height. For example, the width plus height of a window pane measuring 24" x 36" is 60".						
To 40" width plus height	BG@.580	Ea	—	19.70	19.70	33.50
41" to 50" width plus height	BG@.640	Ea	—	21.70	21.70	36.90
51" to 60" width plus height	BG@.710	Ea	—	24.10	24.10	41.00
61" to 70" width plus height	BG@.900	Ea	—	30.60	30.60	52.00
81" to 90" width plus height	BG@1.00	Ea	—	34.00	34.00	57.80
Over 91" width plus height	BG@1.10	Ea	—	37.40	37.40	63.60
Vinyl glazing bead.						
7/32" x 6', brown	—	Ea	1.83	—	1.83	—
7/32" x 6', gray	—	Ea	1.83	—	1.83	—
7/32" x 6', white	—	Ea	1.83	—	1.83	—
Glazing compound, 33®, DAP®. For face-glazing wood or metal window frames. Weather resistant. Smooth and ready mixed.						
1/2 pint	—	Ea	3.74	—	3.74	—
Pint	—	Ea	4.98	—	4.98	—
Quart	—	Ea	6.60	—	6.60	—
Gallon	—	Ea	22.00	—	22.00	—
Glazing compound, For face-glazing metal sash. Requires no painting. Gray.						
Quart	—	Ea	6.60	—	6.60	—
8 oz	—	Ea	3.74	—	3.74	—
Latex glazing compound, DAP®. Resists cracking, sagging, and chalking.						
10.3-ounce tube	—	Ea	5.59	—	5.59	—
Glazing points. Secures glass to sash before glazing putty is applied. Push-T style.						
Bundle	—	Ea	2.01	—	2.01	—

	Craft@Hrs	Unit	Material	Labor	Total	Sell
Wheel glass cutter.						
Steel wheel	—	Ea	3.90	—	3.90	—

Glazing tools. 2-in-1 glazing tool has a slotted V-blade on one end to apply a smooth strip of putty. The other end has a heavy-duty 1-1/4" chisel edge to remove old putty and drive push-style glazing points.

	Craft@Hrs	Unit	Material	Labor	Total	Sell
2-in-1 glazing tool	—	Ea	5.50	—	5.50	—
Pushmate for glazing points	—	Ea	11.50	—	11.50	—
Acrylic sheet cutter	—	Ea	3.50	—	3.50	—
6" nipping/running pliers	—	Ea	13.75	—	13.75	—

Glass Substitutes

Acrylic sheet.

	Craft@Hrs	Unit	Material	Labor	Total	Sell
.093", 11" x 14"	—	Ea	4.32	—	4.32	—
.093", 18" x 24"	—	Ea	9.98	—	9.98	—
.093", 20" x 32"	—	Ea	15.30	—	15.30	—
.093", 24" x 48"	—	Ea	26.50	—	26.50	—
.093", 30" x 36"	—	Ea	24.00	—	24.00	—
.093", 30" x 60"	—	Ea	43.60	—	43.60	—
.093", 36" x 48"	—	Ea	35.70	—	35.70	—
.118", 36" x 72"	—	Ea	67.30	—	67.30	—
.230", 48" x 96"	—	Ea	157.00	—	157.00	—
.236", 18" x 24"	—	Ea	20.40	—	20.40	—
.236", 24" x 48"	—	Ea	59.10	—	59.10	—
.236", 30" x 36"	—	Ea	56.10	—	56.10	—

Fabback acrylic mirror sheet. Ideal for ceilings, doors, children's rooms.

	Craft@Hrs	Unit	Material	Labor	Total	Sell
.118", 24" x 48"	—	Ea	59.40	—	59.40	—

Polycarbonate sheet Plexiglas®, Lexan® XL-10. Impact-strength and long service life. UV-resistant surface. Suitable for vertical-glazing and overhead applications.

	Craft@Hrs	Unit	Material	Labor	Total	Sell
.093" thick, 18" x 24"	—	Ea	20.40	—	20.40	—
.093" thick, 30" x 36"	—	Ea	38.90	—	38.90	—
.093" thick, 32" x 44"	—	Ea	43.10	—	43.10	—
.093" thick, 36" x 48"	—	Ea	93.00	—	93.00	—
.093" thick, 36" x 72"	—	Ea	102.00	—	102.00	—

White silicon plastic adhesive.

	Craft@Hrs	Unit	Material	Labor	Total	Sell
White adhesive	—	Ea	8.50	—	8.50	—

Window Components

Sliding window track lock with steel locking plate. For horizontal or vertical metal sliding windows.

	Craft@Hrs	Unit	Material	Labor	Total	Sell
Aluminum key lock	BC@.250	Ea	8.95	9.00	17.95	30.50
Aluminum lever lock	BC@.250	Ea	3.35	9.00	12.35	21.00
Aluminum sash lock	BC@.250	Ea	3.54	9.00	12.54	21.30
White lever lock	BC@.250	Ea	3.34	9.00	12.34	21.00

Sliding window thumbscrew bar lock. Anti-lift-out bar. Securely locks window in closed or ventilating position.

	Craft@Hrs	Unit	Material	Labor	Total	Sell
Reversible, aluminum	BC@.250	Ea	4.75	9.00	13.75	23.40
Reversible, white	BC@.250	Ea	5.07	9.00	14.07	23.90

Wood window flip lock. Mounts on vertical track. Stops double-hung window in vent position. Wedge flips out of the way to allow opening. Pack of two.

	Craft@Hrs	Unit	Material	Labor	Total	Sell
Brass	BC@.250	Ea	2.52	9.00	11.52	19.60

	Craft@Hrs	Unit	Material	Labor	Total	Sell
Window weatherstrip spring brass seal sets. Peel 'N Stick®, per linear foot per side.						
Double-hung window, head and sill	BC@.008	LF	1.65	.29	1.94	3.30
Add to remove and replace window sash	BC@.250	Ea	—	9.00	9.00	15.30
Casement window, head and sill	BC@.008	LF	.85	.29	1.14	1.94
Gray or black pile gasket	BC@.008	LF	.65	.29	.94	1.60
Gray or black pile gasket with adhesive	BC@.006	LF	.95	.22	1.17	1.99

Specialty Windows and Skylights

Window well, ScapeWEL. Meets code requirements for basement egress window. Snaps into place on-site. For new construction and remodeling, attaches directly to foundation. Two-tier. 41" projection from foundation. 48" high.

	Craft@Hrs	Unit	Material	Labor	Total	Sell
42" wide window, 3' window buck	B1@1.50	Ea	911.00	48.80	959.80	1,630.00
Polycarbonate well cover	—	Ea	407.00	—	407.00	—
54" wide window, 4' window buck	B1@1.50	Ea	841.00	48.80	889.80	1,510.00
Polycarbonate well cover	—	Ea	429.00	—	429.00	—
66" wide window, 5' window buck	B1@1.50	Ea	1,140.00	48.80	1,188.80	2,020.00
Polycarbonate well cover	—	Ea	477.00	—	477.00	—

Tubular skylights with install kit. Acrylic dome on the roof reflects sunlight down a light shaft, providing natural light for bathrooms and hallways. Use a 10"-diameter tube for rooms to 150 square feet and a 14"-diameter tube for rooms to 300 square feet. Use the composition flashing kit for asphalt shingle, slate, or wood shake roofs. Aluminum flashing kit is suitable for other surfaces with pitch from level to 12 in 12. Adjustable tubes make it easy to maneuver around obstacles. Kit contains material to complete a 48" installation from roof surface to interior ceiling surface. Fits between 16" or 24" on center rafters.

	Craft@Hrs	Unit	Material	Labor	Total	Sell
10" tube, aluminum flash kit	B1@3.00	Ea	203.00	97.60	300.60	511.00
10" tube, composition flash kit	B1@3.00	Ea	168.00	97.60	265.60	452.00
10" tube, tile or flat roof	B1@3.00	Ea	206.00	97.60	303.60	516.00
14" tube, aluminum flash kit	B1@3.00	Ea	305.00	97.60	402.60	684.00
14" tube, composition flash kit	B1@3.00	Ea	254.00	97.60	351.60	598.00
14" tube, tile or flat roof	B1@3.00	Ea	309.00	97.60	406.60	691.00

Remove an existing poly dome skylight. No salvage of materials. Includes disposal.

	Craft@Hrs	Unit	Material	Labor	Total	Sell
To 10 square feet	BL@2.65	Ea	—	77.00	77.00	131.00
Add per square foot over 10 square feet	BL@.210	SF	—	6.10	6.10	10.40

Cut roof opening for a skylight. Cut shingles and sheathing, install L-flashing. Self-flashing flush-mount skylights can be installed directly on a roof with a pitch of 3 in 12 or greater. Curb-mounted skylights require a roof curb. Add the cost of roof framing (doubled rafters and two headers) if required. Add the cost of framing a light well through the attic, if required. Add the cost of cutting and finishing a hole in the ceiling (including doubled joists and two headers) if required.

	Craft@Hrs	Unit	Material	Labor	Total	Sell
Measure and cut roof opening for skylight	BC@3.50	Ea	—	126.00	126.00	214.00
Frame roof curb	BC@2.00	Ea	22.50	72.00	94.50	161.00
Curb flashing for skylight	BC@1.25	Ea	87.50	45.00	132.50	225.00

Fixed flush-mount skylights, Velux. Low-E insulating laminated and tempered glass. Wood interior frame. Exterior aluminum cladding covers gaskets and sealants. Pre-mounted brackets allows adjustment for high or low profile roofing materials

	Craft@Hrs	Unit	Material	Labor	Total	Sell
15-5/16" x 46-3/8"	B1@3.42	Ea	218.00	111.00	329.00	559.00
21-1/2" x 27-1/2"	B1@3.42	Ea	180.00	111.00	291.00	495.00
21-1/2" x 38-1/2"	B1@3.42	Ea	226.00	111.00	337.00	573.00
21-1/2" x 46-3/8"	B1@3.42	Ea	251.00	111.00	362.00	615.00
30-5/8" x 38-1/2", Type 74	B1@3.42	Ea	268.00	111.00	379.00	644.00
30-5/8" x 38-1/2", Type 75	B1@3.42	Ea	217.00	111.00	328.00	558.00
30-5/8" x 46-7/8"	B1@3.42	Ea	306.00	111.00	417.00	709.00
30-5/8" x 55"	B1@3.42	Ea	342.00	111.00	453.00	770.00

	Craft@Hrs	Unit	Material	Labor	Total	Sell
Fixed curb-mount skylights, Velux. Low-E insulating laminated and tempered glass. Adapts to curb size and roofing material. Anodized aluminum exterior frame with gasket. UV-resistant ABS inner frame.						
22-1/2" x 22-1/2"	B1@3.00	Ea	160.00	97.60	257.60	438.00
22-1/2" x 46-1/2"	B1@3.42	Ea	210.00	111.00	321.00	546.00
22-1/2" x 46-1/2", impact glass	B1@3.42	Ea	327.00	111.00	438.00	745.00
46-1/2" x 46-1/2"	B1@3.42	Ea	348.00	111.00	459.00	780.00
Venting flush-mount glass skylights, Velux. Low-E tempered insulating glass. Wood interior frame. Aluminum cladding. Thru-screen allows skylight operation without removing screen. Stainless steel scissor operator allows easy opening for maximum ventilation. Removable, integrated insect screen.						
21-1/2" x 27-1/2"	B1@3.00	Ea	419.00	97.60	516.60	878.00
21-1/2" x 38-1/2"	B1@3.42	Ea	455.00	111.00	566.00	962.00
21-1/2" x 46-3/8"	B1@3.42	Ea	492.00	111.00	603.00	1,030.00
30-1/2" x 39"	B1@3.52	Ea	497.00	115.00	612.00	1,040.00
30-1/2" x 55-1/2"	B1@3.52	Ea	600.00	115.00	715.00	1,220.00
44-3/4" x 46-3/8	B1@3.52	Ea	665.00	115.00	780.00	1,330.00
Venting flush-mount laminated glass skylights, Velux. Low-E tempered insulating glass. Wood interior frame. Aluminum cladding. Thru-screen allows skylight operation without removing screen. Stainless steel scissor operator allows easy opening for maximum ventilation. Removable, integrated insect screen.						
21-1/2" x 27-1/2"	B1@3.00	Ea	443.00	97.60	540.60	919.00
21-1/2" x 38-1/2"	B1@3.42	Ea	490.00	111.00	601.00	1,020.00
21-1/2" x 46-3/8"	B1@3.42	Ea	534.00	111.00	645.00	1,100.00
30-5/8" x 38-1/2"	B1@3.42	Ea	562.00	111.00	673.00	1,140.00
30-5/8" x 46-3/8", Type 75	B1@3.52	Ea	296.00	115.00	411.00	699.00
30-5/8" x 46-3/8", Type 74	B1@3.52	Ea	536.00	115.00	651.00	1,110.00
30-5/8" x 55"	B1@3.52	Ea	553.00	115.00	668.00	1,140.00
44-3/8" x 46-3/8"	B1@3.52	Ea	752.00	115.00	867.00	1,470.00
Telescoping control rod, 6' to10'	—	Ea	45.60	—	45.60	—
Fixed skylights. Folded corners. Watertight seal. 360-degree thermal break frame.						
2' x 2', acrylic	B1@3.00	Ea	111.00	97.60	208.60	355.00
2' x 2', glass	B1@3.00	Ea	115.00	97.60	212.60	361.00
2' x 4', acrylic	B1@3.42	Ea	140.00	111.00	251.00	427.00
2' x 4', glass	B1@3.42	Ea	152.00	111.00	263.00	447.00
4' x 4', glass	B1@3.42	Ea	250.00	111.00	361.00	614.00

Doors

9

The front entry door is the most important door in any home. It's usually visible from the street and always makes a first impression on visitors. Whether new or old, the front door should both look good and complement the style of the house.

If only cosmetic repairs are needed to an original door, that's the place to start. If the door is original, it's probably in perfect harmony with the exterior décor. Repairing the door saves the time and expense of finding exactly the right replacement door and hauling away the old door. Finding historically correct replacement doors for older homes can be difficult, and they're likely to be expensive. But if the door is already a replacement, then replacing it again probably won't do any harm.

How can you tell if the entry door is original? Replacement doors can be beautiful, but seldom harmonize with the home décor. For example, a light-colored oak flush door (popular in the 1960s) installed on a home built in the 1920s is an obvious replacement. Hanging a 21st century door doesn't modernize a 19th century home.

Repair or Replace

Balance the cost of repair against the cost of buying and hanging a new door of similar quality. Obviously, repair makes sense when the cost is substantially less than the cost of replacement. Figure 9-1 shows extra-wide French doors set with thick, beveled glass. These doors were very dirty and the bottoms scarred and damaged. Brushing them with detergent removed the grime, and the damage on the bottoms was taken care of by simply covering them with brass kickplates. Two hours of work and a $50 investment salvaged doors that might cost thousands to duplicate.

Figure 9-1
Repaired French doors

The most common problem with entry doors is a deteriorated finish. Peeling paint or alligatored varnish will turn an otherwise attractive door into an eyesore. A bad finish alone isn't an adequate reason to replace an entry door. If the door has a varnish finish that has darkened, wash it before you make a decision about stripping the finish. If the door has been waxed, the finish may be nothing more than dirty. Wax collects dirt and darkens as it ages. That makes the door ugly. But dirty wax also helps protect the finish below. You may find a perfect gem under that wax.

Scrub a small area with a powerful detergent such as Formula 409® or Fantastic®. If you see the color changing dramatically, dirt and wax buildup could be the source of the problem. In that case, it's worth cleaning the entire door. Refinishing may not be needed.

Refinishing an Entry Door

If a door has to be refinished, a heat gun is the best tool for stripping the finish. Heat softens the old paint or varnish, making it easy to scrape off. Stripping like this is easier if you lay the door flat across two sawhorses. Once the paint is off, you can apply either paint or varnish. If it's a quality wood door, varnishing will bring out the beauty of the wood.

If you elect to varnish, begin by washing the door with acetone to remove residue that escaped the heat gun. Residue won't show through paint, but it will show through varnish. When using acetone, wear rubber or neoprene gloves to avoid skin irritation, and work in a well-ventilated area to reduce the risk from fumes.

Pour some acetone into a jar with a screw-on lid. Dip a piece of extra-fine steel wool in the acetone (synthetic steel wool may be an even better choice). Then gently scrub the door surface. Work only on a few square inches at a time. Don't scrub hard; let the acetone do the work. You'll see residue dissolve and the original wood surface appear. After a few minutes of scrubbing, the steel wool will get clogged with old paint or varnish. Rinse the steel wool in acetone. Then continue scrubbing. Be sure to keep the jar of acetone capped between dips. Acetone evaporates very quickly.

Before applying varnish, examine the wood tone. Do you like the color? If it's too light, you might want to stain it. If it's too dark, consider bleaching. Light sanding may also lighten the tone a little. Expect to spend about six hours on a door that requires paint stripping, acetone wash and a coat of varnish. But when you're done, the door will probably be as good as new.

If the door is damaged, painting it may be a better choice. Paint will hide repairs that varnish and stain won't. Varnish will work over minor repairs, such as nail holes filled with wood putty, but larger repairs, like a filled-in hole from an old deadbolt, will show through varnish. Wood putty doesn't absorb stain the same way wood does.

If staining the door is essential, buy colored wood putty that matches the stain selected. The patch will still be visible, but not as apparent.

Antiquing

An antique finish is a painted-on wood grain. When done correctly, the results can be surprisingly believable. You have to look very closely to notice that the finish isn't actually wood grain. Antiquing kits come with complete instructions and all the materials needed, including a special tool for making the fake wood grain.

Antiquing is time-consuming. Any antiqued surface requires three coats: a base coat, the antique wood-grain coat, and a finish coat of varnish. Each coat has to dry overnight before the next coat is applied. Figure about three hours to antique an entry door, not including any other repairs.

Antique finishes are fragile and not recommended for high traffic surfaces. But if the door is too damaged for varnish and painting isn't an option, antiquing will replicate the look and feel of wood grain.

Repairing Holes

Holes are common in older doors. They range from small nail holes, such as used to hang holiday decorations, to abandoned bore holes as big as your fist. Holes of up to about 1/2" can be filled with wood putty. From a foot or two away, the repair won't be noticeable. Repairing larger holes requires a different strategy.

Don't consider a door ruined just because an amateur butchered it trying to install a security lock — or a thief smashed it trying to defeat the same lock. Large holes are fairly easy to fix if you know how. Even a ragged hole the size of a softball can be fixed.

Rather than replace the door, buy a gallon of auto body filler, such as Bondo®. It's strong, weatherproof, workable, paintable, and cheap. A gallon goes a long way. Auto body filler is a gooey, clay-like material. The hardener comes in a separate tube. Mix body filler and hardener on a flat surface such as a piece of scrap wood or heavy cardboard. Do the mixing with a putty knife. Be careful not to get Bondo on your skin, as it burns. And keep anything valuable that's nearby well protected, as it's almost impossible to remove once it's hardened. The more hardener you add to the mix, the faster it hardens. Add too much hardener and it sets in seconds.

Body filler is semi-liquid and will ooze and run when applied to a vertical surface. So lay the door down flat across two sawhorses. If the hole you're repairing goes completely through the door, form the underside of the patch with a piece of cardboard. The form material will bond to the door, so if you use wood for a form, you'll end up having to chisel it off. Cardboard will easily peel off the door. Then you can just sand away the residue.

Fill the hole completely with the first layer of body filler. If you have hills and valleys, let the filler harden, then fill in the valleys. In auto body work, three layers are considered the minimum. When repairing doors, two coats will usually do the job.

Use a Surform plane to smooth out hardened body filler. A Surform plane has hundreds of little cutting teeth. Each tooth trims away a little material. You can work at any angle and not have to worry about cutting too deep. The Surform cuts away hills and blobs in seconds, leaving a flat, although rough, surface. When you're done planing, the surface will look like someone clawed the material with fingernails. Use a power sander with coarse sandpaper to smooth out the marks left by the Surform. The result will be a smooth, strong, solid surface.

Figure 9-2

Door has been patched with Bondo®

You'll have to paint this type of repair. It won't take a stain. However, once the door is painted, the repair will be completely invisible. Don't try to bore or drill through the body filler to install a lockset. It's likely to break loose. Bore for a new lockset either above or below the repair, never in exactly the same place.

If you have to install the lock at the same height, turn the door around (outside to inside) so the hinge edge becomes the lock edge and vice versa. Since the hinge edge has never been bored for a lock, you'll have clean wood to work with. The problem, of course, is that the hinge edge has been mortised for hinges. Fill the old hinge mortise cuts with body filler and make new mortise cuts for hinges on the other edge. This adds about a half hour to the task.

Body filler comes in handy when replacing a mortise lock with a modern cylinder lockset. You can buy a lock conversion kit for this. The kit includes an escutcheon plate that fastens over and covers the old mortise hole. The new cylinder lockset mounts in the plate. Unfortunately, these plates aren't very strong and don't offer much protection against forced entry. Although it's more work to plug the mortise void with body filler and bore for a new circular lockset, the finished product is stronger and far more attractive.

Figure 9-2 shows a door repaired with Bondo® after it was kicked in. The old mortise lock burst out, leaving a huge, ragged hole. Figure 9-3 shows the same door, flipped inside to out, painted, and hung with new hinges and lockset. The complete job, including repair, painting, and hanging, took approximately four hours and was much less expensive than a new door.

Figure 9-3

Repaired door, painted and installed

Rebuilding an Entry Door

Most wood doors are assembled from panels. Figure 9-4 shows an exploded view of door panels. The top and bottom rails are each one piece, and each side is one panel. Each divider between panels is a single piece. The more panels a door has, the more pieces there are. All these pieces are held together with glue. Weather and moisture cause the glue to deteriorate, usually along the lower half of the door.

If an entry door is coming apart but is otherwise worthy of repair, you'll have to disassemble the door and rebuild it. Estimate the time required at about six hours for most doors. The door will be off the hinges for a day or two, so you'll need to board up the opening.

Rebuilding a door with bad joints is a lot like repairing an old piece of furniture that has come unglued. First, take it completely apart. Pry the joints apart carefully, working from the edges to minimize tool marks on the more visible door parts. Most joints will come apart easily. All joints have small, interlocking pieces that are glued together. Try not to break these joints. If a joint can't be pried apart easily, look for small nails driven into the joint at right angles. Remove these nails very carefully to avoid damaging the joint.

Outside rails are the hardest to get apart. That's where the strength of the door is — or rather, was. Once you get the outside rails apart, the center panels and rails will usually fall right out.

Usually, most of the glue in failing joints has turned to powder. In order for the new glue to form a tight bond in the joint, you'll need to thoroughly clean out all the old, powdery glue. Wipe the visible powder away; then remove any remaining glue residue with a sanding disc. You may need to do a little hand sanding where the sanding disc can't reach.

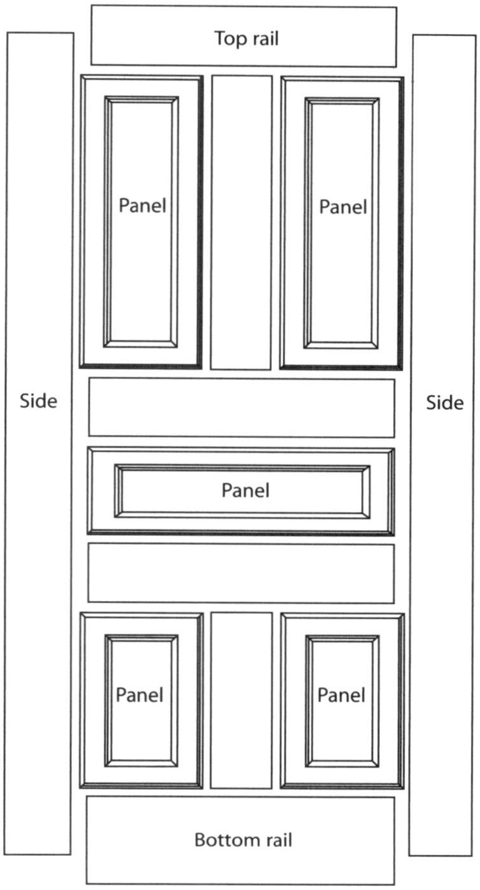

Figure 9-4
Diagram of a panel door

Reassembling the Door

When you've removed the old glue, reglue and reassemble the door. Lay the door flat and make sure that all the corners are straight and true. Once the glue has set, you won't be able to straighten anything. Use a good carpenter's glue and bar clamps to hold the joints firmly in place. Be careful to wipe up any glue drips — once dry, these are hard to remove. Leave the door undisturbed overnight.

If any joints were damaged in the disassembly process, add strength by driving screws through the joint. Use long, thin 8" to 10" wood screws. Working from the edge of the door, pre-drill a hole through the weak joint. The screw should be long enough to go through one rail and into the next. Dip the screw in glue and drive it snug. Since you're driving from the edge of the door, the screw heads will only show when the door is open. Adding screws will substantially strengthen the door.

Framing and Replacing Exterior Doors

Entry doors usually measure 6'8" high by 3' wide by 1^3/$_4$" thick. Side and rear doors should be at least 2'6" wide. Rough opening height should be the height of the door plus 2^1/$_4$" above the finish floor. The width of the rough opening should be the width of the door plus 2^1/$_2$". For a 3' door opening, make the header out of two lengths of 2" x 6" set with two 16d nails driven into the studs on each side. If the stud space on each side of the door isn't accessible, toenail the header to studs. Cripple studs (door bucks) support the header on each side of the opening.

If you're framing a new door opening, install a prehung door assembly complete with jamb, hinges and threshold. Prehung doors cost a little more than slab doors, but reduce the labor cost.

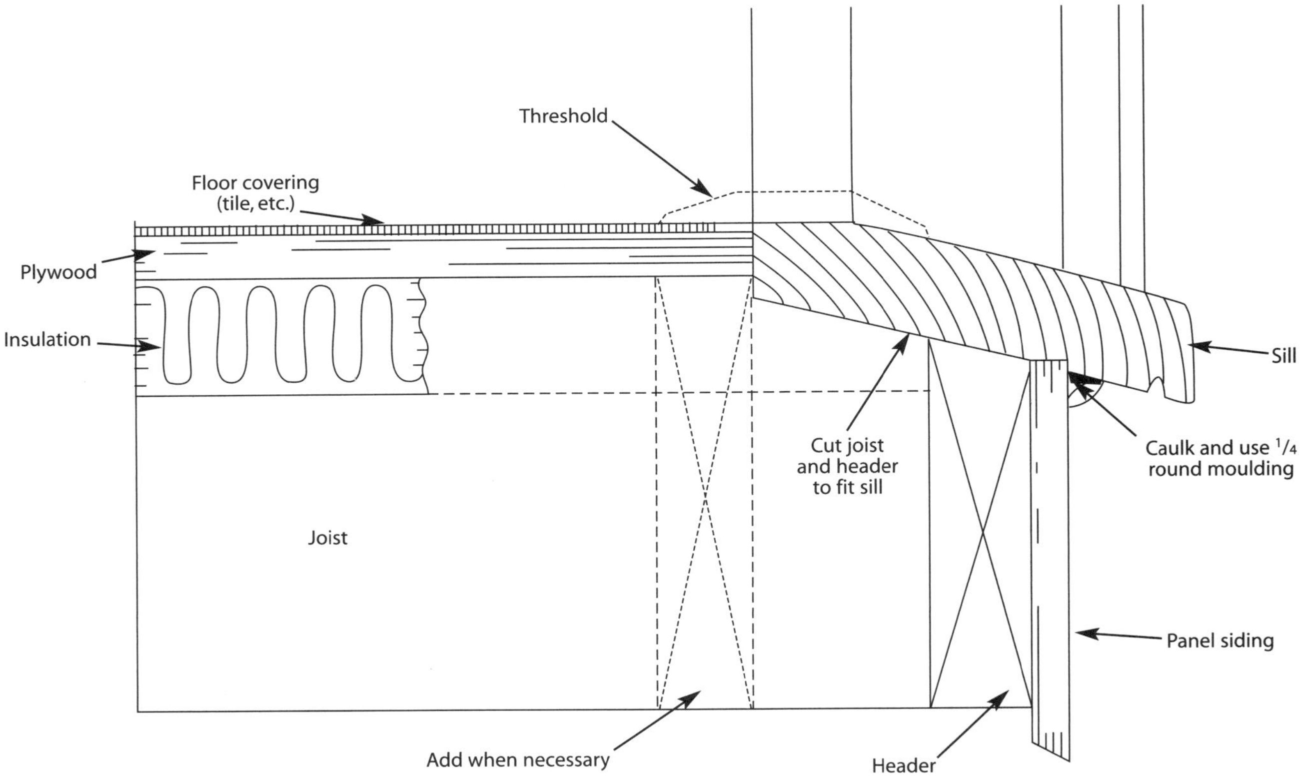

Figure 9-5
Door installation at sill

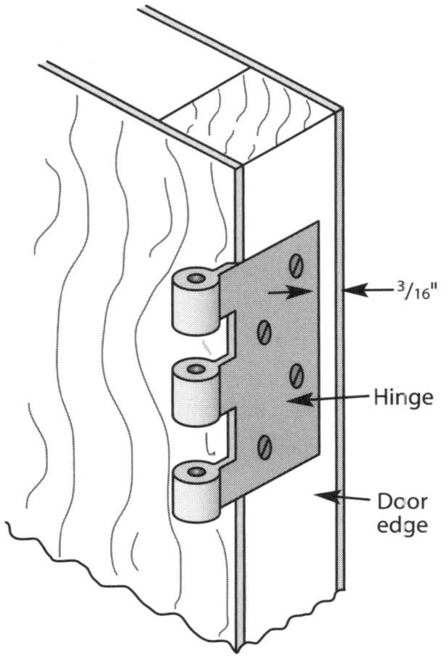

Figure 9-6
Installation of door hinges

In a new installation, you may have to trim floor joists and the joist header to receive the sill. The top of the sill should be the same height as the finish floor so the threshold can be installed over the joint. See Figure 9-5. Shim the sill when necessary so it's fully supported by floor framing. When joists run parallel to the sill, install headers and a short support member at the edge of the sill. Install the threshold over the joint by nailing with finishing nails.

When the opening is ready for installation, apply a ribbon of caulking sealer down each side and above the door opening. This caulk should seal the gap between the wall and the door casing. Set the door frame in the opening and secure it by nailing through the side and head casing. Nail the hinge side first.

Exterior doors should have a solid core. Hollow-core doors offer little security and warp too much during cold weather to use on the exterior. If a door has to be trimmed to fit the frame, cut on the hinge edge. Mortise hinges into the edge of the door with about 3/16" or 1/4" back spacing. See Figure 9-6. Use 3 1/2" by 3 1/2" loose-pin hinges on exterior doors that swing in. Use non-removable pins on out-swinging doors. Use three hinges (rather than two) on solid core doors to minimize warping.

Figure 9-7 shows standard door clearances. Measure the opening width and plane the edge for proper side clearances. Slightly bevel each door edge that fits against the stop. If necessary, square and trim the top of the door for proper fit. In cold climates, exterior doors should have weather seals.

Interior Doors

Most interior doors get heavy use. But they don't have many moving parts and tend to be durable. Sticking or binding with the jamb and failure to latch are the most common problems. Sticking or binding will rarely be the result of heavy use or abuse. The most common cause is movement in the building frame. Settling at the foundation, shrinkage of framing lumber or heavy floor loads can warp the frame out of square and cause a door to bind. In most cases, minor adjustments will solve the problem.

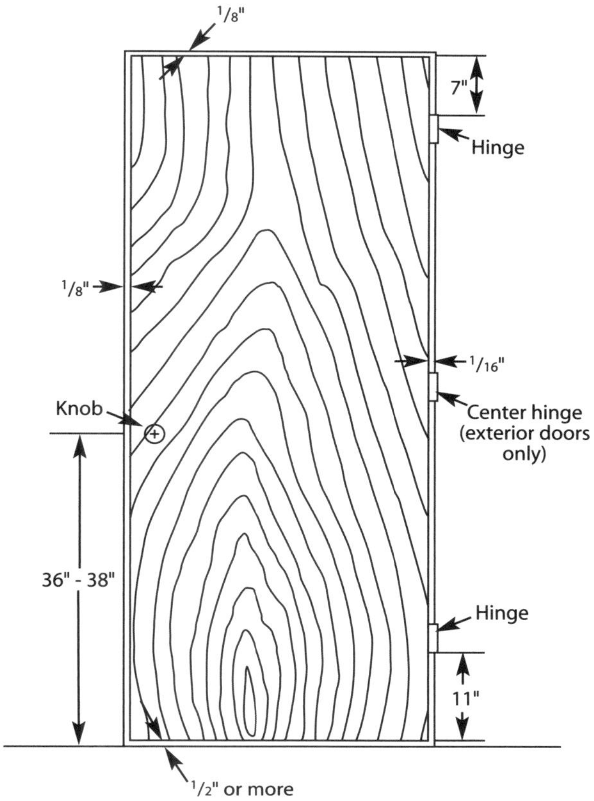

Figure 9-7
Door clearances

To remedy a binding door, first find the point of contact. You'll usually see scuff marks where the door and jamb are rubbing together. Here are some simple repairs.

- *Binding on the top or lock edge.* Use a hand plane to remove 1/8" of door where you see scuff marks. Some paint touch-up will be needed where the door edge was sanded or planed off.
- *Binding on the hinge side.* The hinges may be routed too deeply, causing the door to "bounce" open if the latch doesn't catch. Loosen the hinge leaf enough to add a shim under the leaf. Then re-tighten the leaf screws.
- *Scraping at the bottom edge.* The frame may be out of square, or a thicker floor cover has been installed. Plane or saw off enough door edge to provide floor clearance.
- *The latch doesn't close — interior door.* Remove the strike plate and adjust the position slightly or shim it out slightly. To replace the latch, you may need longer screws. Before replacing the old screws, try inserting a filler (such as a wood matchstick) into the old screw hole. Then drive the screw in the same hole. If the latch is secure, you're done.

Doors will also bind when they absorb excess moisture during humid weather. This is a symptom of the door not being completely sealed at the edges. Sealing the edges with paint or varnish should solve the problem.

Making Changes

Adding or moving a door can make navigating a home more convenient. Traffic through a room should be at the edges, not through the middle. Many older homes have doors centered in a wall, which cuts the wall space in two and makes furniture arrangement difficult. Moving a door opening from the middle of a wall space to the corner of the room will usually add livability to a home. See Figure 9-8. In some cases, simply closing up a doorway is the best choice. If you relocate a door to a corner of the room, be sure the new door swings against the nearest wall (rather than into the open room) and away from the light switch (so the switch is just inside the doorway). Also keep in mind that entire walls (affecting each adjoining room) may need to be re-drywalled when relocating a door. If it's a loadbearing wall, you'll need to allow room for installation of the header. Be sure you and your customer are aware of the additional costs and complexities involved before deciding to relocate a door.

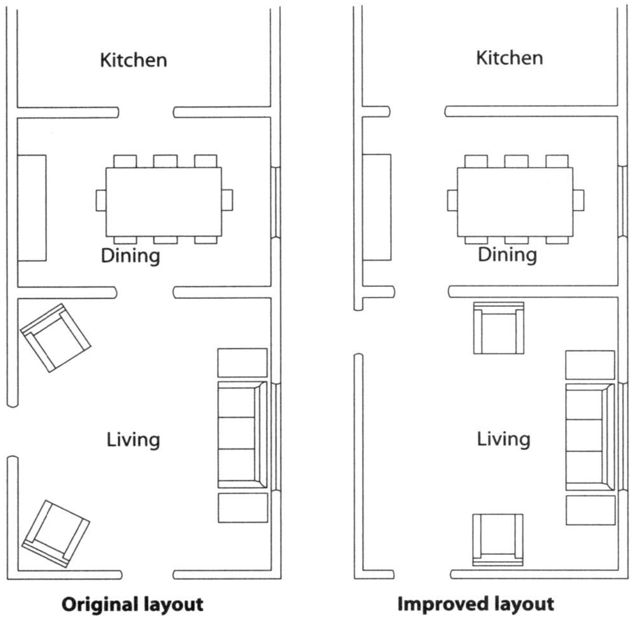

Figure 9-8

Relocation of doors to direct traffic to one side of the rooms

Rough opening dimensions for interior doors are the same as rough opening dimensions for exterior doors. The opening width is the door width plus 2^1/2". The opening height is the door height plus 2^1/4" above the finished floor. The head jamb and two side jambs are the same width as the overall wall thickness where wood casing is used. Where metal casing is used with drywall (Figure 9-9), the jamb width is the same as the stud depth.

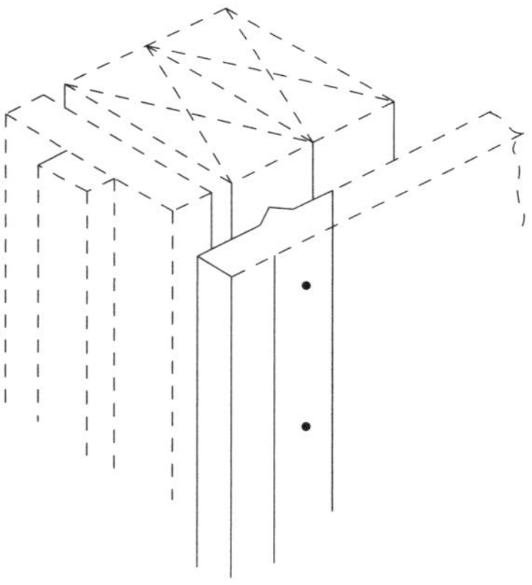

Figure 9-9

Metal casing used with drywall

Installation goes faster if you hang the door in the jamb before setting the frame in the wall. Let the door act as a jig around which you assemble the jambs. To save time, buy jambs in precut sets. The jamb should be about ^{1}/8" wider than the thickness of the finished wall. That's probably 4^9/16" for a 2" x 4" stud wall, or 6^9/16" for a 2" x 6" stud wall. The jamb height is ^{3}/8" less than the rough opening height. Fit the door to the frame, using the clearances shown earlier in Figure 9-7. Bevel the edges slightly where the door meets the stops. Cutting and assembling the jamb will take about a half hour.

Before installing the door, temporarily set the narrow wood strips used as stops. Stops are usually ^{7}/16" thick and from 1^1/2" to 2^1/4" wide. See Figure 9-10. Cut a mitered joint where the stops meet at the head and side jambs. A 45-degree bevel cut 1" to 1^1/2" above the finish floor (as in Figure 9-10) will eliminate a dirt pocket and make cleaning easier. This is called a sanitary stop.

Rout or mortise the hinges into the edge of the door with a ^{3}/16" to ^{1}/4" back spacing. Adjust the hinge, if necessary, so screws penetrate into solid wood. For interior doors, use two 3" x 3" loose-pin hinges. If a router isn't available, mark the hinge outline and depth of cut. Then remove just enough wood with a chisel so that the surface of the hinge is flush with the remainder of the door edge.

Place the door in the opening. Block the door in place with proper clearances all around. Then mark the location of the door hinges on the jamb. Remove the door from the opening and rout or chisel the jamb to the thickness of the hinge half. Install the hinge halves on the jamb.

Setting hinges on the door and jamb will require about an hour if you're using hand tools. If a router and door jig are available, reduce that time by about half.

Place the door in the opening and insert the hinge pins. Plumb and fasten the hinge side of the frame first. Drive shingle wedge sets between the side jamb and the rough door buck to plumb the jamb. Place a wedge set opposite the latch, each hinge and at intermediate locations. Nail the jamb with pairs of 8d nails at each wedge. Then fasten the opposite jamb the same way. When the door jamb is secure, cut off the shingle wedges flush with the wall. Allow about a half hour to set the door and jamb in the frame and nail the assembly in place.

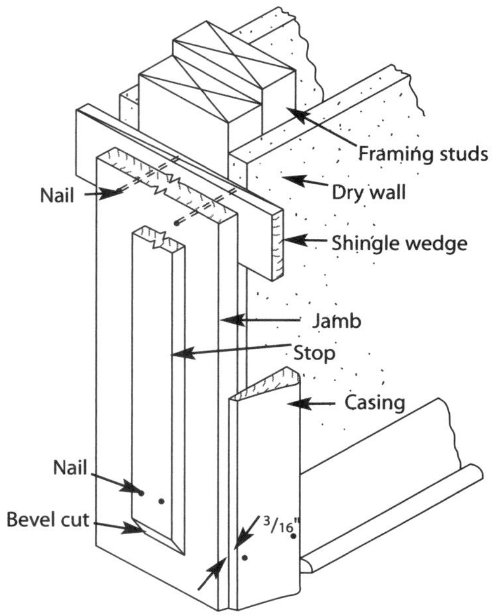

Figure 9-10

Installation of sanitary stop

Door Locksets

Many doors come pre-bored for the lockset. There are four types of locksets:

1. Decorative entry keyed locksets;
2. Common keyed locks;
3. Privacy locksets with an inside lock button and a safety slot that allows opening from the outside; and
4. Passage latch sets with no lock.

There are also dummy knobs that merely attach to the face of the door.

After the lockset has been installed, mark the location of the latch on the jamb. Do this with the door nearly closed. Then mark an outline for the strike plate at this position. Rout the jamb so the strike plate will be flush with the face of the jamb. See Figure 9-11. Installing the lockset and setting the strike will take about 15 minutes, assuming the door is already bored for a lockset. If you have to drill the door for a cylinder lock, allow 45 minutes. Routing for a mortise lockset will take over an hour.

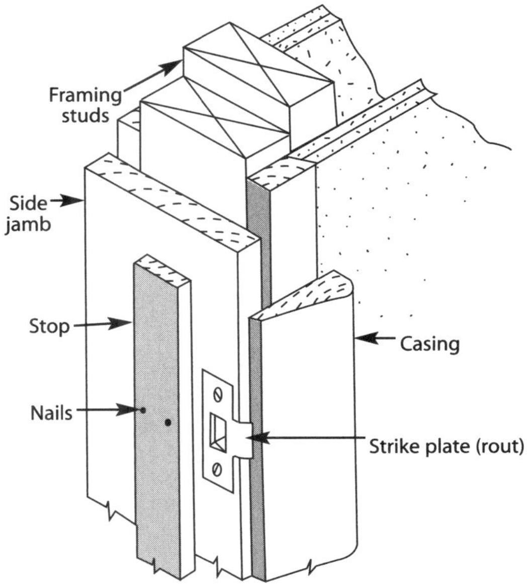

Figure 9-11

Installation of strike plate

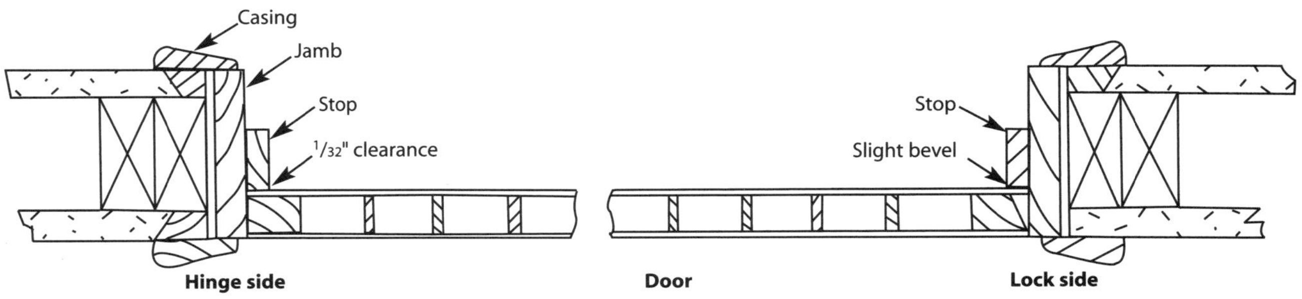

Figure 9-12

Location of stops

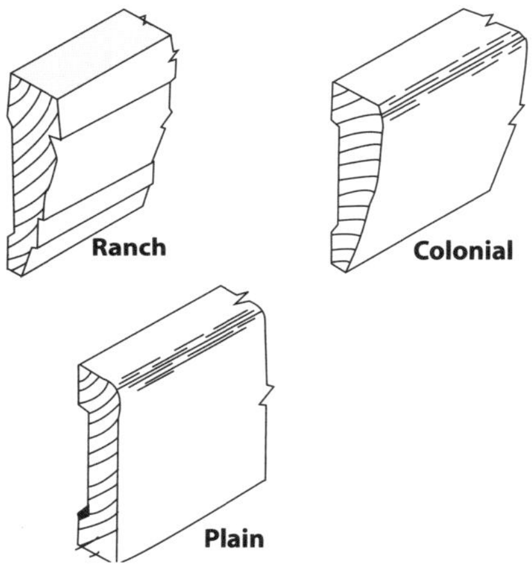

Figure 9-13

Styles of door casings

The stops, which were temporarily tacked in place, can now be permanently installed. Nail the stop on the lock side first, setting it against the door face when the door is fully latched. Nail the stops with $1\frac{1}{2}$"-long finishing nails or brads spaced in pairs about 16" apart. The stop at the hinge side of the door should have a clearance of $^1/_{32}$" (Figure 9-12). Allow 15 minutes to set the stops in place.

Setting the Trim

Casing is the trim set around a door or window opening. Most casing is from $^1/_2$" to $^3/_4$" thick and from $2\frac{1}{4}$" to $3\frac{1}{2}$" wide. Figure 9-13 shows three popular casing profiles.

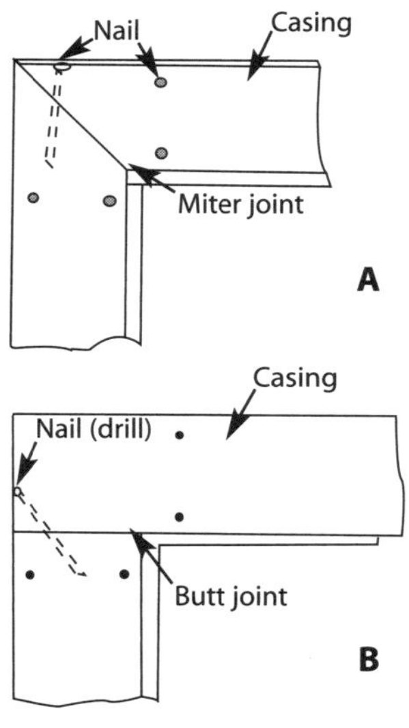

Figure 9-14

Casing joints

Set the casing back about $^3/_{16}$" from the face of the jamb. (Refer back to Figure 9-10.) Nail casing with 6d or 7d finishing nails, depending on the thickness of the casing. Space nails in pairs about 16" apart. Casings with molded profiles need a mitered joint where the head and side casings join (Figure 9-14 A). Rectangular casings can be butt-joined (Figure 9-14 B). With metal casing, fit the casing over the drywall, position the sheet properly, then nail through the drywall and casing into the stud behind. See Figure 9-15. Use the same type of nails and spacing as for drywall alone. Setting casing on both sides of the door will take about an hour.

Hanging a Slab Door

In round numbers, allow about two hours for routine assembly and hanging of a slab door in a rough wall opening. Here's a breakdown:

- Cut and fit the jamb, 30 minutes
- Set hinges on the door and jamb, 15 minutes

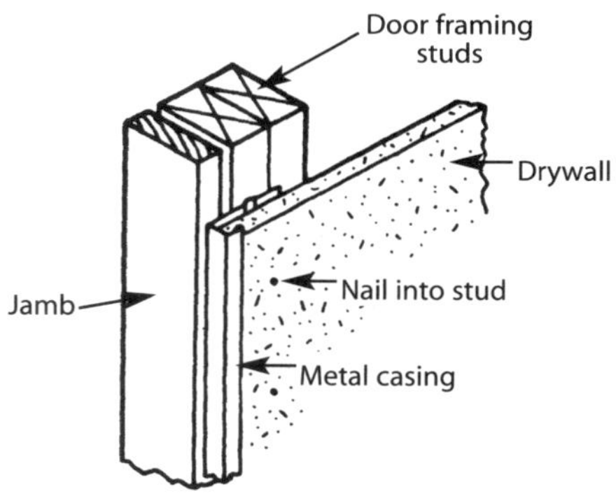

Figure 9-15
Metal casing

- ❖ Install jambs and hang the door, 15 minutes
- ❖ Install the lockset and strike, 15 minutes
- ❖ Set door stop on the jamb, 15 minutes
- ❖ Set casing on two sides, 30 minutes
- ❖ Add 15 minutes for a French door
- ❖ Add 30 minutes for setting a threshold

An experienced finish carpenter using a pneumatic nailer, router and door jig may be able to cut this time in half.

But as mentioned at the start of Chapter 1, your typical door replacement job may be anything but routine.

- ❖ Work starts with removing the old door. That's probably not too hard — once you break through several coats of accumulated paint on the hinges. One whack with a hammer should work wonders.
- ❖ Check the opening for plumb and square. Allow an extra hour or more if you have to align the framing. If the wall is way out of plumb, you may have a serious structural problem that requires reframing the entire wall.
- ❖ If hinge butts are loose, stripped out or poorly aligned, the next step may be installing a blind Dutchman on the jamb where the original hinges were attached.
- ❖ If the jamb is badly chewed up from years of neglect and abuse, you'll have to remove and replace the entire frame.
- ❖ In any case, you'll have to refinish the door and repair cosmetic defects in the interior and exterior wall finish.
- ❖ Finally, you have to haul away the old door and debris.

The pages that follow show a labor estimate of 1.45 hours for hanging a new exterior slab door. That estimate includes setting three hinges on the door and jamb, fitting and hanging the door in an existing cased opening, checking the operation and making minor adjustments as required to the stop, jamb or strike. Add the cost of removing the old door, reframing the opening, repairs to the existing jamb, resetting the latch and strike, and replacing interior casing or exterior brick mold, if required.

The labor estimate for hanging an interior slab door is 1.15 hours and includes setting two hinges on the door and jamb and hanging the door in an existing cased opening. Add the cost of removing the old door, reframing the opening, repairs to the existing jamb, resetting the latch and strike, and replacing casing on two sides, if required.

Prehung doors are a better choice for most jobs and offer obvious advantages.

Prehung Doors

Prehung doors (door, hinges, jamb and casing) cost a little more than slab doors (the door alone) but save money on labor because they require less cutting and fitting. Prehung doors come with the hinges attached, casing set, jamb assembled and door bored for a lockset. Exterior prehung doors usually include an adjustable threshold. The main jamb is attached to the door and is installed from one side of the opening. The other half of the jamb installs from the other side of the opening. When assembled, the two halves fit snugly together. Split jamb prehung doors have a tongue-and-groove joint behind the jamb stop that adjusts to fit the actual wall thickness. The casing will extend about 2" beyond the door edge at both the head and side jambs.

Prehung doors are easy to install. Allow about 45 minutes for an interior prehung door and 60 minutes for an exterior prehung door, assuming the opening is plumb, square and flush. But the door you order has to be exactly the door you need. It's impractical to make changes at the jobsite. All of the following are set at the mill:

- Door width and height
- Left or right hinge
- Inward or outward swing
- Jamb depth that matches the finished wall thickness
- Bore diameter and setback.

> Identify the "hinge" of a prehung door by opening it away from you. Right-hand (or right-hinge) doors swing out of the way to the right when fully open. Left-hand doors swing to the left. Inward swing is most common for exterior doors.

Closet Doors

Old homes tend to have larger rooms, higher ceilings and completely inadequate closet space, at least by 21st century standards. Even if there's enough closet space, it may not be well arranged for good use. Adding or rearranging closet space is a popular home improvement project.

One easy closet improvement is to replace a conventional single closet door with doors that open the full width of the closet. You'll need to remove the wall finish and enough studs to provide the opening needed for accordion, bypass or bi-fold doors (Figure 9-16). If the closet wall is a loadbearing wall, install a header across the new opening: two 2 x 8s for a 5' opening; two 2 x 10s for a 6' opening; or two 2 x 12s for an 8' opening. No header is needed for non-loadbearing walls. Frame the opening the same as for any new door opening. The rough opening should be $2^1/_2$" wider than the door or set of doors.

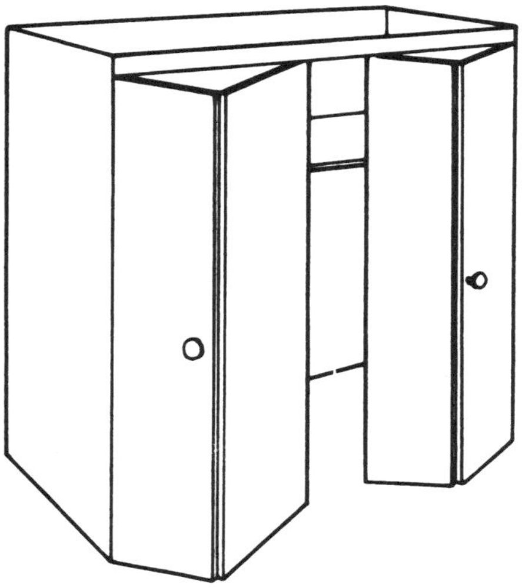

Figure 9-16
Double-hinged door set for full-width opening closet

You can also make closets more useful by adding or altering shelves and clothes rods. The usual closet has one rod with a single shelf above. If hanging space is limited, install a second clothes rod about halfway between the existing rod and the floor. This type of space is good for children's clothing, shirts and folded pants. Support a new shelf or rod with 1" x 4" cleats nailed to the closet end walls. Nail these cleats with three 6d nails at each end of the closet and at the intermediate stud. Shelf ends can rest directly on these cleats. Attach clothes rods to the same cleats that support the shelf. A long closet rod may sag in the middle when supported with nothing but end cleats. In that case, add steel support brackets fastened to studs at the back of the closet.

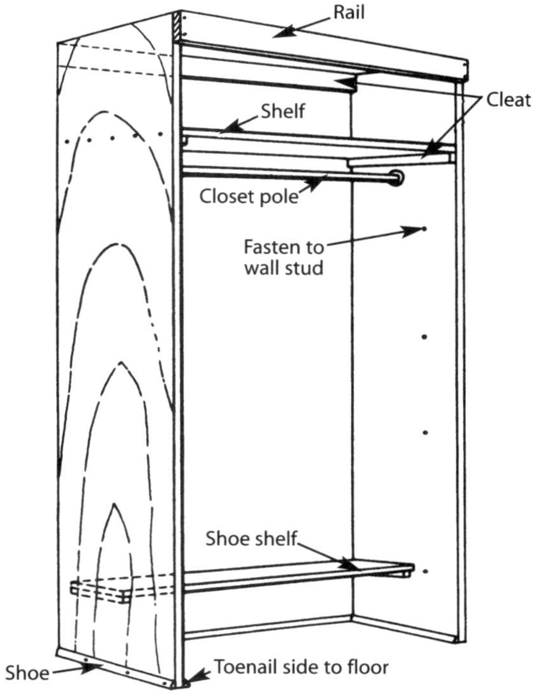

Figure 9-17
Wardrobe closet

New Closet Space

A plywood wardrobe cabinet is an economical and practical alternative to building a conventional closet with studs, drywall and casing. Figure 9-17 shows a simple plywood wardrobe built against a wall. For extra convenience, install additional shelves, drawers and doors.

Build wardrobe closets from $^5/_8$" or $^3/_4$" plywood or particleboard supported on cleats. Use a 1" x 4" top rail and back cleat. Fasten the cleat to the back wall. Fasten the sidewalls to a wall stud. Toenail base shoe molding to the floor to hold the bottoms of the sidewalls in place. Add shelves and closet poles as needed. Then enclose the space with bi-fold doors.

Plan for a coat closet near both the front and rear entrances. There should be a cleaning closet in the work area, and a linen closet in the bedroom area. Each bedroom needs a closet. In most real estate listings, you can't refer to a room as a "bedroom" unless it has a closet. If bedrooms are large, it's a simple matter to build a closet across one end of the room. Build partition walls with 2 x 3 or 2 x 4 framing lumber covered with gypsum board. Provide a cased opening for the closet doors.

In a small house, look for wasted space at the end of a hallway or at a wall offset. If the front door opens directly into the living room, consider building a coat closet beside or in front of the door to form an entry (Figure 9-18). In a story-and-a-half house, build closets in attic space where headroom is too limited for occupancy.

Closets used for hanging clothes should be about 24" deep. Shallower closets can be useful with appropriate hanging fixtures. Deeper closets need rollout hanging rods. To make the best use of closet space, plan for a full-front opening.

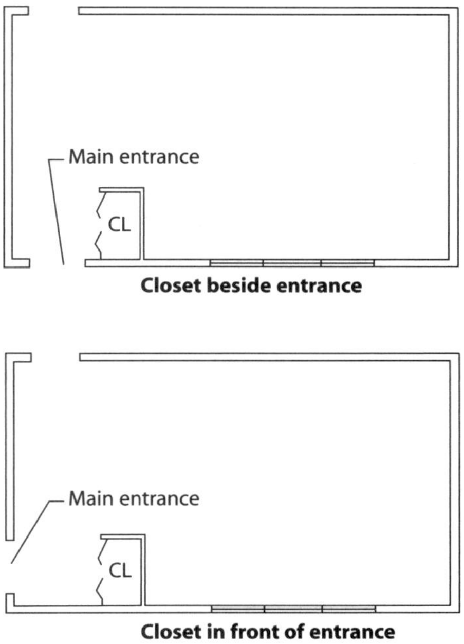

Figure 9-18
Entry formed by coat closet

Hollow core door

Solid core door

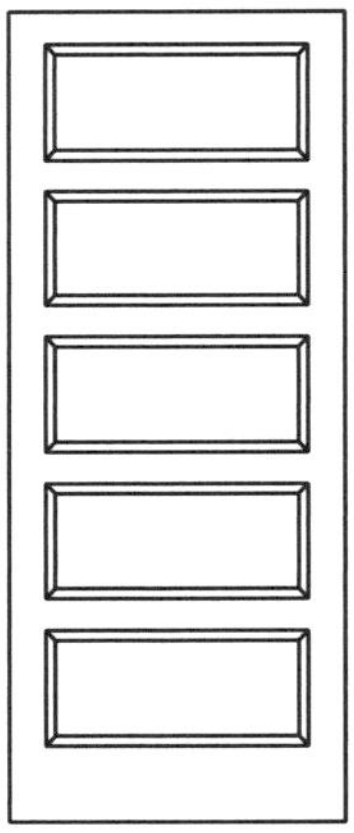

French door, 5 lites

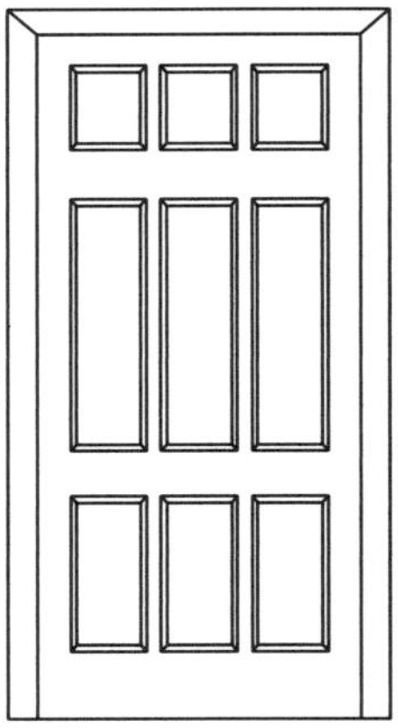

Colonial 9-panel door

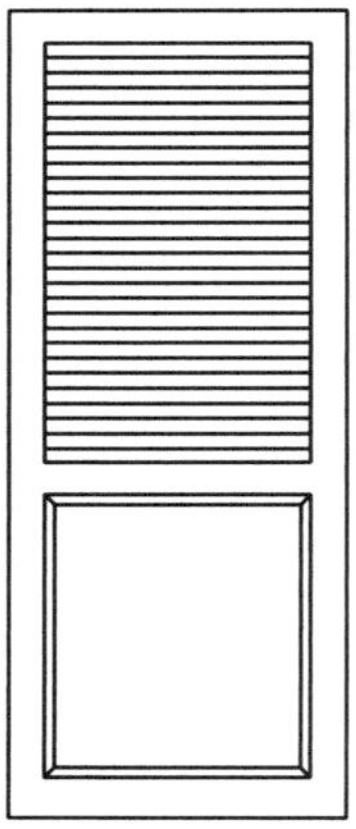

Half louver door

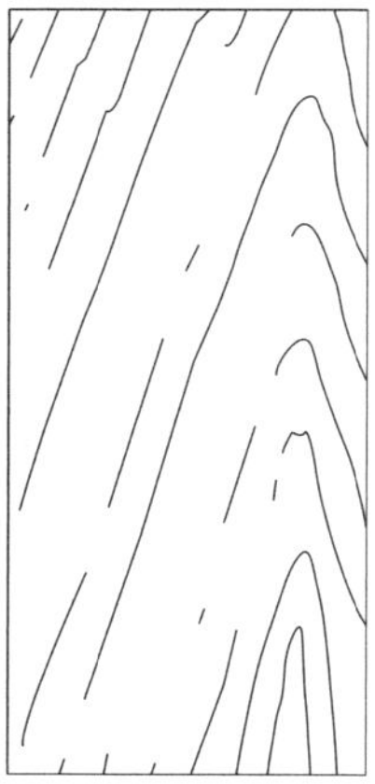

Flush door

Door Terms

- Interior doors are usually 1 3/8" thick and have a "hollow" core. A wood frame is filled with a matrix of corrugated paper and covered with a thin veneer.
- Exterior doors are usually 1 3/4" thick and solid core — again a thin veneer laid over a frame, but with some other material filling the door cavity. This filler could be foam or composition board or even lumber. Other popular exterior door choices include solid wood, steel and fiberglass.
- French doors are mostly glass, with one to as many as fifteen panes set in wood sash.
- Insulated French doors have dual layers of glass, usually with a dividing grille installed between layers.
- Colonial doors have raised decorative wood panels that resemble doors popular in colonial America. Six or nine panels are most common.
- Flush doors are smooth and flat, with no decorative treatment.
- Louver doors include wood louvers that allow air to circulate but obscure vision, an advantage for enclosing a closet, pantry or water heater.
- Grilles (grids) are laid over the glass in a single-pane French door to create the illusion of multiple glass panes.
- Paint grade doors are usually primed at the factory and are designed to be painted after installation.
- Unfinished doors can be stained or given a clear coating, such as urethane, to enhance the beauty of the wood.

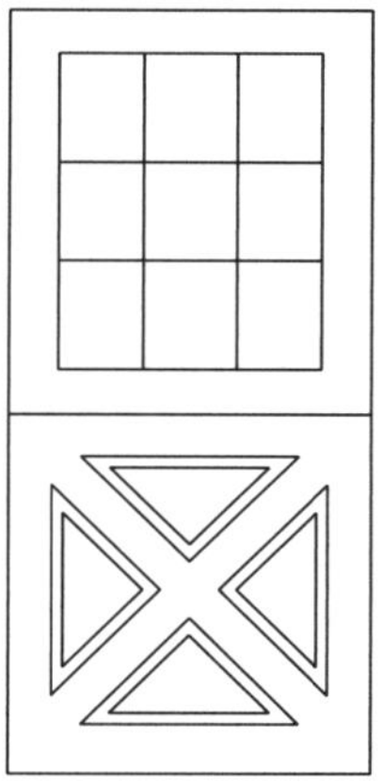

Crossbuck door

Bar doors

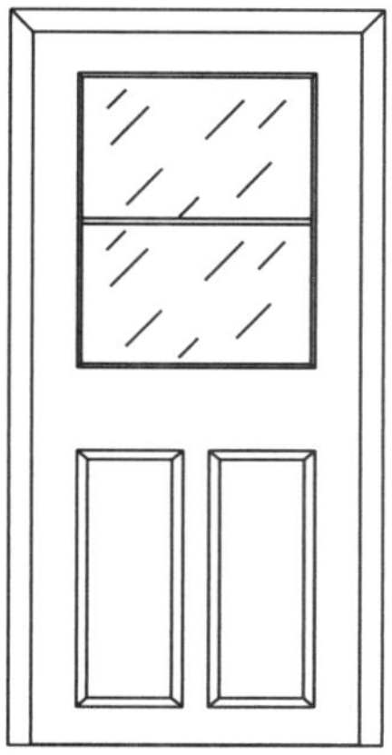

Ventilating door

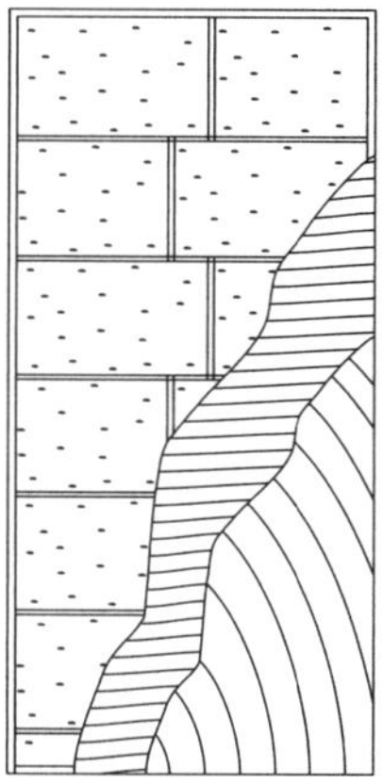

Fire door

- ❖ Entry doors are decorative exterior doors, often with glass panels set in a border of brass caming.
- ❖ Slab doors are the door alone, as distinguished from prehung doors which are sold with jamb, hinges, trim and threshold attached.
- ❖ Prehung doors are sold as a package with the door, jamb, hinges, and trim already assembled.
- ❖ Brick mold (also called stucco mold) is the casing around each side and head on the exterior side of an exterior door. Prehung doors are sold either with or without brick mold.
- ❖ Crossbuck doors have a raised X design in the bottom half of the door.
- ❖ Bar (café) doors are familiar to everyone who has seen a Western movie.
- ❖ Ventilating doors include a window that can be opened.
- ❖ Fire doors carry a fire rating and are designed to meet building code requirements for specialized occupancies, such as hotel rooms or dormitories. Code also requires that fire doors be used between a house and an attached garage, and that an automatic door closer is installed.
- ❖ Bi-fold doors on closets fold in half when open, revealing nearly the full interior.
- ❖ Bypass closet doors slide left or right, never revealing more than one-half of the interior.

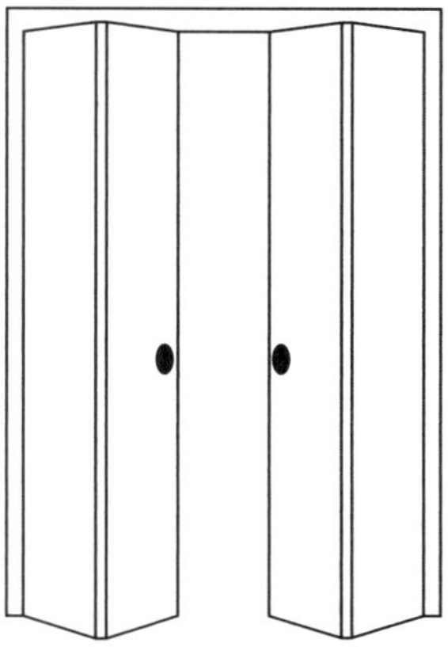

Bi-fold doors

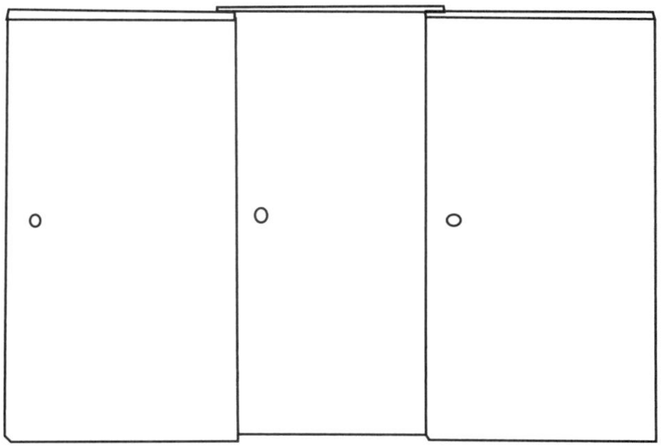

Bypass doors

Security door

3/4 oval lite door

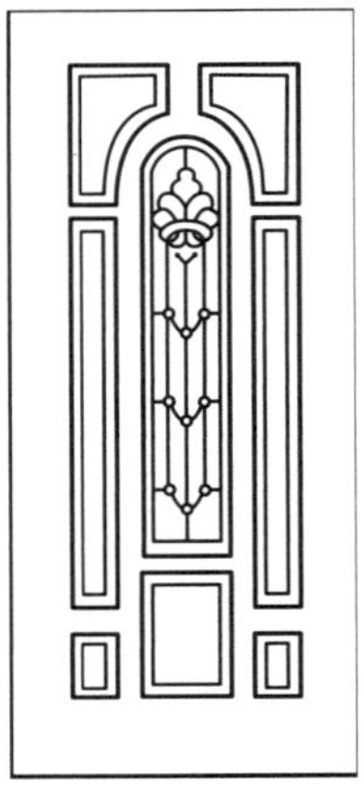

Arch lite panel door

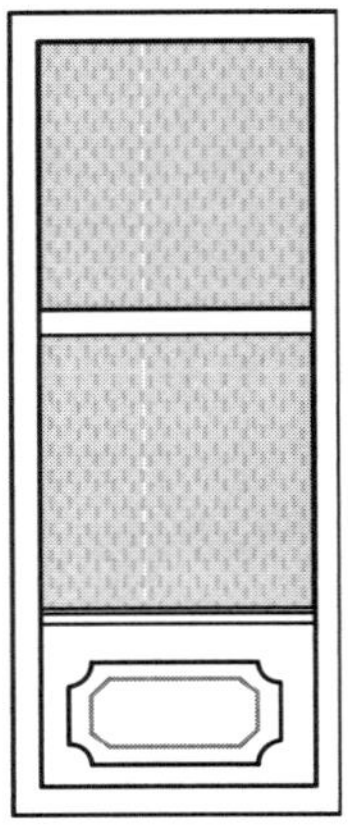

Self-storing storm door

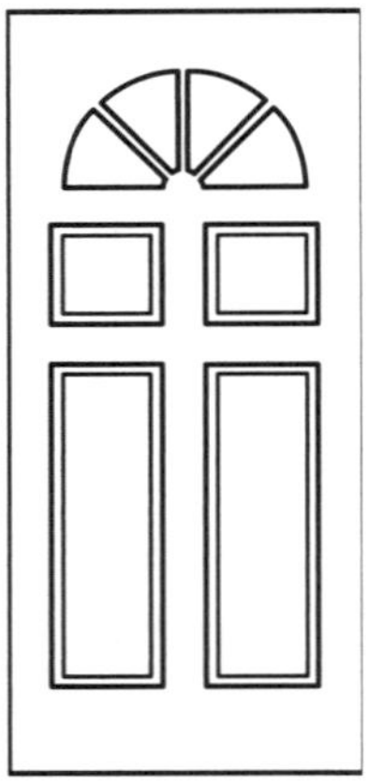

Fan lite panel door

- Hardboard doors are made of composition wood veneer, the least expensive material available.
- Casing is the trim around each side and head of an interior door. Casing covers the joint between jamb and wall.
- Finger joint trim is made from two or more lengths of wood joined together with a finger-like joint that yields nearly as much strength as trim made of a single, solid piece.
- Hardwood doors are finished with a thin layer of true hardwood veneer.
- Lauan doors are finished with a veneer of Philippine hardwood from the mahogany family.
- Stiles and rails form the framework of a flush door. Stiles are vertical and every door has two — a lock stile and the hinge stile. Rails run horizontally at the top and bottom of the door.
- Molded face doors have decorative molding applied to at least one side.

	Craft@Hrs	Unit	Material	Labor	Total	Sell
Exterior Doors						
Cut wall and frame exterior door opening, install slab door and trim. Includes temporary wall bracing and closure, header, trimmer studs, exterior flush slab door ($56), frame, threshold, three hinges, stop, brick mold, casing, cylinder lockset ($80.00), and weatherstrip. Add the cost of debris removal, painting, and floor and wall finish if needed.						
36" x 80" slab exterior door	BC@10.7	Ea	470.00	385.00	855.00	1,450.00
Remove and replace single exterior slab door and trim. Includes removing the existing door, trim and jamb, aligning the rough frame, installing exterior flush slab door ($60), frame, threshold, three hinges, stop, brick mold, casing, cylinder lockset ($75.00), and weatherstrip. Add the cost of debris removal and painting.						
Single door, stud wall	BC@6.40	Ea	425.00	230.00	655.00	1,110.00
Single door, masonry wall	BC@7.35	Ea	430.00	265.00	695.00	1,180.00
Remove and replace double exterior slab doors and trim. Includes removing the existing doors, trim and jamb, aligning the rough frame, installing two exterior 6-panel slab doors ($114), frame, threshold, six hinges, stop, brick mold, casing, cylinder lockset ($78.80), two surface slide bolts, and weatherstrip.						
Double door, stud wall, per pair	BC@8.75	Ea	520.00	315.00	835.00	1,420.00
Double door, masonry wall, per pair	BC@9.75	Ea	525.00	351.00	876.00	1,490.00
Remove, clean, refinish and reinstall exterior slab door. Includes removing door finish on all six sides with heat gun and solvent, filling small voids, sanding and refinishing with varnish or stain and sealer, and rehanging the door.						
Single door	PT@6.00	Ea	16.40	221.00	237.40	404.00
Double door	PT@11.0	Ea	27.20	404.00	431.20	733.00
Refinish jamb and trim, per opening	PT@2.00	Ea	16.40	73.50	89.90	153.00
Add for antique finish, per door	PT@1.50	Ea	21.80	55.20	77.00	131.00
Door repairs.						
Remove set of three hinge pins	BC@.067	Ea	—	2.41	2.41	4.10
Install set of three hinge pins	BC@.091	Ea	—	3.28	3.28	5.58
Patch 6" hole with body filler	BC@4.00	Ea	24.10	144.00	168.10	286.00
Align strike plate with latch bolt	BC@.500	Ea	—	18.00	18.00	30.60
Plane 1/8" off door edge	BC@.250	Ea	—	9.00	9.00	15.30
Loosen and reset one hinge butt	BC@.333	Ea	—	12.00	12.00	20.40
Touch-up painting	PT@.250	Ea	2.25	9.19	11.44	19.40
Install exterior slab door in an existing framed and cased opening. Includes minor repair to the jamb (blind Dutchman), three hinges, flush slab door ($60.00), stop, cylinder lockset ($80.00), and weatherstrip.						
36" x 80" exterior door	BC@3.10	Ea	215.00	112.00	327.00	556.00
Remove and replace exterior door casing, jamb and stop. Includes frame, stop, and casing on one side only. Per door opening.						
Door opening to 36" x 80"	BC@1.50	Ea	122.00	54.00	176.00	299.00
Remove and replace exterior slab door, on existing hinges. Includes exterior 6-panel slab door ($56.90) and cylinder lockset ($78.80).						
36" x 80" exterior door	BC@1.90	Ea	153.00	68.40	221.40	376.00

	Craft@Hrs	Unit	Material	Labor	Total	Sell

Wood Exterior Slab Doors

Flush hardboard exterior doors. Solid core. 1-3/4" thick. Primed and ready to paint. Labor includes setting three hinges on the door and jamb and hanging the door in an existing cased opening.

	Craft@Hrs	Unit	Material	Labor	Total	Sell
30" x 80"	BC@1.45	Ea	53.00	52.20	105.20	179.00
32" x 80"	BC@1.45	Ea	53.00	52.20	105.20	179.00
34" x 80"	BC@1.45	Ea	53.00	52.20	105.20	179.00
36" x 80"	BC@1.45	Ea	53.00	52.20	105.20	179.00

Flush hardwood exterior doors. Solid core. 1-3/4" thick. Labor includes setting three hinges on the door and jamb and hanging the door in an existing cased opening.

	Craft@Hrs	Unit	Material	Labor	Total	Sell
28" x 80"	BC@1.45	Ea	60.20	52.20	112.40	191.00
30" x 80"	BC@1.45	Ea	78.10	52.20	130.30	222.00
32" x 80"	BC@1.45	Ea	83.40	52.20	135.60	231.00
36" x 80"	BC@1.45	Ea	55.10	52.20	107.30	182.00
36" x 84"	BC@1.45	Ea	94.40	52.20	146.60	249.00
30" x 80", diamond light	BC@1.45	Ea	91.40	52.20	143.60	244.00
30" x 80", square light	BC@1.45	Ea	91.40	52.20	143.60	244.00
32" x 80", square light	BC@1.45	Ea	90.10	52.20	142.30	242.00
36" x 80", diamond light	BC@1.45	Ea	98.50	52.20	150.70	256.00
36" x 80", square light	BC@1.45	Ea	98.50	52.20	150.70	256.00

Flush birch exterior doors. Solid core.1-3/4" thick. Stainable. Labor includes setting three hinges on the door and jamb and hanging the door in an existing cased opening.

	Craft@Hrs	Unit	Material	Labor	Total	Sell
32" x 80"	BC@1.45	Ea	53.00	52.20	105.20	179.00
34" x 80"	BC@1.45	Ea	53.00	52.20	105.20	179.00
36" x 80"	BC@1.45	Ea	53.00	52.20	105.20	179.00

Flush lauan exterior doors. Solid core. 1-3/4" thick. Stainable. Labor includes setting three hinges on the door and jamb and hanging the door in an existing cased opening.

	Craft@Hrs	Unit	Material	Labor	Total	Sell
30" x 80"	BC@1.45	Ea	53.00	52.20	105.20	179.00
30" x 84"	BC@1.45	Ea	53.00	52.20	105.20	179.00
32" x 80"	BC@1.45	Ea	57.10	52.20	109.30	186.00
36" x 80"	BC@1.45	Ea	57.10	52.20	109.30	186.00

Flush fire-rated doors. 1-3/4" thick. 20-minute fire rating. Labor includes setting three hinges on the door and jamb and hanging the door in an existing cased opening.

	Craft@Hrs	Unit	Material	Labor	Total	Sell
32" x 80", hardboard veneer	BC@1.45	Ea	53.00	52.20	105.20	179.00
36" x 80", hardboard veneer	BC@1.45	Ea	53.00	52.20	105.20	179.00
32" x 80", birch veneer	BC@1.45	Ea	65.30	52.20	117.50	200.00
36" x 80", birch veneer	BC@1.45	Ea	65.30	52.20	117.50	200.00

Red oak exterior doors. Wood stave core. 1-3/4" thick. Labor includes setting three hinges on the door and jamb and hanging the door in an existing cased opening.

	Craft@Hrs	Unit	Material	Labor	Total	Sell
30" x 80"	BC@1.45	Ea	62.00	52.20	114.20	194.00
32" x 80"	BC@1.45	Ea	62.00	52.20	114.20	194.00
34" x 80"	BC@1.45	Ea	68.00	52.20	120.20	204.00
36" x 80"	BC@1.45	Ea	68.00	52.20	120.20	204.00

6-panel fir exterior doors. 1-3/4" thick. Beveled, hip-raised panels. Doweled stile and rail joints. 4-1/2" wide stiles. Unfinished clear fir. Ready to paint or stain. Labor includes setting three hinges on the door and jamb and hanging the door in an existing cased opening.

	Craft@Hrs	Unit	Material	Labor	Total	Sell
30" x 80"	BC@1.45	Ea	160.00	52.20	212.20	361.00
32" x 80"	BC@1.45	Ea	160.00	52.20	212.20	361.00
36" x 80"	BC@1.45	Ea	160.00	52.20	212.20	361.00

	Craft@Hrs	Unit	Material	Labor	Total	Sell
6-panel Douglas fir entry doors. 1-3/4" thick. Solid core. Labor includes setting three hinges on the door and jamb and hanging the door in an existing cased opening.						
30" x 80"	BC@1.45	Ea	127.00	52.20	179.20	305.00
32" x 80"	BC@1.45	Ea	129.00	52.20	181.20	308.00
36" x 80"	BC@1.45	Ea	131.00	52.20	183.20	311.00
Mahogany entry doors. 1-3/4" thick. Beveled tempered glass lite. Unfinished solid wood. Ready to paint or stain. Labor includes setting three hinges on the door and jamb and hanging the door in an existing cased opening.						
30" x 80", arch lite	BC@1.45	Ea	420.00	52.20	472.20	803.00
30" x 80", moon lite	BC@1.45	Ea	420.00	52.20	472.20	803.00
30" x 80", oval lite	BC@1.45	Ea	438.00	52.20	490.20	833.00
4-panel 4-lite fir exterior doors. Vertical grain Douglas fir. 1-3/4" thick. 3/4" thick panels. Unfinished. Veneered and doweled construction. 1/8" tempered safety glass. Labor includes setting three hinges on the door and jamb and hanging the door in an existing cased opening.						
36" x 80"	BC@1.70	Ea	187.00	61.20	248.20	422.00
2-panel 9-lite entry doors. 1-3/4" thick. Single-pane glass. Solid core with doweled joints. Unfinished. Suitable for stain or paint. Labor includes setting three hinges on the door and jamb and hanging the door in an existing cased opening.						
30" x 80"	BC@1.70	Ea	190.00	61.20	251.20	427.00
32" x 80"	BC@1.70	Ea	190.00	61.20	251.20	427.00
36" x 80"	BC@1.70	Ea	190.00	61.20	251.20	427.00
1-lite clear fir entry doors. 1-3/4" thick. Clear fir surface ready to stain or paint. Insulated glass. Labor includes setting three hinges on the door and jamb and hanging the door in an existing cased opening.						
32" x 80"	BC@1.70	Ea	178.00	61.20	239.20	407.00
36" x 80"	BC@1.70	Ea	180.00	61.20	241.20	410.00
Fir 10-lite exterior doors. 1-3/4" thick. Unfinished. Labor includes setting three hinges on the door and jamb and hanging the door in an existing cased opening.						
30" x 80" x 1-3/4"	BC@1.70	Ea	178.00	61.20	239.20	407.00
32" x 80" x 1-3/4"	BC@1.70	Ea	178.00	61.20	239.20	407.00
36" x 80" x 1-3/4"	BC@1.70	Ea	178.00	61.20	239.20	407.00
Fir 15-lite entry doors. Single-pane glass. Ready for stain or paint. 1-3/4" thick. Doweled stile and rail joints. Labor includes setting three hinges on the door and jamb and hanging the door in an existing cased opening.						
30" x 80" x 1-3/4"	BC@1.70	Ea	178.00	61.20	239.20	407.00
32" x 80" x 1-3/4"	BC@1.70	Ea	178.00	61.20	239.20	407.00
36" x 80" x 1-3/4"	BC@1.70	Ea	178.00	61.20	239.20	407.00
Fir fan lite entry doors. Vertical grain Douglas fir. 1-3/4" thick. 3/4" thick panels. Unfinished. Veneer and doweled construction. 1/8" tempered safety glass. Labor includes setting three hinges on the door and jamb and hanging the door in an existing cased opening.						
32" x 80"	BC@1.45	Ea	201.00	52.20	253.20	430.00
30" x 80"	BC@1.45	Ea	193.00	52.20	245.20	417.00
Deco fan lite entry doors. 1-3/4" thick. 3/4" thick panels. Unfinished solid wood. Ready to paint or stain. Beveled insulated glass. Labor includes setting three hinges on the door and jamb and hanging the door in an existing cased opening.						
36" x 80"	BC@1.45	Ea	213.00	52.20	265.20	451.00

	Craft@Hrs	Unit	Material	Labor	Total	Sell
Flush oak entry doors. 1-3/4" thick. Prefinished. Labor includes setting three hinges on the door and jamb and hanging the door in an existing cased opening.						
36" x 80"	BC@1.45	Ea	882.00	52.20	934.20	1,590.00
Reversible hollow metal fire doors.						
34" x 80"	BC@.721	Ea	142.00	26.00	168.00	286.00
36" x 80"	BC@.721	Ea	162.00	26.00	188.00	320.00
Knock down steel door frames. Bored for (3) 4-1/2" hinges, 4-7/8" strike plate and 2-3/4" backset. 90-minute fire label manufactured to N.Y.C. M.E.A. procedures. Hollow core. For drywall and masonry applications. Galvanized steel.						
30" x 80"	BC@.944	Ea	71.40	34.00	105.40	179.00
32" x 80"	BC@.944	Ea	71.40	34.00	105.40	179.00
34" x 80"	BC@.944	Ea	71.40	34.00	105.40	179.00
36" x 80"	BC@.944	Ea	71.40	34.00	105.40	179.00
36" x 84"	BC@.944	Ea	71.40	34.00	105.40	179.00
90-minute fire-rated door frame parts.						
4-7/8" x 80" legs, pack of 2	BC@.778	Ea	65.30	28.00	93.30	159.00
4-7/8" x 36" head	BC@.166	Ea	17.30	5.98	23.28	39.60
Steel cellar doors. Durable heavy gauge steel.						
57" x 45" x 24-1/2"	BC@1.00	Ea	336.00	36.00	372.00	632.00
63" x 49" x 22"	BC@1.00	Ea	342.00	36.00	378.00	643.00
71" x 53" x 19-1/2"	BC@1.00	Ea	356.00	36.00	392.00	666.00

Door Trim

	Craft@Hrs	Unit	Material	Labor	Total	Sell
Install door trim.						
Door casing, two sides, hand nailed	BC@.500	Ea	—	18.00	18.00	30.60
Door casing, two sides, power nailed	BC@.333	Ea	—	12.00	12.00	20.40
Install mail slot in door	BC@.529	Ea	—	19.00	19.00	32.30
T-astragal molding. Covers the vertical joint where a pair of doors meets.						
1-1/4" x 2" x 7', pine	BC@.529	Ea	24.80	19.00	43.80	74.50
Remove and replace 80" long astragal	BC@.680	Ea	—	24.50	24.50	41.70
Primed exterior door jamb. Primed finger joint pine. 1-1/4" thick. Set for a 3' door consists of two 6'8" legs and one 3' head.						
4-5/8" wide, per 6'8" leg	BC@.204	Ea	31.20	7.35	38.55	65.50
5-1/4" wide, per 6'8" leg	BC@.204	Ea	37.00	7.35	44.35	75.40
6-9/16" wide, per 6'8" leg	BC@.204	Ea	23.60	7.35	30.95	52.60
4-5/8" wide, per set	BC@.500	Ea	56.80	18.00	74.80	127.00
5-1/4" wide, per set	BC@.500	Ea	37.00	18.00	55.00	93.50
6-9/16" wide, per set	BC@.500	Ea	36.00	18.00	54.00	91.80
Primed kerfed exterior door jamb. 1-1/4" thick. Set for a 3' door consists of two 6'8" legs and one 3' head.						
4-5/8" x 7' leg	BC@.204	Ea	24.00	7.35	31.35	53.30
4-5/8" x 3' head	BC@.092	Ea	11.40	3.31	14.71	25.00
6-5/8" x 7' leg	BC@.204	Ea	25.00	7.35	32.35	55.00
6-5/8" x 3' head	BC@.092	Ea	12.50	3.31	15.81	26.90

	Craft@Hrs	Unit	Material	Labor	Total	Sell
Aluminum thermal door sill. With vinyl insert. 36" x 5-5/8", mill finish.						
Install 30" to 36" aluminum threshold	BC@.518	Ea	15.50	18.70	34.20	58.10
Install 56" to 72" aluminum threshold	BC@.623	Ea	31.50	22.40	53.90	91.60
Remove and replace 30" to 36" threshold	BC@.850	Ea	15.50	30.60	46.10	78.40
Remove and replace 56" to 72" threshold	BC@1.20	Ea	31.50	43.20	74.70	127.00
Oak high boy threshold. Fits standard door sizes.						
5/8" x 3-1/2" x 36"	BC@.518	Ea	14.50	18.70	33.20	56.40
Stop molding. Nails to face of door jamb to prevent door from swinging through. Solid pine.						
3/8" x 1-1/4"	BC@.015	LF	.77	.54	1.31	2.23
7/16" x 3/4"	BC@.015	LF	.56	.54	1.10	1.87
7/16" x 1-3/8"	BC@.015	LF	.77	.54	1.31	2.23
7/16" x 1-5/8"	BC@.015	LF	.88	.54	1.42	2.41
7/16" x 2-1/8"	BC@.015	LF	1.36	.54	1.90	3.23
Solid pine ranch casing.						
9/16" x 2-1/4"	BC@.015	LF	.73	.54	1.27	2.16
9/16" x 2-1/4" x 7'	BC@.105	Ea	5.19	3.78	8.97	15.20
11/16" x 3-1/2"	BC@.015	LF	1.81	.54	2.35	4.00
11/16" x 2-1/2" x 7'	BC@.105	Ea	9.21	3.78	12.99	22.10
Primed finger joint ranch door casing.						
9/16" x 3-1/4"	BC@.015	LF	.91	.54	1.45	2.47
11/16" x 2-1/4"	BC@.015	LF	.59	.54	1.13	1.92
11/16" x 2-1/4" x 7'	BC@.105	Ea	4.14	3.78	7.92	13.50
11/16" x 2-1/2"	BC@.015	LF	1.06	.54	1.60	2.72
11/16" x 2-1/2" x 7'	BC@.105	Ea	7.43	3.78	11.21	19.10
Colonial door casing. Solid pine.						
1/2" x 2-1/4"	BC@.015	LF	1.32	.54	1.30	2.21
9/16" x 2-1/4" x 7'	BC@.105	Ea	7.93	3.78	11.71	19.90
9/16" x 2-1/4"	BC@.015	LF	1.13	.54	1.67	2.84
11/16" x 2-1/2"	BC@.015	LF	2.82	.54	3.36	5.71
11/16" x 2-1/2" x 7'	BC@.105	Ea	9.39	3.78	13.17	22.40
11/16" x 3-1/2"	BC@.015	LF	2.15	.54	2.69	4.57
2-1/2", pre-mitered, 3' opening	BC@.250	Ea	23.10	9.00	32.10	54.60
Fluted pine casing and rosette set. Includes 1 header piece, 2 side pieces, 2 rosettes and 2 base blocks.						
Set	BC@.250	Ea	38.40	9.00	47.40	80.60
Brick mold set. Trim for the exterior side of an exterior door. Two legs and one head. Primed finger joint pine. 36" x 80" door. 1-1/4" x 2".						
Finger joint pine	BC@.250	Ea	36.50	9.00	45.50	77.40
Pressure treated pine	BC@.250	Ea	23.20	9.00	32.20	54.70
Remove and replace brick mold	BC@.350	Ea	—	12.60	12.60	21.40
Brick mold.						
1-3/16" x 2", hemlock	BC@.015	LF	1.86	.54	2.40	4.08
1-1/4" x 2", fir	BC@.015	LF	1.25	.54	1.79	3.04
1-1/4" x 2", solid pine	BC@.015	LF	2.54	.54	3.08	5.24

	Craft@Hrs	Unit	Material	Labor	Total	Sell
Stucco molding. Redwood.						
13/16" x 1-1/2"	BC@.015	LF	1.10	.54	1.64	2.79
1" x 1-3/8"	BC@.015	LF	1.10	.54	1.64	2.79
Remove door hinge butt. Remove four screws per 3-1/2" hinge butt half. Per hinge.						
Door or jamb half hinge butt only	BC@.089	Ea	—	3.20	3.20	5.44
Door and jamb half hinge butts	BC@.135	Ea	—	4.86	4.86	8.26
Remove and reset door hinge in existing mortise	BC@.275	Ea	—	9.90	9.90	16.80
Door hinges. Solid brass. Mortise type. Non-rising removable tip. Includes screws. Installation labor is included with cost of the door.						
3" x 3", square corners 3 pk	—	Ea	11.20	—	11.20	—
3-1/2" x 3-1/2", round corners	—	Ea	13.80	—	13.80	—
3-1/2" x 3-1/2", square corners	—	Ea	14.10	—	14.10	—
4" x 4", round corners 3 pk	—	Ea	15.30	—	15.30	—
4" x 4", square corners 3 pk	—	Ea	15.30	—	15.30	—

Door Weatherstrip

	Craft@Hrs	Unit	Material	Labor	Total	Sell
Remove and replace vinyl door weatherstrip.						
36" x 80" exterior door	BC@.589	Ea	12.60	21.20	33.80	57.50
Aluminum and vinyl adjustable door weatherstrip. Seals out drafts and moisture. Fits top and sides of 36" x 80" door. Slotted holes for easy alignment.						
3/4" x 17', brown or white	BC@.381	Ea	12.60	13.70	26.30	44.70
3/4" x 17', grey	BC@.381	Ea	11.70	13.70	25.40	43.20
Gray foam weatherstrip. Self-adhesive. Resilient. Open-cell. Compresses flat for a tight seal. Sticks to any dry clean surface. Width by thickness by length.						
1/4" x 1/2" x 17'	BC@.250	Ea	2.53	9.00	11.53	19.60
3/8" x 1/2" x 17'	BC@.250	Ea	2.80	9.00	11.80	20.10
Aluminum and vinyl door bottom. Heavy gauge aluminum with vinyl sweep. Seals out drafts, rain and dust. Screws included. 36" x 1-5/8".						
Brown	BC@.250	Ea	6.61	9.00	15.61	26.50
Brushed chrome	BC@.250	Ea	10.20	9.00	19.20	32.60
Drip cap door bottom seal. Drip cap diverts rain away. Seals out drafts, dust and insects. 36" long.						
1-1/2", brite gold	BC@.250	Ea	11.20	9.00	20.20	34.30
Magnetic weatherstrip. Provides airtight, refrigerator-like seal. Hinge compression tear strip adjusts after installation.						
36" long	BC@.250	Ea	13.70	9.00	22.70	38.60
Flex-O-Matic door sweep. As the door opens, the spring-action door sweep rises to clear the carpet. When the door is closed, the door sweep automatically lowers to firmly press against floor or carpet. Keeps out dirt and dust.						
Brite gold	BC@.250	Ea	5.07	9.00	14.07	23.90

	Craft@Hrs	Unit	Material	Labor	Total	Sell
Prehung Wood Exterior Doors						
Cut wall and frame exterior door opening, install prehung door. Includes cutting stud wall, temporary wall bracing and closure, header, trimmer studs, exterior 6-panel prehung steel door ($142) with frame, threshold, hinges, stop, brick mold, casing, weatherstrip and cylinder lockset ($80.00). Add the cost of debris removal, painting, floor and wall finish if needed.						
36" x 80" slab exterior door	BC@8.20	Ea	265.00	295.00	560.00	952.00
Remove door and frame and replace with an exterior prehung steel door. Includes removing the existing door, trim and jamb, aligning the rough frame, installing one exterior 6-panel prehung steel door ($142) with frame, threshold, hinges, stop, brick mold, casing, weatherstrip and cylinder lockset ($80.00). Add the cost of debris removal and painting.						
Single door, stud wall	BC@3.90	Ea	225.00	140.00	365.00	621.00
Single door, masonry wall	BC@4.85	Ea	225.00	175.00	400.00	680.00
Remove double doors and frame and replace with double exterior prehung steel doors. Includes removing an existing double door, trim and jamb, aligning the rough frame, installing double exterior 6-panel prehung steel doors ($305), with frame, threshold, hinges, stop, brick mold, casing, and weatherstrip. Add cylinder lockset ($80.00) and surface slide bolts ($15).						
Double door, stud wall, per pair	BC@6.25	Ea	400.00	225.00	625.00	1,060.00
Double door, masonry wall, per pair	BC@7.25	Ea	400.00	261.00	661.00	1,120.00
Flush hardwood prehung exterior doors. 1-3/4" door. Solid particleboard core. Lauan veneer. Weatherstripped. Includes oak sill and threshold with vinyl insert. Finger joint 4-9/16" jamb and finger joint brick mold. 2-1/8" bore for lockset with 2-3/8" backset. Labor includes hanging the door in an existing framed opening.						
32" x 80"	BC@1.00	Ea	101.00	36.00	137.00	233.00
36" x 80"	BC@1.00	Ea	101.00	36.00	137.00	233.00
Flush birch prehung exterior doors. 1-3/4" door. Solid particleboard core. Weatherstripped. Includes oak sill and threshold. Finger joint 4-9/16" jamb and finger joint brick mold. 2-1/8" bore for lockset with 2-3/8" backset. Labor includes hanging the door in an existing framed opening.						
36" x 80"	BC@1.00	Ea	118.00	36.00	154.00	262.00
Flush hardboard prehung exterior doors. Primed. Solid core. Weatherstripped. Includes sill and threshold. Finger joint 4-9/16" jamb and finger joint brick mold. 2-1/8" bore for lockset with 2-3/8" backset. Labor includes hanging the door in an existing framed opening.						
32" x 80"	BC@1.00	Ea	101.00	36.00	137.00	233.00
36" x 80"	BC@1.00	Ea	101.00	36.00	137.00	233.00
Fan lite prehung hemlock exterior doors. Weatherstripped. Includes sill and threshold. Finger joint 4-9/16" jamb and finger joint brick mold. 2-1/8" bore for lockset with 2-3/8" backset. Labor includes hanging the door in an existing framed opening.						
36" x 80"	BC@1.00	Ea	309.00	36.00	345.00	587.00
9-lite prehung fir exterior doors. Inward swing. No brick mold. Solid core. Weatherstripped. Includes sill and threshold. Finger joint 4-9/16" jamb and finger joint brick mold. 2-1/8" bore for lockset with 2-3/8" backset. Labor includes hanging the door in an existing framed opening.						
36" x 80"	BC@1.00	Ea	240.00	36.00	276.00	469.00
Prehung hardboard fire-rated doors. 20-minute fire rating. 80" high. Labor includes hanging the door in an existing framed opening.						
30" x 80"	BC@1.00	Ea	152.00	36.00	188.00	320.00
32" x 80"	BC@1.00	Ea	158.00	36.00	194.00	330.00
36" x 80"	BC@1.00	Ea	161.00	36.00	197.00	335.00

	Craft@Hrs	Unit	Material	Labor	Total	Sell

Prehung Steel Exterior Doors

Utility flush prehung steel exterior doors. No brick mold. 26 gauge galvanized steel. Foam core. Compression weatherstrip. Triple fin sweep. Non-thermal threshold. Single bore.

	Craft@Hrs	Unit	Material	Labor	Total	Sell
32" x 80"	BC@1.00	Ea	118.00	36.00	154.00	262.00
36" x 80"	BC@1.00	Ea	118.00	36.00	154.00	262.00

Premium flush prehung steel exterior doors. No brick mold. 24 gauge galvanized steel. Foam core. Compression weatherstrip. Triple fin sweep. Thermal threshold.

	Craft@Hrs	Unit	Material	Labor	Total	Sell
30" x 80"	BC@1.00	Ea	142.00	36.00	178.00	303.00
32" x 80"	BC@1.00	Ea	142.00	36.00	178.00	303.00
36" x 80"	BC@1.00	Ea	142.00	36.00	178.00	303.00

6-panel prehung steel exterior doors. With brick mold. Inward swing. Non-thermal threshold. Bored for lockset with 2-3/4" backset. Includes compression weatherstrip and thermal-break threshold.

	Craft@Hrs	Unit	Material	Labor	Total	Sell
32" x 80"	BC@1.00	Ea	118.00	36.00	154.00	262.00
36" x 80"	BC@1.00	Ea	118.00	36.00	154.00	262.00

6-panel premium prehung steel exterior doors. 4-9/16" jamb. Adjustable thermal-break sill. Inward swing. 24 gauge galvanized steel. Bored for 2-3/8" lockset. 12" lock block. Impact-resistant laminated glass. Polyurethane core. Triple sweep and compression weatherstripping. Factory primed.

	Craft@Hrs	Unit	Material	Labor	Total	Sell
32" x 80", no brick mold	BC@1.00	Ea	121.00	36.00	157.00	267.00
32" x 80", with brick mold	BC@1.00	Ea	142.00	36.00	178.00	303.00
36" x 80", no brick mold	BC@1.00	Ea	121.00	36.00	157.00	267.00
36" x 80", with brick mold	BC@1.00	Ea	142.00	36.00	178.00	303.00
72" x 80", no brick mold	BC@1.50	Ea	294.00	54.00	348.00	592.00

6-panel prehung steel exterior doors. 24 gauge galvanized steel. Foam core. Caming matches hinge and sill finishes. Jamb guard security plate. No brick mold. Fi XEd sill. 4-9/16" primed jamb. Triple sweep compression weatherstripping. Factory primed. Thermal-break threshold. 20-minute fire rating. 10-year limited warranty.

	Craft@Hrs	Unit	Material	Labor	Total	Sell
30" x 80"	BC@1.00	Ea	121.00	36.00	157.00	267.00
32" x 80"	BC@1.00	Ea	121.00	36.00	157.00	267.00
36" x 80"	BC@1.00	Ea	121.00	36.00	157.00	267.00

6-panel prehung steel entry door with sidelites. Two 10"-wide sidelites. Sidelites have 5 divided lites.

	Craft@Hrs	Unit	Material	Labor	Total	Sell
36" x 80"	BC@2.00	Ea	438.00	72.00	510.00	867.00

Fan lite premium prehung steel exterior doors. With brick mold. 24 gauge galvanized steel. Foam core. Fi XEd sill. Compression weatherstripping. Factory primed. Thermal-break threshold.

	Craft@Hrs	Unit	Material	Labor	Total	Sell
32" x 80", 4-9/16" jamb	BC@1.00	Ea	192.00	36.00	228.00	388.00
32" x 80", 6-9/16" jamb	BC@1.00	Ea	192.00	36.00	228.00	388.00
36" x 80", 4-9/16" jamb	BC@1.00	Ea	192.00	36.00	228.00	388.00
36" x 80", 6-9/16" jamb	BC@1.00	Ea	192.00	36.00	228.00	388.00

Half lite prehung steel exterior doors. 4-9/16" jamb. 22" x 36" light. Adjustable sill. With brick mold.

	Craft@Hrs	Unit	Material	Labor	Total	Sell
32" x 80"	BC@1.00	Ea	274.00	36.00	310.00	527.00
36" x 80"	BC@1.00	Ea	274.00	36.00	310.00	527.00

Half lite prehung steel exterior doors with mini-blind. 22" x 36" light. Adjustable sill. 24 gauge galvanized steel. Non-yellowing, white, warp-resistant, paintable, high-performance light frames. Brick mold applied.

	Craft@Hrs	Unit	Material	Labor	Total	Sell
32" x 80", 4-9/16" jamb	BC@1.00	Ea	266.00	36.00	302.00	513.00
36" x 80", 4-9/16" jamb	BC@1.00	Ea	266.00	36.00	302.00	513.00

	Craft@Hrs	Unit	Material	Labor	Total	Sell
Full lite prehung steel exterior doors. 22" x 64" light. 4-9/16" jamb. Adjustable sill. No brick mold. 24 gauge galvanized steel.						
32" x 80"	BC@1.00	Ea	234.00	36.00	270.00	459.00
36" x 80"	BC@1.00	Ea	234.00	36.00	270.00	459.00
2-lite, 4-panel prehung steel exterior doors. 4-9/16" jamb. Adjustable sill. Ready-to-install door and jamb system. 24 gauge galvanized steel. Non-yellowing, white, warp-resistant, paintable, high-performance light frames. Brick mold applied.						
30" x 80"	BC@1.00	Ea	195.00	36.00	231.00	393.00
32" x 80"	BC@1.00	Ea	195.00	36.00	231.00	393.00
36" x 80"	BC@1.00	Ea	195.00	36.00	231.00	393.00
9-lite prehung steel exterior doors. Insulating glass with internal 9-lite grille. Adjustable sill. Ready-to-install door and jamb system. 24 gauge galvanized steel. Non-yellowing, white, warp-resistant, paintable, high-performance light frames. Brick mold applied.						
32" x 80", 4-9/16" jamb	BC@1.00	Ea	203.00	36.00	239.00	406.00
32" x 80", 6-9/16" jamb	BC@1.00	Ea	207.00	36.00	243.00	413.00
11-lite prehung steel exterior doors. 24 gauge, high-profile, deep embossed steel skins. Impact-resistant glass. Double-bored for deadbolt and doorknob. Brass adjustable threshold. 4-9/16" primed frame. No brick mold.						
32" x 80" left	BC@1.00	Ea	241.00	36.00	277.00	471.00
15-lite prehung steel exterior doors. Multi-paned insulated glass. 24 gauge hot-dipped galvanized steel. Polyurethane foam core. Compression weatherstrip. Thermal-break construction. No brick mold. Adjustable seal. Aluminum sill with composite adjustable threshold.						
32" x 80"	BC@1.00	Ea	234.00	36.00	270.00	459.00
36" x 80"	BC@1.00	Ea	234.00	36.00	270.00	459.00
Flush ventlite prehung steel exterior doors. No brick mold. 4-9/16" primed jamb. 24 gauge galvanized steel. Foam core. Magnetic weatherstrip. Triple-seal sweep. Thermal-break threshold. Bored for lockset and deadbolt. 20-minute fire rating. 10-year limited warranty.						
30" x 80"	BC@1.00	Ea	234.00	36.00	270.00	459.00
32" x 80"	BC@1.00	Ea	234.00	36.00	270.00	459.00
36" x 80"	BC@1.00	Ea	234.00	36.00	270.00	459.00

Decorative Steel Entry Doors

	Craft@Hrs	Unit	Material	Labor	Total	Sell
Decorative 3/4 oval lite prehung steel entry doors. 24 gauge galvanized steel. Non-yellowing, white, warp-resistant, paintable, high-performance light frames. Decorative impact-resistant glass. Rot-resistant frame. Polyurethane insulated core. With brick mold. Adjustable sill.						
36" x 80", brass caming, 4-9/16" jamb	BC@1.00	Ea	305.00	36.00	341.00	580.00
36" x 80", zinc caming, 4-9/16" jamb	BC@1.00	Ea	305.00	36.00	341.00	580.00
36" x 80", zinc caming, 6-9/16" jamb	BC@1.00	Ea	323.00	36.00	359.00	610.00
Camber top prehung steel entry doors. 4-9/16" jamb.						
36" x 80", with brick mold	BC@1.00	Ea	277.00	36.00	313.00	532.00
36" x 80", no brick mold	BC@1.00	Ea	277.00	36.00	313.00	532.00

	Craft@Hrs	Unit	Material	Labor	Total	Sell
Decorative central arch lite prehung steel entry door. Decorative laminated glass with brass caming to match hinge and sill finish. 24 gauge galvanized steel. Polyurethane core. Adjustable thermal-break sill. Triple sweep compression weatherstripping. Non-yellowing and warp-resistant light frame. Factory primed.						
36" x 80", with brick mold, 4-9/16" jamb	BC@1.00	Ea	304.00	36.00	340.00	578.00
36" x 80", with no brick mold, 4-9/16" jamb	BC@1.00	Ea	294.00	36.00	330.00	561.00
Decorative fan lite prehung steel entry doors. Polyurethane core. Adjustable thermal-break sill with composite saddle. 24 gauge galvanized steel. Laminated glass. Brass caming matches hinge and sill. Triple sweep and compression weatherstrip. Non-yellowing and warp-resistant light frame. Factory primed. Includes brick mold.						
32" x 80"	BC@1.00	Ea	224.00	36.00	260.00	442.00
36" x 80"	BC@1.00	Ea	224.00	36.00	260.00	442.00
Decorative half lite prehung steel entry doors. 22" x 36" laminated safety glass. Brass caming matches sill and hinges. 24 gauge galvanized insulated steel.						
36" x 80", no brick mold	BC@1.00	Ea	283.00	36.00	319.00	542.00
36" x 80", with brick mold	BC@1.00	Ea	283.00	36.00	319.00	542.00

Fiberglass Prehung Entry Doors

	Craft@Hrs	Unit	Material	Labor	Total	Sell
6-panel fiberglass prehung exterior doors. Polyurethane core. PVC stiles and rails. Double bore. 4-9/16" prefinished jambs. Adjustable thermal-break brass threshold. Weatherstripped. With brick mold.						
32" x 80"	BC@1.00	Ea	171.00	36.00	207.00	352.00
36" x 80"	BC@1.00	Ea	171.00	36.00	207.00	352.00
Fan lite prehung light oak fiberglass entry doors. With brick mold. Factory prefinished. Triple pane insulated glass with brass caming. Adjustable brass thermal-break threshold and fully weatherstripped. 4-9/16" jamb.						
36" x 80"	BC@1.00	Ea	376.00	36.00	412.00	700.00
3/4 oval lite prehung fiberglass entry doors. With brick mold. Ready to finish.						
36" x 80", 4-9/16" jamb	BC@1.00	Ea	336.00	36.00	372.00	632.00
36" x 80", 6-9/16" jamb	BC@1.00	Ea	336.00	36.00	372.00	632.00
3/4 oval lite prehung light oak fiberglass doors. With brick mold. Prefinished. Polyurethane core. Triple pane insulated glass. Glue chip and clear 30" x 18" beveled glass surrounded by brass caming. Double bore. Extended wood lock block. 4-9/16" primed jamb. Thermal-break brass threshold with adjustable oak cap.						
36" x 80", 4-9/16" jamb	BC@1.00	Ea	478.00	36.00	514.00	874.00
36" x 80", 6-9/16" jamb	BC@1.00	Ea	512.00	36.00	548.00	932.00
Full height oval lite prehung medium oak fiberglass entry doors.						
36" x 80", 4-9/16" jamb	BC@1.00	Ea	517.00	36.00	553.00	940.00
36" x 80", 6-9/16" jamb	BC@1.00	Ea	532.00	36.00	568.00	966.00

	Craft@Hrs	Unit	Material	Labor	Total	Sell
Center arch lite medium oak prehung fiberglass doors. Factory prefinished. Triple pane insulated glass with brass caming. Adjustable brass thermal-break threshold and fully weatherstripped. 4-9/16" jamb except as noted. PVC stiles and rails.						
36" x 80", no brick mold	BC@1.00	Ea	427.00	36.00	463.00	787.00
36" x 80", with brick mold	BC@1.00	Ea	427.00	36.00	463.00	787.00
36" x 80", with brick mold, 6-9/16" jamb	BC@1.00	Ea	448.00	36.00	484.00	823.00
Fan lite prehung fiberglass entry doors. With brick mold. Ready to finish. Insulated glass. Adjustable thermal-break brass threshold. 4-9/16" jamb except as noted.						
32" x 80"	BC@1.00	Ea	254.00	36.00	290.00	493.00
36" x 80"	BC@1.00	Ea	254.00	36.00	290.00	493.00
36" x 80", 6-9/16" jamb	BC@1.00	Ea	295.00	36.00	331.00	563.00
9-lite fiberglass prehung exterior doors. Inward swing. With brick mold. Smooth face.						
32" x 80"	BC@1.00	Ea	244.00	36.00	280.00	476.00
36" x 80"	BC@1.00	Ea	244.00	36.00	280.00	476.00
15-lite smooth prehung fiberglass entry doors. No brick mold. Factory prefinished. Left or right hinge. Triple pane, insulated glass with brass caming. Adjustable brass thermal-break threshold and fully weatherstripped. 4-9/16" jamb.						
32" x 80"	BC@1.00	Ea	254.00	36.00	290.00	493.00
32" x 80"	BC@1.00	Ea	254.00	36.00	290.00	493.00
Full lite prehung smooth fiberglass doors. With brick mold. Inward swing.						
36" x 80", 4-9/16" jamb	BC@1.00	Ea	802.00	36.00	838.00	1,420.00
36" x 80", 6-9/16" jamb	BC@1.00	Ea	802.00	36.00	838.00	1,420.00
Smooth prehung fiberglass ventlite exterior doors. Ready to finish. No brick mold.						
32" x 80"	BC@1.00	Ea	254.00	36.00	290.00	493.00
36" x 80"	BC@1.00	Ea	254.00	36.00	290.00	493.00

Entry Locksets

	Craft@Hrs	Unit	Material	Labor	Total	Sell
Door lockset repairs.						
Adjust strike plate to align with latch bolt	BC@.200	Ea	—	7.20	7.20	12.20
Remove cylinder or deadlock	BC@.137	Ea	—	4.93	4.93	8.38
Reinstall same cylinder or deadlock	BC@.183	Ea	—	6.59	6.59	11.20
Install cylinder lock and strike in pre-bored door	BC@.250	Ea	—	9.00	9.00	15.30
Bore door, install cylinder lock, strike	BC@.750	Ea	—	27.00	27.00	45.90
Remove and replace cylinder or deadlock in existing bore	BC@.300	Ea	—	10.80	10.80	18.40
Remove and replace combination cylinder lock and deadbolt in existing bore	BC@1.52	Ea	—	54.70	54.70	93.00
Remove mortise lock	BC@.146	Ea	—	5.26	5.26	8.94
Install mortise lock	BC@1.31	Ea	—	47.20	47.20	80.20
Remove and replace mortise lock	BC@1.45	Ea	—	52.20	52.20	88.70
Keyed entry locksets. Front or back door. Keyed exterior, turn button interior. Labor cost assumes the door is already bored for a lockset.						
Antique brass	BC@.250	Ea	18.90	9.00	27.90	47.40
Polished brass	BC@.250	Ea	14.20	9.00	23.20	39.40
Satin chrome	BC@.250	Ea	14.80	9.00	23.80	40.50

	Craft@Hrs	Unit	Material	Labor	Total	Sell
Keyed entry locksets. Meets ANSI Grade 2 standards for residential security. One-piece knob. Triple option faceplate fits square, radius-round, and drive-in door preps. 10-year finish warranty, lifetime mechanical warranty. Labor cost assumes the door is already bored for a lockset.						
Antique brass	BC@.250	Ea	27.50	9.00	36.50	62.10
Antique pewter	BC@.250	Ea	30.00	9.00	39.00	66.30
Bright brass	BC@.250	Ea	24.50	9.00	33.50	57.00
Satin chrome	BC@.250	Ea	23.40	9.00	32.40	55.10
Lever locksets. Dual torque springs eliminate wobble. Labor cost assumes the door is already bored for a lockset.						
Bright brass	BC@.250	Ea	40.80	9.00	49.80	84.70
Satin nickel	BC@.250	Ea	46.80	9.00	55.80	94.90
Egg knob entry locksets. Labor cost assumes the door is already bored for a lockset.						
Satin nickel finish	BC@.250	Ea	51.10	9.00	60.10	102.00
Egg knob entry locksets — solid forged brass. 5-pin cylinder. Full lip strike. Adjustable backset. Labor cost assumes the door is already bored for a lockset.						
Polished brass	BC@.250	Ea	81.60	9.00	90.60	154.00
Colonial knob entry locksets. Labor cost assumes the door is already bored for a lockset and frame is already bored for a strike plate.						
Polished brass	BC@.250	Ea	73.00	9.00	82.00	139.00
Double cylinder entry door handle set. Forged brass handle with thumb latch. Separate Grade 1 security double cylinder deadbolt. Labor cost assumes the door is already bored for a lockset, but frame is not.						
Satin nickel finish	BC@.820	Ea	92.10	29.50	121.60	207.00
Bronze finish	BC@.820	Ea	112.00	29.50	141.50	241.00
Deadbolt and lockset combo. Includes keyed alike entry knob and Grade 2 security deadbolt with 4 keys. Labor cost assumes the door is already bored for a lockset.						
Brass, single cylinder	BC@.750	Ea	47.90	27.00	74.90	127.00
Brass, double cylinder	BC@.750	Ea	58.10	27.00	85.10	145.00
Deadbolts. All metal and brass. ANSI Grade 2 rating. 1" throw deadbolt. Includes 2 keys. Polished brass.						
Single cylinder	BC@.500	Ea	38.40	18.00	56.40	95.90
Double cylinder	BC@.500	Ea	48.90	18.00	66.90	114.00

Screen Doors

	Craft@Hrs	Unit	Material	Labor	Total	Sell
Install wood screen door. Complete with three hinge butts, pneumatic door closer, latch lock and guard. Labor cost includes adjustments to existing screen door frame.						
Per screen door	BC@2.17	Ea	—	78.10	78.10	133.00
Install aluminum or vinyl screen door. Complete with pneumatic door closer, chain and spring restrainer and latch lock.						
Per screen door	BC@1.35	Ea	—	48.60	48.60	82.60
Remove and replace wood screen door. Using existing hinge butts. With pneumatic door closer, latch lock and guard. Labor cost assumes new door is an exact replacement for the old door.						
Per screen door	BC@1.00	Ea	—	36.00	36.00	61.20
Remove and replace aluminum or vinyl screen door. Complete with pneumatic door closer, chain and spring restrainer and latch lock.						
Per screen door	BC@1.97	Ea	—	70.90	70.90	121.00

	Craft@Hrs	Unit	Material	Labor	Total	Sell
Remove wood and replace with aluminum or vinyl screen door. With pneumatic door closer, chain spring retainer and latch lock.						
Per screen door	BC@1.70	Ea	—	61.20	61.20	104.00
Screen door hardware.						
Remove and replace pneumatic closer	BC@.296	Ea	—	10.70	10.70	18.20
Remove and replace latch lock	BC@.185	Ea	—	6.66	6.66	11.30
Remove and replace chain spring retainer	BC@.203	Ea	—	7.31	7.31	12.40
Remove and replace door bottom wiper	BC@.187	Ea	—	6.73	6.73	11.40
Wood screen doors. 3-3/4"-wide stile and rail with 7-1/4" bottom rail. Removable screen. Mortise and tenon solid construction. Wide center push bar. Ready to finish.						
32" x 80"	BC@2.17	Ea	75.50	78.10	153.60	261.00
36" x 80"	BC@2.17	Ea	75.50	78.10	153.60	261.00
T-style wood screen doors. 3-3/4"-wide stile and rail with 7-1/4" bottom rail. Removable screen. Mortise and tenon solid construction. All joints sealed with water-resistant adhesive. Wide center push bar.						
30" x 80"	BC@2.17	Ea	20.40	78.10	98.50	167.00
32" x 80"	BC@2.17	Ea	20.40	78.10	98.50	167.00
36" x 80"	BC@2.17	Ea	20.40	78.10	98.50	167.00
Metal screen doors. Includes hardware and fiberglass screening. 1-piece steel frame. Heavy-duty pneumatic closer. 5" kickplate.						
32" x 80"	BC@1.35	Ea	44.90	48.60	93.50	159.00
36" x 80"	BC@1.35	Ea	44.90	48.60	93.50	159.00
Five-bar solid vinyl screen doors. Accepts standard screen door hardware.						
32" x 80"	BC@1.35	Ea	90.80	48.60	139.40	237.00
36" x 80"	BC@1.35	Ea	90.80	48.60	139.40	237.00
Solid pine screen doors. Right or left side installation. 1-1/4" thick. Unfinished. Hardware included.						
32" x 80"	BC@1.35	Ea	180.00	48.60	228.60	389.00
36" x 80"	BC@1.35	Ea	180.00	48.60	228.60	389.00
Steel security screen doors. Class I rated to withstand 300 pounds. All welded steel construction for strength and durability. Double lockbox with extension plate. Baked-on powder coating.						
36" x 80"	BC@1.35	Ea	122.00	48.60	170.60	290.00
Replacement aluminum screen doors. Includes all hardware. 2-piece protective grille. 3" x 1-1/4" heavy-duty rails.						
36" wide	BC@1.35	Ea	90.30	48.60	138.90	236.00
Replacement roll-formed screen doors. 7/8" 1-piece unitized roll-formed frame, mechanically secured corners, epoxy-painted finish over electro-galvanizing, 5" heavy-duty kickplate.						
32" x 80", gray	BC@1.35	Ea	63.20	48.60	111.80	190.00
36" x 80", white	BC@1.35	Ea	95.90	48.60	144.50	246.00
36" x 80", bronze	BC@1.35	Ea	94.90	48.60	143.50	244.00
Replacement steel screen doors. Powder-coated finish. With hardware and closer.						
36" x 80"	BC@1.35	Ea	85.70	48.60	134.30	228.00

	Craft@Hrs	Unit	Material	Labor	Total	Sell
Storm Doors						
Remove existing storm door.						
Wood or aluminum	BC@.353	Ea	—	12.70	12.70	21.60
Remove and replace storm door	BC@1.80	Ea	—	64.80	64.80	110.00
Self-storing aluminum storm and screen doors. Pneumatic closer and sweep. Tempered safety glass. Maintenance-free finish. Push-button hardware. 1" x 2-1/8" frame size. Bronze or white. Includes screen.						
30" x 80" x 1-1/4"	BC@1.45	Ea	94.90	52.20	147.10	250.00
32" x 80" x 1-1/4"	BC@1.45	Ea	94.90	52.20	147.10	250.00
36" x 80" x 1-1/4"	BC@1.45	Ea	94.90	52.20	147.10	250.00
Self-storing vinyl-covered wood storm doors. White vinyl-covered wood core. Push-button handle set. Self-storing window and screen. Single black closer.						
30" x 80"	BC@1.45	Ea	98.90	52.20	151.10	257.00
32" x 80"	BC@1.45	Ea	98.90	52.20	151.10	257.00
34" x 80"	BC@1.45	Ea	98.90	52.20	151.10	257.00
36" x 80"	BC@1.45	Ea	98.90	52.20	151.10	257.00
Store-in-Door™ storm doors. 1-1/2"-thick polypropylene frame. Reversible hinge. Slide window or screen, completely concealed when not in use. Brass-finished keylock hardware system with separate color-matched deadbolt. Triple-fin door sweep. 2 color-matched closers. Full-length piano hinge. Flexible weather seal. White.						
30" x 80", crossbuck style	BC@1.45	Ea	267.00	52.20	319.20	543.00
32" x 80", crossbuck style	BC@1.45	Ea	267.00	52.20	319.20	543.00
36" x 80", crossbuck style	BC@1.45	Ea	267.00	52.20	319.20	543.00
30" x 80", traditional style	BC@1.45	Ea	267.00	52.20	319.20	543.00
36" x 80", traditional style	BC@1.45	Ea	267.00	52.20	319.20	543.00
Aluminum and wood storm doors. Aluminum over wood core. Self-storing window and screen. Black lever handle set. Separate deadbolt. Single black closer. White finish. Traditional panel style.						
30" x 80"	BC@1.45	Ea	150.00	52.20	202.20	344.00
32" x 80"	BC@1.45	Ea	150.00	52.20	202.20	344.00
36" x 80"	BC@1.45	Ea	150.00	52.20	202.20	344.00
Triple-track storm doors. Low-maintenance aluminum over solid wood core. Ventilates from top, bottom or both. Solid brass exterior handle. Color-matched interior handle. Separate deadbolt. 80" high. White, almond or bronze.						
30" x 80"	BC@1.45	Ea	203.00	52.20	255.20	434.00
32" x 80"	BC@1.45	Ea	203.00	52.20	255.20	434.00
34" x 80"	BC@1.45	Ea	203.00	52.20	255.20	434.00
36" x 80"	BC@1.45	Ea	203.00	52.20	255.20	434.00
All-season storm doors. 1-1/4" thick frame. 1-piece construction with window molding built in. Self-storing tempered safety glass and screen for easy ventilation. Includes solid-brass lockset with deadbolt security. Brass-finished sweep for tight threshold seal. Color-matched screw covers, and two closers. Reversible hinge opens to 180 degrees. White, almond or sandstone.						
32" x 80"	BC@1.45	Ea	217.00	52.20	269.20	458.00
36" x 80"	BC@1.45	Ea	217.00	52.20	269.20	458.00
All-season triple-track storm doors. Ventilates from top or bottom, or both. 1" composite frame. Brass handle set with deadbolt security. Wood-grain finish. Color-matched screws. 1-piece construction with window molding built in. Self-storing window and screen. White, almond or sandstone.						
32" x 80"	BC@1.45	Ea	221.00	52.20	273.20	464.00
36" x 80"	BC@1.45	Ea	221.00	52.20	273.20	464.00

	Craft@Hrs	Unit	Material	Labor	Total	Sell
Colonial triple-track storm doors. 1-piece composite frame. 12-lite grille. Vents from top, bottom, or both. Solid-brass handle set with deadbolt security. Brass-finished sweep ensures tight seal across the entire threshold. Wood-grain finish. White, almond or sandstone.						
32" x 80"	BC@1.45	Ea	280.00	52.20	332.20	565.00
36" x 80"	BC@1.45	Ea	284.00	52.20	336.20	572.00
Full-view brass all-season storm doors. Water-diverting rain cap. Brass-finished, color-matched sweep. One closer. Window and insect screen snap in and out. Reversible. Opens 180 degrees when needed. Separate deadbolt for added security. Tempered glass. Heavy-gauge one-piece 1" aluminum construction with reinforced corners and foam insulation.						
36" x 80", green	BC@1.45	Ea	359.00	52.20	411.20	699.00
36" x 80", white	BC@1.45	Ea	359.00	52.20	411.20	699.00
Full-view aluminum storm doors. 1-1/2"-thick heavy-gauge aluminum frame with foam insulation. Extra perimeter weatherstripping and water-diverting rain cap. Solid brass handle set and double-throw deadbolt. Brass-finished sweep. Full-length, piano-style hinge. Reversible hinge for left- or right-side entry door handle. Two heavy-duty, color-matched closers. Solid brass perimeter locks. Insect screen. White, brown, green or almond.						
30" x 80"	BC@1.45	Ea	293.00	52.20	345.20	587.00
32" x 80"	BC@1.45	Ea	293.00	52.20	345.20	587.00
36" x 80"	BC@1.45	Ea	293.00	52.20	345.20	587.00
Full-view woodcore storm doors. 1-1/2-inch thick, heavy-gauge aluminum frame over foam insulation. Perimeter weatherstripping and water-diverting rain cap. Full-length, piano-style hinge. Reversible hinge for left- or right-side entry door handle. Two heavy-duty closers.						
30" x 80", White	BC@1.45	Ea	293.00	52.20	345.20	587.00
32" x 80", White	BC@1.45	Ea	293.00	52.20	345.20	587.00
36" x 80", White	BC@1.45	Ea	293.00	52.20	345.20	587.00

Security Doors

	Craft@Hrs	Unit	Material	Labor	Total	Sell
Security storm doors. 16 gauge steel. Brass deadbolt lock. Tempered safety glass. Interchangeable screen panel. Heavy-gauge all-welded steel frame with mitered top. Double weatherstripped jamb and sweep. Tamper-resistant hinges with concealed interior screws. Heavy-duty pneumatic closer and safety wind chain. Prehung with reversible hinge.						
30" x 80"	BC@1.45	Ea	336.00	52.20	388.20	660.00
32" x 80"	BC@1.45	Ea	336.00	52.20	388.20	660.00
36" x 80"	BC@1.45	Ea	336.00	52.20	388.20	660.00
Class II security doors. Rated to withstand 400 pounds. Double lockbox with extension plate. Decorative geometric ornaments. All welded steel construction. Expanded, galvanized metal screen. 3/4" square frame design.						
32" x 80"	BC@1.45	Ea	118.00	52.20	170.20	289.00
36" x 80"	BC@1.45	Ea	120.00	52.20	172.20	293.00
Heavy-duty security door with screen. Right- or left-hand exterior mounting. All welded steel construction. Baked-on powder coating. Double lockbox with tamper-resistant extension plate. Uses standard 2-3/8" backset lock (not included). Galvanized perforated metal security screen. 1" x 1" door frame. 1" x 1-1/2" jambs, prehung on hinge side.						
32" x 80"	BC@1.45	Ea	146.00	52.20	198.20	337.00
36" x 80"	BC@1.45	Ea	148.00	52.20	200.20	340.00
Paradise-style security door. Five semi-concealed hinges. Heavy-duty deadbolt. All-steel frame.						
30" x 80"	BC@1.45	Ea	441.00	52.20	493.20	838.00
32" x 80"	BC@1.45	Ea	441.00	52.20	493.20	838.00
36" x 80"	BC@1.45	Ea	441.00	52.20	493.20	838.00
Folding security gates. Riveted 3/4" top and bottom channel. Gate pivots and folds flat to one side. Provides air circulation and positive visibility.						
48" max width, 79" high	BC@1.25	Ea	111.00	45.00	156.00	265.00

	Craft@Hrs	Unit	Material	Labor	Total	Sell

Interior Slab Doors

Cut wall and frame interior door opening, install slab door and trim. Includes header, trimmer studs, interior flush lauan slab door ($24.00), jamb, two hinges stop, casing two sides, and passage lockset ($28). Add the cost of debris removal, painting, and floor and wall finish, as needed.

	Craft@Hrs	Unit	Material	Labor	Total	Sell
32" to 36" slab interior door	BC@6.30	Ea	168.00	227.00	395.00	672.00

Install interior slab door in an existing framed and cased opening. Includes minor repair to the jamb (blind Dutchman), two hinges, interior flush lauan slab door ($24.00), stop, and passage lockset ($28).

	Craft@Hrs	Unit	Material	Labor	Total	Sell
28" to 36" x 80" door	BC@2.93	Ea	89.30	106.00	195.30	332.00

Remove and replace interior door casing, jamb and stop. Includes jamb, stop, and casing one side only. Per door opening.

	Craft@Hrs	Unit	Material	Labor	Total	Sell
Door opening to 36" x 80"	BC@1.20	Ea	72.00	43.20	115.20	196.00

Remove and replace interior slab door, on existing hinges. Includes interior flush lauan slab door ($26.90), and passage lockset ($28).

	Craft@Hrs	Unit	Material	Labor	Total	Sell
36" x 80" exterior door	BC@2.00	Ea	56.00	72.00	128.00	218.00

Install interior slab door in an existing cased opening. Includes interior flush lauan slab door ($25.00), two hinges and stop. Add the cost of additional work to the jamb (if needed), installing the lockset and strike, setting or adjusting the stop, and setting the door casing.

	Craft@Hrs	Unit	Material	Labor	Total	Sell
36" x 80" interior door	BC@1.15	Ea	53.00	41.40	94.40	160.00

Flush hardboard interior doors. Hollow core. Bored for lockset. Ready to paint or stain. 1-3/8" thick.

	Craft@Hrs	Unit	Material	Labor	Total	Sell
24" x 80"	BC@1.15	Ea	26.50	41.40	67.90	115.00
28" x 80"	BC@1.15	Ea	28.60	41.40	70.00	119.00
30" x 80"	BC@1.15	Ea	30.60	41.40	72.00	122.00
32" x 80"	BC@1.15	Ea	32.60	41.40	74.00	126.00
36" x 80"	BC@1.15	Ea	34.70	41.40	76.10	129.00

Flush birch interior doors. Hollow core. 1-3/8" thick. Wood veneer. Stainable and paintable. All wood stile and rail.

	Craft@Hrs	Unit	Material	Labor	Total	Sell
18" x 80"	BC@1.15	Ea	30.60	41.40	72.00	122.00
24" x 80"	BC@1.15	Ea	30.60	41.40	72.00	122.00
28" x 80"	BC@1.15	Ea	32.60	41.40	74.00	126.00
30" x 80"	BC@1.15	Ea	34.70	41.40	76.10	129.00
32" x 80"	BC@1.15	Ea	36.70	41.40	78.10	133.00
36" x 80"	BC@1.15	Ea	39.10	41.40	80.50	137.00

Flush lauan interior doors. Hollow core. 1-3/8" thick. Wood veneer. Stainable and paintable. All wood stile and rail.

	Craft@Hrs	Unit	Material	Labor	Total	Sell
18" x 80"	BC@1.15	Ea	17.30	41.40	58.70	99.80
24" x 80"	BC@1.15	Ea	22.40	41.40	63.80	108.00
28" x 80"	BC@1.15	Ea	22.40	41.40	63.80	108.00
30" x 80"	BC@1.15	Ea	22.40	41.40	63.80	108.00
32" x 80"	BC@1.15	Ea	23.50	41.40	64.90	110.00
36" x 80"	BC@1.15	Ea	24.50	41.40	65.90	112.00

Flush oak interior doors. Hollow core. 1-3/8" thick. Wood veneer. Stainable and paintable. All wood stile and rail.

	Craft@Hrs	Unit	Material	Labor	Total	Sell
18" x 80"	BC@1.15	Ea	29.00	41.40	70.40	120.00
24" x 80"	BC@1.15	Ea	32.60	41.40	74.00	126.00
28" x 80"	BC@1.15	Ea	33.70	41.40	75.10	128.00
30" x 80"	BC@1.15	Ea	38.80	41.40	80.20	136.00
32" x 80"	BC@1.15	Ea	39.80	41.40	81.20	138.00
36" x 80"	BC@1.15	Ea	42.80	41.40	84.20	143.00

	Craft@Hrs	Unit	Material	Labor	Total	Sell
Flush hardboard solid core interior doors. Primed. 1-3/8" thick.						
30" x 80"	BC@1.15	Ea	46.90	41.40	88.30	150.00
32" x 80"	BC@1.15	Ea	49.00	41.40	90.40	154.00
36" x 80"	BC@1.15	Ea	51.00	41.40	92.40	157.00
Flush lauan solid core interior doors. 1-3/8" thick. Stainable.						
28" x 80"	BC@1.15	Ea	56.10	41.40	97.50	166.00
32" x 80"	BC@1.15	Ea	56.10	41.40	97.50	166.00
36" x 80"	BC@1.15	Ea	56.10	41.40	97.50	166.00
30" x 84"	BC@1.15	Ea	51.00	41.40	92.40	157.00
32" x 84"	BC@1.15	Ea	53.00	41.40	94.40	160.00
34" x 84"	BC@1.15	Ea	57.10	41.40	98.50	167.00
36" x 84"	BC@1.15	Ea	57.10	41.40	98.50	167.00
Flush birch solid core interior doors. 1-3/8" thick. Solid core.						
24" x 80"	BC@1.15	Ea	78.50	41.40	119.90	204.00
32" x 80"	BC@1.15	Ea	79.00	41.40	120.40	205.00
36" x 80"	BC@1.15	Ea	79.50	41.40	120.90	206.00
Flush hardboard interior doors with lite. 1-3/8" thick.						
30" x 80" x 1-3/8"	BC@1.15	Ea	109.00	41.40	150.40	256.00
32" x 80" x 1-3/8"	BC@1.15	Ea	109.00	41.40	150.40	256.00
Hardboard 6-panel colonist interior doors. Hollow core. Raised panels. 1-3/8" thick. Masonite high-density fiberboard with CraftMaster door facings. Embossed simulated woodgrain panels. Primed.						
18" x 80"	BC@1.15	Ea	19.40	41.40	60.80	103.00
24" x 80"	BC@1.15	Ea	19.40	41.40	60.80	103.00
28" x 80"	BC@1.15	Ea	22.40	41.40	63.80	108.00
30" x 80"	BC@1.15	Ea	22.40	41.40	63.80	108.00
32" x 80"	BC@1.15	Ea	22.40	41.40	63.80	108.00
36" x 80"	BC@1.15	Ea	22.40	41.40	63.80	108.00
Hardboard 6-panel molded face interior doors. 1-3/8" thick. Molded and primed hardboard. Embossed simulated woodgrain. Hollow core. Ready to paint. Bored for lockset.						
24" x 80"	BC@1.15	Ea	26.50	41.40	67.90	115.00
28" x 80"	BC@1.15	Ea	26.50	41.40	67.90	115.00
30" x 80"	BC@1.15	Ea	26.50	41.40	67.90	115.00
32" x 80"	BC@1.15	Ea	26.50	41.40	67.90	115.00
36" x 80"	BC@1.15	Ea	26.50	41.40	67.90	115.00
Lauan 6-panel molded face interior doors. Bored for latchset. 1-3/8" thick.						
24" x 80"	BC@1.15	Ea	25.50	41.40	66.90	114.00
28" x 80"	BC@1.15	Ea	26.50	41.40	67.90	115.00
30" x 80"	BC@1.15	Ea	26.50	41.40	67.90	115.00
32" x 80"	BC@1.15	Ea	26.50	41.40	67.90	115.00
36" x 80"	BC@1.15	Ea	26.50	41.40	67.90	115.00

	Craft@Hrs	Unit	Material	Labor	Total	Sell
Pine 6-panel molded face interior doors. Radiata pine veneer. Primed. Hollow core. 1-3/8" thick. 3/4" double hip raised panels. Unfinished. Ready to paint, stain, or varnish.						
24" x 80"	BC@1.15	Ea	50.00	41.40	91.40	155.00
30" x 80"	BC@1.15	Ea	53.80	41.40	95.20	162.00
32" x 80"	BC@1.15	Ea	54.60	41.40	96.00	163.00
36" x 80"	BC@1.15	Ea	56.40	41.40	97.80	166.00
Pine 6-panel stile and rail interior doors. Ready to paint or stain. 1-3/8" thick.						
24" x 80"	BC@1.15	Ea	95.90	41.40	137.30	233.00
28" x 80"	BC@1.15	Ea	123.00	41.40	164.40	279.00
30" x 80"	BC@1.15	Ea	101.00	41.40	142.40	242.00
32" x 80"	BC@1.15	Ea	103.00	41.40	144.40	245.00
36" x 80"	BC@1.15	Ea	107.00	41.40	148.40	252.00
Fir 1-panel stile and rail interior doors. 1-3/8" thick.						
28" x 80"	BC@1.15	Ea	188.00	41.40	229.40	390.00
30" x 80"	BC@1.15	Ea	189.00	41.40	230.40	392.00
32" x 80"	BC@1.15	Ea	190.00	41.40	231.40	393.00
Fir 3-panel stile and rail interior doors. 1-3/8" thick.						
28" x 80"	BC@1.15	Ea	229.00	41.40	270.40	460.00
30" x 80"	BC@1.15	Ea	229.00	41.40	270.40	460.00
32" x 80"	BC@1.15	Ea	230.00	41.40	271.40	461.00
Fir 6-panel stile and rail interior doors. Select grade. 1-3/8" thick.						
24" x 80"	BC@1.15	Ea	138.00	41.40	179.40	305.00
28" x 80"	BC@1.15	Ea	143.00	41.40	184.40	313.00
30" x 80"	BC@1.15	Ea	150.00	41.40	191.40	325.00
32" x 80"	BC@1.15	Ea	153.00	41.40	194.40	330.00
36" x 80"	BC@1.15	Ea	157.00	41.40	198.40	337.00
Oak 6-panel stile and rail interior doors. Red oak. Double bevel hip raised panels. 1-3/8" thick.						
24" x 80"	BC@1.15	Ea	75.00	41.40	116.40	198.00
28" x 80"	BC@1.15	Ea	80.60	41.40	122.00	207.00
30" x 80"	BC@1.15	Ea	86.60	41.40	128.00	218.00
32" x 80"	BC@1.15	Ea	91.70	41.40	133.10	226.00
36" x 80"	BC@1.15	Ea	96.80	41.40	138.20	235.00
French stile and rail interior doors. Clear pine. Ready to paint or stain. 1-3/8" thick. Pre-masked tempered glass.						
24" x 80", 10 lite	BC@1.40	Ea	120.00	50.40	170.40	290.00
28" x 80", 10 lite	BC@1.40	Ea	129.00	50.40	179.40	305.00
30" x 80", 15 lite	BC@1.40	Ea	124.00	50.40	174.40	296.00
32" x 80", 15 lite	BC@1.40	Ea	128.00	50.40	178.40	303.00
36" x 80", 15 lite	BC@1.40	Ea	132.00	50.40	182.40	310.00
Half louver stile and rail interior doors. 1-3/8" thick.						
24" x 80"	BC@1.15	Ea	163.00	41.40	204.40	347.00
28" x 80"	BC@1.15	Ea	170.00	41.40	211.40	359.00
30" x 80"	BC@1.15	Ea	174.00	41.40	215.40	366.00
32" x 80"	BC@1.15	Ea	181.00	41.40	222.40	378.00
36" x 80"	BC@1.15	Ea	185.00	41.40	226.40	385.00

	Craft@Hrs	Unit	Material	Labor	Total	Sell
Full louver cafe (bar) doors. Per pair of doors. With hardware.						
24" wide, 42" high	BC@1.15	Ea	79.10	41.40	120.50	205.00
30" wide, 42" high	BC@1.15	Ea	86.60	41.40	128.00	218.00
32" wide, 42" high	BC@1.15	Ea	95.10	41.40	136.50	232.00
36" wide, 42" high	BC@1.15	Ea	97.30	41.40	138.70	236.00
Stile and rail cafe doors. Per pair of doors. With hardware.						
30" x 42"	BC@1.15	Ea	51.00	41.40	92.40	157.00
32" x 42"	BC@1.15	Ea	54.00	41.40	95.40	162.00
36" x 42"	BC@1.15	Ea	56.00	41.40	97.40	166.00

Interior Door Jambs

	Craft@Hrs	Unit	Material	Labor	Total	Sell
Finger joint pine interior jamb set. Unfinished. Two 6'8" legs and one 3' head.						
4-9/16" x 11/16"	BC@.500	Ea	27.90	18.00	45.90	78.00
5-1/4" x 11/16"	BC@.500	Ea	30.00	18.00	48.00	81.60
Solid clear pine interior jamb. Varnish or stain. Jamb set includes two 6'8" legs and one 3' head.						
11/16" x 4-9/16" x 7' leg	BC@.210	Ea	10.20	7.56	17.76	30.20
11/16" x 4-9/16" x 3' head	BC@.090	Ea	7.14	3.24	10.38	17.60
11/16" x 4-9/16" jamb set	BC@.500	Ea	25.50	18.00	43.50	74.00
Door jamb set with hinges.						
80" x 4-9/16", per set	BC@.500	Ea	45.60	18.00	63.60	108.00
80" x 6-9/16", per set	BC@.500	Ea	51.00	18.00	69.00	117.00

Prehung Interior Doors

Cut wall and frame interior door opening, install prehung door and trim. Includes header, trimmer studs, interior flush prehung door ($60.00) with jamb, two hinges, stop, casing two sides, and passage lockset ($30). Add the cost of debris removal, painting, floor and wall finish, as needed.

	Craft@Hrs	Unit	Material	Labor	Total	Sell
32" to 36" prehung interior door	BC@5.08	Ea	120.00	183.00	303.00	515.00

Install interior prehung door in an existing wall opening. Includes labor setting the door, casing, jamb and stops. Add the cost of setting the lockset.

	Craft@Hrs	Unit	Material	Labor	Total	Sell
32" to 36" x 80" interior door	BC@.750	Ea	—	27.00	27.00	45.90
Flush prehung hardboard interior doors. 1-3/8" thick. Hollow core. Primed. Ready to paint.						
24" x 80"	BC@.750	Ea	56.10	27.00	83.10	141.00
28" x 80"	BC@.750	Ea	63.20	27.00	90.20	153.00
30" x 80"	BC@.750	Ea	65.30	27.00	92.30	157.00
32" x 80"	BC@.750	Ea	67.30	27.00	94.30	160.00
36" x 80"	BC@.750	Ea	69.40	27.00	96.40	164.00

	Craft@Hrs	Unit	Material	Labor	Total	Sell
Flush prehung lauan interior doors. 1-3/8" thick. 4-9/16" jamb. Bored for lockset.						
18" x 78"	BC@.750	Ea	55.00	27.00	82.00	139.00
24" x 78"	BC@.750	Ea	56.10	27.00	83.10	141.00
28" x 78"	BC@.750	Ea	57.10	27.00	84.10	143.00
30" x 78"	BC@.750	Ea	58.10	27.00	85.10	145.00
32" x 78"	BC@.750	Ea	60.20	27.00	87.20	148.00
36" x 78"	BC@.750	Ea	62.20	27.00	89.20	152.00
24" x 80"	BC@.750	Ea	56.10	27.00	83.10	141.00
28" x 80"	BC@.750	Ea	57.10	27.00	84.10	143.00
30" x 80"	BC@.750	Ea	58.10	27.00	85.10	145.00
32" x 80"	BC@.750	Ea	60.20	27.00	87.20	148.00
36" x 80"	BC@.750	Ea	62.20	27.00	89.20	152.00
Flush prehung lauan interior doors. 1-3/8" thick. Hollow core. Bored for 2-1/8" lockset with 2-3/8" backset. Casing and lockset sold separately. Wood stile construction. Can be stained or finished naturally. Two brass hinges. Paint grade pine split jambs and 2-1/4" casings. 4" to 4-5/8" adjustable jamb.						
24" x 80"	BC@.750	Ea	69.40	27.00	96.40	164.00
28" x 80"	BC@.750	Ea	70.40	27.00	97.40	166.00
30" x 80"	BC@.750	Ea	74.50	27.00	101.50	173.00
32" x 80"	BC@.750	Ea	78.50	27.00	105.50	179.00
36" x 80"	BC@.750	Ea	80.60	27.00	107.60	183.00
Flush prehung birch interior doors. Stain grade. 1-3/8" thick. Wood stiles. Bored for lockset. Multi-finger jointed door jamb 11/16" x 4-9/16". 3 brass finish hinges 3-1/2" x 3-1/2".						
24" x 80"	BC@.750	Ea	72.40	27.00	99.40	169.00
28" x 80"	BC@.750	Ea	74.50	27.00	101.50	173.00
30" x 80"	BC@.750	Ea	78.50	27.00	105.50	179.00
32" x 80"	BC@.750	Ea	80.60	27.00	107.60	183.00
36" x 80"	BC@.750	Ea	81.60	27.00	108.60	185.00
Flush prehung oak interior doors. Stain grade. 1-3/8" thick. Wood stiles. Bored for lockset. Multi-finger jointed door jamb 11/16" x 4-9/16". 3 brass finish hinges 3-1/2" x 3-1/2".						
24" x 80"	BC@.750	Ea	65.30	27.00	92.30	157.00
28" x 80"	BC@.750	Ea	68.30	27.00	95.30	162.00
30" x 80"	BC@.750	Ea	70.40	27.00	97.40	166.00
32" x 80"	BC@.750	Ea	90.80	27.00	117.80	200.00
36" x 80"	BC@.750	Ea	93.80	27.00	120.80	205.00
Flush prehung red oak interior doors. Hollow core. 1-3/8" thick. Stain grade.						
24" x 80"	BC@.750	Ea	70.70	27.00	97.70	166.00
28" x 80"	BC@.750	Ea	73.70	27.00	100.70	171.00
30" x 80"	BC@.750	Ea	75.80	27.00	102.80	175.00
32" x 80"	BC@.750	Ea	78.20	27.00	105.20	179.00
36" x 80"	BC@.750	Ea	80.80	27.00	107.80	183.00
Interior prehung lauan heater closet doors.						
24" x 60"	BC@.750	Ea	72.60	27.00	99.60	169.00

	Craft@Hrs	Unit	Material	Labor	Total	Sell
6-panel colonial prehung interior doors. 11/16" x 4-9/16" jamb. Bored for lockset. Casing and hardware sold separately. Hollow core. 1-3/8" thick.						
18" x 80", 3-panel	BC@.750	Ea	54.10	27.00	81.10	138.00
24" x 80"	BC@.750	Ea	51.00	27.00	78.00	133.00
28" x 80"	BC@.750	Ea	51.00	27.00	78.00	133.00
30" x 80"	BC@.750	Ea	51.00	27.00	78.00	133.00
32" x 80"	BC@.750	Ea	51.00	27.00	78.00	133.00
36" x 80"	BC@.750	Ea	51.00	27.00	78.00	133.00
6-panel colonist prehung interior doors. 6-9/16" jamb. Hollow core.						
24" x 80"	BC@.750	Ea	102.00	27.00	129.00	219.00
30" x 80"	BC@.750	Ea	110.00	27.00	137.00	233.00
32" x 80"	BC@.750	Ea	112.00	27.00	139.00	236.00
36" x 80"	BC@.750	Ea	115.00	27.00	142.00	241.00
6-panel colonist prehung interior doors. Ready to paint or stain. Split flush jambs. Colonial casing attached. Bored for lockset.						
18" x 78"	BC@.750	Ea	70.40	27.00	97.40	166.00
24" x 78"	BC@.750	Ea	70.40	27.00	97.40	166.00
28" x 78"	BC@.750	Ea	72.40	27.00	99.40	169.00
30" x 78"	BC@.750	Ea	73.40	27.00	100.40	171.00
32" x 78"	BC@.750	Ea	76.50	27.00	103.50	176.00
36" x 78"	BC@.750	Ea	78.50	27.00	105.50	179.00
18" x 80"	BC@.750	Ea	70.40	27.00	97.40	166.00
24" x 80"	BC@.750	Ea	70.40	27.00	97.40	166.00
28" x 80"	BC@.750	Ea	72.40	27.00	99.40	169.00
30" x 80"	BC@.750	Ea	73.40	27.00	100.40	171.00
32" x 80"	BC@.750	Ea	76.50	27.00	103.50	176.00
36" x 80"	BC@.750	Ea	78.50	27.00	105.50	179.00
6-panel hardboard prehung interior doors. Wood stiles. Bored for lockset. Primed finger joint 11/16" x 4-9/16" jamb. 3 brass finish hinges 3-1/2" x 3-1/2". Hollow core.						
24" x 80"	BC@.750	Ea	54.10	27.00	81.10	138.00
28" x 80"	BC@.750	Ea	56.10	27.00	83.10	141.00
30" x 80"	BC@.750	Ea	67.30	27.00	94.30	160.00
32" x 80"	BC@.750	Ea	70.40	27.00	97.40	166.00
36" x 80"	BC@.750	Ea	71.40	27.00	98.40	167.00
6-panel hardboard prehung interior doors. 4-9/16" adjustable split jambs. Hollow core. Paint grade casing applied. Pre-drilled for lockset. Ready to paint or gel stain.						
24" x 80"	BC@.750	Ea	64.30	27.00	91.30	155.00
28" x 80"	BC@.750	Ea	66.30	27.00	93.30	159.00
30" x 80"	BC@.750	Ea	67.30	27.00	94.30	160.00
32" x 80"	BC@.750	Ea	70.40	27.00	97.40	166.00
36" x 80"	BC@.750	Ea	71.40	27.00	98.40	167.00
6-panel hardboard prehung double interior doors. 1-3/8" thick. Flat jamb.						
48" x 80"	BC@1.25	Ea	158.00	45.00	203.00	345.00
60" x 80"	BC@1.25	Ea	165.00	45.00	210.00	357.00
6-panel prehung fire-rated interior doors. 4-9/16" jamb. 20-minute fire rating.						
32" x 80"	BC@.750	Ea	197.00	27.00	224.00	381.00
36" x 80"	BC@.750	Ea	201.00	27.00	228.00	388.00

	Craft@Hrs	Unit	Material	Labor	Total	Sell
6-panel oak prehung interior doors. Hollow core. 1-3/8" thick.						
24" x 80"	BC@.750	Ea	147.00	27.00	174.00	296.00
28" x 80"	BC@.750	Ea	174.00	27.00	201.00	342.00
30" x 80"	BC@.750	Ea	171.00	27.00	198.00	337.00
32" x 80"	BC@.750	Ea	164.00	27.00	191.00	325.00
36" x 80"	BC@.750	Ea	169.00	27.00	196.00	333.00
6-panel prehung interior double doors. Bored for lockset. Ready to paint or stain. 1-3/8" thick. 4-9/16" jambs.						
4'0" x 6'8"	BC@1.50	Ea	171.00	54.00	225.00	383.00
6'0" x 6'8"	BC@1.50	Ea	178.00	54.00	232.00	394.00
6-panel hemlock prehung interior doors. Bored for lockset. Ready to paint or stain. 1-3/8" thick. 4-9/16" jambs.						
24" x 80"	BC@.750	Ea	182.00	27.00	209.00	355.00
28" x 80"	BC@.750	Ea	185.00	27.00	212.00	360.00
30" x 80"	BC@.750	Ea	188.00	27.00	215.00	366.00
32" x 80"	BC@.750	Ea	192.00	27.00	219.00	372.00
36" x 80"	BC@.750	Ea	197.00	27.00	224.00	381.00
2-panel knotty alder prehung interior doors. Bored for lockset. 1-3/8" thick.						
24" x 80"	BC@.750	Ea	202.00	27.00	229.00	389.00
28" x 80"	BC@.750	Ea	207.00	27.00	234.00	398.00
30" x 80"	BC@.750	Ea	212.00	27.00	239.00	406.00
32" x 80"	BC@.750	Ea	217.00	27.00	244.00	415.00
36" x 80"	BC@.750	Ea	222.00	27.00	249.00	423.00
Pine full louver prehung interior doors. Ready to install. Provides ventilation while maintaining privacy. 11/16" x 4-9/16" clear face pine door jamb. 1-3/8" thick. Includes (3) 3-1/2" x 3-1/2" hinges with brass finish. Bored for lockset.						
24" x 80"	BC@.750	Ea	135.00	27.00	162.00	275.00
28" x 80"	BC@.750	Ea	152.00	27.00	179.00	304.00
30" x 80"	BC@.750	Ea	152.00	27.00	179.00	304.00
32" x 80"	BC@.750	Ea	158.00	27.00	185.00	315.00
36" x 80"	BC@.750	Ea	160.00	27.00	187.00	318.00
Pine louver over louver prehung interior doors. Bored for lockset. 1-3/8" thick.						
32" x 80"	BC@.750	Ea	207.00	27.00	234.00	398.00
Prehung pantry doors. Finger joint flat jamb. Bored for lockset. 1-3/8" thick.						
28" x 80"	BC@.750	Ea	235.00	27.00	262.00	445.00
30" x 80"	BC@.750	Ea	235.00	27.00	262.00	445.00
10-lite wood French prehung double interior doors. Pre-masked tempered glass for easy finishing. True divided light design. Ponderosa pine. 1-3/8" thick.						
48" x 80", unfinished	BC@1.25	Ea	342.00	45.00	387.00	658.00
60" x 80", unfinished	BC@1.25	Ea	364.00	45.00	409.00	695.00
48" x 80", primed	BC@1.25	Ea	291.00	45.00	336.00	571.00
60" x 80", primed	BC@1.25	Ea	290.00	45.00	335.00	570.00
15-lite wood French prehung double interior doors. Pre-masked tempered glass for easy finishing. True divided light design. Ponderosa pine. 1-3/8" thick.						
48" x 80"	BC@1.25	Ea	288.00	45.00	333.00	566.00
60" x 80"	BC@1.25	Ea	288.00	45.00	333.00	566.00

	Craft@Hrs	Unit	Material	Labor	Total	Sell
Oak French prehung double interior doors. Bored for lockset. 1-3/8" thick.						
48" x 80"	BC@1.25	Ea	427.00	45.00	472.00	802.00
60" x 80"	BC@1.25	Ea	488.00	45.00	533.00	906.00
72" x 80"	BC@1.25	Ea	510.00	45.00	555.00	944.00

Interior Latchsets

	Craft@Hrs	Unit	Material	Labor	Total	Sell
Knob passage latch.						
Dummy, brass	BC@.250	Ea	10.20	9.00	19.20	32.60
Dummy, satin nickel	BC@.250	Ea	12.20	9.00	21.20	36.00
Hall or closet, brass	BC@.250	Ea	23.40	9.00	32.40	55.10
Hall or closet, satin nickel	BC@.250	Ea	31.10	9.00	40.10	68.20
Egg knob privacy latchset.						
Bright brass	BC@.250	Ea	34.00	9.00	43.00	73.10
Satin nickel	BC@.250	Ea	36.20	9.00	45.20	76.80
Privacy lever latchset. Both levers locked or unlocked by turn button. Outside lever can be unlocked by emergency key (included).						
Antique brass	BC@.250	Ea	15.00	9.00	24.00	40.80
Polished brass	BC@.250	Ea	15.00	9.00	24.00	40.80
Satin nickel	BC@.250	Ea	15.00	9.00	24.00	40.80

Door Hardware

	Craft@Hrs	Unit	Material	Labor	Total	Sell
Door knockers. Polished brass.						
3-1/2" x 6", knocker and eye viewer	BC@.250	Ea	25.00	9.00	34.00	57.80
6-1/2" x 3-1/2", colonial, ornate	BC@.250	Ea	50.00	9.00	59.00	100.00
Door pull plates.						
4" x 16", satin chrome	BC@.333	Ea	26.00	12.00	38.00	64.60
Door push plates.						
3" x 12", bright brass	BC@.333	Ea	12.20	12.00	24.20	41.10
3-1/2" x 15" anodized aluminum	BC@.333	Ea	11.20	12.00	23.20	39.40
3-1/2" x 15", solid brass	BC@.333	Ea	17.30	12.00	29.30	49.80
Door kick plates.						
6" x 30", anodized aluminum	BC@.333	Ea	18.30	12.00	30.30	51.50
6" x 30", lifetime brass finish	BC@.333	Ea	38.50	12.00	50.50	85.90
6" x 30", mirror brass finish	BC@.333	Ea	51.20	12.00	63.20	107.00
6" x 34", polished brass finish	BC@.333	Ea	52.00	12.00	64.00	109.00
8" x 34", .050" stainless	BC@.333	Ea	28.00	12.00	40.00	68.00
8" x 34", anodized aluminum	BC@.333	Ea	33.30	12.00	45.30	77.00
8" x 34", lifetime brass finish	BC@.333	Ea	45.90	12.00	57.90	98.40
8" x 34", satin nickel finish	BC@.333	Ea	45.30	12.00	57.30	97.40
8" x 34", magnetic, brass finish	BC@.333	Ea	47.80	12.00	59.80	102.00
8" x 34", polished brass finish	BC@.333	Ea	63.50	12.00	75.50	128.00
Door stops.						
3" spring wall stop	BC@.084	Ea	2.13	3.02	5.15	8.76
2" wall stop, bright brass	BC@.084	Ea	4.00	3.02	7.02	11.90
2" wall stop, satin stainless	BC@.084	Ea	4.50	3.02	7.52	12.80
4" kick down stop, aluminum	BC@.120	Ea	4.50	4.32	8.82	15.00
4" kick down stop, bright brass	BC@.120	Ea	4.50	4.32	8.82	15.00
4" kick down door holder, brass	BC@.166	Ea	17.00	5.98	22.98	39.10
Floor-mounted door bumper	BC@.166	Ea	8.00	5.98	13.98	23.80

	Craft@Hrs	Unit	Material	Labor	Total	Sell
Door peep sights. Brass with optical lens.						
160-degree, 1/2" diameter	BC@.300	Ea	6.00	10.80	16.80	28.60
190-degree, 1/2" diameter	BC@.300	Ea	9.00	10.80	19.80	33.70
160-degree, 1" diameter	BC@.300	Ea	12.00	10.80	22.80	38.80
House numbers. Brass or black coated.						
4" high, brass	BC@.103	Ea	3.03	3.71	6.74	11.50
4" high, stainless steel	BC@.103	Ea	7.00	3.71	10.71	18.20
5" high, bronze	BC@.103	Ea	7.50	3.71	11.21	19.10

Patio Doors

Install a sliding or hinged patio door. Setting a door in an existing framed opening.						
To 6' wide x 8' high	B1@3.58	Ea	—	116.00	116.00	197.00
Over 6' wide x 8' high to 12' x 8'	B1@4.60	Ea	—	150.00	150.00	255.00
Remove and replace a sliding or hinged patio door. Includes removing the existing door and frame down to the rough opening, trim the siding as needed, caulk the opening, shim and nail the door in place, set exterior trim. Add the cost of painting, debris removal, wall and floor finish, as required.						
To 6' wide x 8' high	B1@5.50	Ea	—	179.00	179.00	304.00
Over 6' wide x 8' high to 12' x 8'	B1@6.95	Ea	—	226.00	226.00	384.00
10-lite swinging patio double doors. White finish. Dual-glazed with clear tempered glass. Fully weatherstripped. Pre-drilled lock-bore.						
6'0" wide x 6'8" high	B1@3.58	Ea	637.00	116.00	753.00	1,280.00
Vinyl-clad hinged patio double doors, Frenchwood®, Andersen. White exterior. Perma-Shield® low-maintenance finish. Clear pine interior. 1-lite. 3-point locking system. Adjustable hinges with ball-bearing pivots. Low-E2 tempered insulating glass. Mortise and tenon joints.						
6'0" x 6'8"	B1@3.58	Ea	1,400.00	116.00	1,516.00	2,580.00
Center hinge 15-lite aluminum patio double doors. Sized for replacement. Unfinished clear pine interior. White extruded aluminum-clad exterior. Weatherstripped frame. Self-draining sill. Bored for lockset. 3/4" Low-E insulated safety glass.						
6'0" x 6'8"	B1@3.58	Ea	909.00	116.00	1,025.00	1,740.00
Aluminum inward swing 1-lite patio double doors. Low-E insulated safety glass. Clear wood interior ready to paint or stain.						
6'0" x 6'8"	B1@3.58	Ea	744.00	116.00	860.00	1,460.00
Aluminum inward swing 10-lite patio double doors. Unfinished interior. Low-E2 argon-filled insulated glass. Thermal-break sill. Bored for lockset. With head and foot bolts.						
6'0" x 6'8"	B1@3.58	Ea	764.00	116.00	880.00	1,500.00
Vinyl-clad gliding patio doors, Andersen. White. Tempered 1" Low-E insulating glass. Anodized aluminum track.						
6'0" x 6'8"	B1@3.58	Ea	703.00	116.00	819.00	1,390.00
Vinyl-clad gliding patio doors. Pine core. Dual-pane Low-E tempered insulating glass. Stainless-steel tracks. Ball-bearing rollers. Combination weatherstrip and interlock.						
6'0" x 6'8"	B1@3.58	Ea	1,020.00	116.00	1,136.00	1,930.00
10-lite sliding patio doors. Solid pine core. Weatherstripped on 4 sides of operating panels. Butcher block 4-1/2"-wide stiles and 9-1/2" bottom rail. 4-9/16" jamb width. White. Actual dimensions: 71-1/4" wide x 79-1/2" high. Pre-drilled lock-bore.						
6'0" wide x 6'8" high	B1@3.58	Ea	623.00	116.00	739.00	1,260.00

	Craft@Hrs	Unit	Material	Labor	Total	Sell
Aluminum sliding patio doors. Unfinished interior. White aluminum-clad exterior. With lockset. Insulating Low-E2 glass.						
6'0" x 6'8", 1-lite	B1@3.58	Ea	611.00	116.00	727.00	1,240.00
6'0" x 6'8", 15-lite look	B1@3.58	Ea	611.00	116.00	727.00	1,240.00
Aluminum sliding patio doors. With screen. Single glazed. Reversible, self-aligning ball bearing rollers. Zinc finish lock. Dimensions nominal width by height. Actual height and width are 1/2" less.						
5' x 6'8", mill finish	B1@3.58	Ea	205.00	116.00	321.00	546.00
6' x 6'8" mill finish	B1@3.58	Ea	383.00	116.00	499.00	848.00
6' x 6'8" white finish	B1@3.58	Ea	401.00	116.00	517.00	879.00
6' x 6'8" bronze finish	B1@3.58	Ea	405.00	116.00	521.00	886.00
6' x 6'8" white finish, solar gray glass	B1@3.58	Ea	475.00	116.00	591.00	1,000.00
8' x 8'0", white finish	B1@4.60	Ea	386.00	150.00	536.00	911.00
Vinyl sliding patio doors. Single glazed. By opening size. Actual width and height are 1/2" less.						
5'0" x 6'8", with grille	B1@3.58	Ea	479.00	116.00	595.00	1,010.00
6'0" x 6'8"	B1@3.58	Ea	479.00	116.00	595.00	1,010.00
6'0" x 6'8", with grille	B1@3.58	Ea	518.00	116.00	634.00	1,080.00
8'0" x 6'8"	B1@3.58	Ea	599.00	116.00	715.00	1,220.00
8'0" x 6'8", with grille	B1@3.58	Ea	639.00	116.00	755.00	1,280.00
1-lite swinging prehung steel patio double doors. Polyurethane core. High-performance weatherstrip. Adjustable thermal-break sill. Non-yellowing frame and grille. Inward swing.						
6'0" x 6'8"	B1@3.58	Ea	343.00	116.00	459.00	780.00
10-lite swinging prehung steel patio double doors. 24 gauge galvanized steel. Double bored for deadbolt. Compression weatherstripping. Aluminum T-astragal. Single glazed.						
6'0" x 6'8"	B1@3.58	Ea	294.00	116.00	410.00	697.00
10-lite swinging steel patio double doors. Inward swing. Insulated Low-E glass.						
5'0" x 6'8"	B1@3.58	Ea	451.00	116.00	567.00	964.00
6'0" x 6'8"	B1@3.58	Ea	441.00	116.00	557.00	947.00
10-lite venting steel patio double doors. Inward swing. Insulated glass.						
6'0" x 6'8"	B1@3.58	Ea	570.00	116.00	686.00	1,170.00
8'0" x 6'8"	B1@3.58	Ea	689.00	116.00	805.00	1,370.00
15-lite swinging steel patio double doors. Inward swing.						
5'0" x 6'8", insulated glass	B1@3.58	Ea	355.00	116.00	471.00	801.00
6'0" x 6'8", Low-E insulated glass	B1@3.58	Ea	469.00	116.00	585.00	995.00
6'0" x 6'8", single glazed	B1@3.58	Ea	353.00	116.00	469.00	797.00
15-lite prehung swinging fiberglass patio double doors. Outward swing. Insulated glass. Adjustable mill finish thermal-break threshold.						
6'0" x 6'8"	B1@3.58	Ea	662.00	116.00	778.00	1,320.00

	Craft@Hrs	Unit	Material	Labor	Total	Sell

Patio Door Replacement Parts

	Craft@Hrs	Unit	Material	Labor	Total	Sell
Replacement sliding patio door screen and track.						
2'6" x 6'8" screen	BC@.335	Ea	132.00	12.10	144.10	245.00
3'0" x 6'8" screen	BC@.335	Ea	134.00	12.10	146.10	248.00
Replacement sliding patio door screens. Adjustable 78-1/2" to 80" height. Self-latching. Fiberglass screening. Steel frame. Powder-coated finish.						
30" wide, bronze	BC@.335	Ea	47.90	12.10	60.00	102.00
36" wide, bronze	BC@.335	Ea	47.90	12.10	60.00	102.00
48" wide, gray	BC@.335	Ea	53.00	12.10	65.10	111.00
Patio door hardware set, Frenchwood® and Perma-Shield®, Andersen. Fits Andersen patio doors.						
Bright brass, gliding doors	—	Ea	198.00	—	198.00	—
Bright brass, hinged doors	—	Ea	225.00	—	225.00	—
Metro, stone, hinged doors	—	Ea	60.00	—	60.00	—
Metro, white, gliding doors	—	Ea	60.00	—	60.00	—
Sliding aluminum patio door hardware kit.						
6'0" x 6'8", bronze	BC@.450	Ea	50.00	16.20	66.20	113.00
6'0" x 6'8", white	BC@.450	Ea	50.00	16.20	66.20	113.00
Sliding patio door replacement handle set.						
White	BC@.250	Ea	35.00	9.00	44.00	74.80
Sliding patio door replacement locks.						
Auxiliary footlock with screws	BC@.250	Ea	33.70	9.00	42.70	72.60
Reachout deadlock, white	BC@.250	Ea	32.00	9.00	41.00	69.70
Sliding patio door replacement rollers.						
Lower screen roller, per pair	BC@.450	Ea	12.20	16.20	28.40	48.30
Steel tandem rollers	BC@.450	Ea	12.20	16.20	28.40	48.30
Patio door window grilles. For 6'0" x 6' 8" door.						
15-lite	—	Ea	65.00	—	65.00	—
Patio door jamb extension kit.						
6-9/16", white	—	Ea	40.00	—	40.00	—
Swinging patio door threshold.						
6'0", oak	BC@.500	Ea	35.70	18.00	53.70	91.30
Cat patio pet door.						
6" x 6", mill finish	BC@.800	Ea	134.00	28.80	162.80	277.00
Pet door. Mill finish aluminum.						
Small	BC@.800	Ea	24.30	28.80	53.10	90.30
Medium	BC@.800	Ea	34.70	28.80	63.50	108.00
Large	BC@.800	Ea	43.50	28.80	72.30	123.00
Extra Large	BC@.800	Ea	55.10	28.80	83.90	143.00
Super Large	BC@.800	Ea	86.70	28.80	115.50	196.00

	Craft@Hrs	Unit	Material	Labor	Total	Sell
Door Closers						
Remove and replace door closers.						
Interior or exterior door	BC@.540	Ea	—	19.40	19.40	33.00
Pneumatic door closers. For wood or metal out-swing doors to 90 degrees. Adjustable closing speed. Internal spring keeps rod from bending. 2-hole end plug for adjusting latching power. Bronze, black or white.						
Closer	BC@.462	Ea	9.46	16.60	26.06	44.30
Light-duty door closers. For free-swinging doors up to 85 pounds and 30" wide. Parallel arm mount. Adjustable closing speed. 125-degree maximum opening angle. Brown or ivory.						
Non hold-open	BC@.462	Ea	30.70	16.60	47.30	80.40
Hold-open	BC@.462	Ea	30.70	16.60	47.30	80.40
Medium-duty door closers. Adjustable closing and latching speeds. Designed for free-swinging residential and light commercial doors up to 140 pounds. Brown or ivory.						
Closer	BC@.462	Ea	33.60	16.60	50.20	85.30
Heavy-duty door closers.						
No cover, painted steel finish	BC@.462	Ea	55.10	16.60	71.70	122.00
No cover, aluminum finish	BC@.462	Ea	55.10	16.60	71.70	122.00
With cover, aluminum finish	BC@.462	Ea	76.50	16.60	93.10	158.00
Commercial-grade hydraulic door closers. For free-swinging doors up to 176 pounds. Includes brackets for three possible mounting applications. Adjustable back check. Adjustable closing and latching speeds. Bronze or silver.						
Closer	BC@.462	Ea	53.90	16.60	70.50	120.00
Adjustable hinge pin door closers. Includes screws.						
Brass	BC@.250	Ea	13.20	9.00	22.20	37.70
Accordion Folding Doors						
Accordion folding doors. Solid PVC slats with flexible hinges. Vertical embossed surface. Includes hardware. Fully assembled. 32" x 80". Hung in a cased opening.						
Light-weight, white	BC@.750	Ea	25.00	27.00	52.00	88.40
Standard-weight, gray	BC@.750	Ea	35.30	27.00	62.30	106.00
Standard-weight, white	BC@.750	Ea	36.70	27.00	63.70	108.00
Premium accordion folding doors. Vertical embossed surface. Preassembled. Prefinished. Adjustable width and height. Flexible vinyl hinges. Hung in a cased opening.						
8" x 80" section	BC@.750	Ea	26.00	27.00	53.00	90.10
32" x 80"	BC@.750	Ea	62.20	27.00	89.20	152.00
Folding door lock.						
Door lock	BC@.250	Ea	8.83	9.00	17.83	30.30

	Craft@Hrs	Unit	Material	Labor	Total	Sell

Bypass Closet Doors

White bypass closet doors. Prefinished white vinyl-covered hardboard panels instead of mirrors. White steel frame. Jump-proof bottom track. 1-1/2" dual race ball bearing wheels. Injection-molded top guides. Dimensions are width by height. Per pair of doors.

	Craft@Hrs	Unit	Material	Labor	Total	Sell
48" x 80"	BC@.750	Ea	57.10	27.00	84.10	143.00
60" x 80"	BC@1.00	Ea	66.30	36.00	102.30	174.00
72" x 80"	BC@1.00	Ea	79.60	36.00	115.60	197.00
48" x 96"	BC@.750	Ea	65.30	27.00	92.30	157.00
60" x 96"	BC@1.00	Ea	81.60	36.00	117.60	200.00
72" x 96"	BC@1.00	Ea	95.90	36.00	131.90	224.00

Economy mirrored bypass closet doors. Gold finished steel. Plate mirror with double strength glass. Bottom rail with jump-proof track. Dimensions are width by height. Per pair of doors.

	Craft@Hrs	Unit	Material	Labor	Total	Sell
48" x 80"	BC@.750	Ea	68.30	27.00	95.30	162.00
60" x 80"	BC@1.00	Ea	88.70	36.00	124.70	212.00
72" x 80"	BC@1.00	Ea	103.00	36.00	139.00	236.00
96" x 80"	BC@1.25	Ea	177.00	45.00	222.00	377.00

Good quality mirrored bypass closet doors. White frame. Safety-backed mirror. Dimensions are width by height. Per pair of doors.

	Craft@Hrs	Unit	Material	Labor	Total	Sell
48" x 80", white	BC@.750	Ea	95.90	27.00	122.90	209.00
60" x 80", white	BC@1.00	Ea	114.00	36.00	150.00	255.00
72" x 80", white	BC@1.00	Ea	135.00	36.00	171.00	291.00
96" x 80", white	BC@1.25	Ea	222.00	45.00	267.00	454.00
48" x 80", brushed nickel	BC@.750	Ea	123.00	27.00	150.00	255.00
60" x 80", brushed nickel	BC@1.00	Ea	143.00	36.00	179.00	304.00
72" x 80", brushed nickel	BC@1.00	Ea	166.00	36.00	202.00	343.00
48" x 96", white	BC@1.00	Ea	144.00	36.00	180.00	306.00
60" x 96", white	BC@1.00	Ea	161.00	36.00	197.00	335.00
72" x 96", white	BC@1.25	Ea	196.00	45.00	241.00	410.00
96" x 96", white	BC@1.25	Ea	271.00	45.00	316.00	537.00

Better quality mirrored bypass closet doors. White steel frame. Select 3mm plate mirror. Jump-proof bottom track. 1-1/2" dual race ball bearing wheels. Injection-molded top guides. Dimensions are width by height. Per pair of doors.

	Craft@Hrs	Unit	Material	Labor	Total	Sell
48" x 80"	BC@.750	Ea	95.90	27.00	122.90	209.00
60" x 80"	BC@1.00	Ea	113.00	36.00	149.00	253.00
72" x 80"	BC@1.00	Ea	135.00	36.00	171.00	291.00
96" x 80"	BC@1.25	Ea	199.00	45.00	244.00	415.00
48" x 96"	BC@1.00	Ea	133.00	36.00	169.00	287.00
60" x 96"	BC@1.00	Ea	154.00	36.00	190.00	323.00
72" x 96"	BC@1.25	Ea	174.00	45.00	219.00	372.00
96" x 96"	BC@1.25	Ea	253.00	45.00	298.00	507.00

Beveled mirror bypass closet doors. 1/2" bevel on both sides of glass. Safety-backed mirror. Includes hardware. Dimensions are width by height. Per pair of doors.

	Craft@Hrs	Unit	Material	Labor	Total	Sell
48" x 80"	BC@.750	Ea	131.00	27.00	158.00	269.00
60" x 80"	BC@1.00	Ea	152.00	36.00	188.00	320.00
72" x 80"	BC@1.00	Ea	174.00	36.00	210.00	357.00
96" x 80"	BC@1.25	Ea	220.00	45.00	265.00	451.00

	Craft@Hrs	Unit	Material	Labor	Total	Sell
Premium mirrored bypass closet doors. Bright gold aluminum frame with integrated handle. Select quality 3mm plate mirror and heavy-duty 1-1/2" wide commercial-grade tubular side frame molding. Ultraglide® 2-1/4" deep felt-lined top channel, jump-proof bottom track and precision 1-1/2" dual race ball bearing wheels. Dimensions are width by height. Per pair of doors.						
48" x 81"	BC@.750	Ea	200.00	27.00	227.00	386.00
60" x 81"	BC@1.00	Ea	225.00	36.00	261.00	444.00
72" x 81"	BC@1.00	Ea	250.00	36.00	286.00	486.00
Bypass door accessories.						
Bumper	BC@.175	Ea	2.52	6.30	8.82	15.00
Carpet riser	BC@.175	Ea	2.00	6.30	8.30	14.10
Guide	BC@.175	Ea	2.05	6.30	8.35	14.20

Bi-Fold Closet Doors

	Craft@Hrs	Unit	Material	Labor	Total	Sell
Lauan bi-fold flush closet doors. 1-3/8" thick. 2 prehinged panels. Ready to paint, stain, or varnish.						
24" x 80"	BC@.700	Ea	31.60	25.20	56.80	96.60
30" x 80"	BC@.700	Ea	36.70	25.20	61.90	105.00
32" x 80"	BC@.700	Ea	39.80	25.20	65.00	111.00
36" x 80"	BC@.700	Ea	41.80	25.20	67.00	114.00
Primed hardboard bi-fold flush closet doors. 1-3/8" thick. 2 prehinged panels.						
24" x 80"	BC@.700	Ea	31.60	25.20	56.80	96.60
30" x 80"	BC@.700	Ea	36.70	25.20	61.90	105.00
32" x 80"	BC@.700	Ea	39.00	25.20	64.20	109.00
36" x 80"	BC@.700	Ea	41.80	25.20	67.00	114.00
Clear mahogany bi-fold flush closet doors. 1-1/8" thick. 2 prehinged panels. Unfinished.						
24" x 80"	BC@.700	Ea	49.10	25.20	74.30	126.00
30" x 80"	BC@.700	Ea	57.80	25.20	83.00	141.00
32" x 80"	BC@.700	Ea	61.80	25.20	87.00	148.00
36" x 80"	BC@.700	Ea	63.40	25.20	88.60	151.00
2-door colonist molded face bi-fold closet doors. 1-3/8" thick. Primed. Two panels.						
24" x 78"	BC@.700	Ea	37.70	25.20	62.90	107.00
30" x 78"	BC@.700	Ea	41.80	25.20	67.00	114.00
32" x 78"	BC@.700	Ea	46.90	25.20	72.10	123.00
36" x 78"	BC@.700	Ea	48.90	25.20	74.10	126.00
4-panel birch bi-fold flush closet doors. Wood stile and rail construction. 1-3/8" thick. Open left or right. Includes track and hardware.						
24" x 80"	BC@.700	Ea	31.60	25.20	56.80	96.60
30" x 80"	BC@.700	Ea	36.70	25.20	61.90	105.00
36" x 80"	BC@.700	Ea	41.80	25.20	67.00	114.00
48" x 80"	BC@.800	Ea	57.50	28.80	86.30	147.00
4-panel lauan flush bi-fold closet doors. Wood stile and rail construction. 1-3/8" thick. Open left or right. Includes track and hardware.						
48" x 80"	BC@.800	Ea	57.10	28.80	85.90	146.00
4-panel red oak flush bi-fold closet doors. Wood stile and rail construction. 1-3/8" thick. Open left or right. Includes track and hardware.						
24" x 80"	BC@.700	Ea	34.70	25.20	59.90	102.00
30" x 80"	BC@.700	Ea	39.80	25.20	65.00	111.00
36" x 80"	BC@.700	Ea	44.90	25.20	70.10	119.00
48" x 80"	BC@.800	Ea	67.30	28.80	96.10	163.00

	Craft@Hrs	Unit	Material	Labor	Total	Sell
4-panel molded face bi-fold closet doors. 1-3/8" thick. Primed. Two doors.						
24" x 80"	BC@.700	Ea	39.60	25.20	64.80	110.00
30" x 80"	BC@.700	Ea	44.30	25.20	69.50	118.00
36" x 80"	BC@.700	Ea	49.90	25.20	75.10	128.00
6-panel molded face bi-fold closet doors. 1-3/8" thick. Unfinished. Molded and primed hardboard. Two doors.						
24" x 80"	BC@.700	Ea	37.70	25.20	62.90	107.00
30" x 80"	BC@.700	Ea	41.80	25.20	67.00	114.00
32" x 80"	BC@.700	Ea	46.90	25.20	72.10	123.00
36" x 80"	BC@.700	Ea	48.90	25.20	74.10	126.00
48" x 80"	BC@.800	Ea	70.40	28.80	99.20	169.00
60" x 80"	BC@.900	Ea	80.60	32.40	113.00	192.00
72" x 80"	BC@1.00	Ea	96.90	36.00	132.90	226.00
Colonial 6-panel pine bi-fold closet doors. 1-3/8" thick. Includes track and hardware. Unfinished.						
24" x 78"	BC@.700	Ea	82.60	25.20	107.80	183.00
30" x 78"	BC@.700	Ea	87.70	25.20	112.90	192.00
36" x 78"	BC@.700	Ea	92.80	25.20	118.00	201.00
24" x 80"	BC@.700	Ea	79.10	25.20	104.30	177.00
30" x 80"	BC@.700	Ea	89.10	25.20	114.30	194.00
32" x 80"	BC@.700	Ea	91.00	25.20	116.20	198.00
36" x 80"	BC@.700	Ea	97.90	25.20	123.10	209.00
Clear pine 2-panel colonial bi-fold closet doors. Clear pine. Recessed rails. Unfinished. Open left or right. Includes all hardware and track. Two hinged panels.						
24" x 80"	BC@.700	Ea	130.00	25.20	155.20	264.00
30" x 80"	BC@.700	Ea	136.00	25.20	161.20	274.00
32" x 80"	BC@.700	Ea	143.00	25.20	168.20	286.00
36" x 80"	BC@.700	Ea	150.00	25.20	175.20	298.00
6-panel oak bi-fold closet doors. Includes track and hardware. 1-1/8" thick.						
24" x 80"	BC@.700	Ea	140.00	25.20	165.20	281.00
30" x 80"	BC@.700	Ea	150.00	25.20	175.20	298.00
36" x 80"	BC@.700	Ea	160.00	25.20	185.20	315.00
Metal louver over louver bi-fold doors. Includes track and hardware. 1-3/8" thick. Two prehinged panels.						
24" x 80"	BC@.700	Ea	56.10	25.20	81.30	138.00
30" x 80"	BC@.700	Ea	64.30	25.20	89.50	152.00
36" x 80"	BC@.700	Ea	73.40	25.20	98.60	168.00
Oak louver over louver bi-fold doors. Includes track and hardware. 1-3/8" thick. Two hinged panels.						
24" x 96"	BC@.700	Ea	65.30	25.20	90.50	154.00
30" x 96"	BC@.700	Ea	75.50	25.20	100.70	171.00
36" x 96"	BC@.700	Ea	75.50	25.20	100.70	171.00
Pine louver over louver bi-fold doors. 1-1/8" thick ponderosa pine. 3-11/16" wide rails. 1-1/4" wide stiles. Two prehinged panels. Unfinished.						
24" x 80"	BC@.700	Ea	54.30	25.20	79.50	135.00
30" x 80"	BC@.700	Ea	59.70	25.20	84.90	144.00
32" x 80"	BC@.700	Ea	65.20	25.20	90.40	154.00
36" x 80"	BC@.700	Ea	70.60	25.20	95.80	163.00

	Craft@Hrs	Unit	Material	Labor	Total	Sell
Louver over panel white bi-fold doors. With track and hardware. Two hinged panels.						
24" x 80"	BC@.700	Ea	49.00	25.20	74.20	126.00
30" x 80"	BC@.700	Ea	53.90	25.20	79.10	134.00
36" x 80"	BC@.700	Ea	58.80	25.20	84.00	143.00
Pine louver over panel bi-fold doors. Recessed rails. Open left or right. 1/2" swing space. Two prehinged panels. Includes track and hardware. Unfinished stain grade.						
24" x 80"	BC@.700	Ea	45.00	25.20	70.20	119.00
30" x 80"	BC@.700	Ea	60.00	25.20	85.20	145.00
36" x 80"	BC@.700	Ea	65.00	25.20	90.20	153.00
Bi-fold door hardware.						
Bi-fold hinge	BC@.175	Ea	3.34	6.30	9.64	16.40
Carpet riser	BC@.175	Ea	1.99	6.30	8.29	14.10
Hardware set	BC@.175	Ea	7.34	6.30	13.64	23.20
48" track	—	Ea	23.30	—	23.30	—
60" track	—	Ea	24.40	—	24.40	—
72" track	—	Ea	27.10	—	27.10	—

Mirrored Bi-Fold Doors

	Craft@Hrs	Unit	Material	Labor	Total	Sell
Framed bi-fold mirror doors. White frame. Top-hung with single wheel top hanger. Snap-in bottom guide and fascia. Low-rise bottom track. Non-binding door operation. Reversible top and bottom track. 3mm safety-backed mirror. Two hinged panels.						
24" x 80"	BC@.700	Ea	89.00	25.20	114.20	194.00
30" x 80"	BC@.700	Ea	100.00	25.20	125.20	213.00
36" x 80"	BC@.700	Ea	110.00	25.20	135.20	230.00
Bevel edge bi-fold mirror doors. 1/2" vertical bevel with bearing hinges. White finished steel frame with safety-backed mirror. Includes hardware. Full access to closet opening. Reversible high- and low-profile top and bottom tracks. Two hinged panels.						
24" x 80"	BC@.700	Ea	105.00	25.20	130.20	221.00
30" x 80"	BC@.700	Ea	113.00	25.20	138.20	235.00
36" x 80"	BC@.700	Ea	96.40	25.20	121.60	207.00
Frameless bi-fold mirror doors. Frameless style maximizes the mirror surface. Beveled edge glass. All preassembled hinges. Safety-reinforced mirror backing. Two hinged panels. Includes hardware.						
24" x 80"	BC@.700	Ea	100.00	25.20	125.20	213.00
30" x 80"	BC@.700	Ea	110.00	25.20	135.20	230.00
36" x 80"	BC@.700	Ea	120.00	25.20	145.20	247.00
Chrome bi-fold mirror doors. Two hinged panels. Includes hardware.						
24" x 80"	BC@.700	Ea	152.00	25.20	177.20	301.00
30" x 80"	BC@.700	Ea	161.00	25.20	186.20	317.00
36" x 80"	BC@.700	Ea	172.00	25.20	197.20	335.00

	Craft@Hrs	Unit	Material	Labor	Total	Sell
Pocket Door Hardware						
Pocket door frame and track. Fully assembled frame. Designed to hold 150-pound standard 80" door. Door sold separately.						
24" x 80"	BC@1.00	Ea	74.50	36.00	110.50	188.00
28" x 80"	BC@1.00	Ea	76.50	36.00	112.50	191.00
30" x 80"	BC@1.00	Ea	80.60	36.00	116.60	198.00
32" x 80"	BC@1.00	Ea	83.60	36.00	119.60	203.00
36" x 80"	BC@1.00	Ea	85.70	36.00	121.70	207.00
Universal pocket door jamb kit.						
Clear pine	—	Ea	34.70	—	34.70	—
Pocket door edge pulls.						
2-1/8" round brass	BC@.300	Ea	4.06	10.80	14.86	25.30
2-1/8" round brass dummy	BC@.300	Ea	5.96	10.80	16.76	28.50
1-3/8" x 3" oval brass	BC@.300	Ea	3.76	10.80	14.56	24.80
1-3/8" x 3" rectangular brass	BC@.300	Ea	3.76	10.80	14.56	24.80
3/4" round brass	BC@.300	Ea	2.59	10.80	13.39	22.80
Brass plate	BC@.300	Ea	4.74	10.80	15.54	26.40
Ball bearing pocket door hangers.						
Hanger	BC@.175	Ea	9.38	6.30	15.68	26.70
Converging pocket door kit.						
Door kit	—	Ea	4.06	—	4.06	—
Pocket door hangers.						
3/8" offset, one wheel	BC@.175	Ea	2.97	6.30	9.27	15.80
1/16" offset, one wheel	BC@.175	Ea	2.97	6.30	9.27	15.80
7/8" offset, two wheel	BC@.175	Ea	4.32	6.30	10.62	18.10
2-door bypass pocket door track hardware.						
48" width	BC@.350	Ea	11.00	12.60	23.60	40.10
60" width	BC@.350	Ea	13.50	12.60	26.10	44.40
72" width	BC@.350	Ea	16.00	12.60	28.60	48.60
Mobile Home Doors and Windows						
Exterior mobile home combo doors. Inswinging door with outswinging storm and screen door.						
32" x 72"	BC@1.00	Ea	397.00	36.00	433.00	736.00
32" x 76"	BC@1.00	Ea	397.00	36.00	433.00	736.00
34" x 76"	BC@1.00	Ea	397.00	36.00	433.00	736.00
Exterior mobile home doors. Left hinge. Outswinging.						
32" x 72"	BC@1.00	Ea	202.00	36.00	238.00	405.00
32" x 76"	BC@1.00	Ea	219.00	36.00	255.00	434.00
34" x 76"	BC@1.00	Ea	224.00	36.00	260.00	442.00

	Craft@Hrs	Unit	Material	Labor	Total	Sell
Garage Doors						
Install overhead rollup garage door sections only. Assemble and install rollup door. Excludes trolley and springs.						
4 sections high, 8' to 9' wide	B1@2.73	Set	—	88.80	88.80	151.00
4 sections high, 12' to 16' wide	B1@4.27	Set	—	139.00	139.00	236.00
6 or 7 sections high, 12' to 16' wide	B1@5.19	Set	—	169.00	169.00	287.00
Install overhead rollup garage door. Assemble and install rollup door. Includes trolley and torsion spring counter-balance installed on wood frame building.						
4 sections high, 8' to 9' wide	B1@6.46	Ea	—	210.00	210.00	357.00
4 sections high, 12' to 16' wide	B1@8.33	Ea	—	271.00	271.00	461.00
6 or 7 sections high, 12' to 16' wide	B1@9.60	Ea	—	312.00	312.00	530.00
Remove and replace overhead rollup garage door sections. 4-section-high door. No painting included.						
One bottom section, 8' to 9' wide	B1@2.20	Ea	—	71.60	71.60	122.00
One bottom section, 12' to 16' wide	B1@3.00	Ea	—	97.60	97.60	166.00
Remove and replace overhead rollup garage door sections. 6- or 7-section-high door. No painting included.						
One bottom section, 12' to 16' wide	B1@3.84	Ea	—	125.00	125.00	213.00
Install overhead rollup garage door torsion spring counter-balance. Per set of two torsion springs.						
4 section door 8' to 9' wide	B1@1.02	Ea	—	33.20	33.20	56.40
4 section door 12' to 16' wide	B1@1.33	Ea	—	43.30	43.30	73.60
6 or 7 section door 12' to 16' wide	B1@2.01	Ea	—	65.40	65.40	111.00
Remove and replace torsion springs on overhead garage door. Remove set of two old torsion springs and replace with set of two new torsion springs.						
4 section door 8' to 9' wide	B1@1.10	Ea	—	35.80	35.80	60.90
4 section door 12' to 16' wide	B1@1.40	Ea	—	45.50	45.50	77.40
6 or 7 section door 12' to 16' wide	B1@2.10	Ea	—	68.30	68.30	116.00
Install trollies for overhead rollup garage door. Installation on a frame building. Includes trollies but no torsion springs.						
4 section door, 8' to 16' wide	B1@2.70	Ea	—	87.80	87.80	149.00
6 to 7 section door, 12' to 16' wide	B1@3.32	Ea	—	108.00	108.00	184.00
Install garage door bottom seal.						
8' to 9' wide	B1@.143	Ea	—	4.65	4.65	7.91
12' wide	B1@.204	Ea	—	6.64	6.64	11.30
16' wide	B1@.259	Ea	—	8.43	8.43	14.30
Steel rollup garage door. Torsion springs. 2-layer insulated construction. 7/8" insulation. Steel frame with overlapped section joints. With hardware and lock.						
8' x 7'	B1@6.46	Ea	294.00	210.00	504.00	857.00
9' x 7'	B1@6.46	Ea	320.00	210.00	530.00	901.00
16' x 7'	B1@8.33	Ea	544.00	271.00	815.00	1,390.00
Raised panel steel garage door. 2-layer construction. 7/8" thick polystyrene insulation.						
8' x 7'	B1@6.46	Ea	300.00	210.00	510.00	867.00
9' x 7'	B1@6.46	Ea	320.00	210.00	530.00	901.00
16' x 7'	B1@8.33	Ea	544.00	271.00	815.00	1,390.00

	Craft@Hrs	Unit	Material	Labor	Total	Sell
Bonded steel garage door. 3-layer steel exterior. 2"-thick insulation. With torsion springs, hardware and lock.						
8' x 7'	B1@6.46	Ea	505.00	210.00	715.00	1,220.00
8' x 7', molding face	B1@6.46	Ea	528.00	210.00	738.00	1,250.00
9' x 7'	B1@6.46	Ea	539.00	210.00	749.00	1,270.00
9' x 7', molding face	B1@6.46	Ea	550.00	210.00	760.00	1,290.00
16' x 7'	B1@6.46	Ea	843.00	210.00	1,053.00	1,790.00
16' x 7', molding face	B1@8.33	Ea	955.00	271.00	1,226.00	2,080.00
8' x 6'6", with glass window	B1@6.46	Ea	793.00	210.00	1,003.00	1,710.00
2-layer steel garage door. 7/8"-thick insulation. With torsion springs, hardware and lock.						
16' x 7', raised panel face	B1@8.33	Ea	544.00	271.00	815.00	1,390.00
Non-insulated steel garage door. With rollup hardware and lock.						
8' x 7'	B1@6.46	Ea	229.00	210.00	439.00	746.00
9' x 7'	B1@6.46	Ea	249.00	210.00	459.00	780.00
6' x 7'	B1@8.33	Ea	447.00	271.00	718.00	1,220.00
Wood garage doors. Unfinished, jamb hinge, with hardware. panel uninsulated hardboard door						
9' wide x 7' high	B1@4.42	Ea	309.00	144.00	453.00	770.00
16' wide x 7' high	B1@5.90	Ea	535.00	192.00	727.00	1,240.00
Styrofoam core hardboard door						
9' wide x 7' high	B1@4.42	Ea	332.00	144.00	476.00	809.00
16' wide x 7' high	B1@5.90	Ea	586.00	192.00	778.00	1,320.00
Redwood door, raised panels						
9' wide x 7' high	B1@4.42	Ea	615.00	144.00	759.00	1,290.00
16' wide x 7' high	B1@5.90	Ea	1,360.00	192.00	1,552.00	2,640.00
Hardboard door with two small lites						
9' wide x 7' high	B1@4.42	Ea	340.00	144.00	484.00	823.00
16' wide x 7' high	B1@5.90	Ea	615.00	192.00	807.00	1,370.00
Hardboard door with sunburst lites						
9' wide x 7' high	B1@4.42	Ea	865.00	144.00	1,009.00	1,720.00
16' wide x 7' high	B1@5.90	Ea	954.00	192.00	1,146.00	1,950.00
Hardboard door with cathedral-style lites						
9' wide x 7' high	B1@4.42	Ea	435.00	144.00	579.00	984.00
16' wide x 7' high	B1@5.90	Ea	919.00	192.00	1,111.00	1,890.00
Garage door jamb hinge. With carriage bolts and nuts.						
No. 1, 14 gauge	BC@.333	Ea	4.68	12.00	16.68	28.40
No. 2, 14 gauge	BC@.333	Ea	4.58	12.00	16.58	28.20
No. 3, 14 gauge	BC@.333	Ea	4.58	12.00	16.58	28.20
Steel garage door weather seal. Cellular vinyl.						
7/16" x 2" x 7'	BC@.400	Ea	7.23	14.40	21.63	36.80
7/16" x 2" x 9'	BC@.500	Ea	8.52	18.00	26.52	45.10
7/16" x 2" x 16'	BC@.750	Ea	19.60	27.00	46.60	79.20

	Craft@Hrs	Unit	Material	Labor	Total	Sell
Garage door hardware.						
Adjustable kicker	BC@.250	Ea	10.70	9.00	19.70	33.50
Door cantilever bolts	BC@.250	Ea	3.90	9.00	12.90	21.90
Garage door handle	BC@.500	Ea	4.28	18.00	22.28	37.90
Garage door wall pushbutton	BC@.250	Ea	8.45	9.00	17.45	29.70
Garage door rubber bumpers	BC@.250	Ea	3.88	9.00	12.88	21.90
16' Door truss rod kit	BC@.500	Ea	29.30	18.00	47.30	80.40
9' Door hardware kit	BC@1.00	Ea	93.40	36.00	129.40	220.00
16' Door hardware kit	BC@1.00	Ea	73.40	36.00	109.40	186.00
Containment C-hook	—	Ea	3.18	—	3.18	—
Spring connector C-hook	—	Ea	5.60	—	5.60	—
Double door low headroom kit	BC@.500	Ea	66.80	18.00	84.80	144.00
Track hanger kit	—	Ea	18.90	—	18.90	—
Garage door replacement hardware.						
3" replacement pulley	BC@.500	Ea	3.32	18.00	21.32	36.20
4" replacement pulley	BC@.500	Ea	6.12	18.00	24.12	41.00
7' lift cable set	BC@.500	Ea	4.08	18.00	22.08	37.50
8' lift cable set	BC@.500	Ea	4.85	18.00	22.85	38.80
Safety lift cable set	BC@.500	Ea	4.69	18.00	22.69	38.60
Window insert	BC@.250	Ea	51.00	9.00	60.00	102.00
Track, bolt and flange nuts	—	Ea	2.14	—	2.14	—
Replacement nylon rollers	BC@.500	Ea	2.55	18.00	20.55	34.90
G-top bracket	—	Ea	2.04	—	2.04	—
Lift handles	BC@.250	Ea	2.14	9.00	11.14	18.90

Garage door openers. Radio controlled, electric, single or double door, includes typical electric hookup to existing adjacent 110 volt outlet. 1/2 HP opener, wireless receiver and two multiple frequency transmitters.

	Craft@Hrs	Unit	Material	Labor	Total	Sell
Chain drive glide opener,	BC@3.90	Ea	233.00	140.00	373.00	634.00
Screw drive glide opener	BC@3.90	Ea	268.00	140.00	408.00	694.00
Add for additional transmitter	—	Ea	35.70	—	35.70	—
Add for operator reinforcement kit	—	Ea	26.00	—	26.00	—
Add for low headroom kit	—	Ea	72.10	—	72.10	—
Add for garage door monitor	—	Ea	27.00	—	27.00	—
Add for monitor add-on sensor	—	Ea	16.00	—	16.00	—
Add for 8' chain drive extension kit	—	Ea	35.60	—	35.60	—
Add for 8' belt drive extension kit	—	Ea	52.80	—	52.80	—
Add for remote light control	—	Ea	22.20	—	22.20	—
Add for wireless keyless entry control	—	Ea	42.40	—	42.40	—
Add for motion detecting wall control	—	Ea	23.00	—	23.00	—
Add for opener surge protector	—	Ea	17.10	—	17.10	—
Add for replacement safety beam kit	—	Ea	35.80	—	35.80	—
Garage door single extension spring.						
70 pounds	BC@.500	Ea	20.30	18.00	38.30	65.10
110 pounds	BC@.500	Ea	23.70	18.00	41.70	70.90
120 pounds	BC@.500	Ea	24.20	18.00	42.20	71.70
130 pounds	BC@.500	Ea	25.80	18.00	43.80	74.50
140 pounds	BC@.500	Ea	28.50	18.00	46.50	79.10
150 pounds	BC@.500	Ea	31.10	18.00	49.10	83.50
160 pounds	BC@.500	Ea	31.90	18.00	49.90	84.80
U-bolt spring anchor	BC@.100	Ea	11.20	3.60	14.80	25.20
8-link spring anchor chain	BC@.100	Ea	7.13	3.60	10.73	18.20

	Craft@Hrs	Unit	Material	Labor	Total	Sell

Emergency Exit Hardware

Install rim panic bar on metal door. Includes measure and mark, drill holes, cut bar to length, assemble bar, set strike plate.

	Craft@Hrs	Unit	Material	Labor	Total	Sell
Per bar, no exit bolts	BC@.685	Ea	239.00	24.70	263.70	448.00

Touch bar exit hardware. For use with 1-3/4"-thick hollow metal and wood doors. Extruded anodized aluminum with stainless steel spring. 3/4" throw stainless steel latch bolts. Strike for 5/8" stop; shim for 1/2" stop; 2-3/4" backset, 4-1/4" minimum stile. Includes standard mounting with sheet metal and machine screws. Recommended mounting height: 40-5/16" from ceiling to finished floor. 36" maximum width. Rim-type cylinders with Schlage C keyway. Includes measure and mark, drill holes, assemble, set strike plate.

	Craft@Hrs	Unit	Material	Labor	Total	Sell
Install, single door	BC@1.45	Ea	141.00	52.20	193.20	328.00
Install, per set of double doors	BC@2.97	Ea	281.00	107.00	388.00	660.00
Remove, single door	BC@.369	Ea	—	13.30	13.30	22.60
Remove, double doors and mortise lock	BC@1.38	Ea	—	49.70	49.70	84.50

Basement doors, Bilco. 12 gauge primed steel, center opening basement doors. Costs include assembly and installation hardware. Add the cost of excavation, concrete, masonry, anchor placement and finish painting.

	Craft@Hrs	Unit	Material	Labor	Total	Sell
Doors (overall dimensions)						
19-1/2" H, 55" W, 72" L	BC@3.41	Ea	436.00	123.00	559.00	950.00
22" H, 51" W, 64" L	BC@3.41	Ea	398.00	123.00	521.00	886.00
30" H, 47" W, 58" L	BC@3.41	Ea	391.00	123.00	514.00	874.00
52" H, 51" W, 43-1/4" L	BC@3.41	Ea	467.00	123.00	590.00	1,000.00
Door extensions (available for 19-1/2" H, 55" W, 72" L door only)						
6" deep	BC@1.71	Ea	194.00	61.60	255.60	435.00
12" deep	BC@1.71	Ea	253.00	61.60	314.60	535.00
18" deep	BC@1.71	Ea	316.00	61.60	377.60	642.00
24" deep	BC@1.71	Ea	384.00	61.60	445.60	758.00
Basement door stair stringers, steel, pre-cut for 2" x 10" wood treads (without treads)						
32" to 39" stair height	BC@1.71	Ea	91.40	61.60	153.00	260.00
48" to 55" stair height	BC@1.71	Ea	115.00	61.60	176.60	300.00
56" to 64" stair height	BC@1.71	Ea	128.00	61.60	189.60	322.00
65" to 72" stair height	BC@1.71	Ea	144.00	61.60	205.60	350.00
73" to 78" stair height	BC@1.71	Ea	198.00	61.60	259.60	441.00
81" to 88" stair height	BC@1.71	Ea	215.00	61.60	276.60	470.00
89" to 97" stair height	BC@1.71	Ea	230.00	61.60	291.60	496.00

Walls and Ceilings 10

Before about 1950, most homes were built with interior walls and ceilings finished in lath and plaster. During the 1950s, gypsum wallboard became generally available and most owners and industry professionals switched over. While plaster and gypsum wallboard are equally durable (when protected from moisture and impact), and both make attractive wall and ceiling coverings, gypsum wallboard is preferred because it's so much less expensive to install. Lath and plaster is very labor-intensive, and because it requires days or weeks to dry out, it tends to delay construction. Today, it's unlikely you'll be called on to do a lath and plaster job; but you may have to make repairs on one.

Nearly all plaster develops hairline cracks. These cracks and small holes must be patched with patching plaster, joint compound or spackle before repainting. Plan on covering or replacing the surface if plaster is uneven, bulging, loose, or there are many large holes. If there are only a few large holes, you can patch them. If there are more than a few, it's easier to replace the entire wall. Loose or bulging plaster is a sign of water damage. When you find damage caused by water, your first step will be eliminating the source of moisture. Otherwise, the repair will be worthless.

Gypsum wallboard is usually called drywall, though you'll hear the same material referred to as wallboard, gypboard, Sheetrock or simply rock. The term drywall is a little misleading. First, drywall isn't just for walls. It adapts equally well to installation on ceilings. Second, drywall is dry only by comparison with plaster, which goes on wet. The cement (sometimes called mud) used to cover drywall joints is mixed with water and needs at least a few hours to dry completely.

Drywall does have its problems. Unlike plaster walls and ceilings, it can develop nail pops — nail heads that work loose and rise above the surface. Also, the paper tape used to cover panel joints can delaminate and curl if not bedded correctly. Loose tape is a sign of a poor drywall job.

Choices for Renewing Wall and Ceiling Cover

1. Patch and smooth what's there. This is the least expensive choice. But it's not practical when walls or ceilings are seriously damaged.

2. Install furring strips as a base for new wall cover, or apply new wall and ceiling cover directly over the old. This option works when the existing surface is flat and firm. I don't generally recommend this, as the wall will be

Minimum Material Thickness for On-Center Spacing of Studs and Joists		
Material	16" OC	24" OC
Drywall	$^3/_8$"	$^1/_2$"
Plywood Paneling	$^1/_4$"	$^3/_8$"
Hardboard Paneling	$^1/_4$"	—
Tongue-&-Groove Plank	$^3/_8$"	$^1/_2$"

Note 1. New wall cover set with nails or adhesive directly on existing wall cover can be any thickness because it's supported continuously on the existing surface.
Note 2. $^1/_4$" plywood or hardboard set on framing members 16" on center may be slightly wavy unless applied over at least $^3/_8$" drywall.

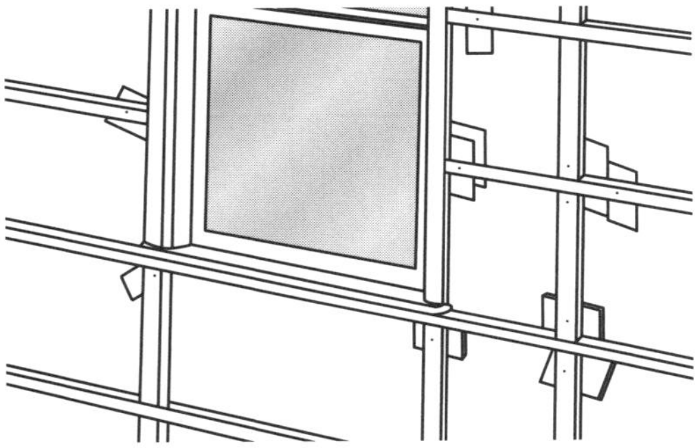

Figure 10-1
Shingle shims behind furring to produce a smooth vertical surface

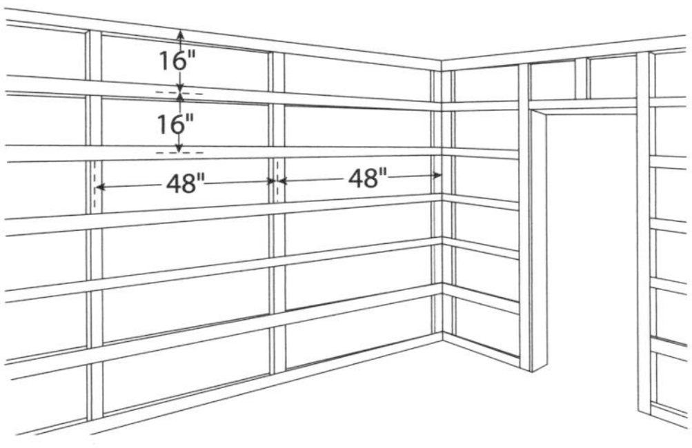

Figure 10-2
Application of horizontal furring to interior wall

significantly thicker, requiring moldings and casings to be redone, and new trim around wall penetrations. Plus, setting furring strips requires a great deal of work, and can result in an inferior job. Generally it's easier to tear off the existing wall cover and replace it, but you may come across a situation where this may be the best choice — perhaps when it's just one wall in a room and there are no openings. It's a lot less messy.

If new wall cover is going to be nailed over old or set on furring strips, the first step will be finding the studs. They're usually installed 16" on center and at the edge of doors and windows. Where you see a nail head in drywall or baseboard, assume there's a stud under the nail. If stud locations aren't obvious, put a stud finder to work. Stud finders locate dense material in the wall cavity — usually a stud. Mark the center of each stud at the top and bottom of the wall. Then snap a chalkline between the two marks.

Check walls for flatness by holding the straight side of a 2 x 4 against the surface. Furring will correct a bulging or indented wall. Mark wall locations that are uneven. Also check for true vertical alignment by holding a large carpenter's level against a straight 2 x 4 held against the wall. Use shingles to shim out furring and create a flat, vertical surface. See Figure 10-1.

Figure 10-2 shows 1" x 2" furring strips set 16" on center. Nail the furring at each stud. Remove existing base trim, window casings, and door casings. Apply furring around all the openings. Be sure there will be a vertical furring strip under each vertical joint in the new wall cover.

3. Your third choice for renewing wall and ceiling cover is to strip off what's covering the studs and joists and start over. The result is a wall or ceiling that's as good as new. In fact, it *is* new. This is by far the best option for seriously damaged walls, although stripping off the existing wall covering is messy, and it creates a disposal problem.

Before applying the new wall cover, be sure studs are aligned. Run a string across the front of the studs, holding it taut. The studs should all touch the string equally. Those that don't touch the string, or push the string out, are out of line with the rest and need to be re-aligned or replaced.

Hanging Drywall

You can hang drywall either with the long edges horizontal (Figure 10-3A) or vertical (Figure 10-3B). Hanging the drywall horizontally reduces the number of vertical joints at the middle

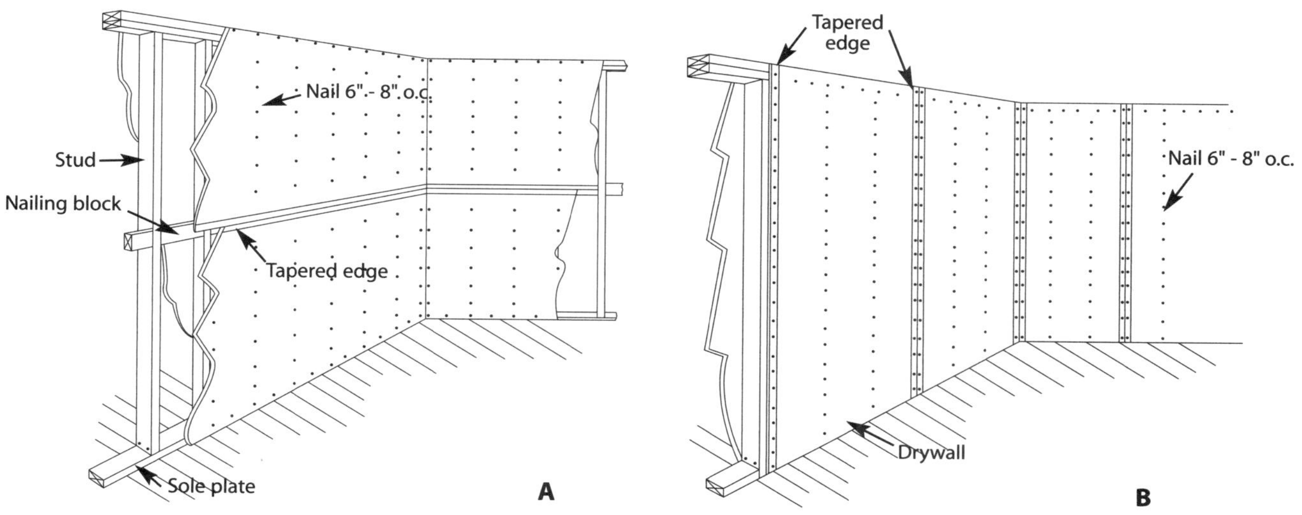

Figure 10-3

*Installing drywall on walls: **A**, horizontal application; **B** vertical application*

of walls. If you install drywall horizontally, be sure all joints are supported at the ends. If you install drywall vertically, you won't need nailing blocks — the stud acts as the nailing block. All drywall joints must be supported, or they'll crack. Whether hung vertically or horizontally, proper fastening is essential. If you have a drywall screw gun, drive screws every 12" on ceiling joists and every 16" on wall studs. Adjust the gun so it sets the screw just slightly into the board without breaking the paper. Some drywall specialists set the board initially with a few nails driven at the edges and then secure the board in place with screws.

If you nail the board in place, drive a nail every 8" along wall studs and every 6" along ceiling joists. Use 4d ring shank nails on 3/8" board and 5d ring shank nails on 1/2" board. Keep nails at least 3/8" away from the edges of the board. Use a drywall hammer with a slightly convex head that leaves a dimple at each nail location. Be careful not to break the surface of the paper. See Figure 10-4. You'll cover the dimple later with drywall joint compound (mud).

Also be aware of where you're driving fasteners. There may be pipe runs through the studs, and copper pipe is thin, soft and easily punctured with a drywall nail or a screw. Copper pipes and wire are supposed to be protected by metal plates to prevent punctures. But don't count on it. Take care when working around pocket door frames. Don't drive any nails or screw points into the pocket door cavity. When in doubt, use shorter nails or screws.

Drywall is heavy. If you're hanging full 4' x 8' sheets on ceilings, you'll need a crew of at least two, and three would be better. One person working alone can drywall ceilings if it's an emergency, but it's not recommended. Start with the ceiling panels. Center the panel edges on the ceiling joist. Plan the layout so that any cut edges will wind up in the corners, and not

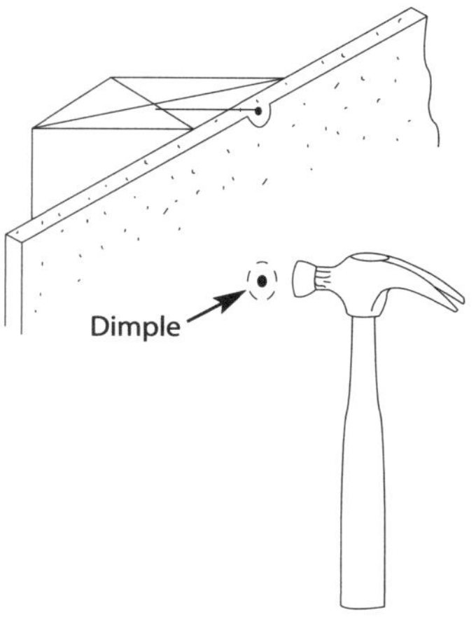

Figure 10-4

Drive nails in "dimple" fashion

in the middle of the ceiling. All joints must center on a joist. If there's no ceiling joist where you need one, you'll have to add one. Don't cut the drywall sheets to make the joint center on a joist — the seam will show.

When the ceiling is done, begin hanging drywall on the walls. The standard ceiling height is 97 1/8" between the floor and the bottom of the ceiling joists. Subtracting 1/2" for the thickness of ceiling panels leaves a 96 5/8" wall height. Two panel widths total 96". Hang wall panels 5/8" above the finished floor for a snug fit at the ceiling. The 5/8" at the floor will be covered with baseboard. Use a drywall foot lift to hold the panel 5/8" above the floor while driving nails or screws.

Regular core 1/2" drywall panels are recommended for single-layer wall application in new residential construction. 3/8" panels are recommended for ceilings in residential repair and remodeling in single or double-layers. Type X drywall is designed to meet requirements for fire safety. Greenboard is water-resistant for use behind tile. Brownboard is designed for exterior sheathing or soffits.

Joint compound comes in both powder and pre-mixed forms. Home improvement specialists generally use pre-mixed compound. Dry mix has to be used right away after adding water. Pre-mixed cement will last for weeks if kept in a sealed container. Regardless of which you use, the mix should have a soft, putty-like consistency that spreads easily with a trowel or wide putty knife. Mud that runs off the knife is too thin.

Most drywall has a tapered, or beveled edge. Joint compound and tape fill this recess, leaving a smooth, flat surface. If a sheet has been cut to fit, the edge won't be tapered. You can tape a square edge the same way you tape beveled edges, but the joint compound will rise slightly above the finished surface. The extra depth won't be as obvious if you feather out the joint compound at least 4" beyond the joint, but one located in the middle of the ceiling is going to be noticeable. In the corners, a taped cut edge won't show. Taping and feathering cut joints will slow your job down to a crawl. Avoid this situation whenever possible.

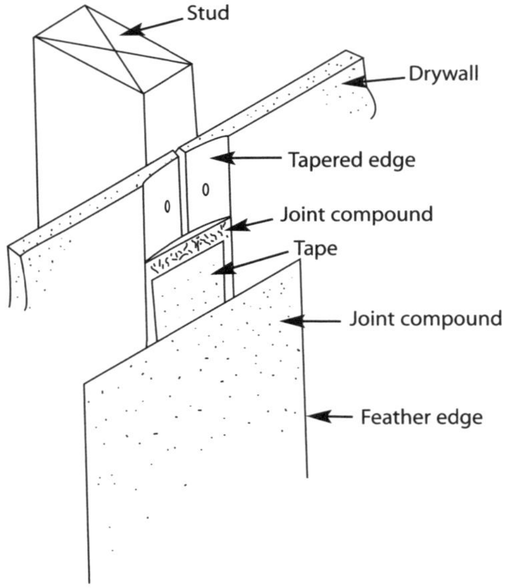

Figure 10-5
Detail of joint treatment

Taping and finishing joints takes three or four days. Figure 10-5 illustrates the taping sequence. Paper joint tape is the most economical, about $2.50 per thousand square feet of board hung. Fiberglass tape costs more, about $6.00 per thousand square feet of board. But self-adhesive fiberglass joint mesh needs no embedding coat and is more durable than paper tape. It flexes rather than curling or tearing if the joint moves.

Taping and Finishing Drywall

1. Start with the ceiling and work down the walls. If you've selected paper tape rather than fiberglass, spread joint compound over panel edges with a 5"-wide taping knife. Don't skimp on the mud. If you're using self-adhesive fiberglass tape, press tape over the joints and skip to step three below.

2. Press the paper tape into the mud with a drywall knife, not your hand. Then smooth the surface with the knife. Press hard enough to force joint compound through the small perforations in the tape, if the tape has perforations. But don't press too hard. Some mud should remain under the tape. When you're done taping and embedding the tape, let everything dry overnight before beginning the finish coats.

3. Cover the tape with cement, feathering the outer edges at least 2" on each side of the paper tape. Feather an additional 2" when covering a cut joint, as there's no taper. Then let the cement dry overnight.

4. When dry, sand lightly. A pole sander speeds this work. Then apply a thin second finish coat, feathering the edges a little past the edge of the prior coat. Use drywall topping compound designed for finish coats to create a smooth joint that's easy to sand. You can buy "all-purpose" compound, which can be used both for bedding and topping. But it's better to use bedding compound for bedding, and topping compound for finishing. Use a wider drywall knife for this finish coat, up to 12" wide. To save sanding time, keep this finish coat as smooth as possible. For top quality work, apply a third coat of mud after the second coat has dried and been sanded.

5. When the last coat of cement is dry, sand smooth.

6. Fill all nail and screw dimples with at least one coat of joint compound. Sand the surface after each coat is dry.

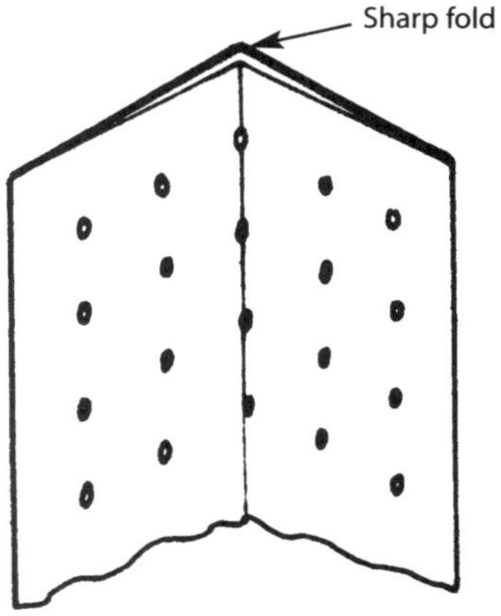

Figure 10-6
Corner tape

Use folded perforated tape on interior corners. See Figure 10-6. Fold tape down the center to form a right angle. Tape designed for this purpose already has a crease down the middle. Then apply cement on each side of the corner and press the tape in place with the putty knife. Use a drywall corner knife to embed the tape on both intersecting walls at the same time. Finish the corner with a coat of joint compound. Smooth the cement on both surfaces of a corner at the same time with a corner knife. Let the corner dry overnight and then sand the surface smooth.

For exterior corners, use metal or plastic drywall corner bead. This makes a more durable corner, able to withstand impacts that are likely at external corners. Also apply paper drywall tape over the edge of the metal bead. Nail or screw outside corners to the board every 8" and finish with joint compound. When you're finished applying tape, bead and mud at both internal and external corners, you should have a 4" strip of mud on each side of every corner. Don't worry if this strip isn't smooth. Sanding and more finishing will follow.

Drywall mud shrinks as it dries. So apply a little more than actually needed to make a smooth finish, especially over fastener heads. Bear in mind that drywall sanding is very messy! It creates huge, choking clouds of dust. Wear a dust mask to avoid inhaling this stuff. In an unoccupied house, the mess isn't as much of a problem. However, if you're working in an occupied home, you'll need to control the dust. Homeowners don't appreciate having everything in their home covered with a thick layer of white powder. You'll either need to seal off your work area with plastic sheeting, or use a dustless sanding system, such as a wet sanding sponge.

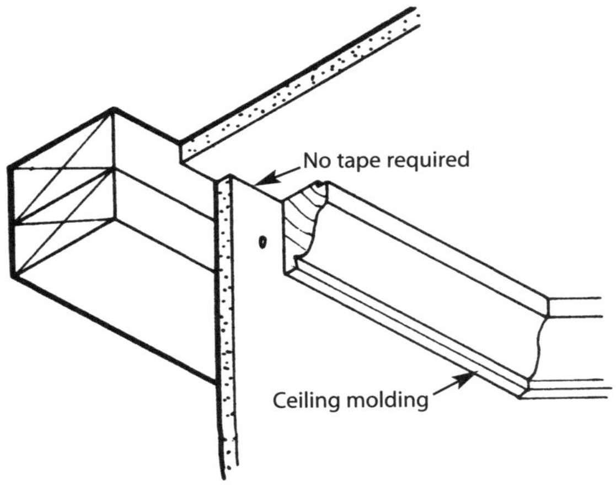

Figure 10-7
Ceiling molding

If you plan to install crown molding, no taping is needed where walls meet the ceiling. See Figure 10-7. Set the trim with 8d finishing nails spaced 12" to 16" apart. Be sure to nail into the top wall plate.

Drywall Ceilings

You can apply new drywall directly over old plaster or on furring strips nailed over an uneven plaster ceiling. Applying furring strips on a ceiling won't create the problems that furring would on a wall. But furring out the ceiling will usually be more work than tearing down the ceiling cover and starting over. Use 2" x 2" or 2" x 3" furring strips nailed perpendicular to the joists. Space these furring strips 16" on center for $^{3}/_{8}$" drywall or 24" on center for $^{1}/_{2}$" drywall. Nail the furring strips with two 8d nails at each joist. Stagger board end joints. Be sure board edges end on a joist or furring strip. Don't jam the boards tightly together. It's best if there's only light contact at each edge.

Hanging ceiling panels is easier with a drywall lift that allows precise positioning while leaving two hands free for driving nails or screws. If you don't have a drywall lift, cut two braces like the one shown in Figure 10-8. Make them slightly longer than the ceiling height. Nail or screw the drywall to all supporting members, spacing the fasteners 7" to 8" apart. If you use nails, select 5d ($1^{1}/_{4}$") ring-shank nails for $^{1}/_{2}$" drywall and 4d (1") ring shank nails for $^{3}/_{8}$" drywall. Again, don't break the surface of the paper when driving fasteners. Finish ceiling joints the same way you finish wall joints.

Textured Finishes for Drywall

Wall finish is usually smooth to make cleaning easier. But the ceiling finish may be textured, usually with some form of joint compound. Texture hides ridges and bumps in the ceiling and improves acoustics by eliminating echo off the ceiling. But textured ceilings are also an admission that there's something to hide. Many owners don't like textured ceilings and know that texture is used to hide defects. The first thing they'll say when they see a textured ceiling is, "What's wrong with your ceiling?"

Orange peel texture consists of thinned joint compound applied with a long-nap paint roller. In an emergency, you can make ceiling texture by thinning out joint compound with water until it reaches a consistency similar to that of paint. But it's better to buy mix that's specifically made for texturing. It's much easier than trying to make your own. To ensure proper consistency, try applying some mixture to a scrap piece of drywall held upright. Adjust the consistency as necessary by adding

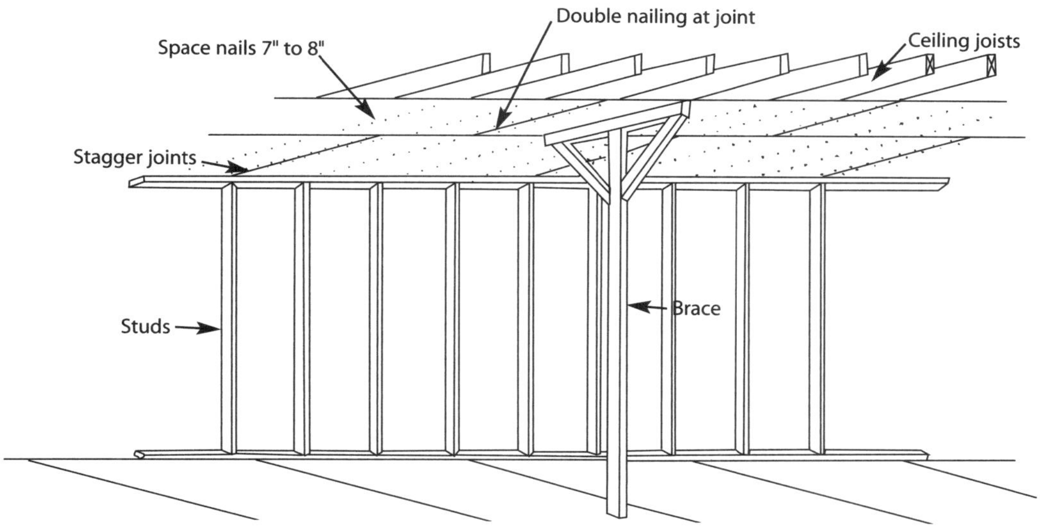

Figure 10-8
Installing drywall on ceiling

more water or joint compound to the bucket of mixture. When you've got the mixture just right, roll it onto the ceiling or wall. Keep the rolling pattern uniform so the texture appears to have a grain. When the mixture dries, avoid the temptation to sand the surface. The texture is very fragile. Sanding can knock off too much of the desirable surface.

Spatter finish is done with a compressor-operated spatter gun that shoots globules of thinned drywall mud on the ceiling at random. Scrape overspray off the walls. Other mixes are available to create different effects. You can get nearly the same spatter effect by dipping a stiff-bristle brush in thinned drywall mud and slinging mud on the ceiling with a snap of the wrist. This takes practice. Don't count on getting this right on the first attempt. Control the size of the spatters by making the mix thinner or thicker. Obviously, this is messy work. But it's an effective technique when you have to match only a few square feet of spatter-finished ceiling. For minor patching, you can also buy spatter finish in an aerosol can.

Knockdown finish uses the same technique — spatter blown or snapped on the ceiling. But the mix should be stiffer so spatters are between the size of a pea and a grape. Let the globules dry for a few minutes. Then knock the tops off with a masonry trowel. Work the trowel in all directions to avoid creating an obvious grain in the texture.

Skip trowel or imperial texture is like a knockdown finish, only more so. Apply mud to the ceiling or wall in a random pattern. Then smooth out what's there, leaving irregular patterns of texture in some areas and no texture in others. When done, it should look like Spanish stucco.

Veneer plaster is used in one or two $^1/_8$" coats over a veneer plaster base such as blueboard. Blueboard is similar to drywall, with a paper surface designed to bond well with the veneer plaster. Apply enough plaster to trowel a smooth, even finish over the entire surface. It's a lot of work, but veneer plaster hides imperfections and joints better than regular plaster, and provides a hard coat that protects the paper surface below.

Cottage cheese or popcorn texture is applied with a compressor, hopper and applicator gun. Although popcorn ceiling texture isn't currently in fashion, and you're not likely to be installing it, you may still be called on to do a repair job. Aerosol sprays are available to match the existing cottage cheese texture if you're only patching a small area.

No matter what finish you apply, the job isn't done until the surface has been primed and painted.

Repairing Drywall

Fill nail holes and small cracks in boards by applying a smooth coat of drywall compound. Let it dry. Then sand the surface smooth. Repairing larger holes in drywall isn't as easy. There's nothing but wall cavity behind a full penetration of the board. Drywall compound will fall into holes wider than about $^1/_2$". Cover larger cracks and small holes with self-adhesive fiberglass tape. Then press stiff drywall compound into the mesh. When the first coat is dry, apply a finish coat. With a patch like this, feather the drywall compound 12" on each side of the crack to avoid leaving an obvious ridge. Again, don't count on getting this right the first time.

Holes larger than a golf ball need some type of backing to hold the drywall mud until it sets. You can buy a drywall repair kit with clips that support drywall cut to cover nearly any size hole. If these drywall clips create lumps or otherwise don't work for you, make a patch kit with cardboard, string and a short length of dowel. Cut a piece of stiff cardboard slightly larger than the hole. Loop a short length of string through the center of the cardboard patch. Then fold the cardboard in half and insert it into the cavity. Pull the string tight, flattening the cardboard against the cavity side of the board and closing off the hole. Tie the loose end of the string around a short dowel laid across the hole. Then apply a coat of drywall compound over the hole and against the cardboard backing. Leave the patch slightly concave. When dry, cut the string and remove the dowel.

Many experienced drywall experts use neither clips nor cardboard. Instead, they cut a piece of scrap wood that will fit through the hole and extend about 2" to either side. They screw this in place with drywall screws on either side of the hole. This puts a firm foundation behind a portion of the hole. Then they cut a piece of drywall to fit the hole and screw it to the scrap wood. Once it's in place, they lay lengths of self-adhesive fiberglass drywall tape over the patch so it laps several inches onto firm wallboard. Then they apply a finish coat of joint compound and feather out several inches beyond the patch. When dry, they sand the patch smooth. Once primed and painted, there should be no evidence of the repair.

A hole more than 12" across is probably too large for a cardboard-backed patch. Instead, mark and cut out a rectangular section of wallboard all the way to the middle of the studs at both sides. Cut two nailing blocks to fit horizontally between the studs. Insert the blocks into the cutout and toenail them at the top and bottom of the rectangular cutout. Then cut drywall to fit in the cutout. Tape and finish the perimeter of the cutout as with any drywall joint.

Wood Paneling

Plywood paneling is sold in many grains and species. Hardboard imprinted with a wood grain pattern is generally less expensive. Better hardboard paneling has a realistic wood grain pattern. Both plywood and hardboard paneling are sold with a hard, plastic finish that's easily wiped clean. Hardboard is also available with vinyl coatings in many patterns and colors, including some that have the appearance of ceramic tile.

Wood paneling should be delivered to the site a few days before application. Panels need time to adapt to room temperature and humidity before you put them up. Stack panels in the room separated by full length furring strips so air can circulate to panel faces and backs. Figure 10-9.

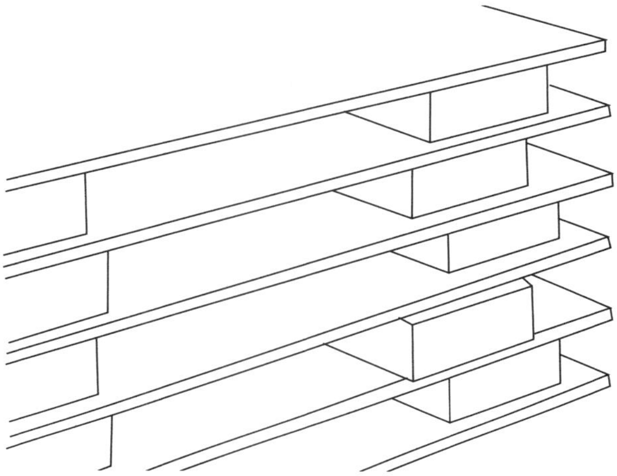

Figure 10-9

Stacking panels for conditioning to room environment prior to use

Always start a panel application with a truly vertical edge. If a corner is straight and vertical, butt the first panel into that corner. Cut subsequent panels so they lap on studs. If you don't have a vertical corner, tack a panel perfectly vertical and 2" from the starting corner. Use an art compass to scribe the outline of the corner on the panel edge. See Figure 10-10. Cut the panel along this line and move it into the corner. Butt the next panel against the first, being careful to keep the long edges truly vertical. Use the same art compass to scribe a line for panel top edges.

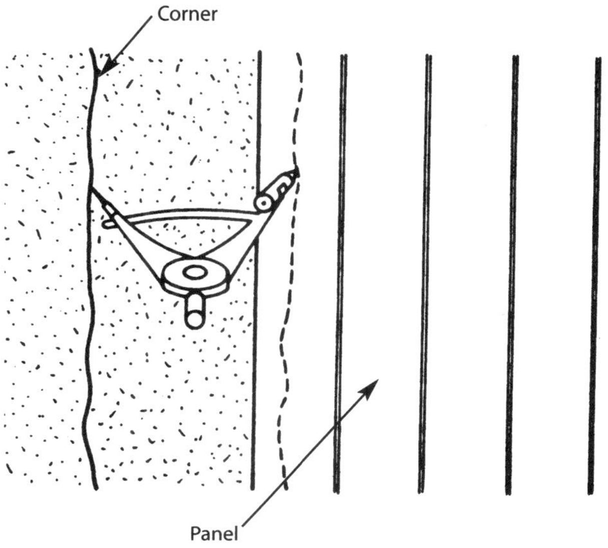

Figure 10-10

Scribing of cut at panel edge to provide exact fit in a corner or at ceiling

Fasten the panels with nails or adhesive. Adhesive saves filling nail holes on the panel surface. Use adhesive that provides "work time" before forming a tight bond. That makes it easier to adjust panels for a good fit. If panels are nailed, use small finishing nails (brads). Use $1^1/_2$"-long brads for $^1/_4$"- or $^3/_8$"-thick materials. Drive a brad each 8" to 10" along edges and at intermediate supports. Most panels are grooved to simulate hardboard panels. Drive brads in these grooves. Set brads slightly below the surface with a nail set. Many vendors of prefinished paneling also sell matching nails that require no putty to fill nail holes. Other vendors sell wood-filler putty to match their panels.

Hardwood Paneling

Most hardwood paneling is 8" wide or less. Hardwood paneling needs several days to adapt to room temperature and moisture conditions before being applied. Most paneling is applied with the long edges running vertically. But rustic patterns may be applied horizontally or diagonally to achieve a special effect.

Nail vertical paneling to horizontal furring strips or to nailing blocks set between studs. Use 1½" to 2" finishing or casing nails. Blind nail through the tongue on narrow strips. For 8" boards, face nail near each edge.

Ceiling Tile

Tile attached to the ceiling is usually 12" x 12". Suspended ceiling panels are usually 2' x 2' or 2' x 4'. Ceiling tile can be set with adhesive if the surface is smooth, level and firm. Dab a small spot of adhesive at the center and at each corner of the tile. Edge-matched tile can be stapled if the backing is wood.

You can set tile on furring strips to cover unsightly defects. But it's usually faster, cheaper and results in a better job if you tear off the existing cover and start over. If you want to try setting tile over the existing ceiling, use 1" x 3" or 1" x 4" furring strips where ceiling joists are 16" or 24" on center. Fasten the furring with two 7d or 8d nails at each joist. Where trusses or ceiling joists are spaced up to 48" apart, fasten 2" x 2" or 2" x 3" furring strips with two 10d nails at each joist. The furring should be a low-density wood, such as a soft pine, if tile is to be stapled to the furring.

Lay furring strips from the center of the room to the edges. Find the center by snapping chalklines from opposite corners. The ceiling center is where the diagonal lines cross. Place the first furring strip at the room center and at a right angle to the joists. Run parallel furring strips 12" each to both edges of the room. See Figure 10-11A. Edge courses on opposite walls should be equal in width. Plan spacing perpendicular to joists the same way. End courses should also be equal in width. Install tile the same way, working from the center to the edges. Set edge tile last so you get a close fit. Ceiling tile usually has a tongue on two adjacent edges and grooves on the other edges. Keep the tongue edges on the open side so they can be stapled to furring strips. Attach edge tile on the groove side with finishing nails or adhesive. Use one staple at each furring strip on the leading edge and two staples along the side, as in Figure 10-11B. Drive a small finishing nail or use adhesive to set edge tile against the wall.

A

B

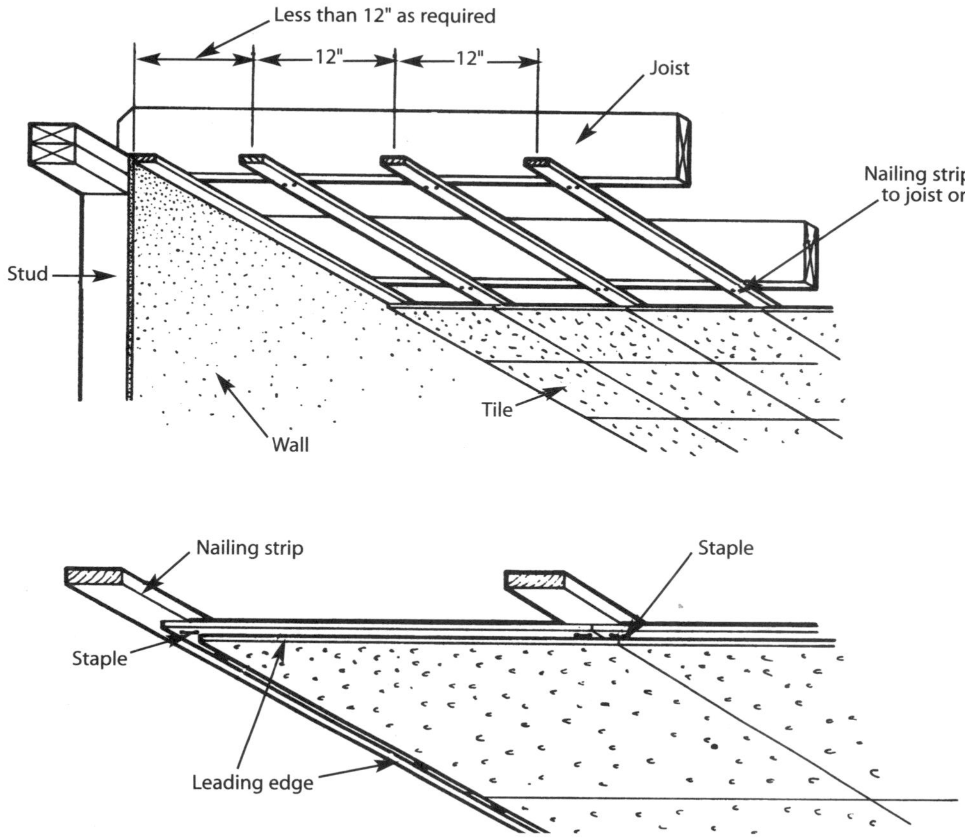

Figure 10-11

*Ceiling tile installation: **A**, nailing strip location; **B**, stapling*

Be careful not to soil the tile surface. Grease will leave permanent stains on ceiling tile. Professional tile installers rub cornmeal between their palms to keep their hands oil-free.

Suspended Ceilings

These are fine for basements, or other informal areas. Using them in "formal" rooms, like a living room or dining room, creates a "low-income" effect that may not be exactly what your client intends. Suspended ceilings cover imperfections, lower the ceiling to a more practical height, and add a plenum for running new electrical, plumbing, and HVAC lines.

The ceiling grid is suspended from wires or straps attached to joists. Panels drop into the completed grid. Ceiling height can be any level. Hanger wires may be only 2" or 3" long if the primary purpose is to cover fractured plaster. In earthquake zones, seismic bracing may be required by the building code. Your building department will have more information on this.

Interior Trim

Many older homes have trim styles no longer available at building material dealers. Matching trim exactly may require expensive custom fabrication. Try to remove trim in salvage condition so it can be re-installed. If trim is damaged or if you have to move doors or windows, it may be easier to replace all the trim in the room rather than try to match existing trim.

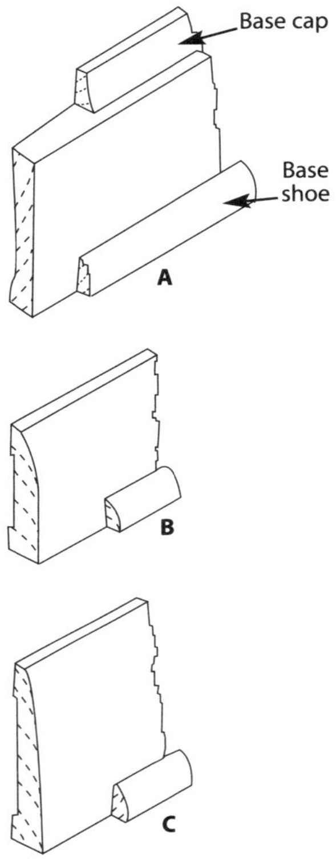

Figure 10-12
*Baseboard: **A**, two-piece; **B**, narrow; **C**, medium width*

Keep in mind that trim work requires a very high level of carpentry skill. Trim needs to be essentially perfect: sloppy joints and visible nail heads won't do. Don't ask a rough carpenter to do trim work — the results will be a disappointment If trim is going to be painted, select a trim made of extruded polymer, ponderosa pine or northern white pine, or primed MDF. Highly decorative cast trim is another good choice if trim will be painted. Most natural finish trim in modern homes is pine or oak. These woods can be very attractive if they're nicely finished.

Casing

Casing is the interior edge trim for door and window openings. Modern casing patterns vary in width from $2^1/4$" to $3^1/2$" and in thickness from $^1/2$" to $^3/4$". Install casing about $^3/16$" back from the face of the door or window jamb. Nail with 6d or 7d casing or finishing nails, depending on the thickness of the casing. Space nails in pairs about 16" apart, nailing to both jambs and framing. Casing with molded forms requires mitered joints, while rectangular casing can be butt-joined.

Baseboard

Finish the joint between the wall and floor with baseboard. Figure 10-12 shows several sizes and forms of baseboard. Two-piece base consists of a baseboard topped with a small base cap. The cap covers any gap caused by irregularities in the wall finish. Base shoe is nailed into the subfloor and covers irregularities in the finished floor. Walls made of drywall seldom need a base cap. Carpeted floors hide variations in the floor and make base shoe unnecessary.

Install square-edged baseboard with butt joints at inside corners and mitered joints at outside corners. Nail at each stud with two 8d finishing nails. Molded base, base cap, and base shoe require a coped joint at inside corners and a mitered joint at outside corners. See Figure 10-13.

Other Molding

Ceiling molding may be strictly decorative or may be used to hide the joint where the wall and ceiling meet. Use crown molding to cover the gap where wood paneling meets the ceiling. Attach crown molding with finishing nails driven into upper wall plates. Wide crown molding should be nailed both to the wall plate and the ceiling joists.

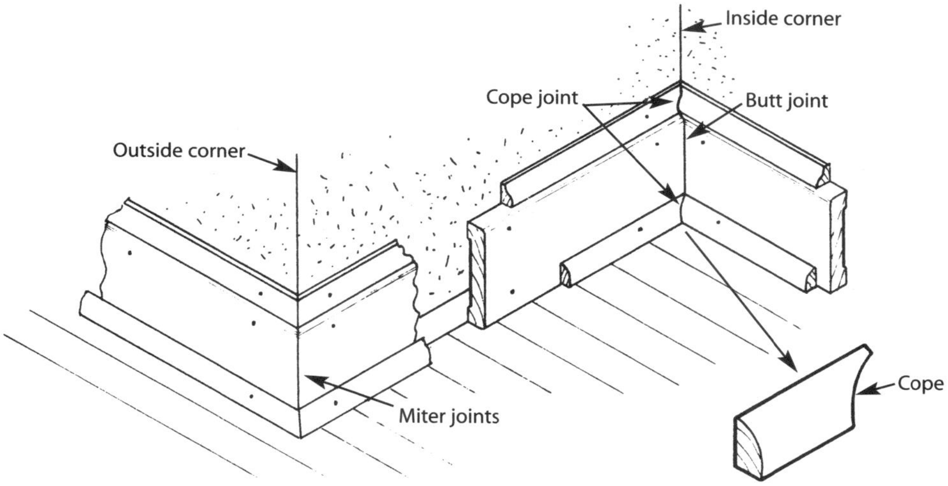

Figure 10-13
Installation of base molding

Removing Partitions

Modern taste favors more open space in homes. For example, new homes often have the family room and kitchen combined as one open area. Many older homes have a formal dining room or kitchen dining area enclosed by four walls and with a door that can be closed. Removing a nonbearing wall (partition) can add livability to an older home.

Nonbearing partitions support neither the roof nor a floor above. Breaking out a partition is only a cosmetic change. If wall cover is plaster or drywall, there's no salvage value in the partition. But save the trim, if possible. You may need it later.

Many partition walls include plumbing, electric or HVAC lines. Plan how those will be handled before you begin demolition.

When the partition is gone, there will be a strip of exposed floor, ceiling and wall where bottom, top and side edges of the partition had been. Finish the ceiling and walls with strips of drywall, tape, joint compound and paint. Filling the strip in the floor isn't as easy. Usually the best you can do is level the surface and cover the area with carpet or vinyl. With oak strip floors, it's possible to patch holes by weaving in new oak strips. However, this is a tricky job, and the entire floor may have to be refinished to get a perfect match of colors.

Removing a loadbearing partition requires the same patching of walls, ceiling, and floor. But you also have to add support for ceiling joists. If there's attic space above the partition, install a support beam above the ceiling joists. If it's a loadbearing wall on the first floor of a two-story house, it's holding up the second floor. If you remove it, the upstairs rooms could collapse onto the first floor. You'll need a large beam and posts to carry the weight that the wall was carrying. In this case, the beam will have to be below the ceiling joists, since the ceiling joists of

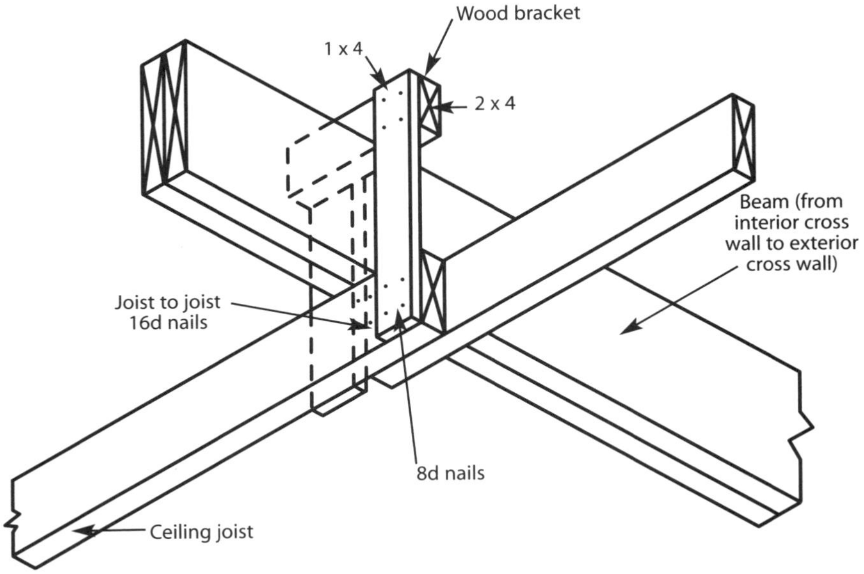

Figure 10-14
Framing for flush ceiling with wood brackets

the first floor are also the floor joists of the second floor. Be sure both ends of the beam are well supported on a bearing wall or post that is supported by the foundation. Support joists with metal framing anchors or wood brackets, as illustrated in Figure 10-14. To eliminate the need for temporary support, install the new beam before demolition begins.

If there's no attic space for a concealed beam, substitute an exposed beam at least 6'8" above the floor. Support ceiling joists temporarily with jacks and blocking while the partition is demolished and until the new beam is in place.

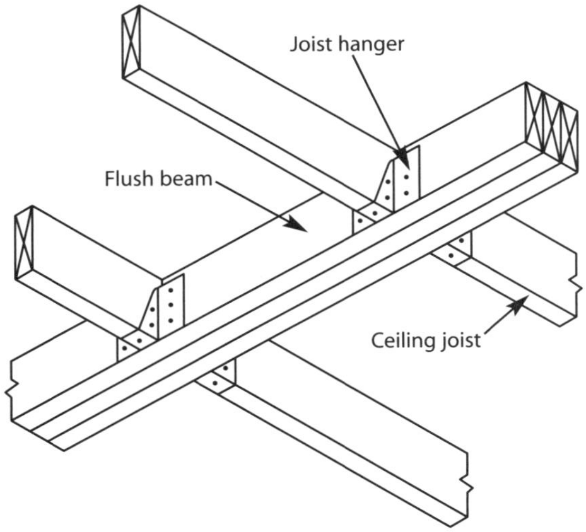

Figure 10-15
Flush beam with joist hangers

Figure 10-15 illustrates another option if an exposed beam is objectionable and there's no attic space for a concealed beam. Existing joists are supported by a new beam inserted where the top of the bearing wall had been. Place temporary joist supports on both sides of the bearing wall. Then remove the bearing wall and cut the joists as needed. Insert the new beam and install a hanger for each joist. Posts will also be needed to support this new beam.

The size of the beam required will vary with the span, load and lumber grade. Beam sizing like this is work for a civil engineer. In some communities, you'll need the approval of a licensed engineer before a permit is issued. Your building department or lumber yard probably has span tables for beams and load tables for posts that cover the most common residential situations. You probably won't need an engineer unless you're spanning a huge opening. For example, licensed contractors are often allowed to do simple engineering like this for houses up to 3,000 square feet. Houses this size and smaller are fairly straightforward. Larger houses and commercial buildings are more likely to present more complex engineering problems. When in doubt, make the framing much stronger than necessary. Building departments never have a problem with this.

Adding Partitions

Partition walls support nothing but their own weight and can be framed from 2" x 3" lumber, though 2" x 4" studs and plates are more common. The first step is to install the top plate. If ceiling joists are perpendicular to the partition, nail the top plate to each joist using 16d nails. If ceiling joists are parallel to the top plate and the partition is not directly under a joist, install solid blocking between joists. Blocks should be no more than 24" on center. See Figure 10-16. Nail the top plate to the blocks.

To be sure the new partition will be vertical, hold a plumb line along the side of the top plate at several points. Mark where the plumb bob touches the floor. Nail the sole plate to the floor joist at that position. If there's no joist where needed, nail solid blocking between joists. Blocks should be no more than 24" on center. Cut studs to fit snugly between the plates every 16" on center. Stud lengths may vary, so measure for each stud. Toenail the studs to the plates using 8d nails.

If you have enough space, assemble the wall on the floor and tilt it into place. Nail the top plate to the studs first. Tilt the assembled wall into place. Then toenail the studs to the bottom plate as described above.

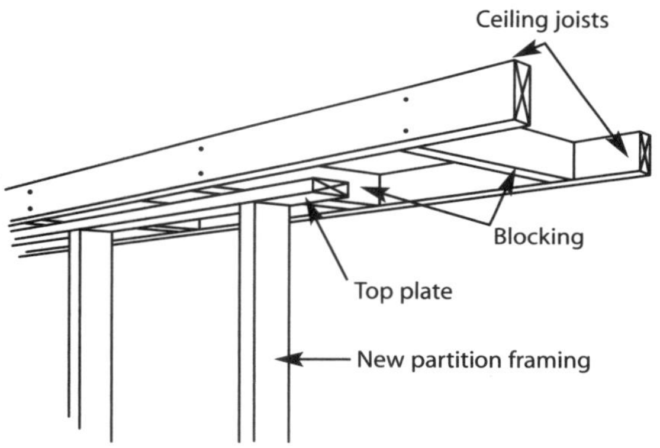

Figure 10-16

Blocking between joists to which the top plate of a new partition is nailed

	Craft@Hrs	Unit	Material	Labor	Total	Sell

Drywall

Ceiling and wall demolition. Knock down with hand tools at heights to 9' and handle debris to a trash bin on site. Building structure to remain. Includes the cost of breaking out old ceiling or wall cover, pulling or driving the old fasteners and cleaning up the debris. Add the cost of hauling debris off the site and dump fees. These figures assume demolition with a crowbar. Knock a hole with a crowbar, hook the crowbar in the hole, pull, and get out of the way! Ceilings come down fast. Add extra time if you're planning to save the floor, and put down floor protection. Plaster will fall in big chunks that can gouge holes in the floor. Figures in parentheses show the volume and weight of materials after demolition.

	Craft@Hrs	Unit	Material	Labor	Total	Sell
Plaster on ceiling (175 to 200 SF per CY and 8 pounds per SF)						
Lath and plaster only	BL@.010	SF	—	.29	.29	.49
Lath, plaster and furring	BL@.015	SF	—	.44	.44	.75
Suspended lath and plaster	BL@.010	SF	—	.29	.29	.49
Plaster on walls (150 SF per CY and 8 pounds per SF)						
Lath and plaster only	BL@.011	SF	—	.32	.32	.54
Lath, plaster and furring	BL@.015	SF	—	.44	.44	.75

Remove drywall. Includes the cost of breaking out old board, pulling or driving the old fasteners and cleaning up the debris. Add the cost of hauling debris off the site and dump fees.

	Craft@Hrs	Unit	Material	Labor	Total	Sell
Remove full panels on ceilings	BL@.009	SF	—	.26	.26	.44
Remove full panels on walls	BL@.008	SF	—	.23	.23	.39
Remove full panels and furring strips	BL@.028	SF	—	.81	.81	1.38
Remove and install drywall, no joint treatment per square foot	BC@.022	SF	—	.79	.79	1.34
Add for job setup, per room	BC@.200	Ea	—	7.20	7.20	12.20

Repair drywall. Includes joint tape, three coats of joint compound and finish sanding.

	Craft@Hrs	Unit	Material	Labor	Total	Sell
Cut out section, to 4' x 4'	D1@.288	Ea	—	10.10	10.10	17.20
Remove and replace section, to 4' x 4'	D1@.840	Ea	6.08	29.40	35.48	60.30
Taping and finishing only, ceilings	DT@.010	SF	.08	.35	.43	.73
Taping and finishing only, walls	DT@.008	SF	.08	.28	.36	.61
Tape and finish cracks, ceilings	DT@.060	LF	.13	2.12	2.25	3.83
Tape and finish cracks, walls	DT@.050	LF	.13	1.77	1.90	3.23

Patch hole in drywall. Cut back drywall, insert backing in wall cavity, apply three coats of joint compound, sand smooth.

	Craft@Hrs	Unit	Material	Labor	Total	Sell
Per patch	BC@1.00	Ea	4.42	36.00	40.42	68.70

Regular core drywall. Tapered edges. Cut ends. Labor includes cutting board around electrical boxes and obstacles, installing on wall studs or ceiling joists 8' to 12' above floor level, joint tape, three coats of joint compound and finish sanding. Material includes 1/2 gallon of premixed joint compound per 100 square feet, 38 linear feet of 2" perforated joint tape per 100 square feet, 1/2 pound of drywall screws per 100 square feet and 10% waste.

	Craft@Hrs	Unit	Material	Labor	Total	Sell
3/8" x 4' x 8', ceilings	D1@.023	SF	.33	.80	1.13	1.92
3/8" x 4' x 8', walls	D1@.017	SF	.33	.59	.92	1.56
1/2" x 4' x 8', ceilings	D1@.024	SF	.38	.84	1.22	2.07
1/2" x 4' x 8', walls	D1@.018	SF	.38	.63	1.01	1.72
1/2" x 4' x 9', ceilings	D1@.024	SF	.32	.84	1.16	1.97
1/2" x 4' x 9', walls	D1@.018	SF	.32	.63	.95	1.62
1/2" x 4' x 10', ceilings	D1@.024	SF	.33	.84	1.17	1.99
1/2" x 4' x 10', walls	D1@.018	SF	.33	.63	.96	1.63
1/2" x 4' x 12', ceilings	D1@.024	SF	.32	.84	1.16	1.97
1/2" x 4' x 12', walls	D1@.018	SF	.32	.63	.95	1.62
1/2" x 54" x 12', ceilings	D1@.024	SF	.44	.84	1.28	2.18
1/2" x 54" x 12', walls	D1@.018	SF	.44	.63	1.07	1.82

	Craft@Hrs	Unit	Material	Labor	Total	Sell
1/2" x 4' x 12', ceilings, sag resistant	D1@.024	SF	.44	.84	1.28	2.18
1/2" x 4' x 12', soffit board	D1@.035	SF	.52	1.22	1.74	2.96
1/2" panels on ceilings, no tape or finish	D1@.012	SF	.29	.42	.71	1.21
1/2" panels on walls, no tape or finish	D1@.008	SF	.29	.28	.57	.97
Arches, soffits, recesses, columns, angles	D1@.035	SF	—	1.22	1.22	2.07

Firecode Type X drywall. Fire-resistant gypsum core. Labor includes cutting board around electrical boxes and obstacles, installing on wall studs or ceiling joists 8' to 12' above floor level, joint tape, three coats of joint compound and finish sanding. Material includes 1/2 gallon of premixed joint compound per 100 square feet, 38 linear feet of 2" perforated joint tape per 100 square feet, 1/2 pound of drywall screws per 100 square feet and 10% waste.

	Craft@Hrs	Unit	Material	Labor	Total	Sell
5/8" x 4' x 6', ceilings	D1@.025	SF	.44	.87	1.31	2.23
5/8" x 4' x 6', walls	D1@.019	SF	.44	.66	1.10	1.87
5/8" x 4' x 8', ceilings	D1@.025	SF	.43	.87	1.30	2.21
5/8" x 4' x 8', walls	D1@.019	SF	.43	.66	1.09	1.85
5/8" x 4' x 9', ceilings	D1@.025	SF	.38	.87	1.25	2.13
5/8" x 4' x 9', walls	D1@.019	SF	.38	.66	1.04	1.77
5/8" x 4' x 10', ceilings	D1@.025	SF	.37	.87	1.24	2.11
5/8" x 4' x 10', walls	D1@.019	SF	.37	.66	1.03	1.75
5/8" x 4' x 12', ceilings	D1@.025	SF	.38	.87	1.25	2.13
5/8" x 4' x 12', walls	D1@.019	SF	.38	.66	1.04	1.77
5/8" panels on ceilings, no tape or finish	D1@.013	SF	.30	.45	.75	1.28
5/8" panels on walls, no tape or finish	D1@.009	SF	.30	.31	.61	1.04

Water-resistant drywall. Moisture-resistant gypsum core and paper as a base for tile or plastic-faced wall panels in bathrooms and kitchens. Not designed for use in high-moisture areas such as tub and shower surrounds. Includes cutting board around obstacles and 10% waste but no taping or finishing.

	Craft@Hrs	Unit	Material	Labor	Total	Sell
1/2" x 4' x 8'	D1@.008	SF	.31	.28	.59	1.00
1/2" x 4' x 12'	D1@.008	SF	.31	.28	.59	1.00
5/8" x 4' x 8'	D1@.009	SF	.43	.31	.74	1.26
5/8" x 4' x 12'	D1@.009	SF	.43	.31	.74	1.26

Mold- and mildew-resistant drywall. DensArmor has a glass mat backing. DensArmor Plus has glass mat on both front and back. Includes cutting board around obstacles and 10% waste but no taping or finishing.

	Craft@Hrs	Unit	Material	Labor	Total	Sell
1/2" x 4' x 8' DensArmor	D1@.008	SF	.39	.28	.67	1.14
1/2" x 4' x 8' DensArmor Plus	D1@.008	SF	.39	.28	.67	1.14
1/2" x 4' x 12' DensArmor Plus	D1@.009	SF	.39	.31	.70	1.19
5/8" x 4' x 8' DensArmor Plus	D1@.009	SF	.47	.31	.78	1.33
5/8" x 4' x 12' DensArmor Plus	D1@.009	SF	.47	.31	.78	1.33

Moisture- and fire-resistant drywall. For use under masonry veneer, aluminum, steel and vinyl siding, wood and mineral shingles, and stucco. Includes cutting board around obstacles and 10% waste, but no taping or finishing. Used to meet fire code in some types of multi-unit residential applications.

	Craft@Hrs	Unit	Material	Labor	Total	Sell
1/2" x 2' x 8'	D1@.008	SF	.35	.28	.63	1.07
5/8" x 2' x 8'	D1@.009	SF	.46	.31	.77	1.31

Sound-deadening drywall. For use as a base layer under drywall finish. Interior applications only. Must be kept dry. Includes cutting board around electrical boxes and obstacles, hanging on wall studs or ceiling joists 8' to 12' above floor level and 10% waste.

	Craft@Hrs	Unit	Material	Labor	Total	Sell
1/2" x 4' x 8', ceilings	D1@.012	SF	.35	.42	.77	1.31
1/2" x 4' x 8', walls	D1@.008	SF	.35	.28	.63	1.07

	Craft@Hrs	Unit	Material	Labor	Total	Sell
Flexible drywall. For covering curved surfaces. Includes cutting board around electrical boxes and obstacles, hanging board on wall studs or ceiling joists 8' to 12' above floor level, joint tape, three coats of joint compound and finish sanding. Material includes 1/2 gallon of premixed joint compound per 100 square feet, 38 linear feet of 2" perforated joint tape per 100 square feet, 1/2 pound of drywall screws per 100 square feet and 10% waste.						
1/4" x 4' x 8', ceilings	D1@.035	SF	.35	1.22	1.57	2.67
1/4" x 4' x 8', walls	D1@.035	SF	.35	1.22	1.57	2.67
Drywall repair sheet. Includes taping and finishing.						
1/2" x 16" x 16", greenboard	D1@1.00	Ea	7.56	35.00	42.56	72.40
1/2" x 24" x 24", regular	D1@1.00	Ea	4.59	35.00	39.59	67.30
5/8" x 16" x 16", regular	D1@1.00	Ea	5.51	35.00	40.51	68.90
Drywall texture. Finish applied to gypsum wallboard.						
Orange peel, rolled on, one coat	PT@.006	SF	.15	.22	.37	.63
Spatter finish, one coat	PT@.007	SF	.16	.26	.42	.71
Knockdown finish, one coat	PT@.008	SF	.16	.29	.45	.77
Skip trowel finish, one coat	PT@.009	SF	.13	.33	.46	.78
Smooth finish veneer plaster, one coat	PR@.018	SF	.14	.66	.80	1.36
Wall and ceiling spray texture, unaggregated. Creates spatter, knock down and orange peel texture. Powder mixes with water. Can be spray applied over drywall, concrete or plaster.						
40-pound bag, walls and ceilings	—	Ea	12.65	—	12.65	—
50-pound bag, walls and ceilings	—	Ea	13.10	—	13.10	—
Wall and ceiling spray texture, aggregated. Polystyrene aggregate mixes with water for spray application. Produces a white, simulated acoustical ceiling finish. Coarser mix conceals minor surface defects better. For gypsum, plaster and concrete ceilings. One pound of dry mix covers 8 square feet.						
Regular, 40-pound bag	PT@1.60	Ea	12.20	58.80	71.00	121.00
Medium, 32-pound bag	PT@1.28	Ea	10.50	47.10	57.60	97.90
Medium, 40-pound bag	PT@1.60	Ea	14.60	58.80	73.40	125.00
Coarse, 40-pound bag	PT@1.60	Ea	11.20	58.80	70.00	119.00
Spray texture touch-up kit. Hand-operated sprayer for small repairs and touch-up work. Pump body adjusts to spray orange peel, spatter, or knockdown textures. Reusable and refillable. Kit covers 15 square feet. Refill covers 10 square feet for orange peel and medium spatter or 6 square feet for heavy spatter/knockdown finish.						
Spray texture touch-up kit	PT@.250	Ea	26.00	9.19	35.19	59.80
Premixed refill kit	PT@.167	Ea	10.40	6.14	16.54	28.10
Popcorn ceiling texture. Finish applied to gypsum ceiling board. Applied with a compressed air hopper gun. Add the cost of a compressor.						
Blown on polystyrene texture	PT@.005	SF	.34	.18	.52	.88
Pneumatic hopper gun, purchase	—	Ea	73.70	—	73.70	—
Shop-type compressor rental, per day	—	Ea	60.00	—	60.00	—
Popcorn ceiling spray mix. Polystyrene. Mix with water and spray. 14-pound bag.						
Bag, titanium white	PT@1.00	Ea	13.20	36.80	50.00	85.00
Popcorn ceiling texture touch-up kit. Hand-operated sprayer for small repair and touch-up work. Pump body sprays acoustic popcorn texture. Reusable and refillable. Includes texture with polystyrene chip material for matching. Covers two square feet.						
Touch-up kit	PT@.500	Ea	15.30	18.40	33.70	57.30
Pack of 2 dry refills	PT@1.00	Ea	6.00	36.80	42.80	72.80

	Craft@Hrs	Unit	Material	Labor	Total	Sell
Popcorn ceiling patch. For repair of popcorn or acoustical rough-textured ceilings. Blends with most existing textures to cover holes, scuffs and scars. Apply with a paint roller or putty knife. One gallon covers 200 square feet.						
14 pound spray	PT@.250	Ea	9.15	9.19	18.34	31.20
5 gallons	PT@3.50	Ea	24.70	129.00	153.70	261.00
Acousti-Tex™ ceiling texture, Homax®. Mix one 20-ounce bag with one gallon of paint. Apply with a brush or roller. Creates a simulated cottage cheese ceiling texture. Covers cracks, flaws and scrapes.						
20 ounce bag	—	Ea	6.54	—	6.54	—
Ceiling spray texture. Premixed. Bottle attaches to spray texture hopper gun and covers 200 square feet.						
2.2 liter bottle	PT@1.00	Ea	15.20	36.80	52.00	88.40
Textured ceiling finish aerosol. 10-ounce aerosol can. Repair for small patches and cracks. Water cleanup.						
Popcorn, white	PT@.167	Ea	20.70	6.14	26.84	45.60
Knockdown or heavy spatter	PT@.167	Ea	25.30	6.14	31.44	53.40
Orange peel	PT@.167	Ea	25.30	6.14	31.44	53.40
Imperial veneer plaster base, USG. Rigid plaster base resists sag and sound transmission. Face paper resist plaster slide. Includes 10% waste.						
1/2" x 4' x 8', square edge	BC@.012	SF	.31	.43	.74	1.26
1/2" x 4' x 12', square edge	BC@.012	SF	.37	.43	.80	1.36
5/8" x 4' x 8', fire-resistant Type X	BC@.013	SF	.41	.47	.88	1.50
5/8" x 4' x 12', fire-resistant Type X	BC@.013	SF	.45	.47	.92	1.56
Cameo Veneer Plaster Base Board, Georgia-Pacific. High-suction face paper. Use as a base for 1- or 2-coat Cameo Veneer Plaster. Includes 10% waste.						
3/8" x 4' x 8', tapered edge	BC@.012	SF	.31	.43	.74	1.26
1/2" x 4' x 12'	BC@.012	SF	.31	.43	.74	1.26
DensShield® Tile Backer, Georgia-Pacific. Glass mat facing acrylic coating that blocks moisture. For walls, floors, countertops, showers, laboratories. Includes 10% waste.						
1/4" x 4' x 4'	TL@.042	SF	.69	1.41	2.10	3.57
1/2" x 32" x 5'	TL@.042	Ea	10.80	1.41	12.21	20.80
Durock® Cement Board, USG. Tile base for tub and shower areas, underlayment for tile on floors and countertops. Use the smooth side for ceramic tile mastic and the rough side for thin-set mortar and Portland cement. Hang with Durock® screws or galvanized roofing nails. Includes taping and sealing seams with Durock tape and 10% waste.						
5/16" x 4' x 4'	TL@.037	SF	.71	1.24	1.95	3.32
1/2" x 3' x 5'	TL@.042	SF	.71	1.41	2.12	3.60
1/2" x 4' x 8', interior	TL@.042	SF	.90	1.41	2.31	3.93
5/8" x 3' x 5'	TL@.047	SF	.97	1.58	2.55	4.34
5/8" x 4' x 8', interior	TL@.047	SF	.97	1.58	2.55	4.34

Drywall Joint Compound

	Craft@Hrs	Unit	Material	Labor	Total	Sell
Lightweight ready-mix joint compound. For embedding drywall joint tape and finishing drywall joints. One-half gallon finishes three 4' x 8' drywall panels. Installation cost is included in the cost of hanging drywall.						
3.5 gallon carton	—	Ea	8.10	—	8.10	—
4.5 gallon carton, tinted	—	Ea	12.20	—	12.20	—
4.5 gallon pail	—	Ea	13.70	—	13.70	—

	Craft@Hrs	Unit	Material	Labor	Total	Sell

All-purpose ready-mix joint compound. For embedding joint tape, finishing drywall joints, repairing small cracks and holes in drywall and plaster surfaces, and simple hand-applied texturing. One-half gallon finishes three 4' x 8' drywall panels. Installation cost is included in the cost of hanging drywall.

	Craft@Hrs	Unit	Material	Labor	Total	Sell
3 pound, 1 quart tub	—	Ea	3.41	—	3.41	—
12 pound pail	—	Ea	6.50	—	6.50	—
48 pound carton	—	Ea	7.29	—	7.29	—
50 pound carton	—	Ea	9.63	—	9.63	—
61.7 pound carton	—	Ea	11.60	—	11.60	—
61.7 pound pail	—	Ea	13.70	—	13.70	—

Lightweight setting-type joint compound powder. 20-, 45-, and 90-minute setting ranges allow one-day drywall joint finishing with next-day texturing. After a coat has set, apply another coat. No need to wait for each coat to dry completely. Weighs 25% less than all-purpose compound. Sands as easily as ready-mix joint compound. Ideal for patching drywall and plaster surfaces. 18-pound bag.

	Craft@Hrs	Unit	Material	Labor	Total	Sell
20-minute set time	—	Ea	10.20	—	10.20	—
45-minute set time	—	Ea	10.20	—	10.20	—
90-minute set time	—	Ea	10.20	—	10.20	—

Ready-mix joint topping compound. For filling, leveling and finishing coats over drywall joints. Sands easier and faster than all-purpose compound, and weighs 35% less. Not suitable for embedding joint tape, skim coating, or texturing.

	Craft@Hrs	Unit	Material	Labor	Total	Sell
3.5 gallon carton	—	Ea	8.71	—	8.71	—

Drywall ready-mix primer. Primes new drywall panel walls and ceilings before texturing or painting. Brush, roller, or spray application. Dries white in 30 minutes. Topcoat in 1 hour. 200 square feet per gallon.

	Craft@Hrs	Unit	Material	Labor	Total	Sell
5 gallon pail	—	Ea	56.10	—	56.10	—

Tuf-Tex hopper gun respray texture. For use with hopper gun or trailer-mounted spray rig. Approximately 500 square feet per bag.

	Craft@Hrs	Unit	Material	Labor	Total	Sell
40 pound bag	—	Ea	16.60	—	16.60	—

Plaster of Paris. For patching interior drywall or plaster walls and ceilings. Sets hard in 30 minutes.

	Craft@Hrs	Unit	Material	Labor	Total	Sell
25 pounds	—	Ea	16.30	—	16.30	—

One coat veneer plaster. High resistance to cracking, nail-popping, impact and abrasion failure. Mill-mixed plaster components help assure uniform installation performance and finished job quality. Can be applied directly to concrete block.

	Craft@Hrs	Unit	Material	Labor	Total	Sell
50 pounds	—	Ea	13.30	—	13.30	—

Drywall Joint Tape

Drywall joint tape. Center-creased paper tape. Six 4' x 8' drywall panels require about 75 linear feet of tape. Tape installation cost is included in the cost of hanging drywall. 2-1/16" wide. Perforated.

	Craft@Hrs	Unit	Material	Labor	Total	Sell
250' roll	—	Ea	1.79	—	1.79	—
500' roll	—	Ea	3.26	—	3.26	—

Fiberglass drywall tape. For drywall joints and repairs, veneer plastering, stucco and tile backer board. Self-adhesive 100% fiberglass mesh. Apply directly to drywall joint to eliminate embedding coat. 1-7/8" wide. Six 4' x 8' drywall panels require about 75 linear feet of tape. Cost of joint tape application is included in the Drywall section above.

	Craft@Hrs	Unit	Material	Labor	Total	Sell
50' long, white	—	Ea	2.35	—	2.35	—
150' long, white	—	Ea	4.59	—	4.59	—
150' long, yellow	—	Ea	4.18	—	4.18	—
300' long, white	—	Ea	6.68	—	6.68	—
300' long, yellow	—	Ea	6.58	—	6.58	—
500' long, white	—	Ea	10.50	—	10.50	—
500' long, yellow	—	Ea	10.20	—	10.20	—

	Craft@Hrs	Unit	Material	Labor	Total	Sell

Drywall Bead

Galvanized bullnose corner bead. All-metal galvanized steel reinforcement for protecting external drywall corners. Rounded 3/4" radius outside corner on 1/2" or 5/8" drywall. Attached with nails or crimping tool. Finish with USG joint compounds.

	Craft@Hrs	Unit	Material	Labor	Total	Sell
3/4" x 8' long	DI@.128	Ea	4.36	4.43	8.79	14.90
3/4" x 10' long	DI@.128	Ea	7.79	4.43	12.22	20.80

Plastic bullnose corner bead. For use where smooth or rounded corners are specified. Will not dent or rust. Paint and drywall mud adhere well. Finish compatible with drywall mud and all paint finishes.

	Craft@Hrs	Unit	Material	Labor	Total	Sell
3/4" x 8'	DI@.128	Ea	3.67	4.43	8.10	13.80

135-degree open-angle bullnose drywall arch. Plastic. For hard-to-do arches. No shiner. Wide flange.

	Craft@Hrs	Unit	Material	Labor	Total	Sell
3/4" x 8'	DI@.128	Ea	6.10	4.43	10.53	17.90

Bullnose open plastic corner. Used where three exterior corners intersect. Treated for maximum joint adhesion and rust protection. Wide flange.

	Craft@Hrs	Unit	Material	Labor	Total	Sell
3-way	DI@.160	Ea	1.01	5.54	6.55	11.10

Paper-faced metal bullnose outside corner, tape-on. Smooth, rounded radius outside corner for 1/2" or 5/8" drywall. Paper tape cover is laminated to a rust-resistant metal profile for good adhesion of joint compound. Held in place with joint compound instead of nails to bond the bead to the drywall surface. Bead is covered with joint compound.

	Craft@Hrs	Unit	Material	Labor	Total	Sell
3/4" x 8' long	DI@.128	Ea	2.62	4.43	7.05	12.00
9/16" x 13/16" x 10'	DI@.160	Ea	2.24	5.54	7.78	13.20
11/16" x 15/16" x 8'	DI@.216	Ea	2.19	7.48	9.67	16.40
1-1/16" x 1-1/16" x 8'	DI@.216	Ea	2.35	7.48	9.83	16.70
1-1/16" x 1-1/16" x 10'	DI@.270	Ea	2.93	9.35	12.28	20.90

Galvanized bullnose corner bead. Protects corners from damage with a more rounded edge for curves. Textured flanges allow for better bonding.

	Craft@Hrs	Unit	Material	Labor	Total	Sell
1-1/4" x 8'	DI@.128	Ea	1.86	4.43	6.29	10.70
1-1/4" x 9'	DI@.144	Ea	2.18	4.99	7.17	12.20
1-1/4" x 10'	DI@.160	Ea	2.02	5.54	7.56	12.90

Beadex® paper-faced metal L-shaped tape-on trim. Edge reinforcement for drywall panels where panels abut other materials such as suspended ceilings, beams, plaster, masonry, concrete, door or window jambs. Paper tape covering laminated to an L-shaped rust-resistant metal profile provides excellent adhesion of joint compound, texture and paint. Applied using joint compound instead of nails to bond trim to the drywall surface. Trim is finished with joint compound. For use with 1/2" drywall.

	Craft@Hrs	Unit	Material	Labor	Total	Sell
1/2" x 8'	DI@.160	Ea	2.42	5.54	7.96	13.50
1/2" x 10'	DI@.160	Ea	2.52	5.54	8.06	13.70

Paper-faced metal inside corner bead. Tape-on trim used to protect drywall panel assembly edges, corners and sides. Forms a true, inner (90-degree) corner. No fasteners required, adheres to compound. Eliminates edge cracking, nail pops and chipping.

	Craft@Hrs	Unit	Material	Labor	Total	Sell
8' long	DI@.216	Ea	2.49	7.48	9.97	16.90

Drywall flex corner bead. For use where curve or arch is formed. Pre-slotted flange allows for perfect arches. Tapered flanges are perforated and striated for adhesion. No need to tape the flange. Finish is compatible with drywall mud and all paint finishes.

	Craft@Hrs	Unit	Material	Labor	Total	Sell
1-1/4" x 10'	DI@.270	Ea	3.34	9.35	12.69	21.60

	Craft@Hrs	Unit	Material	Labor	Total	Sell
Beadex® tape-on flexible metal corner. Provides straight, strong inside or outside corners of any angle. Paper tape is 2-1/16" wide laminated to two 7/16" wide galvanized steel strips with a 1/16" gap between the strips. The strength of steel with the superior bond of joint compound on paper. Strong, chip-resistant, smooth finish. Applied with joint compound instead of nails. Guaranteed against edge cracking. Bead is finished with joint compound.						
25' roll	DI@.675	Ea	3.56	23.40	26.96	45.80
Drywall metal J-bead. Protects drywall edges around doors and windows. Textured flanges allow for better bonding.						
1/2" x 8'	DI@.216	Ea	1.90	7.48	9.38	15.90
1/2" x 10'	DI@.270	Ea	2.32	9.35	11.67	19.80
Drywall plastic J-bead. For capping vertical and horizontal edges of raw drywall. Helps keep moisture out of board. Finish is compatible with drywall mud and paint finishes.						
1/2" x 10'	DI@.270	Ea	1.68	9.35	11.03	18.80
5/8" x 10'	DI@.270	Ea	1.93	9.35	11.28	19.20
Drywall metal J-trim. Galvanized steel. Provides maximum protection and neat finished edges where panels join window and door jambs, and at internal angles. Install with nails or screws.						
1/2" x 8'	DI@.216	Ea	2.73	7.48	10.21	17.40
1/2" x 10'	DI@.270	Ea	3.83	9.35	13.18	22.40
5/8" x 8'	DI@.216	Ea	2.68	7.48	10.16	17.30
5/8" x 10'	DI@.270	Ea	2.71	9.35	12.06	20.50
Vinyl drywall outside corner bead. Rust- and impact-resistant. Tapered legs are perforated and striated for better adhesion. No need to tape flange. Prevents electrolysis. Finish compatible with drywall mud and all paint finishes.						
1-1/4" x 8' outside corner	DI@.216	Ea	1.81	7.48	9.29	15.80
1-1/4" x 10' outside corner	DI@.270	Ea	2.30	9.35	11.65	19.80
1/2" x 10' zip corner	DI@.270	Ea	2.75	9.35	12.10	20.60
10' bullnose corner	DI@.270	Ea	2.47	9.35	11.82	20.10
10' arch bullnose	DI@.270	Ea	4.03	9.35	13.38	22.70
1/2" x 8' J-trim	DI@.216	Ea	1.46	7.48	8.94	15.20
5/8" x 8' J-trim	DI@.216	Ea	1.46	7.48	8.94	15.20
Drywall L-trim. Metal covered with paper. Trims drywall where it abuts with other building components such as suspended ceilings, beams, plaster, masonry, or concrete walls, and untrimmed door or window jambs.						
1/2" x 8'	DI@.216	Ea	3.90	7.48	11.38	19.30
1/2" x 10'	DI@.270	Ea	5.52	9.35	14.87	25.30
Utility access door. For plumbing and electrical access. 28-gauge galvanized steel. Piano hinge. Pressure fit door. Drywall frame with nailing flange.						
14" x 14"	DI@.500	Ea	15.20	17.30	32.50	55.30
Drywall repair kit. Includes 1-pound bag of setting compound, 5' of drywall, fiberglass drywall tape, 120-grit sandpaper, plastic spreader, four drywall repair clips, and eight drywall screws.						
Kit	—	Ea	13.50	—	13.50	—
Drywall repair patch. No-rust aluminum plate provides rigid backing. Fiberglass mesh eliminates the need to embed tape.						
4" x 4"	—	Ea	3.62	—	3.62	—
8" x 8"	—	Ea	5.08	—	5.08	—
Crack repair mesh, Quik-Tape™. Self-adhesive 100% fiberglass mesh tape. Provides tensile strength and dimensional stability.						
6" wide, 25' long	—	Ea	3.43	—	3.43	—

	Craft@Hrs	Unit	Material	Labor	Total	Sell
Drywall repair clips. For repair of holes over 2". Suitable for both 1/2" and 5/8" thick drywall. Clips hold and reinforce the new piece of drywall sized to fit the hole being repaired.						
Kit with 6 clips	—	Ea	3.65	—	3.65	—
Tile backer tape. Pressure sensitive, anti-corrosion coating.						
2" x 50'	—	Ea	4.05	—	4.05	—
Flexible metal corner tape. For inside and outside corners.						
2" x 25'	DI@.175	Ea	5.20	6.06	11.26	19.10
2" x 100'	DI@.700	Ea	12.20	24.20	36.40	61.90
Corner bead spray adhesive, 3M Company. Quick grab so corner bead can be joined without delay. Longer open time allows assembly flexibility. Covers up to 1,200 linear feet per can. Tinted adhesive helps provide better wall coverage in hard-to-see areas.						
Adhesive	—	Ea	6.50	—	6.50	—

Drywall Fasteners

	Craft@Hrs	Unit	Material	Labor	Total	Sell
Drywall screws, Twinfast. Sharp, non-walking point. Dual lead. Each 100 square feet of board requires approximately 170 screws (1/2 pound). The cost of fasteners needed to hang drywall is included in the previous section, Drywall.						
#6 x 1", 5 pounds, 2,000 screws	—	Ea	22.40	—	22.40	—
#6 x 1-1/4", 5 pounds, 1,700 screws	—	Ea	22.40	—	22.40	—
#6 x 1-5/8", 5 pounds, 1,550 screws	—	Ea	22.40	—	22.40	—
Drywall nails. Approximately 300 nails per pound. Each 100 square feet of board requires approximately 200 nails (2/3 of a pound). The cost of fasteners needed to hang drywall is included in the previous section, Drywall.						
1-3/8", 30 pounds	—	Ea	40.80	—	40.80	—
1-1/2", 1 pound	—	Ea	4.00	—	4.00	—
1-1/2", coated, 1 pound	—	Ea	3.55	—	3.55	—
1-1/2", galvanized, 1 pound	—	Ea	3.92	—	3.92	—

Drywall Tools

	Craft@Hrs	Unit	Material	Labor	Total	Sell
Electronic stud sensor. Detects wood at 3/4" depth.						
Electronic stud sensor	—	Ea	22.00	—	22.00	—
Drywall screwdriver, 6-amp, DeWalt. Helical-cut steel, heat-treated gears. Quiet clutch. Depth-sensitive nosepiece. Tool weight: 2.9 pounds. UL Listed and OSHA approved. 60 inch-pounds of torque.						
0 To 4,000 RPM	—	Ea	101.00	—	101.00	—
Battery-powered drywall screw gun, DuraSpin, Senco. Drives 600 1" to 3" drywall screws an hour. Adjustable nosepiece. Includes spare drive, bits and case.						
18 volts	—	Ea	235.00	—	235.00	—
Drywall to wood collated screws, Senco. Collated pack of 1,000.						
#6 x 1-1/4"	—	Ea	17.30	—	17.30	—
#6 x 1-5/8"	—	Ea	19.50	—	19.50	—
#7 x 2"	—	Ea	22.60	—	22.60	—
Drywall to steel collated screws, Senco. Collated pack of 1,000.						
#6 x 1-1/4", light steel	—	Ea	18.20	—	18.20	—
#6 x 1-1/4", heavy steel	—	Ea	24.40	—	24.40	—

	Craft@Hrs	Unit	Material	Labor	Total	Sell
Phillips-head drywall screwdriver bit, Hilti.						
#1 insert bit, pack of 10	—	Ea	8.04	—	8.04	—
#2 insert bit, pack of 10	—	Ea	9.15	—	9.15	—
Drywall hammer, Estwing. All solid steel construction. Head and handle forged in one piece. Fully polished. Shock-reduction nylon vinyl grip. Crowned and scored face.						
13-1/2" long, 14 ounce	—	Ea	19.40	—	19.40	—
Drywall T-square. Head riveted to blade. Silver anodized aluminum. 1/8" and 1/16" graduations.						
48" blade	—	Ea	11.20	—	11.20	—
Stainless steel curved drywall trowel. Soft handle.						
11" trowel	—	Ea	20.20		20.20	—
14" trowel	—	Ea	21.90		21.90	—
Drywall corner trowel. Blade set at 80-degree angle that flexes to 90 degrees during use.						
Hardwood handle	—	Ea	10.50		10.50	—
Single-texture drywall brush. Horsehair plastic, round for swirled or sponged effects on ceilings and walls. Standard threaded hole in center can be used with any standard broom handle.						
2-1/2" trim, 4-3/4" block	—	Ea	13.20	—	13.20	—
Stippling brush	—	Ea	9.10	—	9.10	—
Crow's foot texturing brush. Tampico hemp bristles set in a wooden base. Can be hand held or center-hole pole mounted.						
13" x 9"	—	Ea	24.00	—	24.00	—
Drywall stilts. High-strength aluminum alloy with adjustable heights. Spring wishbone locks securely to prevent legs from sliding. Rubber soles prevent slipping. To 225 pounds.						
18" to 28" lift	—	Ea	280.00	—	280.00	—
Drywall panel lift. Lifts and holds drywall in place on the ceiling.						
Daily rental	—	Ea	25.00	—	25.00	—
Panel lift deposit	—	Ea	300.00	—	300.00	—
Drywall sander kit. For drywall sanding without dust. Includes 6' hose with swivel ends, 2 hose adapters, 1 medium grit sandscreen, 1-piece high-impact polystyrene plastic sanding head, and instructions.						
Drywall sander kit	—	Ea	20.40	—	20.40	—
Drywall sanding fiberglass respirator. Strengthened outer shell helps prevent collapse due to moisture buildup. Soft inner shell for comfort and durability. Adjustable metal nosepiece with foam seal fits variety of face sizes. Heavy-duty pre-stretched straps.						
Respirator	—	Ea	5.50	—	5.50	—
Drywall sanding sponge. For sanding drywall and plaster during spot repair. Resists clogging and buildup. Fine and medium sanding grits in one sponge.						
Small	—	Ea	2.20	—	2.20	—
Large	—	Ea	3.24	—	3.24	—
Plastic drywall hand sander. Lightweight, durable plastic base with multiple sandpaper usage. Use with die-cut or 1/2-sheet standard sandpaper.						
3-1/4" x 9-1/4"	—	Ea	6.10	—	6.10	—

	Craft@Hrs	Unit	Material	Labor	Total	Sell
Pole and hand drywall sander.						
Sander	—	Ea	7.65	—	7.65	—
Sand & Kleen™ dustless drywall sanding system, Magna. 20' of hose to sand 7 drywall panels between water changes. 5-gallon aquair filter, 1-1/4", 2-1/4", and 2-1/2" vacuum adapters.						
No. MT800 dustless sander	—	Ea	62.90	—	62.90	—

4 x 8 Paneling

	Craft@Hrs	Unit	Material	Labor	Total	Sell
Embossed tileboard, ABTCO. Printed hardboard. Including 10% waste.						
Aquatile, 4' x 8' panels, 1/8" thick	BC@.020	SF	1.39	.72	2.11	3.59
Glaztile III, 4' x 8' panels, 1/8" thick	BC@.020	SF	1.23	.72	1.95	3.32
Embossed tileboard. Water-resistant surface. The look of real ceramic tile. 1/8" x 4' x 8' panels. Including 10% waste.						
Alicante	BC@.034	SF	.83	1.22	2.05	3.49
Bisque plain	BC@.034	SF	.66	1.22	1.88	3.20
Blue floral	BC@.034	SF	.95	1.22	2.17	3.69
Florableu	BC@.034	SF	.81	1.22	2.03	3.45
Golden lace	BC@.034	SF	.79	1.22	2.01	3.42
Morning glories	BC@.034	SF	.82	1.22	2.04	3.47
Scored white	BC@.034	SF	.77	1.22	1.99	3.38
Silver quartz	BC@.034	SF	.97	1.22	2.19	3.72
Stone white	BC@.034	SF	.68	1.22	1.90	3.23
Thrifty white	BC@.034	SF	.43	1.22	1.65	2.81
Tuscan marble	BC@.034	SF	.74	1.22	1.96	3.33
White	BC@.034	SF	.54	1.22	1.76	2.99
Primed fiberboard paneling. Medium-density fiberboard 4' x 8' panels. Per square foot including 10% waste. 3/16" thick. White.						
Beaded	BC@.020	SF	.69	.72	1.41	2.40
Beaded, grooved 3" on center	BC@.020	SF	.71	.72	1.43	2.43
Paneling, Georgia-Pacific. Durable wood fiber backing. Simulated woodgrain face. Can be used in basements. 4' x 8' panels. Per square foot, including 10% waste.						
Blanc polare lauan, 5/32"	BC@.020	SF	.79	.72	1.51	2.57
Moonlight, 1/8"	BC@.020	SF	.54	.72	1.26	2.14
Mount Vernon paneling, Georgia-Pacific. Simulated wood grain face on medium-density fiberboard backing. Can be used in basements. 4' x 8' panels. Per square foot, including 10% waste.						
Kimberly oak, 1/8"	BC@.020	SF	.38	.72	1.10	1.87
Jubilee® beaded paneling, Georgia-Pacific. 3/16" thick. Ready to paint. Simulated woodgrain and decorative finishes. Acryglas® topcoat. 4' x 8' panels. Per square foot including 10% waste.						
Beaded white ice, 3/16"	BC@.020	SF	.80	.72	1.52	2.58
Jubilee®, 3/16"	BC@.020	SF	.77	.72	1.49	2.53
Hardboard paneling. Printed wood grain. 4' x 8' panels. Per square foot, including 10% waste.						
Alaskan oak, 1/8"	BC@.020	SF	.55	.72	1.27	2.16
Beaded finished oak, 5/32"	BC@.020	SF	1.51	.72	2.23	3.79
Charterhouse oak, 1/8"	BC@.020	SF	.81	.72	1.53	2.60
Knotty cedar, 5/32"	BC@.020	SF	.68	.72	1.40	2.38
Norwegian pine, 1/8"	BC@.020	SF	.73	.72	1.45	2.47
Spartan oak, 1/8"	BC@.020	SF	.40	.72	1.12	1.90
Valencia birch, 1/8"	BC@.020	SF	.69	.72	1.41	2.40

	Craft@Hrs	Unit	Material	Labor	Total	Sell
Water resistant, 1/4"	BC@.020	SF	.79	.72	1.51	2.57
Westminster red brick, 1/4"	BC@.020	SF	.98	.72	1.70	2.89
Vintage maple, 1/8"	BC@.020	SF	.67	.72	1.39	2.36
Traditional oak, 1/8"	BC@.020	SF	.80	.72	1.52	2.58
Honey oak, 1/8"	BC@.020	SF	.73	.72	1.45	2.47
Cherry oak, 1/8"	BC@.020	SF	.73	.72	1.45	2.47
Arabella pine, 5/32"	BC@.020	SF	.56	.72	1.28	2.18
Heather oak, 5/32"	BC@.020	SF	.68	.72	1.40	2.38
Juliet cherry, beaded, 5/32"	BC@.020	SF	.73	.72	1.45	2.47
Kristen leather, 5/32"	BC@.020	SF	.87	.72	1.59	2.70

Designer paneling, Fashion Series. 3.2 millimeter backing. Texture of wallpaper. Not recommended for floors, ceilings, and high-moisture areas. 4' x 8' panels. Per square foot, including 10% waste.

Sculptured stripe, 3.2mm thick	BC@.034	SF	.81	1.22	2.03	3.45

Hardwood Paneling

Ann Arbor birch prefinished paneling, Georgia-Pacific. For use in basements and below ground level. Rotary cut face. Plywood backing. Permagard® topcoat finish. 4' x 8' panels. Per square foot, including 10% waste.

5.0mm thick	BC@.034	SF	1.40	1.22	2.62	4.45

Bridgeport Bryant birch paneling, Georgia-Pacific. Offers the beauty of real wood with a durable hardwood plywood backing. Can be used in basements and below ground level. 4' x 8' panels. Per square foot, including 10% waste.

5.0mm thick	BC@.034	SF	1.31	1.22	2.53	4.30

Brookstone birch paneling, Georgia-Pacific. May be used in basements and below ground level. Rotary cut face and prefinished hardwood plywood substrate. Permagard® topcoat finish and unfinished. 4' x 8' panels. Per square foot, including 10% waste.

Unfinished, 5.0mm thick	BC@.034	SF	1.37	1.22	2.59	4.40

Remove tongue-and-groove plank paneling. Remove 1/4" x 8" x 4' to 8' pieces in salvage condition. Per square foot of wall covered.

Working from a ladder	BC@.040	SF	—	1.44	1.44	2.45

Beaded Cape Cod plank. Tongue-and-groove. Medium-density fiberboard backing.

1/4" x 21" x 8'	BC@.040	SF	1.31	1.44	2.75	4.68
9/16" x 8', trim kit	BC@.250	Ea	13.10	9.00	22.10	37.60

Edge V-plank. 4" wide x 8' long. Including 10% waste.

Beaded Adirondack oak, 5/16"	BC@.040	SF	5.04	1.44	6.48	11.00
Beaded knotty pine, 3/4"	BC@.040	SF	1.66	1.44	3.10	5.27
Knotty cedar, 5/16"	BC@.040	SF	1.60	1.44	3.04	5.17

Edge V-plank trim pack. Tongue-and-groove.

Adirondack oak, 5/16" x 6'	BC@.250	Ea	29.30	9.00	38.30	65.10

Pine planking. 4" x 8' x 5/16". Tongue-and-groove. Including 10% waste.

Beaded knotty pine	BC@.040	SF	1.20	1.44	2.64	4.49
Knotty pine 8' trim pack	BC@.250	Ea	14.15	9.00	23.15	39.40
V-groove pine	BC@.040	SF	1.31	1.44	2.75	4.68

Wainscot Paneling

Hardboard wainscot paneling. Paneling covers only the lower 32" of walls. Hardboard with the classic look of fine raised wood. Cost per panel.

	Craft@Hrs	Unit	Material	Labor	Total	Sell
8" x 32"	BC@.035	Ea	5.73	1.26	6.99	11.90
12" x 32", Pack of 2	BC@.105	Ea	23.00	3.78	26.78	45.50
16" x 32", Pack of 2	BC@.140	Ea	23.60	5.04	28.64	48.70
48" x 32", Pack of 2	BC@.420	Ea	50.40	15.10	65.50	111.00
Dual outlet cover	BC@.250	Ea	5.35	9.00	14.35	24.40
Universal outlet cover	BC@.250	Ea	5.35	9.00	14.35	24.40

Birch wainscot paneling, Bedford Village, Georgia-Pacific. Precut to 32" high. Unfinished. Stain to match an existing wainscot room, or customize. Beaded grooves every 1-1/2". 48" wide x 32" high. 5mm thick. Including 10% waste.

	Craft@Hrs	Unit	Material	Labor	Total	Sell
Ann Arbor	BC@.040	SF	1.54	1.44	2.98	5.07
Unfinished	BC@.040	SF	1.66	1.44	3.10	5.27

Pine wainscot paneling. 4" x 32". Tongue-and-groove. Including 10% waste.

	Craft@Hrs	Unit	Material	Labor	Total	Sell
4' x 32" x 1/4", paintable white	BC@.040	Ea	12.40	1.44	13.84	23.50
4' x 32" x 5/16", beaded knotty pine	BC@.040	Ea	14.10	1.44	15.54	26.40
3/4" x 8', knotty pine trim pack	BC@.250	Ea	19.60	9.00	28.60	48.60

Fiberglass Reinforced Plastic Panels

FRP wall and ceiling panels, Structoglas®, Sequentia. For residential construction and remodeling. Won't rust, rot, corrode, stain, dent, peel, or splinter. One side textured. Surface won't support mold, mildew, or other bacterial growth. Impermeable to cooking fumes and grease. Cleans with household detergents. Meets all major model code requirements. Class C flame spread. 4' x 8', .090 thick panels. Including 10% waste.

	Craft@Hrs	Unit	Material	Labor	Total	Sell
Almond	BC@.032	SF	1.14	1.15	2.29	3.89
White	BC@.032	SF	1.14	1.15	2.29	3.89

FRP panel molding. Fiberglass reinforced plastic. White. 10' long.

	Craft@Hrs	Unit	Material	Labor	Total	Sell
Division bar	BC@.160	Ea	3.55	5.76	9.31	15.80
Inside corner	BC@.160	Ea	3.55	5.76	9.31	15.80
Outside corner	BC@.160	Ea	3.55	5.76	9.31	15.80
End cap, 8' long	BC@.160	Ea	2.52	5.76	8.28	14.10

Nylon panel rivets. Non-staining. Maintains moisture seal and provides mechanical fastening for FRP panels when used with silicone sealants. 1/16" minimum grip, 1/2" maximum grip. Pack of 100.

	Craft@Hrs	Unit	Material	Labor	Total	Sell
Almond	—	Ea	14.00	—	14.00	—
White	—	Ea	14.40	—	14.40	—

Ceiling Tile

Acoustical ceiling tile. 12" x 12" tile. Working from a ladder at heights 8' to 12' above floor level.

	Craft@Hrs	Unit	Material	Labor	Total	Sell
Remove nailed or glued tile ceiling, per square foot of ceiling	BC@.009	SF	—	.32	.32	.54
Remove and install glued or stapled tile at various locations, per tile	BC@.100	Ea	—	3.60	3.60	6.12
Remove and install border tile by slipping tile under molding, per tile replaced	BC@.110	Ea	—	3.96	3.96	6.73
Staple tile directly on furring, no molding included, per square foot of tile	BC@.020	SF	—	.72	.72	1.22
Add for job setup, per room	BC@.150	Ea	—	5.40	5.40	9.18

	Craft@Hrs	Unit	Material	Labor	Total	Sell
Staple tile to 1" x 2" furring strips, including strips, per square foot of tile	BC@.042	SF	—	1.51	1.51	2.57
Add for job setup, per room	BC@.150	Ea	—	5.40	5.40	9.18
Set tile directly on ceiling with adhesive, including cove molding, per square foot of tile	BC@.020	SF	—	.72	.72	1.22
Add for job setup, per room	BC@.150	Ea	—	5.40	5.40	9.18
Secure loose tile by nailing or driving screws, per tile	BC@.035	Ea	—	1.26	1.26	2.14

Ceiling tile, Armstrong. Covers cracked or stained drywall ceilings. Fire-retardant. Stapled or set with adhesive.

	Craft@Hrs	Unit	Material	Labor	Total	Sell
12" x 12", Glenwood	BC@.020	Ea	.89	.72	1.61	2.74
12" x 12", Grenoble	BC@.020	Ea	1.01	.72	1.73	2.94
12" x 12", White, washable	BC@.020	Ea	.89	.72	1.61	2.74

Tin design ceiling tile, Armstrong. Hidden seams, durable, vinyl-coated, fire-retardant.

	Craft@Hrs	Unit	Material	Labor	Total	Sell
12" x 12"	BC@.020	Ea	1.81	.72	2.53	4.30

Wood fiber ceiling tile, Advantage Series, USG. Class C panels for non-acoustical and non fire-rated applications. Tongue-and-groove edging hides staples for a smooth, clean look. 12" x 12" x 1/2" with staple flange. Stapled or set with adhesive.

	Craft@Hrs	Unit	Material	Labor	Total	Sell
Lace	BC@.020	Ea	.89	.72	1.61	2.74
Tivoli	BC@.020	Ea	1.01	.72	1.73	2.94
White wood fiber	BC@.020	Ea	.92	.72	1.64	2.79

Acoustical ceiling tile adhesive, 237, Henry. For use on 12" x 12" or 12" x 24" tile. Bonds to concrete, concrete block, drywall, plaster and brick. 30-minute working time. One gallon covers 60 square feet.

	Craft@Hrs	Unit	Material	Labor	Total	Sell
Per square foot	—	SF	.23	—	.23	—

Suspended Ceilings

Remove suspended tile ceiling. (200 to 250 SF of ceiling yields one CY of debris.)

	Craft@Hrs	Unit	Material	Labor	Total	Sell
Including suspended grid	BL@.010	SF	—	.29	.29	.49
Including grid in salvage condition	BL@.019	SF	—	.55	.55	.94

Install suspended ceiling. Working from a ladder at heights 8' to 12' above floor level. No light fixtures included. Suspended from wires attached to wood joists through holes broken in drywall, plaster or fiberboard ceiling.

	Craft@Hrs	Unit	Material	Labor	Total	Sell
24" x 24" ceiling grid with panels	BC@.042	SF	—	1.51	1.51	2.57
24" x 48" ceiling grid with panels	BC@.028	SF	—	1.01	1.01	1.72
48" x 48" ceiling grid with panels	BC@.025	SF	—	.90	.90	1.53
Add for job setup, per room	BC@.183	Ea	—	6.59	6.59	11.20

Remove and replace 24" x 48" ceiling panels on an existing grid.

	Craft@Hrs	Unit	Material	Labor	Total	Sell
No cutting of panels, per panel	BC@.052	Ea	—	1.87	1.87	3.18
10% of panels cut, per panel	BC@.059	Ea	—	2.12	2.12	3.60
20% of panels cut, per panel	BC@.064	Ea	—	2.30	2.30	3.91
30% of panels cut, per panel	BC@.070	Ea	—	2.52	2.52	4.28
40% of panels cut, per panel	BC@.076	Ea	—	2.74	2.74	4.66
50% of panels cut, per panel	BC@.082	Ea	—	2.95	2.95	5.02

	Craft@Hrs	Unit	Material	Labor	Total	Sell
Prelude® suspended ceiling system cross tee, Armstrong. Intermediate-duty performance. Meets all building codes for non-residential construction. Hot-dipped galvanized, rust-resistant steel. Double-web construction. Requires only 2-1/2" clearance.						
2', stab-in, white	BC@.033	Ea	1.41	1.19	2.60	4.42
4', stab-in, white	BC@.033	Ea	2.31	1.19	3.50	5.95
Prelude® fire guard suspended ceiling system cross tee, Armstrong. Fire rated intermediate-duty performance. Meets all building codes for non-residential construction. Hot-dipped galvanized, rust-resistant steel. Double-web construction. Requires only 2-1/2" clearance.						
2', stab-in	BC@.033	Ea	1.77	1.19	2.96	5.03
4', stab-in, white	BC@.033	Ea	2.93	1.19	4.12	7.00
Prelude® suspended ceiling system main beam, Armstrong. Intermediate-duty performance. Hot-dipped galvanized, rust-resistant steel. Double-web construction. PeakForm™ end. 12.6 pound per foot carrying capacity.						
12', white	BC@.167	Ea	5.92	6.01	11.93	20.30
12', fire guard, white	BC@.167	Ea	6.37	6.01	12.38	21.00
Prelude® suspended ceiling system wall molding, Armstrong. Fire rated performance. Hemmed edges. Hot-dipped galvanized, rust-resistant steel, 12' x 7/8" x 7/8", 0.018" thick. Meets all building codes for non-residential construction.						
12', white	BC@.167	Ea	4.45	6.01	10.46	17.80
CeilingMAX™ zero-clearance ceiling tile grid system. Surface mounting for covering old plaster, drywall or paste-up ceilings. No wires to hang or leveling required. Top hanger fastens directly to joists or existing ceiling. Use any 2' x 2' or 4' x 4' acoustic ceiling tile. Eliminates demolition and re-installation. Runner snaps into top hanger to lock entire grid system and ceiling panels. Cross tee is installed perpendicular to top hanger to support ceiling panels. Wall bracket is installed around perimeter of room. Grid snaps out to access plenum above the grid. White vinyl.						
2' cross tee	BC@.033	Ea	1.41	1.19	2.60	4.42
8' runner insert	BC@.333	Ea	5.78	12.00	17.78	30.20
8' top hanger	BC@.333	Ea	6.89	12.00	18.89	32.10
8' wall bracket	BC@.111	Ea	3.52	4.00	7.52	12.80
Drill tip suspension lag screws.						
Package of 100	—	Ea	35.00	—	35.00	—
White rivets.						
Package of 100	—	Ea	3.80	—	3.80	—
Ceiling wire hanging kit. Contains nine 12 gauge support wires and nine J-hook nails. Suspends 256 square feet with main tees supported on 4' centers.						
Ceiling wire kit	BC@1.67	Ea	4.00	60.10	64.10	109.00
Suspended ceiling installation kit, Armstrong. Kit includes everything needed to install 64 square feet of suspended ceiling grid. White.						
2' x 2' grid	BC@1.25	Ea	46.60	45.00	91.60	156.00
2' x 4' grid	BC@1.25	Ea	34.50	45.00	79.50	135.00
Random textured suspended ceiling panel, Armstrong. Square lay-in panel. Recommended for use with Prelude 15/16" exposed-tee grid. Class A fire-resistance. Reduces sound by 55%. 5-year limited warranty.						
2' x 2' x 5/8" thick	BC@.027	Ea	2.49	.97	3.46	5.88
2' x 4' x 5/8" thick	BC@.056	Ea	6.21	2.02	8.23	14.00

	Craft@Hrs	Unit	Material	Labor	Total	Sell
Contractor suspended ceiling panel, Armstrong. Fire-retardant. Acoustical NRC .55. Reduces sound by 55%.						
2' x 2', random textured	BC@.027	Ea	3.43	.97	4.40	7.48
Plain white suspended ceiling panel, Armstrong. Fire-retardant. 10-year limited warranty.						
2' x 4', square edge	BC@.056	Ea	6.05	2.02	8.07	13.70
Classic fine-textured suspended ceiling panel, Armstrong. 3-dimensional look. Reduces sound by 60%. High light reflectance. BioBlock paint inhibits mold and mildew. HumiGuard Plus for extra sag resistance. Fire-retardant.						
2' x 2'	BC@.027	Ea	7.45	.97	8.42	14.30
Acoustical suspended ceiling panel, Armstrong. Square edge. Light reflectance minimum LR 0.80. Reduces sound by 55%.						
2' x 4', random textured	BC@.056	Ea	4.21	2.02	6.23	10.60
Textured suspended ceiling panel, Armstrong. Good for basements to hide plumbing and electrical fixtures while allowing access. Acoustical (NRC .55, CAC min. 35). Fire-retardant. Square edge. 5-year limited warranty.						
2' x 4'	BC@.056	Ea	4.12	2.02	6.14	10.40
Sahara suspended ceiling panel, Armstrong. Regular edge panel. HumiGuard™ Plus and Bioguard™ paint. Class A fire-resistant. 3-dimensional look. Reduces sound by 50%.						
2' x 2'	BC@.027	Ea	5.67	.97	6.64	11.30
HomeStyle™ suspended ceiling panel, Armstrong. Grid-hiding designs. Fire-retardant. 10-year limited warranty.						
2' x 2', Brighton	BC@.027	Ea	3.46	.97	4.43	7.53
2' x 2', Royal oak	BC@.027	Ea	5.24	.97	6.21	10.60
Fiberglass suspended ceiling panel, Armstrong. Square-edge panel with HumiGuard Plus for extra sag resistance for high humidity.						
2' x 4' x 5/8" thick, random fissured	BC@.056	Ea	6.33	2.02	8.35	14.20
2' x 4' x 5/8" thick, Esprit	BC@.056	Ea	6.36	2.02	8.38	14.20
Fiberglass suspended ceiling panel, Armstrong. Acoustical. UL-approved fire-retardant. Light reflectance minimum LR 0.80. Class A surface-burning characteristics for Flame Spread 25 or under. Meets Class A requirements of ASTM E 1264.						
2' x 4', reveal-edge design	BC@.056	Ea	6.81	2.02	8.83	15.00
Fissured fire guard suspended ceiling panel, Armstrong. Square lay-in panel. Recommended for use with Prelude 15/16" exposed-tee grid. Class A fire-resistance. UL approved for fire-rated ceiling assemblies. Reduces sound by 55%.						
2' x 4' x 5/8"	BC@.056	Ea	5.88	2.02	7.90	13.40
Prestige suspended ceiling panel, Armstrong. Architectural detail. Grid-blending pattern. Fire-retardant.						
2' x 2'	BC@.027	Ea	9.81	.97	10.78	18.30
Grenoble suspended ceiling panel, Armstrong. HomeStyle™ ceiling panel. Fire-retardant.						
2' x 4'	BC@.056	Ea	5.99	2.02	8.01	13.60

	Craft@Hrs	Unit	Material	Labor	Total	Sell
RADAR™ Illusion suspended ceiling panel, Professional Series, USG Interiors. Non-directional pattern for fast installation. Made with a perforated, water-felted surface for good sound absorption. Face-scored to create the illusion of a smaller-scaled ceiling system without compromising accessibility and speed of installation. Surface cleans with a soft brush or vacuum.						
2' x 4' x 3/4", slant edge	BC@.056	Ea	6.45	2.02	8.47	14.40
2' x 2' x 5/8", square edge	BC@.027	Ea	2.20	.97	3.17	5.39
2' x 4' x 5/8", square edge	BC@.056	Ea	4.19	2.02	6.21	10.60
2' x 2' x 5/8", shadowline tapered edge	BC@.027	Ea	2.41	.97	3.38	5.75
Fire guard suspended ceiling panel, Armstrong. Fine-textured ceiling panels with high light reflectance and scratch resistance. Washable. Sag resistant. Resists growth of mold and mildew.						
2' x 2' x 1/2", square edge	BC@.056	Ea	1.71	2.02	3.73	6.34
Fifth Avenue suspended ceiling panel, Professional Series, USG Interiors. Medium-textured acoustical panels with directional pattern. Made with a water-felted manufacturing process for good sound absorption. Surface cleans easily with a soft brush or vacuum. For home and light commercial applications. Square edge.						
2' x 2' x 5/8"	BC@.027	Ea	2.27	.97	3.24	5.51
2' x 4' x 5/8"	BC@.056	Ea	4.19	2.02	6.21	10.60
Firecode™ Fifth Avenue suspended ceiling panel, Professional Series, USG Interiors. Medium-textured acoustical panels with directional fissures. Made with a water-felted manufacturing process for good sound absorption. Cleans easily with a soft brush or vacuum. Washable. Class A fire-resistant.						
2' x 4' x 5/8", square edge	BC@.056	Ea	5.31	2.02	7.33	12.50
Sheetrock® gypsum lay-in ceiling panel, Professional Series, USG Interiors. Vinyl-faced panels for interior or exterior applications. Fire-rated construction up to 1-1/2 hours. USDA accepted for food processing areas.						
2' x 4' x 1/2", square edge	BC@.056	Ea	7.03	2.02	9.05	15.40
Stonehurst mineral fiber suspended ceiling panel, Advantage Series, USG Interiors. Square edge design for quick installation.						
2' x 4' x 9/16"	BC@.056	Ea	4.12	2.02	6.14	10.40
Luna™ suspended ceiling panel, Elite Series, USG Interiors. Fine-textured acoustical panels with Climaplus performance to prevent visible sag. High sound absorption and sound blocking. High light reflectance values. Cleans easily with a soft brush or vacuum. High humidity resistance.						
2' x 2' x 3/4", slant edge	BC@.027	Ea	6.69	.97	7.66	13.00
Alpine suspended ceiling panel, USG Interiors. Medium-textured acoustical panels made with a water-felted manufacturing process for good sound absorption. Perforated. Excellent light reflectance. Cleans easily with a soft brush or vacuum.						
2' x 2' x 5/8", slant edge	BC@.027	Ea	3.38	.97	4.35	7.40
Astro textured suspended ceiling panel, Professional Series, USG Interiors. Ceiling panels with textured surface and good sound absorption.						
2' x 2' x 5/8", square edge	BC@.027	Ea	4.84	.97	5.81	9.88
Saville Row suspended ceiling panel, Professional Series, USG Interiors. Medium, natural random texture with a sculptured, upscale appearance at an economy price. Made with a water-felted manufacturing process and perforated for good sound absorption. Excellent light reflectance. Cleans easily with a soft brush or vacuum.						
2' x 2' x 3/4", slant edge	BC@.027	Ea	6.26	.97	7.23	12.30

	Craft@Hrs	Unit	Material	Labor	Total	Sell
Acrylic suspended lighting panel.						
2' x 4', cracked ice, clear	BC@.056	Ea	9.96	2.02	11.98	20.40
2' x 4', cracked ice, white	BC@.056	Ea	9.96	2.02	11.98	20.40
2' x 4', prismatic, clear	BC@.056	Ea	9.96	2.02	11.98	20.40
2' x 4', prismatic, white	BC@.056	Ea	9.96	2.02	11.98	20.40
Styrene suspended lighting panel.						
2' x 2', cracked ice, clear	BC@.027	Ea	3.97	.97	4.94	8.40
2' x 2', cracked ice, white	BC@.027	Ea	3.97	.97	4.94	8.40
2' x 2', prismatic, clear	BC@.027	Ea	3.97	.97	4.94	8.40
2' x 2', prismatic, white	BC@.027	Ea	3.97	.97	4.94	8.40
2' x 4', cracked ice, clear	BC@.056	Ea	7.11	2.02	9.13	15.50
2' x 4', cracked ice, white	BC@.056	Ea	6.07	2.02	8.09	13.80
2' x 4', eggcrate, metallic	BC@.056	Ea	26.50	2.02	28.52	48.50
2' x 4', eggcrate, white	BC@.056	Ea	14.30	2.02	16.32	27.70
2' x 4', flat mist, white	BC@.056	Ea	10.60	2.02	12.62	21.50
2' x 4', prismatic, clear	BC@.056	Ea	7.11	2.02	9.13	15.50
2' x 4', prismatic, white	BC@.056	Ea	7.11	2.02	9.13	15.50
2' x 4', Victorian pattern	BC@.056	Ea	18.60	2.02	20.62	35.10

Fluorescent troffer light fixture. For suspending in a ceiling grid. Labor includes fixture hookup only. Add the cost of rough-in electrical and lamps. 32 watt, T-8.

	Craft@Hrs	Unit	Material	Labor	Total	Sell
2' x 2', 2 light	BE@.800	Ea	41.90	31.20	73.10	124.00
2' x 4', 3 light	BE@.800	Ea	46.90	31.20	78.10	133.00
2' x 4', 4 light	BE@.800	Ea	50.90	31.20	82.10	140.00

Parabolic troffer fluorescent fixture. For suspending in a ceiling grid. Labor includes fixture hookup only. Add the cost of rough-in electrical and lamps. 32 watt, T-8. Contoured louvers reduce glare. Contains electronic ballast for maximum efficiency.

	Craft@Hrs	Unit	Material	Labor	Total	Sell
2' x 4', 3 light	BE@.800	Ea	52.00	31.20	83.20	141.00

U-lamp troffer fluorescent fixture. For suspending in a ceiling grid. Labor includes fixture hookup only. Add the cost of rough-in electrical and lamps. Uses two U6 lamps. Energy-saving ballast. White unibody steel housing, reinforced flat white steel door, and mitered corners. Hinged and mitered doorframe. Comes with cam latches for easy opening. 40 or 34 watt bulbs.

	Craft@Hrs	Unit	Material	Labor	Total	Sell
2' x 2', 2 light	BE@.800	Ea	41.80	31.20	73.00	124.00

Ornamental Wood Molding

Hardwood Victorian base molding. Labor for moldings is best estimated by the piece rather than by the foot. The hard part is cutting lengths precisely to size. The more cuts needed, the more time required. For example, one 12' run of molding requires two cuts — one on each end. Twelve 1' lengths of molding would require 24 cuts, even though it's the same 12 linear feet. The estimates that follow are based on linear feet, but assume that full 8' lengths can be used to good advantage.

	Craft@Hrs	Unit	Material	Labor	Total	Sell
5/8" x 4", per foot	BC@.032	LF	2.61	1.15	3.76	6.39

Beaded baseboard molding. Solid wood. Embossed. Stain or paint to match paneling, cabinets, hardwood floors, or walls. Baseboard connectors provide architectural detail. Protects corners. Traditional period-style detail. White hardwood oak. Per foot.

	Craft@Hrs	Unit	Material	Labor	Total	Sell
3/4" x 6-1/2", Victorian	BC@.032	LF	3.39	1.15	4.54	7.72
Embossed molding, basswood.						
No. 630, 3/4" x 1/2" x 8'	BC@.290	Ea	8.14	10.40	18.54	31.50
Base corner molding. Per corner.						
No. 623, 7/8" x 7/8" x 6-3/4", inside	BC@.050	Ea	2.52	1.80	4.32	7.34
No. 623, 7/8" x 7/8" x 6-3/4", outside	BC@.050	Ea	4.18	1.80	5.98	10.20
Rosette block molding. Per block.						
2-3/4" x 2-3/4" x 1"	BC@.050	Ea	2.34	1.80	4.14	7.04
3-3/4" x 3-3/4" x 11/16"	BC@.050	Ea	2.31	1.80	4.11	6.99
Square rosette molding, solid pine.						
7/8" x 2-1/2" x 2-1/2"	BC@.050	Ea	1.61	1.80	3.41	5.80
7/8" x 3-1/2" x 3-1/2"	BC@.050	Ea	2.34	1.80	4.14	7.04
Plinth block.						
3-3/4" x 5-1/2", solid pine	BC@.050	Ea	5.48	1.80	7.28	12.40
3-3/4" x 5-3/4", solid pine	BC@.050	Ea	5.49	1.80	7.29	12.40
Ornamental crown molding. Per foot.						
3-3/4", dentil, pine	BC@.044	LF	5.39	1.58	6.97	11.80
3-3/4", rope, pine	BC@.044	LF	6.42	1.58	8.00	13.60
4-7/16", embossed	BC@.044	LF	4.73	1.58	6.31	10.70
4-7/8", pine	BC@.044	LF	3.62	1.58	5.20	8.84
5-1/2", dentil, pine	BC@.044	LF	7.89	1.58	9.47	16.10
Embossed decorative hardwood molding.						
Base, 5/8" x 4" x 8'	BC@.128	Ea	20.50	4.61	25.11	42.70
Egg and Dart, 1/2" x 3/4" x 8'	BC@.128	Ea	9.02	4.61	13.63	23.20
Egg and Dart, 5/8" x 1" x 8'	BC@.128	Ea	9.88	4.61	14.49	24.60
Flower, 5/8" x 1" x 8'	BC@.128	Ea	11.30	4.61	15.91	27.00
Flower, 3/4" x 1-15/16" x 8'	BC@.128	Ea	9.41	4.61	14.02	23.80
Greek Key, 3/8" x 3/4" x 8'	BC@.128	Ea	5.32	4.61	9.93	16.90
Leaf, 3/8" x 3/4" x 8'	BC@.128	Ea	5.32	4.61	9.93	16.90
Molding, 3/8" x 3/4" x 8'	BC@.128	Ea	6.65	4.61	11.26	19.10
Repeating Flower, 3/8" x 1-15/16" x 8'	BC@.128	Ea	8.41	4.61	13.02	22.10
Rope, 1/4" x 1" x 8'	BC@.128	Ea	6.64	4.61	11.25	19.10
Rope Flute, 1/2" x 2-1/4" x 7'	BC@.112	Ea	10.90	4.03	14.93	25.40
Rope Flute, 1/2" x 3" x 7'	BC@.112	Ea	16.10	4.03	20.13	34.20
Sunburst, 3/4" x 1-15/16" x 8'	BC@.128	Ea	8.97	4.61	13.58	23.10
Victorian, 5/8" x 3" x 8'	BC@.128	Ea	18.34	4.61	22.95	39.00
Vine, 3/8" x 3/4" x 8'	BC@.128	Ea	5.97	4.61	10.58	18.00
Vine, 3/8" x 1-3/4" x 8'	BC@.128	Ea	9.33	4.61	13.94	23.70

	Craft@Hrs	Unit	Material	Labor	Total	Sell
Primed Finger Joint Molding						
Primed finger joint base molding. Per foot.						
No. S, 7/16" x 3"	BC@.032	LF	.80	1.15	1.95	3.32
No. B, 7/16" x 3"	BC@.032	LF	.80	1.15	1.95	3.32
No. 663, 9/16" x 3-1/4"	BC@.032	LF	1.19	1.15	2.34	3.98
No. 753, 9/16" x 3-1/4"	BC@.032	LF	1.01	1.15	2.16	3.67
No. 750, 9/16" x 4-1/4"	BC@.032	LF	1.62	1.15	2.77	4.71
No. 620, 9/16" x 4-1/4"	BC@.032	LF	1.56	1.15	2.71	4.61
No. LB-11, 9/16" x 4-1/2"	BC@.032	LF	1.37	1.15	2.52	4.28
No. 618, 9/16" x 5-1/4"	BC@.032	LF	1.91	1.15	3.06	5.20
No. 5180, 9/16" x 5-1/4"	BC@.032	LF	1.91	1.15	3.06	5.20
No. 5163, 9/16" x 5-1/4"	BC@.032	LF	1.37	1.15	2.52	4.28
No. A314, 5/8" x 3-1/4"	BC@.032	LF	.80	1.15	1.95	3.32
No. B322, 5/8" x 3-1/4"	BC@.032	LF	.80	1.15	1.95	3.32
Primed finger joint base cap molding. Per foot.						
No. 1207, 11/16" x 7/8"	BC@.018	LF	.85	.65	1.50	2.55
No. 167, 11/16" x 1-1/8"	BC@.018	LF	.81	.65	1.46	2.48
No. 163, 11/16" x 1-3/8"	BC@.018	LF	.81	.65	1.46	2.48
Primed finger joint base shoe molding. Per foot.						
No. 129, 7/16" x 11/16"	BC@.020	LF	.42	.72	1.14	1.94
No. WM126, 1/2" x 3/4"	BC@.020	LF	.42	.72	1.14	1.94
Primed finger joint bead molding. Per foot.						
No. WM75, 9/16" x 1-5/8"	BC@.032	LF	.80	1.15	1.95	3.32
No. WM74, 9/16" x 1-3/4"	BC@.032	LF	.96	1.15	2.11	3.59
Primed finger joint brick molding. Pine.						
No. 180WMMPA, 1-1/4" x 2"	BC@.250	LF	2.26	9.00	11.26	19.10
No. WM180, 1-1/4" x 2" x 7'	BC@.105	LF	1.16	3.78	4.94	8.40
Primed finger joint ranch casing.						
No. C101, 1/2" x 2-1/8" x 7'	BC@.105	Ea	4.93	3.78	8.71	14.80
No. 713, 9/16" x 3-1/4"	BC@.015	LF	.91	.54	1.45	2.47
No. 327, 11/16" x 2-1/4"	BC@.015	LF	.59	.54	1.13	1.92
No. 327, 11/16" x 2-1/4" x 7'	BC@.105	Ea	4.14	3.78	7.92	13.50
No. 315, 11/16" x 2-1/2" x 7'	BC@.105	Ea	7.71	3.78	11.49	19.50
Primed finger joint colonial casing. Pine. Per foot.						
No. WM445, 5/8" x 3-1/4"	BC@.015	LF	1.16	.54	1.70	2.89
No. WM444, 5/8" x 3-1/2"	BC@.015	LF	1.21	.54	1.75	2.98
No. 442, 11/16" x 2-1/4"	BC@.015	LF	1.05	.54	1.59	2.70
Primed finger joint chair rail molding. Per foot.						
No. 297H, 5/8" x 3"	BC@.030	LF	1.35	1.08	2.43	4.13
No. CR-7, 11/16" x 2-5/8"	BC@.030	LF	1.91	1.08	2.99	5.08
No. AIL300, 1-1/16" x 2-3/4"	BC@.030	LF	1.79	1.08	2.87	4.88

	Craft@Hrs	Unit	Material	Labor	Total	Sell
Primed finger joint crown molding. Per foot.						
No. WM53, 9/16" x 2-5/8"	BC@.044	LF	1.22	1.58	2.80	4.76
No. AMC-R-47, 9/16" x 4-5/8"	BC@.044	LF	2.43	1.58	4.01	6.82
No. L47-PFJ, 9/16" x 4-5/8"	BC@.044	LF	2.40	1.58	3.98	6.77
Primed finger joint reversible flute and reed molding. Per foot.						
No. 286-PFJ, 11/16" x 3-1/4"	BC@.023	LF	1.54	.83	2.37	4.03
No. 286-POM, 11/16" x 3-1/4"	BC@.023	LF	2.50	.83	3.33	5.66
Primed finger joint quarter round molding. Per foot.						
No. WM106, 11/16" x 11/16"	BC@.020	LF	.83	.72	1.55	2.64
No. WM105, 3/4" x 3/4"	BC@.020	LF	.84	.72	1.56	2.65
Primed finger joint stool molding. Per foot.						
No. LWM1021, 11/16" x 5-1/4"	BC@.037	LF	3.05	1.33	4.38	7.45
Round edge stop molding. Per foot.						
No. P433FJ, 3/8" x 1-1/4"	BC@.015	LF	.84	.54	1.38	2.35
No. P349PR, 7/16" x 1-1/4"	BC@.015	LF	1.11	.54	1.65	2.81
No. P435PR, 1/2" x 1-5/8"	BC@.015	LF	.89	.54	1.43	2.43

Medium Density Fiberboard (MDF) Molding

	Craft@Hrs	Unit	Material	Labor	Total	Sell
MDF colonial base molding. Medium density fiberboard. Per foot.						
7/16" x 3"	BC@.032	LF	1.09	1.15	2.24	3.81
No. WM-663, 9/16" x 3-1/4"	BC@.032	LF	1.13	1.15	2.28	3.88
No. WM662, 9/16" x 3-1/2"	BC@.032	LF	.94	1.15	2.09	3.55
9/16" x 5-1/8"	BC@.032	LF	1.90	1.15	3.05	5.19
No. 620-MDF, 4-1/4"	BC@.032	LF	1.11	1.15	2.26	3.84
No. L163E, 5-1/8"	BC@.032	LF	1.64	1.15	2.79	4.74
MDF primed base corner. Per corner piece.						
No. MDF409A-711, 7/16" x 2-1/2"	BC@.050	Ea	1.88	1.80	3.68	6.26
No. MDF413A, 7/16" x 2-1/2", 3 step	BC@.050	Ea	1.88	1.80	3.68	6.26
No. MDF410A-711, 7/16" x 3-1/2"	BC@.050	Ea	2.34	1.80	4.14	7.04
No. MDF24A, 5/8" x 6", imperial	BC@.050	Ea	4.04	1.80	5.84	9.93
No. MDF28A, 5/8" x 6-1/2", vintage	BC@.050	Ea	4.04	1.80	5.84	9.93
MDF casing. Medium density fiberboard.						
No. AMH52, 1" x 4"	BC@.015	LF	1.80	.54	2.34	3.98
No. AMH56, 1" x 3-1/2"	BC@.015	Ea	1.39	.54	1.93	3.28
No. C322, 5/8" x 2-1/2", per foot	BC@.015	LF	.85	.54	1.39	2.36
No. 1684, 12mm x 3-3/8", per foot	BC@.015	LF	1.28	.54	1.82	3.09
MDF chair rail. Medium density fiberboard. Per foot.						
9/16" x 3", primed	BC@.030	LF	1.73	1.08	2.81	4.78
11/16" x 2-1/2"	BC@.030	LF	1.43	1.08	2.51	4.27
5/8" x 2-5/8"	BC@.030	LF	1.43	1.08	2.51	4.27

	Craft@Hrs	Unit	Material	Labor	Total	Sell
MDF crown molding. Medium density fiberboard. Per foot.						
No. 45, 9/16" x 5-1/4"	BC@.044	LF	1.83	1.58	3.41	5.80
No. L47, 9/16" x 4-5/8" x 16'	BC@.044	LF	1.82	1.58	3.40	5.78
MDF ultralite crown molding. Per foot.						
No. 40, 11/16" x 4-1/4"	BC@.044	LF	2.54	1.58	4.12	7.00
No. 49, 15mm x 3-5/8"	BC@.044	LF	1.16	1.58	2.74	4.66
No. 47, 15mm x 4-5/8"	BC@.044	LF	1.69	1.58	3.27	5.56
No. 43, 18mm x 7-1/4"	BC@.044	LF	2.68	1.58	4.26	7.24
MDF primed plinth block.						
1" x 3-3/4" x 8"	BC@.050	Ea	2.65	1.80	4.45	7.57
1" x 4-1/8" x 9", Victorian	BC@.050	Ea	3.55	1.80	5.35	9.10
1-1/4" x 2-1/2" x 5-3/4"	BC@.050	Ea	2.43	1.80	4.23	7.19
1-1/4" x 3-1/2" x 6-1/2"	BC@.050	Ea	3.00	1.80	4.80	8.16

Polyurethane Cast Molding

	Craft@Hrs	Unit	Material	Labor	Total	Sell
Cross head polyurethane cast molding.						
6" x 60", dentil	BC@.250	Ea	48.20	9.00	57.20	97.20
6" x 60", Tuscan	BC@.250	Ea	48.20	9.00	57.20	97.20
Frieze polyurethane cast molding, American Molding. Horizontal flat band, usually installed just below cornice molding.						
4-7/8" x 96", rococo	BC@.250	Ea	18.20	9.00	27.20	46.20
7-1/2" x 92-1/4", grand palmetto	BC@.250	Ea	42.80	9.00	51.80	88.10
7-1/2" x 92-1/4", grand Tuscan	BC@.250	Ea	42.80	9.00	51.80	88.10
Polyurethane cast molding.						
4-1/8" x 8', dentil	BC@.200	Ea	17.70	7.20	24.90	42.30
4-1/8" x 8', egg and dart	BC@.200	Ea	20.80	7.20	28.00	47.60
4-1/8" x 8', Grecian	BC@.200	Ea	20.80	7.20	28.00	47.60
5-7/8" x 8', dentil	BC@.200	Ea	27.00	7.20	34.20	58.10
5-7/8" x 8', egg and dart	BC@.200	Ea	27.00	7.20	34.20	58.10
5-7/8" x 8', rope crown	BC@.200	Ea	27.00	7.20	34.20	58.10
Polyurethane corner block.						
4-1/8", inside	BC@.050	Ea	7.25	1.80	9.05	15.40
4-1/8", outside	BC@.050	Ea	7.25	1.80	9.05	15.40
5-7/8", inside	BC@.050	Ea	8.29	1.80	10.09	17.20
5-7/8", outside	BC@.050	Ea	8.29	1.80	10.09	17.20
Rope polyurethane cast molding. Rope molding is a bead or Torus molding carved to imitate rope.						
3-1/2" x 8', casing	BC@.400	Ea	19.70	14.40	34.10	58.00
3-1/2" x 8', chair rail	BC@.300	Ea	19.70	10.80	30.50	51.90
5-1/2" x 8', base	BC@.250	Ea	27.30	9.00	36.30	61.70
Chair rail polyurethane cast molding.						
3-1/2" x 8', dentil	BC@.300	Ea	20.80	10.80	31.60	53.70
3-1/2" x 8', Greek key	BC@.300	Ea	20.80	10.80	31.60	53.70

	Craft@Hrs	Unit	Material	Labor	Total	Sell
Poly molding.						
Bead, brite white, 8'	BC@.128	Ea	4.15	4.61	8.76	14.90
Bead, cherry/mahogany, 8'	BC@.128	Ea	3.45	4.61	8.06	13.70
Bead, cinnamon chestnut, 8'	BC@.128	Ea	3.60	4.61	8.21	14.00
Bed mold, brite white, 8'	BC@.128	Ea	6.01	4.61	10.62	18.10
Bed mold, cinnamon chestnut, 8'	BC@.128	Ea	4.86	4.61	9.47	16.10
Bed mold, cherry/mahogany, 8'	BC@.128	Ea	4.68	4.61	9.29	15.80
Casing and base, brite white, 8'	BC@.184	Ea	6.43	6.63	13.06	22.20
Casing and base, cherry/mahogany, 8'	BC@.184	Ea	6.43	6.63	13.06	22.20
Casing and base, cinnamon chestnut, 8'	BC@.184	Ea	6.86	6.63	13.49	22.90
Outside corner, brite white, 8'	BC@.128	Ea	4.17	4.61	8.78	14.90
Outside corner, cherry/mahogany, 8'	BC@.128	Ea	2.42	4.61	7.03	12.00
Outside corner, cinnamon chestnut, 8'	BC@.128	Ea	2.50	4.61	7.11	12.10
Plywood cap, brite white, 8'	BC@.128	Ea	4.17	4.61	8.78	14.90
Shoe, brite white, 8'	BC@.128	Ea	4.14	4.61	8.75	14.90
Shoe, cherry/mahogany, 8'	BC@.128	Ea	2.42	4.61	7.03	12.00
Shoe, cinnamon chestnut, 8'	BC@.128	Ea	2.50	4.61	7.11	12.10
Stop, brite white, 8'	BC@.200	Ea	3.53	7.20	10.73	18.20
Stop, cinnamon chestnut, 8'	BC@.200	Ea	3.53	7.20	10.73	18.20

White extruded polymer molding. Durable in high traffic areas. Easy to clean. Paintable and stainable. Unaffected by moisture, will not mold, mildew, swell, or rot. Termite proof. No sanding needed. Per linear foot.

	Craft@Hrs	Unit	Material	Labor	Total	Sell
No. 49, 1/2" x 3-5/8", crown	BC@.044	LF	1.23	1.58	2.81	4.78
No. 623, 9/16" x 3-1/4", base	BC@.032	LF	1.30	1.15	2.45	4.17
No. 356, 11/16" x 2-1/4", casing	BC@.050	LF	.99	1.80	2.79	4.74
No. 105, 3/4" x 3/4", quarter round	BC@.032	LF	.77	1.15	1.92	3.26
1-3/8", closet pole	BC@.032	LF	1.53	1.15	2.68	4.56
No. 390, 2-5/8", chair rail	BC@.032	LF	1.84	1.15	2.99	5.08

Prefinished Molding

Embossed polystyrene base molding, Heirloom™, Royal Mouldings. Deep embossed ornate design. Won't crack, warp, split, or bow. Paint process finish. Stainable.

	Craft@Hrs	Unit	Material	Labor	Total	Sell
No. 57A66438V, 5" x 9/16" x 8', Bone white	BC@.250	Ea	25.20	9.00	34.20	58.10
No. 0A66408009, 5" x 9/16" x 8', Clearwood	BC@.250	Ea	20.40	9.00	29.40	50.00
8', Highlands oak	BC@.250	Ea	33.60	9.00	42.60	72.40
8', Imperial oak	BC@.250	Ea	34.00	9.00	43.00	73.10

Colonial base molding, Royal Mouldings. Pre-sealed and pre-sanded. Non-repetitive woodgrain finish for consistent appearance. Stain or paint with any oil or latex paint applied by brush, roller or sprayer. Stain or paint absorbs uniformly without discoloration. Clean by wiping with paint thinner, naphtha or ammonia.

	Craft@Hrs	Unit	Material	Labor	Total	Sell
7/16" x 3-1/4" x 8', Bone white	BC@.250	Ea	8.91	9.00	17.91	30.40
7/16" x 3-1/4" x 8', Clearwood	BC@.250	Ea	8.52	9.00	17.52	29.80

Colonial base paperwrap molding, Royal Collection™, Royal Mouldings.

	Craft@Hrs	Unit	Material	Labor	Total	Sell
7/16" x 3-1/4" x 8', Highlands oak	BC@.250	Ea	11.33	9.00	20.33	34.60
7/16" x 3-1/4" x 8', Imperial oak	BC@.250	Ea	11.33	9.00	20.33	34.60
7/16" x 3-1/4" x 8', Natural maple	BC@.250	Ea	11.33	9.00	20.33	34.60

	Craft@Hrs	Unit	Material	Labor	Total	Sell
Colonial base polystyrene molding, Royal Mouldings. Non-repetitive woodgrain finish. Pre-sealed and pre-sanded. Stain or paint with any oil or latex paint. Stain and paint absorb uniformly.						
7/16" x 3-1/4" x 8', Bone white	BC@.250	Ea	11.40	9.00	20.40	34.70
7/16" x 3-1/4" x 8', Dominion oak	BC@.250	Ea	9.13	9.00	18.13	30.80
Combination base molding, Royal Mouldings. 11/16" x 5-1/4".						
8', Bone white	BC@.250	Ea	18.04	9.00	27.04	46.00
8', Clearwood	BC@.250	Ea	18.50	9.00	27.50	46.80
Ranch base paperwrap molding, Royal Collection™, Royal Mouldings.						
7/16" x 3-1/4" x 8', Clearwood	BC@.250	Ea	7.92	9.00	16.92	28.80
1/2" x 2-7/16" x 8', Dominion oak	BC@.250	Ea	9.00	9.00	18.00	30.60
1/2" x 2-7/16" x 8', Highlands oak	BC@.250	Ea	9.23	9.00	18.23	31.00
1/2" x 2-7/16" x 8', Imperial oak	BC@.250	Ea	9.24	9.00	18.24	31.00
Base shoe molding, Royal Mouldings. Pre-sealed and pre-sanded. Stainable or paintable with any oil or latex paint applied by brush, roller or sprayer. Non-repetitive woodgrain finish for consistent appearance. Stain or paint absorbs uniformly without discoloration. Clean by wiping with paint thinner, naphtha or ammonia.						
3/8" x 11/16" x 8', Clearwood	BC@.128	Ea	4.38	4.61	8.99	15.30
Colonial cap paperwrap molding, Royal Collection™, Royal Mouldings. Non-repetitive woodgrain finish. Pre-sealed and pre-sanded. Stainable or paintable with any oil or latex paint. Stain and paint absorb uniformly.						
9/16" x 1-1/8" x 8', Highlands oak	BC@.250	Ea	7.08	9.00	16.08	27.30
9/16" x 1-1/8" x 8', Imperial oak	BC@.250	Ea	7.10	9.00	16.10	27.40
9/16" x 1-1/8" x 8', Natural maple	BC@.250	Ea	7.11	9.00	16.11	27.40
Colonial casing paperwrap molding, Royal Collection™, Royal Mouldings.						
9/16" x 2-1/4" x 7', Highlands oak	BC@.105	Ea	9.15	3.78	12.93	22.00
9/16" x 2-1/4" x 7', Imperial oak	BC@.105	Ea	9.15	3.78	12.93	22.00
9/16" x 2-1/4" x 7', Natural maple	BC@.105	Ea	9.15	3.78	12.93	22.00
Colonial casing polystyrene molding, Royal Mouldings. Non-repetitive woodgrain finish. Pre-sealed and pre-sanded. Stainable or paintable with any oil or latex paint. Stain and paint absorb uniformly.						
7/16" x 1-9/16" x 7'	BC@.105	Ea	5.52	3.78	9.30	15.80
9/16" x 2-1/4" x 7', Bone white	BC@.105	Ea	6.83	3.78	10.61	18.00
9/16" x 2-1/4" x 7', Dominion oak	BC@.105	Ea	7.05	3.78	10.83	18.40
Colonial casing prefinished molding, Royal Mouldings.						
No. CWPS, 7', "Creations"	BC@.105	Ea	9.80	3.78	13.58	23.10
9/16" x 2-1/4" x 7', "Reflections"	BC@.105	Ea	7.12	3.78	10.90	18.50
Highlands oak, 7', "Heirloom"	BC@.105	Ea	28.10	3.78	31.88	54.20
Imperial oak, 7', "Heirloom"	BC@.105	Ea	28.10	3.78	31.88	54.20

	Craft@Hrs	Unit	Material	Labor	Total	Sell
Economy casing molding, Royal Mouldings. 9/16" x 2-1/4" x 7'.						
Highlands oak	BC@.105	Ea	4.48	3.78	8.26	14.00
Imperial oak	BC@.105	Ea	4.48	3.78	8.26	14.00
Fluted casing embossed polystyrene molding, Heirloom™, Royal Mouldings. Stainable or paintable finish. Deep embossed ornate design. Won't crack, warp, split, or bow. Paint process finish.						
11/16" x 4" x 8', Bone white	BC@.105	Ea	21.20	3.78	24.98	42.50
11/16" x 4" x 8', Clearwood	BC@.105	Ea	24.10	3.78	27.88	47.40
Fluted casing prefinished molding, Royal Mouldings.						
2-1/8", Bone white, 8'	BC@.105	Ea	15.00	3.78	18.78	31.90
3-1/8", Bone white, 8'	BC@.105	Ea	15.00	3.78	18.78	31.90
Ranch casing molding, Clearwood PS™, Royal Mouldings. Pre-sealed and pre-sanded. Stainable or paintable with any oil or latex paint applied by brush, roller, or sprayer. Non-repetitive woodgrain finish for consistent appearance. Stain or paint absorbs uniformly without discoloration. Clean by wiping with paint thinner, naphtha, or ammonia.						
9/16" x 2-1/4" x 7', Bone white	BC@.105	Ea	5.61	3.78	9.39	16.00
Ranch casing molding, Royal Mouldings. Pre-sealed and pre-sanded. Stainable or paintable with any oil or latex paint applied by brush, roller or sprayer. Non-repetitive woodgrain finish for consistent appearance. Stain or paint absorbs uniformly without discoloration. Clean by wiping with paint thinner, naphtha or ammonia.						
9/16" x 2-1/4" x 7', Clearwood	BC@.105	Ea	4.99	3.78	8.77	14.90
Ranch casing paperwrap molding, Royal Collection™, Royal Mouldings.						
9/16" x 2-1/4" x 7', Highlands oak	BC@.105	Ea	7.43	3.78	11.21	19.10
9/16" x 2-1/4" x 7', Imperial oak	BC@.105	Ea	7.41	3.78	11.19	19.00
Ranch casing polystyrene molding, Royal Mouldings. Non-repetitive woodgrain finish. Pre-sealed and pre-sanded. Stainable or paintable with any oil or latex paint. Stain and paint absorb uniformly.						
13/32" x 1-9/16" x 7'	BC@.105	Ea	5.19	3.78	8.97	15.20
9/16" x 2-1/4" x 7', Dominion oak	BC@.105	Ea	6.90	3.78	10.68	18.20
5/8" x 1-5/8" x 7', Clearwood	BC@.105	Ea	7.29	3.78	11.07	18.80

	Craft@Hrs	Unit	Material	Labor	Total	Sell
Reversible polystyrene casing, Decorator Lykewood® RB3, Royal Mouldings. Won't crack, warp, split, or bow. Paint process finish.						
1" x 3-1/2" x 7', Bone white	BC@.105	Ea	12.70	3.78	16.48	28.00
1" x 3-1/2" x 7', Clearwood	BC@.105	Ea	12.90	3.78	16.68	28.40
Chair rail embossed polystyrene molding, Heirloom™, Royal Mouldings. Stainable or paintable finish. Deep embossed ornate design. Won't crack, warp, split, or bow. Paint process finish.						
4-1/2" x 1-1/4" x 8', Bone white	BC@.300	Ea	20.40	10.80	31.20	53.00
4-1/2" x 1-1/4" x 8', Clearwood	BC@.300	Ea	20.20	10.80	31.00	52.70
Chair rail embossed prefinished molding, Heirloom™, Royal Mouldings.						
Highlands oak, 8'	BC@.300	Ea	37.70	10.80	48.50	82.50
Imperial oak, 8'	BC@.300	Ea	37.70	10.80	48.50	82.50
Chair rail molding, Clearwood PS™, Royal Mouldings. Pre-sealed and pre-sanded. Stainable or paintable with any oil or latex paint applied by brush, roller, or sprayer. Non-repetitive woodgrain finish for consistent appearance. Stain or paint absorbs uniformly without discoloration. Clean by wiping with paint thinner, naphtha, or ammonia.						
11/16" x 2-5/8" x 8', Bone white	BC@.300	Ea	10.10	10.80	20.90	35.50
Chair rail paperwrap molding, Royal Collection™, Royal Mouldings.						
11/16" x 2-5/8" x 8', Highlands oak	BC@.300	Ea	10.40	10.80	21.20	36.00
11/16" x 2-5/8" x 8', Imperial oak	BC@.300	Ea	10.40	10.80	21.20	36.00
11/16" x 2-5/8" x 8', Natural maple	BC@.300	Ea	10.40	10.80	21.20	36.00
Chair rail polystyrene molding, Royal Mouldings. Non-repetitive woodgrain finish. Pre-sealed and pre-sanded. Stainable or paintable with any oil or latex paint. Stain and paint absorb uniformly.						
11/16" x 2-5/8" x 8', Bone white	BC@.300	Ea	9.13	10.80	19.93	33.90
11/16" x 2-5/8" x 8', Clearwood	BC@.300	Ea	7.47	10.80	18.27	31.10
Chair rail prefinished molding, Reflections™, Royal Mouldings.						
11/16" x 2-5/8" x 8', White marble	BC@.300	Ea	14.30	10.80	25.10	42.70
Inside corner and cove molding, Royal Mouldings. Pre-sealed and pre-sanded. Stainable or paintable with any oil or latex paint applied by brush, roller or sprayer. Non-repetitive woodgrain finish for consistent appearance. Stain or paint absorbs uniformly without discoloration. Clean by wiping with paint thinner, naphtha or ammonia.						
11/16" x 11/16" x 8', Clearwood	BC@.300	Ea	3.22	10.80	14.02	23.80
Inside corner and cove paperwrap molding, Royal Collection™, Royal Mouldings.						
5/16" x 15/16" x 8', Highlands oak	BC@.300	Ea	4.68	10.80	15.48	26.30
5/16" x 15/16" x 8', Imperial oak	BC@.300	Ea	4.68	10.80	15.48	26.30
5/16" x 15/16" x 8', Natural maple	BC@.300	Ea	4.61	10.80	15.41	26.20

	Craft@Hrs	Unit	Material	Labor	Total	Sell
Inside corner and cove prefinished molding, Reflections™, Royal Mouldings.						
11/16" x 11/16" x 8',						
White marble	BC@.300	Ea	3.76	10.80	14.56	24.80
Inside corner molding, Clearwood PS™, Royal Mouldings. Pre-sealed and pre-sanded. Stainable or paintable with any oil or latex paint applied by brush, roller, or sprayer. Non-repetitive woodgrain finish for consistent appearance. Stain or paint absorbs uniformly without discoloration. Clean by wiping with paint thinner, naphtha, or ammonia.						
5/16" x 1" x 8',						
Bone white	BC@.300	Ea	3.36	10.80	14.16	24.10
Inside corner polystyrene molding, Royal Mouldings. Non-repetitive woodgrain finish. Pre-sealed and pre-sanded. Stainable or paintable with any oil or latex paint. Stain and paint absorb uniformly.						
5/16" x 1" x 8',						
Dominion oak	BC@.300	Ea	3.75	10.80	14.55	24.70
Inside corner prefinished molding, Royal Mouldings.						
Natural pine	BC@.300	Ea	3.85	10.80	14.65	24.90
Perfect corner, bone white	BC@.300	Ea	5.75	10.80	16.55	28.10
Outside corner molding, Royal Mouldings. Pre-sealed and pre-sanded. Stainable or paintable with any oil or latex paint applied by brush, roller or sprayer. Non-repetitive woodgrain finish for consistent appearance. Stain or paint absorbs uniformly without discoloration. Clean by wiping with paint thinner, naphtha or ammonia.						
11/16" x 11/16" x 8',						
Clearwood	BC@.300	Ea	4.05	10.80	14.85	25.20
7/8" x 7/8" x 8',						
Bone white	BC@.300	Ea	3.75	10.80	14.55	24.70
1" x 1" x 8',						
Bone white	BC@.300	Ea	5.90	10.80	16.70	28.40
1-1/8" x 1-1/8" x 8',						
Clearwood	BC@.300	Ea	7.48	10.80	18.28	31.10
Outside corner polystyrene molding, Royal Mouldings. Non-repetitive woodgrain finish. Pre-sealed and pre-sanded. Stainable or paintable with any oil or latex paint. Stain and paint absorb uniformly.						
7/8" x 7/8" x 8',						
Dominion oak	BC@.300	Ea	4.36	10.80	15.16	25.80
Outside corner prefinished molding, Reflections™, Royal Mouldings.						
7/8" x 7/8" x 8', white marble	BC@.300	Ea	4.10	10.80	14.90	25.30
8', Perfect corner, bone white	BC@.300	Ea	5.75	10.80	16.55	28.10
Colonial outside corner paperwrap molding, Royal Collection™, Royal Mouldings.						
1" x 1" x 8', Highlands	BC@.300	Ea	6.03	10.80	16.83	28.60
1" x 1" x 8', Imperial oak	BC@.300	Ea	6.03	10.80	16.83	28.60
1" x 1" x 8', Natural maple	BC@.300	Ea	6.02	10.80	16.82	28.60
Colonial outside corner polystyrene molding, Royal Mouldings. Non-repetitive woodgrain finish. Pre-sealed and pre-sanded. Stainable or paintable with any oil or latex paint. Stain and paint absorb uniformly.						
1" x 1" x 8', Dominion oak	BC@.300	Ea	6.11	10.80	16.91	28.70

	Craft@Hrs	Unit	Material	Labor	Total	Sell

Oak counter edge, Royal Mouldings. Molding specially shaped for edge of tile countertops. Provides durable protection from everyday impacts that can chip ceramics. Wood adds a warm, natural accent to tile.

	Craft@Hrs	Unit	Material	Labor	Total	Sell
13/16" x 1-3/4" x 5', Oak trim	BC@.300	Ea	16.90	10.80	27.70	47.10
13/16" x 1-3/4" x 8', Oak trim	BC@.480	Ea	26.80	17.30	44.10	75.00

Cove molding, Royal Mouldings. Pre-sealed and pre-sanded. Stainable or paintable with any oil or latex paint applied by brush, roller or sprayer. Non-repetitive woodgrain finish for consistent appearance. Stain or paint absorbs uniformly without discoloration. Clean by wiping with paint thinner, naphtha or ammonia.

	Craft@Hrs	Unit	Material	Labor	Total	Sell
9/16" x 1-5/8" x 8', Clearwood	BC@.300	Ea	5.25	10.80	16.05	27.30

Spring cove molding, Clearwood PS™, Royal Mouldings. Pre-sealed and pre-sanded. Stainable or paintable with any oil or latex paint applied by brush, roller, or sprayer. Non-repetitive woodgrain finish for consistent appearance. Stain or paint absorbs uniformly without discoloration. Clean by wiping with paint thinner, naphtha, or ammonia.

	Craft@Hrs	Unit	Material	Labor	Total	Sell
9/16" x 1-5/8" x 8', Bone white	BC@.300	Ea	5.87	10.80	16.67	28.30

Spring cove polystyrene molding, Royal Mouldings. Non-repetitive woodgrain finish. Pre-sealed and pre-sanded. Stainable or paintable with any oil or latex paint. Stain and paint absorb uniformly.

	Craft@Hrs	Unit	Material	Labor	Total	Sell
9/16" x 1-5/8" x 8', Dominion oak	BC@.300	Ea	5.78	10.80	16.58	28.20

Colonial crown paperwrap molding, Royal Collection™, Royal Mouldings.

	Craft@Hrs	Unit	Material	Labor	Total	Sell
9/16" x 3-5/8" x 8', Highlands oak	BC@.352	Ea	15.50	12.70	28.20	47.90
9/16" x 3-5/8" x 8', Imperial oak	BC@.352	Ea	13.50	12.70	26.20	44.50
11/16" x 3-5/8" x 8', Natural maple	BC@.352	Ea	15.90	12.70	28.60	48.60

Crown embossed polystyrene molding, Heirloom™, Royal Mouldings. Stainable or paintable finish. Deep embossed ornate design. Won't crack, warp, split, or bow. Paint process finish.

	Craft@Hrs	Unit	Material	Labor	Total	Sell
5" x 11/16" x 8', Bone white	BC@.352	Ea	19.40	12.70	32.10	54.60
5" x 11/16" x 8', Clearwood	BC@.352	Ea	21.70	12.70	34.40	58.50

Crown polystyrene molding, Royal Mouldings. Non-repetitive woodgrain finish. Pre-sealed and pre-sanded. Stainable or paintable with any oil or latex paint. Stain and paint absorb uniformly.

	Craft@Hrs	Unit	Material	Labor	Total	Sell
11/16" x 3-5/8" x 8', Bone white	BC@.350	Ea	17.90	12.60	30.50	51.90

Crown prefinished molding, Heirloom™, Royal Mouldings.

	Craft@Hrs	Unit	Material	Labor	Total	Sell
8', Highlands oak	BC@.352	Ea	32.70	12.70	45.40	77.20
8', Oak	BC@.352	Ea	32.70	12.70	45.40	77.20

Crown prefinished molding, Royal Mouldings.

	Craft@Hrs	Unit	Material	Labor	Total	Sell
11/16" x 3-5/8" x 8', Clearwood	BC@.352	Ea	9.20	12.70	21.90	37.20
8', Bone white, smooth	BC@.352	Ea	18.10	12.70	30.80	52.40

	Craft@Hrs	Unit	Material	Labor	Total	Sell
Lattice molding, Royal Mouldings. Pre-sealed and pre-sanded. Stainable or paintable with any oil or latex paint applied by brush, roller or sprayer. Non-repetitive woodgrain finish for consistent appearance. Stain or paint absorbs uniformly without discoloration. Clean by wiping with paint thinner, naphtha or ammonia.						
5/32" x 1-1/8" x 8', Bone white	BC@.250	Ea	3.85	9.00	12.85	21.80
1/4" x 1-1/8" x 8', Clearwood	BC@.250	Ea	3.66	9.00	12.66	21.50
1/4" x 1-3/4" x 8', Clearwood	BC@.250	Ea	4.17	9.00	13.17	22.40
Lattice polystyrene molding, Royal Mouldings. Non-repetitive woodgrain finish. Pre-sealed and pre-sanded. Stainable or paintable with any oil or latex paint. Stain and paint absorb uniformly.						
5/32" x 1-1/8" x 8', Dominion oak	BC@.250	Ea	4.02	9.00	13.02	22.10
Parting stop molding, Royal Mouldings. Pre-sealed and pre-sanded. Stainable or paintable with any oil or latex paint applied by brush, roller or sprayer. Non-repetitive woodgrain finish for consistent appearance. Stain or paint absorbs uniformly without discoloration. Clean by wiping with paint thinner, naphtha or ammonia.						
1/2" x 3/4" x 8', Clearwood	BC@.200	Ea	3.70	7.20	10.90	18.50
Colonial stop molding, Royal Mouldings. Pre-sealed and pre-sanded. Stainable or paintable with any oil or latex paint applied by brush, roller or sprayer. Non-repetitive woodgrain finish for consistent appearance. Stain or paint absorbs uniformly without discoloration. Clean by wiping with paint thinner, naphtha or ammonia.						
3/8" x 1-1/4" x 7', Clearwood	BC@.175	Ea	3.50	6.30	9.80	16.70
Colonial stop paperwrap molding, Royal Collection™, Royal Mouldings.						
3/8" x 1-1/4" x 7', Highlands oak	BC@.175	Ea	4.78	6.30	11.08	18.80
3/8" x 1-1/4" x 7', Imperial oak	BC@.175	Ea	4.78	6.30	11.08	18.80
3/8" x 1-1/4" x 7', Natural maple	BC@.175	Ea	5.37	6.30	11.67	19.80
Colonial stop polystyrene molding, Royal Mouldings. Non-repetitive woodgrain finish. Pre-sealed and pre-sanded. Stainable or paintable with any oil or latex paint. Stain and paint absorb uniformly.						
3/8" x 1-1/4" x 7', Bone white	BC@.175	Ea	4.68	6.30	10.98	18.70
3/8" x 1-1/4" x 7', Dominion oak	BC@.175	Ea	4.68	6.30	10.98	18.70
3/8" x 1-1/4" x 7', Washed oak	BC@.175	Ea	4.68	6.30	10.98	18.70
Ranch stop molding, Royal Mouldings. Pre-sealed and pre-sanded. Stainable or paintable with any oil or latex paint applied by brush, roller or sprayer. Non-repetitive woodgrain finish for consistent appearance. Stain or paint absorbs uniformly without discoloration. Clean by wiping with paint thinner, naphtha or ammonia.						
3/8" x 1-1/4" x 7', Clearwood	BC@.175	Ea	3.34	6.30	9.64	16.40
Ranch stop polystyrene molding, Royal Mouldings. Non-repetitive woodgrain finish. Pre-sealed and pre-sanded. Stainable or paintable with any oil or latex paint. Stain and paint absorb uniformly.						
3/8" x 1-1/4" x 7', Bone white	BC@.175	Ea	4.66	6.30	10.96	18.60

	Craft@Hrs	Unit	Material	Labor	Total	Sell
Vinyl tileboard molding, prefinished rigid divider, Royal Mouldings. 1/8" to 4mm insert.						
8', white	BC@.200	Ea	2.57	7.20	9.77	16.60
8', white	BC@.200	Ea	2.54	7.20	9.74	16.60
8', white	BC@.200	Ea	2.65	7.20	9.85	16.70
8', white	BC@.200	Ea	2.65	7.20	9.85	16.70

Interior Partition Walls

Remove non-bearing drywall stud walls. Includes allowance for plates, blocking and wall cover. Per SF of wall area demolished, measured one side. Add the cost of patching at floor, wall and ceiling joints, hauling away debris, and dump fees. No salvage of material assumed.

	Craft@Hrs	Unit	Material	Labor	Total	Sell
2" x 3" framing, drywall both sides	BL@.022	SF	—	.64	.64	1.09
2" x 4" framing, drywall both sides	BL@.026	SF	—	.76	.76	1.29
2" x 6" framing, drywall both sides	BL@.034	SF	—	.99	.99	1.68

Remove plastered stud walls. Includes allowance for plates, blocking and wall cover. Per SF of wall area demolished, measured one side. Add the cost of patching at floor, wall and ceiling joints, hauling away debris, and dump fees. No salvage of material assumed.

	Craft@Hrs	Unit	Material	Labor	Total	Sell
2" x 3" framing, lath and plaster both sides	BL@.038	SF	—	1.10	1.10	1.87
2" x 4" framing, lath and plaster both sides	BL@.042	SF	—	1.22	1.22	2.07

Add 2" x 4" interior stud partition with 1/2" drywall both sides. Add the cost of patching, finishing and trim.

	Craft@Hrs	Unit	Material	Labor	Total	Sell
Cost per square foot of wall	B1@.064	SF	.93	2.08	3.01	5.12
Cost per running foot, 8' high walls	B1@.512	LF	7.45	16.70	24.15	41.10

Add 2" x 4" interior stud partition with 5/8" drywall both sides. Includes drywall taping. Add the cost of patching, finishing and trim.

	Craft@Hrs	Unit	Material	Labor	Total	Sell
Cost per square foot of wall	B1@.068	SF	1.05	2.21	3.26	5.54
Cost per running foot, 8' high walls	B1@.544	LF	8.41	17.70	26.11	44.40

Add 2" x 6" interior stud partition with 1/2" drywall both sides. Add the cost of patching, finishing and trim.

	Craft@Hrs	Unit	Material	Labor	Total	Sell
Cost per square foot of wall	B1@.072	SF	1.20	2.34	3.54	6.02
Cost per running foot, 8' high walls	B1@.576	LF	9.58	18.70	28.28	48.10

Add 2" x 6" interior stud partition with 5/8" drywall both sides. Add the cost of patching, finishing and trim.

	Craft@Hrs	Unit	Material	Labor	Total	Sell
Cost per square foot of wall	B1@.076	SF	1.32	2.47	3.79	6.44
Cost per running foot, 8' high walls	B1@.608	LF	10.50	19.80	30.30	51.50

Floors and Tile

11

Some types of floor cover are more durable than others. Vinyl lasts longer than carpet. Ceramic tile lasts longer than wood block or strip flooring. But no floor material has a life expectancy equal to that of the house itself. That makes flooring a popular focus in home improvement work.

All floor cover requires a base that's structurally sound, clean, level (to $^1/_4$" per 10' span) and dry (moisture content of the subfloor should not exceed 13 percent). Concrete makes a good base, assuming the surface is smooth and incorporates a good vapor barrier. Untempered hardboard, plywood, and particleboard also make a good base for flooring. Use either $^1/_4$"- or $^3/_8$"-thick sheets. Underlayment needs a $^1/_{32}$" gap at the edges and the ends to allow for expansion. Underlayment-grade plywood has a sanded, C-plugged or better face. If moisture isn't a problem, use interior type plywood. Otherwise use either exterior or interior grade plywood with exterior glue. Trowel on a smooth coat of cement-based underlay to prepare nearly any floor surface for resilient flooring.

Repair, Recover or Remove?

Adhesive used to secure resilient flooring tends to deteriorate when moisture comes up through the subfloor. If resilient tile comes loose, try resetting the tile in new adhesive that's designed for use below grade. If the resilient tile is cracked, broken or has chipped edges, it's usually better to install new flooring. Matching new tile with old tile isn't practical. Resilient tile changes color with age. But it may not be necessary to remove the old surface when installing new. If the old surface is scarred, stained, abraded or has been embossed by the weight of furniture, apply a liquid leveler, or trowel on a cement-based underlayment to smooth the surface. Then install the new floor cover. If unevenness in the underlayment is showing through, remove the old surface and do some leveling before installing the new floor cover. Remove resilient tile if the new floor cover is also to be resilient tile.

If a wood floor is smooth and free of large cracks, refinishing may put the floor back in like-new condition. Most wood flooring can be sanded and refinished several times. Softwood flooring with no subfloor is an exception. Even one sanding might weaken the floor too much. Plywood block flooring can sometimes be sanded and refinished. Thin wood flooring and wood flooring with wide cracks usually has to be replaced — any patch would be obvious.

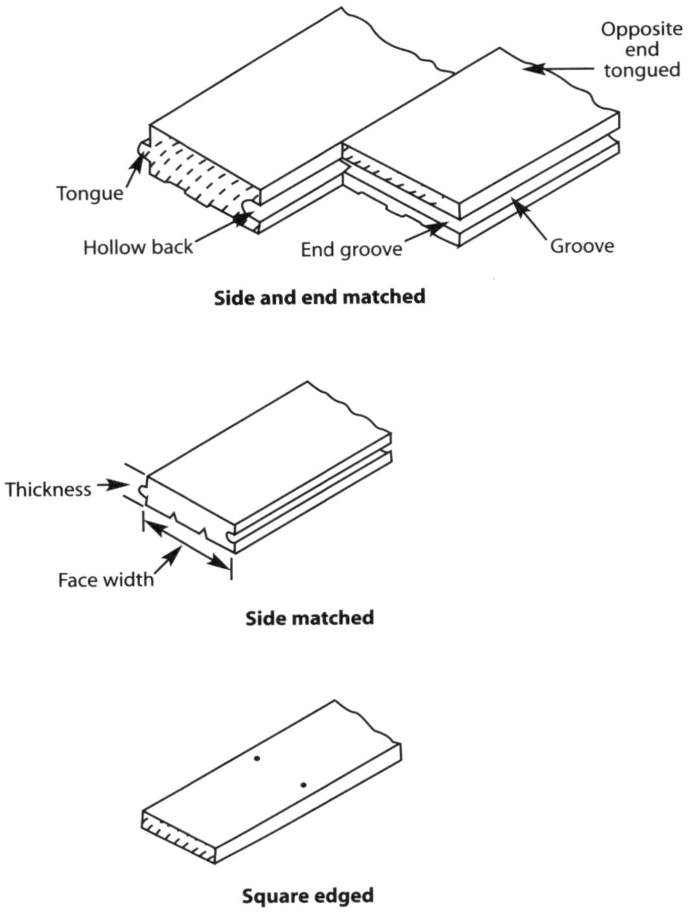

Figure 11-1
Strip flooring

Wood flooring, sheet vinyl with resilient backing, and carpeting can be installed directly over an existing hardwood floor, assuming any voids have been filled and the surface isn't loose. Shrinkage cracks are more common where boards are wide. Be sure to check for boards that are buckling, cupping or cracking due to moisture. If there's a moisture problem, solve that before you lay new flooring.

You can install laminated wood flooring over ceramic, wood or resilient flooring, so long as the surface is firm and dry.

New Wood Flooring

Hardwood flooring is available in tongue-and-groove strips and blocks (parquet). Some thinner patterns of strip flooring are square-edged. See Figure 11-1. The most common hardwood strip flooring is $^{25}/_{32}$" thick by $2^{1}/_{4}$" wide and has a hollow back. Strips are random lengths and vary from 2' to 16' long. The face is slightly wider than the bottom so joints will be tight on the surface.

Softwood flooring is also available in strips and blocks. Most softwood strip flooring has tongue-and-groove edges, although some types are end matched. Softwood flooring costs less than most hardwood species, but it's also less wear-resistant and shows surface abrasions more readily. Use softwood flooring in light traffic areas such as closets. No matter which type of flooring you select, give the material a few days to reach the moisture content of the room where it will be installed.

Strip flooring is normally laid at right angles to the floor joists. When new strip flooring is installed over old, lay the new strips at right angles to the old, no matter what direction the floor joists run. Use 8d flooring nails for $^{25}/_{32}$" thick flooring, 6d flooring nails for $^{1}/_{2}$" flooring, and 4d casing nails for $^{3}/_{8}$" flooring. Some manufacturers recommend ring-shank or screw-shank nails. To help prevent splitting the tongue, use flooring brads with blunted points.

Begin installation of tongue-and-groove flooring by placing the first strip $^{1}/_{2}$" to $^{5}/_{8}$" away from the wall. That allows for expansion and prevents buckling when the moisture content increases. Nail straight down through the face of the first strip, as in Figure 11-2. The nail should be close enough to the wall to be covered by the base or shoe molding. Try to nail into a joist if the new flooring is laid at right

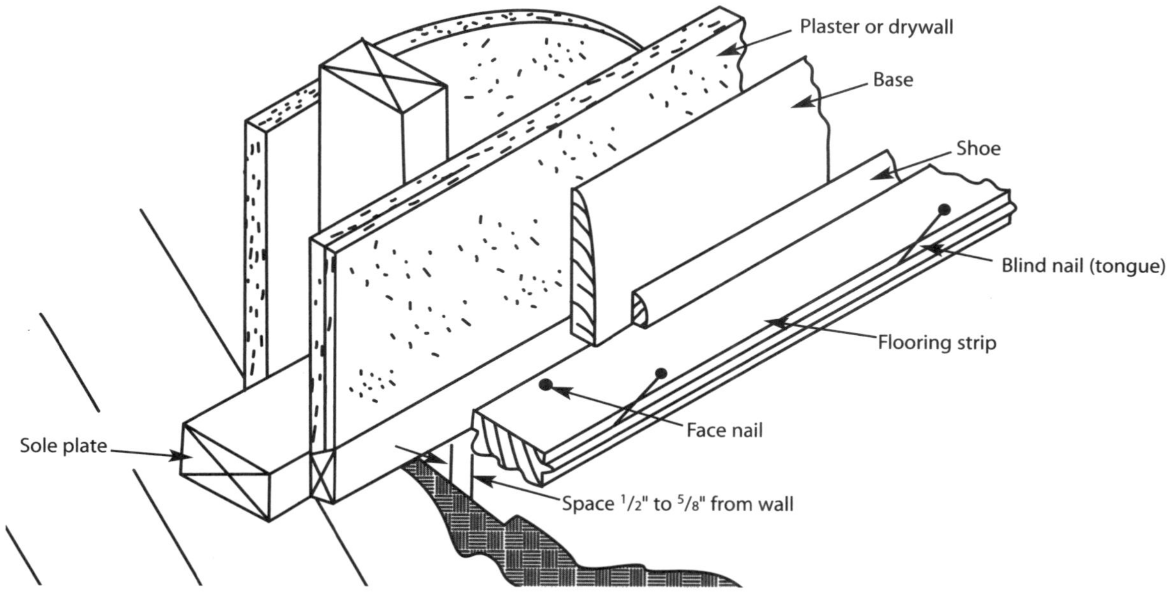

Figure 11-2
Installation of first strip of flooring

angles to the joists. Drive a second nail through the tongue of this first strip. All other strips are nailed through the tongue only. Drive these nails at an angle of 45 to 50 degrees. But leave the head just above the surface to avoid damaging the strip with your hammer. Use a large nail set to drive nails the last quarter inch. You can lay the nail set flat against the flooring when setting these nails, see Figure 11-3.

Stagger the end joints of strip flooring so butts are separated in adjacent courses. Install each new strip tightly against the previously-installed strip. Use shorter strips and crooked strips at the end of courses or in closets. Leave a 1/2" to 5/8" space between the last course of flooring and the wall, just as with the first course. Face-nail the last course where the base or shoe will cover the nail head.

Square-edged strip flooring must be installed over a substantial subfloor and should be face-nailed. Other than that, the installation procedure is the same as for matched (tongue-and-groove) flooring. Wood strip flooring is always nailed.

Parquet tile is made from narrow wood slats formed into a square. Parquet block flooring can be applied with adhesive over a concrete floor protected from moisture with a vapor barrier. Spread adhesive on the slab or underlayment with a notched trowel. Then lay parquet in the adhesive. If you elect to nail parquet flooring to wood underlayment, nail through the tongue, the same as with wood strip flooring. Minimize problems associated with shrinkage and swelling by changing the grain direction of alternate blocks.

You can install particleboard tile over underlayment the same way you install parquet tile — except particleboard tile shouldn't be installed directly over concrete.

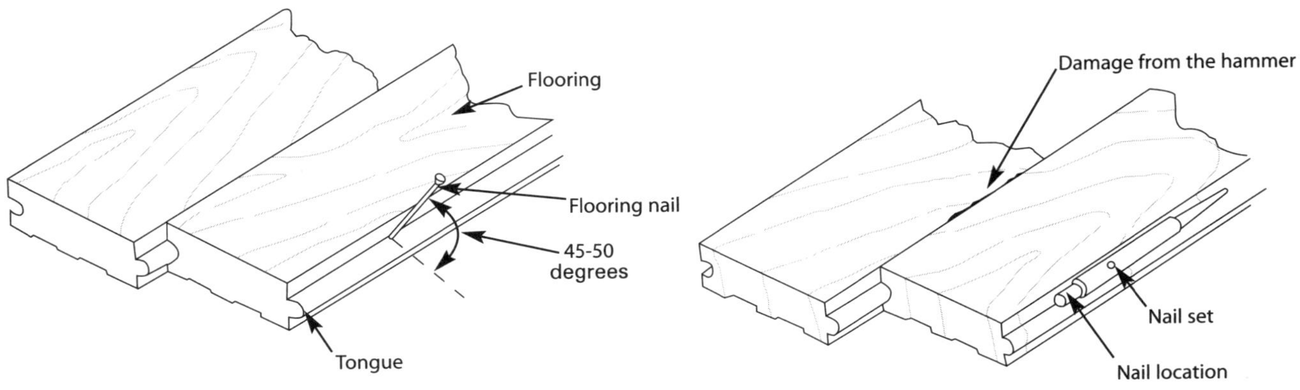

Figure 11-3
Nailing strip flooring

Follow the manufacturer's installation instructions. Particleboard tile is usually 9" x 9" and 3/8" thick, with tongue-and-groove edges. The back is often marked with small saw kerfs to stabilize the tile and provide a better grip for adhesive.

Laminate Wood Flooring

Laminate flooring strips are made from layers of wood and finished with a hard synthetic surface. Pergo® is one popular name. Most laminate flooring is loose lay; neither nails nor adhesive are used. Instead, the flooring floats on a cushioning material designed to reduce noise from foot traffic. Laminate flooring can be installed over nearly any firm, flat flooring material. Install strips parallel to the longest wall in the room. Keep the strips about 1/4" away from the side wall and end wall so the floor can expand with changes in temperature and moisture. Cut laminate flooring with the finish side down, using a carbide-tip blade.

Lay the tongue side of the first strip against the wall. Continue laying boards along that wall, fitting ends snug against the previous board. Use spacers to maintain a 1/4" gap between the flooring and the wall. Avoid short lengths of flooring at the end of a course. If the last board in any course is less than 8", trim that amount off the first board in the course and move the entire course down by that distance. Second and later courses lock into the previous course. Stagger end joints in adjacent courses. Finish the job with base molding that covers the 1/4" gap at side and end walls.

Resilient Flooring

Sheet vinyl with resilient backing smoothes out minor surface imperfections. Some sheet vinyl is designated *loose lay* and doesn't require adhesive. But use double-faced tape at joints and around edges to keep the covering in place. Manufacturers recommend spreading adhesive under all parts of the sheet for most products. Others can be bonded at the perimeter and seams only. Minimize the number of joints needed by using wider sheets — some sheet vinyl comes in widths up to 15'.

Both resilient sheet flooring and resilient tile require a smooth surface for proper adhesive bonding. You can repair an irregular surface with an embossing leveler or a masonry leveling compound. When the surface is dry, spread adhesive with a notched trowel, following the adhesive manufacturer's instructions. Lay the tile so joints don't coincide with the joints in the underlayment.

Seamless flooring, consisting of resin chips combined with a urethane binder, can be applied over any stable base, including old floor tile. Apply this liquid in several coats, allowing each coat to dry. A complete application may take several days, depending on the brand. You can repair a seamless floor by applying another coat. Damaged spots are easy to patch by adding more chips and binder.

Cork Tile

Cork is a natural sound absorber and insulator. It is quiet underfoot, and can last for decades when properly maintained. Cork will expand and contract based on humidity, although to a lesser degree than wood. Cork tiles should be given time to acclimate to the environment before installation. Remove tiles from their packaging and store them in the room where they will be installed for at least 48 hours prior to installation.

Most manufacturers recommend using a water-based contact cement adhesive for cork installation. Cork is porous, allowing the water in the adhesive to evaporate and create a strong bond. It's a good idea to test for proper adhesion before proceeding with the installation. Excessive moisture can damage cork flooring. For kitchen, bathroom or other high-risk applications, follow the manufacturer's guidelines for sealing cork floors with urethane or floor wax.

Granite and Marble Tile

Common granite and marble surface finishes include *polished*, *honed*, and *flamed*. A polished surface is highly reflective, and is best suited for low-traffic areas. A honed surface has a duller, more slip-resistant finish that's less likely to show scratches. Flamed tiles have a deeply textured surface that's useful for applications requiring additional slip-resistance.

Marble is softer and more porous than granite, so it's more susceptible to scratches, but it can be repolished when necessary. Marble is also susceptible to damage from alcohols, oils and acids commonly found in the home. A penetrative sealer is generally recommended when installing marble in high-risk areas such as kitchens and bathrooms.

Colors and grains will vary from tile to tile and batch to batch. To assure the installation will have a uniform look, be sure you have enough material to finish a complete area. Consider using tiles randomly from different boxes when allocating material for the job. This will more-evenly distribute irregularities, and may actually create a more homogeneous look.

Flooring Over Concrete

A concrete floor that stays dry in all seasons probably has a good vapor barrier under the slab. If the surface is also smooth and level, nearly any type of resilient flooring or carpet can be installed directly over the slab. If a basement slab is both uneven and moist to the touch, one remedy is to lay a vapor barrier over the existing slab, then cover the entire surface with a 2"- to 3"-thick concrete topping. Another approach is to lay a good-quality vapor barrier directly on the slab, then anchor furring strips or sleepers to the slab with concrete nails or shot fasteners. You can then install hardwood strip flooring directly over the sleepers. See Figure 11-4. For tile or sheet vinyl, nail underlayment or plywood to the sleepers before you install the finish floor.

Ceramic Tile

Ceramic tile can be set in either mortar *(thin-set or thick set)* or applied with adhesive. Adhesive is more convenient because no mixing is required, though cleanup takes a little longer. Tile is set on backerboard, cement board reinforced with polymer-coated glass mesh. Common names are Durock®, WonderBoard®, RhinoBoard® and Hardibacker®. For floors and counters, set the backerboard in adhesive on 3/4" exterior grade plywood. For walls, affix backerboard to the studs with cement board screws every 8". On ceilings, drive cement board screws every 6". Regular drywall screws don't have enough holding strength for use on backerboard. Cover panel joints with fiberglass mesh and joint cement. One side of backerboard is rough for use with tile in thin-set mortar. The other side is smooth for use with tile adhesive.

Ceramic tile sizes range from 1" square mosaic to 12" x 12" and even larger. Mosaic tile are usually sold in 12" x 12" squares held together with a mesh backing. The most popular tile size for walls and counters is 4 1/4" x 4 1/4".

Avoid using tile with a bright glaze finish on floors. A highly-reflective finish tends to be slippery and offers less resistance to wear. Vitrified porcelain tiles are hard to cut accurately with a tile cutter and may require a circular ceramic wet saw. You also have to apply adhesive to both the tile and the floor when you're installing porcelain tile.

Ceramic Tile Definitions

- Field tiles make up most of the job, the "field".
- Border tiles are trim pieces set around the edge of the field.
- Listello tiles have a decorative design different from field tile and are generally used on the edge of the field, like the frame of a picture.
- Rope tiles, as you might expect, have a rope design, usually in raised relief, and are used on the border.

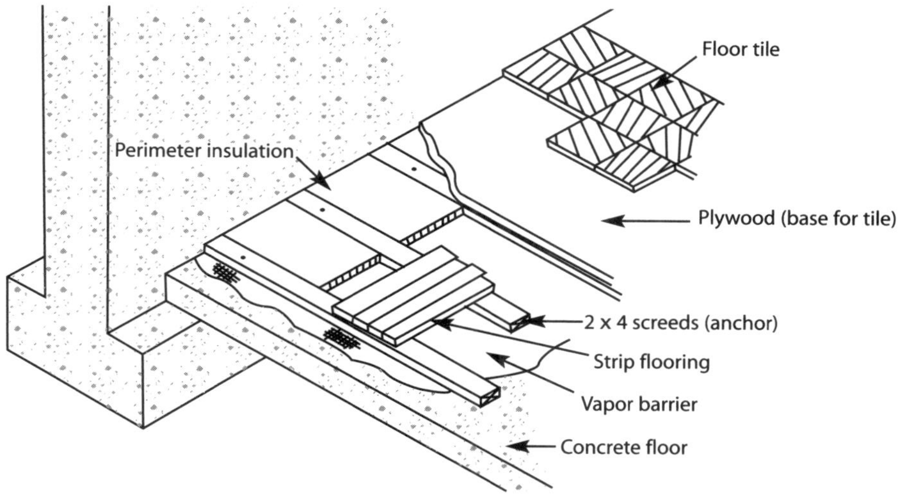

Figure 11-4
Installation of wood floors in a basement

Tips on Ceramic Tile

Most ceramic tile carries a PEI (Porcelain Enamel Institute) wear rating:

- *Class 1*, no foot traffic. Interior residential and commercial walls only.
- *Class 2*, light traffic. Interior residential and commercial walls and residential bathroom floors.
- *Class 3*, light to moderate traffic. Residential floors, countertops, and walls.
- *Class 4*, moderate to heavy traffic. Residential, medium commercial and light institutional floors and walls.
- *Class 5*, heavy to extra-heavy traffic. Residential, commercial and institutional floors and walls.

Indoor vs. Outdoor Tile

Tile that absorbs water will crack when exposed to freezing and thawing. Tile with an absorption rating of 3 percent or less is usually considered acceptable for outdoor use. That includes vitrified and porcelain ceramic tiles. Outdoor tile is very dense and doesn't break easily. Use thin-set mortar with a latex admix.

Matching Styles and Batches

Tile colors and glazes can vary from batch to batch. To make matching easier, many tile manufacturers emboss batch numbers into the back of each tile.

	Craft@Hrs	Unit	Material	Labor	Total	Sell
Remove flooring. Per 100 square feet of floor. Using a long handled spudding spade. Includes removing baseboard as required. Debris piled on site. No salvage of materials or furniture moving included.						
Remove nailed hardwood floor and prepare surface for new flooring	BF@3.22	CSF	—	105.00	105.00	179.00
Remove glued hardwood floor and prepare surface for new flooring	BF@5.58	CSF	—	183.00	183.00	311.00
Remove sheet vinyl set in adhesive	BF@.620	CSF	—	20.29	20.29	34.70
Remove sheet vinyl and adhesive residue for new flooring	BF@1.37	CSF	—	44.84	44.84	76.50
Remove loose lay sheet flooring	BF@.138	CSF	—	4.52	4.52	7.72
Remove resilient tile, no floor prep	BF@.777	CSF	—	25.43	25.43	43.40
Remove resilient tile and adhesive residue for new flooring	BF@1.53	CSF	—	50.08	50.08	85.50
Remove vinyl cove base	BF@2.50	CSF	—	81.80	81.80	139.00
Remove rubber-back glue-down carpet, no scraping of adhesive included	BF@.217	CSF	—	7.10	7.10	12.10
Remove rubber-back glue-down carpet, and prepare for new flooring	BF@.967	CSF	—	31.65	31.65	54.10
Remove ceramic tile, air hammer	BL@2.90	CSF	—	84.25	84.25	141.00
Remove ceramic tile and tile backer for new flooring	BL@5.80	CSF	—	168.50	168.50	282.00
Wood Strip and Plank Flooring						
Bruce oak solid hardwood strip flooring. These estimates assume work is done by skilled tradespeople, but not flooring specialists. Highly skilled flooring specialists may be able to install strip and plank flooring in 50% less time. Prefinished solid oak strip. Eased edge, square ends. 2-1/4" wide x 3/4" thick. Includes 10% cutting waste.						
Gunstock (western)	BF@.050	SF	4.40	1.64	6.04	10.30
Gunstock (eastern)	BF@.050	SF	4.06	1.64	5.70	9.69
Natural (western)	BF@.050	SF	4.40	1.64	6.04	10.30
Natural (eastern)	BF@.050	SF	4.06	1.64	5.70	9.69
Butterscotch	BF@.050	SF	6.61	1.64	8.25	14.00
Marsh	BF@.050	SF	3.05	1.64	4.69	7.97
Bruce hardwood molding. Prefinished.						
3'11" reducer	BF@.160	Ea	20.20	5.24	25.44	43.20
3'11" stair nose	BF@.160	Ea	20.30	5.24	25.54	43.40
3'11" T-molding	BF@.160	Ea	20.30	5.24	25.54	43.40
7'10" quarter round	BF@.320	Ea	15.10	10.50	25.60	43.50
Bruce touch-up kit. Includes two touch-up markers, three filler sticks and 1/2 ounce of Dura Luster urethane.						
Kit	—	Ea	21.20	—	21.20	—
Northwoods hardwood strip flooring, Harris-Tarkett. These estimates assume work is done by skilled tradespeople, but not flooring specialists. Highly skilled flooring specialists may be able to install strip and plank flooring in 50% less time. Alumide® enhanced urethane finish resists stains and cleans easily. Precise tongue-and-groove for precise fit. Can be sanded and refinished up to three times. Nail-down installation only, on or above grade. 3/4" thick x 2-1/4" wide. Random lengths. Includes 10% cutting waste.						
Oak Toffee	BF@.050	SF	4.38	1.64	6.02	10.20
Red Oak natural	BF@.050	SF	4.38	1.64	6.02	10.20

	Craft@Hrs	Unit	Material	Labor	Total	Sell
Northwoods hardwood molding. Prefinished toffee.						
7'10" quarter round	BF@.320	Ea	15.20	10.50	25.70	43.70
3'11" reducer strip, carpet to hard surface	BF@.160	Ea	20.30	5.24	25.54	43.40
3'11" stair nose	BF@.160	Ea	20.30	5.24	25.54	43.40
3'11" T-molding	BF@.160	Ea	20.30	5.24	25.54	43.40

Rosewood hardwood strip flooring. These estimates assume work is done by skilled tradespeople, but not flooring specialists. Highly skilled flooring specialists may be able to install strip and plank flooring in 50% less time. Prefinished one-piece strips. Nail or staple to wood subfloor. 3/4" thick. Random lengths except as noted. Includes 10% cutting waste.

	Craft@Hrs	Unit	Material	Labor	Total	Sell
2-1/4" wide	BF@.050	SF	4.70	1.64	6.34	10.80
2-1/4" wide, 7' long	BF@.048	SF	7.19	1.57	8.76	14.90
3-1/2" wide, 7' long	BF@.047	SF	6.89	1.54	8.43	14.30
4-1/2" wide, select	BF@.047	SF	8.92	1.54	10.46	17.80
5-1/2" wide, premium	BF@.045	SF	9.07	1.47	10.54	17.90
5-1/2" wide, select	BF@.045	SF	10.70	1.47	12.17	20.70
5-1/2" wide, select, 6' long	BF@.045	SF	11.06	1.47	12.53	21.30

Rosewood hardwood molding. 72" long. Prefinished.

	Craft@Hrs	Unit	Material	Labor	Total	Sell
Grooved reducer	BF@.360	Ea	24.50	11.80	36.30	61.70
Quarter round	BF@.360	Ea	19.20	11.80	31.00	52.70
Stair nosing	BF@.360	Ea	54.00	11.80	65.80	112.00
T-molding	BF@.360	Ea	45.40	11.80	57.20	97.20

Maple hardwood flooring. These estimates assume work is done by skilled tradespeople, but not flooring specialists. Highly skilled flooring specialists may be able to install strip and plank flooring in 50% less time. Light colored with less distinct grain. Extremely hard. 3/4" thick x 2-1/4" wide. Random lengths. Includes 10% cutting waste. Prefinished.

	Craft@Hrs	Unit	Material	Labor	Total	Sell
Ginger	BF@.050	Ea	9.89	1.64	11.53	19.60
Natural	BF@.050	Ea	7.58	1.64	9.22	15.70

Maple hardwood molding. Prefinished.

	Craft@Hrs	Unit	Material	Labor	Total	Sell
3'11" floor to carpet reducer strip	BF@.160	Ea	18.40	5.24	23.64	40.20
3'11" reducer strip	BF@.160	Ea	18.40	5.24	23.64	40.20
3'11" stair nose	BF@.160	Ea	18.40	5.24	23.64	40.20
3'11" T-molding	BF@.160	Ea	18.40	5.24	23.64	40.20

Longstrip hardwood flooring. These estimates assume work is done by skilled tradespeople, but not flooring specialists. Highly skilled flooring specialists may be able to install strip and plank flooring in 50% less time. Random lengths. Cost per square foot including 10% cutting waste. Prefinished.

	Craft@Hrs	Unit	Material	Labor	Total	Sell
3/4" x 3-1/4", Medium brown teak	BF@.047	SF	3.40	1.54	4.94	8.40
3/8" x 5", Rustic antique oak	BF@.045	SF	5.66	1.47	7.13	12.10

Pre-finished natural flooring strips. These estimates assume work is done by skilled tradespeople, but not flooring specialists. Highly skilled flooring specialists may be able to install strip and plank flooring in 50% less time. 3/4" thick x 2-1/4" wide. Includes 10% cutting waste.

	Craft@Hrs	Unit	Material	Labor	Total	Sell
Natural strips	BF@.050	SF	3.68	1.64	5.32	9.04

	Craft@Hrs	Unit	Material	Labor	Total	Sell
Vanguard TapTight hardwood plank flooring, Harris-Tarkett. These estimates assume work is done by skilled tradespeople, but not flooring specialists. Highly skilled flooring specialists may be able to install strip and plank flooring in 50% less time. Factory pre-glued tongue-and-groove. Aluminide®-enhanced urethane finish. Thick top layer can be sanded and refinished up to 3 times. Cross-directional construction for added strength and stability. Floating, glue-down or staple-down installation on grade or above grade. 9/16" x 7.5" x 4' planks. Case covers 13.25 square feet with 10% end-cutting waste.						
Heritage Maple	BF@.045	SF	4.39	1.47	5.86	9.96
Oak Wheat	BF@.045	SF	4.47	1.47	5.94	10.10
Red Oak natural	BF@.045	SF	4.47	1.47	5.94	10.10
TapTight hardwood moldings. Prefinished.						
3'11" carpet reducer	BF@.160	Ea	19.10	5.24	24.34	41.40
3'11" stair nose	BF@.160	Ea	13.70	5.24	18.94	32.20
3'11" T-molding	BF@.160	Ea	19.10	5.24	24.34	41.40
7'10" quarter round	BF@.160	Ea	19.10	5.24	24.34	41.40
Vanguard TapTight floating floor installation kit, Harris-Tarkett. Includes pull tool, spacers and tapping block.						
Kit	—	Ea	17.80	—	17.80	—
Bamboo plank flooring. These estimates assume work is done by skilled tradespeople, but not flooring specialists. Highly skilled flooring specialists may be able to install strip and plank flooring in 50% less time. 5/8" thick x 3-5/8" wide. Includes 10% waste. Prefinished.						
Natural	BF@.050	SF	5.11	1.64	6.75	11.50
Spice	BF@.050	SF	4.86	1.64	6.50	11.10
Bamboo flooring molding. Prefinished.						
Quarter round, 78"	BF@.360	Ea	11.00	11.80	22.80	38.80
Reducer, 47"	BF@.240	Ea	22.00	7.86	29.86	50.80
Stair nose, 47"	BF@.240	Ea	33.20	7.86	41.06	69.80
T-molding, 47"	BF@.240	Ea	21.00	7.86	28.86	49.10
Wood plank flooring. These estimates assume work is done by skilled tradespeople, but not flooring specialists. Highly skilled flooring specialists may be able to install strip and plank flooring in 50% less time. Prefinished. 3/8" thick x 3" wide. Includes 10% cutting waste.						
Butterscotch	BF@.050	SF	3.30	1.64	4.94	8.40
Gunstock	BF@.050	SF	3.30	1.64	4.94	8.40
Rustic oak plank flooring. These estimates assume work is done by skilled tradespeople, but not flooring specialists. Highly skilled flooring specialists may be able to install strip and plank flooring in 50% less time. Prefinished. 3/4" thick x 2-1/4" wide. Includes 10% cutting waste.						
Toffee	BF@.050	SF	3.25	1.64	4.89	8.31
Umber	BF@.050	SF	3.25	1.64	4.89	8.31
Heart pine plank flooring. These estimates assume work is done by skilled tradespeople, but not flooring specialists. Highly skilled flooring specialists may be able to install strip and plank flooring in 50% less time. Century-old heart pine timber milled to 3/4" thick. Tongue-and-groove. Random lengths. Includes 10% cutting waste. Prefinished.						
3/4" x 2-1/4" x 7'	BF@.050	SF	5.53	1.64	7.17	12.20
3/4" x 5-1/2", Premium	BF@.045	SF	7.20	1.47	8.67	14.70
3/4" x 5-1/2", Select	BF@.045	SF	7.97	1.47	9.44	16.00

	Craft@Hrs	Unit	Material	Labor	Total	Sell
Heart pine molding. 72" long. Prefinished.						
Baby threshold	BF@.360	Ea	21.20	11.80	33.00	56.10
Grooved reducer	BF@.360	Ea	21.80	11.80	33.60	57.10
Quarter round	BF@.360	Ea	11.10	11.80	22.90	38.90
Stair nose	BF@.360	Ea	44.40	11.80	56.20	95.50
T-molding	BF@.360	Ea	34.00	11.80	45.80	77.90
Threshold	BF@.360	Ea	34.00	11.80	45.80	77.90

Unfinished red oak strip flooring, Harris-Tarkett. These estimates assume work is done by skilled tradespeople, but not flooring specialists. Highly skilled flooring specialists may be able to install strip and plank flooring in 50% less time. Random lengths. Nested bundle covers 17.6 square feet including 10% end-cutting waste. Add the cost of finishing.

	Craft@Hrs	Unit	Material	Labor	Total	Sell
3/4" x 2-1/4", Select	BF@.050	SF	2.55	1.64	4.19	7.12
3/4" x 2-1/4", #1 Common	BF@.050	SF	2.33	1.64	3.97	6.75
3/4" x 2-1/4", #2 Common	BF@.050	SF	2.01	1.64	3.65	6.21
3/4" x 3-1/4", Select	BF@.047	SF	3.31	1.54	4.85	8.25
3/4" x 3-1/4", #1 Common	BF@.047	SF	2.44	1.54	3.98	6.77
3/4" x 3-1/4", #2 Common	BF@.047	SF	1.83	1.54	3.37	5.73

Unfinished white oak strip flooring, Harris-Tarkett. These estimates assume work is done by skilled tradespeople, but not flooring specialists. Highly skilled flooring specialists may be able to install strip and plank flooring in 50% less time. Random lengths. Nested bundle covers 17.6 square feet including 10% end-cutting waste.

	Craft@Hrs	Unit	Material	Labor	Total	Sell
3/4" x 2-1/4", Select	BF@.050	SF	2.69	1.64	4.33	7.36
3/4" x 2-1/4", #1 Common	BF@.050	SF	2.50	1.64	4.14	7.04
Unfinished oak molding.						
3/4" x 2" x 39", reducer	BF@.160	Ea	10.70	5.24	15.94	27.10
3/4" x 3/4" x 78", quarter round	BF@.360	Ea	12.40	11.80	24.20	41.10
3/4" x 78", stair nose	BF@.360	Ea	24.40	11.80	36.20	61.50

Unfinished maple strip flooring. These estimates assume work is done by skilled tradespeople, but not flooring specialists. Highly skilled flooring specialists may be able to install strip and plank flooring in 50% less time. Random lengths. Including 10% end-cutting waste.

	Craft@Hrs	Unit	Material	Labor	Total	Sell
3/4" x 2-1/4", 1st Grade	BF@.050	SF	4.04	1.64	5.68	9.66
Acrylic wood filler. Fills nicks and dents in hardwood flooring.						
3 ounce tube	—	Ea	5.22	—	5.22	—
Floor installation tools. Heavy duty.						
Pull bar	—	Ea	7.83	—	7.83	—
Tapping block	—	Ea	7.82	—	7.82	—

Flooring installation clamp. Use to install most wood and laminate flooring. Adjustable cinching system. Recommended use: every 3' to 5'.

	Craft@Hrs	Unit	Material	Labor	Total	Sell
18' maximum span	—	Ea	45.00	—	45.00	—

Sand hardwood floor. Sanding floors requires skill. An inexperienced operator can ruin a hardwood floor with a power sander. Using a vibrating or drum sander. Per 100 square feet of floor. No furniture moving included. Add the cost of sandpaper and equipment.

	Craft@Hrs	Unit	Material	Labor	Total	Sell
Sand only, 3 passes	BF@1.08	CSF	—	35.30	35.30	60.00
Sand and scrape, 4 passes	BF@1.31	CSF	—	42.90	42.90	72.90
Small room or closet, 3 passes	BF@2.80	CSF	—	91.60	91.60	156.00
Small room or closet, 4 passes	BF@3.40	CSF	—	111.00	111.00	189.00

	Craft@Hrs	Unit	Material	Labor	Total	Sell

Wood Parquet Flooring

Hevea parquet flooring, OakCrest Products. Solid hardwood. Locking tongue-and-groove on all five sides with Tru-Square™ edges. 7.5mm thick. Wire back. UV-cured polyurethane finish, no-wax and stain-resistant. Can be sanded and refinished several times over the life of the floor. Pack of 10 covers 9 square feet with 10% waste. Add the cost of adhesive.

	Craft@Hrs	Unit	Material	Labor	Total	Sell
12" x 12"	BC@.055	SF	2.00	1.98	3.98	6.77

Bruce oak parquet flooring. 12" x 12" x 5/16". Pack of 10 covers 9 square feet with 10% waste. Add the cost of adhesive.

	Craft@Hrs	Unit	Material	Labor	Total	Sell
Chestnut	BF@.055	SF	2.26	1.80	4.06	6.90
Desert	BF@.055	SF	2.26	1.80	4.06	6.90

Parquet floor tile. Urethane finish. 12" x 12" x 5/16". Case of 25 covers 22.5 square feet with 10% waste. Add the cost of adhesive.

	Craft@Hrs	Unit	Material	Labor	Total	Sell
Mellow	BF@.055	SF	1.37	1.80	3.17	5.39
Sorrento Chestnut	BF@.055	SF	1.37	1.80	3.17	5.39
Sorrento Quartz	BF@.055	SF	1.37	1.80	3.17	5.39

Bruce oak self-stick parquet flooring. Dura-Luster® urethane finish. 12" x 12" x 7/16". Pack of 10 covers 9 square feet with 10% waste.

	Craft@Hrs	Unit	Material	Labor	Total	Sell
Chestnut	BF@.050	SF	2.42	1.64	4.06	6.90
Desert	BF@.050	SF	2.42	1.64	4.06	6.90

Hardwood moldings for Bruce parquet.

	Craft@Hrs	Unit	Material	Labor	Total	Sell
3'11" reducer	BF@.160	Ea	22.60	5.24	27.84	47.30
3'11" stair nose	BF@.160	Ea	22.60	5.24	27.84	47.30
3'11" T-molding	BF@.160	Ea	22.60	5.24	27.84	47.30
7'10" quarter round	BF@.360	Ea	17.00	11.80	28.80	49.00

Wood Flooring Adhesives

Adhesive remover, Jasco®. Nonflammable and water rinseable. Softens old adhesive in 5 to 15 minutes. One gallon removes adhesive from 100 to 200 square feet. Semi-paste clings to vertical surfaces. Use with neoprene or rubber gloves and adequate ventilation.

	Craft@Hrs	Unit	Material	Labor	Total	Sell
Quart	—	Ea	13.20	—	13.20	—
Gallon	—	Ea	42.80	—	43.70	—

Adhesive primer, Jasco®. For use on plywood, concrete, and glasscrete. Primes floors before applying adhesive. Prevents dry-out when adhesive is drawn below the surface. Ensures adhesion of self-stick floor tile. Water cleanup. Covers 300 to 500 square feet per gallon.

	Craft@Hrs	Unit	Material	Labor	Total	Sell
Quart	—	Ea	4.01	—	4.01	—
Gallon	—	Ea	11.20	—	11.20	—

Wood flooring adhesive. Synthetic latex-based adhesive for use on engineered laminated wood plank and wood parquet flooring. May be used over interior grade plywood subflooring and concrete on and above grade. Not recommended for foam-backed, solid wood flooring or high-pressure laminate flooring (e.g., Wilsonart, Pergo, Formica, etc.). Excellent early grab and develops good water resistance. Covers 40 to 50 square feet per gallon when applied with a 1/8" x 1/8" x 1/8" square-notch trowel for parquet or a 1/4" x 1/4" x 1/4" V-notch trowel for laminated plank.

	Craft@Hrs	Unit	Material	Labor	Total	Sell
Gallon	—	Ea	60.30	—	60.30	—
3.5 gallons	—	Ea	202.00	—	202.00	—
Pro trowel	—	Ea	16.40	—	16.40	—

	Craft@Hrs	Unit	Material	Labor	Total	Sell

Laminate Plank Flooring

Pergo Presto laminate wood flooring. Tongue-and-groove joints lock together for a tight fit. Guaranteed secure joint. Case covers 17.59 square feet before end-cutting waste.

	Craft@Hrs	Unit	Material	Labor	Total	Sell
Beech	BF@.051	SF	2.72	1.67	4.39	7.46
Antique Cherry	BF@.051	SF	2.72	1.67	4.39	7.46
Covington Oak	BF@.051	SF	1.81	1.67	3.48	5.92
Manitoba Maple	BF@.051	SF	2.72	1.67	4.39	7.46

Pergo floor molding.

	Craft@Hrs	Unit	Material	Labor	Total	Sell
Four-in-one kit, 94"	BF@.320	Ea	30.60	10.50	41.10	69.90
Wall base, 94"	BF@.320	Ea	11.10	10.50	21.60	36.70

Pergo flooring installation spacers. For maintaining proper spacing along walls and fixed objects. Minimizes movement of floor during installation. Package covers about 48 square feet.

	Craft@Hrs	Unit	Material	Labor	Total	Sell
Pack of 48	—	Ea	8.65	—	8.65	—

SilentStep™ laminate flooring underlayment, Pergo®. Sound reduction for laminate flooring. Flexible. Cuts without leaving debris or dust. Roll covers 100 square feet.

	Craft@Hrs	Unit	Material	Labor	Total	Sell
Flooring underlayment	BF@.010	SF	.64	.33	.97	1.65

Soundbloc™ underlayment foam, Pergo®. Evens out minor irregularities in subfloor. Provides sound and heat insulation. Makes floor comfortable underfoot. Roll covers 100 square feet.

	Craft@Hrs	Unit	Material	Labor	Total	Sell
Foam underlayment	BF@.006	SF	.28	.20	.48	.82

Underlayment. Vapor barrier and sound reducing underlayment in one. For concrete subfloors. 8" band of film along edges for overlap. Alkaline resistant. Roll covers 400 square feet.

	Craft@Hrs	Unit	Material	Labor	Total	Sell
Foam and film pack	—	SF	.25	—	.25	—

Pergo laminate glue. Penetrates the core material to create a super-strong joint. Pergo warranty is valid only if Pergo glue is used for installation.

	Craft@Hrs	Unit	Material	Labor	Total	Sell
16-ounce tube	—	Ea	7.02	—	7.02	—

Laminate flooring sealant, Pergo®. Acrylic-based sealing compound for moisture resistance. For use in bathrooms, kitchens, laundry rooms and other wet areas. Formulated to maintain high level of flexibility and elasticity.

	Craft@Hrs	Unit	Material	Labor	Total	Sell
11-ounce tube	—	Ea	9.15	—	9.15	—

Finishing putty, Pergo®. Water-based putty formulated to fill nail and screw holes in Pergo moldings and wall base. Repairs small chips and dents in the laminate layer.

	Craft@Hrs	Unit	Material	Labor	Total	Sell
Tube	—	Ea	10.20	—	10.20	—

Pergo® universal installation kit. Contains underseal tapping block, 48 installation spacers, pull bar and glue scraper.

	Craft@Hrs	Unit	Material	Labor	Total	Sell
Kit	—	Ea	24.00	—	24.00	—

	Craft@Hrs	Unit	Material	Labor	Total	Sell
Hampton TrafficMaster laminate flooring. Engineered "longstrip" wood floors with premium oak veneers. Cross-ply, multiple layer construction for long-term dimensional stability and structural integrity. Each plank is 7.5" wide, 72" long, 7/16" thick.						
Raintree Acacia , 14 SF per case	BF@.051	SF	1.52	1.67	3.19	5.42
Estate Maple, 25.3 SF per case	BF@.051	SF	2.03	1.67	3.70	6.29
Ainsley Oak, 26.8 SF per case	BF@.051	SF	.69	1.67	2.36	4.01
Silverhill Oak, 19.9 SF per case	BF@.051	SF	1.83	1.67	3.50	5.95
Black Hills Hickory, 19.9 SF per case	BF@.051	SF	2.32	1.67	3.99	6.78
Hampton floor molding.						
3'11" carpet reducer	BF@.160	Ea	15.50	5.24	20.74	35.30
3'11" reducer	BF@.160	Ea	15.50	5.24	20.74	35.30
3'11" stair nose	BF@.160	Ea	12.90	5.24	18.14	30.80
3'11" T-molding	BF@.160	Ea	15.00	5.24	20.24	34.40
7'10" quarter round	BF@.320	Ea	9.13	10.50	19.63	33.40
7'10" wall base	BF@.320	Ea	12.40	10.50	22.90	38.90
Legends strip laminate flooring. Case covers 20 square feet before cutting waste.						
Kempas	BF@.051	SF	2.80	1.67	4.47	7.60
Brazilian cherry	BF@.051	SF	2.80	1.67	4.47	7.60
Wheat	BF@.051	SF	2.80	1.67	4.47	7.60
Cognac	BF@.051	SF	2.80	1.67	4.47	7.60
Coffee	BF@.051	SF	2.80	1.67	4.47	7.60
Legends molding. Molding used to transition from laminate flooring to carpet or at the base of a wall.						
3'11" reducer	BF@.160	Ea	13.00	5.24	18.24	31.00
3'11" stair nose	BF@.160	Ea	20.10	5.24	25.34	43.10
3'11" T-molding	BF@.160	Ea	15.50	5.24	20.74	35.30
8' quarter round	BF@.320	Ea	7.73	10.50	18.23	31.00
8' wall base	BF@.320	Ea	15.60	10.50	26.10	44.40
Wood and laminate flooring underlayment, Shaw HardSurfaces. Roll covers 50 square feet.						
3-in-1 underlayment	BF@.010	SF	.77	.33	1.10	1.87
Foam underlayment, Shaw HardSurfaces. Evens out minor irregularities in subfloor. Provides sound and heat insulation. Makes floor comfortable underfoot. Roll covers 100 square feet.						
2-in-1 underlayment	BF@.006	SF	.27	.20	.47	.80
Moisturbloc™ polyethylene film. For installation over concrete subfloors. Lasts for the life of the floor. 6 mil polyethylene film with increased alkaline resistance and 0.2% HALS additive. 120 square foot roll.						
Poly film	BF@.006	SF	.13	.20	.33	.56
Laminate flooring installation system. Adjustable cinching system pulls laminate floor together and provides a perfect, tight seam.						
Cinch system	—	Ea	22.40	—	22.40	—

Floor Surface Prep

	Craft@Hrs	Unit	Material	Labor	Total	Sell
Prepare surface for new flooring. Per 100 square feet of floor.						
Scrape concrete floor to remove adhesive residue	BF@.750	CSF	—	24.50	24.50	41.70
Apply skim-coat to level surface	BF@.563	CSF	72.00	18.40	90.40	154.00

	Craft@Hrs	Unit	Material	Labor	Total	Sell
Adhesive remover. Removes latex, acrylic and pressure-sensitive adhesives from concrete, wood, sheet vinyl, tile, ceramic and terrazzo substrates. Highly concentrated cleaning solution.						
Quart	—	Ea	13.20	—	13.20	—
Sentinel 747 flooring adhesive remover. For cleaning cutback, emulsion and outdoor flooring mastics. Coverage is 75 to 150 square feet per gallon.						
Gallon	—	Ea	21.40	—	21.40	—
Henry 550 embossing leveler. Levels embossed areas in existing resilient flooring prior to installing new resilient flooring. One-part latex polymer resin. No mixing required. Can be used directly over urethane wear surfaces with no special preparation. Eliminates the need to remove the old floor. Prevents pattern show-through. Coverage: approximately 54 square feet per quart over heavy embossed floors or 108 square feet per quart over lightly embossed floors. Working time 15 to 20 minutes.						
Quart	—	Ea	27.50	—	27.50	—
Liquid underlayment. For residential use as an embossing leveler. Can be used directly over CleanSweep surfaces with no special preparation. Eliminates the need to remove a damaged floor. Prevents pattern show-through. Coverage is approximately 216 square feet per gallon over heavily embossed floors or 432 square feet per gallon over lightly embossed floors. Working time is 15 to 20 minutes.						
Gallon	—	Ea	81.60	—	81.60	—
Latex primer and additive, S-185, Armstrong. For priming wood and concrete before installing resilient floors. Improves adhesion. Fast-drying, ready for adhesive in 60 minutes. Gallon covers 400 square feet.						
Quart	—	Ea	5.75	—	5.75	—
Gallon	—	Ea	10.50	—	10.50	—
Henry 336 floor primer. Improves adhesion on dry, dusty and porous substrates. Bonds to concrete, poured-in-place gypsum subfloors, wood and wood underlayments, radiant-heated subfloors where the surface temperature does not exceed 85 degrees F (29 degrees C), wall surfaces such as wood, plaster, drywall, and masonry.						
Quart	—	Ea	5.99	—	5.99	—
Gallon	—	Ea	11.80	—	11.80	—
Fast-setting patch and underlayment, S-184, Armstrong. Cement-based smooth finish under resilient flooring. Ready for adhesive in 60 minutes. Can be used to skim coat over old cutback adhesive.						
3 pounds	—	Ea	7.55	—	7.55	—
10 pounds	—	Ea	15.70	—	15.70	—
25 pounds	—	Ea	18.30	—	18.30	—
Cement-based underlayment, Armstrong. Fast-setting cement-based patch and underlayment for use over wood and concrete in resilient flooring application. Ready for most flooring in 1-2 hours. 10-pound bag covers 10 square feet at 1/8" thickness.						
10 pounds	—	Ea	11.80	—	11.80	—
40 pounds	—	Ea	22.50	—	22.50	—
Henry 331 patch. Ready-to-use gypsum-based patch – just add water. Fills voids and levels floors from particle thin to 1" thick in a single application. Non-shrinking. Fast-setting.						
25 pounds	—	Ea	7.93	—	7.93	—

	Craft@Hrs	Unit	Material	Labor	Total	Sell

Henry 547 universal underlayment and floor patch. Use as a patch, underlayment or embossing leveler. For patching, leveling and skim coating; covering existing "cutback" adhesive residue; and leveling the embossing of existing resilient flooring. Sets for most flooring in 1 to 2 hours. Excellent fill characteristics allows for installations from as little as a featheredge to depths of 1" thick. Non-sanded formulation allows product to be troweled to a smooth finish. Compressive strength exceeds 5,000 PSI. Non-shrinking formulation. Bonds to concrete, wood substrates, steel, stainless steel, brass, lead, ceramic, terrazzo, marble, existing "cutback" adhesive residue and existing resilient flooring (as an embossing leveler). Metal substrates must be clean and abraded. 10-year limited warranty. One pound covers one square foot at 1/8" thick.

	Craft@Hrs	Unit	Material	Labor	Total	Sell
3 pounds	—	Ea	6.60	—	6.60	—
10 pounds	—	Ea	13.20	—	13.20	—
25 pounds	—	Ea	22.40	—	22.40	—

Henry 546 feather edge additive. Improves the bond and enhances flexural strength when used as a mix instead of water with Henry 547. Extends the pot life for added working time. Not to be used when installing over asphalt "cutback" adhesive residue.

	Craft@Hrs	Unit	Material	Labor	Total	Sell
Gallon	—	Ea	17.20	—	17.20	—

Skimcoat white cover patch and underlayment. Covers and levels defects in concrete, wood, metal, vinyl, ceramic tile and other subfloors. Smooth-spreading, quick-setting and fast-drying. High compressive strength and adhesion-resistance. Non-staining. Non-alkaline, adhesive-friendly. Non-shrinking, dust-free in thick or thin applications. Use on virtually any flooring surface from particle thin to over 1" thick. One pound covers one square foot at 1/8" thick.

	Craft@Hrs	Unit	Material	Labor	Total	Sell
10 pounds	—	Ea	14.70	—	14.70	—
25 pounds	—	Ea	22.40	—	22.40	—
35 pounds	—	Ea	39.30	—	39.30	—
50 pounds	—	Ea	53.90	—	53.90	—

Sheet Vinyl Flooring

Medley™ sheet vinyl flooring, Armstrong. CleanSweep® surface. ToughGuard® durability — guaranteed not to rip, tear or gouge. 15-year warranty. 6' wide roll. 0.08" thick. Urethane no-wax wear layer. Includes 10% waste.

	Craft@Hrs	Unit	Material	Labor	Total	Sell
Chalk White	BF@.300	SY	18.50	9.82	28.32	48.10
Edinburgh White	BF@.300	SY	19.10	9.82	28.92	49.20
Sage Stone	BF@.300	SY	20.20	9.82	30.02	51.00

Caspian II™ sheet vinyl flooring, Armstrong. CleanSweep® surface. ToughGuard® durability — guaranteed not to rip, tear or gouge. 15-year limited warranty. 12' wide rolls. 0.07" thick. Urethane no-wax wear layer. 3.8 Performance Appearance Rating. Includes 10% waste.

	Craft@Hrs	Unit	Material	Labor	Total	Sell
Modular Stone Beige	BF@.300	SY	11.40	9.82	21.22	36.10
Resonna Walnut	BF@.300	SY	11.40	9.82	21.22	36.10
River Slate	BF@.300	SY	11.40	9.82	21.22	36.10
Corner Dot, Taupe	BF@.300	SY	16.90	9.82	26.72	45.40
Decker Station, Honey Spice	BF@.300	SY	15.90	9.82	25.72	43.70
Mt. Bethel Park, Moss	BF@.300	SY	15.90	9.82	25.72	43.70
Shavano Park, Spice	BF@.300	SY	15.90	9.82	25.72	43.70
Walkway, Green	BF@.300	SY	15.90	9.82	25.72	43.70
Walkway, Taupe	BF@.300	SY	15.90	9.82	25.72	43.70
Walkway, Warm Beige	BF@.300	SY	15.90	9.82	25.72	43.70

	Craft@Hrs	Unit	Material	Labor	Total	Sell
Themes sheet vinyl flooring, Armstrong. Easy to clean. ToughGuard® durability. Guaranteed not to rip, tear or gouge. 10-year limited warranty. Urethane no-wax wear layer. 12' wide rolls. 0.08" thick. 2.9 Performance Appearance Rating. Includes adhesive and 10% waste.						
Most colors and designs	BF@.300	SY	9.58	9.82	19.40	33.00
Sandstone	BF@.300	SY	9.89	9.82	19.71	33.50
Marble Wisp, White Crystal	BF@.300	SY	9.89	9.82	19.71	33.50
Prescott, Natural White	BF@.300	SY	9.89	9.82	19.71	33.50
Sundial™ sheet vinyl flooring, Armstrong. Urethane wear layer. Modified loose lay. No adhesive needed. 10-year warranty. 12' wide rolls. 0.098" thick. Includes 10% waste.						
Most colors and designs	BF@.300	SY	11.40	9.82	21.22	36.10
Chestnut Corner, Rust	BF@.300	SY	10.90	9.82	20.72	35.20
Myrtlewood, Light Oak	BF@.300	SY	12.20	9.82	22.02	37.40
Quinault, Light Limestone	BF@.300	SY	13.40	9.82	23.22	39.50
Sentinel sheet vinyl flooring, Armstrong. 5-year limited warranty. 12' wide rolls. Includes adhesive and 10% waste.						
Most colors and designs	BF@.300	SY	9.00	9.82	18.82	32.00
Darker colors, intricate designs	BF@.300	SY	9.00	9.82	18.82	32.00
Metro™ sheet vinyl flooring, Armstrong. ToughGuard® durability. Guaranteed not to rip, tear or gouge. 5-year limited warranty. 12' wide rolls. 0.045" thick. Vinyl no-wax wear layer. 2.3 Performance Appearance Rating. Includes adhesive and 10% waste.						
Adobe Tan	BF@.300	SY	8.17	9.82	17.99	30.60
Bluelake	BF@.300	SY	7.25	9.82	17.07	29.00
Cocoa	BF@.300	SY	8.15	9.82	17.97	30.50
Media Heights, Sandstone	BF@.300	SY	8.71	9.82	18.53	31.50
Navy	BF@.300	SY	10.40	9.82	20.22	34.40
Sand	BF@.300	SY	8.17	9.82	17.99	30.60
Cambray™ sheet vinyl flooring, Armstrong. Vinyl ToughGuard® durability. FHA spec. Guaranteed not to rip, tear or gouge. 5-year limited warranty. 12' wide rolls. 1.7 Performance Appearance Rating. Includes 10% waste.						
Aegean Mist, White	BF@.300	SY	7.89	9.82	17.71	30.10
Country Flair, Rose	BF@.300	SY	7.95	9.82	17.77	30.20
Hampton House, White	BF@.300	SY	8.24	9.82	18.06	30.70
Middlesex, Toast	BF@.300	SY	7.88	9.82	17.70	30.10
Royelle® sheet vinyl flooring, Armstrong. No-wax surface. Easy to install. 12' wide rolls. One-year limited warranty. 0.05" thick. Vinyl no-wax wear layer. 1.3 Performance Appearance Rating. Includes 10% waste.						
Augusta, Dark Oak	BF@.300	SY	5.21	9.82	15.03	25.60
Kaley's Korner, Seashell	BF@.300	SY	5.21	9.82	15.03	25.60
Sheffley, Black and White	BF@.300	SY	4.95	9.82	14.77	25.10

	Craft@Hrs	Unit	Material	Labor	Total	Sell

Vinyl Floor Tile

Metro Series™ vinyl floor tile, Armstrong. 12" x 12". Resists stains, scratches, scuffs and indentations. 5-year warranty. 0.045" thick. Vinyl no-wax wear layer. 2.3 Performance Appearance Rating. Per square foot, including adhesive and 10% waste.

	Craft@Hrs	Unit	Material	Labor	Total	Sell
Avlana, Golden Mosaic	BF@.023	SF	.93	.75	1.68	2.86
Salina, Beige	BF@.023	SF	.81	.75	1.56	2.65
Salina, Gray	BF@.023	SF	.88	.75	1.63	2.77
Toledo, Slate Blue	BF@.023	SF	.88	.75	1.63	2.77
Tetherow Provincial, Cherry	BF@.023	SF	.88	.75	1.63	2.77

Elegant Images™ Series vinyl floor tile, Armstrong. 12" x 12". ToughGuard® durability. Guaranteed not to rip, tear or gouge. Resists stains, scratches, scuffs and indentations. 25-year warranty. Per square foot, including adhesive and 10% waste.

	Craft@Hrs	Unit	Material	Labor	Total	Sell
Cornwall, Burnt Almond	BF@.023	SF	2.35	.75	3.10	5.27
Modern, Rose/Green Inset	BF@.023	SF	1.67	.75	2.42	4.11

Vernay™ Series vinyl floor tile, Armstrong. 12" x 12". ToughGuard® durability. Guaranteed not to rip, tear or gouge. Resists stains, scratches, scuffs and indentations. 25-year warranty. Self-adhesive. Per square foot, including 10% waste.

	Craft@Hrs	Unit	Material	Labor	Total	Sell
Allenbury, Shadow Blue	BF@.021	SF	.70	.69	1.39	2.36
Classic Marble, Blue	BF@.021	SF	.79	.69	1.48	2.52
Englewood, Oak	BF@.021	SF	.77	.69	1.46	2.48
Floral Focus, Rose Petal	BF@.021	SF	.78	.69	1.47	2.50
Montelena, Cinnamon	BF@.021	SF	.78	.69	1.47	2.50
Montelena, White	BF@.021	SF	.77	.69	1.46	2.48
Snapshot, Bianco White	BF@.021	SF	.87	.69	1.56	2.65

Chelsea Collection™ vinyl floor tile, Armstrong. 12" x 12". ToughGuard® durability. Guaranteed not to rip, tear or gouge. Resists stains, scratches, scuffs and indentations. 10-year warranty. Self-adhesive. Per square foot, including 10% waste.

	Craft@Hrs	Unit	Material	Labor	Total	Sell
Marble Beauty	BF@.021	SF	2.65	.69	3.34	5.68

Themes Collection™ vinyl floor tile, Armstrong. 12" x 12". ToughGuard® durability. Guaranteed not to rip, tear or gouge. Resists stains, scratches, scuffs and indentations. 10-year warranty. Per square foot, including adhesive and 10% waste.

	Craft@Hrs	Unit	Material	Labor	Total	Sell
Center Piece, Black	BF@.023	SF	1.17	.75	1.92	3.26
Charmayne, Sandstone	BF@.023	SF	1.17	.75	1.92	3.26
Classic Mosaic, Deep Blue/Natural	BF@.023	SF	1.17	.75	1.92	3.26
Classic Mosaic, Burgundy/Natural	BF@.023	SF	1.17	.75	1.92	3.26
Classic Mosaic, Hunter Green/Almond	BF@.023	SF	1.22	.75	1.97	3.35
Colebrook, Fieldstone	BF@.023	SF	1.17	.75	1.92	3.26
Midway, Beige	BF@.023	SF	1.17	.75	1.92	3.26
Midway, Bisque	BF@.023	SF	1.17	.75	1.92	3.26
Naples, Black	BF@.023	SF	1.17	.75	1.92	3.26
Naples, Green	BF@.023	SF	1.17	.75	1.92	3.26
Naples, Natural	BF@.023	SF	1.15	.75	1.90	3.23
Senegal, Light Beige	BF@.023	SF	1.09	.75	1.84	3.13
Stonehaven, Orange Bisque	BF@.023	SF	1.17	.75	1.92	3.26
Tripoli, Blue/Natural	BF@.023	SF	1.17	.75	1.92	3.26

	Craft@Hrs	Unit	Material	Labor	Total	Sell
Safety Zone vinyl floor tile, Armstrong. Slip-retardant vinyl composition tile for use where slips and falls are a concern. Meets or exceeds ADA slip-retardant performance ranges. Low profile for easy maintenance. Styled for all commercial interiors. Per square foot, including adhesive and 10% waste.						
Earth Stone	BF@.023	SF	2.30	.75	3.05	5.19
Slate Black	BF@.023	SF	2.30	.75	3.05	5.19
Stone Beige	BF@.023	SF	2.30	.75	3.05	5.19
Weathered Alabaster	BF@.023	SF	2.30	.75	3.05	5.19
Stylistik® II vinyl floor tile, Armstrong. 12" x 12". Resists stains, scratches, scuffs and indentations. 5-year warranty. 0.065" thick. Vinyl no-wax wear layer. Self-adhesive. Per square foot, including 10% waste.						
Bayville, Bisque	BF@.021	SF	1.26	.69	1.95	3.32
Criswood, Russet Oak	BF@.021	SF	1.16	.69	1.85	3.15
Criswood, Vintage Oak	BF@.021	SF	1.07	.69	1.76	2.99
Firenza, Sapphire	BF@.021	SF	1.12	.69	1.81	3.08
Gladstone, Blue/Natural	BF@.021	SF	1.26	.69	1.95	3.32
Oakland, Dusty Brick	BF@.021	SF	1.13	.69	1.82	3.09
Pequea Park, Canyon Multi	BF@.021	SF	1.10	.69	1.79	3.04
Terramora, Black	BF@.021	SF	1.13	.69	1.82	3.09
Terramora, White	BF@.021	SF	1.08	.69	1.77	3.01
Chesapeake Collection™ vinyl floor tile, Armstrong. 12" x 12". 25-year warranty. 0.08" thick. Urethane no-wax wear layer. Per square foot, including adhesive and 10% waste.						
Leesport, Sienna/Gray	BF@.023	SF	2.14	.75	2.89	4.91
Millport, Wheat	BF@.023	SF	2.06	.75	2.81	4.78
Steinway, White Essence	BF@.023	SF	1.81	.75	2.56	4.35
Natural Images™ vinyl floor tile, Armstrong. 12" x 12". ToughGuard® durability. 25-year warranty. Per square foot including adhesive and 10% waste						
Stonegate, Antique Ivory	BF@.023	SF	2.82	.75	3.57	6.07
Stonegate, Blue Slate	BF@.023	SF	2.85	.75	3.60	6.12
Harbour Collection™ vinyl floor tile, Armstrong. 12" x 12". ToughGuard® durability. Guaranteed not to rip, tear or gouge. Resists stains, scratches, scuffs and indentations. 25-year warranty. Per square foot including adhesive and 10% waste.						
Andora, Spring Green	BF@.023	SF	1.88	.75	2.63	4.47
Arrington, Dusty Green	BF@.023	SF	1.88	.75	2.63	4.47
Arrington, Trail Beige	BF@.023	SF	1.88	.75	2.63	4.47
Matheson Park, Emerald	BF@.023	SF	1.88	.75	2.63	4.47
Matheson Park, Sapphire	BF@.023	SF	1.77	.75	2.52	4.28
Place n Press floor tile, Armstrong. 12" x 12". No-wax easy-to-clean finish. Household gauge. Self-adhesive. Per square foot, including 10% waste.						
Lockeport, Slate Blue	BF@.021	SF	.53	.69	1.22	2.07
Parkson, Light Oak	BF@.021	SF	.53	.69	1.22	2.07
Excelon® Imperial Texture vinyl composition tile, Armstrong. 1/8"-thick tile. 12" x 12". Color pattern goes through the tile. Commercial traffic rated. Per square foot, including 10% waste and adhesive.						
Blue Cloud	BF@.023	SF	.81	.75	1.56	2.65
Blue/Gray	BF@.023	SF	.81	.75	1.56	2.65
Caribbean Blue	BF@.023	SF	.81	.75	1.56	2.65
Cherry Red	BF@.023	SF	.81	.75	1.56	2.65
Classic Black	BF@.023	SF	.81	.75	1.56	2.65
Classic White	BF@.023	SF	.81	.75	1.56	2.65
Cool White	BF@.023	SF	.81	.75	1.56	2.65
Cottage Tan	BF@.023	SF	.81	.75	1.56	2.65

	Craft@Hrs	Unit	Material	Labor	Total	Sell
Fortress White	BF@.023	SF	.81	.75	1.56	2.65
Hazelnut	BF@.023	SF	.76	.75	1.51	2.57
Marina Blue	BF@.023	SF	.81	.75	1.56	2.65
Pearl White	BF@.023	SF	.76	.75	1.51	2.57
Sandrift White	BF@.023	SF	.81	.75	1.56	2.65
Sea Green	BF@.023	SF	.81	.75	1.56	2.65
Shelter White	BF@.023	SF	.81	.75	1.56	2.65
Sterling	BF@.023	SF	.81	.75	1.56	2.65
Teal	BF@.023	SF	.81	.75	1.56	2.65

Excelon® Civic Square vinyl composition tile, Armstrong. Color pattern goes through the tile. 1/8"-thick tile. 12" x 12". Per square foot, including adhesive and 10% waste.

	Craft@Hrs	Unit	Material	Labor	Total	Sell
Oyster White	BF@.023	SF	.70	.75	1.45	2.47
Stone Tan	BF@.023	SF	.70	.75	1.45	2.47

Rubber studded tile. 12" x 12". Highly resilient to reduce fatigue. 10-year wear warranty. Slip-resistant. Sound absorbent. High-abrasion resistance. Recommended in kitchens and active use rooms.

	Craft@Hrs	Unit	Material	Labor	Total	Sell
Black	BF@.025	SF	9.70	.82	10.52	17.90
Indigo	BF@.025	SF	9.70	.82	10.52	17.90
Taupe	BF@.025	SF	9.70	.82	10.52	17.90

Vinyl Flooring Adhesive

Sheet flooring adhesive, Armstrong. For residential felt-backed floors. Water-based rubber-resin. Can be used on all grade levels of concrete, existing resilient floors, ceramic, terrazzo, marble, and wood floors. Solvent-free and low odor. Non-staining. Long working time. Gallon covers 350 to 400 square feet.

	Craft@Hrs	Unit	Material	Labor	Total	Sell
Quart	—	Ea	8.70	—	8.70	—
Gallon	—	Ea	19.40	—	19.40	—

Henry 430 clear thin-spread vinyl tile adhesive. For installing vinyl composition tile (VCT). 24-hour working time. Dries clear so chalk lines show through. Moisture and alkali resistance allows installation above or below grade. Bonds to concrete, existing asphalt "cutback" adhesive residue, underlayments, wood substrates, terrazzo, clean and abraded steel, stainless steel, aluminum, lead, copper, brass and bronze, and existing resilient flooring. Gallon covers 350 to 400 square feet.

	Craft@Hrs	Unit	Material	Labor	Total	Sell
Quart	—	Ea	5.54	—	5.54	—
Gallon	—	Ea	16.30	—	16.30	—
4 gallons	—	Ea	47.30	—	47.30	—

Resilient tile adhesive, Armstrong. For Excelon®, dry-back vinyl, and residential dry-back tile. Used on all grade levels of concrete, ceramic, terrazzo, marble and polymeric poured floors. Water-based. Up to 6-hour working time. Low odor and solvent-free. Gallon covers 350 to 400 square feet.

	Craft@Hrs	Unit	Material	Labor	Total	Sell
Quart	—	Ea	6.53	—	6.53	—
Gallon	—	Ea	17.30	—	17.30	—
4 gallons	—	Ea	66.30	—	66.30	—

CX-941 Rubber Adhesive. Gallon covers 60 square feet.

	Craft@Hrs	Unit	Material	Labor	Total	Sell
2 gallon bucket	—	Ea	137.00	—	137.00	—

Henry 440 cove base adhesive. Wet-set adhesive for rubber and vinyl cove base. Aggressive initial grab prevents slip during installation. Cleans up with water.

	Craft@Hrs	Unit	Material	Labor	Total	Sell
11 ounces	—	Ea	3.97	—	3.97	—
1 quart	—	Ea	7.64	—	7.64	—
30 ounces	—	Ea	7.33	—	7.33	—
1 gallon	—	Ea	18.00	—	18.00	—

	Craft@Hrs	Unit	Material	Labor	Total	Sell
Acrylic tape, Cal-Flor. Double-face. For Armstrong Sundial sheet vinyl flooring. Allows flooring to be repositioned during installation.						
2" x 50' roll	—	Ea	16.20	—	16.20	—
Low-gloss seam coating kit, S-564, Armstrong. Protects seams from dirt and wear. Includes seam cleaner, coating, deglosser, and applicator. Covers approximately 100 linear feet per kit.						
Seam coat kit	—	Ea	10.70	—	10.70	—
High-gloss seam coating kit, S-595, Armstrong. For use on high-gloss residential sheet vinyl floors. Includes coating, accelerator, and cleaner. Coverage is 250 linear feet.						
3 ounces	—	Ea	13.90	—	13.90	—

Vinyl Flooring Tools

	Craft@Hrs	Unit	Material	Labor	Total	Sell
Extendable roller. Extension tube with firm-grip handle. Extends from 17" to 27". For rolling vinyl floor covering, carpet, wall covering and cove base.						
7-1/2" wide	—	Ea	38.00	—	38.00	—
Margin notch trowel. 2" x 6" flexible steel trowel for patch jobs and hard-to-reach areas.						
1/4" x 1/4" x 1/4" square notch	—	Ea	4.16	—	4.16	—
5/16" x 1/4" x 1/16" V-notch	—	Ea	4.06	—	4.06	—
Notched trowels. Large, foam comfort grip. Flexible spring steel blade.						
1/8" x 1/8" V-notch	—	Ea	8.32	—	8.32	—
3/16" x 1/4" U-notch	—	Ea	13.50	—	13.50	—
1/2" x 1/4" U-notch	—	Ea	13.50	—	13.50	—
3/8" x 1/4" U-notch	—	Ea	13.50	—	13.50	—
2-3/4" x 8" square notch	—	Ea	2.08	—	2.08	—
Underlay finish trowel. Constructed of welded tool steel. Comfortable hardwood handle.						
2" x 5" margin trowel	—	Ea	4.07	—	4.07	—
6" pointing trowel	—	Ea	3.96	—	3.96	—
Chalk line. 100" self-chalking line with aluminum casing and plumb bob.						
100" chalk line	—	Ea	7.11	—	7.11	—
Chalk refill, 8 oz. plastic bottle	—	Ea	1.94	—	1.94	—
White rubber mallet. Will not mark coping or tile. Rubber head securely mounted on hickory handle.						
Mallet	—	Ea	18.20	—	18.20	—
Vinyl tile cutter. Cuts vinyl tiles to 12" x 12". Hardened steel blade with die-cast aluminum base. Cushioned rubber handle. Ball bearings at stress points with "guillotine" action. Built-in measuring gauge. Works equally well with peel-and-stick tiles.						
Tile cutter	—	Ea	51.00	—	51.00	—
Cove base adhesive nozzle. Fits all cove base caulking guns. Spreads adhesive evenly for cove base.						
1-3/4" long, 3" wide, 4-1/2" high	—	Ea	3.03	—	3.03	—
S-153 Scribing felt, Armstrong. For transferring the pattern of the floor to the flooring material.						
Per square yard	—	Ea	.22	—	.22	—
477 square yard roll	—	Ea	102.00	—	102.00	—

	Craft@Hrs	Unit	Material	Labor	Total	Sell

Flooring Transition Strips

Vinyl flooring transition reducer strip, Roppe. Beveled edge for resilient flooring material. Butting edge is 1/8".

	Craft@Hrs	Unit	Material	Labor	Total	Sell
3' long, black	BF@.250	Ea	1.70	8.18	9.88	16.80
3' long, brown	BF@.250	Ea	1.70	8.18	9.88	16.80
3' long, gray	BF@.250	Ea	1.70	8.18	9.88	16.80

Carpet-to-tile transition strip, Roppe. Joins 1/4" carpet to 1/8" tile.

	Craft@Hrs	Unit	Material	Labor	Total	Sell
4' long, black	BF@.250	Ea	5.33	8.18	13.51	23.00
4' long, brown	BF@.250	Ea	5.31	8.18	13.49	22.90
4' long, gray	BF@.250	Ea	5.32	8.18	13.50	23.00

Oak-to-ceramic tile transition strip.

	Craft@Hrs	Unit	Material	Labor	Total	Sell
3', high pile	BF@.250	Ea	15.50	8.18	23.68	40.30
3', laminate	BF@.250	Ea	15.50	8.18	23.68	40.30
3', vinyl	BF@.250	Ea	15.50	8.18	23.68	40.30
6', high pile	BF@.500	Ea	22.40	16.40	38.80	66.00
6', laminate	BF@.500	Ea	22.40	16.40	38.80	66.00
6', vinyl	BF@.500	Ea	22.40	16.40	38.80	66.00

Cove Base for Vinyl Flooring

Dryback vinyl wall base. Resists scratches, scuffs and whitening. Flexible. 0.080 gauge. Per linear foot.

	Craft@Hrs	Unit	Material	Labor	Total	Sell
4", brown	BF@.025	LF	.61	.82	1.43	2.43
4", black	BF@.025	LF	.61	.82	1.43	2.43
4", fawn	BF@.025	LF	.61	.82	1.43	2.43
4", snow	BF@.025	LF	.67	.82	1.49	2.53
4", almond	BF@.025	LF	.67	.82	1.49	2.53
4", gray	BF@.025	LF	.67	.82	1.49	2.53

Thermoplastic dryback wall base, Roppe. Type TP thermoplastic rubber. More flexible than vinyl due to its rubber content. Thickness hides many wall irregularities. Good stability. Rib back for positive adhesion. Per foot.

	Craft@Hrs	Unit	Material	Labor	Total	Sell
2-1/2" x 120' x 1/8", almond	BF@.025	LF	.52	.82	1.34	2.28
2-1/2" x 120' x 1/8", black	BF@.025	LF	.49	.82	1.31	2.23
2-1/2" x 120' x 1/8", brown	BF@.025	LF	.49	.82	1.31	2.23
2-1/2" x 120' x 1/8", dark gray	BF@.025	LF	.52	.82	1.34	2.28
2-1/2" x 120' x 1/8", fawn	BF@.025	LF	.52	.82	1.34	2.28
2-1/2" x 120' x 1/8", hunter green	BF@.025	LF	.62	.82	1.44	2.45
2-1/2" x 120' x 1/8", white	BF@.025	LF	.59	.82	1.41	2.40
4" x 120' x 1/8", almond	BF@.025	LF	.59	.82	1.41	2.40
4" x 120' x 1/8", black	BF@.025	LF	.59	.82	1.41	2.40
4" x 120' x 1/8", brown	BF@.025	LF	.59	.82	1.41	2.40
4" x 120' x 1/8", fawn	BF@.025	LF	.62	.82	1.44	2.45
4" x 120' x 1/8", hunter green	BF@.025	LF	.52	.82	1.34	2.28
4" x 120' x 1/8", white	BF@.025	LF	.61	.82	1.43	2.43

Rubber dryback wall base. 100% synthetic rubber. Extremely flexible. Won't shrink, grab or separate. Resists scuffing, gouging and most chemicals. Low gloss stain finish. Flexible with shade control. 1/8" thick. Per foot.

	Craft@Hrs	Unit	Material	Labor	Total	Sell
2-1/2", almond	BF@.025	LF	.73	.82	1.55	2.64
2-1/2", black	BF@.025	LF	.72	.82	1.54	2.62
2-1/2", brown	BF@.025	LF	.72	.82	1.54	2.62
2-1/2", dark gray	BF@.025	LF	.72	.82	1.54	2.62
2-1/2", fawn	BF@.025	LF	.72	.82	1.54	2.62
2-1/2", white	BF@.025	LF	.72	.82	1.54	2.62
4", almond	BF@.025	LF	.80	.82	1.62	2.75

	Craft@Hrs	Unit	Material	Labor	Total	Sell
4", black	BF@.025	LF	.80	.82	1.62	2.75
4", brown	BF@.025	LF	.80	.82	1.62	2.75
4", dark gray	BF@.025	LF	.81	.82	1.63	2.77
4", fawn	BF@.025	LF	.80	.82	1.62	2.75
4", white	BF@.025	LF	.81	.82	1.63	2.77

Self-stick wall base, Roppe. Full-cover adhesive for sure-stick installation. Top lip design for tight fit. Special coating resists scuffs and scratches. 1/8" thick. Per foot.

	Craft@Hrs	Unit	Material	Labor	Total	Sell
2-1/2", almond	BF@.016	LF	.81	.52	1.33	2.26
2-1/2", black	BF@.016	LF	.82	.52	1.34	2.28
2-1/2", brown	BF@.016	LF	.82	.52	1.34	2.28
2-1/2", dark gray	BF@.016	LF	.82	.52	1.34	2.28
2-1/2", snow	BF@.016	LF	.82	.52	1.34	2.28
4", almond	BF@.016	LF	.97	.52	1.49	2.53
4", black	BF@.016	LF	.97	.52	1.49	2.53
4", brown	BF@.016	LF	.97	.52	1.49	2.53
4", dark gray	BF@.016	LF	.97	.52	1.49	2.53
4", snow	BF@.016	LF	.97	.52	1.49	2.53

Residential self-stick vinyl wall base, Armstrong. Per foot.

	Craft@Hrs	Unit	Material	Labor	Total	Sell
4", almond high gloss	BF@.016	LF	1.19	.52	1.71	2.91
4", architectural white	BF@.016	LF	1.20	.52	1.72	2.92
4", black high gloss	BF@.016	LF	1.19	.52	1.71	2.91
4", gray	BF@.016	LF	1.20	.52	1.72	2.92
4", walnut high gloss	BF@.016	LF	1.19	.52	1.71	2.91
4", white high gloss	BF@.016	LF	1.19	.52	1.71	2.91

Cork Wall and Floor Cover

Natural cork tile, Boone International. Case of 18 tiles covers 16.2 square feet with 10% waste.

	Craft@Hrs	Unit	Material	Labor	Total	Sell
12" x 12" x 3/8"	BF@.030	SF	1.79	.98	2.77	4.71

Cork roll.

	Craft@Hrs	Unit	Material	Labor	Total	Sell
4' wide, 1/4" thick, per linear foot	—	Ea	2.41	—	2.41	—
4' x 50' x 1/4" roll	—	Ea	120.00	—	120.00	—

Ceramic Tile Installation

Ceramic tile adhesive. Premixed Type 1, Acryl-4000. Gallon covers 70 square feet applied with #1 trowel (3/16" x 5/32" V-notch), 50 square feet applied with #2 trowel (3-/16" x 1/4" V-notch) or 40 square feet applied with #3 trowel (1/4" x 1/4" square-notch).

	Craft@Hrs	Unit	Material	Labor	Total	Sell
1/2 pint	—	Ea	3.02	—	3.02	—
1 gallon	—	Ea	13.20	—	13.20	—
3-1/2 gallons	—	Ea	36.70	—	36.70	—
Tile adhesive applied with #2 trowel	—	SF	.21	—	.21	—

Thin-set tile mortar. 50 pound bag covers 100 square feet applied with #1 trowel (1/4" x 1/4" square-notch), 80 square feet applied with #2 trowel (1/4" x 3/8" square-notch), 45 square feet applied with #3 trowel (1/2" x 1/2" square-notch). Cost per 50 pound bag.

	Craft@Hrs	Unit	Material	Labor	Total	Sell
FlexBond®, gray	—	Ea	29.60	—	29.60	—
Marble and granite mix	—	Ea	22.20	—	22.20	—
MasterBlend™, white	—	Ea	14.30	—	14.30	—
Standard, gray	—	Ea	5.93	—	5.93	—
MasterBlend™ applied with #3 trowel	—	SF	.32	—	.32	—

	Craft@Hrs	Unit	Material	Labor	Total	Sell
Tile grout. Polymer-modified dry grout for joints 1/16" to 1/2". Coverage varies with tile and joint size. Cost per 25 pound bag.						
Sanded, dark colors	—	Ea	14.20	—	14.20	—
Sanded, light colors	—	Ea	13.60	—	13.60	—
Unsanded	—	Ea	14.20	—	14.20	—
Setting counter, wall and floor tile in adhesive. Costs are for adhesive and grout. Add the cost of surface preparation (backerboard) and the cost of tile.						
Countertops						
Ceramic mosaic field tile	TL@.203	SF	.28	6.83	7.11	12.10
4-1/4" x 4-1/4" to 6" x 6" glazed field tile	TL@.180	SF	.28	6.06	6.34	10.80
Countertop trim pieces and edge tile	TL@.180	LF	.05	6.06	6.11	10.40
Floors						
Ceramic mosaic field tile	TL@.121	SF	.28	4.07	4.35	7.40
4-1/4" x 4-1/4" to 6" x 6" glazed field tile	TL@.110	SF	.28	3.70	3.98	6.77
8" x 8" and larger field tile	TL@.100	SF	.28	3.36	3.64	6.19
Floor trim pieces and edge tile	TL@.200	LF	.05	6.73	6.78	11.50
Walls						
Ceramic mosaic field tile	TL@.143	SF	.28	4.81	5.09	8.65
4-1/4" x 4-1/4" to 6" x 6" glazed field tile	TL@.131	SF	.28	4.41	4.69	7.97
8" x 8" and larger field tile	TL@.120	SF	.28	4.04	4.32	7.34
Wall trim pieces and edge tile	TL@.200	LF	.05	6.73	6.78	11.50
Setting counter, wall and floor tile in thin-set mortar. Costs are for mortar and grout. Add the cost of surface preparation (backerboard) and tile.						
Countertops						
Ceramic mosaic field tile	TL@.300	SF	.35	10.10	10.45	17.80
4-1/4" x 4-1/4" to 6" x 6" glazed field tile	TL@.250	SF	.35	8.41	8.76	14.90
Countertop trim pieces and edge tile	TL@.400	LF	.06	13.50	13.56	23.10
Floors						
Ceramic mosaic field tile	TL@.238	SF	.35	8.01	8.36	14.20
4-1/4" x 4-1/4" to 6" x 6" field tile	TL@.210	SF	.35	7.06	7.41	12.60
8" x 8" and larger field tile	TL@.200	SF	.35	6.73	7.08	12.00
Quarry or paver tile	TL@.167	SF	.35	5.62	5.97	10.10
Marble, granite or stone tile	TL@.354	SF	.35	11.90	12.25	20.80
Floor trim pieces and edge tile	TL@.210	LF	.06	7.06	7.12	12.10
Walls						
Ceramic mosaic field tile	TL@.315	SF	.35	10.60	10.95	18.60
4-1/4" x 4-1/4" to 6" x 6" glazed field tile	TL@.270	SF	.35	9.08	9.43	16.00
8" x 8" and larger field tile	TL@.260	SF	.35	8.75	9.10	15.50
Wall trim pieces and edge tile	TL@.270	LF	.06	9.08	9.14	15.50

	Craft@Hrs	Unit	Material	Labor	Total	Sell

Tile backerboard. (Durock™ or Wonderboard™). Water-resistant underlayment for ceramic tile on floors, walls, countertops, and other interior wet areas. Material cost for 100 square feet (CSF) of board includes waterproofing membrane, the backerboard (10% waste,) 50 pounds of job-mixed latex-fortified mortar for the joints and surface skim coat, 75 linear feet of fiberglass joint tape, and 200 1-1/4" backerboard screws. For scheduling purposes, estimate that a crew of 2 can install, tape and apply the skim coat on the following quantity of backerboard in an 8-hour day: countertops 180 SF, floors 525 SF, and walls 350 SF. Use $250.00 as a minimum charge for this type work.

	Craft@Hrs	Unit	Material	Labor	Total	Sell
Using 15 lb. felt waterproofing, per 400 SF, roll	—	Ea	13.70	—	13.70	—
Using 1/4" or 1/2" backerboard, per 100 SF		CSF	66.60	—	66.60	—
Using 50 lbs. of latex-fortified mortar each 100 SF, per 50 lb. sack	—	Ea	14.70	—	14.70	—
Using 75 LF of backerboard tape for each 100 SF, per 50 LF roll	—	Ea	4.13	—	4.13	—
Using 200 backerboard screws for each 100 SF, per pack of 200 screws	—	Ea	4.15	—	4.15	—
Backerboard with waterproofing, mortar, tape and screws.						
Countertops	T1@.090	SF	1.15	2.82	3.97	6.75
Floors	T1@.030	SF	1.15	.94	2.09	3.55
Walls	T1@.045	SF	1.15	1.41	2.56	4.35

Wonderboard® backerboard. Underlayment for tile floors in kitchens and baths and for countertops. 1/4" thickness reduces subfloor modifications to adjacent floors, thresholds, carpets and cabinets. 1/2" thickness matches up with surrounding 1/2" (12 mm) drywall without the need to use shims or spacers. Easy to score, snap, cut and nail. For installation cost, see Tile backerboard. Includes 10% waste.

	Craft@Hrs	Unit	Material	Labor	Total	Sell
1/4" x 3' x 5'	—	SF	.60	—	.60	—
1/2" x 3' x 5'	—	SF	.67	—	.67	—

RhinoBoard™ backerboard. Lightweight fiber-cement backerboard for interior floors and countertops. Used under ceramic tile, natural stone, wood, vinyl, and other resilient flooring. Easy to score, snap, cut and nail. For installation costs, see Tile backerboard. Includes 10% waste.

	Craft@Hrs	Unit	Material	Labor	Total	Sell
1/4" x 3' x 5'	—	SF	.78	—	.78	—

Hardibacker® backerboard, James Hardie. Water-resistant, cement-based underlayment for floors, walls and countertops. Bonds with mastic or latex-modified thin-set tile adhesive. Non-abrasive to enamel or porcelain surfaces. Easy to score and snap. For installation costs, see Tile backerboard. Includes 10% waste.

	Craft@Hrs	Unit	Material	Labor	Total	Sell
1/4" x 3' x 5'	—	SF	.66	—	.66	—
1/2" x 3' x 5'	—	SF	.77	—	.77	—
1/2" x 4' x 8'	—	SF	.66	—	.66	—

Aqua-Tough™ tile underlayment, USG Interiors. Underlayment for ceramic tile, resilient flooring, laminated or hardwood flooring in both wet and dry areas. For installation costs, see Tile backerboard. Includes 10% waste.

	Craft@Hrs	Unit	Material	Labor	Total	Sell
1/4" x 4' x 4'	—	SF	.53	—	.53	—

8" x 8" Ceramic Floor Tile

8" ceramic floor tile. Glazed ceramic residential indoor tile. Class 3, light-to-moderate traffic. Case of 25 tiles covers 10 square feet, including 10% waste. See Ceramic Tile Installation for tile setting costs, including tile backer, thin-set mortar or adhesive and grout.

	Craft@Hrs	Unit	Material	Labor	Total	Sell
Ivory	—	SF	1.18	—	1.18	—
White	—	SF	1.18	—	1.18	—

	Craft@Hrs	Unit	Material	Labor	Total	Sell
8" Ragione ceramic floor tile. Satin finish for ease of maintenance. Subtle shading provides a warm, natural, earthy feel. Stained finish makes tile durable, practical and easy to maintain. Case of 25 tiles covers 10 square feet, including 10% waste. See Ceramic Tile Installation for tile setting costs, including tile backer, thin-set mortar or adhesive and grout.						
Almond	—	SF	.97	—	.97	—
White	—	SF	.97	—	.97	—
8" ceramic floor tile. Case of 36 tiles covers 15 square feet, including 10% waste. See Ceramic Tile Installation for tile setting costs, including tile backer, thin-set mortar or adhesive and grout.						
Castle Gray	—	SF	1.19	—	1.19	—
8" Natura® ceramic floor tile. A natural stone-look ceramic tile with warm earth tones. Suitable for all interior residential and light commercial applications. Case of 25 tiles covers 10 square feet, including 10% waste. See Ceramic Tile Installation for tile setting costs, including tile backer, thin-set mortar or adhesive and grout.						
Stone	—	SF	2.02	—	2.02	—
8" x 8" ceramic floor tile. Class 3, light-to-moderate traffic for bathrooms, kitchens, foyers, dining rooms, and recreation areas. Case of 30 tiles covers 12.92 square feet, including 10% waste. See Ceramic Tile Installation for tile setting costs, including tile backer, thin-set mortar or adhesive and grout.						
Pietra, Alabaster	—	SF	1.01	—	1.01	—
Pietra, Gem	—	SF	1.01	—	1.01	—
8" glazed ceramic floor tile. Extra durable matte glaze. Non-slip finish. Exceptional wear and scratch resistance. Case of 25 tiles covers 10 square feet, including 10% waste. See Ceramic Tile Installation for tile setting costs, including tile backer, thin-set mortar or adhesive and grout.						
Blue	—	SF	2.53	—	2.53	—
White	—	SF	2.53	—	2.53	—

12" x 12" Ceramic Floor Tile

	Craft@Hrs	Unit	Material	Labor	Total	Sell
12" x 12" ceramic residential floor tile. Class 3, light-to-moderate traffic. For bathrooms, kitchens, foyers, dining rooms and recreation areas. Case of 15 tiles covers 15 square feet, including 10% waste. See Ceramic Tile Installation for tile setting costs, including tile backer, thin-set mortar or adhesive and grout.						
Carrara, Beige	—	SF	1.40	—	1.40	—
Carrara, Gray	—	SF	1.34	—	1.34	—
Pietra, Alabaster	—	SF	1.08	—	1.08	—
Pietra, Gem	—	SF	1.29	—	1.29	—
Saltillo, Monterey	—	SF	1.08	—	1.08	—
12" x 12" glazed porcelain residential and commercial floor tile, American Marazzi. For residential, commercial and light institutional applications, including restaurant dining rooms, shopping malls, offices, lobbies, showrooms and corridors. Class 4, moderate-to-heavy traffic. Case of 15 tiles covers 15 square feet, including 10% waste. This product does not meet the minimum coefficient of friction to be considered a slip-resistant tile. See Ceramic Tile Installation for tile setting costs, including tile backer, thin-set mortar or adhesive and grout.						
Rapolano, Noce	—	SF	1.43	—	1.43	—
Salento Sabbia, Beige	—	SF	1.61	—	1.61	—
Salento Sabbia, Gray	—	SF	1.13	—	1.13	—
Vermont, Caledonia Porcelain	—	SF	1.89	—	1.89	—
12" x 12" ceramic floor tile. For all residential applications. Class 3, light-to-moderate traffic, including bathrooms, kitchens, foyers, dining rooms and recreation areas. Case of 15 tiles covers 15 square feet, including 10% waste. This product does not meet the minimum coefficient of friction to be considered a slip-resistant tile. See Ceramic Tile Installation for tile setting costs, including tile backer, thin-set mortar or adhesive and grout.						
Verona, Beige	—	SF	1.61	—	1.62	—
Verona, White	—	SF	1.41	—	1.41	—

	Craft@Hrs	Unit	Material	Labor	Total	Sell

12" x 12" ceramic floor tile, Dal-Tile. Field tile. For interior floors, walls, and counter tops. Case of 11 tiles covers 10 square feet, including 10% waste. See Ceramic Tile Installation for tile setting costs, including tile backer, thin-set mortar or adhesive and grout.

	Craft@Hrs	Unit	Material	Labor	Total	Sell
Dakota™, Rushmore Gray	—	SF	1.80	—	1.80	—
French Quarter®, Cobblestone	—	SF	1.44	—	1.44	—
French Quarter®, Mardi Gras	—	SF	2.70	—	2.70	—
Gold Rush®, California Sand	—	SF	1.31	—	1.31	—
Gold Rush®, Golden Nugget	—	SF	1.94	—	1.94	—
Gold Rush®, Goldust	—	SF	2.50	—	2.50	—
Edgefield™, Beige	—	SF	1.59	—	1.59	—

12" x 12" ceramic floor and wall tile, Eliane Ceramic Tile. Marbleized look floor ceramic designed for bathrooms. Can be used as wall coverings. Rated PEI 4 and Mohs scale 3. Eleven tiles cover 10 square feet, including 10% waste. See Ceramic Tile Installation for tile setting costs, including tile backer, thin-set mortar or adhesive and grout.

	Craft@Hrs	Unit	Material	Labor	Total	Sell
Autumn, White	—	SF	1.80	—	1.80	—
Illusione, Beige	—	SF	1.86	—	1.86	—
Illusione, Caramel	—	SF	1.95	—	1.95	—
Illusione, Ice	—	SF	1.85	—	1.85	—

Large Ceramic Floor Tile

16" x 16" ceramic floor tile, Megatrade. Smooth stone finish. PEI rating of 4. Indoor use only. Case of 9 tiles covers 15.5 square feet, including 10% waste. See Ceramic Tile Installation for tile setting costs, including tile backer, thin-set mortar or adhesive and grout.

	Craft@Hrs	Unit	Material	Labor	Total	Sell
Ivory	—	SF	4.04	—	4.04	—
White	—	SF	4.04	—	4.04	—

Fossile Cotto 16" x 16" ceramic floor tile, Eliane Ceramic Tile. Rated PEI 4 and Mohs scale 5. Indoor. Suitable for offices, living rooms, bedrooms, bathrooms, halls and kitchens. Rustic look. Can also be used as wall covering. Case of 9 tiles covers 16 square feet, including 10% waste. See Ceramic Tile Installation for tile setting costs, including tile backer, thin-set mortar or adhesive and grout.

	Craft@Hrs	Unit	Material	Labor	Total	Sell
Rustic look	—	SF	2.25	—	2.25	—

Illusione 16" x 16" ceramic floor tile, Eliane Ceramic Tile. Rated PEI 4 and Mohs scale 3. Marbleized look. Designed for bathrooms. Can also be used as wall coverings. Case of 9 tiles covers 14.4 square feet, including 10% waste. See Ceramic Tile Installation for tile setting costs, including tile backer, thin-set mortar or adhesive and grout.

	Craft@Hrs	Unit	Material	Labor	Total	Sell
Beige	—	SF	3.11	—	3.11	—
Caramel	—	SF	3.11	—	3.11	—
Ice	—	SF	3.11	—	3.11	—

16" x 16" ceramic residential and medium commercial floor tile, American Marazzi. PEI class 4, moderate-to-heavy traffic. For residential, medium commercial and light institutional applications, including restaurant dining rooms, shopping malls, offices, lobbies, showrooms and corridors. Case of 8 tiles covers 13.77 square feet, including 10% waste. Does not meet the minimum coefficient of friction to be considered a slip-resistant tile. See Ceramic Tile Installation for tile setting costs, including tile backer, thin-set mortar or adhesive and grout.

	Craft@Hrs	Unit	Material	Labor	Total	Sell
Island Sand	—	SF	1.26	—	1.26	—
Montagna Cortina	—	SF	4.80	—	4.80	—

	Craft@Hrs	Unit	Material	Labor	Total	Sell
16" x 16" glazed porcelain residential and medium commercial floor tile, American Marazzi. PEI Class 4, moderate-to-heavy traffic. For residential, medium commercial and light institutional applications, including restaurant dining rooms, shopping malls, offices, lobbies, showrooms and corridors. Case of 9 tiles covers 15.5 square feet. Does not meet the minimum coefficient of friction to be considered a slip-resistant tile. See Ceramic Tile Installation for tile setting costs, including tile backer, thin-set mortar or adhesive and grout.						
Romance, Almond	—	SF	1.68	—	1.68	—

Natural Stone and Slate Flooring

	Craft@Hrs	Unit	Material	Labor	Total	Sell
Botticino Americana marble floor tile. Highly polished genuine marble. See Ceramic Tile Installation for tile setting costs, including tile backer, thin-set mortar or adhesive and grout.						
12" x 12"	—	Ea	6.80	—	6.80	—
Grigio Venato marble tile. Gray with natural shade variations. For interior and exterior floors and walls. See Ceramic Tile Installation for tile setting costs, including tile backer, thin-set mortar or adhesive and grout.						
12" x 12"	—	Ea	2.20	—	2.20	—
Coral Venato marble tile. Polished marble with soft hues of rose and gray to match all decors. For interior and exterior floors, wall fireplaces and bathroom countertops. Gloss finish. High durability. See Ceramic Tile Installation for tile setting costs, including tile backer, thin-set mortar or adhesive and grout.						
12" x 12"	—	Ea	1.98	—	1.98	—
Maron Venato marble tile. For floors, fireplaces, and countertops. Residential and commercial applications. High sheen finish. See Ceramic Tile Installation for tile setting costs, including tile backer, thin-set mortar or adhesive and grout.						
12" x 12"	—	Ea	3.40	—	3.40	—
White Gioia marble floor tile. First-quality Italian marble. Each tile has unique polished surface. Residential and commercial applications. Rich classic look. High sheen finish. See Ceramic Tile Installation for tile setting costs, including tile backer, thin-set mortar or adhesive and grout.						
12" x 12"	—	Ea	5.09	—	5.09	—
Nero Black marble floor tile. Tile size is approximate. See Ceramic Tile Installation for tile setting costs, including tile backer, thin-set mortar or adhesive and grout.						
12" x 12"	—	Ea	3.71	—	3.71	—
Marble floor tile, Bestview Ltd. 12" x 12" and standard thickness of 10mm. Suitable for both interior and exterior wall and floor applications. Highly polished surface and beveled on the edges. See Ceramic Tile Installation for tile setting costs, including tile backer, thin-set mortar or adhesive and grout.						
Botticino Beige	—	Ea	4.35	—	4.35	—
Emperador Lite	—	Ea	4.01	—	4.01	—
Cream Jade	—	Ea	1.86	—	1.86	—
Travertine stone tile, Emser International. Rustic look with natural color variations. For floors or fireplaces. See Ceramic Tile Installation for tile setting costs, including tile backer, thin-set mortar or adhesive and grout.						
12" x 12", Dorato, regular	—	Ea	4.55	—	4.55	—
12" x 12", Coliseum, honed	—	Ea	6.57	—	6.57	—
12" x 12", Medium beige	—	Ea	3.54	—	3.54	—

	Craft@Hrs	Unit	Material	Labor	Total	Sell
Travertine floor tile. See Ceramic Tile Installation for tile setting costs, including tile backer, thin-set mortar or adhesive and grout.						
12" x 12", Noce	—	Ea	7.58	—	7.58	—
12" x 12", Rojo	—	Ea	6.57	—	6.57	—
16" x 16", Desert Sand	—	Ea	9.09	—	9.09	—
18" x 18", Desert Sand	—	Ea	11.60	—	11.60	—
Granite floor tile. For bathrooms, entryways, fireplaces or countertops. See Ceramic Tile Installation for tile setting costs, including tile backer, thin-set mortar or adhesive and grout.						
12" x 12", Mauve	—	Ea	4.55	—	4.55	—
12" x 12", Absolute Black	—	Ea	10.10	—	10.10	—
12" x 12", Mystique	—	Ea	3.54	—	3.54	—
12" x 12", Pacific Gray	—	Ea	8.08	—	8.08	—
12" x 12", Sonoma Montecruz	—	Ea	3.54	—	3.54	—
Natural granite tile. Pattern and color varies. For interior, exterior walls, heavy-duty interior floors and polished, honed, or flamed surfaces. Slightly eased edges. 3/8" thick. See Ceramic Tile Installation for tile setting costs, including tile backer, thin-set mortar or adhesive and grout.						
12" x 12", Black	—	Ea	11.10	—	11.10	—
12" x 12", Balmoral Red	—	Ea	11.10	—	11.10	—
Natural gauged slate floor tile. Rustic earth tones. Indoor or outdoor use. See Ceramic Tile Installation for tile setting costs, including tile backer, thin-set mortar or adhesive and grout.						
12" x 12", Autumn	—	Ea	2.20	—	2.20	—
12" x 12", Earth	—	Ea	1.97	—	1.97	—
12" x 12", Golden White Quartz	—	Ea	3.05	—	3.05	—
16" x 16", Earth	—	Ea	4.95	—	4.95	—
Natural stone floor tile. See Ceramic Tile Installation for tile setting costs, including tile backer, thin-set mortar or adhesive and grout.						
12" x 12", Absolute Black	—	Ea	6.06	—	6.06	—
12" x 12", Caffe Mist	—	Ea	3.54	—	3.54	—
12" x 12", Malaga Beige	—	Ea	5.56	—	5.56	—
12" x 12", Panetella Brown	—	Ea	3.54	—	3.54	—

Paver Tile

	Craft@Hrs	Unit	Material	Labor	Total	Sell
Saltillo floor tile. 12" x 12". Refined natural Saltillo clay tile. Slight rustic appearance with subtle color variations. Rounded edges. Indoor or outdoor use. See Ceramic Tile Installation for tile setting costs, including tile backer, thin-set mortar or adhesive and grout.						
Handcrafted, regular	—	Ea	.68	—	.68	—
Handcrafted, super	—	Ea	.85	—	.85	—
Oil sealed	—	Ea	1.31	—	1.31	—
Quarrybasics® Mayflower Red quarry tile, U.S. Ceramic Tile. Indoor or outdoor use. 6" x 6" x 1/2" thick. Relieved edges. Four tiles per square foot before cutting waste and breakage. Low absorption. See Ceramic Tile Installation for tile setting costs, including tile backer, thin-set mortar or adhesive and grout.						
6" x 6" x 1/2" flat tile	—	SF	.43	—	.43	—
6" bullnose trim	—	Ea	1.72	—	1.72	—
6" bullnose corner	—	Ea	2.51	—	2.51	—
6" round top cove base	—	Ea	1.63	—	1.63	—

	Craft@Hrs	Unit	Material	Labor	Total	Sell

Brick paver, Magnolia Brick and Tile. 4" x 8" x 1/2". Magnolia Brick Pavers install like ceramic tile, but look and perform like real brick floors. Case of pavers covers approximately 10 square feet. See Ceramic Tile Installation for tile setting costs, including tile backer, thin-set mortar or adhesive and grout.

	Craft@Hrs	Unit	Material	Labor	Total	Sell
Old Kentucky	—	SF	1.77	—	1.77	—
Old South	—	SF	2.07	—	2.07	—
Savannah	—	SF	2.08	—	2.08	—
St. Louis	—	SF	2.08	—	2.08	—

Tumbled Marble Tile Flooring

4" x 4" tumbled stone tile, Jeffrey Court. Case of 9 tiles covers 1 square foot. See Ceramic Tile Installation for tile setting costs, including tile backer, thin-set mortar or adhesive and grout.

	Craft@Hrs	Unit	Material	Labor	Total	Sell
Botticino	—	SF	5.41	—	5.41	—
Light Travertine	—	SF	5.41	—	5.41	—
Roso Perlina	—	SF	5.42	—	5.42	—
Travertino Noce	—	SF	5.42	—	5.42	—

6" x 6" tumbled slate tile, Jeffrey Court. Case of 4 tiles covers 1 square foot. See Ceramic Tile Installation for tile setting costs, including either thin-set mortar or adhesive and grout.

	Craft@Hrs	Unit	Material	Labor	Total	Sell
Sequoia	—	SF	5.01	—	5.01	—

Tumbled marble listello tile, Westminster Ceramics. For walls or floors. Stone marble from Italy. Per linear foot of tile before waste and breakage. See Ceramic Tile Installation for tile setting costs, including either thin-set mortar or adhesive and grout.

	Craft@Hrs	Unit	Material	Labor	Total	Sell
3" x 12", Chiaro Rosso	—	Ea	5.19	—	5.19	—
3" x 12", Chiaro Rosso Verde	—	Ea	4.94	—	5.94	—
3" x 12", Noce Chiaro	—	Ea	5.14	—	5.14	—
4" x 12", Bottichino, Scabas	—	Ea	5.74	—	5.74	—
4" x 12", Chiaro Rosso	—	Ea	5.76	—	5.76	—
4" x 12", Chiaro Rosso Verde	—	Ea	5.78	—	5.78	—
4" x 12", Noce	—	Ea	5.66	—	5.66	—

Ceramic Tile Threshold, Saddle

Carrera sill and saddle. 5/8" thick.

	Craft@Hrs	Unit	Material	Labor	Total	Sell
4" x 24", double Hollywood saddle	TL@.687	Ea	9.99	23.10	33.09	56.30
5" x 30", Hollywood saddle	TL@.687	Ea	20.20	23.10	43.30	73.60
5" x 36", double Hollywood saddle	TL@.687	Ea	30.00	23.10	53.10	90.30
5" x 37", sill	TL@.687	Ea	15.90	23.10	39.00	66.30
5" x 73", sill	TL@.687	Ea	29.90	23.10	53.00	90.10
6" x 37", sill	TL@.687	Ea	20.60	23.10	43.70	74.30

Marble threshold. A natural marble threshold provides a transition from one flooring surface to another or from one room to another.

	Craft@Hrs	Unit	Material	Labor	Total	Sell
36" x 2" x 3/8", gray	TL@.687	Ea	6.53	23.10	29.63	50.40
36" x 2" x 3/8", white	TL@.687	Ea	8.41	23.10	31.51	53.60
36" x 2" x 5/8", white	TL@.687	Ea	9.64	23.10	32.74	55.70
36" x 4" x 3/8", white	TL@.687	Ea	8.61	23.10	31.71	53.90
36" x 4" x 5/8", white, gray veins	TL@.687	Ea	9.41	23.10	32.51	55.30

	Craft@Hrs	Unit	Material	Labor	Total	Sell
Ceramic threshold. Off-white.						
1/4" x 36"	TL@.515	Ea	15.40	17.30	32.70	55.60
1/4" x 72"	TL@.730	Ea	34.20	24.60	58.80	100.00
1/2" x 36"	TL@.515	Ea	17.00	17.30	34.30	58.30
1/2" x 72"	TL@.730	Ea	35.50	24.60	60.10	102.00
3/4" x 36"	TL@.515	Ea	27.40	17.30	44.70	76.00
3/4" x 72"	TL@.730	Ea	49.40	24.60	74.00	126.00
Ceramic window sill.						
4" x 37" x 3/4"	TL@.687	Ea	15.00	23.10	38.10	64.80
4" x 54" x 3/4"	TL@.687	Ea	26.10	23.10	49.20	83.60
4" x 74" x 3/4"	TL@.730	Ea	35.90	24.60	60.50	103.00
5" x 37" x 3/4"	TL@.687	Ea	22.60	23.10	45.70	77.70
5" x 54" x 3/4"	TL@.687	Ea	32.00	23.10	55.10	93.70
5" x 74" x 3/4"	TL@.730	Ea	37.60	24.60	62.20	106.00
6" x 37" x 3/4"	TL@.687	Ea	26.20	23.10	49.30	83.80
6" x 54" x 3/4"	TL@.687	Ea	39.60	23.10	62.70	107.00
6" x 74" x 3/4"	TL@.730	Ea	47.60	24.60	72.20	123.00

Ceramic Field Wall Tile

4-1/4" x 4-1/4" ceramic wall tile, U.S. Ceramic Tile. For full-wall interior residential, commercial, and institutional applications. Matte glaze suitable for residential countertops and light-duty floors. Designer's choice glaze. A case of 80 tiles covers 10 square feet, and a case of 120 tiles covers 15 square feet. Add for cutting waste and breakage. See Ceramic Tile Installation for tile setting costs, including tile backer, thin-set mortar or adhesive and grout.

	Craft@Hrs	Unit	Material	Labor	Total	Sell
Black	—	SF	3.43	—	3.43	—
Burgundy	—	SF	2.97	—	2.97	—
Crystal Bone	—	SF	2.51	—	2.51	—
Crystal White	—	SF	2.51	—	2.51	—
Fawn Beige, bright glaze	—	SF	1.72	—	1.72	—
Flaxseed	—	SF	1.72	—	1.72	—
Gold Dust, bright glaze	—	SF	1.72	—	1.72	—
Granite	—	SF	1.80	—	1.80	—
Kelly	—	SF	2.97	—	2.97	—
Medium Blue, bright glaze	—	SF	1.72	—	1.72	—
Moonstone Bone, matte glaze	—	SF	1.71	—	1.71	—
Navy	—	SF	3.59	—	3.59	—
Pink	—	SF	1.62	—	1.62	—
Royal Cobalt	—	SF	2.97	—	2.97	—
Snow White, bright glaze	—	SF	1.43	—	1.43	—
Sterling, bright glaze	—	SF	1.72	—	1.72	—
White, matte glaze	—	SF	1.71	—	1.71	—
Windrift White, matte glaze	—	SF	2.20	—	2.20	—
Yellow	—	SF	1.72	—	1.72	—

4-1/4" x 4-1/4" ceramic single bullnose tile, U.S. Ceramic Tile. Per trim piece. Add for cutting waste and breakage. See Ceramic Tile Installation for tile setting costs, including tile backer, thin-set mortar or adhesive and grout.

	Craft@Hrs	Unit	Material	Labor	Total	Sell
Fawn Beige	—	Ea	.93	—	.93	—
Flaxseed	—	Ea	.91	—	.91	—
Gold Dust, bright glaze	—	Ea	1.06	—	1.06	—
Granite	—	Ea	.73	—	.73	—
Medium Blue, bright glaze	—	Ea	.93	—	.93	—
Moonstone, matte glaze	—	Ea	.93	—	.93	—
Pink	—	Ea	.93	—	.93	—

	Craft@Hrs	Unit	Material	Labor	Total	Sell
Sterling	—	Ea	.92	—	.92	—
White, matte glaze	—	Ea	.91	—	.91	—
Yellow	—	Ea	.89	—	.89	—

4-1/4" x 4-1/4" ceramic bullnose corner tile, U.S. Ceramic Tile. Per trim piece. Add for cutting waste and breakage. See Ceramic Tile Installation for tile setting costs, including tile backer, thin-set mortar or adhesive and grout.

	Craft@Hrs	Unit	Material	Labor	Total	Sell
Crystal White	—	Ea	1.75	—	1.75	—
Fawn Beige	—	Ea	1.53	—	1.53	—
Flaxseed	—	Ea	1.53	—	1.53	—
Gold Dust, bright glaze	—	Ea	1.62	—	1.62	—
Granite	—	Ea	1.53	—	1.53	—
Medium Blue, bright glaze	—	Ea	1.53	—	1.53	—
Moonstone, matte glaze	—	Ea	1.49	—	1.49	—
Pink	—	Ea	1.53	—	1.53	—
Sterling	—	Ea	1.53	—	1.53	—
White, matte glaze	—	Ea	3.35	—	3.35	—
Yellow	—	Ea	1.50	—	1.50	—

Ceramic sink rail, U.S. Ceramic Tile. Per trim piece. Add for cutting waste and breakage. See Ceramic Tile Installation for tile setting costs, including tile backer, thin-set mortar or adhesive and grout.

	Craft@Hrs	Unit	Material	Labor	Total	Sell
1-5/8" x 2-1/4", corner, Moonstone	—	Ea	6.70	—	6.70	—
2" x 2", corner, Bone, bright glaze	—	Ea	3.59	—	3.59	—
2" x 6", Bone, bright glaze	—	Ea	1.82	—	1.82	—
2" x 6", Moonstone	—	Ea	2.00	—	2.00	—
2" x 6", White, matte glaze	—	Ea	2.02	—	2.02	—

2" x 2" ceramic surface cap corner, U.S. Ceramic Tile. Per trim piece. Add for cutting waste and breakage. See Ceramic Tile Installation for tile setting costs, including tile backer, thin-set mortar or adhesive and grout.

	Craft@Hrs	Unit	Material	Labor	Total	Sell
Black	—	Ea	.96	—	.96	—
Bone, bright glaze	—	Ea	.68	—	.68	—
Burgundy	—	Ea	.98	—	.98	—
Fawn Beige	—	Ea	.68	—	.68	—
Flaxseed	—	Ea	.70	—	.70	—
Gold Dust, bright glaze	—	Ea	.70	—	.70	—
Granite	—	Ea	.69	—	.69	—
Kelly	—	Ea	.95	—	.95	—
Medium Blue, bright glaze	—	Ea	.67	—	.67	—
Moonstone, matte glaze	—	Ea	1.45	—	1.45	—
Navy	—	Ea	.94	—	.94	—
Pink	—	Ea	.48	—	.48	—
Sterling, bright glaze	—	Ea	.70	—	.70	—
White, bright glaze	—	Ea	.55	—	.55	—
White, matte glaze	—	Ea	1.45	—	1.45	—
Yellow	—	Ea	.65	—	.65	—

2" x 6" ceramic surface cap, U.S. Ceramic Tile. Per trim piece. Add for cutting waste and breakage. See Ceramic Tile Installation for tile setting costs, including tile backer, thin-set mortar or adhesive and grout.

	Craft@Hrs	Unit	Material	Labor	Total	Sell
Black, Designer's Choice glaze	—	Ea	1.06	—	1.06	—
Bone, bright glaze	—	Ea	.79	—	.79	—
Burgundy, Designer's Choice glaze	—	Ea	1.05	—	1.05	—
Fawn Beige	—	Ea	.74	—	.74	—
Flaxseed	—	Ea	.80	—	.80	—
Gold Dust, bright glaze	—	Ea	.73	—	.73	—
Granite	—	Ea	.53	—	.53	—

	Craft@Hrs	Unit	Material	Labor	Total	Sell
Kelly, Designer's Choice glaze	—	Ea	1.05	—	1.05	—
Medium Blue, bright glaze	—	Ea	.74	—	.74	—
Moonstone, matte glaze	—	Ea	.71	—	.71	—
Navy, Designer's Choice glaze	—	Ea	1.02	—	1.02	—
Pink	—	Ea	.71	—	.71	—
White, matte glaze	—	Ea	.71	—	.71	—
Yellow	—	Ea	.74	—	.74	—

Gold Rush® glazed ceramic tile, Dal-Tile. Glazed paver with unique textured surface. Durable and easy to maintain. Ideal for floors, walls and countertops. Provides good slip, wear and stain resistance. No sealers or waxes needed. One case covers 11 square feet.

	Craft@Hrs	Unit	Material	Labor	Total	Sell
6" x 6", Goldust	—	SF	2.80	—	2.80	—
6" x 12", Goldust	—	SF	3.30	—	3.30	—

6" x 6" ceramic wall tile, U.S. Ceramic Tile. For full-wall interior residential, commercial, and institutional applications. Bright glaze suitable for residential countertops and light-duty floors. Case of 50 tiles covers 12.5 square feet. Add for cutting waste and breakage. See Ceramic Tile Installation for tile setting costs, including tile backer, thin-set mortar or adhesive and grout.

	Craft@Hrs	Unit	Material	Labor	Total	Sell
Biscuit	—	SF	1.94	—	1.94	—
White	—	SF	1.94	—	1.94	—

6" x 6" ceramic residential and commercial wall tile, American Marazzi. For interior residential and commercial wall applications and for residential bathroom floor applications only. PEI Class 2, light traffic. Case of 48 tiles covers 12 square feet.

	Craft@Hrs	Unit	Material	Labor	Total	Sell
Explorer Atlantis, 6" x 6", per SF	—	SF	2.81	—	2.81	—
2" x 2" angle, each	—	Ea	2.57	—	2.57	—
2" x 6" single bullnose, each	—	Ea	2.35	—	2.35	—
Explorer Columbia, 6" x 6", per SF	—	SF	2.19	—	2.19	—
2" x 2" angle, each	—	Ea	2.49	—	2.49	—
2" x 6" single bullnose, each	—	Ea	2.24	—	2.24	—
Explorer Gemini, 6" x 6", per SF	—	SF	2.19	—	2.19	—
2" x 2" angle, each	—	Ea	2.49	—	2.49	—
2" x 6" single bullnose, each	—	Ea	1.17	—	1.17	—
Explorer Voyager, 6" x 6", per SF	—	SF	2.34	—	2.34	—
2" x 2" angle, each	—	Ea	2.49	—	2.49	—
2" x 6" single bullnose, each	—	Ea	2.35	—	2.35	—

6" x 8" ceramic wall tile, U.S. Ceramic Tile. For full-wall interior residential, commercial, and institutional applications. Matte glaze suitable for residential countertops and light-duty floors.

	Craft@Hrs	Unit	Material	Labor	Total	Sell
Beige	—	SF	2.29	—	2.29	—
Biscuit, bright glaze	—	SF	1.68	—	1.68	—
Blush	—	SF	2.29	—	2.29	—
Sierra Gray, bright glaze	—	SF	1.99	—	1.99	—
Sierra Sand, bright glaze	—	SF	1.99	—	1.99	—
White, bright glaze	—	SF	1.53	—	1.53	—

Tile edge protection, Schluter Systems. Finishes and protects the edges of tiled surfaces. Provides a smooth transition between surfaces of the same height or different heights. Per 8'2-1/2" length.

	Craft@Hrs	Unit	Material	Labor	Total	Sell
1/4", aluminum/bright brass	—	Ea	16.06	—	16.06	—
5/16", aluminum	—	Ea	8.31	—	8.31	—
5/16", aluminum/brass, round edge	—	Ea	18.50	—	18.50	—
5/16", aluminum/chrome, round edge	—	Ea	17.90	—	17.90	—
5/16", aluminum/gold	—	Ea	13.10	—	13.10	—
3/8", aluminum	—	Ea	8.88	—	8.88	—
3/8", aluminum/brass, round edge	—	Ea	20.80	—	20.80	—

	Craft@Hrs	Unit	Material	Labor	Total	Sell
3/8", aluminum/gold	—	Ea	14.40	—	14.40	—
3/8", brass	—	Ea	27.80	—	27.80	—
3/8", solid brass	—	Ea	15.10	—	15.10	—
1/2", aluminum	—	Ea	9.66	—	9.66	—

Tile reducer strip, Schluter Systems. Protective edge profile with a sloped transition. Use for higher tiled surfaces to lower carpet, vinyl or wood.

	Craft@Hrs	Unit	Material	Labor	Total	Sell
5/16" x 4', aluminum/bright gold	—	Ea	6.30	—	6.30	—
5/16" x 8', aluminum/bright brass	—	Ea	13.50	—	13.50	—
5/16" x 8', anodized aluminum	—	Ea	14.30	—	14.30	—
3/8" x 4', aluminum/bright brass	—	Ea	18.10	—	18.10	—
3/8" x 8', anodized aluminum	—	Ea	14.40	—	14.40	—
3/8" x 4', aluminum/bright gold	—	Ea	6.70	—	6.70	—
3/8" x 4', brass	—	Ea	21.70	—	21.70	—

Mosaic Tile

Octagon mosaic dot ceramic tile, Dal-Tile. Face-mounted 12" x 12" sheets. One sheet covers one square foot before cutting waste and breakage. See Ceramic Tile Installation for tile setting costs, including tile backer, thin-set mortar or grout.

	Craft@Hrs	Unit	Material	Labor	Total	Sell
2", burgundy matte with white	—	SF	2.82	—	2.82	—
2", white with black	—	SF	2.83	—	2.83	—
2", white with white	—	SF	2.83	—	2.83	—

Octagon mosaic dot ceramic tile. Face-mounted 12" x 12" sheets. One sheet covers one square foot before cutting waste and breakage. See Ceramic Tile Installation for tile setting costs, including tile backer, thin-set mortar or grout.

	Craft@Hrs	Unit	Material	Labor	Total	Sell
2-1/4", white with black	—	SF	2.33	—	2.33	—
2-1/4", white with green	—	SF	2.11	—	2.11	—
2-1/4", white with white	—	SF	2.44	—	2.44	—

Accent mosaic ceramic tile, Dal-Tile. Face-mounted 12" x 12" sheets. One sheet covers one square foot before cutting waste and breakage. See Ceramic Tile Installation for tile setting costs, including tile backer, thin-set mortar or grout.

	Craft@Hrs	Unit	Material	Labor	Total	Sell
2" x 2", mottled sandalwood	—	SF	7.90	—	7.90	—

Ceramic Tile Fixtures

Ceramic tile corner shelf, Professional, Lenape. Genuine porcelain. Mastic installation.

	Craft@Hrs	Unit	Material	Labor	Total	Sell
4-1/2" x 4-1/2" x 2-1/4", bone	TL@.500	Ea	9.17	16.80	25.97	44.10
4-1/2" x 4-1/2" x 2-1/4", white	TL@.500	Ea	9.17	16.80	25.97	44.10
7-1/4" x 7-1/4" x 3", bone	TL@.500	Ea	20.40	16.80	37.20	63.20
7-1/4" x 7-1/4" x 3", white	TL@.500	Ea	19.30	16.80	36.10	61.40

Ceramic tile soap dish, Dal-Tile. Mastic installation.

	Craft@Hrs	Unit	Material	Labor	Total	Sell
4" x 6", almond	TL@.250	Ea	9.68	8.41	18.09	30.80
4" x 6", white	TL@.250	Ea	9.67	8.41	18.08	30.70

Ceramic tile toilet tissue paper holder, Professional, Lenape. 4-1/4" x 6", Genuine porcelain. Mastic installation.

	Craft@Hrs	Unit	Material	Labor	Total	Sell
Bone	BC@.300	Ea	6.62	10.80	17.42	29.60
White	BC@.300	Ea	5.60	10.80	16.40	27.90

	Craft@Hrs	Unit	Material	Labor	Total	Sell
Ceramic tile toothbrush and tumbler holder, Professional, Lenape. 4-1/4" x 4-1/4". Genuine porcelain. Mastic installation.						
Bone	TL@.250	Ea	5.60	8.41	14.01	23.80
White	TL@.250	Ea	5.60	8.41	14.01	23.80
Ceramic tile 24" towel bar assembly, Professional, Lenape. Includes bar and posts. High-fired porcelain. Coordinates with most ceramic tile. Mastic installation.						
Bone	BC@.400	Ea	9.17	14.40	23.57	40.10
White	BC@.400	Ea	9.17	14.40	23.57	40.10
5-piece ceramic tile porcelain bath fixture set, Lenape. Includes soap dish, toothbrush holder, tub soap dish, toilet paper holder and 24" towel bar and posts. Genuine porcelain. Made in USA. Lifetime warranty. Mastic installation.						
Bone	BC@.500	Ea	25.00	18.00	43.00	73.10
White	BC@.500	Ea	25.00	18.00	43.00	73.10
SunTouch® ceramic tile floor warming system, Watts Radiant. Electric floor warming for tile and stone floors. Dual-insulated heating cable woven into thin tile warming mats that install in thin-set mortar below tile and stone. All mats are 120 volts AC and deliver 12 watts per square foot. All sizes are 30" wide by 4', 6', 8', and 10' long. Warms the floor and keeps it dry. Adds warmth to the floor, as well as the entire room. The SunTouch® system consists of an electric mat (or mats) and a control (either programmable or non-programmable). The system comes with an in-depth installation manual and installation video. The mat installs easily — just roll it out on the subfloor or backerboard and secure it with the double-sided tape provided, or just staple it down. Then apply the mortar.						
4' x 30" mat	—	Ea	132.00	—	132.00	—
6' x 30" mat	—	Ea	150.00	—	150.00	—
8' x 30" mat	—	Ea	193.00	—	193.00	—
10' x 30" mat	—	Ea	202.00	—	202.00	—
Floor warming system thermostat, Watts Radiant. Multi-function thermostat. Senses the floor temperature via a remote floor sensor. Includes a large digital display showing floor temperature, comfort temperature, and set-back temperature. Built-in 15 amp relay can control up to 150 square feet of SunTouch (15 amp load). With ground-fault circuit interrupter. Select both comfort setting and set-back setting at the touch of a button.						
Non-programmable	—	Ea	101.00	—	101.00	—
Programmable	—	Ea	142.00	—	134.00	—

Tile Adhesive and Grout

	Craft@Hrs	Unit	Material	Labor	Total	Sell
Semco anti-fracture tile membrane. Protects tile installations from minor substrate movement. Provides crack resistance. Can set tile directly over membrane. Easy-to-use, one-step system. Covers 40 to 45 square feet per gallon when applied with a 3/16" x 1/4" V-notch trowel.						
1 Gallon	—	Ea	46.90	—	46.90	—
RedGard™ waterproofing and anti-fracture membrane, Custom® Building Products. Elastomeric waterproofing and anti-fracture membrane for interior or exterior commercial and residential tile and stone installations. Can also be used as a slab-on-grade moisture barrier under resilient or wood flooring. Applied with roller, trowel or sprayer. Reduces crack transmission in ceramic tile and stone floors.						
Gallon	—	Ea	46.90	—	46.90	—
VersaBond® bonding mortar, Custom® Building Products. All-purpose tile mortar for walls and floors. Trowels smoothly and cures fast, even in cold weather. Just add water and mix; no additives needed.						
Gray, 50 pounds	—	Ea	14.30	—	14.30	—
White, 50 pounds	—	Ea	14.80	—	14.80	—

	Craft@Hrs	Unit	Material	Labor	Total	Sell
Medium-bed mortar, Custom® Building Products. For installing large tile over 12" x 12" tile or natural stone. To 3/4" thick on horizontal applications. Meets ANSI A118.4.						
Gray, 50 pounds	—	Ea	12.00	—	12.00	—
CustomBlend® thin-set mortar, Custom® Building Products. Economical bonding adhesive for most basic job tile installations. Good for floor installations and Saltillo or clay pavers. Offers good open time, pot life and adjustment time to apply and adjust tiles. Interior or exterior use. Mix with acrylic mortar admix for best results when setting dense tile or setting over hard-to-bond-to surfaces such as laminates, existing ceramic tile and plywood.						
Gray, 50 pounds	—	Ea	5.99	—	5.99	—
White, 50 pounds	—	Ea	8.00	—	8.00	—
100 percent solids epoxy mortar, Custom® Building Products. For use where high chemical resistance is required, as in commercial kitchens, dairies, and animal hospitals. Pure, 3-component epoxy mortar with high bond and compressive strengths. Provides exceptional resistance to chemicals and staining. Water washable. Also for installation of true green marble.						
5-gallon pail	—	Ea	71.10	—	71.10	—
FlexBond® premium flexible bonding mortar, Custom® Building Products. For tiling over surfaces subject to minor movement and hard-to-bond-to surfaces such as laminates, ceramic tile and plywood. Allows tiling over cracks up to 1/16" (2 mm) without repairs. Just add water and mix. No additives needed.						
Gray, 50 pounds	—	Ea	28.50	—	28.50	—
White, 50 pounds	—	Ea	31.60	—	31.60	—
CustomFloat® bedding mortar, Custom® Building Products. A lightweight Portland-cement-based, pre-blended mortar for use as a bedding or brown coat. Very low shrinkage, exceptional bond strength. For floors, walls and countertops. Excellent for overhead work such as showers, arches and coves.						
50 pounds	—	Ea	15.30	—	15.30	—
Porcelain mortar, Custom® Building Products. For setting impervious porcelain, glass mosaics and large tile to concrete. Also used over radiant heating systems.						
50 pounds	—	Ea	27.50	—	27.50	—
Marble and granite mortar mix, Custom® Building Products. For setting marble, granite, other natural stone, large modular tile, Saltillo pavers and other tiles requiring a medium-bed mortar. Can be applied up to 3/4" (19 mm) thick. Polymer-modified for high bond strength. Interior or exterior.						
White, 50 pounds	—	Ea	23.20	—	23.20	—
Marble and granite stone-setting adhesive, Custom® Building Products. Formulated specifically for setting moisture-sensitive stones such as green and black marble and granite. Fast-setting, self-drying technology inhibits warping and staining that can occur with traditional thin-set mortars. Grout in two hours. No additives required. Interior use only.						
12.5 pounds	—	Ea	14.20	—	14.20	—
Glass block mortar mix, Custom® Building Products. For installing all types of glass block. Suitable for new installations or to repair old mortar joints. Pre-mixed.						
White, 50 pounds	—	Ea	12.20	—	12.20	—
PremiumPlus® thin-set mortar, Custom® Building Products. Extra high bond strength for most basic job tile installations. Offers maximum open time, pot life and extra long working time.						
Gray, 50 pounds	—	Ea	13.30	—	13.30	—
White, 50 pounds	—	Ea	15.40	—	15.40	—
Tile repair mortar, Custom® Building Products. A polymer-modified mortar with maximum bond strength. Use to replace broken interior or exterior tile. Ready to grout in 24 hours.						
White, 1 pound	—	Ea	4.44	—	4.44	—

	Craft@Hrs	Unit	Material	Labor	Total	Sell
Underwater tile repair mortar, Custom® Building Products. Replace broken tile underwater without draining fountains, pools or spas. Mix with water for immediate use.						
1 pound	—	Ea	8.85	—	8.85	—
Acrylic mortar admix, Custom® Building Products. Greatly increases bond strength, water-resistance and impact resistance of non-polymer modified thin-sets, mortars and grouts. Use with PremiumPlus®, MasterBlend™ and CustomBlend® thin-sets, Portland cement mortar mixes and non-polymer modified grouts. Recommended to use with non-polymer modified thin-set mortars when setting vitreous, dense tile or setting over hard-to-bond-to surfaces such as laminates, existing ceramic tile and plywood.						
3 quarts	—	Ea	9.17	—	9.17	—
2-1/2 gallons	—	Ea	20.40	—	20.40	—
AFM® anti-fracture membrane kit, ProtectoWrap. Peel-and-stick 40 mm membrane. For indoor use under thin-set tile, stone and marble for crack suppression. Includes primer, knife, and instruction sheet.						
12" x 25' roll, per square foot	—	SF	1.51	—	1.51	—
5' wide roll, per square foot	—	SF	1.27	—	1.27	—
Ceramic tile adhesive caulk, Tile Perfect. For tile, tubs, showers, and countertops. Mold and mildew resistant. Remains flexible and paintable.						
5.5 ounce tube	—	Ea	4.05	—	4.05	—
White dry tile grout (non-sanded), Custom® Building Products. Portland-cement-based grout for interior/exterior floor and wall tiles with narrow joints up to 1/8" (3 mm). For soft-glazed, high-glazed and marble tile subject to scratching by sanded grouts.						
1 pound	—	Ea	2.00	—	2.00	—
5 pounds	—	Ea	5.07	—	5.07	—
10 pounds	—	Ea	6.95	—	6.95	—
25 pounds	—	Ea	9.66	—	9.66	—
Grout colorant, Custom® Building Products. Renews or changes grout color, interior or exterior. Seals grout joints and evens color. Water-based and fade-resistant. Available in 10 popular Polyblend® grout colors.						
1/2 pint, most colors	—	Ea	12.20	—	12.20	—
Saltillo grout mix, Custom® Building Products. Pre-blended mix of Portland cement, silica sand and pigments. For any interior or exterior tile with wide grout joints, 1/2" to 1-1/4". Recommend mixing with acrylic mortar admix diluted 1:1 with water for higher strength.						
Gray, tan or red, 50 pounds	—	Ea	9.15	—	9.15	—
Sanded tile grout, Weco. Provides a smooth, colorfast joint for wall and floor installations.						
Most colors, 25 pounds	—	Ea	9.79	—	9.79	—
Polyblend® ceramic tile caulk, Custom® Building Products. Fills and repairs joints around sinks, tubs, showers and countertops. Siliconate acrylic formula.						
Sanded, various colors, 5.5 ounces	—	Ea	3.75	—	3.75	—
High-gloss finish/sealer, Miracle Sealants Co. Medium- to high-gloss interior surface sealer for Saltillo tile, terra cotta tile, unglazed ceramic tile, brick, slate and concrete. One gallon covers 500 to 1000 square feet.						
Quart	—	Ea	10.20	—	10.20	—
Gallon	—	Ea	25.50	—	25.50	—
Sealer's Choice 15 Gold, Aqua Mix. Water-based sealer leaves a no-sheen natural look. Maximum stain protection in food preparation areas, on porous tile and natural stone. One gallon covers 300 to 1500 square feet.						
Pint	—	Ea	35.00	—	35.00	—
Quart	—	Ea	50.00	—	50.00	—
Gallon	—	Ea	136.00	—	136.00	—

	Craft@Hrs	Unit	Material	Labor	Total	Sell
Tile gloss and seal. Two coats protect from grease, food and oil. For unglazed Mexican paver, terrazzo, flagstone, and Saltillo. Clear interior and exterior tile sealer. Non-yellowing, non-flammable, low odor. Dries in 8 to 24 hours. Water clean up. Wax can be applied over product. One gallon covers 125 to 300 square feet.						
Gallon	—	Ea	35.70	—	35.70	—
2.5 gallons	—	Ea	89.30	—	89.30	—
Grout sealer. Water-based sealer resists water, oil and acid-based contaminants. Repels food, dirt and grease. Does not change the appearance of grout. Inhibits mildew and bacteria. For interior and exterior use.						
Pint	—	Ea	10.10	—	10.10	—
Ditra detaching membrane, Schluter Systems. Polyethylene membrane with a grid structure of square, cut-back cavities on a base of anchoring fleece. Forms an uncoupling, waterproofing, and vapor pressure equalization layer. Installed in binding mortar; spread with a notched trowel. Apply tile over Ditra using thin-set mortar.						
Detaching membrane	—	SF	1.75	—	1.75	—

Carpet

Includes sweeping a prepared surface, setting carpet tack strip, rebond pad, unrolling 12'- to 15'-wide carpet, measuring, marking, cutting, trimming one edge, hot melt tape on seams and disposal of debris.

	Craft@Hrs	Unit	Material	Labor	Total	Sell
Minimum quality olefin, 25 to 35 ounce, with 1/2" (3 lb. density) pad	BF@.150	SY	18.30	4.91	23.21	39.50
Standard quality polyester or olefin, 35 to 50 ounce, with 1/2" (6 lb. density) pad	BF@.160	SY	24.30	5.24	29.54	50.20
Better quality nylon, 50 (plus) ounce, with 1/2" (6 lb. density) pad	BF@.170	SY	36.40	5.56	41.96	71.30
Wool carpet, 36 ounce medium traffic	BF@.300	SY	42.50	9.82	52.32	88.90
Wool carpet, 42 ounce heavy traffic	BF@.300	SY	54.70	9.82	64.52	110.00
Sisal carpet, no pad	BF@.130	SY	12.10	4.25	16.35	27.80
Foam-backed olefin carpet, no pad	BF@.130	SY	8.45	4.25	12.70	21.60
Add for typical furniture moving	BF@.060	SY	—	1.96	1.96	3.33
Add for waterfall (box steps) stairways	BF@.190	Step	—	6.22	6.22	10.60
Add for wrapped steps (open riser), sewn	BF@.250	Step	—	8.18	8.18	13.90
Add for steps with sewn edge on one side	BF@.310	Step	—	10.10	10.10	17.20
Add for circular stair steps, typical cost	BF@.440	Step	—	14.40	14.40	24.50
Add for carpet edge binding	BF@.062	LF	—	2.03	2.03	3.45
Rebond 1/2" thick 4-pound pad only		SY	2.44	—	2.44	—
Rebond 1/2" thick 6-pound pad only		SY	4.64	—	4.64	—
Carpet tile, 22" x 23", with adhesive	BF@.022	SF	3.20	.72	3.92	6.66
Peel-and-stick carpet tile, 12" x 12"	BF@.020	SF	1.12	.65	1.77	3.01
Indoor/outdoor carpet, 16 ounce olefin	BF@.025	SF	.56	.82	1.38	2.35

Kitchens 12

Living area in a home can be divided into three categories: the private area (bedrooms and bathrooms), the relaxing area (living room, dining room, den) and the work area (kitchen and laundry).

This work area is more efficient when the kitchen is like the hub of the wheel. The dining area should be only a few steps away. The entrance off the garage should be nearby in another direction to make unloading groceries easier. The laundry room should be another spoke off the hub. Anyone preparing meals in the kitchen should be able to monitor the washer and dryer. Storage space (or a pantry) should be close at hand. But remember that a kitchen is a work area. Traffic from the rear entry or laundry shouldn't pass through the kitchen work triangle — range-to-refrigerator-to-sink.

In the 1960s and 1970s, designers favored small kitchens on the theory that smaller work triangles saved steps and reduced fatigue. The proliferation of appliances at the end of the 20th century made small kitchens seem cramped. Modern kitchen counters need space for juicers, extractors, grinders, mixers, toasters, dispensers, coffee makers, microwave ovens, electric pots, grills, griddles and more.

Avoid setting a door that swings *into* a kitchen. Doors interfere with use of appliances, cabinets and countertops. If a door is essential, consider a pocket or folding door. Provide enough glass area to make the kitchen a light and cheerful place. Time spent in this work area is more enjoyable when there's a window over the kitchen sink with a view. To make food preparation easier for outdoor living, install a large sliding window over the sink or counter. Extend the window sill 14" beyond the window to form a level shelf for passage of plates or a food tray between interior and exterior.

Every kitchen is also a family meeting place. Consider combining work with pleasure: Merge the kitchen (work area) with the family room (relaxing area) by removing a partition or by adding extra living space. Figure 12-1 shows a kitchen and family room combined with an island cook center that doubles as a breakfast bar for casual dining.

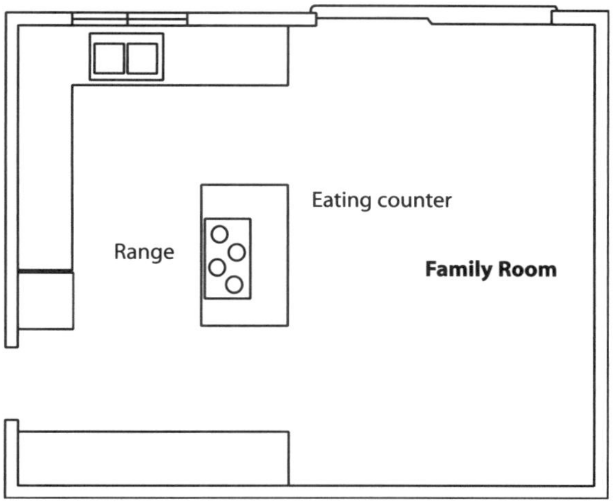

Figure 12-1
Island counter dividing kitchen and family room

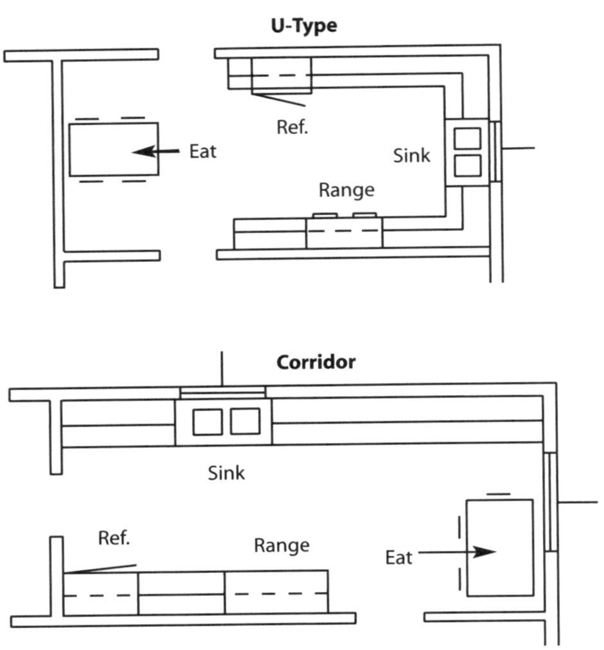

Figure 12-2

Kitchen arrangements

Food usually moves from the refrigerator to the sink to the range. That's the work triangle. Figure 12-2 shows four common kitchen floor plans: "U", "L", corridor and sidewall kitchens. The work triangle is smallest in "U" and corridor kitchens. Use a sidewall kitchen when space is very limited. The "L" arrangement is best for a square kitchen that includes an eating area with a table.

Cabinets and Counters

Here's a yardstick to judge whether there's enough cabinet and counter space in a home. Figures are in linear feet of counter or cabinet face per thousand square feet of floor in the living area (everything under the roof except the garage, soffit, patio and porch). Allow a little more counter and cabinet space in a small home (under 1,200 square feet). A very large home (over 4,000 square feet) may need less space than indicated below. Measure linear feet of counter at the front or back edge, whichever is shorter.

- *Counter.* 8 linear feet per 1,000 square feet of living area
- *Base cabinet.* 7 linear feet per 1,000 square feet of living area
- *Wall cabinet.* 6 linear feet per 1,000 square feet of living area

Notes: Base cabinet length assumes one sink base cabinet. Wall cabinets can include vertical storage cabinets (such as for trays and cookie sheets) above a built-in oven. Count utility cabinets (6' high or more) as both base and wall cabinet. Credit a corner rotating shelf (Lazy Susan) base cabinet with 2 linear feet of base cabinet, regardless of the actual width.

- *Drawer base cabinet.* one per 10 linear feet of base cabinet
- *Counter width and height.* 25" wide x 36" high
- *Dishwasher.* 24" wide x 35" high
- *Trash compactor.* 15" wide x 35" high

Modern kitchens include a work desk with connections for electric service, telephone, computer and TV. Minimum desk width is about 3', with a book shelf (for cook books) above the desk. Minimum desk cutout (seating area) is 24". Install a 14"- to 20"-wide drawer base cabinet at the right of the cutout. Desk depth can be 25", the same as normal counter depth. Desk height should be 28". Provide task lighting above the desk.

Refreshing the Look

Kitchen cabinets seldom wear out before they go out of style. If kitchen cabinet space is adequate and well arranged, refinishing cabinets and adding new hardware may be enough. Replace roller catches with magnetic hardware. The magnet attaches to the cabinet interior and the complementary metal plate fits on the door interior. New knobs or pulls (handles) complete the makeover.

For a more modern look, replace cabinet doors and drawer fronts on the existing cabinet frames. Consider flush doors with concealed (European) hinges, as in Figure 12-3. Cabinet doors fit edge to edge for a continuous panel effect. Finger slots at the edge of doors and drawers make cabinet knobs and pulls unnecessary.

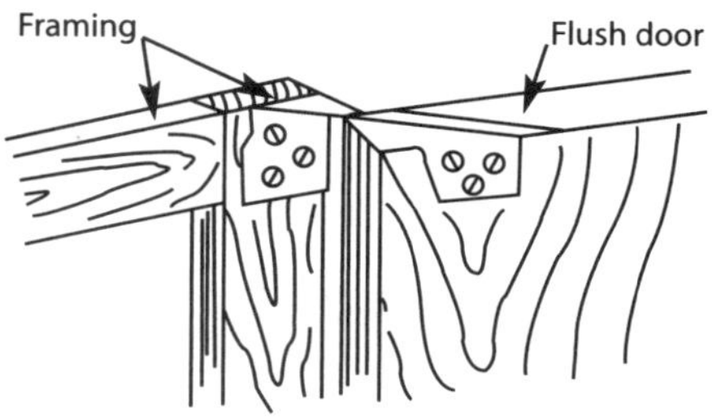

Figure 12-3
Concealed hinge used with flush cabinet door

Countertops are subject to considerable wear and often need to be replaced before the cabinets. Installing a simulated stone countertop will modernize any kitchen. Figure 12-4 shows standard counter and cabinet dimensions. Counter height is usually 36". Counter depth is usually 25".

Cabinet Costs

Cabinets are like furniture. Prices vary widely. For example, a drawer with hardwood rails, plywood bottom and dovetail joints will cost considerably more than a particleboard drawer with stapled butt joints. Better-quality custom and semi-custom cabinets are made from 3/4" or 1/2" furniture-grade plywood covered with hardwood veneer. Drawers have full-extension roller hardware. Less-expensive cabinets are made from 3/8" or 1/2" particleboard, usually with a melamine coating.

"Ready to assemble" cabinets cost the least and are sold primarily in do-it-yourself outlets. Options are limited and units require about an hour of assembly before they are ready to install.

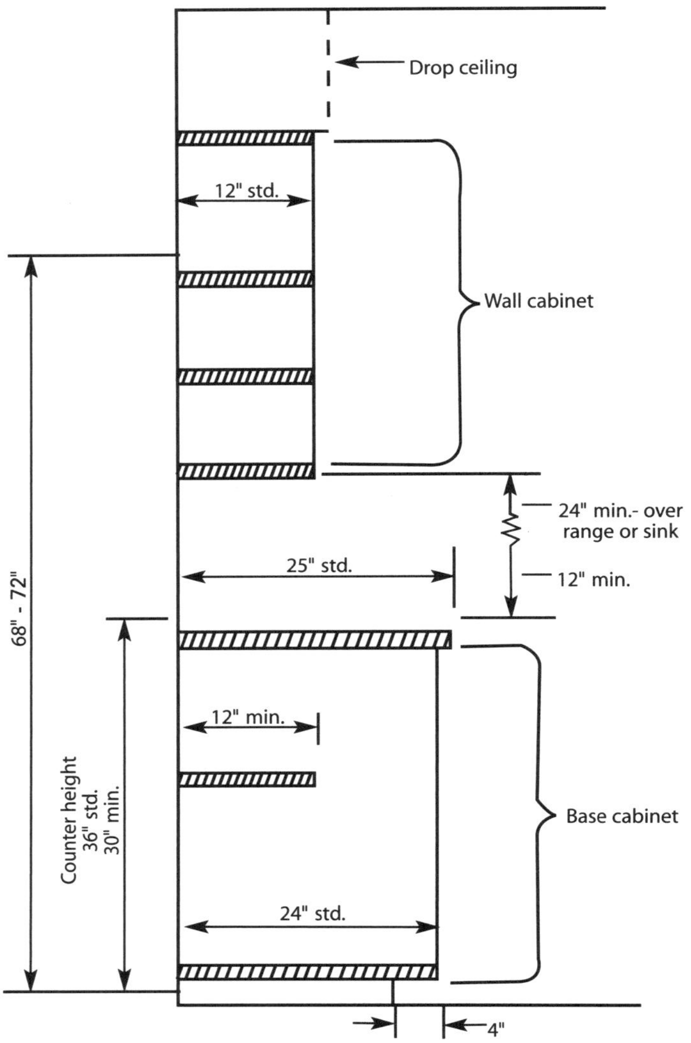

Figure 12-4

Kitchen cabinet standards

Cabinet costs listed in this chapter are based on good quality stock units such as those offered by American Woodmark, IXL (Triangle Pacific), Kabinart, Merillat, Mills Pride, Prestige, and Thomasville. Semi-custom cabinets will cost about 50 percent more. Manufacturers of semi-custom cabinets include Brandom, Decora, Diamond, Kemper, KraftMaid, Schrock, Shenandoah and Yorktowne. Costs for true custom cabinets will be about double the figures listed later in this chapter. Vendors of true custom cabinets include Crystal, J.H. Brubaker, DBS, Fieldstone, Neff, Omega, Plato, Poggenpohl, Rutt, Snaidero, SieMatic, Studio Becker, Wood-Mode, and any cabinet shop in your community.

The advantage of custom-made cabinets is flexibility in style, finish, size and design. For example, custom cabinets can be made in any width or height to fit any kitchen. Stock cabinets come in widths that increment 3 inches at a time, such as 15", 18", 21" and 24". Most installations will require a filler strip to extend the line of cabinets to exactly the right length. The disadvantage of custom cabinets is price — about double the cost of stock cabinets. Incidentally, nearly all cabinets are priced with the screws, hinges, rails and guides needed to finish the job. But door and drawer pulls and knobs are generally left to the discretion of the homeowner.

Cabinet prices in this chapter reflect what most home improvement contractors install — stock cabinets with flat panel faces, picture frame molding or a simple design. Doors and drawers with grooves, raised panels, bead or elaborate molding will cost more. Full overlay doors (installed with Euro hinges mortised into the interior cabinet wall) hide nearly all the cabinet frame and will cost more than traditional doors with hinges mounted on the exterior. Doors and drawers set inside the cabinet face frame *(inset)* will cost even more than full overlay doors and drawers.

About 60 percent of the cost of most cabinets is the wood itself. So you can expect to pay more for cabinets made with exotic wood veneers such as cherry, hickory, alder, redwood or teak on a plywood base. Oak, birch, maple and pine are the most common wood species used for cabinets. The least expensive cabinets have a melamine or plastic laminate surface on a particleboard core. Cabinets surfaced in stainless steel are among the most expensive. Some older homes have steel cabinets that were popular in the middle of the 20th century. Steel cabinets are still made, but their primary use is in hospitals and laboratories rather than in homes.

Wood cabinets are stained to bring out the warmth of the wood grain and then sealed for moisture protection. Install unfinished cabinets if your client wants to match existing woodwork in the home. But be prepared to do a lot of sanding and hand rubbing to produce a finish equal to the best of stock cabinets. Many cabinet vendors offer special custom glazes and layered finishes at extra cost — usually about 10 percent more than standard stained cabinets.

Kitchen Remodeling Checklist

- ❐ *Cabinets* — Soffits, countertops, size of backsplash, accommodating a dishwasher?
- ❐ *Ceiling* — Repaired, acoustic tile, suspended ceiling, drywall
- ❐ *Electrical* — See the Kitchen Electrical Service Checklist for new outlets needed
- ❐ *Floor* — Underlayment, vinyl tile, sheet vinyl, ceramic or clay tile
- ❐ *Heating and cooling* — Ducts or radiators relocated
- ❐ *Doors* — Change in location or swing. New trim
- ❐ *Plumbing* — New venting, gas line relocated, change sink location, dishwasher
- ❐ *Structural changes* — Partitions removed or moved
- ❐ *Wall repairs* — Ceramic tile, paint, wallpaper, paneling
- ❐ *Windows* — Relocated, increase or decrease in size. Storm windows
- ❐ *Ventilation* — Kitchen exhaust hood, ducted or ductless

Kitchen Appliances

Most home improvement contractors avoid reselling kitchen appliances that only need to be plugged in. Every homeowner knows where to get a good deal on a refrigerator. Instead of quoting prices, include a cash allowance for appliances in the contract price. For example, your contract might include the following language:

> *This agreement includes an allowance of $ _______ for the following appliances:*
>
> ____
>
> *Owner agrees to bear the cost of appliances that exceed this allowance. Contractor agrees to furnish and install all gas, water, drain, vent and electrical lines required for operation of these appliances. Contractor is not responsible for installation, service or maintenance of these appliances.*

Built-in appliances, such as garbage disposers, cooktops and wall ovens are an exception to the rule on appliances. These are fixtures, a part of the home itself, and should be included in your bid as a courtesy to the owner. Be sure to specify in the contract the brand and model the customer chose so that's what you price.

You'll never see an old kitchen with adequate electrical service. Upgrading the electrical service is a prime reason for remodeling most kitchens. Consider the following checklist when planning extra runs from the electrical service panel. Note that some building codes require as many as three ground-fault receptacles in the kitchen. If you're adding base cabinets, figure which electrical outlets have to be moved.

Kitchen Electrical Service Checklist

- ❐ Ceiling fixture
- ❐ Ceiling paddle fan
- ❐ Clock in soffit
- ❐ Dishwasher
- ❐ Disposer
- ❐ Electric range
- ❐ Hood over the stove
- ❐ Soffit lighting (fluorescent strip)
- ❐ Light fixture recessed over the sink
- ❐ Light fixture over the desk
- ❐ Microwave oven
- ❐ Oven
- ❐ Refrigerator
- ❐ Small appliance outlets
- ❐ Three ground-fault receptacles
- ❐ Trash compactor
- ❐ TV
- ❐ Wall or ceiling exhaust fan
- ❐ Wall switches

Other Cost Considerations

Extensive kitchen remodeling will usually require HVAC work such as moving duct, registers, grilles, hydronic piping or radiators. Consider also the cost of patching walls or ceilings after HVAC materials have been moved or added. Figure a half-day of work ($150) as the minimum charge for drywall hanging, taping and finishing.

Kitchen Demolition

Remove cabinets. Per linear foot of cabinet face or back (whichever is longer). The cost of removing base cabinets includes the cost of removing the countertop. Add the cost of moving plumbing, electrical and HVAC lines. These figures assume debris is piled on site. No salvage of materials or fixture moving is included.

	Craft@Hrs	Unit	Material	Labor	Total	Sell
Wall cabinets, wood	BL@.250	LF	—	7.26	7.26	12.30
Wall cabinets, metal	BL@.350	LF	—	10.20	10.20	17.30
Base cabinets, wood	BL@.400	LF	—	11.60	11.60	19.70
Base cabinets, metal	BL@.550	LF	—	16.00	16.00	27.20
Remove cabinet door only	BC@.250	Ea	—	9.00	9.00	15.30
Remove and replace cabinet door	BC@.580	Ea	—	20.90	20.90	35.50

Remove kitchen fixtures. These figures include turning off the water, disconnecting the drain, capping the lines, disconnecting faucets and fittings, removing the sink from the base cabinet and piling debris on site. No salvage value assumed. Reduce these costs by 25% if countertop and base cabinets are also being demolished. If a new floor surface has been laid over the original floor, there may be no way to remove undercounter appliances without damaging either the floor or the countertop.

	Craft@Hrs	Unit	Material	Labor	Total	Sell
Self-rimming sink	BL@1.00	Ea	—	29.10	29.10	49.50
Ledge-type or huddee ring sink	BL@1.65	Ea	—	47.90	47.90	81.40
Self-rimming sink with disposer	BL@1.50	Ea	—	43.60	43.60	74.10
Ledge-type or huddee ring sink with disposer	BL@2.00	Ea	—	58.10	58.10	98.80
Garbage disposer	BL@.500	Ea	—	14.50	14.50	24.70
Undercounter dishwasher	BL@1.00	Ea	—	29.10	29.10	49.50
Undercounter trash compactor	BL@.800	Ea	—	23.20	23.20	39.40

Remove ceiling cover. Per square foot of ceiling. Add the cost of removing electrical fixtures. Debris piled on site. No salvage of materials included except as noted.

	Craft@Hrs	Unit	Material	Labor	Total	Sell
Plaster ceiling						
Including lath and furring	BL@.025	SF	—	.73	.73	1.24
Including suspended grid	BL@.020	SF	—	.58	.58	.99
Acoustic tile ceiling						
Including suspended grid	BL@.010	SF	—	.29	.29	.49
Including grid in salvage condition	BL@.019	SF	—	.55	.55	.94
Tile glued or stapled to ceiling	BL@.015	SF	—	.44	.44	.75
Tile on strip furring, including furring	BL@.025	SF	—	.73	.73	1.24
Drywall ceiling						
Nailed or attached with screws to joists	BL@.010	SF	—	.29	.29	.49
Dropped drywall ceiling on wood or metal grid, including grid	BL@.022	SF	—	.64	.64	1.09
Drywall on plaster and lath, including lath and plaster	BL@.041	SF	—	1.19	1.19	2.02

Remove floor cover. Per square yard of floor cover removed. Debris piled on site. No salvage of materials included.

	Craft@Hrs	Unit	Material	Labor	Total	Sell
Ceramic tile, pneumatic breaker	BL@.263	SY	—	7.64	7.64	13.00
Hardwood, nailed	BL@.290	SY	—	8.42	8.42	14.30
Hardwood, glued	BL@.503	SY	—	14.60	14.60	24.80
Linoleum or sheet vinyl	BL@.056	SY	—	1.63	1.63	2.77
Resilient tile	BL@.300	SY	—	8.72	8.72	14.80
Terrazzo	BL@.286	SY	—	8.31	8.31	14.10
Carpet on tack strip	BL@.028	SY	—	.81	.81	1.38
Bonded floor, scraped	BL@.135	SY	—	3.92	3.92	6.66

	Craft@Hrs	Unit	Material	Labor	Total	Sell
Remove wall cover. Per square foot of wall. Debris piled on site.						
Drywall nailed or attached with screws to joists	BL@.010	SF	—	.29	.29	.49
Drywall on plaster and lath, including lath and plaster	BL@.041	SF	—	1.19	1.19	2.02
Plywood or insulation board	BL@.018	SF	—	.52	.52	.88
Plaster and lath	BL@.025	SF	—	.73	.73	1.24
Beaded partitioning dado	BL@.025	SF	—	.73	.73	1.24
Soffits. These figures assume a 12" wide by 12" high soffit or drop ceiling (as in Figure 12-4) installed above wall cabinets. Per linear foot of soffit measured on the longest side. Demolition figures assume debris is piled on site. Figure a half-day of work ($150) as the minimum charge for drywall or plaster finish.						
Remove wood frame soffit with drywall or plaster finish	BL@.059	LF	—	1.71	1.71	2.91
Install wood frame soffit with drywall finish	BC@.471	LF	3.24	17.00	20.24	34.40
Install wood frame soffit with rock lath and plaster finish	BC@.600	LF	2.98	21.60	24.58	41.80
Illuminated soffit with ceiling strips, battens and luminous ceiling panel	BC@.318	LF	4.61	11.50	16.11	27.40
Undercounter refrigerator	BL@1.00	Ea	—	29.10	29.10	49.50
Pendent, wall or recessed light fixture	BL@.590	Ea	—	17.10	17.10	29.10

Kitchen Cabinets

For detailed coverage of custom-built cabinets, see *National Framing & Finish Carpentry Estimator*, http://CraftsmanSiteLicense.com.

	Craft@Hrs	Unit	Material	Labor	Total	Sell
Unfinished oak cabinet door only. Ready to stain. With hardware. Labor cost assumes exact fit replacement door with standard hinges.						
10" x 22"	BC@.250	Ea	20.00	9.00	29.00	49.30
10" x 28"	BC@.250	Ea	21.80	9.00	30.80	52.40
13" x 13"	BC@.250	Ea	16.60	9.00	25.60	43.50
13" x 22"	BC@.250	Ea	21.80	9.00	30.80	52.40
13" x 28"	BC@.250	Ea	24.40	9.00	33.40	56.80
14-1/2" x 22"	BC@.250	Ea	23.20	9.00	32.20	54.70
14" x 28"	BC@.250	Ea	24.60	9.00	33.60	57.10
16" x 13"	BC@.250	Ea	19.30	9.00	28.30	48.10
16" x 22"	BC@.250	Ea	20.60	9.00	29.60	50.30
16" x 28"	BC@.250	Ea	23.05	9.00	32.05	54.50
22" x 22"	BC@.250	Ea	31.90	9.00	40.90	69.50
22" x 28"	BC@.250	Ea	31.90	9.00	40.90	69.50
Unfinished oak cabinet drawer front only. Ready to stain. With hardware.						
10" x 5-3/4"	BC@.250	Ea	7.08	9.00	16.08	27.30
13" x 5-3/4"	BC@.250	Ea	7.43	9.00	16.43	27.90
16" x 5-3/4"	BC@.250	Ea	8.28	9.00	17.28	29.40
22" x 5-3/4"	BC@.250	Ea	10.90	9.00	19.90	33.80
Unfinished oak cabinet door trim. Ready to stain.						
3" x 30" filler strip	BC@.500	Ea	9.14	18.00	27.14	46.10
3" x 36" valance	BC@.250	Ea	17.20	9.00	26.20	44.50
4-1/2" x 96" toe kick	BC@.750	Ea	8.16	27.00	35.16	59.80
Unfinished oak cabinet end panel. Ready to stain.						
11" x 30", wall cabinet	BC@.200	Ea	7.14	7.20	14.34	24.40
24" x 34-1/2", base cabinet	BC@.200	Ea	11.22	7.20	18.42	31.30
24" x 84", utility cabinet	BC@.200	Ea	21.40	7.20	28.60	48.60

Unfinished Birch Kitchen Cabinets

	Craft@Hrs	Unit	Material	Labor	Total	Sell
Unfinished birch base cabinets. 34-1/2" high, 24" deep. Birch face frame and door frame. Ready to stain. With hardware.						
12" wide, 1 door	BC@.640	Ea	42.60	23.00	65.60	112.00
15" wide, 1 door	BC@.640	Ea	57.20	23.00	80.20	136.00
18" wide, 1 door	BC@.770	Ea	65.50	27.70	93.20	158.00
18" wide, 3 drawer	BC@.770	Ea	70.20	27.70	97.90	166.00
24" wide, 2 doors	BC@.770	Ea	78.50	27.70	106.20	181.00
24" wide, 3 drawer	BC@.910	Ea	109.00	32.80	141.80	241.00
30" wide, 2 doors	BC@.910	Ea	90.50	32.80	123.30	210.00
36" wide, 2 doors	BC@.910	Ea	102.00	32.80	134.80	229.00
36" wide, blind corner base, 1 door	BC@.910	Ea	138.00	32.80	170.80	290.00
36" wide, sink base, 1 door	BC@.770	Ea	97.30	27.70	125.00	213.00
48" wide, sink base, 2 doors	BC@.770	Ea	119.00	27.70	146.70	249.00
60" wide, sink base, 2 doors	BC@.910	Ea	141.00	32.80	173.80	295.00
Unfinished birch wall cabinets. 12" deep. Birch face frame and door frame. Ready to stain. With hardware.						
12" wide x 30" high, 1 door	BC@.640	Ea	28.00	23.00	51.00	86.70
15" wide x 30" high, 1 door	BC@.640	Ea	37.80	23.00	60.80	103.00
18" wide x 30" high, 1 door	BC@.770	Ea	44.00	27.70	71.70	122.00
24" wide x 30" high, corner, 1 door	BC@.770	Ea	86.80	27.70	114.50	195.00
24" wide x 30" high, 2 doors	BC@.770	Ea	56.70	27.70	84.40	143.00
30" wide x 30" high, 2 doors	BC@.910	Ea	64.10	32.80	96.90	165.00
36" wide x 30" high, 2 doors	BC@.910	Ea	78.10	32.80	110.90	189.00
30" wide x 15" high, 2 doors	BC@.770	Ea	48.10	27.70	75.80	129.00
36" wide x 15" high, 2 doors	BC@.770	Ea	52.70	27.70	80.40	137.00
42" wide x 21" high, 2 doors	BC@.910	Ea	66.30	32.80	99.10	168.00
54" wide x 24" high, laundry, 2 doors	BC@.910	Ea	86.30	32.80	119.10	202.00
Unfinished birch utility cabinet. 2 doors and 2 drawers. 24" deep. Birch face frame and door frame. Ready to stain. With hardware.						
18" wide x 84" high	BC@.910	Ea	183.00	32.80	215.80	367.00
Unfinished birch cabinet end panels.						
12" wide x 30" high	BC@.200	Ea	5.16	7.20	12.36	21.00
24" wide x 35" high	BC@.200	Ea	7.63	7.20	14.83	25.20
24" wide x 84" high	BC@.200	Ea	18.50	7.20	25.70	43.70
Unfinished birch filler strip. Ready to stain.						
3" x 32"	BC@.500	Ea	3.60	18.00	21.60	36.70

	Craft@Hrs	Unit	Material	Labor	Total	Sell

Unfinished Red Oak Kitchen Cabinets

Unfinished red oak base cabinets. Red oak face frames and flat panel. Picture frame doors. 34-1/2" high, 24" deep. Ready to stain. With hardware.

	Craft@Hrs	Unit	Material	Labor	Total	Sell
12" wide, 1 door	BC@.640	Ea	65.20	23.00	88.20	150.00
15" wide, 1 door	BC@.640	Ea	69.50	23.00	92.50	157.00
18" wide, 1 door	BC@.770	Ea	83.10	27.70	110.80	188.00
18" wide 3-drawer base	BC@.770	Ea	132.00	27.70	159.70	271.00
24" wide, 2 doors	BC@.770	Ea	94.40	27.70	122.10	208.00
24" wide 3-drawer base	BC@.770	Ea	147.00	27.70	174.70	297.00
30" wide, 2 doors	BC@.910	Ea	130.00	32.80	162.80	277.00
36" Lazy Susan corner base	BC@.910	Ea	210.00	32.80	242.80	413.00
36" wide, 2 doors	BC@.910	Ea	142.00	32.80	174.80	297.00
36" wide blind base	BC@.910	Ea	135.00	32.80	167.80	285.00
36" wide corner base	BC@.910	Ea	168.00	32.80	200.80	341.00
36" wide sink base, 1 door	BC@.770	Ea	113.00	27.70	140.70	239.00
60" wide sink base, 2 doors	BC@.910	Ea	169.00	32.80	201.80	343.00

Unfinished red oak wall cabinets. Red oak face frames and flat panel. 12" deep. Ready to stain. With hardware.

	Craft@Hrs	Unit	Material	Labor	Total	Sell
12" wide x 30" high, 1 door	BC@.640	Ea	44.70	23.00	67.70	115.00
15" wide x 30" high, 1 door	BC@.640	Ea	51.60	23.00	74.60	127.00
18" wide x 30" high, 1 door	BC@.770	Ea	51.00	27.70	78.70	134.00
21" wide x 30" high, 1 door	BC@.770	Ea	68.50	27.70	96.20	164.00
24" wide x 30" high, 2 doors	BC@.770	Ea	79.80	27.70	107.50	183.00
24" wide x 30" high, diagonal corner	BC@.770	Ea	113.00	27.70	140.70	239.00
30" wide x 30" high, 2 doors	BC@.910	Ea	91.10	32.80	123.90	211.00
36" wide x 30" high, 2 doors	BC@.910	Ea	113.00	32.80	145.80	248.00
30" wide x 15" high, 2 doors	BC@.770	Ea	58.50	27.70	86.20	147.00
33" wide x 15" high, 2 doors	BC@.770	Ea	66.30	27.70	94.00	160.00
36" wide x 15" high, 2 doors	BC@.770	Ea	70.80	27.70	98.50	167.00
54" wide x 24" high, 2 doors, laundry	BC@.910	Ea	113.00	32.80	145.80	248.00

Unfinished red oak utility cabinet. Red oak face frames and flat panel. Ready to stain. With hardware.

	Craft@Hrs	Unit	Material	Labor	Total	Sell
18" wide x 84" high, 24" deep, 1 door	BC@.910	Ea	203.00	32.80	235.80	401.00
24" wide x 84" high, 18" deep, 1 door	BC@.910	Ea	214.00	32.80	246.80	420.00

Unfinished red oak cabinet end panel. Ready to stain.

	Craft@Hrs	Unit	Material	Labor	Total	Sell
11-3/8" wide x 30" high	BC@.200	Ea	10.30	7.20	17.50	29.80
23" wide x 34-1/2" high	BC@.200	Ea	15.30	7.20	22.50	38.30

Unfinished red oak filler strip. Ready to stain.

	Craft@Hrs	Unit	Material	Labor	Total	Sell
3" x 32"	BC@.500	Ea	7.75	18.00	25.75	43.80
6" x 32"	BC@.500	Ea	11.50	18.00	29.50	50.20

Unfinished Natural Oak Kitchen Cabinets

Unfinished natural oak base cabinets. 34-1/2" high, 24" deep. Solid oak face frame and door frame. Ready to stain. With hardware.

	Craft@Hrs	Unit	Material	Labor	Total	Sell
12" wide, 1 door	BC@.640	Ea	68.80	23.00	91.80	156.00
15" wide, 1 door	BC@.640	Ea	75.80	23.00	98.80	168.00
15" wide, 3-drawer base	BC@.770	Ea	112.00	27.70	139.70	237.00
18" wide, 1 door	BC@.770	Ea	79.70	27.70	107.40	183.00
18" wide, 3-drawer base	BC@.770	Ea	123.00	27.70	150.70	256.00
21" wide, 1 door	BC@.770	Ea	103.00	27.70	130.70	222.00
24" wide, 2 doors	BC@.770	Ea	102.00	27.70	129.70	220.00
24" wide, 3-drawer base	BC@.910	Ea	117.00	32.80	149.80	255.00
30" wide, 2 doors	BC@.910	Ea	138.00	32.80	170.80	290.00

	Craft@Hrs	Unit	Material	Labor	Total	Sell
36" wide, 2 doors	BC@.910	Ea	174.00	32.80	206.80	352.00
36" wide, Lazy Susan base	BC@.910	Ea	244.00	32.80	276.80	471.00
36" wide sink base, 1 door	BC@.770	Ea	115.00	27.70	142.70	243.00
48" wide sink base, 2 doors, oak	BC@.910	Ea	183.00	32.80	215.80	367.00
60" wide sink base, 2 doors, oak	BC@.910	Ea	216.00	32.80	248.80	423.00
Corner sink base	BC@.770	Ea	147.00	27.70	174.70	297.00
Unfinished natural oak wall cabinets. 12" deep. Solid oak face frame and door frame. Ready to stain. With hardware.						
12" wide x 30" high, 1 door	BC@.640	Ea	40.60	23.00	63.60	108.00
15" wide x 30" high, 1 door	BC@.640	Ea	48.60	23.00	71.60	122.00
18" wide x 30" high, 1 door	BC@.770	Ea	54.60	27.70	82.30	140.00
18" wide x 36" high, 1 door	BC@.910	Ea	86.60	32.80	119.40	203.00
24" wide x 30" high diagonal corner	BC@.770	Ea	98.50	27.70	126.20	215.00
24" wide x 36" high, 2 doors	BC@.910	Ea	86.90	32.80	119.70	203.00
30" wide x 36" high, 2 doors	BC@.910	Ea	79.50	32.80	112.30	191.00
24" wide x 15" high, 2 doors	BC@.640	Ea	68.50	23.00	91.50	156.00
30" wide x 15" high, 2 doors	BC@.770	Ea	62.20	27.70	89.90	153.00
33" wide x 15" high, 2 doors	BC@.770	Ea	65.70	27.70	93.40	159.00
36" wide x 15" high, 2 doors	BC@.770	Ea	68.40	27.70	96.10	163.00
36" wide blind corner	BC@.910	Ea	108.00	32.80	140.80	239.00
Unfinished natural oak utility cabinet. 2 doors and 2 drawers. 24" deep. Solid oak face frame and door frame. Ready to stain. With hardware.						
18" wide x 84" high	BC@.910	Ea	193.00	32.80	225.80	384.00
Unfinished natural oak washer-dryer wall cabinet. Oak face frame and door frame. Ready to stain. With hardware. 12" deep.						
24" high x 54" wide, 2 doors	BC@.910	Ea	108.00	32.80	140.80	239.00
Unfinished natural oak cabinet end panel. Ready to stain.						
12" wide x 15" high	BC@.200	Ea	4.38	7.20	11.58	19.70
12" wide x 24" high	BC@.200	Ea	6.84	7.20	14.04	23.90
12" wide x 30" high	BC@.200	Ea	5.61	7.20	12.81	21.80
23" wide x 84" high	BC@.200	Ea	32.70	7.20	39.90	67.80
Unfinished natural oak trim. Ready to stain.						
3" x 32" filler strip	BC@.500	Ea	7.08	18.00	25.08	42.60
6" x 32" filler strip	BC@.500	Ea	14.70	18.00	32.70	55.60
36" x 4" valance	BC@.250	Ea	15.70	9.00	24.70	42.00
36" x 8" valance	BC@.250	Ea	32.20	9.00	41.20	70.00

	Craft@Hrs	Unit	Material	Labor	Total	Sell

Finished Kitchen Cabinets

For detailed coverage of finished custom-built cabinets, see *National Framing & Finish Carpentry Estimator*, http://CraftsmanSiteLicense.com.

Arctic white finished base cabinets. Pre-assembled. 3/4"-thick medium-density fiberboard. Door face and drawer fronts vacuum-formed with white thermofoil. Door back is white melamine laminate. 34-1/2" high x 24" deep. With hardware.

	Craft@Hrs	Unit	Material	Labor	Total	Sell
12" wide, 1 door	BC@.640	Ea	118.00	23.00	141.00	240.00
15" wide, 1 door	BC@.640	Ea	122.00	23.00	145.00	247.00
15" wide, 3-drawer base	BC@.770	Ea	195.00	27.70	222.70	379.00
18" wide, 1 door	BC@.770	Ea	133.00	27.70	160.70	273.00
18" wide, 3-drawer base	BC@.770	Ea	224.00	27.70	251.70	428.00
24" wide, 2 doors	BC@.770	Ea	157.00	27.70	184.70	314.00
30" wide, 2 doors	BC@.910	Ea	201.00	32.80	233.80	397.00
36" wide, 2 doors	BC@.910	Ea	247.00	32.80	279.80	476.00
36" wide, easy reach corner	BC@.910	Ea	264.00	32.80	296.80	505.00
36" wide, sink base	BC@.770	Ea	179.00	27.70	206.70	351.00
45" wide, blind corner base	BC@.910	Ea	221.00	32.80	253.80	431.00

Arctic white finished wall cabinets. Pre-assembled. 3/4"-thick medium-density fiberboard. Door face vacuum formed with white thermofoil. Door back is white melamine laminate. 12" deep. With hardware.

	Craft@Hrs	Unit	Material	Labor	Total	Sell
12" wide x 30" high	BC@.640	Ea	91.00	23.00	114.00	194.00
15" wide x 30" high	BC@.640	Ea	98.90	23.00	121.90	207.00
18" wide x 30" high	BC@.770	Ea	104.00	27.70	131.70	224.00
24" wide x 30" high	BC@.770	Ea	119.00	27.70	146.70	249.00
24" wide x 30" high, corner angle cabinet	BC@.770	Ea	191.00	27.70	218.70	372.00
27" wide x 30" high, blind corner cabinet	BC@.910	Ea	155.00	32.80	187.80	319.00
30" wide x 15" high	BC@.770	Ea	100.00	27.70	127.70	217.00
30" wide x 18" high	BC@.770	Ea	119.00	27.70	146.70	249.00
30" wide x 30" high	BC@.910	Ea	148.00	32.80	180.80	307.00
36" wide x 15" high	BC@.910	Ea	113.00	32.80	145.80	248.00
36" wide x 30" high	BC@.910	Ea	165.00	32.80	197.80	336.00

Arctic white finished utility cabinet. Pre-assembled. 3/4"-thick medium-density fiberboard. Door face vacuum formed with white thermofoil. Door back is white melamine laminate. With hardware.

	Craft@Hrs	Unit	Material	Labor	Total	Sell
18" wide x 48" high x 24" deep	BC@.910	Ea	186.00	32.80	218.80	372.00

Arctic white cabinet trim.

	Craft@Hrs	Unit	Material	Labor	Total	Sell
Crown mold, 8' long	BC@.750	Ea	48.30	27.00	75.30	128.00
Filler strip, 3" x 30"	BC@.500	Ea	16.50	18.00	34.50	58.70
Toe kick, 8' long	BC@.750	Ea	21.20	27.00	48.20	81.90

Oak finished base cabinets. Pre-assembled. Raised veneer center panel framed with 3/4"-thick solid oak. Arched top rail. Drawer fronts are 3/4"-thick solid oak. 34-1/2" high, 24" deep. With hardware.

	Craft@Hrs	Unit	Material	Labor	Total	Sell
12" wide, 1 door	BC@.640	Ea	131.00	23.00	154.00	262.00
15" wide, 1 door	BC@.640	Ea	140.00	23.00	163.00	277.00
15" wide, 3-drawer base	BC@.640	Ea	200.00	23.00	223.00	379.00
18" wide, 1 door	BC@.770	Ea	143.00	27.70	170.70	290.00
18" wide, 3-drawer base	BC@.770	Ea	237.00	27.70	264.70	450.00
24" wide, 2 doors	BC@.770	Ea	173.00	27.70	200.70	341.00
30" wide, 2 doors	BC@.910	Ea	217.00	32.80	249.80	425.00
36" wide, 2 doors	BC@.910	Ea	260.00	32.80	292.80	498.00
36" wide, corner base	BC@.770	Ea	186.00	27.70	213.70	363.00
36" wide, easy-reach base	BC@.910	Ea	277.00	32.80	309.80	527.00
45" wide, blind corner base	BC@.910	Ea	235.00	32.80	267.80	455.00

	Craft@Hrs	Unit	Material	Labor	Total	Sell
Oak finished wall cabinets. Pre-assembled. Raised veneer center panel framed with 3/4"-thick solid oak. 12" deep. Arched top rail. With hardware.						
12" wide x 30" high, 1 door	BC@.640	Ea	97.10	23.00	120.10	204.00
15" wide x 30" high, 1 door	BC@.640	Ea	135.00	23.00	158.00	269.00
27" wide x 30" high, corner	BC@.770	Ea	152.00	27.70	179.70	305.00
30" wide x 15" high, 2 doors	BC@.770	Ea	105.00	27.70	132.70	226.00
30" wide x 18" high, 2 doors	BC@.770	Ea	122.00	27.70	149.70	254.00
30" wide x 30" high, 2 doors	BC@.910	Ea	156.00	32.80	188.80	321.00
36" wide x 15" high, 2 doors	BC@.910	Ea	124.00	32.80	156.80	267.00
36" wide x 30" high, 2 doors	BC@.910	Ea	177.00	32.80	209.80	357.00
Oak finished utility top cabinet. Pre-assembled. Raised veneer center panel framed with 3/4"-thick solid oak. 24" deep. Two doors. Arched top rail. With hardware.						
18" wide x 48" high	BC@.910	Ea	209.00	32.80	241.80	411.00
Finished oak cabinet trim.						
Crown mold, 8' long	BC@.750	Ea	54.00	27.00	81.00	138.00
Toe kick, 8' long	BC@.750	Ea	20.80	27.00	47.80	81.30
Universal fill strip, 3" x 30"	BC@.500	Ea	16.40	18.00	34.40	58.50
Maple finished base cabinets. Pre-assembled. Veneer arched raised center panel framed with 2"-wide, 3/4"-thick solid maple. Base cabinet doors have a square raised center panel. Drawer fronts are 3/4"-thick solid maple. 34-1/2" high x 24" deep. With hardware.						
12" wide, 1 door	BC@.640	Ea	140.00	23.00	163.00	277.00
15" wide, 1 door	BC@.640	Ea	148.00	23.00	171.00	291.00
15" wide, 3-drawer base	BC@.640	Ea	210.00	23.00	233.00	396.00
18" wide, 1 door	BC@.770	Ea	158.00	27.70	185.70	316.00
18" wide, 3-drawer base	BC@.770	Ea	223.00	27.70	250.70	426.00
24" wide, 2 doors	BC@.770	Ea	188.00	27.70	215.70	367.00
30" wide, 2 doors	BC@.910	Ea	225.00	32.80	257.80	438.00
36" sink base	BC@.770	Ea	216.00	27.70	243.70	414.00
36" wide, 2 doors	BC@.910	Ea	244.00	32.80	276.80	471.00
45" wide blind base	BC@.910	Ea	248.00	32.80	280.80	477.00
Maple finished wall cabinets. Pre-assembled. Veneer arched raised center panel framed with 2"-wide, 3/4"-thick solid maple. The top rail has a gentle arch. 12" deep. With hardware.						
12" wide x 30" high, 1 door	BC@.640	Ea	102.00	23.00	125.00	213.00
15" wide x 30" high, 1 door	BC@.640	Ea	112.00	23.00	135.00	230.00
18" wide x 30" high, 1 door	BC@.770	Ea	120.00	27.70	147.70	251.00
24" wide x 30" high, 1 door	BC@.770	Ea	146.00	27.70	173.70	295.00
24" wide x 30" high, angle corner	BC@.770	Ea	186.00	27.70	213.70	363.00
27" wide x 30" high, blind corner	BC@.770	Ea	156.00	27.70	183.70	312.00
30" wide x 30" high, 2 doors	BC@.910	Ea	163.00	32.80	195.80	333.00
36" wide x 30" high, 2 doors	BC@.910	Ea	185.00	32.80	217.80	370.00
36" wide x 30" high, easy reach corner	BC@.910	Ea	283.00	32.80	315.80	537.00
30" wide x 15" high, 2 doors	BC@.770	Ea	115.00	27.70	142.70	243.00
36" wide x 15" high, 2 doors	BC@.770	Ea	138.00	27.70	165.70	282.00
30" wide x 18" high, 2 doors	BC@.770	Ea	128.00	27.70	155.70	265.00
Maple finished utility top cabinet. Pre-assembled. Veneer arched raised center panel framed with 2"-wide, 3/4"-thick solid maple. 24" deep. With hardware.						
18" x 48" high	BC@.910	Ea	220.00	32.80	252.80	430.00

	Craft@Hrs	Unit	Material	Labor	Total	Sell
Maple finished cabinet trim.						
Crown molding, 8' long	BC@.750	Ea	54.60	27.00	81.60	139.00
Toe kick, 8' long	BC@.750	Ea	21.30	27.00	48.30	82.10
Universal filler, 3" x 30"	BC@.500	Ea	16.60	18.00	34.60	58.80

Custom Countertops

Laminated plastic countertops. Custom-fabricated straight, U- or L-shaped tops (such as Formica, Textolite or Wilsonart) on a particleboard base. 22" to 25" wide. Cost per LF of longest edge (either back or front) in solid colors with 3-1/2" to 4" backsplash.

	Craft@Hrs	Unit	Material	Labor	Total	Sell
Accent color front edge	B1@.181	LF	29.80	5.89	35.69	60.70
Full wrap (180 degree) front edge	B1@.181	LF	23.40	5.89	29.29	49.80
Inlaid front and back edge	B1@.181	LF	35.20	5.89	41.09	69.90
Rolled drip edge (post formed)	B1@.181	LF	21.80	5.89	27.69	47.10
Square edge, separate backsplash	B1@.181	LF	15.40	5.89	21.29	36.20
Additional costs for laminated countertops						
Add for contour end splash	—	Ea	26.70	—	26.70	—
Add for drilling 3 plumbing fixture holes	—	LS	11.00	—	11.00	—
Add for half round corner	—	Ea	31.70	—	31.70	—
Add for mitered corners	—	Ea	22.00	—	22.00	—
Add for quarter round corner	—	Ea	22.10	—	22.10	—
Add for seamless tops	—	Ea	31.90	—	31.90	—
Add for sink, range or chop block cutout	—	Ea	8.60	—	8.60	—
Add for square end splash	—	Ea	18.60	—	18.60	—
Add for textures, patterns	—	%	20.0	—	—	—

Solid surface countertops. Many brands available, including Corian, Wilsonart and Avenite. Tops are cut from acrylic or polyester-acrylic sheets measuring 1/2" or 3/4" thick, 30" to 36" wide and 12' long. The estimates below include cutting, polishing, sink and range cutouts and a flat polish edge. Color groups are usually offered with distinct edge treatments. Group 1 colors are light shades with little texture and a matte finish. Group 2 colors include darker colors and stone textures. Group 3 colors are extra dark or bright and include high-gloss surfaces. Solid surface countertops are set directly on cabinets with silicone adhesive.

	Craft@Hrs	Unit	Material	Labor	Total	Sell
Color Group 1, Basic edge	B1@.128	SF	50.50	4.16	54.66	92.90
Color Group 1, Custom edge	B1@.128	SF	58.20	4.16	62.36	106.00
Color Group 1, Premium edge	B1@.128	SF	54.00	4.16	58.16	98.90
Color Group 2, Basic edge	B1@.128	SF	56.20	4.16	60.36	103.00
Color Group 2, Custom edge	B1@.128	SF	65.30	4.16	69.46	118.00
Color Group 2, Premium edge	B1@.128	SF	61.20	4.16	65.36	111.00
Color Group 3, Basic edge	B1@.128	SF	65.30	4.16	69.46	118.00
Color Group 3, Custom edge	B1@.128	SF	74.40	4.16	78.56	134.00
Color Group 3, Premium edge	B1@.128	SF	70.40	4.16	74.56	127.00
Add for delivery (typical)	—	Ea	152.00	—	152.00	—
Add for integral sink, one basin	—	Ea	354.00	—	354.00	—
Add for integral sink, two basins	—	Ea	455.00	—	455.00	—
Add for radius or beveled edge	—	LF	7.00	—	7.00	—

	Craft@Hrs	Unit	Material	Labor	Total	Sell

Granite countertops. Granite countertop work is generally done by a specialist. Custom made to order from a template. 22" to 25" wide. Per square foot of top and backsplash overall dimension, including measuring, cutting, and flat-edge polishing. Add the cost of edge detail as itemized below and jobsite delivery. 3/4" granite weighs about 30 pounds per square foot. 1-1/4" granite weighs about 45 pounds per square foot. 3/4" tops should be set on 3/4" plywood and set in silicon adhesive every two feet. 1-1/4" granite tops can be set directly on base cabinets with silicon adhesive.

	Craft@Hrs	Unit	Material	Labor	Total	Sell
Most 3/4" (2cm) granite tops	B1@.128	SF	40.20	4.16	44.36	75.40
Most 1-1/4" (3cm) granite tops	B1@.128	SF	52.50	4.16	56.66	96.30
Add for 1/2" beveled edge	—	LF	7.21	—	7.21	—
Add for backsplash seaming and edging	—	LF	13.20	—	13.20	—
Add for cooktop overmount cutout	—	Ea	133.00	—	133.00	—
Add for delivery (typical)	—	Ea	156.00	—	156.00	—
Add for eased edge	—	LF	5.15	—	5.15	—
Add for electrical outlet cutout	—	Ea	36.30	—	36.30	—
Add for faucet holes	—	Ea	27.10	—	27.10	—
Add for full bullnose edge	—	LF	20.60	—	20.60	—
Add for half bullnose edge	—	LF	10.30	—	10.30	—
Add to remove existing non-ceramic top	BL@.135	SF	—	3.92	3.92	6.66
Add to remove existing ceramic tile top	BL@.230	SF	—	6.68	6.68	11.40
Add for rounded corners	—	Ea	31.10	—	31.10	—
Add for undermount sink, polished	—	Ea	309.00	—	309.00	—

Engineered stone countertops. Custom made from a template. Quartz particles with acrylic or epoxy binder. Trade names include Crystalite, Silestone and Cambria. Per square foot of top and backsplash surface.

	Craft@Hrs	Unit	Material	Labor	Total	Sell
Small light chips, 3/4" square edge	B1@.128	SF	50.50	4.16	54.66	92.90
Large dark chips, 3/4" square edge	B1@.128	SF	56.70	4.16	60.86	103.00
Add for 1-1/2" bullnose edge	—	LF	49.50	—	49.50	—
Add for 1-1/2" ogee edge	—	LF	86.50	—	86.50	—
Add for 3/4" bullnose edge	—	LF	25.90	—	25.90	—
Add for 3/4" ogee edge	—	LF	33.90	—	33.90	—
Add for delivery (typical)	—	Ea	156.00	—	156.00	—
Add for electrical outlet cutout	—	Ea	30.90	—	30.90	—
Add for radius corner or end	—	Ea	191.00	—	191.00	—
Add to remove existing laminated top	BL@.135	Ea	—	3.92	3.92	6.66
Add for sink cutout	—	Ea	259.00	—	259.00	—

Stock Kitchen Countertops

Post-formed double-radius plastic laminate countertop. Standard laminate. 3" backsplash. 25" width. Substrate is 3/4" industrial grade particleboard.

	Craft@Hrs	Unit	Material	Labor	Total	Sell
5' long	BC@1.00	Ea	58.70	36.00	94.70	161.00
6' long	BC@1.20	Ea	68.90	43.20	112.10	191.00
6' long, left-hand miter	BC@1.20	Ea	68.90	43.20	112.10	191.00
6' long, right-hand miter	BC@1.20	Ea	68.90	43.20	112.10	191.00
8' long	BC@1.60	Ea	90.30	57.60	147.90	251.00
10' long	BC@2.00	Ea	113.00	72.00	185.00	315.00
10' long, left-hand miter	BC@1.60	Ea	113.00	57.60	170.60	290.00
10' long, right-hand miter	BC@1.60	Ea	113.00	57.60	170.60	290.00
End cap kit	BC@.250	Ea	12.60	9.00	21.60	36.70
End splash	BC@.250	Ea	14.20	9.00	23.20	39.40

	Craft@Hrs	Unit	Material	Labor	Total	Sell
Post-formed double-radius premium plastic laminate countertop. 3" backsplash. 25" width. Substrate is 3/4" industrial grade particleboard. Exceeds ANSI A161.2 standard for countertops.						
5' long	BC@1.00	Ea	58.08	36.00	94.08	160.00
6' long	BC@1.20	Ea	69..40	43.20	112.20	191.00
6' long, left-hand miter	BC@1.20	Ea	75.70	43.20	118.90	202.00
6' long, right-hand miter	BC@1.20	Ea	75.70	43.20	118.90	202.00
8' long	BC@1.60	Ea	94.70	57.60	152.30	259.00
10' long	BC@2.00	Ea	116.00	72.00	188.00	320.00
10' long, left-hand miter	BC@2.00	Ea	121.00	72.00	193.00	328.00
10' long, right-hand miter	BC@2.00	Ea	121.00	72.00	193.00	328.00
End cap kit	BC@.250	Ea	12.60	9.00	21.60	36.70
End splash	BC@.250	Ea	14.20	9.00	23.20	39.40
Miter bolt kit. Joins 2 mitered tops. Includes 4 miter bolts, packet of glue, and instructions.						
Kit	—	Ea	5.05	—	5.05	—

Cabinet Knobs, Pulls and Hinges

	Craft@Hrs	Unit	Material	Labor	Total	Sell
Round ceramic knob. Quality porcelain. Classic style. Fasteners included.						
1", white	BC@.060	Ea	1.72	2.16	3.88	6.60
1-1/4", white	BC@.060	Ea	1.82	2.16	3.98	6.77
1-3/8", white	BC@.060	Ea	1.91	2.16	4.07	6.92
1-1/2", white	BC@.060	Ea	2.03	2.16	4.19	7.12
1-1/2", white with apple decal	BC@.060	Ea	2.23	2.16	4.39	7.46
1-1/2", white with pear decal	BC@.060	Ea	2.45	2.16	4.61	7.84
1-1/2", white with plum decal	BC@.060	Ea	2.45	2.16	4.61	7.84
Classic cabinet knob. 30mm diameter.						
Black nickel	BC@.060	Ea	3.20	2.16	5.36	9.11
Brushed red antique	BC@.060	Ea	3.39	2.16	5.55	9.44
Matte blue	BC@.060	Ea	3.25	2.16	5.41	9.20
Matte red	BC@.060	Ea	3.25	2.16	5.41	9.20
Pearl gold	BC@.060	Ea	3.07	2.16	5.23	8.89
Satin chrome	BC@.060	Ea	3.09	2.16	5.25	8.93
Brushed antique satin cabinet knob. Striking brushed; lacquered antique finish. 30mm diameter. Fasteners included.						
Bronze	BC@.060	Ea	3.05	2.16	5.21	8.86
Copper, red	BC@.060	Ea	2.93	2.16	5.09	8.65
Nickel	BC@.060	Ea	2.93	2.16	5.09	8.65
Decorative star cabinet knob.						
Black	BC@.060	Ea	3.20	2.16	5.36	9.11
Nickel	BC@.060	Ea	4.26	2.16	6.42	10.90
Pewter	BC@.060	Ea	3.21	2.16	5.37	9.13
Round cabinet knob. Lacquered black chrome finish. Fasteners included.						
1-1/4" diameter	BC@.060	Ea	3.05	2.16	5.21	8.86

	Craft@Hrs	Unit	Material	Labor	Total	Sell
Round knob.						
1-3/8", almond	BC@.060	Ea	1.06	2.16	3.22	5.47
1-3/8", black	BC@.060	Ea	1.00	2.16	3.16	5.37
1-3/8", green	BC@.060	Ea	1.00	2.16	3.16	5.37
1-3/8", speckled black	BC@.060	Ea	1.00	2.16	3.16	5.37
Satin nickel knob. Lacquered, brushed satin finish. Traditional styling. Fasteners included.						
1-1/4", round	BC@.060	Ea	3.05	2.16	5.21	8.86
1-3/4" x 1-1/16", oval	BC@.060	Ea	1.93	2.16	4.09	6.95
Brass-plated cabinet knob. Decorative knob. Lacquered; polished brass plate. Classic style. Fasteners included.						
1" diameter	BC@.060	Ea	3.05	2.16	5.21	8.86
1-1/4" diameter	BC@.060	Ea	3.05	2.16	5.21	8.86
1-1/4" diameter, ringed design	BC@.060	Ea	3.05	2.16	5.21	8.86
Brass-plated knob with insert. Decorative knob. Lacquered; polished brass plate. Decorative beveled edge. Fasteners included. 1-1/4" diameter.						
Frost maple insert	BC@.060	Ea	2.34	2.16	4.50	7.65
High-gloss white ceramic insert	BC@.060	Ea	2.13	2.16	4.29	7.29
Medium oak wood insert	BC@.060	Ea	2.07	2.16	4.23	7.19
Natural wood round knob. Unfinished solid birch. Pre-sanded. Perfect for refinishing. Fasteners included. Limited lifetime warranty.						
1" diameter	BC@.060	Ea	.93	2.16	3.09	5.25
1-1/4" diameter	BC@.060	Ea	.96	2.16	3.12	5.30
1-1/2" diameter	BC@.060	Ea	1.19	2.16	3.35	5.70
1-3/4" diameter	BC@.060	Ea	1.42	2.16	3.58	6.09
2" diameter	BC@.060	Ea	1.61	2.16	3.77	6.41
Circle swirl knob. Distressed finish. Fasteners included. Limited lifetime warranty.						
4/5" x 1-1/3", black finish	BC@.060	Ea	3.53	2.16	5.69	9.67
33mm, rustic pewter patina finish	BC@.060	Ea	3.56	2.16	5.72	9.72
Solid brass and chrome knob. Polished brass with bright chrome for a contemporary look. High-quality lacquered solid brass. Classic style. Fasteners included.						
1-1/4", 2 tone	BC@.060	Ea	5.13	2.16	7.29	12.40
Solid brass knob. High-quality lacquered solid brass. Fasteners included.						
1-3/16" x 3/4", oval	BC@.060	Ea	4.07	2.16	6.23	10.60
1-1/4", round	BC@.060	Ea	4.07	2.16	6.23	10.60
1-3/8", round	BC@.060	Ea	5.09	2.16	7.25	12.30
1-1/2", round, rope design	BC@.060	Ea	4.41	2.16	6.57	11.20
Small forked branch knob.						
Black	BC@.060	Ea	3.42	2.16	5.58	9.49
Antique brass target knob. Round. Provincial ringed design with antiqued finish. Fasteners included.						
1-1/8" diameter	BC@.060	Ea	1.20	2.16	3.36	5.71
1-5/16" diameter	BC@.060	Ea	1.41	2.16	3.57	6.07

	Craft@Hrs	Unit	Material	Labor	Total	Sell
Round backplate. For knobs.						
1-1/4", chrome	BC@.020	Ea	1.71	.72	2.43	4.13
1-1/4", polished brass	BC@.020	Ea	1.72	.72	2.44	4.15
Birdcage knob. 50mm diameter.						
Antique pewter	BC@.060	Ea	3.03	2.16	5.19	8.82
Flat black	BC@.060	Ea	3.03	2.16	5.19	8.82
Birdcage bail pull. 96mm post spacing.						
Antique pewter	BC@.120	Ea	7.25	4.32	11.57	19.70
Flat black	BC@.120	Ea	7.25	4.32	11.57	19.70
Brass pull with oak insert. Medium oak insert. Traditional die-cast spoon-foot design. Matching knob sold separately. Fasteners included. 3" post spacing.						
Oak and antique brass	BC@.120	Ea	3.31	4.32	7.63	13.00
Oak and polished brass	BC@.120	Ea	3.31	4.32	7.63	13.00
Birdcage wire pull. 3" post spacing.						
Flat black	BC@.120	Ea	4.05	4.32	8.37	14.20
Pewter	BC@.120	Ea	4.05	4.32	8.37	14.20
Bow pull. Versatile design. Classic style for kitchen or bath. Fasteners included. 96mm post spacing.						
Lacquered black nickel	BC@.120	Ea	2.64	4.32	6.96	11.80
Lacquered brass	BC@.120	Ea	2.70	4.32	7.02	11.90
Fusilli pull. Shallow S design. Fasteners included. Limited lifetime warranty. 96mm post spacing.						
Black nickel	BC@.120	Ea	3.84	4.32	8.16	13.90
Bright chrome	BC@.120	Ea	4.07	4.32	8.39	14.30
Brushed bronze antique	BC@.120	Ea	4.07	4.32	8.39	14.30
Brushed satin antique red	BC@.120	Ea	4.61	4.32	8.93	15.20
Brushed satin chrome	BC@.120	Ea	4.07	4.32	8.39	14.30
Brushed satin nickel	BC@.120	Ea	4.07	4.32	8.39	14.30
Flat black	BC@.120	Ea	4.07	4.32	8.39	14.30
Pearl gold	BC@.120	Ea	4.74	4.32	9.06	15.40
Satin gold	BC@.120	Ea	4.74	4.32	9.06	15.40
White decorative cabinet pull. Durable epoxy coating. Classic style for kitchen or bath. Fasteners included. 3" post spacing.						
Minaret	BC@.120	Ea	2.54	4.32	6.86	11.70
Spoon-foot	BC@.120	Ea	2.54	4.32	6.86	11.70

	Craft@Hrs	Unit	Material	Labor	Total	Sell
Chrome decorative cabinet pull. Fasteners included. Limited lifetime warranty. 3" post spacing.						
Tapered bow,						
bright lacquered, polished chrome	BC@.120	Ea	2.01	4.32	6.33	10.80
Die-cast, spoon-foot,						
bright polished chrome	BC@.120	Ea	2.44	4.32	6.76	11.50
Ornate beaded,						
lacquered, polished chrome	BC@.120	Ea	2.54	4.32	6.86	11.70
Satin nickel decorative cabinet pull. Lacquered, brushed satin nickel finish. Fasteners included. 3" post spacing.						
Die-cast spoon-foot	BC@.120	Ea	3.05	4.32	7.37	12.50
Fan-design foot	BC@.120	Ea	2.64	4.32	6.96	11.80
Brass-plated pull with ceramic insert. High-gloss non-porous ceramic insert. Traditional die-cast spoon-foot design. 3" post spacing.						
Almond insert	BC@.120	Ea	3.11	4.32	7.43	12.60
White insert	BC@.120	Ea	3.05	4.32	7.37	12.50
S-swirl pull. Distressed die-cast rust-resistant finish. Fasteners included. Limited lifetime warranty. 3" post spacing.						
Pewter, small	BC@.120	Ea	4.15	4.32	8.47	14.40
Solid brass wire pull. 96mm post spacing.						
2-tone	BC@.120	Ea	5.25	4.32	9.57	16.30
Brass-plated spoon-foot pull. Lacquered, polished brass finish. Traditional die-cast spoon-foot design. Fasteners included. 3" post spacing.						
Smooth	BC@.120	Ea	2.86	4.32	7.18	12.20
Wire cabinet pull. Fasteners included. By post spacing.						
3", black nickel	BC@.120	Ea	2.55	4.32	6.87	11.70
3", flat black	BC@.120	Ea	2.07	4.32	6.39	10.90
3", polished chrome	BC@.120	Ea	2.80	4.32	7.12	12.10
3", satin chrome	BC@.120	Ea	2.83	4.32	7.15	12.20
3", solid brass	BC@.120	Ea	2.70	4.32	7.02	11.90
3", white	BC@.120	Ea	2.03	4.32	6.35	10.80
3-1/2", satin chrome	BC@.120	Ea	3.02	4.32	7.34	12.50
3-1/2", white	BC@.120	Ea	1.93	4.32	6.25	10.60
4", white	BC@.120	Ea	2.03	4.32	6.35	10.80
35mm Euro face frame overlay cabinet hinge. For face frame cabinets with 5/8" overlay cabinet doors. Nickel-plated stamped steel cup and zinc die-cast mounting plate. Three-way adjustable. Door opening angle 100 degrees. Fasteners included. Labor assumes cabinet is already cut for this type of hinge.						
Per hinge	BC@.250	Ea	7.13	9.00	16.13	27.40
Pack of 5	BC@1.25	Ea	18.35	45.00	63.35	108.00

	Craft@Hrs	Unit	Material	Labor	Total	Sell
Semi-concealed cabinet hinge. For face frame cabinets using 3/8" inset doors. Fasteners included. Pack of two.						
Antique brass	BC@.250	Ea	2.19	9.00	11.19	19.00
Chrome	BC@.250	Ea	2.52	9.00	11.52	19.60
Polished brass	BC@.250	Ea	2.19	9.00	11.19	19.00
Semi-concealed self-closing cabinet hinge. For face frame cabinets using 3/8" inset doors. Wrought steel. Self-closing spring. Fasteners included.						
Antique brass	BC@.250	Ea	2.83	9.00	11.83	20.10
Black nickel	BC@.250	Ea	2.88	9.00	11.88	20.20
Chrome	BC@.250	Ea	2.85	9.00	11.85	20.10
Polished brass	BC@.250	Ea	2.84	9.00	11.84	20.10
White	BC@.250	Ea	2.85	9.00	11.85	20.10
White, pack of 20	BC@5.00	Ea	19.00	180.00	199.00	338.00

Stainless Steel Sinks

	Craft@Hrs	Unit	Material	Labor	Total	Sell
Stainless steel single bowl sink, Neptune Series, Elkay. Heavy-gauge stainless steel. Four-hole faucet drilling. Faucet and accessories sold separately. Bright satin finish. 25" long x 22" wide.						
5-1/2" deep	P1@.600	Ea	40.90	21.10	62.00	105.00
6" deep	P1@.600	Ea	45.20	21.10	66.30	113.00
7" deep	P1@.600	Ea	68.30	21.10	89.40	152.00
8" deep	P1@.600	Ea	100.00	21.10	121.10	206.00
Stainless steel double bowl sink, Neptune Series, Elkay. Heavy-gauge nickel bearing stainless steel. Full undercoating to deaden sound and prevent condensation. Exclusive hand-mounting fasteners. 3-1/2" drain openings. Brilliant finish with highlights. 33" x 22". Four-hole faucet drilling. Faucet and accessories sold separately.						
5-1/2" deep, double bowl	P1@.600	Ea	35.40	21.10	56.50	96.10
6" deep, double bowl	P1@.600	Ea	62.80	21.10	83.90	143.00
8" deep, double bowl	P1@.600	Ea	101.00	21.10	122.10	208.00
7" deep, small and large bowls	P1@.600	Ea	109.00	21.10	130.10	221.00
Stainless steel single bowl sink, Signature Series, Elkay. Drop-in single bowl sink. Made from nickel bearing type 302 stainless steel. 3" radius vertical corners. 8-1/4"-deep bowl with mirror-finished interior deck. Machine-ground satin-finished bowls with satin highlighted outer rim edge and bowl radius. Fully undercoated for sound deadening. Self-rimming installation. Four holes. Drain openings are 3-1/2".						
25" x 22", 20 gauge	P1@.600	Ea	111.00	21.10	132.10	225.00
25" x 22", 18 gauge	P1@.600	Ea	173.00	21.10	194.10	330.00
33" x 22", 18 gauge	P1@.600	Ea	203.00	21.10	224.10	381.00
Stainless steel double bowl sink, Signature Series, Elkay. Drop-in double bowl sink. Self-rimming installation. 3" radius vertical coved corners. Machine-ground satin-finished bowls and satin highlighted outer rim edge and bowl radius. Mirror-finished interior deck. Sound Guard undercoat prevents condensation and ensures quiet operation. Four holes. 3-1/2" drain openings, 33" long x 22" wide, 8-1/4" deep.						
20 gauge	P1@.600	Ea	203.00	21.10	224.10	381.00
18 gauge	P1@.600	Ea	240.00	21.10	261.10	444.00
Offset appearance	P1@.600	Ea	193.00	21.10	214.10	364.00
Dual level, 7-1/2" deep	P1@.600	Ea	214.40	21.10	235.50	400.00
Stainless steel kitchen sink, Kohler. 20 gauge with sound deadening. Undercoating reduces disposal noise, minimizes condensation and maintains water temperature. Stain- and corrosion-resistant. Self-rimming, four-hole installation. Faucet sold separately.						
25" x 22", single bowl	P1@.600	Ea	146.00	21.10	167.10	284.00
33" x 22", double bowl	P1@.600	Ea	183.00	21.10	204.10	347.00

	Craft@Hrs	Unit	Material	Labor	Total	Sell
Stainless steel double bowl sink, Staccato, Kohler. 8" basin depth. Double equal basins. Four-hole punched. 18 gauge. Faucet sold separately.						
33" x 22"	P1@.600	Ea	254.00	21.10	275.10	468.00
Stainless steel kitchen sink, Banner. Standard-gauge stainless steel. Rounded corners. Self-rimming design. Includes sink clips. Faucet, sprayer, and strainers sold separately. Five-year limited warranty.						
25" x 22", 7" depth, single bowl	P1@.600	Ea	77.40	21.10	98.50	167.00
33" x 22", 6" depth, double buffed bowl	P1@.600	Ea	50.60	21.10	71.70	122.00
33" x 22", 7" depth, double bowl	P1@.600	Ea	70.10	21.10	91.20	155.00
33" x 22", 8" depth, double bowl	P1@.600	Ea	100.00	21.10	121.10	206.00
Sink mounting clips and screws, Elkay. U-channel. For use with Elkay Signature and Neptune sinks.						
Clips	—	Ea	4.06	—	4.06	—

Enameled Steel Sinks

	Craft@Hrs	Unit	Material	Labor	Total	Sell
Charleston double bowl kitchen sink. White enameled steel with four faucet holes. Self-rimming installation. 33" x 22" x 7-1/2".						
White	P1@.750	Ea	104.00	26.40	130.40	222.00

Cast Iron Sinks

	Craft@Hrs	Unit	Material	Labor	Total	Sell
Dumont double bowl cast iron sink, Eljer. White enameled cast iron. 33" x 22" x 8" deep.						
Self-rimming	P1@.750	Ea	174.00	26.40	200.40	341.00
Tile-in installation	P1@.750	Ea	246.00	26.40	272.40	463.00
Mayfield cast iron single bowl sink, Kohler. Self-rimming. 25" x 22" x 8". Four-hole drilling.						
Almond	P1@.750	Ea	250.00	26.40	276.40	470.00
White	P1@.750	Ea	225.00	26.40	251.40	427.00
Efficiency™ double bowl cast iron sink, Kohler. Self-rimming installation. Offset faucet ledge to maximize basin area. 33" x 22" x 6-3/4".						
White	P1@.750	Ea	199.00	26.40	225.40	383.00
Ashland cast iron double bowl sink, Kohler. 33" x 22". Four-hole faucet drilling. Large and medium double basins. Self-rimming.						
White, 7" deep	P1@.750	Ea	178.00	26.40	204.40	347.00

	Craft@Hrs	Unit	Material	Labor	Total	Sell
Hartland™ double bowl cast iron sink, Kohler. Unique design lines. Four-hole. Self-rimming. Deep, glossy color. 33" long x 22" wide x 7-3/4" deep.						
Almond	P1@.750	Ea	356.00	26.40	382.40	650.00
Bisque	P1@.750	Ea	254.00	26.40	280.40	477.00
White	P1@.750	Ea	295.00	26.40	321.40	546.00
Brookfield™ double bowl cast iron sink, Kohler. Deep, glossy color. 33" long x 22" wide. Faucet, sprayer, and strainers sold separately. Self-rimming. Four holes.						
White, 6-3/4" deep	P1@.750	Ea	249.00	26.40	275.40	468.00
White, 8" deep	P1@.750	Ea	246.00	26.40	272.40	463.00
Cast iron double bowl sink. 33" x 22". Four-hole faucet drilling. 9" deep. Interior corners are nearly square. Large and medium bowls.						
White, tile-in	P1@.750	Ea	290.00	26.40	316.40	538.00
White, under counter	P1@.750	Ea	275.00	26.40	301.40	512.00

Composite Sinks

	Craft@Hrs	Unit	Material	Labor	Total	Sell
Silhouette polycast single bowl sink, American Standard. Porcelain enameled Americast. Lighter than cast iron. Silicone sealant supplied. Four faucet holes. Faucet, sprayer, and strainer sold separately. Self-rimming, tile edge or under-the-counter installation. 25" x 22" x 9-1/2" deep.						
Bisque	P1@.750	Ea	178.00	26.40	204.40	347.00
Black	P1@.750	Ea	221.00	26.40	247.40	421.00
Bone	P1@.750	Ea	170.00	26.40	196.40	334.00
White heat	P1@.750	Ea	141.00	26.40	167.40	285.00
Silhouette polycast double bowl sink, American Standard. Porcelain enameled Americast. Lighter than cast iron. Silicone sealant supplied. Four faucet holes. Faucet, sprayer, and strainer sold separately. Self-rimming, tile edge or under-the-counter installation. 33" x 22" x 9-1/2" deep. Insulating material holds water temperature and reduces noise of running water and garbage disposal. Lead-free porcelain enamel finish and scratch-resistant. Faucets, sprayers and strainers sold separately.						
Bisque	P1@.750	Ea	233.00	26.40	259.40	441.00
Bone	P1@.750	Ea	262.00	26.40	288.40	490.00
White heat	P1@.750	Ea	200.00	26.40	226.40	385.00
Single bowl composite sink. Four-hole faucet drilling. Scratch- and stain-resistant. Acrylic, self-rimming design. Faucets, drains and accessories sold separately.						
25" x 22", black	P1@.600	Ea	157.00	21.10	178.10	303.00
Double bowl composite sink, CorStone. Large bowl stretches from the front of the sink to the back rim. A large, 8-1/2"-deep offset bowl in front of the faucet deck is separated by a lowered dam that helps avoid chipping and banging of dishes. For 33" x 22" sink cut-out. Faucets, drains and accessories sold separately. Equal bowls.						
Kendall, white	P1@.750	Ea	111.00	26.40	137.40	234.00
Kennesaw, biscuit	P1@.750	Ea	177.00	26.40	203.40	346.00
Hi/lo double bowl composite sink. 33" x 22" x 8-1/2" deep. Four-hole faucet drilling. Scratch- and stain-resistant. Acrylic, self-rimming design. Faucets, drains and accessories sold separately.						
White	P1@.750	Ea	202.00	26.40	228.40	388.00
Greenwich double bowl composite sink, CorStone. 33" x 22" x 8-1/2" deep. Four-hole faucet drilling. Scratch- and stain-resistant. Acrylic, self-rimming design. Faucets, drains and accessories sold separately. Equal bowls.						
White	P1@.750	Ea	150.00	26.40	176.40	300.00

	Craft@Hrs	Unit	Material	Labor	Total	Sell
Providence 60/40 bowl composite sink, CorStone. 10" main bowl depth. 33" x 22" sink for use with disposals. Easy installation and easy cleaning.						
Royston white	P1@.750	Ea	204.00	26.40	230.40	392.00
Toccoa black	P1@.750	Ea	223.00	26.40	249.40	424.00
Inverness single bowl composite sink, International Thermocast. 25" x 22" x 9" deep. Cutout dimensions 23" x 20" with 1" radius corners. Rich, high gloss finish that's easy to clean with non-abrasive cleaners. The beauty of porcelain with the strength of cast iron, yet lighter weight. Insulated construction to absorb waste disposal noise. Large basins for more workspace. Resists stain, rust, oxidation, chipping, and scratches. Surface can be polished with ordinary liquid sink, countertop, or auto polish. Drop-in ready. Faucets, drains and accessories sold separately.						
Inverness, white, 4" deck	P1@.750	Ea	121.00	26.40	147.40	251.00
Inverness, bone, 4" deck	P1@.750	Ea	148.00	26.40	174.40	296.00
Wellington, white, 5" deck	P1@.750	Ea	124.00	26.40	150.40	256.00
Beaumont double bowl composite kitchen sink, International Thermocast. Equal bowls. The beauty of porcelain with the strength of cast iron, yet lighter weight. Insulated construction to absorb waste disposal noise. Large basins for more workspace. Resists stain, rust, oxidation, chipping, and scratches. Surface can be polished with ordinary liquid sink, countertop, or auto polish. Drop-in ready. No mounting clips required. Certified by the National Association of Home Builders Research Center. 33" x 22" x 8-1/4" deep.						
Bone	P1@.750	Ea	134.00	26.40	160.40	273.00
White	P1@.750	Ea	102.00	26.40	128.40	218.00
Breckenridge double bowl composite sink, International Thermocast. 33" x 22". Large bowl, 9" deep. Small bowl, 7" deep. Cutout dimensions 31" x 20" with 1" radius corners. Drains positioned to maximize workspace. Insulated construction to absorb waste disposal noise. Resists stain, rust, oxidation, chipping, and scratches. Drop-in ready. No mounting clips required.						
Black	P1@.750	Ea	213.00	26.40	239.40	407.00
White	P1@.750	Ea	172.00	26.40	198.40	337.00
Newport double bowl composite sink, International Thermocast. 33" x 22" x 9" deep. Equal bowls. 4-1/4" deck. Four-hole faucet drilling. Self-rimming. Faucets, drains and accessories sold separately.						
Biscuit	P1@.750	Ea	161.00	26.40	187.40	319.00
Bone	P1@.750	Ea	172.00	26.40	198.40	337.00
White	P1@.750	Ea	162.00	26.40	188.40	320.00
Cambridge offset double bowl composite sink, International Thermocast. 33" x 22". Large bowl 10-1/2" deep. Small bowl 8-1/2" deep. Self-rimming. Four-hole faucet drilling. Cutout dimensions 31" x 20" with 1" radius corners. Insulated construction to absorb waste disposal noise. Faucets, drains and accessories sold separately.						
Black	P1@.750	Ea	215.00	26.40	241.40	410.00
White	P1@.750	Ea	162.00	26.40	188.40	320.00

Pull-Out Kitchen Faucets

Kitchen faucet, single-handle, pull-out spout. Deck mount. One- or three-hole installation. Loop handle. Labor includes fasten in place only, and assumes sink is already in place. Installation times will be reduced by half if installing faucet before sink is in place. Add the cost of plumbing rough-in as required. With soap dispenser.						
Chrome finish	P1@1.10	Ea	102.00	38.70	140.70	239.00
Kitchen faucet, single-handle, pull-out spout. Deck mount. One- or three-hole installation. Loop handle. Ceramic disc cartridge. 20-year warranty. Labor includes fasten in place only, and assumes sink is already in place. Installation times will be reduced by half if installing faucet before sink is in place. Add the cost of plumbing rough-in as required.						
Polished chrome finish	P1@1.00	Ea	150.00	35.20	185.20	315.00

	Craft@Hrs	Unit	Material	Labor	Total	Sell
Kitchen faucet, single-handle, pull-out spout, Reliant+™, American Standard. Deck mount. One- or three-hole installation. Lever handle. Ceramic disc cartridge. Spray or aerated stream. 4-1/2" spout. ADA compliant. Labor includes fasten in place only, and assumes sink is already in place. Installation times will be reduced by half if installing faucet before sink is in place. Add the cost of plumbing rough-in as required.						
Bone finish	P1@1.00	Ea	138.00	35.20	173.20	294.00
Chrome with brass accents	P1@1.00	Ea	130.00	35.20	165.20	281.00
Stainless steel	P1@1.00	Ea	193.00	35.20	228.20	388.00
Velvet finish	P1@1.00	Ea	173.00	35.20	208.20	354.00
White finish	P1@1.00	Ea	120.00	35.20	155.20	264.00
Kitchen faucet, single-handle, pull-out spout, Signature, Delta. Deck mount. One- or three-hole installation. Lever handle. Spray or aerated stream. Arched spout provides 9" of reach and 6-1/2" of height. Touch-Clean® wand makes it easy to wipe away residue and mineral deposits, ensuring proper spraying. Lifetime limited warranty. Labor includes fasten in place only, and assumes sink is already in place. Installation times will be reduced by half if installing faucet before sink is in place. Add the cost of plumbing rough-in as required.						
Chrome	P1@1.00	Ea	205.00	35.20	240.20	408.00
Stainless steel	P1@1.00	Ea	313.00	35.20	348.20	592.00
White	P1@1.00	Ea	287.00	35.20	322.20	548.00
Kitchen faucet, single-handle, pull-out hi-arc spout, Glacier Extensa®, Moen. Deck mount. One- or three-hole installation. Lever handle. Spray or aerated stream. Washerless cartridge. IntuiTouch™ wand. Lifetime limited warranty. Labor includes fasten in place only, and assumes sink is already in place. Installation times will be reduced by half if installing faucet before sink is in place. Add the cost of plumbing rough-in as required.						
Chrome finish	P1@1.00	Ea	282.00	35.20	317.20	539.00
Matte black finish	P1@1.00	Ea	388.00	35.20	423.20	719.00
Sand finish	P1@1.00	Ea	299.00	35.20	334.20	568.00
Stainless steel finish	P1@1.00	Ea	328.00	35.20	363.20	617.00
White finish	P1@1.00	Ea	94.00	35.20	129.20	220.00
Kitchen faucet, single-handle, pull-out spout, Colonnade, Moen. Deck mount. One- or three-hole installation. Lever handle. Integral vacuum breaker. Labor includes fasten in place only, and assumes sink is already in place. Installation times will be reduced by half if installing faucet before sink is in place. Add the cost of plumbing rough-in as required.						
Chrome finish	P1@1.00	Ea	282.00	35.20	317.20	539.00
White finish	P1@1.00	Ea	294.00	35.20	329.20	560.00
Kitchen faucet, single-handle, pull-out spout, Genesis, Price Pfister. Deck mount. One- or three-hole installation. Loop handle. Spray or aerated stream. Ceramic disc cartridge. Pforever Pfaucet™ limited lifetime warranty. Labor includes fasten in place only, and assumes sink is already in place. Installation times will be reduced by half if installing faucet before sink is in place. Add the cost of plumbing rough-in as required.						
Stainless steel	P1@1.00	Ea	201.00	35.20	236.20	402.00
White finish	P1@1.00	Ea	198.00	35.20	233.20	396.00
Kitchen faucet, single-handle, pull-out spout, Forte®, Kohler. Labor includes fasten in place only and assumes sink is already in place. Installation times will be reduced by half if installing faucet before sink is in place. Add the cost of plumbing rough-in as required.						
Polished chrome	P1@1.00	Ea	176.00	35.20	211.20	359.00
Stainless steel	P1@1.00	Ea	208.00	35.20	243.20	413.00

	Craft@Hrs	Unit	Material	Labor	Total	Sell
Kitchen faucet, high arch single-handle, Forte®, Kohler. Remote valve and matching deck-mounted spray head. Labor includes fasten in place only and assumes sink is already in place. Installation times will be reduced by half if installing faucet before sink is in place. Add the cost of plumbing rough-in as required.						
Polished chrome	P1@1.10	Ea	175.00	38.70	213.70	363.00
Brushed nickel	P1@1.10	Ea	230.00	38.70	268.70	457.00
Polished chrome with soap dispenser	P1@1.20	Ea	177.00	42.20	219.20	373.00
Brushed nickel with soap dispenser	P1@1.20	Ea	202.00	42.20	244.20	415.00

Two-Handle Kitchen Faucets

	Craft@Hrs	Unit	Material	Labor	Total	Sell
Wall-mount kitchen faucet, two-handle, American Standard. 8" center set. Porcelain lever handles. Lifetime limited warranty on function and finish. Labor includes fasten in place only. Add the cost of plumbing rough-in as required.						
Chrome finish	P1@1.00	Ea	170.00	35.20	205.20	349.00
Kitchen faucet, two-handle. Deck mount. 8" center set. Labor includes fasten in place only, and assumes sink is already in place. Installation times will be reduced by half if installing faucet before sink is in place. Add the cost of plumbing rough-in as required.						
Chrome finish	P1@1.00	Ea	75.00	35.20	110.20	187.00
Kitchen faucet, two-handle, Savannah, Price Pfister. Solid brass construction. Ceramic disc valving. High-arc spout. Dual porcelain handles. Lifetime warranty. Labor assumes sink is already in place. Installation times will be reduced by half if installing faucet before sink is in place.						
No spray, chrome	P1@1.00	Ea	119.00	35.20	154.20	262.00
No spray, gooseneck spout, chrome	P1@1.00	Ea	132.00	35.20	167.20	284.00
With spray, chrome	P1@1.10	Ea	154.00	38.70	192.70	328.00
With spray, stainless steel	P1@1.10	Ea	239.00	38.70	277.70	472.00
Kitchen faucet, two-handle, deck spray, Waterfall, Delta. Deck-mount. Chrome finish. 3-hole installation. Washerless design. Lifetime function and finish warranty. Labor includes fasten in place only, and assumes sink is already in place. Installation times will be reduced by half if installing faucet before sink is in place. Add the cost of plumbing rough-in as required.						
Chrome	P1@1.10	Ea	184.00	38.70	222.70	379.00
White	P1@1.10	Ea	207.00	38.70	245.70	418.00
Kitchen faucet, two-handle, deck spray, Fairfax, Kohler. Deck mount. Labor includes fasten in place only, and assumes sink is already in place. Installation times will be reduced by half if installing faucet before sink is in place. Add the cost of plumbing rough-in as required.						
Polished chrome finish	P1@1.10	Ea	70.00	38.70	108.70	185.00
Kitchen faucet, two-handle, Glacier Bay. Deck mount. 8" center set. Four-hole installation. Acrylic handles. 12-year warranty. Chrome finish. Labor includes fasten in place only, and assumes sink is already in place. Installation times will be reduced by half if installing faucet before sink is in place. Add the cost of plumbing rough-in as required.						
With deck spray	P1@1.10	Ea	50.00	38.70	88.70	151.00
Without spray	P1@1.00	Ea	40.00	35.20	75.20	128.00

Kitchen faucet, two-handle, with spray, Stanadyne. Labor includes fasten in place only, and assumes sink is already in place. Installation times will be reduced by half if installing faucet before sink is in place. Add the cost of plumbing rough-in as required. Chrome finish.

	Craft@Hrs	Unit	Material	Labor	Total	Sell
Acrylic handles	P1@1.10	Ea	85.70	38.70	124.40	211.00
Chrome handles	P1@1.10	Ea	85.70	38.70	124.40	211.00
High arc	P1@1.10	Ea	120.00	38.70	158.70	270.00
Brass accent	P1@1.10	Ea	128.00	38.70	166.70	283.00

Kitchen faucet, two-handle, J A Manufacturing. Deck-mount. 8" center set. Builder's grade. Labor includes fasten in place only, and assumes sink is already in place. Installation times will be reduced by half if installing faucet before sink is in place. Add the cost of plumbing rough-in as required.

	Craft@Hrs	Unit	Material	Labor	Total	Sell
Brushed nickel, 8" hi-rise, with spray	P1@1.10	Ea	89.80	38.70	128.50	218.00
Chrome finish	P1@1.00	Ea	61.00	35.20	96.20	164.00

Kitchen faucet, two-handle, Brass Craft. Deck mount. 8" center set. Labor includes fasten in place only, and assumes sink is already in place. Installation times will be reduced by half if installing faucet before sink is in place. Add the cost of plumbing rough-in as required.

	Craft@Hrs	Unit	Material	Labor	Total	Sell
Chrome finish	P1@1.00	Ea	17.70	35.20	52.90	89.90

Kitchen faucet, two-handle, American Standard. Deck-mount. 8" center set. Four-hole installation. Ceramic disc cartridge. Matching handles and deck spray. Cast brass waterways. 1/2" brass supply shanks. Labor includes fasten in place only, and assumes sink is already in place. Installation times will be reduced by half if installing faucet before sink is in place. Add the cost of plumbing rough-in as required.

	Craft@Hrs	Unit	Material	Labor	Total	Sell
Without spray, chrome	P1@1.00	Ea	83.40	35.20	118.60	202.00
Without spray, satin nickel	P1@1.00	Ea	105.00	35.20	140.20	238.00
With spray, chrome	P1@1.10	Ea	99.60	38.70	138.30	235.00
With spray, satin nickel	P1@1.10	Ea	127.00	38.70	165.70	282.00

Kitchen Sink Accessories

Kitchen sink accessories, American Standard.

	Craft@Hrs	Unit	Material	Labor	Total	Sell
Undercounter mounting kit	—	Ea	12.90	—	12.90	—
Faucet lift block	—	Ea	7.11	—	7.11	—
Sink mount clips, pack of 14	P1@.050	Ea	7.94	1.76	9.70	16.50
Extra-long sink clip set	P1@.050	Ea	5.09	1.76	6.85	11.60
J-channel installation clip set	P1@.050	Ea	3.71	1.76	5.47	9.30

Sink frame, Vance Industries. Watertight seal. No caulk bead. Prevents edge of sink from chipping, cracking or dimpling. 32" long x 21" wide.

	Craft@Hrs	Unit	Material	Labor	Total	Sell
Stainless steel	—	Ea	39.00	—	39.00	—
Stainless steel, S4	—	Ea	45.00	—	45.00	—
White	—	Ea	29.00	—	29.00	—

Air gap, LDR Industries.

	Craft@Hrs	Unit	Material	Labor	Total	Sell
Body only, stainless	P1@.400	Ea	3.43	14.10	17.53	29.80
Cover only, stainless	P1@.400	Ea	3.17	14.10	17.27	29.40
Chrome	P1@.400	Ea	4.86	14.10	18.96	32.20
White finish	P1@.400	Ea	4.86	14.10	18.96	32.20

Air gap cover, LDR Industries.

	Craft@Hrs	Unit	Material	Labor	Total	Sell
Almond finish	P1@.090	Ea	2.82	3.16	5.98	10.20
Chrome finish	P1@.090	Ea	1.73	3.16	4.89	8.31
White finish	P1@.090	Ea	2.47	3.16	5.63	9.57

	Craft@Hrs	Unit	Material	Labor	Total	Sell
Faucet hole cover, LDR Industries. Blanks off deck faucet sink holes not being used.						
Almond finish	P1@.090	Ea	3.00	3.16	6.16	10.50
Chrome finish	P1@.090	Ea	1.50	3.16	4.66	7.92
Chrome finish, nipple type	P1@.090	Ea	2.98	3.16	6.14	10.40
Stainless steel, bolt down	P1@.090	Ea	4.50	3.16	7.66	13.00
White finish	P1@.090	Ea	2.78	3.16	5.94	10.10
Deluxe sink-top soap dispenser. Liquid soap or lotion dispenser. Pump action.						
Nickel finish	P1@.250	Ea	124.00	8.79	132.79	226.00
Black finish	P1@.250	Ea	71.10	8.79	79.89	136.00
Chrome finish	P1@.250	Ea	134.00	8.79	142.79	243.00
Stainless steel finish	P1@.250	Ea	33.40	8.79	42.19	71.70
Plastic sink-top soap dispenser. Plastic. Liquid soap or lotion dispenser. Pump action. Can be refilled from above the sink.						
Almond finish	P1@.250	Ea	23.40	8.79	32.19	54.70
Chrome finish	P1@.250	Ea	105.00	8.79	113.79	193.00
White finish	P1@.250	Ea	23.40	8.79	32.19	54.70

Sink Baskets and Strainer Inserts

	Craft@Hrs	Unit	Material	Labor	Total	Sell
Kitchen sink strainer insert, Jameco.						
Clip on, stainless steel	P1@.070	Ea	3.04	2.46	5.50	9.35
Lock spin, stainless steel	P1@.070	Ea	3.22	2.46	5.68	9.66
Long clip, stainless steel	P1@.070	Ea	3.11	2.46	5.57	9.47
Plastic	P1@.070	Ea	2.02	2.46	4.48	7.62
Plastic basket and strainer insert, LDR Industries. Designer finish. Fits all standard kitchen sinks.						
Almond	P1@.250	Ea	4.90	8.79	13.69	23.30
Almond, garbage disposal	P1@.250	Ea	7.17	8.79	15.96	27.10
Stainless steel	P1@.250	Ea	8.14	8.79	16.93	28.80
White	P1@.250	Ea	6.10	8.79	14.89	25.30
White, garbage disposal	P1@.250	Ea	7.12	8.79	15.91	27.00
Kitchen sink basket and strainer insert, Spin-N-Lock, Jameco. Premium grade stainless steel. Positive lock-spin metal seal basket closure.						
Stainless steel	P1@.250	Ea	14.80	8.79	23.59	40.10
Decorative kitchen sink basket and strainer insert, Jameco. Chip- and stain-resistant Celcon insert and basket. Stainless steel double cup one-piece underbody. Recommended for cast iron, stainless steel and composite sinks. Watertight ball-lok mechanism basket closure.						
White	P1@.250	Ea	20.90	8.79	29.69	50.50
Kitchen sink basket and strainer insert, Jameco. Ball-lok basket closure for a watertight seal. Metal underbody double-cup construction. Chip- and stain-resistant. Celcon basket and strainer insert. Threaded bushing allows color insert to be changed easily.						
Bone finish	P1@.250	Ea	20.50	8.79	29.29	49.80
Kitchen sink basket and strainer insert, Jameco. Polished brass finish. Heavy-duty chrome-plated solid brass body. Short clip basket closure.						
Polished brass	P1@.250	Ea	30.60	8.79	39.39	67.00
Satin nickel	P1@.250	Ea	30.60	8.79	39.39	67.00

	Craft@Hrs	Unit	Material	Labor	Total	Sell
Kitchen sink basket and strainer insert, Plumb Shop. Brass locknut and slip nut. Stainless steel basket. Solid cast brass body. Recommended for Corian sinks.						
Chrome rim	P1@.250	Ea	20.40	8.79	29.19	49.60
Kitchen sink basket strainer washer, Danco. 1/8" thick rubber.						
1-3/4" x 1-3/8"	P1@.070	Ea	1.32	2.46	3.78	6.43
4-3/8" x 3-7/16"	P1@.070	Ea	1.42	2.46	3.88	6.60
Sink strainer shank nut. 4-3/8" outside diameter. Die cast.						
Shank nut, 3-11/32" inside diameter	P1@.070	Ea	1.50	2.46	3.96	6.73
Strainer nut, 3-3/8" inside diameter	P1@.070	Ea	1.00	2.46	3.46	5.88
Flat sink drain stopper. Universal fit. Rubber.						
Stopper	—	Ea	1.25	—	1.25	—
New style drain stopper, Danco. 5/16" stud fits Rapid Fit drains. For old style plastic and brass drain assemblies. Polished brass. 1-11/16" long						
3/8" diameter thread	—	Ea	7.12	—	7.12	—

Sink-Mounted Hot Water Dispensers

	Craft@Hrs	Unit	Material	Labor	Total	Sell
Hot water dispenser, In-Sink-Erator. Non-swivel high-spout design. 2/3 gallon capacity. 60 cups per hour. Self-closing valve. Adjustable thermostat. Rugged, low-profile stainless steel spout features an integral copper valve body and tank for dependable performance and easy installation. One-year in-home warranty.						
Dispenser	P1@1.25	Ea	213.00	44.00	257.00	437.00
Deluxe instant hot water dispenser, In-Sink-Erator. Near-boiling 190-degree Fahrenheit water instantly. Chrome-plated faucet completely insulated from 190-degree water. Provides 60 cups of 190-degree water per hour. Snap-action adjustable from 140 to 200 degrees, factory preset to 190 degrees. Die-molded, high-efficiency, expanded polystyrene thermal barrier surrounds tank for low heat loss. Adjustable thermostat. Convenient drain, thermally fused to prevent tank damage from dry start-up or loss of water. Tank is always open to atmospheric pressure and requires no relief valve. Instant self-closing valve. Traditional twist-handle actuation. One-year full parts and labor warranty.						
European design, chrome	P1@1.25	Ea	381.00	44.00	425.00	723.00
Standard	P1@1.25	Ea	440.00	44.00	484.00	823.00
White	P1@1.25	Ea	495.00	44.00	539.00	916.00

Utility Sinks, Laundry Tubs

	Craft@Hrs	Unit	Material	Labor	Total	Sell
All-in-one standard utility tub kit, American Shower & Bath. Made of polypropylene. Includes one standard utility tub, two 3/8" OD x 20" long faucet risers, one drain installation kit, one drain stopper, one faucet assembly, four leg levelers and Teflon® tape. Limited one-year warranty.						
19-gallon capacity	P1@1.75	Ea	90.80	61.50	152.30	259.00
Heavy-duty utility laundry tub kit, American Shower & Bath. Made of extra-strength structural foam. Includes ASB® heavy-duty utility tub, 2 Brass Craft® B1-20A SpeediPlumb Plus® connectors. 3/8" OD x 20" long PEX faucet risers. ASB® drain installation kit. Drain stopper. ASB® faucet assembly, 4 leg levelers and Teflon® tape. Limited one-year warranty.						
Laundry tub	P1@1.75	Ea	114.00	61.50	175.50	298.00
Utilatub® polypropylene single bowl laundry tub, E.L. Mustee. Sturdy, one-piece molded tub. Smooth white satin finish. Easy to clean and stain-resistant. Self-draining soap and storage shelves. Includes leakproof 1-1/2" integral molded drain assembly and floor mounting hardware. Ribbed underbody for extra strength. 13" deep, 20-gallon capacity, 23" wide.						
Laundry tub	P1@1.75	Ea	37.10	61.50	98.60	168.00

	Craft@Hrs	Unit	Material	Labor	Total	Sell
Double bowl laundry tub. Sturdy, one-piece molded tub. Smooth white surface. Heavy gauge steel legs include levelers. 13" deep, 19-gallon capacity per tub, 40" wide. Leakproof 1-1/2" integral molded drain, hooks up to standard 1-1/2" S- or P-trap. Accommodates single- or dual-handle faucet with 4" or 8" centers. Meets or exceeds codes ANSI Z124.6, IAPMO listed						
Laundry tub	P1@1.75	Ea	173.00	61.50	234.50	399.00
Drop-in utility sink, American Shower & Bath. 14.5-gallon capacity ABS utility sink. Countertop installation with self-rimming design. Built-in self-draining storage shelf and soap dish. High-gloss stain-resistant finish. Easy to clean.						
Utility sink	P1@1.75	Ea	71.40	61.50	132.90	226.00
Stainless steel utility bowl sink. 20 gauge, stainless steel. 3 faucet hole drillings. 10" extra deep bowl. 15" x 15"						
Utility or island top sink	P1@.600	Ea	153.00	21.10	174.10	296.00
Laundry tub hose. Dispenses water from washing machine discharge hose into laundry tub. Eliminates draping washing machine hose over back/side wall of laundry tub and conceals washer hose. Mounts to back or side wall of laundry tub, includes mounting hardware.						
Hose	—	Ea	10.40	—	10.40	—
Laundry tub overflow tube. Maintains water level in laundry tub and prevents overflows. Replaces drain plug and eliminates reaching into water. 12-3/8" high; may be cut to alter height. Fits all drain openings.						
Overflow tube	—	Ea	6.50	—	6.50	—

Laundry Tub Faucets

	Craft@Hrs	Unit	Material	Labor	Total	Sell
Laundry faucet, two-handle, Brass Craft. Deck-mount. 4" center set. Labor includes fasten in place only. Add the cost of connections and plumbing rough-in as required.						
Rough brass finish	P1@.900	Ea	51.00	31.60	82.60	140.00
Laundry faucet, two-handle, Glacier Bay. Deck-mount. 4" center set. Drip-free washerless cartridge. Standard 3/4" hose thread on end of spout. 12-year limited warranty. Labor includes fasten in place only. Add the cost of connections and plumbing rough-in as required.						
Chrome finish	P1@.900	Ea	26.20	31.60	57.80	98.30
Laundry faucet, two-handle. 1/2" connections. Mini lever handles. Labor includes fasten in place only. Add the cost of connections and plumbing rough-in as required.						
Chrome	P1@.900	Ea	56.40	31.60	88.00	150.00
Laundry faucet, two-handle, wall mount, Price Pfister. Oakland blade handles with flanges. Handicap accessible. Labor includes fasten in place only. Add the cost of connections and plumbing rough-in as required.						
Satin chrome finish	P1@.900	Ea	57.50	31.60	89.10	151.00

Bar Sinks

	Craft@Hrs	Unit	Material	Labor	Total	Sell
Stainless steel bar sink. Single bowl. Standard-gauge stainless steel. Self-rimming design. Buffed finish. 2" drain. Faucet and strainer sold separately.						
15" x 15" x 5-1/2" deep	P1@.600	Ea	100.00	21.10	121.10	206.00

	Craft@Hrs	Unit	Material	Labor	Total	Sell
Stainless steel bar sink, Neptune Series, Elkay. Deep hospitality drop-in single bowl. Medium thickness, 23 gauge nickel-bearing stainless steel. Buffed finish with machine-ground satin-finished bowl. Sound deadened with pads. Self-rimming installation. Square corners. Two faucet holes with 4" spacing. One-hand mounting fasteners. Includes strainer.						
15" x 15", 5" deep	P1@.600	Ea	61.20	21.10	82.30	140.00
Stainless steel bar sink, Signature Series, Elkay. Drop-in single bowl. Self-rimming installation. Made from nickel bearing Type 302 20-gauge stainless steel. Mirror-finished interior deck. Machine-ground satin-finished bowl. Highlighted outer-rim edge and bowl radius. Satin highlighted sink rim. Fully undercoated for sound deadening. Medium thickness. 3" radius vertical coved corners. Two faucet holes with 4" spacing. Complete with strainer.						
15" x 15" x 7-1/2" deep	P1@.600	Ea	95.00	21.10	116.10	197.00
Silhouette polycast island or bar sink, American Standard. 18" x 18" x 9" deep. Porcelain enameled Americast. Lighter than cast iron. Silicone sealant supplied.						
White Heat	P1@.750	Ea	150.00	26.40	176.40	300.00
Composite bar sink. Replaces 15" x 15" bar sinks. Large enough bowl to use as a kitchen, prep or utility sink. Full 3-1/2" drain accepts standard kitchen sized basket strainers. Faucet and drain sold separately.						
White	P1@.750	Ea	60.20	26.40	86.60	147.00
Composite entertainment sink. Serves as a prep, bar or island sink. Accommodates all garbage disposals in its full depth of 8". Use as a stand-alone or as a "point-of-use" sink in a multi-sink kitchen.						
18" x 18" x 8", white	P1@.750	Ea	116.00	26.40	142.40	242.00
Manchester composite bar sink, International Thermocast. 16" x 16" x 7" deep. Cutout dimensions 14.5" x 14.5" with 1" radius corners. Rich high-gloss finish. Resists stain, rust, oxidation, chipping, and scratches. Drop-in ready for quick and easy installation. No mounting clips required. Faucets, drains and accessories sold separately.						
White	P1@.750	Ea	60.20	26.40	86.60	147.00
Junior basket strainer insert. For laundry tubs, bar sinks and mobile homes. Includes junior basket strainer, flat grid strainer and wash tray plug.						
Stainless steel	P1@.070	Ea	8.00	2.46	10.46	17.80

Bar Faucets

	Craft@Hrs	Unit	Material	Labor	Total	Sell
Bar faucet, single-handle, Waterfall. Deck-mount. 4" center set. Two-hole application. Solid brass and stainless steel construction. Designed to work on bar, preparation and laundry sinks. Labor includes fasten in place only. Add the cost of plumbing rough-in and connection as required.						
Chrome finish	P1@1.00	Ea	100.00	35.20	135.20	230.00
Stainless steel	P1@1.00	Ea	155.00	35.20	190.20	323.00
Bar faucet, two-handle. Deck-mount. 4" center set. Porcelain handles. Labor includes fasten in place only. Add the cost of plumbing rough-in as required.						
Brass finish	P1@.906	Ea	150.00	31.90	181.90	309.00
Chrome finish	P1@.906	Ea	100.00	31.90	131.90	224.00
Bar faucet, Cadet Series, American Standard. Deck-mount. Labor includes fasten in place only. Add the cost of plumbing rough-in as required.						
Chrome finish	P1@.906	Ea	50.00	31.90	81.90	139.00

	Craft@Hrs	Unit	Material	Labor	Total	Sell
Bar faucet, two-handle. Deck-mount. 4" center set. Drip-free washerless cartridge. Solid brass waterways. 12-year limited warranty. Labor includes fasten in place only. Add the cost of plumbing rough-in as required.						
Chrome finish	P1@.906	Ea	45.00	31.90	76.90	131.00
Bar faucet, two-handle, J A Manufacturing. Deck-mount. 4" center set. Builder's Grade. Labor includes fasten in place only. Add the cost of connections and plumbing rough-in as required.						
Chrome finish, kit	P1@.906	Ea	55.00	31.90	86.90	148.00
Chrome finish, lever handles	P1@.906	Ea	45.00	31.90	76.90	131.00

Garbage Disposers

	Craft@Hrs	Unit	Material	Labor	Total	Sell
1/3-horsepower food waste disposer, In-Sink-Erator. Plastic 26-ounce grind chamber. Galvanized steel grinding elements with two stainless steel 360-degree swivel lugs. Stainless steel sink flange with positive seal one-piece plastic stopper. Overload protector manual reset. Permanently lubricated upper and lower bearings. Quick Lock™ mounting for fast, easy installation. Cushioned anti-splash baffle. Dishwasher drain connection. One-year parts and in-home service warranty.						
Badger 1®	P1@1.00	Ea	86.30	35.20	121.50	207.00
1/2-horsepower food waste disposer, In-Sink-Erator. Galvanized steel grinding elements with two stainless steel 360-degree swivel lugs. Plastic grind chamber. Permanently lubricated upper and lower bearings. Quick Lock™ mounting for fast, easy installation. Stainless steel sink flange with positive seal one-piece plastic stopper. Cushioned anti-splash baffle. Overload protector manual reset. Two-year parts and in-home service warranty.						
Badger 5®	P1@1.00	Ea	102.00	35.20	137.20	233.00
5/8-horsepower food waste disposer, In-Sink-Erator. Plastic 26-ounce grinding chamber. Insulated outer shell. Galvanized steel grinding elements with two stainless steel 360-degree swivel legs. Stainless steel sink flange with positive seal one-piece plastic stopper. Dishwasher drain connection. Permanently lubricated upper and lower bearings. Heavy-duty induction motor. Overload protector manual reset. Stainless steel mounting. Three-year parts and in-home service warranty. Cushioned anti-splash baffle. Quick Lock™ mounting for fast, easy installation.						
Badger 5® Plus	P1@1.00	Ea	147.00	35.20	182.20	310.00
3/4-horsepower heavy-duty food waste disposer, In-Sink-Erator. 26-ounce plastic grind chamber. Insulated outer shell. Stainless steel grinding elements with two stainless steel 360-degree swivel lugs. Stainless steel sink flange with positive seal one-piece stainless steel stopper. Dishwasher drain connection. Permanently lubricated upper and lower bearings. Overload protector manual reset. Cushioned anti-splash baffle. Quick Lock™ mounting for fast, easy installation. Four-year parts and in-home service warranty.						
Disposer	P1@1.00	Ea	186.00	35.20	221.20	376.00
3/4-horsepower septic system food waste disposer, In-Sink-Erator. Corrosion-proof grind chamber. Stainless steel grinding elements with two stainless steel 360-degree swivel lugs. Permanently lubricated upper and lower bearings. Quick Lock™ mounting for fast, easy installation. Stainless steel sink flange with positive seal one-piece plastic stopper. Overload protector manual reset. Dishwasher drain connection. Replaceable Bio-Charge™ additive cartridge. 26-ounce grind chamber capacity. Three-year parts and in-home service warranty.						
Disposer	P1@1.00	Ea	264.00	35.20	299.20	509.00
3/4-horsepower batch feed disposer, In-Sink-Erator. Extra-large capacity. Locking stopper controls operation. Five-year in-home warranty. Full stainless steel construction. Labor does not include electrical connection or switch.						
Disposer	P1@1.00	Ea	389.00	35.20	424.20	721.00

	Craft@Hrs	Unit	Material	Labor	Total	Sell
1-horsepower food waste disposer, In-Sink-Erator. Auto-reverse grind system helps prevent jams, extends product life and provides extra sanitation. Secondary sound baffle. Stainless steel 50-ounce grind chamber. Stainless steel grinding elements with two stainless steel 360-degree swivel lugs. Corrosion protection shield. One-piece polished stainless steel stopper. Dishwasher drain connection. Permanently lubricated upper and lower bearings. Overload protector manual reset. Cushioned anti-splash baffle. Quick Lock™ mounting for fast, easy installation. Seven-year parts and in-home service warranty.						
Disposer	P1@1.00	Ea	346.00	35.20	381.20	648.00
Garbage disposer sink stopper, In-Sink-Erator. Economy-style sink stopper. Fits all disposers made by In-Sink-Erator. Genuine In-Sink-Erator replacement part.						
Plastic	—	Ea	7.13	—	7.13	—
Polished brass	—	Ea	30.40	—	30.40	—
Stainless steel	—	Ea	11.20	—	11.20	—
Stainless steel, almond	—	Ea	16.70	—	16.70	—
Stainless steel, white	—	Ea	16.70	—	16.70	—
Garbage disposer cord kit, In-Sink-Erator. Power cord UL-approved to resist moisture and safely connect disposer to a plug-in outlet. Includes wire nuts and retaining clamp.						
Kit	—	Ea	12.20	—	12.20	—
Disposer mounting assembly, In-Sink-Erator. Replaces existing disposer mounts. Quick Lock™ mounting assembly.						
Kit	—	Ea	16.30	—	16.30	—
Dishwasher connector, In-Sink-Erator. Connects dishwasher drain to the disposer. Standard size.						
Hose	—	Ea	9.17	—	9.17	—
Sink disposer control switch, In-Sink-Erator. Convenient and stylish new way to control disposer. Can be installed on sink or countertop. Wall-mount power module. 3' grounded power cord and 6' of air tubing.						
Chrome or white	—	Ea	62.40	—	62.40	—
Deluxe disposer mounting gasket, In-Sink-Erator.						
Deluxe gasket A	—	Ea	10.40	—	10.40	—
Standard gasket B	—	Ea	7.99	—	7.99	—
Removable disposer sound baffle, In-Sink-Erator. Muffles disposer noise. Prevents food from splashing up in sink.						
Baffle	—	Ea	7.27	—	7.27	—
Septic disposer replacement cartridge, In-Sink-Erator. Bio-charge 16-ounce enzyme replacement cartridge for septic disposer system. Accelerates digesting action in the septic tank. Lasts up to 4 months.						
Cartridge	—	Ea	15.30	—	15.30	—
Disposer sink flange, In-Sink-Erator. Adds a custom flair.						
Almond	P1@.185	Ea	15.00	6.50	21.50	36.60
Polished brass	P1@.185	Ea	26.30	6.50	32.80	55.80
Stainless steel	P1@.185	Ea	15.00	6.50	21.50	36.60
White	P1@.185	Ea	16.00	6.50	22.50	38.30

	Craft@Hrs	Unit	Material	Labor	Total	Sell
Dishwashers						
1-cycle dishwasher. Two wash levels with normal wash cycle (dial). Heated dry (rocker switch). Hot pre-wash option. Standard upper rack. Deluxe lower rack.						
Black	P1@2.40	Ea	300.00	84.40	384.40	653.00
4-cycle dishwasher, Maytag. Space for 12 place settings. High-pressure, high-water-flow wash system. Three wash arms for spray action above and below each rack. Energy Star® rated. Exceeds federal energy-efficiency standards. Microprocessor controls with angled touch pad. Sound-silencing package.						
Bisque	P1@2.40	Ea	350.00	84.40	434.40	738.00
White	P1@2.40	Ea	350.00	84.40	434.40	738.00
2-level dishwasher, GE. Nautilus dishwasher with the PowerScrub wash system and two wash arms. Touchpad controls. HotStart™ GE QuietPower™ motor.						
White or black	P1@2.40	Ea	350.00	84.40	434.40	738.00
5-cycle, 11-key touch pad dishwasher, Maytag. Tall tub for higher capacity. Space for 14 place settings. Adjustable upper rack allows custom configurations. High-pressure, high-water-flow wash system. Three wash arms. Finer filtration. Energy Star® rated. Exceeds federal energy-efficiency standards. Microprocessor controls with angled touch pad.						
Black or white	P1@2.40	Ea	400.00	84.40	484.40	823.00
13-cycle stainless steel interior dishwasher, Jenn-Air. Tall-tub design.						
Dishwasher	P1@2.40	Ea	675.00	84.40	759.40	1,290.00
Dishwasher Accessories						
Dishwasher connector.						
60" long, stainless steel braid	P1@.185	Ea	21.40	6.50	27.90	47.40
3/8" x 3/8" x 60", poly braid	P1@.150	Ea	18.10	5.27	23.37	39.70
72" long, stainless steel braid	P1@.185	Ea	26.00	6.50	32.50	55.30
Dishwasher connector, compression inlet, Speedi-Plumb™, Plumb Shop. 125 PSI maximum operating pressure. 40 to 140 degree Fahrenheit temperature range. CSA, UL and IAPMO listed.						
3/8" x 3/8" x 48" compression	P1@.200	Ea	14.20	7.03	21.23	36.10
Garbage-disposal-to-dishwasher drain connector, Fernco. Connects dishwasher hose to sink drains and garbage disposals. Engineered for years of leak-free and trouble-free service.						
Model No. PDWC-100	P1@.250	Ea	3.29	8.79	12.08	20.50

	Craft@Hrs	Unit	Material	Labor	Total	Sell

Cooktops

30" electric cooktop. Four Insta-Heat™ radiant elements. Element-in-use and hot surface indicator lights. Frameless glass-ceramic surface. Dual-Choice™ element provides a large and small element in one location. Color-coordinated knobs with metallic accents pull off for cleaning. Overall dimensions 21" deep x 29-15/16" wide x 4-1/8" high. Frameless smooth black top. Labor cost assumes countertop is cut correctly and electrical rough-in is complete.

	Craft@Hrs	Unit	Material	Labor	Total	Sell
MEC7430WW, White	BE@.903	Ea	641.00	35.30	676.30	1,150.00

30" gas cooktop, Maytag. Four sealed surface gas burners. Two 9,200/9,100 BTU burners, one 12,500/10,500 BTU power boost burner, one 5,000/4,000 BTU simmer burner. Porcelain-on-steel surface. Pilotless electronic ignition with flame chamber. Color coordinated easy-grasp pull-off control knobs with metallic accents. Four Duraclean lift-off burner caps. Continuous Duraclean cast-iron burner grates. Labor cost assumes countertop is cut correctly and gas rough-in is complete.

	Craft@Hrs	Unit	Material	Labor	Total	Sell
No. MGC7430W, white	P1@1.25	Ea	629.00	44.00	673.00	1,140.00
No. MGC7430WB, black	P1@1.25	Ea	629.00	44.00	673.00	1,140.00

30" electric ceramic glass cooktop. Five tri-ring (12", 9" and 6") and ribbon heating elements. Frameless design with upfront controls. Overall dimensions: 20-7/8" deep x 29-3/4" wide x 3-1/4" high. Labor cost assumes countertop is cut correctly and electrical rough-in is complete.

	Craft@Hrs	Unit	Material	Labor	Total	Sell
Stainless	BE@.903	Ea	1,240.00	35.30	1,275.30	2,170.00

Wall Ovens

30" electric double wall oven. Two self-cleaning ovens with adjustable cleaning levels. Precision electronic controls. Favorite setting, cook and hold setting, adjustable keep warm setting. Delayed cook/clean options.

	Craft@Hrs	Unit	Material	Labor	Total	Sell
Choice of finishes	BE@1.67	Ea	1,520.00	65.20	1,585.20	2,690.00

27" single wall oven. Preheat signal, auto oven shut-off, self-cleaning, control lock, delay bake option and TrueTemp System. Clear view window pattern.

	Craft@Hrs	Unit	Material	Labor	Total	Sell
Black	BE@1.67	Ea	917.00	65.20	982.20	1,670.00

Mounted Microwave Ovens

1.4 cubic foot over-the-range microwave oven. 950 watts. Cooking complete reminder, defrost, delay start, extend cook function, interior light, one-touch cooking, popcorn feature, vent/exhaust fan. 29.75" wide x 29.75" high x 15" deep. Labor cost assumes the opening has been cut and electrical rough-in is complete.

	Craft@Hrs	Unit	Material	Labor	Total	Sell
White	BE@.903	Ea	215.00	35.30	250.30	426.00
Black	BE@.903	Ea	215.00	35.30	250.30	426.00
Stainless	BE@.903	Ea	215.00	35.30	250.30	426.00

1.3 cubic foot Advantium™ microwave oven, GE. Can be installed over a countertop, range, or cooktop and plugs into a standard 120-volt/15-amp outlet. Halogen cooktop lighting. Microhood two-speed, 300 CFM venting system. 15.75" deep x 16.4" high x 29.88" wide. 900 watts. Labor cost assumes the opening has been cut and electrical rough-in is complete.

	Craft@Hrs	Unit	Material	Labor	Total	Sell
SA1201RSS, stainless and black	BE@.903	Ea	958.00	35.30	993.30	1,690.00

Range Hoods

	Craft@Hrs	Unit	Material	Labor	Total	Sell
Economy ducted range hood. Under-cabinet mounted 190 CFM (cubic feet per minute) exhaust. 6.0 sones. 7" duct connector. 18" deep x 6" high. Washable charcoal filter with replaceable charcoal filter pad. Uses 75-watt incandescent lamp. Polymeric lens. Add the cost of ducting and electrical connection.						
24" or 30" wide, enamel finish	SW@1.42	Ea	94.30	57.40	151.70	258.00
24" or 30" wide, stainless steel	SW@1.42	Ea	139.00	57.40	196.40	334.00
36" wide, enamel finish	SW@1.42	Ea	94.30	57.40	151.70	258.00
36" wide, stainless steel	SW@1.42	Ea	139.00	57.40	196.40	334.00
42" wide, enamel finish	SW@1.42	Ea	97.60	57.40	155.00	264.00
42" wide, stainless steel	SW@1.42	Ea	142.00	57.40	199.40	339.00
Add for ducted hood with a duct damper	—	Ea	15.60	—	15.60	—
Standard quality ducted range hood. Under-cabinet mounted 190 CFM (cubic feet per minute) 3-speed blower. 3.5 sones. Either vertical or horizontal 3-1/4" x 10" duct, 7" round duct. Powder-coated 22 gauge steel. 18" deep x 6" high. Uses one 40-watt incandescent lamp. Add the cost of ducting and electrical connection.						
24" or 30" wide, ducted, black	SW@1.42	Ea	135.00	57.40	192.40	327.00
24" or 30" wide, ducted, almond	SW@1.42	Ea	135.00	57.40	192.40	327.00
24" or 30" wide, ducted, stainless	SW@1.42	Ea	239.00	57.40	296.40	504.00
36" wide, ducted, black	SW@1.42	Ea	150.00	57.40	207.40	353.00
36" wide, ducted, almond	SW@1.42	Ea	150.00	57.40	207.40	353.00
36" wide, ducted, stainless	SW@1.42	Ea	219.00	57.40	276.40	470.00
42" wide, ducted, black	SW@1.42	Ea	160.00	57.40	217.40	370.00
42" wide, ducted, almond	SW@1.42	Ea	160.00	57.40	217.40	370.00
42" wide, ducted, stainless	SW@1.42	Ea	251.00	57.40	308.40	524.00
Add for ductless range hood	—	Ea	51.00	—	51.00	—
Better quality range hood, Allure III NuTone. Under-cabinet mounted 400 CFM (cubic feet per minute). 0.4 sones at 100 CFM. Teflon-coated bottom pan. Dual halogen lamps. Sensor detects excessive heat and adjusts vent speed. 20" deep x 7-1/4" high. Four ducting options: 3 1/4" X 10" horizontal or vertical vent, 7" round vertical vent or non-ducted. Add the cost of ducting and electrical connection.						
30" wide, polyester finish	SW@1.42	Ea	508.00	57.40	565.40	961.00
30" wide, stainless steel finish	SW@1.42	Ea	555.00	57.40	612.40	1,040.00
36" wide, polyester finish	SW@1.42	Ea	480.00	57.40	537.40	914.00
36" wide, stainless steel finish	SW@1.42	Ea	571.00	57.40	628.40	1,070.00
42" wide, polyester finish	SW@1.42	Ea	528.00	57.40	585.40	995.00
42" wide, stainless steel finish	SW@1.42	Ea	587.00	57.40	644.40	1,100.00
Add for filter for ducted application	—	Ea	51.00	—	51.00	—
Add for filter for non-ducted application	—	Ea	27.30	—	27.30	—
Wall-mount range hood, Ballista NuTone. Stainless steel finish with hood-mounted blower. Three-speed control. Dual 20-watt halogen bulbs. Adjusts to fit under 8' to 10' ceiling. Stainless steel grease filter. Built-in backdraft damper. 20" deep, 33" to 54" high. Add the cost of ducting and electrical connection.						
30" wide, 450 CFM	SW@2.50	Ea	1,680.00	101.00	1,781.00	3,030.00
36" wide, 450 CFM	SW@2.50	Ea	1,700.00	101.00	1,801.00	3,060.00
48" wide, 900 CFM	SW@2.50	Ea	1,900.00	101.00	2,001.00	3,400.00
Ceiling-hung range hood, Provisa NuTone. Stainless steel finish with hood-mounted blower. Three-speed control. Four 20-watt halogen bulbs. Telescopic flue adjusts to fit under 8' to 9' ceiling. Stainless steel grease filter. Built-in backdraft damper. 27" deep, 33" high. Add the cost of ducting and electrical connection.						
40" wide, 900 CFM	SW@3.50	Ea	3,250.00	142.00	3,392.00	5,770.00

	Craft@Hrs	Unit	Material	Labor	Total	Sell
Custom range hood power package. Exhaust system components installed in a custom-built range hood 24" to 30" above the cooling surface. Stainless steel. With variable speed control, filters and halogen or incandescent lighting. Bulbs not included. No ductwork included. Add the cost of a decorative enclosure, ducting and electrical connection.						
700 CFM, dual blowers, 7" duct, 1.5 to 4.6 sones						
30" wide x 22" deep	SW@1.42	Ea	539.00	57.40	596.40	1,010.00
32" wide x 22" deep	SW@1.42	Ea	586.00	57.40	643.40	1,090.00
36" wide x 28" deep	SW@1.42	Ea	666.00	57.40	723.40	1,230.00
42" wide x 28" deep	SW@1.42	Ea	734.00	57.40	791.40	1,350.00
48" wide x 28" deep	SW@1.42	Ea	831.00	57.40	888.40	1,510.00
1,300 CFM, dual blowers, twin 7" ducts, 1.5 to 4.6 sones						
30" wide x 22" deep	SW@1.42	Ea	874.00	57.40	931.40	1,580.00
32" wide x 22" deep	SW@1.42	Ea	962.00	57.40	1,019.40	1,730.00
36" wide x 28" deep	SW@1.42	Ea	1,080.00	57.40	1,137.40	1,930.00
42" wide x 28" deep	SW@1.42	Ea	1,190.00	57.40	1,247.40	2,120.00
48" wide x 28" deep	SW@1.42	Ea	1,340.00	57.40	1,397.40	2,380.00
Range splash plate, Broan. Easy to install. Mounts behind range with screws. Keeps kitchen walls clean.						
30" x 24", reversible white or almond	SW@.165	Ea	32.00	6.67	38.67	65.70
30" x 24", stainless steel	SW@.165	Ea	84.00	6.67	90.67	154.00
36" x 24", reversible white or almond	SW@.165	Ea	34.70	6.67	41.37	70.30
36" x 24", stainless steel	SW@.165	Ea	91.20	6.67	97.87	166.00

Bathrooms

13

There are two types of bathroom jobs. The first simply makes better use of the available space. The second is adding space, either by enlarging an existing bathroom or by adding an entirely new bathroom. Both types will be expensive when calculated on the basis of cost per square foot of floor. But real estate professionals agree that money invested in bathrooms is generally money wisely spent. A home with three or four bedrooms and only one and a half bathrooms has a major defect worth fixing. That's especially true when the home is in a neighborhood where most homes have two or more bathrooms.

Adding a Bathroom

Older homes seldom have enough bathrooms. But the bathrooms in those homes are often much larger than the modern 5' by 7' standard for a three-fixture bath (toilet, lavatory and tub). With a large bathroom, remodeling options will be obvious: extra storage cabinets, a double-bowl vanity, more mirrors, a lighted dressing table, separate tub and shower units, or even a bathroom divided into compartments. If an entirely new bathroom is needed, the most obvious issue will be where to find the space. Keep in mind that a bathroom addition outside the existing perimeter wall is seldom a good choice of location.

Generally, bedrooms and bathrooms go together. Access to any new bathroom should be directly off either a bedroom or a central hall connecting bedrooms. Avoid any floor plan that requires walking through another room to get to the new bathroom. An exception would be if the new bathroom is a two-fixture half-bath (powder room) designed to serve a relaxation area. A good bathroom plan minimizes plumbing and electrical runs and locates plumbing fixtures on one wall. This arrangement allows the fixtures to share a common waste line and roof vent.

Counting and Naming Bathrooms

Here's an explanation of what determines a bath type, if you're not familiar with bathroom naming conventions.

Full bath — Three or more fixtures including at least one lavatory, toilet and bathtub. Expect to see a showerhead over the tub.

Three-quarter bath — Three fixtures including a lavatory, toilet and stall shower (but no tub).

Half bath — Two fixtures: a lavatory and toilet.

Powder room — A two-fixture bathroom located near the front entrance.

Mud room — Same as a powder room but located by a back door or laundry.

Notes: When adding up the bathrooms in a house, round three-quarter baths up to a full bath if there are other partial bathrooms. You're sure to confuse someone if you claim a house has two and a quarter baths (full bath, three-quarter bath and half-bath). Instead, call it two and a half baths. Likewise, ignore the benefit of a second lavatory or a bidet in a bathroom. Those fixtures get no extra credit when it comes to what counts in a bathroom.

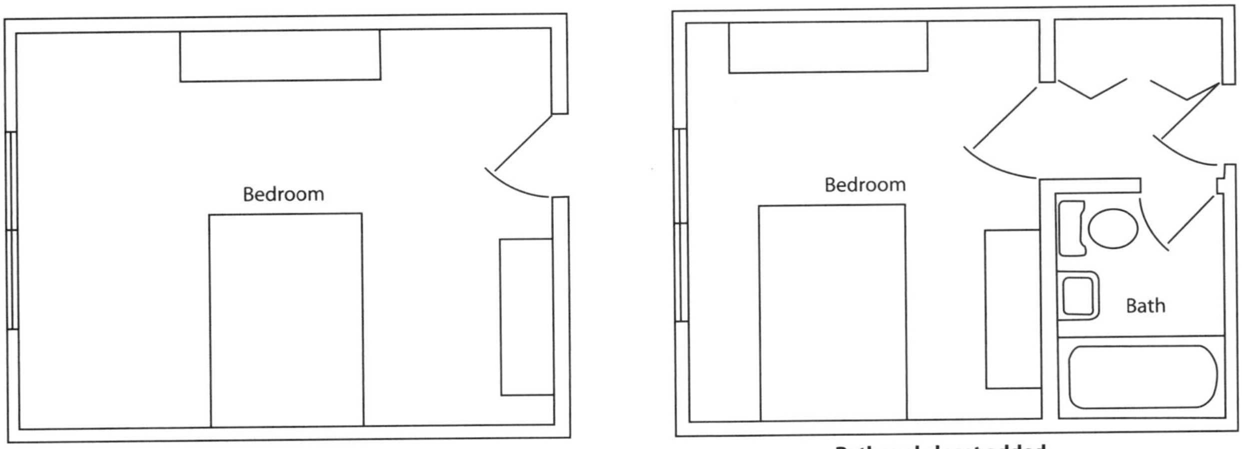

Figure 13-1

Portion of a large bedroom used for a bath addition

Sometimes you'll see homes where a pantry, large closet or open area under a stairway has been converted to a bathroom. Depending on the location, converting space like this into a bathroom can be either a good choice or a poor choice. Don't convert a space into a bathroom simply because two fixtures happen to fit there. A bathroom in an entry or off the kitchen, totally removed from the bedroom area, is only good for a half bath or powder room. There's little practical advantage to a three-fixture bath that's remote from bedrooms.

Finding Space for a Bathroom

You may find bedrooms in older homes that have a walk-in closet that's at least 5' by 7'. If a wall below that closet includes a plumbing drain, waste and vent lines, that closet is an ideal candidate for conversion to a three-fixture bathroom. Any bedroom that measures at least 16' in one direction is a good candidate to be partitioned for a new bathroom addition. See Figure 13-1. If all the bedrooms are small, consider converting the smallest bedroom into a bathroom and adding a new bedroom outside a perimeter wall.

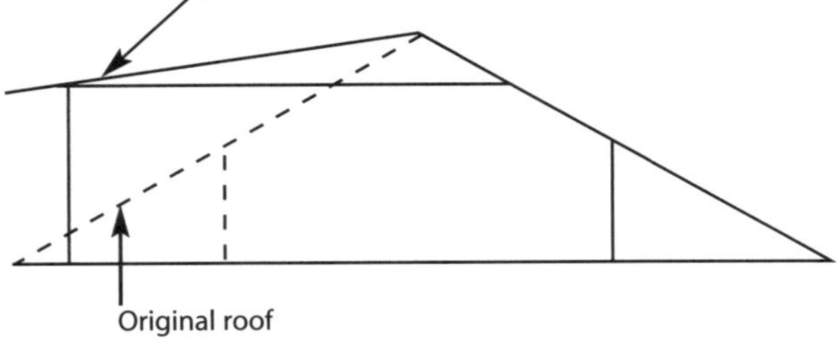

Figure 13-2

Shed dormer used for additional space

In a one-and-one-half-story house, there's sometimes enough space under the shed dormer for another bathroom. See Figure 13-2. If you choose this area, remember that the plumbing wall in the new bathroom requires a wall directly below for pipe runs. Water supply lines to the new bathroom can be copper or poly and will snake through nearly any wall or ceiling cavity. Vent stack and waste lines aren't so flexible and may be 4" in diameter. Hiding a 4" drain pipe in an existing wall isn't easy. Any line that can't be enclosed in an existing wall will need to be framed into an enclosure within a room.

Other options may be converting the basement or garage into living space, or building on an exterior addition to the home. Each comes with its own challenges.

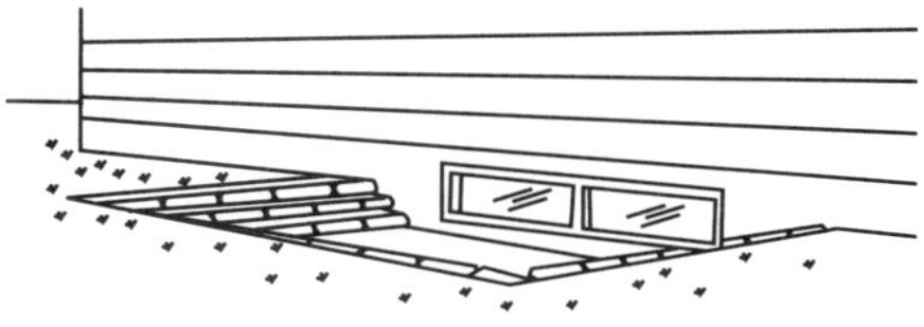

Figure 13-3
Basement window areaway with sloped sides

Basements: An unfinished basement offers good potential for expansion. It could include the bathroom you're adding, or be converted into a bedroom to replace the one you converted into the new bathroom. Building codes allow basements to be converted into habitable rooms if certain conditions are met. A habitable room is any space used for living, sleeping, eating, or cooking. To qualify as a habitable room, most codes require that the ground level outside the basement be no more than 48" above the basement floor, and the average ceiling height inside has to be at least 7'6". But bathrooms aren't considered habitable rooms under the code. So the ground level outside presents no obstacle, and the minimum ceiling height for a bathroom or other non-habitable room can usually be as low as 6'9". Your local building department will have the exact requirements for your location.

Although basements offer good space for conversions, they come with potential problems. The basement can be a difficult area in which to work. Many modifications will involve either tearing up the basement floor, or cutting through the concrete basement wall.

One obvious problem when adding a bathroom in the basement is handling sewage waste lines. If a basement floor is below the sewer main, install a sewage sump with a pump controlled by a float switch.

Another big disadvantage for living space is the lack of natural light and a view to the outside. You can add natural light with area walls or an areaway sloped so sunlight can reach basement windows. See Figure 13-3. If the slope of the lot offers an opportunity for drainage, consider creating a small sunken garden adjacent to one basement wall, as shown in Figure 13-4.

Your building code may require a fire exit in a converted basement. A basement window large enough to serve as a fire exit may meet the code requirements. Otherwise, add an areaway basement door like the one in Figure 13-5. An exterior

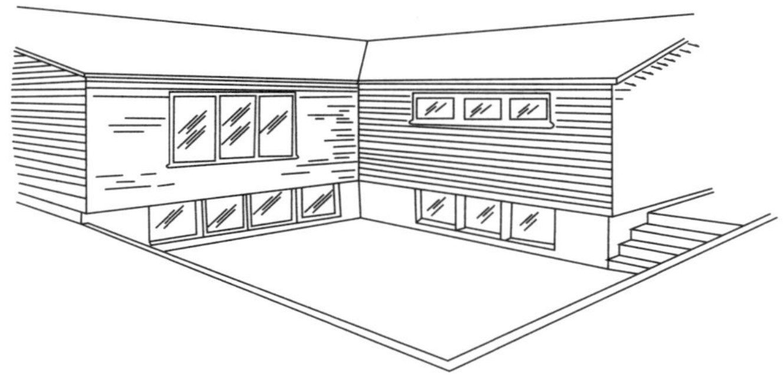

Figure 13-4
Sunken garden forming large basement window areaway

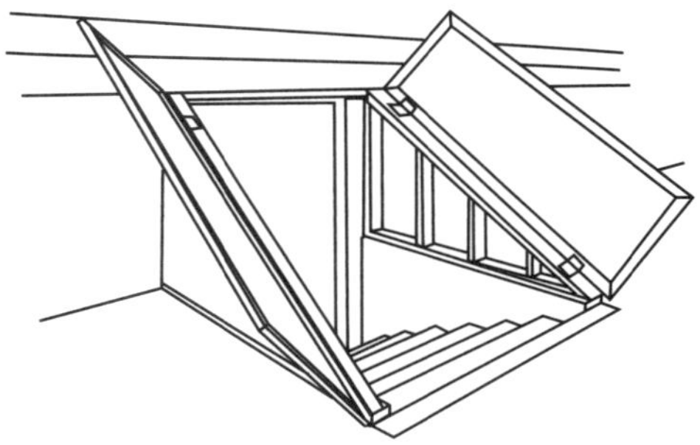

Figure 13-5
Areaway type basement entrance

entrance makes the basement a convenient place for storing lawn furniture and garden equipment. If possible, locate the basement door under the cover of a breezeway or porch roof to protect it from rain or snow.

Garage: If local building codes and zoning ordinances permit garage conversions, a garage attached to the house is another obvious candidate for expansion, especially if the garage is under the main roof. Considerations for the job include adding partition walls and an interior wall finish, and running ductwork for heating and air conditioning. Weigh the gain in living area against the loss of garage space and its convenience. Garages are usually close to the kitchen to save steps when unloading groceries. Even if you only use part of the garage for your addition, you may still lose that direct-access convenience.

Cutting chases in a garage floor for the plumbing drain and supply lines seldom presents a serious problem. But get expert advice on recompacting the soil once access lines are in place. Fill that isn't properly compacted will have expansion characteristics different from the undisturbed soil. The result can be heaving or settling of the subsoil and an uneven floor in the converted garage.

Exterior Additions: Most communities have minimum setback requirements. No construction is allowed within a certain distance of lot lines and the street. The setback may be different on the two sides of the house. For example, the setback could be 5' on one side of the house and 10' on the other. If all the homes in a community have a similar setback, it's safe to assume those homes are built on the setback line. A hundred years ago, builders weren't so careful about their use of space. By modern standards, they squandered available land. You may have more possibilities for expansion when you're dealing with an older home. However, if setbacks restrict you from adding onto the front or side of a house, the only alternative may be to encroach on the back yard.

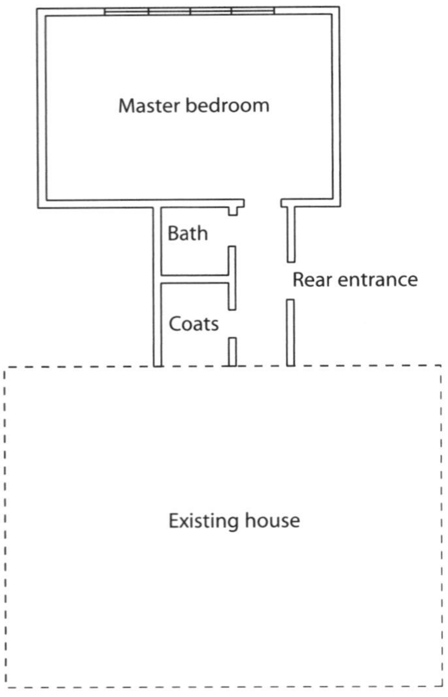

Figure 13-6

Addition using the satellite concept

When planning exterior additions, carefully consider how to meet the needs of the owner while maintaining the overall design of the home. If the owner wants a larger bathroom, you could, as mentioned earlier, convert an existing small bedroom into a bathroom and make the exterior addition a larger new bedroom. Or partition off part of the existing living room for use as bedroom and bathroom. Then make the addition a new family room or a larger living room. In any case, the addition should be constructed to maintain the style of the house. Roofing, siding and windows should match the existing materials to preserve continuity.

The addition won't have a common foundation with the original structure. That puts stress on the point where the existing roof, wall, floor and ceiling join the addition. Cracking along the joint between the original house and the addition can be managed with control joints that allow for independent movement on each side of the joint. To minimize the length of that joint, consider building the addition as a satellite to the main house. See Figure 13-6. Connect the main house to the addition with a corridor that includes the bathroom, a new entry and a closet. Notice that satellite additions have a large perimeter in proportion to the enclosed space. Heating and cooling

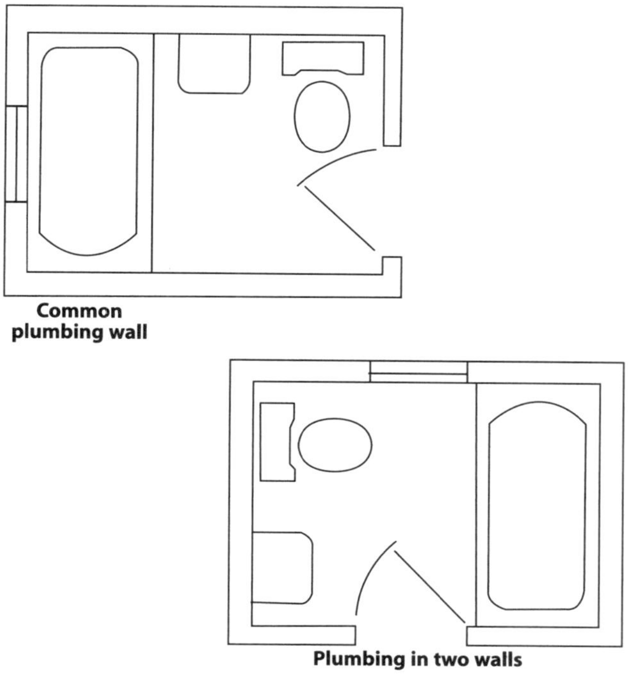

Figure 13-7

Minimum size bathroom (5' by 7')

Figure 13-8

Moderate size bathroom (8' by 8')

this type of addition can be a problem. Be sure to provide good insulation and energy-efficient windows. Chapter 3 includes costs for framing the tie-in between the existing structure and the addition.

Bathroom Floor Plans

Whether adding a new bathroom or just restyling an existing bathroom, it's important to observe some basic design rules. The minimum size for a three-fixture bathroom is 5' by 7' (Figure 13-7). Increasing that to 8' by 8' allows space for linen storage, cleaning equipment and supplies (Figure 13-8).

Figure 13-9 shows a bath divided into two compartments. That makes the bathroom suitable for use by two people, while preserving privacy. The minimum floor space for the toilet portion of a two-compartment bath is 3' by 5', assuming no tub. If a tub is included, you'll need a minimum of 5' by 5'. The lavatory compartment can be 4' x 5'. The interior dimensions of a compact two-compartment bathroom could be as little as 5' by 9' if you have a connecting door that slides or folds rather than swings. If a swinging door is planned, as in Figure 13-9, a floor area measuring 8' by 10'6" or 5' by 12' creates a more comfortable bathroom.

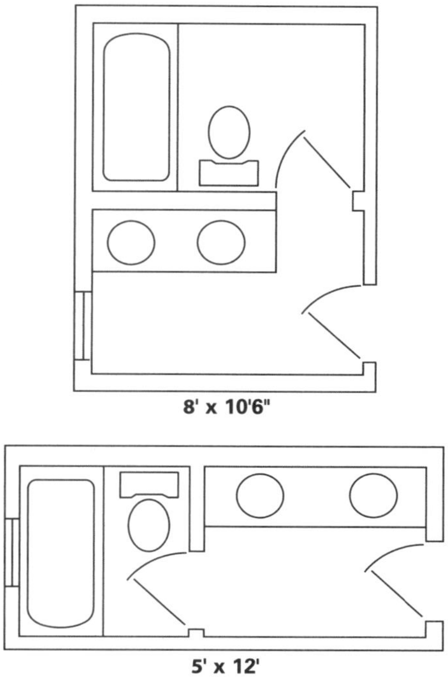

Figure 13-9

Compartmented bathroom

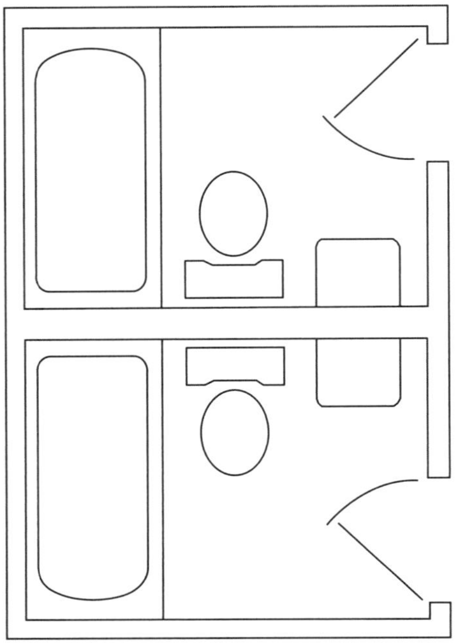

Figure 13-10

Two bathrooms with economical back-to-back arrangement

If two new bathrooms are needed on the same floor, install the plumbing fixtures back to back, as in Figure 13-10, to minimize plumbing runs. Bathrooms built on both floors of a two-story house are most economical when plumbing stacks are aligned vertically in the same wall.

Notice that all bathroom doors in Figures 13-7, 13-8, 13-9 and 13-10 open *into* the bathroom. True, that robs the bathroom of otherwise usable space, but it's better than opening a bathroom door into a hallway or bedroom. Bathroom doors are among the most used in the house and tend to open at unexpected times. A bathroom door that opens into a hallway is a constant hazard. Also, bathroom doors are generally closed when the bathroom is in use. That reclaims some of the space a swinging door would otherwise obstruct in a bathroom. Be sure to consider the swing area of the door when placing fixtures.

Every bathroom needs ventilation to prevent the accumulation of moisture and mold. If the bathroom is on an exterior wall, you should provide a window with an opening equivalent to 10 percent of the bathroom floor area. If the bathroom has only interior walls, the building code requires a fan with ducting to the exterior. Putting both the fan and light on the same switch guarantees the fan will be running when the bathroom is in use, although your customer may prefer independent switches. Don't put any switches or fixtures within arm's reach of someone using the tub or shower.

Cast iron and steel bathtubs are heavy, even when they're not filled with water. Fifty gallons of water in a tub adds another 400 pounds to the floor load. Floor joists in older homes are seldom designed to take such a concentrated load. When adding a tub, reinforce the floor joists and add cross bridging between joists. You may also need to double the studs. Even a shower stall can stress floor joists that weren't designed to support bathroom fixtures. If you plan to install ceramic tile on walls, floors or vanity tops, add extra framing to the floor and wall to support the additional weight. Ceramic tile is heavy and requires rock-solid support to resist cracking.

Bathroom Space Minimums

- 5' x 4' — smallest two-fixture bathroom
- 5' x 7' — smallest three-fixture bathroom
- 12" — from the center of a toilet to the end of a tub
- 15" — from the center of a toilet to the side of a tub
- 15" — from the center of a toilet to an adjacent wall or shower stall
- 15" — from the center of a toilet to the center of an adjacent lavatory

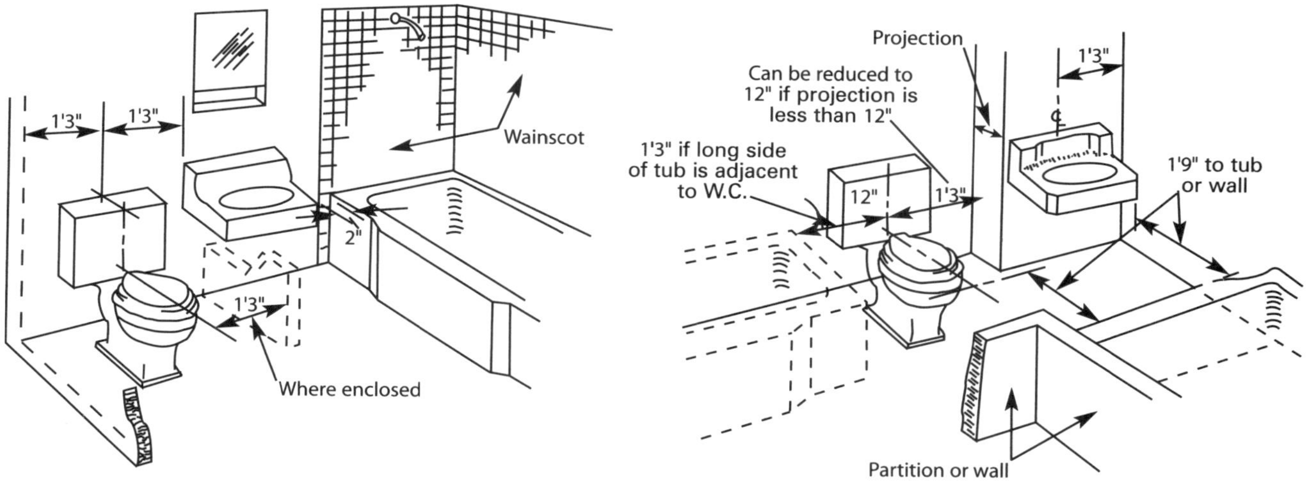

Figure 13-11

Recommended dimensions for fixture spacing

- 15" — from the center of a lavatory to an adjacent wall or shower stall
- 30" — distance between two lavatories mounted in the same countertop
- 32" to 34" — usual height for the top of vanities and lavatories
- 32" x 32" — width and depth of a square shower stall
- 30" x 60" — width and length of standard tubs
- 26" — height of the toilet paper roll holder above the floor
- 21" — depth of standing space while washing hands at a lavatory
- 36" — depth of standing space to open a sliding shower door
- 44" — depth of standing space to open a hinged shower door

Some building codes set minimum bathroom dimensions. As usual, the code is the last word. Figure 13-11 shows recommended minimum distances between and around bathroom fixtures. Remember that these are *minimums*, and more space is usually better.

	Craft@Hrs	Unit	Material	Labor	Total	Sell

Plumbing Repairs

Repair a compression faucet. Turn valves off. Remove packing nut. Remove valve stem and cap. Remove seat washer from stem. Pry out old washer. Install new washer. Install stem. Tighten main gland valves if necessary. Turn valve on. Test faucet.

	Craft@Hrs	Unit	Material	Labor	Total	Sell
Per valve repaired	P1@.350	Ea	—	12.30	12.30	20.90

Remove and replace fixture gasket. Turn valve off. Loosen locknut. Remove spud connection. Remove gasket or washer. Clean spud seat. Install new gasket or washer. Assemble spud connection. Tighten locknut. Reassemble fixture and inspect connection.

	Craft@Hrs	Unit	Material	Labor	Total	Sell
Per gasket replaced	P1@.120	Ea	—	4.22	4.22	7.17

Remove and replace toilet fill valve. Turn water off. Empty tank. Remove valve nut and disconnect supply tube. Install new toilet fill valve. Tighten nut. Attach supply tube. Turn water on. Check operation.

	Craft@Hrs	Unit	Material	Labor	Total	Sell
Per fill valve	P1@.300	Ea	—	10.50	10.50	17.90

Remove and replace lavatory trap. Unscrew trap plug from drain. Disconnect union. Remove and inspect trap. Replace trap. Reconnect union and tighten plug. Test operation.

	Craft@Hrs	Unit	Material	Labor	Total	Sell
Per trap	P1@.155	Ea	—	5.45	5.45	9.27

Remove and replace basin strainer and P-trap. Disconnect strainer from wall nipple and packing nut. Remove strainer lock nut. Remove strainer washer and trap. Clean out trap and strainer. Install strainer nut, packing nut and connect trap to wall nipple. Test operation.

	Craft@Hrs	Unit	Material	Labor	Total	Sell
Per strainer and trap	P1@.446	Ea	—	15.70	15.70	26.70

Remove and replace showerhead. Remove showerhead from 1/2" threaded pipe. Inspect as necessary. Replace and test showerhead.

	Craft@Hrs	Unit	Material	Labor	Total	Sell
Per showerhead	P1@.108	Ea	—	3.80	3.80	6.46

Reset toilet bowl. Turn off angle valve. Remove packing nuts. Remove flange bolts. Remove bowl from flange. Replace gasket or ring. Clean bowl foundation. Reset bowl and connect flange nuts. Replace packing nuts. Turn on water and check operation.

	Craft@Hrs	Unit	Material	Labor	Total	Sell
Per bowl reset	P1@1.27	Ea	—	44.70	44.70	76.00

Bathtubs

Remove bathtub. No salvage of materials assumed. Add for repairs to wall and floor surfaces. Debris piled on site. Labor cost does not include clearing of area or removal of wall finish.

	Craft@Hrs	Unit	Material	Labor	Total	Sell
Remove tub	P1@.722	Ea	—	25.40	25.40	43.20

Enameled steel bathtub. One-piece construction for recess installations. Diagonal brace. Sound-deadening foundation pad. Straight tiling edges. Full wall flange. Slip-resistant bottom. 40-gallon capacity. 60" x 30" x 15-1/4". Labor includes setting and fastening in place only. Add for water supply and drain connections as required. White.

	Craft@Hrs	Unit	Material	Labor	Total	Sell
Left-hand drain	P1@2.50	Ea	128.00	87.90	215.90	367.00
Right-hand drain	P1@2.50	Ea	128.00	87.90	215.90	367.00

Molded bathtub. Clear, molded color. Structural ribs for added strength. Slip-resistant bottom. Available with shower wall set. Faucet sold separately. 60" x 42" x 18". Resistant to chipping, cracking and peeling. Easy-to-clean high-gloss surface. Durable material with color molded throughout fixture. Labor includes setting and fastening in place only. Add for water supply and drain connections as required. White.

	Craft@Hrs	Unit	Material	Labor	Total	Sell
Left-hand drain	P1@2.50	Ea	220.00	87.90	307.90	523.00
Right-hand drain	P1@2.50	Ea	220.00	87.90	307.90	523.00

	Craft@Hrs	Unit	Material	Labor	Total	Sell
Acrylic soaking tub with skirt, Lasco Bathware. Slip-resistant textured bottom. Integral skirt and tiling flange. Labor includes setting and fastening in place only. Add for water supply and drain connections as required. 60" x 30" x 17". White.						
Left-hand drain	P1@2.50	Ea	300.00	87.90	387.90	659.00
Right-hand drain	P1@2.50	Ea	300.00	87.90	387.90	659.00
Oval acrylic soaking tub. Universal drain location. Slip-resistant textured bottom. Integral armrests. Island configuration. Comfort sloped backrest. Pre-leveled base. 60" x 42" x 23". Labor includes setting and fastening in place only. Add for water supply and drain connections as required.						
Ariel I, white	P1@2.50	Ea	688.00	87.90	775.90	1,320.00
Princeton recessed bathtub, American Standard. Integral lumbar support. Slip-resistant. Faucet sold separately. Glossy porcelain finish. Americast material insulates, holding water temperature and reducing the noise of running water. Half the weight of cast iron for easy installation. 60" x 30" x 14". Labor includes setting and fastening in place only. Add for water supply and drain connections as required.						
Bone, left-hand drain	P1@2.50	Ea	376.00	87.90	463.90	789.00
Bone, right-hand drain	P1@2.50	Ea	376.00	87.90	463.90	789.00
White, left-hand drain	P1@2.50	Ea	284.00	87.90	371.90	632.00
White, right-hand drain	P1@2.50	Ea	284.00	87.90	371.90	632.00
Villager™ bathtub, Kohler. Sloping back and safeguard bottom. Flat front rim provides ideal base for shower enclosure tracks. Faucet, fixture trim and accessories sold separately. Labor includes setting and fastening in place only. Add for water supply and drain connections as required.						
Almond, left-hand drain	P1@2.50	Ea	596.00	87.90	683.90	1,160.00
Almond, right-hand drain	P1@2.50	Ea	596.00	87.90	683.90	1,160.00
Bisque, left-hand drain	P1@2.50	Ea	539.00	87.90	626.90	1,070.00
Bisque, right-hand drain	P1@2.50	Ea	539.00	87.90	626.90	1,070.00
White, left-hand drain	P1@2.50	Ea	376.00	87.90	463.90	789.00
White, right-hand drain	P1@2.50	Ea	376.00	87.90	463.90	789.00

Whirlpool Tubs

	Craft@Hrs	Unit	Material	Labor	Total	Sell
Renaissance whirlpool bath, American Standard. 60" x 32" x 19-3/4". High-gloss acrylic. 1-HP pump. 4 body side jets, 2 lumbar back jets, 1 bubble massage control. Deck-mounted on and off switch. Faucet, fixture trim and accessories sold separately. Labor includes setting and fastening in place only. Add for water supply, drain and electrical connections as required.						
Bone	P1@4.00	Ea	611.00	141.00	752.00	1,280.00
White	P1@4.00	Ea	611.00	141.00	752.00	1,280.00
Cadet Elite whirlpool tub, American Standard. Acrylic with fiberglass reinforcement. Form-fitted backrest. Molded-in arm rests with elbow supports. Includes 1.5-HP pump. Labor includes setting and fastening in place only. Add for water supply, drain and electrical connections as required.						
5' x 36", white	P1@4.00	Ea	613.00	141.00	754.00	1,280.00
5' x 42", white	P1@4.00	Ea	725.00	141.00	866.00	1,470.00
Whirlpool tub, American Standard. Built-in apron model for easy, attractive installation. Jet comfort system featuring 6 interchangeable universal flex-jets. Includes 1.5-HP pump. 60" x 32" x 21-1/2". Labor includes setting and fastening in place only. Add for water supply, drain and electrical connections as required.						
White	P1@4.00	Ea	760.00	141.00	901.00	1,530.00

	Craft@Hrs	Unit	Material	Labor	Total	Sell
Devonshire™ whirlpool tub, Sterling Plumbing. Includes 1.25-HP pump with air switch actuator. 8 color-matched Kohler jets direct-adjustable hydro-massage. Crafted of high-gloss acrylic and backed with a solid layer of fiberglass to ensure durability and resistance to chipping, cracking and flexing. Labor includes setting and fastening in place only. Add for water supply, drain and electrical connections as required.						
5', drop-in	P1@4.00	Ea	810.00	141.00	951.00	1,620.00
5', left-hand drain, white	P1@4.00	Ea	1,010.00	141.00	1,151.00	1,960.00
5', right-hand drain, white	P1@4.00	Ea	1,010.00	141.00	1,151.00	1,960.00
Portrait™ whirlpool tub, Kohler. Traditional styling. 8 flexjets. Adjustable hydro-massage. Drop-in installation. High-gloss acrylic and backed with fiberglass. Labor includes setting and fastening in place only. Add for water supply, drain and electrical connections as required.						
5-1/2', with heater	P1@4.00	Ea	2,730.00	141.00	2,871.00	4,880.00
Elite whirlpool bath EZ-Install™ heater, American Standard. Pre-plumbed for easy installation. Maintains constant water temperature. Labor includes setting and fastening in place only. Add for electrical service and connection.						
EZHeat-100	P1@1.00	Ea	203.00	35.20	238.20	405.00
Whirlpool tub apron, American Standard. Labor includes setting and fastening in place only.						
5', white	P1@.500	Ea	135.00	17.60	152.60	259.00
Renaissance or Cadet, bone	P1@.500	Ea	192.00	17.60	209.60	356.00
Williamsburg J-spout Roman tub deck-mount tub filler, American Standard. Lifetime finish. Will not tarnish. Ceramic disc valving. Mounts on 8" to 12" centers. Labor includes set in place and connection only. Add the cost of plumbing rough-in as required.						
Brushed satin nickel finish	P1@1.00	Ea	320.00	35.20	355.20	604.00
Chrome finish	P1@1.00	Ea	259.00	35.20	294.20	500.00
Satin chrome finish	P1@1.00	Ea	320.00	35.20	355.20	604.00
Innovations® Roman tub trim kit, Delta. Chrome and polished brass finish. Flexible deck- and ledge-mount. Brilliant anti-tarnish finish. Installation kit included.						
18 GPM fill rate	P1@1.00	Ea	200.00	35.20	235.20	400.00

Tub Enclosure Walls

	Craft@Hrs	Unit	Material	Labor	Total	Sell
One-piece tub wall with shelf. Bath and shower wall surround. Seamless. One extra-long molded soap dish. High-gloss finish.						
58" to 61" x 31" x 59", white	B1@3.00	Ea	291.00	97.60	388.60	661.00
One-piece tile finish tub wall. Wall surround. High-gloss finish. With integrated soap dishes and simulated ceramic tile finish.						
58" to 60" x 32" x 58", white	B1@3.00	Ea	249.00	97.60	346.60	589.00
Three-piece bathtub wall kit, Pro-Wall 3™. Heavy-gauge, high-impact co-polymer plastic. 2 clear towel bars, 3 molded shelves. Easy installation with few seams to caulk. Back panel can be cut for window opening. Fits alcoves 57" to 61" across the back and 30" across the ends without trimming. Permanent, waterproof, colorfast panels are easy to clean. Resistant to mold and mildew.						
58" wall height, white	B1@3.00	Ea	183.00	97.60	280.60	477.00

	Craft@Hrs	Unit	Material	Labor	Total	Sell
Three-piece bathtub tile kit. Tub wall surround. Mold- and mildew-resistant. Limited lifetime warranty.						
30" x 60", beige, ceramic tile look	B1@3.00	Ea	432.00	97.60	529.60	900.00
Five-piece tub wall surround kit. High-gloss finish. Shower base sold separately.						
49" to 60" x 32" x 59", white	B1@3.00	Ea	191.00	97.60	288.60	491.00
Five-piece overlap construction design tub surround, Tall Elite. 2 towel bars and 6 shelves. Fits alcoves from 49" to 60-1/2" wide by 28" to 31" deep. Panels are easy to trim for window openings. White high-gloss finish.						
80" high x 60" wide	B1@3.00	Ea	155.00	97.60	252.60	429.00
Bathtub wall surround kit, Elite. Walls only. Tub and shower faucets sold separately. 6 shelves. 2 clear towel bars. Thicker corners provide greater strength and durability.						
58" wall height, bone	B1@3.00	Ea	125.00	97.60	222.60	378.00
58" wall height, white	B1@3.00	Ea	125.00	97.60	222.60	378.00
Bathtub tile wall kit. Rounded corners, bullnose edges, and engraved white grout for ceramic tile appearance. Groutless design. PVC finish resists stains, mold and mildew. Panels can be cut to fit window openings. 3 panels. Tub and shower faucets sold separately.						
30" x 60" x 60"	B1@3.00	Ea	235.00	97.60	332.60	565.00
Bathtub walls, Sterling Plumbing. Resistant to chipping, cracking and peeling. Easy-to-clean high-gloss surface. Durable material with color molded throughout fixture. Labor includes fastening in place only.						
Bone	B1@3.00	Ea	225.00	97.60	322.60	548.00
Fiberglass bath wall window trim.						
White	B1@1.00	Ea	90.00	32.50	122.50	208.00

Bypass Door Tub Enclosures

	Craft@Hrs	Unit	Material	Labor	Total	Sell
Bypass door tub enclosure. 60" wide opening. With track.						
Brass, ellipse glass	BC@1.20	Ea	221.00	43.20	264.20	449.00
Brass, hammertone glass	BC@1.20	Ea	152.00	43.20	195.20	332.00
Brass, hammertone mirror glass	BC@1.20	Ea	167.00	43.20	210.20	357.00
Silver, ellipse glass	BC@1.20	Ea	211.00	43.20	254.20	432.00
Silver, frameless, clear glass	BC@1.20	Ea	363.00	43.20	406.20	691.00
Silver, frameless, rainglass	BC@1.20	Ea	314.00	43.20	357.20	607.00
Silver, fluted glass	BC@1.20	Ea	265.00	43.20	308.20	524.00
Silver, hammertone glass	BC@1.20	Ea	137.00	43.20	180.20	306.00
Silver, hammertone mirror glass	BC@1.20	Ea	157.00	43.20	200.20	340.00
Silver, obscure glass	BC@1.20	Ea	240.00	43.20	283.20	481.00
Silver, rainglass	BC@1.20	Ea	176.00	43.20	219.20	373.00
White, ellipse glass	BC@1.20	Ea	221.00	43.20	264.20	449.00

	Craft@Hrs	Unit	Material	Labor	Total	Sell
Sliding tub enclosure door. 60" wide opening. With track.						
Bright nickel, frameless, 1/4" glass	BC@1.20	Ea	355.00	43.20	398.20	677.00
Bright silver finish, rain glass	BC@1.20	Ea	162.00	43.20	205.20	349.00
Deluxe frameless, radiance glass	BC@1.20	Ea	299.00	43.20	342.20	582.00
Deluxe gold, glue chip glass	BC@1.20	Ea	343.00	43.20	386.20	657.00
Deluxe silver finish, rain glass	BC@1.20	Ea	275.00	43.20	318.20	541.00
Deluxe silver, glue chip glass	BC@1.20	Ea	320.00	43.20	363.20	617.00
Gold finish, rain glass	BC@1.20	Ea	174.00	43.20	217.20	369.00
Silver, frameless, 1/4" clear glass	BC@1.20	Ea	262.00	43.20	305.20	519.00
Silver, frameless, clear glass	BC@1.20	Ea	238.00	43.20	281.20	478.00
Silver, frameless, clear glass	BC@1.20	Ea	226.00	43.20	269.20	458.00
Silver, frameless, radiance glass	BC@1.20	Ea	244.00	43.20	287.20	488.00
Silver finish, obscure glass	BC@1.20	Ea	119.00	43.20	162.20	276.00
Silver finish, swan glass	BC@1.20	Ea	167.00	43.20	210.20	357.00
White finish, rain glass	BC@1.20	Ea	172.00	43.20	215.20	366.00
Framed bypass door tub enclosure, Keystone. Fits 57" to 59" wide by 57-3/8" openings. 6'8" high. EZ Kleen® track system. Factory-installed towel bar. No exposed screws.						
Nickel, obscure glass	BC@1.20	Ea	205.00	43.20	248.20	422.00
Silver frame, 3/8" rain glass	BC@1.20	Ea	253.00	43.20	296.20	504.00
Polished Brass	BC@1.20	Ea	370.00	43.20	413.20	702.00
Frameless bypass door shower enclosure. 54" to 59" wide, 71" high.						
Satin nickel track, 3/8" glass	BC@1.20	Ea	578.00	43.20	621.20	1,060.00
Satin nickel track, clear glass	BC@1.20	Ea	383.00	43.20	426.20	725.00
Silver track, 3/8" clear glass	BC@1.20	Ea	768.00	43.20	811.20	1,380.00
Silver track, clear glass	BC@1.20	Ea	324.00	43.20	367.20	624.00
Silver track, Krystal flute glass	BC@1.20	Ea	383.00	43.20	426.20	725.00
White track, clear glass	BC@1.20	Ea	358.00	43.20	401.20	682.00

Fiberglass Tub and Shower Combinations

	Craft@Hrs	Unit	Material	Labor	Total	Sell
Lascoat™ tub and shower unit. Integral toiletry shelves. Installed acrylic grab bar. Slip-resistant textured bottom. 60" x 30" x 72". White. Smooth wall. Labor includes setting and fastening in place only. Add for water supply and drain connections as required. Be sure any tub and shower unit you select for replacement purposes will fit through existing doorways.						
Left-hand drain	P1@4.75	Ea	335.00	167.00	502.00	853.00
Right-hand drain	P1@4.75	Ea	335.00	167.00	502.00	853.00
Lascoat™ two-piece fiberglass tub and shower unit. BathLock™ front installation system. Integral toiletry shelves and acrylic grab bar. Smooth wall design. Leak-proof right angle joint flanges. Slip-resistant textured bottom. White. 60" x 30" x 72". Labor includes setting and fastening in place only. Add for water supply and drain connections as required.						
Left-hand drain	P1@4.75	Ea	407.00	167.00	574.00	976.00
Right-hand drain	P1@4.75	Ea	407.00	167.00	574.00	976.00
Tub and shower, Lasco Bathware. 4" smooth tile design. Integral soap shelf. Slip-resistant, textured bottom. 60" x 32" x 72". White. Labor includes setting and fastening in place only. Add for water supply and drain connections as required. Be sure any tub and shower unit you select for replacement purposes will fit through existing doorways.						
Left-hand drain	P1@4.75	Ea	370.00	167.00	537.00	913.00
Right-hand drain	P1@4.75	Ea	370.00	167.00	537.00	913.00

	Craft@Hrs	Unit	Material	Labor	Total	Sell
Two-piece tub and shower, Lasco Bathware. Smooth wall tile design. BathLock™ front installation system. Leak-proof right-angle joint flanges. Integral soap shelves. Specially designed to hide seam. 60" x 30" x 72". White. Acrylic grab bar. Labor includes setting and fastening in place only. Add for water supply and drain connections as required.						
Left-hand drain	P1@4.75	Ea	426.00	167.00	593.00	1,010.00
Right-hand drain	P1@4.75	Ea	426.00	167.00	593.00	1,010.00
Three-piece molded fiberglass sectional tub and shower, Lasco Bathware. 60" x 30" x 72". Smooth wall design. BathLock™ front installation system. Leak-proof right-angle joint flanges. Integral toiletry shelves. Acrylic grab bar. Slip-resistant textured bottom. Labor includes setting and fastening in place only. Add for water supply and drain connections as required.						
Left-hand drain	P1@4.75	Ea	417.00	167.00	584.00	993.00
Right-hand drain	P1@4.75	Ea	471.00	167.00	638.00	1,080.00
Two-piece fiberglass tub and shower unit. With towel bar, shelf, and soap dish. Smooth wall design. BathLock™ front installation system. Leak-proof right-angle joint flanges. Slip-resistant textured bottom. Acrylic grab bar. Bone finish. 60" x 30" x 72". Labor includes setting and fastening in place only. Add for water supply and drain connections as required.						
Left-hand drain	P1@4.75	Ea	370.00	167.00	537.00	913.00
Right-hand drain	P1@4.75	Ea	370.00	167.00	537.00	913.00

Shower Stalls

	Craft@Hrs	Unit	Material	Labor	Total	Sell
Demolish stall shower. Remove 3 walls and receptor. Remove supply pipe and insulation. Remove drain with a cutting torch. Stack debris on site.						
Per shower stall	BL@1.55	Ea	—	45.00	45.00	76.50
Shower stall kit, American Shower & Bath. Free-standing shower stall. Includes heavy-gauge walls in one-piece design. Molded soap dish and shampoo holder. 2 hand rails for convenient support. Base with no-caulk drain. Faucet, showerhead and installation hardware included. Add the cost of water supply and drain connection.						
White	P1@2.90	Ea	125.00	102.00	227.00	386.00
One-piece shower stall, Lasco Bathware. Dimpled tile finish. Center drain location. Comfort-molded rear corner seats. Slip-resistant textured bottom. Simulated 4" textured tile design. Integral toiletry shelves. Acrylic grab bar. Space-saver design. Add for water supply, drain, valve and showerhead. White. 72" high.						
32" x 32", smooth finish	P1@2.90	Ea	335.00	102.00	437.00	743.00
36" x 36", 4" tile finish	P1@2.90	Ea	365.00	102.00	467.00	794.00
48" x 35", smooth finish	P1@3.25	Ea	374.00	114.00	488.00	830.00
Lascoat™ two-piece molded sectional shower stall, Lasco Bathware. BathLock™ front installation system. Leak-proof right-angle joint flanges. Central drain. Integral soap shelf. Smooth walls. Slip-resistant textured bottom. Specially designed to hide seam. Add the cost of water supply and drain connection, valve and showerhead. White.						
32" x 32" x 72-3/4"	P1@2.90	Ea	413.00	102.00	515.00	876.00
36" x 36" x 72"	P1@2.90	Ea	413.00	102.00	515.00	876.00
Smooth wall shower stall. Center drain. BathLock™ front installation system. Leak-proof right-angle joint flanges. Integral soap shelf. Slip-resistant textured bottom. Add for water supply, drain, valve and showerhead. White. 72" high.						
36" x 36", 3-piece	P1@2.90	Ea	342.00	102.00	444.00	755.00
38" x 38", 2-piece, neo-angle	P1@2.90	Ea	404.00	102.00	506.00	860.00
48" x 34", 2-piece	P1@3.25	Ea	397.00	114.00	511.00	869.00
48" x 34", 3-piece	P1@3.25	Ea	373.00	114.00	487.00	828.00
60" x 34", 3-piece	P1@3.25	Ea	415.00	114.00	529.00	899.00

	Craft@Hrs	Unit	Material	Labor	Total	Sell
Corner entry shower stall kit, American Shower & Bath. White aluminum frame with clear tempered safety glass. Easy-to-clean walls and corner caddy with four shelves and towel bar. Heavy-duty base. Drain, drain cover and installation hardware included. Add the cost of water supply and drain connection, valve and showerhead. White.						
32" x 32" x 71"	P1@2.90	Ea	366.00	102.00	468.00	796.00
Round shower stall kit. Includes shower stall, door, and drain. Installs directly to either studs or finished walls. Reversible sliding door. Non-slip textured base. Bottle holder and shelves. 33-3/4" x 33-3/4" base measured at walls. 78-1/4" height. Easy to clean. Add for water supply, drain connection, valve and showerhead.						
34" x 34"	P1@2.90	Ea	585.00	102.00	687.00	1,170.00
Neo-angle 38-inch shower stall kit, American Shower & Bath. Includes anodized aluminum frame and tempered safety glass pivot door. Easy-to-clean walls. Corner caddy with 4 shelves and towel bar. Heavy-duty base, drain, drain cover and installation hardware. Add the cost of valve, water supply and drain connection.						
Chrome frame, obscure glass	P1@3.25	Ea	407.00	114.00	521.00	886.00
Chrome frame, rainfall glass	P1@3.25	Ea	508.00	114.00	622.00	1,060.00
Chrome with clear glass	P1@3.25	Ea	653.00	114.00	767.00	1,300.00
Chrome with radiance glass	P1@3.25	Ea	518.00	114.00	632.00	1,070.00
Gold frame, autumn glass	P1@3.25	Ea	491.00	114.00	605.00	1,030.00
Gold frame, rainfall glass	P1@3.25	Ea	516.00	114.00	630.00	1,070.00
Silver frame, obscure glass	P1@3.25	Ea	354.00	114.00	468.00	796.00
White frame, rainfall glass	P1@3.25	Ea	428.00	114.00	542.00	921.00
Hazel three-piece neo-angle shower stall kit. Includes shower stall, door, and drain. Large soap dishes and clear acrylic towel bar. Reversible pivot glass panel door with left and right central opening. Magnetic watertight seal. Drip channel on door. Non-slip textured 38" x 38" base. 74" high. Easy to clean. Add the cost of valve, water supply and drain connection.						
Chrome finish, clear glass	P1@3.25	Ea	361.00	114.00	475.00	808.00
White frame, obscure glass	P1@3.25	Ea	334.00	114.00	448.00	762.00
Durastall® three-piece fiberglass shower stall, E.L. Mustee. Tongue-and-groove interlocking seams. Semi-gloss surface. Colorfast, waterproof, and mold- and mildew-resistant. Seamless corners. Swing doors. Shelf and towel bar molded in. Molded fiberglass panels. Fits Mustee Durabase. Easy to clean. Add the cost of base, water supply and drain connection, valve and showerhead. White.						
32" x 32"	P1@3.25	Ea	243.00	114.00	357.00	607.00
36" x 36"	P1@3.25	Ea	264.00	114.00	378.00	643.00
Durabase® one-piece shower base, E.L. Mustee. Underbody ribbed for added strength. Slip-resistant. Mold- and mildew-resistant. Semi-gloss surface. Labor includes setting and fastening in place only. Add for drain connections as required.						
32" long, 32" wide, square, single threshold	P1@.600	Ea	90.80	21.10	111.90	190.00
36" long, 36" wide, neo-angle corner threshold	P1@.600	Ea	101.00	21.10	122.10	208.00
36" long, 36" wide, square, single threshold	P1@.600	Ea	101.00	21.10	122.10	208.00
38" long, 38" wide, neo-angle corner threshold	P1@.800	Ea	116.00	28.10	144.10	245.00
48" long, 32" wide, rectangular, single threshold	P1@.800	Ea	132.00	28.10	160.10	272.00
48" long, 34" wide, rectangular, single threshold	P1@.800	Ea	142.00	28.10	170.10	289.00

	Craft@Hrs	Unit	Material	Labor	Total	Sell
Lascoat™ shower pan, Lasco Bathware. Lascoat gel coat finish. Center drain location. Slip-resistant textured bottom. Integral nailing flange. 3-year warranty on Lascoat finish. White. 7-1/4" height. Labor includes setting and fastening in place only. Add for drain connections as required.						
32" long, 32" wide	P1@.600	Ea	132.00	21.10	153.10	260.00
36" long, 36" wide	P1@.800	Ea	142.00	28.10	170.10	289.00
42" long, 34" wide	P1@.800	Ea	162.00	28.10	190.10	323.00
60" long, 34" wide	P1@.800	Ea	183.00	28.10	211.10	359.00

Shower Enclosure Doors

	Craft@Hrs	Unit	Material	Labor	Total	Sell
Pivot shower enclosure door, Keystone. Obscure glass. Silver frame. Drip apron stops water drips when opening door. Reversible for left- or right-hand opening. Lifetime warranty. 3" adjustment allowance. Factory-installed handle. No exposed screws.						
23-3/4" to 26-3/4" wide	BG@1.20	Ea	137.00	40.80	177.80	302.00
27-3/4" to 30-3/4" wide	BG@1.20	Ea	114.00	40.80	154.80	263.00
30-3/4" to 33-3/4" wide	BG@1.20	Ea	120.00	40.80	160.80	273.00
33-3/4" to 36-3/4" wide	BG@1.20	Ea	161.00	40.80	201.80	343.00
Pivot shower enclosure door. Silver with rain glass.						
24" wide	BG@1.20	Ea	150.00	40.80	190.80	324.00
27" wide	BG@1.20	Ea	155.00	40.80	195.80	333.00
30" wide	BG@1.20	Ea	160.00	40.80	200.80	341.00
33" wide	BG@1.20	Ea	165.00	40.80	205.80	350.00
Pivot shower enclosure door, Contractors Wardrobe.						
Bright clear, rain glass, 59-1/8" x 69-1/2"	BG@1.20	Ea	306.00	40.80	346.80	590.00
Bright clear, rain glass, 59-1/2" x 63-1/4"	BG@1.20	Ea	183.00	40.80	223.80	380.00
Frameless, bright clear, clear glass, 59-3/8" x 68-7/8"	BG@1.20	Ea	234.00	40.80	274.80	467.00
Frameless, bright gold, clear glass, 59-3/8" x 68-7/8"	BG@1.20	Ea	259.00	40.80	299.80	510.00
Pivot shower enclosure door, Sterling Plumbing. Silver frame.						
24" to 27-1/2"	BG@1.20	Ea	130.00	40.80	170.80	290.00
32" to 35-1/2"	BG@1.20	Ea	155.00	40.80	195.80	333.00
36" to 39-1/2"	BG@1.20	Ea	172.00	40.80	212.80	362.00
42" to 45-1/2"	BG@1.20	Ea	207.00	40.80	247.80	421.00
48" to 51-1/2"	BG@1.20	Ea	233.00	40.80	273.80	465.00
Swinging shower enclosure door, Sterling Plumbing. Standard hinged. Use in either right- or left-handed opening. Height is 64". Silver frame finish. Hammered glass texture. Self-draining, easy-to-clean bottom track. Fits most fiberglass surround units.						
23-1/2" to 25" wide	BG@1.20	Ea	101.00	40.80	141.80	241.00
27" to 28-1/2" wide	BG@1.20	Ea	130.00	40.80	170.80	290.00
31" to 32-1/2" wide	BG@1.20	Ea	151.00	40.80	191.80	326.00
Hinged shower enclosure door. Silver frame.						
24", hammertone glass	BG@1.20	Ea	107.00	40.80	147.80	251.00
31", pebble glass	BG@1.20	Ea	186.00	40.80	226.80	386.00
36", pebble glass	BG@1.20	Ea	208.00	40.80	248.80	423.00

	Craft@Hrs	Unit	Material	Labor	Total	Sell
Bypass door shower enclosure.						
48", silver, hammertone glass	BG@1.20	Ea	155.00	40.80	195.80	333.00
60", brass, hammertone glass	BG@1.20	Ea	180.00	40.80	220.80	375.00
60", silver, hammertone glass	BG@1.20	Ea	206.00	40.80	246.80	420.00
60", silver, rain glass	BG@1.20	Ea	247.00	40.80	287.80	489.00

Single-Control Tub and Shower Faucets

Single-control tub and shower faucet, Price Pfister. Round acrylic mixing valve handle. Labor includes set in place and connection only. Add the cost of plumbing rough-in as required.

Chrome	P1@1.40	Ea	95.00	49.20	144.20	245.00
Chrome, shower only	P1@1.00	Ea	77.80	35.20	113.00	192.00

Single-knob tub and shower faucet set. Pressure balance faucet prevents uncomfortable water temperature changes. Durable chrome finish. Labor includes set in place and connection only. Add the cost of plumbing rough-in as required.

Chrome handle	P1@1.40	Ea	96.10	49.20	145.30	247.00
Chrome or brass handle	P1@1.40	Ea	152.00	49.20	201.20	342.00
Clear handle	P1@1.40	Ea	80.80	49.20	130.00	221.00
Porcelain and brass inserts	P1@1.40	Ea	111.00	49.20	160.20	272.00

Single-handle tub and shower set, Stanadyne. Pressure balanced. Classic styling. Washerless cartridge design. Labor includes set in place and connection only. Add the cost of plumbing rough-in as required.

Polished chrome finish, Villeta	P1@1.40	Ea	100.00	49.20	149.20	254.00
Platinum with chrome accents	P1@1.40	Ea	168.00	49.20	217.20	369.00

Reliant+ lever-handle tub and shower faucet set, American Standard. Pressure balanced. Metal lever handle. Ceramic disc valving resists hard and sandy water. 1/2" direct sweat or IPS FPT. High-temperature limit helps prevent scalding. Labor includes set in place and connection only. Add the cost of plumbing rough-in as required.

Chrome, shower only	P1@1.00	Ea	92.30	35.20	127.50	217.00
Polished chrome	P1@1.40	Ea	92.30	49.20	141.50	241.00

Single-control tub and shower faucet, Glacier Bay. Anti-scald pressure balanced mechanism. Brass showerhead with adjustable spray pattern. Drip-free washerless cartridge design.

Chrome	P1@1.40	Ea	76.20	49.20	125.40	213.00

Genesis single-control tub and shower faucet, Price Pfister. Pforever Pfaucet™ warranty. Covers function and finish for life. Labor includes set in place and connection only. Add the cost of plumbing rough-in as required.

Chrome finish	P1@1.40	Ea	121.00	49.20	170.20	289.00

Single-control tub and shower set, Moen. Pressure balanced. Lever handle. Includes adjustable spray showerhead, arm and flange, and diverter spout with flange. ADA compliant. Labor includes set in place and connection only. Add the cost of plumbing rough-in as required.

Chrome, brass accents	P1@1.40	Ea	167.00	49.20	216.20	368.00
Chrome, chrome handle	P1@1.40	Ea	147.00	49.20	196.20	334.00
Polished brass	P1@1.40	Ea	285.00	49.20	334.20	568.00

Single-control tub and shower faucet, Price Pfister. Pressure balanced. Lever handle. Washerless valves. Adjustable showerhead, arm, and flange. Diverter spout. ADA compliant. Labor includes set in place and connection only. Add the cost of plumbing rough-in as required.

Chrome, brass accent	P1@1.40	Ea	160.00	49.20	209.20	356.00
Chrome, porcelain handle	P1@1.40	Ea	139.00	49.20	188.20	320.00
Nickel, brass accent	P1@1.40	Ea	193.00	49.20	242.20	412.00

	Craft@Hrs	Unit	Material	Labor	Total	Sell
Williamsburg lever-handle tub and shower set, American Standard. Lifetime finish, will not tarnish. Ceramic disc valving, great for hard water. Both porcelain and metal lever inserts. Pressure balanced valve prevents accidental scalding. Lifetime warranty on function and finish. Labor includes set in place and connection only. Add the cost of plumbing rough-in as required.						
Brass and chrome finish	P1@1.40	Ea	173.00	49.20	222.20	378.00
Brass with chrome and brass	P1@1.40	Ea	166.00	49.20	215.20	366.00
Brass with satin and chrome	P1@1.40	Ea	260.00	49.20	309.20	526.00
Brass with velvet and chrome	P1@1.40	Ea	220.00	49.20	269.20	458.00
Brushed satin nickel finish	P1@1.40	Ea	217.00	49.20	266.20	453.00
Chrome finish	P1@1.40	Ea	144.00	49.20	193.20	328.00
Chrome and brass finish	P1@1.40	Ea	215.00	49.20	264.20	449.00
Polished brass finish	P1@1.40	Ea	175.00	49.20	224.20	381.00
Satin chrome finish	P1@1.40	Ea	217.00	49.20	266.20	453.00
Single-control tub and shower faucet. Pressure balanced. Lever handle. Showerhead, arm, and flange. Rite-Temp™ pressure-balancing valves eliminate surges of hot and cold water. Nostalgic design. Polished chrome finish. Labor includes set in place and connection only. Add the cost of plumbing rough-in as required.						
With diverter spout	P1@1.40	Ea	161.00	49.20	210.20	357.00
Lever-handle tub and shower faucet. Individual temperature and volume control. Large, decorative showerhead delivers a good spray pattern. Single-hole mount centerset. Brass. Pressure balanced with Scald-Guard®. Labor includes set in place and connection only. Add the cost of plumbing rough-in as required.						
Chrome finish	P1@1.40	Ea	175.00	49.20	224.20	381.00
Pegasus 1000 series single-handle tub and shower faucet, J A Manufacturing. Pressure balanced. 1/4 turn, anti-scald valve temperature control mechanism. Bell-shaped deluxe head with adjustable arm. Favorite temperature set. Shutoff valve. Labor includes set in place and connection only. Add the cost of plumbing rough-in as required.						
Brushed nickel	P1@1.40	Ea	236.00	49.20	285.20	485.00
Brushed nickel, shower only	P1@1.00	Ea	215.00	35.20	250.20	425.00
Chrome	P1@1.40	Ea	198.00	49.20	247.20	420.00
Chrome, shower only	P1@1.00	Ea	166.00	35.20	201.20	342.00
Innovations single-handle tub and shower faucet, Masco. Pressure balanced. Labor includes set in place and connection only. Add the cost of plumbing rough-in as required.						
Pearl nickel finish	P1@1.40	Ea	222.00	49.20	271.20	461.00
Two-handle tub and shower faucet, Price Pfister. Knob handles. Includes showerhead, flange, and arm. Polished chrome. Labor includes set in place and connection only. Add the cost of plumbing rough-in as required.						
Porcelain handle, shower only	P1@1.40	Ea	165.00	49.20	214.20	364.00
Verve®, metal handle, shower only	P1@1.00	Ea	95.00	35.20	130.20	221.00
Windsor, acrylic handle	P1@1.40	Ea	110.00	49.20	159.20	271.00
Two-handle tub and shower faucet, Masco. Scald-Guard® keeps water temperature within 3 degrees Fahrenheit to prevent sudden temperature changes. Touch-Clean® showerhead. Lifetime warranty. Labor includes set in place and connection only. Add the cost of plumbing rough-in as required.						
Polished brass and chrome finish	P1@1.40	Ea	204.00	49.20	253.20	430.00
Three-handle tub and shower faucet set, Price Pfister. Chrome. Includes showerhead, arm, flanges and tub spout. Labor includes set in place and connection only. Add the cost of plumbing rough-in as required.						
Porcelain cross handles	P1@1.40	Ea	183.00	49.20	232.20	395.00
Chrome handles	P1@1.40	Ea	95.90	49.20	145.10	247.00
Windsor® acrylic handles	P1@1.40	Ea	125.00	49.20	174.20	296.00

	Craft@Hrs	Unit	Material	Labor	Total	Sell
Three-handle tub and shower faucet, Moen. Clear knobs. Includes showerhead, arm, flange, and diverter spout. Full-spray, water-saving showerhead. Reliable washerless cartridge. Labor includes set in place and connection only. Add the cost of plumbing rough-in as required.						
Chrome	P1@1.50	Ea	122.00	52.70	174.70	297.00
Savannah three-handle tub and shower faucet set, Price Pfister. Ceramic disc valving. Porcelain levers. Solid brass construction. Pressure balanced with Scald-Guard®. Labor includes set in place and connection only. Add the cost of plumbing rough-in as required.						
White porcelain lever handles	P1@1.50	Ea	197.00	52.70	249.70	424.00
Add-on shower kit for built-in tub, Brass Craft. Includes water-saver showerhead, diverter spout, adjustable risers, and wall bracket. Labor includes set in place and connection only. Add the cost of plumbing rough-in as required.						
Kit	P1@1.00	Ea	40.00	35.20	75.20	128.00

Bath and Shower Accessories

	Craft@Hrs	Unit	Material	Labor	Total	Sell
Power massage showerhead. 3 massage settings: pulsating, gentle and combination. Great for low pressure. Self-cleaning, water-saving showerhead for continuous flow.						
Chrome	P1@.200	Ea	30.60	7.03	37.63	64.00
Brushed Nickel	P1@.200	Ea	40.80	7.03	47.83	81.30
Sunflower showerhead.						
8", easy clean	P1@.200	Ea	73.40	7.03	80.43	137.00

Toilets (Water Closets)

	Craft@Hrs	Unit	Material	Labor	Total	Sell
Hot and cold bidet. Drilled for hot and cold valves. Labor includes setting and connection only. Add the cost of valves and piping.						
Diplomatico, small, bone	P1@1.50	Ea	80.80	52.70	133.50	227.00
White	P1@1.50	Ea	91.80	52.70	144.50	246.00
Bidet faucet set, three-handle. Fits 5" to 9" center holes. Ceramic disc valves. Labor includes set in place and connection only. Add the cost of plumbing rough-in as required.						
Chrome	P1@1.10	Ea	264.00	38.70	302.70	515.00
Bronze	P1@1.10	Ea	399.00	38.70	437.70	744.00
Toilet To Go kit. 1.6-gallon flush. Includes tank, bowl, seat, wax ring and toilet bolts. Easy to install. Exceeds ANSI standards. Expanded 2" trapway. Large 9" x 8-1/2" water surface. Labor includes setting and connection only. Add the cost of wall stop valve and piping.						
Round, Bel-Air, white	P1@1.10	Ea	79.60	38.70	118.30	201.00
Round, Metro, stylish, white	P1@1.10	Ea	105.00	38.70	143.70	244.00
Elongated, Metro	P1@1.10	Ea	125.00	38.70	163.70	278.00
Cadet compact toilet, American Standard. Fits into space of a round front bowl toilet. 1.6-gallon flush. Sanitary dam. Labor includes setting and connection only. Add the cost of wall stop valve, piping and the seat.						
Elongated bowl	P1@.550	Ea	93.00	19.30	112.30	191.00
Tank	P1@.550	Ea	81.10	19.30	100.40	171.00

	Craft@Hrs	Unit	Material	Labor	Total	Sell
Aragon IV one-piece toilet, Foremost International. 1.6-gallon flush. Round front. Seat included. Labor includes setting and connection only. Add the cost of wall stop valve and piping.						
White	P1@1.10	Ea	108.00	38.70	146.70	249.00
Hamilton one-piece toilet, American Standard. Vitreous china. Low consumption. Elongated siphon action bowl. Includes color-matched Rise and Shine™ plastic seat and cover with lift-off hinge. 2-bolt cap. Labor includes setting and connection only. Add the cost of wall stop valve and piping.						
Bone	P1@1.10	Ea	284.00	38.70	322.70	549.00
Linen	P1@1.10	Ea	304.00	38.70	342.70	583.00
White	P1@1.10	Ea	265.00	38.70	303.70	516.00
Cadet pressure-assist toilet, American Standard. Labor includes setting and connection only. Add the cost of wall stop valve, piping and the seat.						
Round front bowl	P1@.550	Ea	147.00	19.30	166.30	283.00
Tank, 1.6 gallon	P1@.550	Ea	254.00	19.30	273.30	465.00
Champion toilet, American Standard. Labor includes setting and connection only. Add the cost of wall stop valve, piping and the seat.						
16-1/2" toilet bowl, white	P1@.550	Ea	170.00	19.30	189.30	322.00
Toilet tank, white	P1@.550	Ea	128.00	19.30	147.30	250.00
Williamsburg Antiquity one-piece toilet, American Standard. Elongated front model fits in the space of a round toilet. 2" trapway. Seat included. Labor includes setting and connection only. Add the cost of wall stop valve and piping.						
Elongated, white	P1@1.10	Ea	298.00	38.70	336.70	572.00
Gabrielle Comfort Height™ one-piece toilet, Kohler. Contemporary look and superior flush. Rim height of 16-1/8" is consistent with standard height of chair. Vigorous flush is a result of siphon-jet flushing technology and 2" fully glazed trapway. 10" x 9" water surface, which maintains a cleaner bowl. Ingenium™ flushing system. Labor includes setting and connection only. Add the cost of wall stop valve and piping.						
Elongated, white	P1@1.10	Ea	516.00	38.70	554.70	943.00
San Raphael™ toilet, Kohler. One-piece design. Kohler's quietest flushing toilet. 2" fully glazed trapway. Polished chrome trip lever. Includes seat. Labor includes setting and connection only. Add the cost of wall stop valve and piping.						
Elongated, white	P1@1.10	Ea	702.00	38.70	740.70	1,260.00
One-piece wall-mount toilet. Vitreous china. Full glazed trapway. Condensation channel. Flush valve. Direct-fed siphon jet action. 10" x 12" water surface area. 25" x 14-3/4" x 15". Labor includes setting and connection only. Add the cost of a flush valve, carrier and piping.						
Top spud, elongated, white	P1@1.10	Ea	225.00	38.70	263.70	448.00
Sloan Royal flush valve. Water saver. For floor or wall-hung top spud bowls.						
Sloan No. 110	P1@.600	Ea	153.00	21.10	174.10	296.00
Sloan Royal urinal flush valve. Water saver. 1.5-gallon flush. For 3/4" top spud urinals.						
Sloan No. 186	P1@.600	Ea	153.00	21.10	174.10	296.00

	Craft@Hrs	Unit	Material	Labor	Total	Sell
Urinal. One-gallon flush. Labor includes setting and connection only. Add the cost of a flush valve, hanger and piping.						
3/4" top spud, white	P1@1.50	Ea	220.00	52.70	272.70	464.00
Urinal. Vitreous china. Flushing rim. Elongated 14" rim from finished wall. Washout flush action. Includes two wall hangers and outlet connection. Labor includes setting and connection only. Add the cost of a flush valve, hanger and piping.						
3/4" top spud	P1@1.50	Ea	348.00	52.70	400.70	681.00

Toilet Installation and Repair

	Craft@Hrs	Unit	Material	Labor	Total	Sell
Low-Boy toilet supply hook-up kit, Brass Craft. For iron pipe or copper tube stub-outs. Kit includes 1-1/2" female pipe thread by 3/8" OD tube compression by 1/2" male pipe thread union adapter, and Teflon® tape. Includes 9" braided Speedi-Plumb water supply connector.						
Kit	P1@.250	Ea	13.50	8.79	22.29	37.90
Angle valve.						
1/2" with 3/8" x 12" supply tube kit	P1@.250	Ea	7.25	8.79	16.04	27.30
Toilet Supply Line. Braided stainless steel.						
3/8" x 7/8" x 12" supply line	P1@.150	Ea	4.97	5.27	10.24	17.40
Decorative flexible supply stop. 1/2" iron pipe size by 3/8" outside diameter.						
Angle valve, polished brass	P1@.250	Ea	11.00	8.79	19.79	33.60
No-Seep wax toilet bowl gasket. Wax gasket with polyethylene flange fits 3" and 4" lines, closet bowls only. Includes 1/4" x 2-1/4" brass bolt kit with stainless washers and brass nuts.						
With brass bolt kit	P1@.100	Ea	5.20	3.52	8.72	14.80
Cast iron water closet flange. Use on heavyweight plastic or soil pipe. Complete with body, brass ring, bolts and gasket.						
4" diameter, compression	P1@.250	Ea	15.00	8.79	23.79	40.40
Toilet base plate cover. For toilet replacement. Conceals ring around smaller base of new toilet. Plastic. Plates can be stacked to raise toilet. 11" wide.						
Round nose, 22-1/2" long	P1@.080	Ea	9.35	2.81	12.16	20.70
Square nose, 18-3/4" long	P1@.080	Ea	8.35	2.81	11.16	19.00
Leak Sentry™ anti-siphon toilet fill valve, Fluidmaster. Prevents automatic refill of a leaky tank. At most, only one tank of water can escape a toilet. Toilet always remains operational. Bowl refill adjustment prevents wasteful overfilling. Fits most toilets, including 1.6 gallon-per-flush models. Corrosion-resistant plastic and stainless steel.						
1.6 gallon flush	P1@.300	Ea	11.20	10.50	21.70	36.90
Complete toilet tank repair kit, Fluidmaster. Includes flush valve with Adjust-Flush™ flapper, anti-siphon fill valve, Sure-Fit® tank lever, 3 bolts and gasket. Flapper dial regulates volume. Valve height adjusts 9" to 14". Lever trims and bends to fit. Universal gasket.						
Kit	P1@.850	Ea	20.40	29.90	50.30	85.50
One-piece toilet repair kit, Fluidmaster. Replaces many one-piece toilet assemblies. Fewer parts. Ideal for wet or dry inlet tubes.						
Kit	P1@.850	Ea	47.90	29.90	77.80	132.00
Tank-to-bowl bolt kit, Fluidmaster. Two 5/16" x 3" brass bolts with washers and wing nuts.						
Kit	P1@.275	Ea	4.73	9.67	14.40	24.50

	Craft@Hrs	Unit	Material	Labor	Total	Sell

Toilet Seats

Molded wood toilet seat, Bemis Manufacturing. Multi-coat enamel finish. Closed front and cover to fit a regular bowl. Color-matched bumpers and hinges. Top-Tite® hinges with non-corrosive, top-tightening bolts and wing nuts. 1" ring thickness including bumper. Includes lid.

	Craft@Hrs	Unit	Material	Labor	Total	Sell
Round, beige	P1@.250	Ea	10.70	8.79	19.49	33.10
Round, bone	P1@.250	Ea	11.20	8.79	19.99	34.00
Round, gold	P1@.250	Ea	11.20	8.79	19.99	34.00
Round, pink	P1@.250	Ea	11.20	8.79	19.99	34.00
Round, seafoam	P1@.250	Ea	11.20	8.79	19.99	34.00
Round, sky blue	P1@.250	Ea	11.20	8.79	19.99	34.00
Round, silver	P1@.250	Ea	10.70	8.79	19.49	33.10
Round, white	P1@.250	Ea	10.20	8.79	18.99	32.30
Round, white, fits Eljer Emblem	P1@.250	Ea	15.50	8.79	24.29	41.30
Round, black, chrome hinge	P1@.250	Ea	22.80	8.79	31.59	53.70
Round, bone, chrome hinge	P1@.250	Ea	21.40	8.79	30.19	51.30
Round, silver, chrome hinge	P1@.250	Ea	21.80	8.79	30.59	52.00
Round, white, chrome hinge	P1@.250	Ea	20.40	8.79	29.19	49.60

Molded wood open-front toilet seat, Bemis Manufacturing. Multi-coat enamel finish. Color-matched bumpers and hinges. Top-Tite® hinges with non-corrosive, top-tightening bolts and wing nuts. 1" ring thickness including bumper. Includes lid. 3-year warranty.

	Craft@Hrs	Unit	Material	Labor	Total	Sell
Elongated, biscuit	P1@.250	Ea	16.20	8.79	24.99	42.50
Elongated, bone	P1@.250	Ea	16.20	8.79	24.99	42.50
Elongated, white	P1@.250	Ea	14.40	8.79	23.19	39.40
Round, biscuit	P1@.250	Ea	12.00	8.79	20.79	35.30
Round, bone	P1@.250	Ea	10.70	8.79	19.49	33.10
Round, white	P1@.250	Ea	10.50	8.79	19.29	32.80

Plastic open-front toilet seat, Bemis Manufacturing. Open-front with cover. Easy to install top-tightening hinges. White.

	Craft@Hrs	Unit	Material	Labor	Total	Sell
Elongated, no cover	P1@.250	Ea	24.40	8.79	33.19	56.40
Elongated, with cover	P1@.250	Ea	20.40	8.79	29.19	49.60
Round, no cover	P1@.250	Ea	20.40	8.79	29.19	49.60
Round, with cover	P1@.250	Ea	19.70	8.79	28.49	48.40

Molded wood shell design toilet seat, Bemis Manufacturing. Multi-coat enamel finish. Closed front and cover to fit a regular bowl. Color-matched bumpers and hinges. Top-Tite® hinges with non-corrosive, top-tightening bolts and wing nuts. 1" ring thickness including bumper. Includes lid.

	Craft@Hrs	Unit	Material	Labor	Total	Sell
Elongated, bone	P1@.250	Ea	28.50	8.79	37.29	63.40
Round, bone	P1@.250	Ea	22.80	8.79	31.59	53.70

Sculpted wood toilet seat. Durable high-gloss finish. Non-corrosive hinges. Installs without tools.

	Craft@Hrs	Unit	Material	Labor	Total	Sell
Elongated, bone	P1@.250	Ea	27.90	8.79	36.69	62.40
Elongated, white	P1@.250	Ea	27.90	8.79	36.69	62.40
Elongated, ivy, white	P1@.250	Ea	30.40	8.79	39.19	66.60
Round, ivy, bone	P1@.250	Ea	24.30	8.79	33.09	56.30
Round, ivy, white	P1@.250	Ea	24.70	8.79	33.49	56.90
Round, rose, white	P1@.250	Ea	24.00	8.79	32.79	55.70
Round, shell, black	P1@.250	Ea	24.50	8.79	33.29	56.60
Round, wave, bone	P1@.250	Ea	24.50	8.79	33.29	56.60

	Craft@Hrs	Unit	Material	Labor	Total	Sell
Soft decorator toilet seat, Bemis Manufacturing. Cushioned for comfort. High-density padding. Durable antibacterial color-fast vinyl with molded wood core. Dial On® exclusive no-wobble hinges. Installs without tools.						
Round, bone	P1@.250	Ea	17.50	8.79	26.29	44.70
Round, pink	P1@.250	Ea	17.00	8.79	25.79	43.80
Round, sky blue	P1@.250	Ea	17.10	8.79	25.89	44.00
Round, white	P1@.250	Ea	20.80	8.79	29.59	50.30
Round, white, chrome hinges	P1@.250	Ea	22.90	8.79	31.69	53.90
Round, white, embroidered butterfly	P1@.250	Ea	21.40	8.79	30.19	51.30
Round, white, Roman marble	P1@.250	Ea	21.40	8.79	30.19	51.30
Natural maple toilet seat. Durable, long-lasting furniture-grade finish. Decorative chrome-plated brass hinge. Will not warp, crack or split.						
Elongated	P1@.250	Ea	49.00	8.79	57.79	98.20
Elongated with diamond inlay	P1@.250	Ea	53.90	8.79	62.69	107.00
Round	P1@.250	Ea	44.10	8.79	52.89	89.90
Round with diamond inlay	P1@.250	Ea	53.90	8.79	62.69	107.00

Pedestal Lavatories

	Craft@Hrs	Unit	Material	Labor	Total	Sell
Bahama pedestal lavatory. Labor includes setting and connecting only. Add the cost of faucet and accessories.						
Gray	P1@.800	Ea	65.70	28.10	93.80	159.00
Sky rose	P1@.800	Ea	65.70	28.10	93.80	159.00
White	P1@.800	Ea	65.70	28.10	93.80	159.00
Williamsburg pedestal lavatory, American Standard. Elegant turn-of-the-century styling. 24-1/2" x 19" top. Labor includes setting and connecting only. Add the cost of faucet and accessories.						
4" centers, linen	P1@.800	Ea	131.00	28.10	159.10	270.00
4" centers, white	P1@.800	Ea	141.00	28.10	169.10	287.00
8" centers, linen	P1@.800	Ea	141.00	28.10	169.10	287.00
8" centers, white	P1@.800	Ea	136.00	28.10	164.10	279.00
Repertoire pedestal lavatory, American Standard. Labor includes setting and connecting only. Add the cost of faucet and accessories.						
4" centers, white	P1@.800	Ea	177.00	28.10	205.10	349.00
8" centers, linen	P1@.800	Ea	217.00	28.10	245.10	417.00
Seychelle pedestal lavatory, American Standard. Labor includes setting and connecting only. Add the cost of faucet and accessories.						
4" centers, white	P1@.800	Ea	172.00	28.10	200.10	340.00
Memoirs™ pedestal lavatory, Kohler. Traditional styling. Integrates with the Memoirs suite. Labor includes setting and connecting only. Add the cost of faucet and accessories.						
4" centers, white	P1@.800	Ea	203.00	28.10	231.10	393.00
8" centers, white	P1@.800	Ea	205.00	28.10	233.10	396.00
Memoirs™ pedestal lavatory, Kohler. Integral backsplash. 27" x 19-3/8" x 34". Vitreous china. Drilled centers. Traditional styling integrates with the Memoirs Suite. Labor includes setting and connecting only. Add the cost of faucet and accessories.						
4" centers, white, stately design	P1@.800	Ea	389.00	28.10	417.10	709.00

	Craft@Hrs	Unit	Material	Labor	Total	Sell
Memoirs™ pedestal lavatory base, Kohler. Traditional styling integrates with the Memoirs™ suite.						
White	P1@.350	Ea	159.00	12.30	171.30	291.00
Classic pedestal lavatory base.						
Bone	P1@.350	Ea	61.20	12.30	73.50	125.00
Coral	P1@.350	Ea	61.20	12.30	73.50	125.00
Silver	P1@.350	Ea	61.20	12.30	73.50	125.00
White	P1@.350	Ea	61.20	12.30	73.50	125.00
Lavatory leg set, Melard.						
Chrome	P1@.350	Ea	22.00	12.30	34.30	58.30
Standard collection pedestal lavatory metal leg set, American Standard.						
Chrome	P1@.350	Ea	336.00	12.30	348.30	592.00
Satin	P1@.350	Ea	361.00	12.30	373.30	635.00

Lavatory Sinks

	Craft@Hrs	Unit	Material	Labor	Total	Sell
Remove pedestal, counter- or wall-mounted lavatory. Turn off water. Disconnect hot and cold supply lines from the wall valves. Unscrew trap plug to the drain. Disconnect trap unions. Remove trap. Remove basin retaining screws or wall hangers. Remove wash basin.						
Per lavatory basin	P1@.630	Ea	—	22.20	22.20	37.70
Round vitreous china lavatory sink, Eljer Plumbingware. Self-rimming drop-in counter mount. 19" diameter. Front overflow. Labor includes setting only. Add the cost of connection, faucet and accessories.						
4" centers, natural	P1@.850	Ea	39.80	29.90	69.70	118.00
4" centers, white	P1@.850	Ea	39.80	29.90	69.70	118.00
Farmington™ drop-in lavatory sink, Kohler. Self-rimming installation. 19-1/4" x 16-1/4" overall. Cast iron. Labor includes setting only. Add the cost of connection, faucet and accessories.						
4" centers, almond	P1@.850	Ea	90.80	29.90	120.70	205.00
4" centers, white	P1@.850	Ea	90.80	29.90	120.70	205.00
8" centers, white	P1@.850	Ea	100.00	29.90	129.90	221.00
Renaissance drop-in lavatory sink, American Standard. Vitreous china. Self-rimming. Round, 19" overall dimensions. Labor includes setting only. Add the cost of connection, faucet and accessories.						
4" centers, bone	P1@.850	Ea	97.00	29.90	126.90	216.00
4" centers, white	P1@.850	Ea	97.00	29.90	126.90	216.00
Lavatory sink, American Standard. Tapered edges for style. Vitreous china. 19" diameter. Labor includes setting only. Add the cost of connection, faucet and accessories.						
4" centers, linen	P1@.850	Ea	93.10	29.90	123.00	209.00
Drop-in lavatory sink, American Standard. Vitreous china. Tapered edges for style. Self-rimming. Front overflow. Supplied with template and color-matched sealant. 20-3/8" x 17". Labor includes setting only. Add the cost of connection, faucet and accessories.						
4" centers, linen	P1@.850	Ea	87.20	29.90	117.10	199.00
Seychelle drop-in lavatory sink, American Standard. Vitreous china. Self-rimming. Bowl has distinctive sculptured pattern. Front overflow. Labor includes setting only. Add the cost of connection, faucet and accessories.						
4" centers, white	P1@.850	Ea	53.90	29.90	83.80	142.00

	Craft@Hrs	Unit	Material	Labor	Total	Sell
Cadet drop-in lavatory sink, American Standard. European-style interior bowl. Self-rimming oval countertop lavatory. Front overflow. Supplied with template and color-matched sealant. Vitreous china, 20" x 17" oval. Labor includes setting only. Add the cost of connection, faucet and accessories.						
4" centers, bone	P1@.850	Ea	70.40	29.90	100.30	171.00
4" centers, white	P1@.850	Ea	70.40	29.90	100.30	171.00
8" centers, white	P1@.850	Ea	91.00	29.90	120.90	206.00
Standard Collection drop-in lavatory sink, American Standard. Oval self-rimming design includes color-matched caulk for seamless, easy installation. Generous interior bowl with "invisible" front overflow. Nominal dimensions 22-7/8" x 18-1/2". Bowl size 15-3/4" wide by 11-1/4" front to back, 6-1/2" deep. Labor includes setting only. Add the cost of connection, faucet and accessories.						
8" centers, white	P1@.850	Ea	98.90	29.90	128.80	219.00
Williamsburg drop-in lavatory sink, American Standard. European-style interior bowl. Self-rimming oval countertop lavatory. Front overflow. Supplied with template and color-matched sealant. 24-1/2" x 18". Vitreous china. Labor includes setting only. Add the cost of connection, faucet and accessories.						
4" centers, linen	P1@.850	Ea	121.00	29.90	150.90	257.00
4" centers, white	P1@.850	Ea	111.00	29.90	140.90	240.00
8" centers, white	P1@.850	Ea	111.00	29.90	140.90	240.00
Memoirs™ drop-in lavatory sink, Kohler. Vitreous china. Glossy finish. Labor includes setting only. Add the cost of connection, faucet and accessories.						
4" centers, white	P1@.850	Ea	160.00	29.90	189.90	323.00
8" centers, white	P1@.850	Ea	160.00	29.90	189.90	323.00
Oval drop-in lavatory sink. Vitreous china. 20" x 17". Self-rimming. Front overflow. Labor includes setting only. Add the cost of connection, faucet and accessories.						
4" centers, natural	P1@.850	Ea	44.90	29.90	74.80	127.00
4" centers, white	P1@.850	Ea	39.80	29.90	69.70	118.00
Ovalyn II undercounter-mount lavatory sink, American Standard. Vitreous china. Unglazed rim. Rear overflow. Supplied with mounting kit. Labor includes setting only. Add the cost of connection, faucet and accessories.						
Bone, 17" x 14"	P1@.850	Ea	70.30	29.90	100.20	170.00
Linen, 17" x 14"	P1@.850	Ea	72.40	29.90	102.30	174.00
White, 17" x 14"	P1@.850	Ea	60.30	29.90	90.20	153.00
Laurel round enameled steel lavatory sink. Self-rimming counter mount. 19" diameter. Labor includes setting only. Add the cost of connection, faucet and accessories.						
4" centers, natural finish	P1@.850	Ea	31.90	29.90	61.80	105.00
Radiant lavatory sink, Kohler. Vitreous china. Glossy finish. Self-rimming, counter mount. Labor includes setting only. Add the cost of connection, faucet and accessories.						
4" centers, white	P1@.850	Ea	94.80	29.90	124.70	212.00
Seychelle countertop lavatory sink, American Standard. Labor includes setting only. Add the cost of connection, faucet and accessories.						
4" centers, bone	P1@.800	Ea	94.10	28.10	122.20	208.00
Summit acrylic lavatory sink, International Thermocast. 20-3/4" x 18" x 6-3/4" deep. 2-gallon bowl. High-gloss finish. Resists stain, rust, oxidation, chipping, and scratches. Labor includes setting only. Add the cost of connection, faucet and accessories.						
4" centers, bone	P1@.850	Ea	82.20	29.90	112.10	191.00
4" centers, white	P1@.850	Ea	60.20	29.90	90.10	153.00

	Craft@Hrs	Unit	Material	Labor	Total	Sell
Magnolia acrylic lavatory sink, International Thermocast. 22" x 19" x 6-3/4" deep. 3-gallon bowl. High-gloss finish. Resists stain, rust, oxidation, chipping, and scratches. Self-rimming counter mount. Labor includes setting only. Add the cost of connection, faucet and accessories.						
4" centers, white	P1@.850	Ea	81.50	29.90	111.40	189.00
8" centers, white	P1@.850	Ea	84.70	29.90	114.60	195.00
Reminiscence lavatory sink, American Standard. Vitreous china. Vintage-style design with raised motif backsplash. Self-rimming. Front overflow. Supplied with template and color-matched sealant. Labor includes setting only. Add the cost of connection, faucet and accessories.						
8" centers, white	P1@.850	Ea	141.00	29.90	170.90	291.00
Morning above-counter lavatory sink, American Standard. For above-counter installation. Integral faucet deck for single-hole faucet. Center drain outlet. Without overflow. Supplied with template and color-matched sealant. Labor includes setting only. Add the cost of connection, faucet and accessories.						
Single-hole faucet, white	P1@.850	Ea	223.00	29.90	252.90	430.00
Memoirs™ lavatory sink, Kohler. Vitreous china. Complies with Americans with Disabilities Act when installed per requirements of Accessibility Guidelines, Section 4.19 Lavatories & Mirrors. Traditional styling coordinates with the Memoirs Suite. Self-rimming, counter mount. Labor includes setting only. Add the cost of connection, faucet and accessories.						
8" centers, white	P1@.850	Ea	341.00	29.90	370.90	631.00
Miami wall-hung lavatory sink. 19-1/2" x 17-1/4". Labor includes setting only. No area prep work included. Add the cost of connection, faucet and accessories.						
4" centers, white	P1@.850	Ea	39.90	29.90	69.80	119.00
Hydra wall-hung lavatory sink, American Standard. Labor includes setting only. No area prep work included. Add the cost of connection, faucet and accessories.						
4" centers, white	P1@.850	Ea	52.80	29.90	82.70	141.00
Wall-hung lavatory sink. Labor includes setting only. No area prep work included. Add the cost of connection, faucet and accessories.						
4" centers, white	P1@.850	Ea	32.60	29.90	62.50	106.00
Lucerne wall-hung lavatory sink, American Standard. Labor includes setting only. No area prep work included. Add the cost of connection, faucet and accessories.						
4" centers, white	P1@.850	Ea	96.00	29.90	125.90	214.00
Minette corner wall-hung lavatory sink, American Standard. Vitreous china. Front overflow. Furnished with two wall hangers. Labor includes setting only. No area prep work included. Add the cost of connection, faucet and accessories.						
4" centers, white	P1@.850	Ea	212.00	29.90	241.90	411.00
Murro wall-hung lavatory sink, American Standard. Vitreous china. Rear overflow. Recessed self-draining deck. For concealed arm or wall. Universal design. Labor includes setting only. No area prep work included. Add the cost of connection, faucet and accessories.						
4" centers, white	P1@.850	Ea	192.00	29.90	221.90	377.00
Lavatory hanger bracket. Supports wall hung china lavatory basin.						
Lavatory hanger bracket	P1@.250	Ea	1.52	8.79	10.31	17.50

	Craft@Hrs	Unit	Material	Labor	Total	Sell
Single-Handle Bath Faucets, 4" Center Set						
Single-handle lavatory faucet, Glacier Bay. Deck-mount. 4" center set. Chrome finish. Lever handle. Washerless cartridge. Includes mechanical pop-up assembly. Labor includes set in place and connection only. Add the cost of plumbing rough-in as required.						
Chrome with brass accents	P1@.900	Ea	86.40	31.60	118.00	201.00
Polished chrome finish	P1@.900	Ea	34.30	31.60	65.90	112.00
Zinc single-handle lavatory faucet. Deck-mount. 4" center set. Lever handle. Washerless cartridge. With mechanical pop-up assembly. Labor includes set in place and connection only. Add the cost of plumbing rough-in as required.						
Chrome finish, zinc	P1@.900	Ea	60.50	31.60	92.10	157.00
Acrylic knob single-control lavatory faucet. Includes pop-up plug and waste assembly. Washerless cartridge. 4" center set. Water- and energy-saving aerator. Labor includes set in place and connection only. Add the cost of plumbing rough-in as required.						
Chrome	P1@.900	Ea	62.20	31.60	93.80	159.00
Reliant single-handle lavatory faucet, American Standard. Deck-mount. Lever handle. All-metal construction. Ceramic disc valve cartridge with adjustable hot temperature limit safety stop. Includes mechanical pop-up assembly. 2.5-gallons-per-minute flow-restricted aerator. 1/2" male threaded copper connector and nuts. 4" center set. Labor includes set in place and connection only. Add the cost of plumbing rough-in as required.						
Polished chrome	P1@.900	Ea	62.40	31.60	94.00	160.00
Classic single-handle lavatory faucet. Deck-mount. 4" center set. Lever handle. Washerless cartridge. Lifetime warranty. Includes mechanical pop-up assembly. Labor includes set in place and connection only. Add the cost of plumbing rough-in as required.						
Chrome finish	P1@.900	Ea	83.30	31.60	114.90	195.00
Single-handle lavatory faucet. Lever handle. 4" center set. Metal poppet. ADA compliant. Labor includes set in place and connection only. Add the cost of plumbing rough-in as required.						
Chrome	P1@.900	Ea	72.00	31.60	103.60	176.00
Villeta single-handle lavatory faucet. Deck-mount. 4" center set. Lever handle. Includes mechanical pop-up assembly. Lifetime limited warranties against leaks, drips and finish defects. Labor includes set in place and connection only. Add the cost of plumbing rough-in as required.						
Chrome finish	P1@.900	Ea	78.80	31.60	110.40	188.00
Chateau single-control lavatory faucet, Moen. With waste assembly. ADA approved. Washerless cartridge. 4" center set. Includes pop-up assembly. Chrome finish. Labor includes set in place and connection only. Add the cost of plumbing rough-in as required.						
Acrylic knob control	P1@.900	Ea	97.30	31.60	128.90	219.00
Handle control	P1@.900	Ea	63.20	31.60	94.80	161.00
Single-handle lavatory faucet, Delta. 4" center set. Deck-mount. 3-hole installation. Lifetime warranty. ADA compliant. Includes mechanical pop-up assembly. Labor includes set in place and connection only. Add the cost of plumbing rough-in as required.						
Chrome finish	P1@.900	Ea	101.00	31.60	132.60	225.00
Chrome finish, brass accents	P1@.900	Ea	125.00	31.60	156.60	266.00
Polished brass, brass accents	P1@.900	Ea	164.00	31.60	195.60	333.00
Parisa single-handle lavatory faucet, Price Pfister. Solid brass. 4" center set. Single-control ceramic disc cartridge with temperature memory. High arc spout. Use with or without the deckplate for single-hole mounting. ADA approved. Add the cost of plumbing rough-in as required.						
Brushed nickel finish	P1@.900	Ea	90.80	31.60	122.40	208.00
Chrome finish	P1@.900	Ea	59.20	31.60	90.80	154.00

	Craft@Hrs	Unit	Material	Labor	Total	Sell
Lavatory single-handle faucet, Kohler. Deck-mount. 4" center set. Chrome finish. Latch handle. 5" spout. Ceramic disc valve cartridge. Includes mechanical pop-up assembly. Labor includes set in place and connection only. Add the cost of plumbing rough-in as required.						
Polished chrome	P1@.900	Ea	111.00	31.60	142.60	242.00
Infrared-operated hands-free lavatory faucet. 4" center set. Deck-mount. Infrared sensor automatically activates solenoid mixing valve to dispense tempered water. Labor includes set in place and connection only. Add the cost of plumbing rough-in as required.						
Battery operated	P1@.900	Ea	437.00	31.60	468.60	797.00

Two-Handle Bath Faucets

	Craft@Hrs	Unit	Material	Labor	Total	Sell
Two-handle lavatory faucet. Deck-mount. Includes mechanical pop-up assembly, except where noted. 4" center set. Labor includes set in place and connection only. Add the cost of plumbing rough-in as required.						
Cleo, satin chrome and brass	P1@.900	Ea	92.60	31.60	124.20	211.00
Leonardo	P1@.900	Ea	44.90	31.60	76.50	130.00
Lotus, chrome and porcelain	P1@.900	Ea	76.00	31.60	107.60	183.00
Modern, chrome finish	P1@.900	Ea	44.00	31.60	75.60	129.00
Porcelain lever handles	P1@.900	Ea	46.10	31.60	77.70	132.00
Teapot, brushed nickel	P1@.900	Ea	71.40	31.60	103.00	175.00
Without mechanical pop-up	P1@.900	Ea	12.00	31.60	43.60	74.10
Acrylic two-handle lavatory faucet, Glacier Bay. Deck-mount. 4" center set. Chrome finish. Solid brass waterways. Drip-free washerless cartridge design. Labor includes set in place and connection only. Add the cost of plumbing rough-in as required.						
With pop-up drain	P1@.900	Ea	35.90	31.60	67.50	115.00
Without pop-up drain	P1@.900	Ea	25.70	31.60	57.30	97.40
Two-handle lavatory faucet, American Standard. Deck-mount. 4" center set. Chrome finish. Includes mechanical pop-up assembly. Ceramic disc cartridge. 1/2" brass supply shanks for easy installation. Labor includes set in place and connection only. Add the cost of plumbing rough-in as required.						
Acrylic handles	P1@.900	Ea	31.20	31.60	62.80	107.00
Contemporary design	P1@.900	Ea	31.60	31.60	63.20	107.00
Ravenna design	P1@.900	Ea	98.00	31.60	129.60	220.00
Bedford two-handle lavatory faucet, Price Pfister. Deck-mount. 4" center set. Chrome finish. Metal verve handle. Includes brass mechanical pop-up assembly. Metal knobs. Labor includes set in place and connection only. Add the cost of plumbing rough-in as required.						
With mechanical pop-up drain	P1@.900	Ea	52.90	31.60	84.50	144.00
Without pop-up drain	P1@.900	Ea	42.00	31.60	73.60	125.00
Two-lever French handle lavatory faucet. Deck-mount. 4" center set. Drip-free ceramic disc cartridge. Solid brass waterways. Includes brass mechanical pop-up assembly. Labor includes set in place and connection only. Add the cost of plumbing rough-in as required.						
Chrome finish	P1@.900	Ea	52.80	31.60	84.40	143.00
Polished brass	P1@.900	Ea	84.50	31.60	116.10	197.00
Touch-control two-handle lavatory faucet. Deck-mount. 4" center set. Chrome finish. Acrylic handles. Limited lifetime function and finish warranty. Labor includes set in place and connection only. Add the cost of plumbing rough-in as required.						
No drain rod hole	P1@.900	Ea	38.10	31.60	69.70	118.00
With mechanical pop-up drain	P1@.900	Ea	50.90	31.60	82.50	140.00

	Craft@Hrs	Unit	Material	Labor	Total	Sell
Acrylic two-handle lavatory faucet, Delta. Deck-mount. Fits 3-hole sinks with 4" centers. Includes mechanical pop-up assembly. Machined brass cartridges. Washerless design. Lifetime warranty. Labor includes set in place and connection only. Add the cost of plumbing rough-in as required.						
Chrome finish	P1@.900	Ea	52.80	31.60	84.40	143.00
Teapot two-handle lavatory faucet, Glacier Bay. Deck-mount. 4" center set. Teapot style. White ceramic handles. Includes mechanical pop-up assembly. Labor includes set in place and connection only. Add the cost of plumbing rough-in as required.						
Brass finish	P1@.900	Ea	49.40	31.60	81.00	138.00
Chrome finish	P1@.900	Ea	34.70	31.60	66.30	113.00
Two-handle lavatory faucet, Masco. Deck-mount. 4" center set. No pop-up included. Lifetime faucet and finish warranty. Solid brass and stainless steel construction. Labor includes set in place and connection only. Add the cost of plumbing rough-in as required.						
Chrome finish	P1@.900	Ea	86.40	31.60	118.00	201.00
Chrome finish, brass accents	P1@.900	Ea	125.00	31.60	156.60	266.00
Bathroom décor kit with two-handle lavatory faucet. Includes chrome finish deck-mount teapot faucet, 24" towel bar, towel ring, toilet paper holder, robe hook and mounting hardware. Titanium PVD polished brass finish. Labor includes set in place and connection only. Add the cost of plumbing rough-in as required.						
Faucet and bath accessory kit	P1@1.90	Ea	78.40	66.80	145.20	247.00
Gerber two-handle lavatory faucet. Deck-mount. 4" center set. Includes mechanical pop-up drain assembly. Traditional styling. Labor includes set in place and connection only. Add the cost of plumbing rough-in as required.						
Cast brass body	P1@.900	Ea	49.00	31.60	80.60	137.00
Williamsburg two-handle lavatory faucet, American Standard. Deck-mount. 1/4-turn ceramic disc cartridge. 4" center set. Solid brass construction with mechanical pop-up assembly. Labor includes set in place and connection only. Add the cost of plumbing rough-in as required.						
Brass and chrome	P1@.900	Ea	114.00	31.60	145.60	248.00
Chrome with brass accents	P1@.900	Ea	107.00	31.60	138.60	236.00
Chrome, porcelain handles	P1@.900	Ea	85.20	31.60	116.80	199.00
Satin brass finish	P1@.900	Ea	128.00	31.60	159.60	271.00
Satin chrome finish	P1@.900	Ea	130.00	31.60	161.60	275.00
Velvet and chrome finish	P1@.900	Ea	135.00	31.60	166.60	283.00
Williamsburg two-handle lavatory faucet, American Standard. Deck-mount. 8" center set. Chrome finish except as noted. J-spout. 1/4-turn ceramic disc cartridge resists hard and sandy water. Complete with both metal and porcelain lever handles. Solid brass construction with metal pop-up drain. Labor includes set in place and connection only. Add the cost of plumbing rough-in as required.						
Brass accents	P1@.900	Ea	155.00	31.60	186.60	317.00
Brass, porcelain handles	P1@.900	Ea	133.00	31.60	164.60	280.00
Brass, two sets of handles	P1@.900	Ea	106.00	31.60	137.60	234.00
Porcelain lever handles	P1@.900	Ea	107.00	31.60	138.60	236.00
Satin chrome finish	P1@.900	Ea	171.00	31.60	202.60	344.00
Two sets of lever handles	P1@.900	Ea	124.00	31.60	155.60	265.00
J-spout two-handle lavatory faucet, American Standard. Deck-mount. 4" center set. 1/4-turn ceramic disc cartridge. Mechanical pop-up assembly. Labor includes set in place and connection only. Add the cost of plumbing rough-in as required.						
Brass, porcelain handles	P1@.900	Ea	138.00	31.60	169.60	288.00
Chrome, porcelain handles	P1@.900	Ea	90.80	31.60	122.40	208.00

	Craft@Hrs	Unit	Material	Labor	Total	Sell

Savannah two-handle lavatory faucet, Price Pfister. Deck-mount. 4" center set. High-arc swivel spout. Porcelain lever handles. Includes mechanical pop-up assembly. ADA compliant. Labor includes set in place and connection only. Add the cost of plumbing rough-in as required.

Chrome finish	P1@.900	Ea	94.70	31.60	126.30	215.00

Georgetown two-handle lavatory faucet, Price Pfister. Deck-mount. 4" center set. Chrome finish. Includes mechanical pop-up assembly. Labor includes set in place and connection only. Add the cost of plumbing rough-in as required.

Brass accents	P1@.900	Ea	67.30	31.60	98.90	168.00
Nickel, brass accent	P1@.900	Ea	140.00	31.60	171.60	292.00
Porcelain cross handles	P1@.900	Ea	103.00	31.60	134.60	229.00
Porcelain lever handles	P1@.900	Ea	92.30	31.60	123.90	211.00

Georgetown two-handle lavatory faucet, Price Pfister. Deck-mount. 8" center set. Includes mechanical pop-up assembly. TwistPfit™ installation. Pforever Pfaucet™. Labor includes set in place and connection only. Add the cost of plumbing rough-in as required.

Chrome, porcelain cross handle	P1@.900	Ea	161.00	31.60	192.60	327.00
Chrome, porcelain lever handle	P1@.900	Ea	162.00	31.60	193.60	329.00
Nickel, brass accent	P1@.900	Ea	196.00	31.60	227.60	387.00

Two-handle lavatory faucet, Delta. Deck-mount. Includes mechanical pop-up assembly. 4" center set. Labor includes set in place and connection only. Add the cost of plumbing rough-in as required.

Brass	P1@.900	Ea	172.00	31.60	203.60	346.00
Chrome, brass accents	P1@.900	Ea	144.00	31.60	175.60	299.00
Chrome, porcelain handles	P1@.900	Ea	96.00	31.60	127.60	217.00

Traditional two-handle lavatory faucet, Delta. Deck-mount. Brilliance lifetime anti-tarnish chrome finish. Fits 3-hole sinks with 4" centers. Lifetime faucet and finish warranty. Labor includes set in place and connection only. Add the cost of plumbing rough-in as required.

Brass accents	P1@.900	Ea	106.00	31.60	137.60	234.00

Victorian two-handle lavatory faucet, Masco. Deck-mount. Metal lever handle. 4" center 3-hole installation. Labor includes set in place and connection only. Add the cost of plumbing rough-in as required.

Chrome finish	P1@.900	Ea	123.00	31.60	154.60	263.00
Satin nickel finish	P1@.900	Ea	235.00	31.60	266.60	453.00
Venetian bronze finish	P1@.900	Ea	160.00	31.60	191.60	326.00

Monticello two-handle lavatory faucet, Moen. Deck-mount. 4" center set. Chrome finish. Includes mechanical pop-up assembly. Reliable washerless cartridge. Labor includes set in place and connection only. Add the cost of plumbing rough-in as required.

Chrome	P1@.900	Ea	86.40	31.60	118.00	201.00
Chrome, brass handles	P1@.900	Ea	96.00	31.60	127.60	217.00

Platinum two-handle lavatory faucet, Moen. Deck-mount. 4" center set. Includes mechanical pop-up assembly. Labor includes set in place and connection only. Add the cost of plumbing rough-in as required.

Platinum finish, chrome accents	P1@.900	Ea	134.00	31.60	165.60	282.00

Decorator two-handle lavatory faucet, Moen. Metal lever handles. 4" center set. With waste assembly. ADA compliant. Labor includes set in place and connection only. Add the cost of plumbing rough-in as required.

Polished brass	P1@.900	Ea	197.00	31.60	228.60	389.00

Asceri Hi-Arc lavatory faucet, Moen. Deck-mount. 4" center set. Includes mechanical pop-up assembly. Lifetime limited warranty against leaks, drips and finish defects. Labor includes set in place and connection only. Add the cost of plumbing rough-in as required.

Chrome with brass accents	P1@.900	Ea	192.00	31.60	223.60	380.00

	Craft@Hrs	Unit	Material	Labor	Total	Sell
Monticello adjustable-center two-handle lavatory faucet, Moen. Deck-mount. Adjustable center fits 8" to 16" centers. Includes mechanical pop-up assembly. Finish will not scratch or tarnish. Metal construction. Washerless cartridge. Comes with 1/2"-high IPS connections. Labor includes set in place and connection only. Add the cost of plumbing rough-in as required.						
Chrome with brass accents	P1@.900	Ea	159.00	31.60	190.60	324.00
Polished brass	P1@.900	Ea	265.00	31.60	296.60	504.00
Adjustable-center two-handle lavatory faucet, Moen. Deck-mount. Fits 8" to 16" centers. LifeShine™ non-tarnish finish. Lever handles. Includes mechanical pop-up assembly. ADA compliant. Labor includes set in place and connection only. Add the cost of plumbing rough-in as required.						
Chrome	P1@.900	Ea	149.00	31.60	180.60	307.00
Adjustable-center two-handle lavatory faucet, J A Manufacturing. Deck-mount. Fits 6" to 12" centers. Energy-saving aerator. Drip-free ceramic disc cartridge. Double set screws strengthen handles. Labor includes set in place and connection only. Add the cost of plumbing rough-in as required.						
Bronze finish	P1@.900	Ea	137.00	31.60	168.60	287.00
Brushed nickel finish	P1@.900	Ea	181.00	31.60	212.60	361.00
Brushed nickel and chrome	P1@.900	Ea	222.00	31.60	253.60	431.00
Chrome finish	P1@.900	Ea	212.00	31.60	243.60	414.00
Fairfax two-handle lavatory faucet, Kohler. Deck-mount. 8" center set. Lever handles. Includes mechanical pop-up assembly. Labor includes set in place and connection only. Add the cost of plumbing rough-in as required.						
Polished chrome	P1@.900	Ea	159.00	31.60	190.60	324.00
Fixture valve and supply lines. Flexible hose. Cone washer for universal fit.						
3/8" C x 7/8" BC x 12", PVC faucet supply	P1@.250	Ea	6.36	8.79	15.15	25.80
1/2" FIP x 1/2" FIP x 16", PVC faucet supply	P1@.250	Ea	5.42	8.79	14.21	24.20
1/2" FIP x 1/2" FIP x 20", PVC faucet supply	P1@.250	Ea	6.10	8.79	14.89	25.30

Vanity Cabinets Without Top

	Craft@Hrs	Unit	Material	Labor	Total	Sell
Springfield vanity. White finish. No assembly required. Maple finished interior with concealed hinges. Add the cost of countertop, basin and plumbing. 32" high.						
18" wide x 16" deep, 1 door	BC@.400	Ea	68.00	14.40	82.40	140.00
24" wide x 18" deep, 2 doors	BC@.400	Ea	94.40	14.40	108.80	185.00
30" wide x 18" deep, 3 doors	BC@.400	Ea	104.00	14.40	118.40	201.00
Charleston oak vanity. Fully assembled. Maple finished interior with concealed hinges. Includes chrome knobs. Add the cost of countertop, basin and plumbing. 32" high.						
24" wide x 18" deep, 2 doors	BC@.400	Ea	89.80	14.40	104.20	177.00
30" wide x 18" deep, 2 doors	BC@.400	Ea	103.00	14.40	117.40	200.00
Kingston vanity. White finish. Fully assembled. Maple finished interior with concealed hinges. Includes decorative door hardware. Add the cost of countertop, basin and plumbing. 18" deep. 34" high.						
24" wide, 1 door, 2 drawers	BC@.400	Ea	149.00	14.40	163.40	278.00
30" wide, 1 door, 2 drawers	BC@.400	Ea	165.00	14.40	179.40	305.00
36" wide, 2 doors, 2 drawers	BC@.400	Ea	180.00	14.40	194.40	330.00

	Craft@Hrs	Unit	Material	Labor	Total	Sell
Arkansas white vanity cabinet. Fully assembled, ready to install. Front surfaced in rigid thermofoil vinyl. Deluxe raised-panel doors match drawer fronts. Extra deep drawers. Concealed and adjustable steel European-style hinges. Chrome hardware included. Add the cost of countertop, basin and plumbing. 18" deep. 34" high.						
24" wide, 1 door, 2 drawers	BC@.400	Ea	138.00	14.40	152.40	259.00
30" wide, 1 door, 2 drawers	BC@.400	Ea	177.00	14.40	191.40	325.00
36" wide, 2 doors, 2 drawers	BC@.400	Ea	193.00	14.40	207.40	353.00
Monterey vanity. White. Fully assembled. Maple-finished interior with fully-adjustable concealed hinges. Extra-durable, thick plywood drawer box. Countertop, faucet and basin sold separately. 21" deep. 32" high. Add the cost of countertop, basin and plumbing.						
24" wide, 1 door, 2 drawers	BC@.400	Ea	212.00	14.40	226.40	385.00
30" wide, 1 door, 2 drawers	BC@.400	Ea	224.00	14.40	238.40	405.00
36" wide, 1 door, 2 drawers	BC@.400	Ea	264.00	14.40	278.40	473.00
48" wide, 2 doors, 4 drawers	BC@.500	Ea	342.00	18.00	360.00	612.00
60" wide, 2 doors, 4 drawers	BC@.500	Ea	430.00	18.00	448.00	762.00
Monterey oak vanity. Fully assembled. Maple-finished interior with fully-adjustable concealed hinges. Extra-durable, thick plywood drawer box. Add the cost of countertop, basin and plumbing. 21" deep. 32" high.						
24" wide, 1 door, 2 drawers	BC@.400	Ea	211.00	14.40	225.40	383.00
30" wide, 1 door, 2 drawers	BC@.400	Ea	224.00	14.40	238.40	405.00
36" wide, 1 door, 2 drawers	BC@.400	Ea	262.00	14.40	276.40	470.00
48" wide, 1 door, 4 drawers	BC@.500	Ea	341.00	18.00	359.00	610.00
60" wide, 1 door, 4 drawers	BC@.500	Ea	429.00	18.00	447.00	760.00
Monterey maple vanity. Fully assembled, ready to install. Solid maple and veneer door drawers and header. Large-capacity drawers. Solid maple face frame. Concealed adjustable hinges. Add the cost of countertop, basin and plumbing. 21" deep. 32" high.						
24" wide, 1 door, 2 drawers	BC@.400	Ea	237.00	14.40	251.40	427.00
30" wide, 1 door, 2 drawers	BC@.400	Ea	297.00	14.40	311.40	529.00
36" wide, 1 door, 2 drawers	BC@.400	Ea	305.00	14.40	319.40	543.00
48" wide, 1 door, 4 drawers	BC@.500	Ea	353.00	18.00	371.00	631.00
60" wide, 1 door, 4 drawers	BC@.500	Ea	462.00	18.00	480.00	816.00
Danville vanity with bottom drawer. Fully assembled. Durable white finish. Raised arch panel door. Patented deep-storage bottom drawer. Concealed adjustable hinges. 21" deep. 32" high. Add the cost of countertop, basin and plumbing.						
24", 2 doors, 1 drawer	BC@.400	Ea	225.00	14.40	239.40	407.00
30", 1 door, 4 drawers	BC@.400	Ea	273.00	14.40	287.40	489.00
36", 2 doors, 4 drawers	BC@.400	Ea	307.00	14.40	321.40	546.00
48", 2 doors, 7 drawers	BC@.500	Ea	410.00	18.00	428.00	728.00
Virginia bottom drawer oak vanity. Solid oak and oak veneer. No assembly required. Maple-finished interior with concealed hinges. Ball bearing glides on bottom drawer. Add the cost of countertop, basin and plumbing. 21" deep. 32" high.						
24" wide, 1 door, 2 drawers	BC@.400	Ea	230.00	14.40	244.40	415.00
30" wide, 1 door, 3 drawers	BC@.400	Ea	274.00	14.40	288.40	490.00
36" wide, 2 doors, 3 drawers	BC@.400	Ea	305.00	14.40	319.40	543.00
48" wide, 2 doors, 7 drawers	BC@.500	Ea	369.00	18.00	387.00	658.00

Cultured Marble Vanity Tops

	Craft@Hrs	Unit	Material	Labor	Total	Sell
Recessed eclipse oval bowl cultured marble vanity top. Radius edge detail. Includes overflow. Labor includes connecting the top to the drain. Dove white.						
25" wide x 22" deep	P1@.654	Ea	91.80	23.00	114.80	195.00
31" wide x 22" deep	P1@.654	Ea	103.00	23.00	126.00	214.00
37" wide x 22" deep	P1@.654	Ea	122.00	23.00	145.00	247.00
49" wide x 22" deep	P1@.654	Ea	158.00	23.00	181.00	308.00
61" wide x 22" deep, two bowls	P1@1.25	Ea	224.00	44.00	268.00	456.00
22" long side splash	P1@.200	Ea	31.20	7.03	38.23	65.00
Neptune shell marble countertop. Traditional scalloped shell bowl. 3-step edge detail. Includes overflow. Labor includes connecting the top to the drain. Glacier white.						
25" wide x 22" deep	P1@.654	Ea	102.00	23.00	125.00	213.00
31" wide x 22" deep	P1@.654	Ea	112.00	23.00	135.00	230.00
37" wide x 22" deep	P1@.654	Ea	133.00	23.00	156.00	265.00
49" wide x 22" deep	P1@.654	Ea	168.00	23.00	191.00	325.00
22" long side splash	P1@.200	Ea	18.60	7.03	25.63	43.60
Newport cultured marble vanity top. White. Labor includes connecting the top to the drain.						
19" wide x 17" deep	P1@.654	Ea	29.60	23.00	52.60	89.40
25" wide x 19" deep	P1@.654	Ea	55.10	23.00	78.10	133.00
31" wide x 19" deep	P1@.654	Ea	65.30	23.00	88.30	150.00
37" wide x 19" deep	P1@.654	Ea	75.50	23.00	98.50	167.00
19" long side splash	P1@.200	Ea	17.30	7.03	24.33	41.40
Rectangular cultured marble vanity top. Solid cultured marble. Heat- and stain-resistant. Labor includes connecting the top to the drain. Faucet sold separately. 19" deep.						
25" wide, dove white	P1@.654	Ea	56.20	23.00	79.20	135.00
31" wide, dove white	P1@.654	Ea	67.30	23.00	90.30	154.00
37" wide, dove white	P1@.654	Ea	81.60	23.00	104.60	178.00
19" side splash, dove white	P1@.200	Ea	16.60	7.03	23.63	40.20
25" wide, pearl onyx	P1@.654	Ea	102.00	23.00	125.00	213.00
31" wide, pearl onyx	P1@.654	Ea	113.00	23.00	136.00	231.00
37" wide, pearl onyx	P1@.654	Ea	122.00	23.00	145.00	247.00
19" side splash, pearl onyx	P1@.200	Ea	8.34	7.03	15.37	26.10
Rectangular premium cultured marble vanity top. 4" backsplash. Overflow drain. 1" lip on three sides to guard against drips. Labor includes connecting the top to the drain. Solid white. Faucet sold separately.						
25" wide x 22" deep	P1@.654	Ea	83.90	23.00	106.90	182.00
31" wide x 22" deep	P1@.654	Ea	91.70	23.00	114.70	195.00
49" wide x 22" deep	P1@.654	Ea	151.00	23.00	174.00	296.00
22" long side splash	P1@.200	Ea	19.40	7.03	26.43	44.90
Shell bowl cultured marble vanity top. White. Labor includes connecting the top to the drain.						
25" wide x 19" deep	P1@.654	Ea	66.30	23.00	89.30	152.00
31" wide x 19" deep	P1@.654	Ea	77.50	23.00	100.50	171.00
37" wide x 19" deep	P1@.654	Ea	62.70	23.00	85.70	146.00
37" wide x 22" deep, offset left	P1@.654	Ea	123.00	23.00	146.00	248.00

	Craft@Hrs	Unit	Material	Labor	Total	Sell
Shell bowl cultured marble vanity top. White. Labor includes connecting the top to the drain.						
31" wide x 22" deep	P1@.654	Ea	102.00	23.00	125.00	213.00
37" wide x 22" deep	P1@.654	Ea	111.00	23.00	134.00	228.00
49" wide x 22" deep	P1@.654	Ea	141.00	23.00	164.00	279.00
22" long side splash	P1@.200	Ea	15.40	7.03	22.43	38.10
Aspen cultured marble vanity top and bowl. Swirl finish. Labor includes connecting the top to the drain.						
31" wide x 22" deep	P1@.654	Ea	113.00	23.00	136.00	231.00
37" wide x 22" deep	P1@.654	Ea	132.00	23.00	155.00	264.00
49" wide x 22" deep	P1@.654	Ea	169.00	23.00	192.00	326.00
22" long side splash	P1@.200	Ea	18.70	7.03	25.73	43.70
Recessed oval bowl cultured marble vanity top. Pearl onyx. Labor includes connecting the top to the drain.						
31" wide x 22" deep	P1@.654	Ea	139.00	23.00	162.00	275.00
37" wide x 22" deep	P1@.654	Ea	158.00	23.00	181.00	308.00
49" wide x 22" deep	P1@.654	Ea	215.00	23.00	238.00	405.00
22" long side splash	P1@.200	Ea	31.20	7.03	38.23	65.00

Solid Surface Vanity Tops

Solid surface vanity sink top. 22" deep. Oval halo bowl. 1-1/4" premium edge detail.						
25" wide, saddle	BC@.377	Ea	143.00	13.60	156.60	266.00
31" wide, saddle	BC@.468	Ea	148.00	16.90	164.90	280.00
37" wide, saddle	BC@.558	Ea	176.00	20.10	196.10	333.00
49" wide, saddle	BC@.740	Ea	210.00	26.60	236.60	402.00
61" wide, saddle, double bowl	B1@.920	Ea	296.00	29.90	325.90	554.00
22" long side splash, saddle	BC@.332	Ea	21.40	12.00	33.40	56.80
31" wide, wheat	BC@.468	Ea	148.00	16.90	164.90	280.00
37" wide, wheat	BC@.558	Ea	176.00	20.10	196.10	333.00
49" wide, wheat	BC@.740	Ea	210.00	26.60	236.60	402.00
22" long side splash, wheat	BC@.332	Ea	21.40	12.00	33.40	56.80
Granite vanity top. 4" center.						
31" wide, beige	BC@.400	Ea	166.00	14.40	180.40	307.00
37" wide, beige	BC@.400	Ea	186.00	14.40	200.40	341.00

Vanity Cabinets with Countertop and Basin

Raised panel vanity with white marble top. Berkeley series. Solid oak and veneer raised panel doors. 31" high. Frameless construction with no exposed hinges. Maple-finish vanity interior. Assembled. Includes decorative hardware. Faucet sold separately. Labor cost includes setting in place only.

18" wide x 16" deep	BC@.400	Ea	91.50	14.40	105.90	180.00
24" wide x 18" deep	BC@.400	Ea	114.00	14.40	128.40	218.00
30" wide x 18" deep	BC@.400	Ea	148.00	14.40	162.40	276.00

Raised panel vanity with white marble top. Del Mar series. White rigid thermofoil raised panel doors. 31" high. Frameless construction with no exposed hinges. Maple-finish vanity interior. Fully assembled. Includes decorative door hardware. Faucet sold separately. Labor cost includes setting in place only.

18" wide x 16" deep	BC@.400	Ea	91.50	14.40	105.90	180.00
24" wide x 18" deep	BC@.400	Ea	118.00	14.40	132.40	225.00
30" wide x 18" deep	BC@.400	Ea	153.00	14.40	167.40	285.00

	Craft@Hrs	Unit	Material	Labor	Total	Sell
Vanity and cultured marble top. Colorado style. Fully assembled. Faucet sold separately. 31" high. Oak finish. Labor cost includes setting in place only.						
18" wide x 16" deep	BC@.400	Ea	68.30	14.40	82.70	141.00
24" wide x 18" deep	BC@.400	Ea	89.80	14.40	104.20	177.00
30" wide x 18" deep	BC@.400	Ea	120.00	14.40	134.40	228.00
30" wide, slide drawer	BC@.400	Ea	145.00	14.40	159.40	271.00
36" wide, slide drawer	BC@.400	Ea	164.00	14.40	178.40	303.00
White vanity with cultured marble top. Arkansas style. Rigid thermofoil front. Raised square panel door. Concealed 35mm European hinges. Glue and dowel construction. Full overlay frameless doors. Fully assembled. Faucet sold separately. Labor cost includes setting in place only.						
18" wide x 16" deep	BC@.400	Ea	68.30	14.40	82.70	141.00
24" wide x 18" deep	BC@.400	Ea	89.70	14.40	104.10	177.00
30" wide x 18" deep	BC@.400	Ea	120.00	14.40	134.40	228.00
36" wide x 18" deep	BC@.400	Ea	158.00	14.40	172.40	293.00
Vanity cabinet with top. Cultured marble top. Solid doors and face frame. Brass hardware. Faucet sold separately. Labor cost includes setting in place only.						
24" wide x 18" deep x 31" high, oak	BC@.400	Ea	105.00	14.40	119.40	203.00
24" wide x 18" deep x 31" high, white	BC@.400	Ea	105.00	14.40	119.40	203.00
Vanity cabinet with cultured marble top and mirror. 2-door, 1-drawer vanity cabinet. White melamine finish. Fully assembled bottom drawer vanity cabinet. Matching decorative Heritage-style vanity and mirror. Durable European-style cultured marble top. Extended Euro top provides extra surface area for accessories. 21-1/2" wide x 23-1/2" high mirror. Faucet sold separately. Labor cost includes setting in place only.						
24" wide x 13" deep x 31" high	BC@.400	Ea	194.00	14.40	208.40	354.00

Bath Exhaust Fans

	Craft@Hrs	Unit	Material	Labor	Total	Sell
ValueTest™ economy ceiling and wall exhaust fan, NuTone. For baths, utility, and recreation rooms. Non-ducted. For baths up to 45 square feet; other rooms up to 60 square feet. Installs in ceiling or in wall. Plastic duct collar. White polymeric grille. UL listed for use in tub and shower enclosure with GFCI branch circuit wiring. Housing dimensions: 8-1/16" length, 7-3/16" width, 3-7/8" depth. Grille size, 8-11/16" x 9-1/2". 3" duct. 0.75 amp. Labor includes setting and connecting only. Add the cost of wiring and ducting. CFM (cubic feet of air per minute).						
50 CFM, 2.5 sones	BE@1.00	Ea	15.30	39.10	54.40	92.50
70 CFM, 4.0 sones	BE@1.00	Ea	30.50	39.10	69.60	118.00
Economy ceiling or wall bath fan. Steel housing, with permanently lubricated motor and built-in damper. Torsion spring-mounted white polymeric grille. Built-in mounting ears. Housing dimensions: 7-1/4" length, 7-1/2" width, 3-5/8" depth. UL listed to use over bathtubs and showers when connected to a GFCI-protected branch circuit. For bathrooms up to 45 square feet. Labor includes setting and connecting only. Add the cost of wiring and ducting. CFM (cubic feet of air per minute).						
50 CFM, 2.0 sones	BE@1.00	Ea	15.30	39.10	54.40	92.50
Ceiling or wall bath fan, Broan Manufacturing. Steel housing, with permanently lubricated motor and built-in damper. Centrifugal blower wheel. Torsion spring-mounted white polymeric grille. Built-in double steel mounting ears with keyhole slots. Fits 3" round ducts. Housing dimensions: 7-1/4" length, 7-1/2" width, 3-5/8" depth. UL listed for use over bathtubs and showers when connected to a GFCI-protected branch circuit. Labor includes setting and connecting only. Add the cost of wiring and ducting. CFM (cubic feet of air per minute).						
50 CFM, 2.0 sones	BE@1.00	Ea	15.30	39.10	54.40	92.50
70 CFM, 3.0 sones	BE@1.00	Ea	30.50	39.10	69.60	118.00

	Craft@Hrs	Unit	Material	Labor	Total	Sell

Vertical discharge bath fan. Galvanized steel housing, with built-in damper and spin-on white polymeric grille. Polymeric fan blade and duct connectors. Built-in double steel mounting ears with keyhole slots. Fits 8" ducts. UL listed. Labor includes setting and connecting only. Add the cost of wiring and ducting. CFM (cubic feet of air per minute).

	Craft@Hrs	Unit	Material	Labor	Total	Sell
60 CFM, 4.0 sones	BE@1.00	Ea	50.00	39.10	89.10	151.00
180 CFM, 5.0 sones	BE@1.00	Ea	100.00	39.10	139.10	236.00

Vertical discharge bath fan, NuTone. For 75-square-foot bath. Mounts in ceiling, discharges through duct to roof or wall. White polymeric grille. Fits 7" round ducts. Housing dimensions: 6-15/16" diameter x 6-1/2" length. Grille size, 8-11/16" x 9-1/2". UL listed for use in tub/shower when used with GFI branch circuit wiring. Not recommended for kitchen use. Labor includes setting and connecting only. Add the cost of wiring and ducting. CFM (cubic feet of air per minute).

	Craft@Hrs	Unit	Material	Labor	Total	Sell
70 CFM, 2.5 sones	BE@1.00	Ea	60.20	39.10	99.30	169.00

Vertical discharge utility fan, NuTone. Mounts in ceiling. Discharges through duct to roof or wall. Pre-wired motor with plug-in receptacle. Adjustable hanger bars for 16" or 24" on-center joists. Silver anodized aluminum grille. Fits 7" round ducts. Housing dimensions: 11" diameter x 5-5/8" deep. UL listed. Labor includes setting and connecting only. Add the cost of wiring and ducting. CFM (cubic feet of air per minute).

	Craft@Hrs	Unit	Material	Labor	Total	Sell
210 CFM, 6.5 sones	BE@1.00	Ea	86.20	39.10	125.30	213.00

Ceiling or wall bath exhaust fan, NuTone. White polymeric grille. Torsion spring grille mounting requires no tools. Plug-in, permanently lubricated motor. Centrifugal blower wheel. Rugged, 26 gauge galvanized steel housing. Keyhole mounting brackets for quick, accurate installation. Tapered, polymeric duct fitting with built-in backdraft damper. UL listed for use over bathtubs and showers when connected to a GFCI-protected branch circuit. 120 volts, 0.5 amps, 2.5 sones, 80 CFM (HVI-2100 certified) and 4" round duct. Labor includes setting and connecting only. Add the cost of wiring and ducting. CFM (cubic feet of air per minute).

	Craft@Hrs	Unit	Material	Labor	Total	Sell
80 CFM, 2.5 sones	BE@1.00	Ea	73.50	39.10	112.60	191.00

Bath exhaust fan, Broan Manufacturing. Aluminum centrifugal blower wheel powered by 4-pole 1,500-RPM low-sound motor. Fits 4" ducts. White polymeric designer grille, torsion spring-mounted. Housing dimensions: 8" length, 8-1/4" width, 5-3/4" depth. UL listed for use over bathtubs and showers when connected to a GFCI-protected branch circuit. Labor includes setting and connecting only. Add the cost of wiring and ducting. CFM (cubic feet of air per minute).

	Craft@Hrs	Unit	Material	Labor	Total	Sell
110 CFM, 4.0 sones	BE@1.00	Ea	87.20	39.10	126.30	215.00

Bath exhaust fan. Pre-wired outlet box for plug-in motor plate, torsion spring-mounted low-profile grille. Steel housing, damper to eliminate backdrafts. Housing: 9-1/4" x 9-1/2" x 7-5/8". AMCA licensed for air and sound. UL listed to use over tub or shower enclosure with GFCI-protected branch circuit. HVI Certified. 4" duct. Labor includes setting and connecting only. Add the cost of wiring and ducting. CFM (cubic feet of air per minute).

	Craft@Hrs	Unit	Material	Labor	Total	Sell
80 CFM, 2.5 sones	BE@1.00	Ea	100.00	39.10	139.10	236.00
80 CFM, 3.5 sones	BE@1.00	Ea	110.00	39.10	149.10	253.00
90 CFM, 1.5 sones	BE@1.00	Ea	130.00	39.10	169.10	287.00
120 CFM, 2.5 sones	BE@1.00	Ea	160.00	39.10	199.10	338.00
130 CFM, 2.5 sones	BE@1.00	Ea	170.00	39.10	209.10	355.00

Ultra Silent® low-sound bath fan, Broan Manufacturing. Centrifugal blower. 7-5/8" high housing. UL listed for use over bathtubs and showers when connected to a GFCI-protected branch circuit. AMCA licensed for both air and sound. 4" duct. Labor includes setting and connecting only. Add the cost of wiring and ducting. CFM (cubic feet of air per minute).

	Craft@Hrs	Unit	Material	Labor	Total	Sell
80 CFM, 0.6 sones	BE@1.00	Ea	122.00	39.10	161.10	274.00
110 CFM, 1.5 sones	BE@1.00	Ea	156.00	39.10	195.10	332.00

	Craft@Hrs	Unit	Material	Labor	Total	Sell
Thru-the-wall utility fan, Broan Manufacturing. Steel housing with built-in damper and white polymeric grille. Permanently lubricated motor. Rotary on and off switch. Housing dimensions: 14-1/4" long, 14-1/4" wide, 4-1/2" deep. UL listed. Labor includes setting and connecting only. Add the cost of wiring and exterior finish. CFM (cubic feet of air per minute).						
180 CFM, 5.0 sones	BE@1.00	Ea	82.60	39.10	121.70	207.00
Chain-operated wall exhaust fan. Steel housing. Low-profile white polymeric grille. Permanently lubricated motor. Pull-chain opens door and turns on fan. Fits walls from 4-1/2" to 9-1/2" thick. Labor includes setting and connecting only. Add the cost of wiring and ducting. CFM (cubic feet of air per minute).						
8", 250 CFM, 4.5 sones	BE@.850	Ea	22.90	33.20	56.10	95.40
QuieTTest® low-sound bath fan, NuTone. For 105-square-foot bath. Pre-wired outlet box with plug-in receptacle. Adjustable hanger brackets. Fits 4" round ducts. Low-profile white polymeric grille, torsion spring-mounted. Housing dimensions: 9-3/8" length, 11-1/4" width, 7-7/8" depth. Grille size, 14-1/4" x 12-1/16". UL listed for use in tub and shower when used with GFI branch circuit wiring. Not recommended for kitchen use. Labor includes setting and connecting only. Add the cost of wiring and ducting. CFM (cubic feet of air per minute).						
110 CFM, 2.0 sones	BE@1.00	Ea	132.00	39.10	171.10	291.00
QuieTTest® low-sound ceiling blower, NuTone. For 375-square-foot room. Pre-wired outlet box with plug-in receptacle. Rounded, low-profile white polymeric grille with silver anodized trim at each end. Housing dimensions: 14-1/4" length, 10" width, 9" depth. Grille size, 16-1/2" x 12-3/32". Labor includes setting and connecting only. Add the cost of wiring and ducting. CFM (cubic feet of air per minute).						
300 CFM, 4.5 sones	BE@1.00	Ea	151.00	39.10	190.10	323.00

Lighted Bath Exhaust Fans

	Craft@Hrs	Unit	Material	Labor	Total	Sell
ValueTest™ economy bath fan and light, NuTone. Fan and light operate separately or together. Fits 4" ducts. White polymeric grille with break-resistant lens. Uses 100-watt lamp (not included). Housing dimensions: 9" length, 9" width, 5-1/2" depth. Grille size, 10-3/4" x 12-1/8". UL listed for use in tub and shower when used with GFI branch circuit wiring. Labor includes setting and connecting only. Add the cost of wiring and ducting. CFM (cubic feet of air per minute).						
50 CFM, 2.5 Sones, 45 SF bath	BE@1.00	Ea	32.30	39.10	71.40	121.00
70 CFM, 4.5 Sones, 65 SF bath	BE@1.00	Ea	67.20	39.10	106.30	181.00
Decorative globe bath fan and light, NuTone. Decorative glass fixture with a powerful exhaust fan. The fan exhausts through inconspicuous openings in the base. Corrosion-resistant gloss-white finish base with frosted melon glass globe. Can be installed throughout the house in bathrooms, bedrooms or hallways. Uses two standard 60-watt bulbs (sold separately). Labor includes setting and connecting only. Add the cost of wiring and ducting. CFM (cubic feet of air per minute).						
70 CFM, 3.5 sones, white	BE@.850	Ea	107.00	33.20	140.20	238.00
Exhaust Air deluxe bath fan with light, NuTone. Ventilation for baths up to 95 square feet, other rooms up to 125 square feet. 100-watt ceiling and 7-watt night light or energy-saving fluorescent light (lamps not included). Snap-on grille assembly. Housing dimensions: 9" length, 9" width, 6" depth. Polymeric white grille, 15" diameter. UL listed for use in tub and shower when used with GFI branch circuit wiring. Labor includes setting and connecting only. Add the cost of wiring and ducting. CFM (cubic feet of air per minute).						
100 CFM, 3.5 sones	BE@1.00	Ea	132.00	39.10	171.10	291.00
Economy bath exhaust fan and light, Broan Manufacturing. Type IC. Polymeric impeller. Plug-in permanently lubricated motor. 13-watt double twin tube compact fluorescent bulb provides equivalent of 60 watts of incandescent light (bulb not included). White polymeric grille. UL listed for use over bathtubs and showers when connected to GFCI-protected branch circuit. 4" duct. Labor includes setting and connecting only. Add the cost of wiring and ducting. CFM (cubic feet of air per minute).						
50 CFM, 3.5 sones	BE@1.00	Ea	43.80	39.10	82.90	141.00
70 CFM, 3.5 sones	BE@1.00	Ea	68.00	39.10	107.10	182.00

	Craft@Hrs	Unit	Material	Labor	Total	Sell

Decorative bath exhaust fan with light, Broan Manufacturing. Corrosion-resistant finish. Frosted melon glass globe. 60 watt max. Labor includes setting and connecting only. Add the cost of wiring and ducting. CFM (cubic feet of air per minute).

70 CFM, round globe, white	BE@1.00	Ea	95.40	39.10	134.50	229.00

Designer Series bath fan with light, Broan Manufacturing. With light and 7-watt night light. Bulbs not included. Type IC permanently lubricated motor. Centrifugal blower for high performance at low sound levels. Quit damper prevents cold backdrafts. Oak frame with multi-layer moisture protection. Steel housing with adjustable brackets that span to 24". 100-watt light with glass lens and low-profile polished brass finished grille assembly. Housing dimensions: 8" x 8-1/4" x 5-3/4". Labor includes setting and connecting only. Add the cost of wiring and ducting. CFM (cubic feet of air per minute).

80 CFM, 0.6 sones	BE@1.00	Ea	129.00	39.10	168.10	286.00
100 CFM, 3.5 sones	BE@1.00	Ea	134.00	39.10	173.10	294.00

Bath fan with light, Broan Manufacturing. 18-watt twin-tube fluorescent bulb, built-in 7-watt night light (bulbs not included). Type IC balanced centrifugal blower. Permanently lubricated motor. 7-5/8"-high housing. Mounts directly to ceiling joists or spans up to 24" with slide bar brackets. 4" duct. Labor includes setting and connecting only. Add the cost of wiring and ducting. CFM (cubic feet of air per minute).

110 CFM, 1.5 sones	BE@1.00	Ea	199.00	39.10	238.10	405.00
120 CFM, 2.5 sones	BE@1.00	Ea	215.00	39.10	254.10	432.00

Bath Exhaust Fans with Heater

Infrared bulb heater and fan, Broan Manufacturing. 4-point adjustable mounting brackets span up to 24". Uses 250-watt, 120-volt R-40 infrared bulbs (not included). Heater and fan units include 70 CFM (cubic feet of air per minute), 3.5 sones ventilator fan. Damper and duct connector included. Plastic matte-white molded grille and compact housings. Labor includes setting and connecting only. Add the cost of wiring and ducting.

1-bulb heater, 4" duct	BE@1.00	Ea	53.50	39.10	92.60	157.00
2-bulb heater, 7" duct	BE@1.10	Ea	73.00	43.00	116.00	197.00

Surface-mount ceiling bath heater and fan, Broan Manufacturing. 1,250 watts. 4,266 Btus (British thermal units). 120 volts. Chrome alloy wire element for instant heat. Built-in fan. Automatic overheat protection. Low-profile housing. Permanently lubricated motor. Mounts to standard 3-1/4" round or 4" octagonal ceiling electrical box. Satin-finish aluminum grille extends 2-3/4" from ceiling. 11" diameter x 2-3/4" deep. Labor includes setting and connecting only. Add the cost of wiring and ducting.

10.7 amp	BE@.750	Ea	58.40	29.30	87.70	149.00

Designer Series bath heater and exhaust fan, Broan Manufacturing. 1,500-watt fan-forced heater. 120-watt light capacity, 7-watt night light (bulbs not included). Permanently lubricated motor. Polymeric damper prevents cold backdrafts. 4-point adjustable mounting brackets with keyhole slots. Torsion spring grille mounting, no tools needed. Non-glare, light diffusing glass lenses. Fits single gang opening. Suitable for use with insulation. Includes 4-function control unit. 70 CFM (cubic feet of air per minute), 3.5 sones. Labor includes setting and connecting only. Add the cost of wiring and ducting.

4" duct, 100-watt lamp	BE@1.00	Ea	82.70	39.10	121.80	207.00

Heat-A-Lamp® bulb heater and fan, NuTone. Non-IC. Swivel-mount. Torsion spring holds plates firmly to ceiling. 250-watt radiant heat from one R-40 infrared heat lamp. Uses 4" duct. Adjustable socket for ceilings up to 1" thick. Adjustable hanger bars. Automatic reset for thermal protection. White polymeric finish. UL listed. 6-7/8" high. Labor includes setting and connecting only. Add the cost of wiring and ducting.

2.6 amp, 1 lamp, 12-1/2" x 10"	BE@1.50	Ea	60.10	58.60	118.70	202.00
5.0 amp, 2 lamp, 15-3/8" x 11"	BE@1.50	Ea	68.70	58.60	127.30	216.00

	Craft@Hrs	Unit	Material	Labor	Total	Sell
Deluxe Heat-A-Ventlite® heater and ventilator, NuTone. With light and night light. Type IC. 1,500-watt fan-forced heat with 55-watt exhaust motor. White enamel grille. Includes wall switch with separate on-off controls for all functions. Housing dimensions: 13-1/4" diameter x 7-1/2". Frame size, 15-3/8" diameter. 5,118-Btu (British thermal unit) heater. 100-watt light capacity, 7-watt night light (lamps not included). Labor includes setting and connecting only. Add the cost of wiring and ducting. CFM (cubic feet of air per minute).						
70 CFM, 3.5 sones	BE@1.50	Ea	199.00	58.60	257.60	438.00
70 CFM, 4.0 sones	BE@1.50	Ea	167.00	58.60	225.60	384.00

Bath Exhaust Fan Accessories

	Craft@Hrs	Unit	Material	Labor	Total	Sell
Timer switch.						
60-minute timer	BE@.250	Ea	15.30	9.76	25.06	42.60
SensAire® 3-function wall control switch, Broan Manufacturing. Fits single gang opening. Top switch for on-auto-off. Other switches for light and night light. Use with SensAire® fans and lights.						
White	BE@.250	Ea	18.30	9.76	28.06	47.70
Bath vent kit, Deflect-O. UL listed. Supurr-Flex® duct, aluminum roof vent, 3" to 4" increaser and 2 clamps. Fire resistant. Labor includes working around existing drywall, feeding the duct through existing attic space, and cutting through existing roofing.						
4" x 8' flexible duct	BC@2.00	Ea	18.30	72.00	90.30	154.00
Roof vent kit, NuTone. For venting kitchen or bath exhaust fan through slanted roof. Works with both 3" and 4" ducted units. Labor includes working around existing drywall, feeding the duct through existing attic space, and cutting through existing roofing.						
8' length	BC@2.00	Ea	23.40	72.00	95.40	162.00

Mirrors

	Craft@Hrs	Unit	Material	Labor	Total	Sell
Vanity mirror. Polished edges.						
24" wide x 36" high	BC@.331	Ea	25.50	11.90	37.40	63.60
36" wide x 36" high	BC@.331	Ea	35.70	11.90	47.60	80.90
36" wide x 42" high	BC@.331	Ea	40.80	11.90	52.70	89.60
48" wide x 36" high	BC@.331	Ea	45.90	11.90	57.80	98.30
60" wide x 36" high	BC@.331	Ea	60.20	11.90	72.10	123.00
Frosted oval mirror. Clear etched border accents the color of wall behind it. Ready to hang vertically or horizontally.						
23" wide x 29" high	BC@.331	Ea	45.90	11.90	57.80	98.30
24" wide x 36" high	BC@.331	Ea	65.10	11.90	77.00	131.00

Medicine Cabinets

	Craft@Hrs	Unit	Material	Labor	Total	Sell
Stainless steel framed bath cabinet, mirrored swing door. Polystyrene body. Recess- or surface-mount. Magnetic catch. 3 fixed shelves. Fits 14" x 18" wall opening. 4-3/4" deep.						
16" wide x 20" high	BC@.950	Ea	13.20	34.20	47.40	80.60
Stainless steel framed door medicine cabinet. Plastic cabinet. Stainless steel door frame. Surface- or recess-mount. 2 adjustable shelves. Fits 13-1/2" x 23-1/2" wall opening. 4-1/2" deep.						
16-1/8" wide x 26-1/8" high	BC@.950	Ea	25.50	34.20	59.70	101.00

	Craft@Hrs	Unit	Material	Labor	Total	Sell
Frameless swing door medicine cabinet. Polystyrene body. Frameless beveled swing door. Recess- or surface-mount. 1/2" beveled mirror. Magnetic catch. 2 adjustable shelves. Reversible left- or right-hand opening. 4-3/4" deep.						
16" wide x 24" high	BC@.950	Ea	31.20	34.20	65.40	111.00
Frameless beveled swing door medicine cabinet. Polystyrene body with beveled mirror edges. 2 adjustable shelves. Recess- or surface-mount. 4-1/2" deep.						
16" wide x 26" high	BC@.950	Ea	44.90	34.20	79.10	134.00
Swing door medicine cabinet, double frame. 2 fixed shelves. Recess- or surface-mount. All wood construction.						
16" x 26" high, oak	BC@.950	Ea	40.70	34.20	74.90	127.00
16" x 26" high, white	BC@.950	Ea	40.70	34.20	74.90	127.00
Chrome framed lighted sliding door cabinet. White stainless steel cabinet. 2 fixed shelves. Surface-mount. Includes on/off light switch and electrical outlet. Bulbs sold separately. 7-3/4" deep.						
24" wide x 20" high	BC@.900	Ea	45.90	32.40	78.30	133.00
Frameless octagonal swing door medicine cabinet. Polystyrene body.						
16" wide x 24" high	BC@.950	Ea	55.10	34.20	89.30	152.00
Etched glass medicine cabinet. Etched mirror panel. Adjustable shelves.						
24" x 25" high, oak	BC@.950	Ea	65.00	34.20	99.20	169.00
24" x 25" high, white	BC@.950	Ea	65.00	34.20	99.20	169.00
Frameless corner cabinet. 4 fixed shelves. Surface-mount.						
14" wide x 36" high	BC@.950	Ea	90.80	34.20	125.00	213.00
Etched glass bath storage cabinet.						
24" wide x 28" high, oak	BC@.950	Ea	79.20	34.20	113.40	193.00
24" wide x 28" high, white	BC@.950	Ea	81.20	34.20	115.40	196.00
Frameless beveled tri-view mirror medicine cabinet. Wood body. Surface-mount. 2 adjustable shelves. Concealed adjustable hinges.						
24-1/4" wide x 26" high	BC@.950	Ea	90.80	34.20	125.00	213.00
29-1/2" wide x 25-3/8" high	BC@.950	Ea	101.00	34.20	135.20	230.00
36" wide x 29-7/8" high	BC@.950	Ea	132.00	34.20	166.20	283.00
Frameless beveled edge mirrored medicine cabinet. Recess- or surface-mount.						
20" x 26" high	BC@.950	Ea	132.00	34.20	166.20	283.00
24" x 30" high, aluminum	BC@.950	Ea	148.00	34.20	182.20	310.00
Top lighted swing door bath cabinet. Solid hardwood frame. Concealed adjustable hinges. 3 adjustable shelves. 3 bulbs, not included. 24" wide x 36" high.						
Maple, nickel-plated top light	BC@1.25	Ea	133.00	45.00	178.00	303.00
Oak, brass-plated top light	BC@1.25	Ea	120.00	45.00	165.00	281.00
White, chrome-plated top light	BC@1.25	Ea	115.00	45.00	160.00	272.00
Single door hinged medicine cabinet. Beveled front mirror. White aluminum interior. Surface-mounted or recessed.						
24" x 30" high, white	BC@.950	Ea	124.00	34.20	158.20	269.00
Oval beveled edge medicine cabinet. Aluminum body. Surface- or recess-mount. Labor cost for recess mounting includes installation into an existing, correctly framed opening.						
24" x 36" high	BC@.950	Ea	199.00	34.20	233.20	396.00

	Craft@Hrs	Unit	Material	Labor	Total	Sell
Frameless beveled edge bi-view medicine cabinet. Recess- or surface-mount. Adjustable glass shelves. Aluminum body. Labor cost for recess mounting includes installation into an existing, correctly framed opening.						
30" x 26" high, mirrored interior	BC@.950	Ea	174.00	34.20	208.20	354.00
30" x 26" high, white interior	BC@.950	Ea	140.00	34.20	174.20	296.00
36" x 26" high, mirrored interior	BC@.950	Ea	190.00	34.20	224.20	381.00
36" x 26" high, white Interior	BC@.950	Ea	162.00	34.20	196.20	334.00
48" x 26" high, mirrored interior	B1@.950	Ea	275.00	30.90	305.90	520.00
Frameless bi-view medicine cabinet. Recess- or surface-mount. Labor cost for recess mounting includes installation into an existing, correctly framed opening.						
30" x 30" high, silver	BC@.950	Ea	252.00	34.20	286.20	487.00
Tri-view medicine cabinet. 36" high. Decorative crown molding and beveled mirrors. Adjustable shelves. Surface-mount.						
30" wide, maple	BC@.950	Ea	158.00	34.20	192.20	327.00
30" wide, oak	BC@.950	Ea	128.00	34.20	162.20	276.00
30" wide, white	BC@.950	Ea	134.00	34.20	168.20	286.00
36" wide, maple	BC@.950	Ea	191.00	34.20	225.20	383.00
36" wide, oak	BC@.950	Ea	169.00	34.20	203.20	345.00
36" wide, white	BC@.950	Ea	149.00	34.20	183.20	311.00
48" wide, maple	B1@.950	Ea	209.00	30.90	239.90	408.00
48" wide, oak	B1@.950	Ea	176.00	30.90	206.90	352.00
48" wide, white	B1@.950	Ea	177.00	30.90	207.90	353.00
Tri-view medicine cabinet. 3 beveled front mirrors. White aluminum interior. Surface-mount or recessed. Labor cost for recess mounting includes installation into an existing, correctly framed opening.						
30" x 31" high, white	BC@.950	Ea	200.00	34.20	234.20	398.00
36" x 31" high, white	BC@.950	Ea	238.00	34.20	272.20	463.00
48" x 31" high, white	B1@.950	Ea	285.00	30.90	315.90	537.00
Tri-view medicine cabinet with light bar. Recess-or surface-mount. Built-in top light bar. Adjustable shelves. 36" high. Labor cost for recess mounting includes installation into an existing, correctly framed opening.						
30" wide, maple	BC@.950	Ea	196.00	34.20	230.20	391.00
30" wide, oak	BC@.950	Ea	169.00	34.20	203.20	345.00
30" wide, white	BC@.950	Ea	169.00	34.20	203.20	345.00
36" wide, maple	BC@.950	Ea	218.00	34.20	252.20	429.00
36" wide, oak	BC@.950	Ea	196.00	34.20	230.20	391.00
36" wide, white	BC@.950	Ea	125.00	34.20	159.20	271.00
48" wide, oak	B1@.950	Ea	223.00	30.90	253.90	432.00
48" wide, white	B1@.950	Ea	223.00	30.90	253.90	432.00

Medicine Cabinet Light Bars

	Craft@Hrs	Unit	Material	Labor	Total	Sell
Light bar for beveled tri-view cabinet. Frameless. Beveled mirror edges. Bulbs not included. Labor cost does not include electrical rough-in.						
24", 4 bulb	BE@.600	Ea	40.80	23.40	64.20	109.00
30", 5 bulb	BE@.600	Ea	48.20	23.40	71.60	122.00
36", 8 bulb	BE@.600	Ea	60.40	23.40	83.80	142.00
48", 8 bulb	BE@.600	Ea	84.80	23.40	108.20	184.00

	Craft@Hrs	Unit	Material	Labor	Total	Sell
Medicine cabinet light bar. Surface-mount. UL and CSA listed. Bulbs not included. Labor cost does not include electrical rough-in.						
24", oak, 4 bulb	BE@.600	Ea	42.50	23.40	65.90	112.00
24", white, 4 bulb	BE@.600	Ea	42.10	23.40	65.50	111.00
30", oak, 5 bulb	BE@.600	Ea	52.40	23.40	75.80	129.00
30", white, 5 bulb	BE@.600	Ea	52.40	23.40	75.80	129.00
36", oak, 6 bulb	BE@.600	Ea	62.90	23.40	86.30	147.00
36", white, 5 bulb	BE@.600	Ea	63.50	23.40	86.90	148.00

Bathroom Accessories

	Craft@Hrs	Unit	Material	Labor	Total	Sell
Standard collection bath accessories, polished chrome, American Standard. Durable solid brass with chrome finish. Easy installation on tile, wood, plaster or drywall. Concealed mounting with no exposed hardware.						
Robe hook	BC@.280	Ea	16.90	10.10	27.00	45.90
24" shelf with glass	BC@.280	Ea	71.00	10.10	81.10	138.00
Toilet paper holder	BC@.280	Ea	31.10	10.10	41.20	70.00
18" towel bar	BC@.280	Ea	37.20	10.10	47.30	80.40
24" towel bar	BC@.280	Ea	39.00	10.10	49.10	83.50
Towel ring	BC@.280	Ea	85.10	10.10	95.20	162.00
Tumbler/toothbrush holder	BC@.280	Ea	30.20	10.10	40.30	68.50
Standard collection bath accessories, satin chrome, American Standard. Durable solid brass with satin chrome finish. Easy installation on tile, wood, plaster or drywall. Concealed mounting with no exposed hardware.						
Robe hook	BC@.280	Ea	18.62	10.10	28.72	48.80
24" shelf with glass	BC@.280	Ea	88.70	10.10	98.80	168.00
Toilet paper holder	BC@.280	Ea	39.00	10.10	49.10	83.50
18" towel bar	BC@.280	Ea	46.10	10.10	56.20	95.50
24" towel bar	BC@.280	Ea	47.80	10.10	57.90	98.40
Towel ring	BC@.280	Ea	38.10	10.10	48.20	81.90
Williamsburg bath accessories, satin chrome, American Standard. Satin finish with optional chrome accent pieces included. Matches Williamsburg line of faucets and accessories. Solid brass construction. Lifetime warranty. Easy installation. Drill-free anchors. Concealed mounting hardware.						
Robe hook	BC@.280	Ea	15.50	10.10	25.60	43.50
Soap dish	BC@.280	Ea	41.00	10.10	51.10	86.90
Toilet paper holder	BC@.280	Ea	33.10	10.10	43.20	73.40
18" towel bar	BC@.280	Ea	37.40	10.10	47.50	80.80
24" towel bar	BC@.280	Ea	39.50	10.10	49.60	84.30
Towel ring	BC@.280	Ea	32.00	10.10	42.10	71.60
Tumbler/toothbrush holder	BC@.280	Ea	40.20	10.10	50.30	85.50
Williamsburg bath accessories, chrome and brass, American Standard. Matches Williamsburg line of faucets and accessories. Chrome finish with optional brass accent pieces included. Solid brass construction. Easy installation. Drill-free anchors. Concealed mounting hardware.						
Robe hook	BC@.280	Ea	17.00	10.10	27.10	46.10
Soap dish	BC@.280	Ea	43.10	10.10	53.20	90.40
Toilet paper holder	BC@.280	Ea	35.20	10.10	45.30	77.00
18" towel bar	BC@.280	Ea	39.00	10.10	49.10	83.50
24" towel bar	BC@.280	Ea	44.20	10.10	54.30	92.30
Towel ring	BC@.280	Ea	35.20	10.10	45.30	77.00
Tumbler/toothbrush holder	BC@.280	Ea	43.80	10.10	53.90	91.60

	Craft@Hrs	Unit	Material	Labor	Total	Sell
Williamsburg bath accessories, chrome and porcelain, American Standard. Matches Williamsburg line of faucets and accessories. Solid brass construction. Easy installation. Drill-free anchors. Concealed mounting hardware.						
Robe hook	BC@.280	Ea	16.40	10.10	26.50	45.10
Soap dish	BC@.280	Ea	44.40	10.10	54.50	92.70
Toilet paper holder	BC@.280	Ea	34.80	10.10	44.90	76.30
18" towel bar	BC@.280	Ea	41.60	10.10	51.70	87.90
24" towel bar	BC@.280	Ea	42.50	10.10	52.60	89.40
Towel ring	BC@.280	Ea	33.90	10.10	44.00	74.80
Tumbler/toothbrush holder	BC@.280	Ea	44.50	10.10	54.60	92.80
Williamsburg bath accessories, satin chrome and satin brass, American Standard. Matches Williamsburg line of faucets and accessories. Satin finish with optional brass accent pieces included. Solid brass construction. Easy installation. Drill-free anchors. Concealed mounting hardware.						
Robe hook	BC@.280	Ea	16.80	10.10	26.90	45.70
Toilet paper holder	BC@.280	Ea	34.80	10.10	44.90	76.30
18" towel bar	BC@.280	Ea	40.10	10.10	50.20	85.30
24" towel bar	BC@.280	Ea	42.40	10.10	52.50	89.30
Towel ring	BC@.280	Ea	34.50	10.10	44.60	75.80
Williamsburg bath accessories, polished brass, American Standard. Matches Williamsburg line of faucets and accessories. Solid brass construction. Easy installation with drill-free anchors. Concealed mounting hardware.						
Robe hook	BC@.280	Ea	15.50	10.10	25.60	43.50
Toilet paper holder	BC@.280	Ea	35.80	10.10	45.90	78.00
18" towel bar	BC@.280	Ea	42.50	10.10	52.60	89.40
24" towel bar	BC@.280	Ea	43.90	10.10	54.00	91.80
Towel ring	BC@.280	Ea	33.90	10.10	44.00	74.80
Best Value™ Series bath accessories, polished chrome, Franklin Brass. Exposed screw.						
Recessed toilet paper holder	BC@.280	Ea	7.50	10.10	17.60	29.90
Robe hook	BC@.280	Ea	2.25	10.10	12.35	21.00
Soap dish	BC@.280	Ea	3.47	10.10	13.57	23.10
Toilet paper holder	BC@.280	Ea	5.39	10.10	15.49	26.30
Toothbrush holder	BC@.280	Ea	3.47	10.10	13.57	23.10
Toothbrush and tumbler holder	BC@.280	Ea	3.34	10.10	13.44	22.80
18" towel bar	BC@.280	Ea	3.55	10.10	13.65	23.20
24" towel bar	BC@.280	Ea	4.05	10.10	14.15	24.10
30" towel bar	BC@.280	Ea	5.08	10.10	15.18	25.80
Towel bar posts, 1 pair	BC@.280	Ea	4.06	10.10	14.16	24.10
Towel ring	BC@.280	Ea	5.09	10.10	15.19	25.80

	Craft@Hrs	Unit	Material	Labor	Total	Sell
Futura bath accessories, polished chrome, Franklin Brass. Zinc die-cast traditional design with concealed mounting.						
Recessed soap dish, with grab bar	BC@.280	Ea	24.60	10.10	34.70	59.00
Recessed toilet paper holder	BC@.280	Ea	14.30	10.10	24.40	41.50
Robe hook	BC@.280	Ea	4.18	10.10	14.28	24.30
Soap dish	BC@.280	Ea	6.11	10.10	16.21	27.60
Toilet paper holder	BC@.280	Ea	9.17	10.10	19.27	32.80
Toothbrush holder	BC@.280	Ea	6.11	10.10	16.21	27.60
18" towel bar	BC@.280	Ea	9.15	10.10	19.25	32.70
24" towel bar	BC@.280	Ea	12.20	10.10	22.30	37.90
30" towel bar	BC@.280	Ea	14.30	10.10	24.40	41.50
Towel ring	BC@.280	Ea	9.17	10.10	19.27	32.80
Alexandria bath accessories, glazed ceramic and chrome, Bath Unlimited. Wall-mount.						
Glass shelf	BC@.280	Ea	40.80	10.10	50.90	86.50
Lotion dispenser	BC@.280	Ea	10.30	10.10	20.40	34.70
Robe hook	BC@.280	Ea	16.30	10.10	26.40	44.90
Soap dish	BC@.280	Ea	8.21	10.10	18.31	31.10
Soap dish, ceramic	BC@.280	Ea	20.40	10.10	30.50	51.90
Tissue cover	BC@.280	Ea	20.60	10.10	30.70	52.20
Toilet paper holder	BC@.280	Ea	24.50	10.10	34.60	58.80
Toothbrush/tumbler holder	BC@.280	Ea	20.40	10.10	30.50	51.90
Toothbrush holder	BC@.280	Ea	8.20	10.10	18.30	31.10
18" towel bar	BC@.280	Ea	30.50	10.10	40.60	69.00
24" towel bar	BC@.280	Ea	30.60	10.10	40.70	69.20
24" towel bar, double	BC@.280	Ea	40.80	10.10	50.90	86.50
Towel ring	BC@.280	Ea	23.50	10.10	33.60	57.10
College Circle bath accessories, polished chrome, Bath Unlimited.						
Double robe hook	BC@.280	Ea	9.23	10.10	19.33	32.90
Soap dish	BC@.280	Ea	11.50	10.10	21.60	36.70
Toilet paper holder	BC@.280	Ea	15.40	10.10	25.50	43.40
18" towel bar	BC@.280	Ea	16.40	10.10	26.50	45.10
24" towel bar	BC@.280	Ea	18.00	10.10	28.10	47.80
Towel ring	BC@.280	Ea	12.30	10.10	22.40	38.10
Tumbler/toothbrush holder	BC@.280	Ea	12.20	10.10	22.30	37.90

	Craft@Hrs	Unit	Material	Labor	Total	Sell
Dakota bath accessories, black finish, Bath Unlimited. Modern look.						
Robe hook	BC@.280	Ea	8.56	10.10	18.66	31.70
Soap dish	BC@.280	Ea	17.10	10.10	27.20	46.20
Toilet paper holder	BC@.280	Ea	13.80	10.10	23.90	40.60
18" towel bar	BC@.280	Ea	17.10	10.10	27.20	46.20
24" towel bar	BC@.280	Ea	18.80	10.10	28.90	49.10
Towel ring	BC@.280	Ea	13.70	10.10	23.80	40.50
Tumbler/toothbrush holder	BC@.280	Ea	16.30	10.10	26.40	44.90
Greenwich bath accessories, chrome, Bath Unlimited.						
Robe hook	BC@.280	Ea	13.40	10.10	23.50	40.00
Soap dish	BC@.280	Ea	14.40	10.10	24.50	41.70
Tissue holder	BC@.280	Ea	20.40	10.10	30.50	51.90
Toothbrush/tumbler holder	BC@.280	Ea	14.40	10.10	24.50	41.70
18" towel bar	BC@.280	Ea	22.40	10.10	32.50	55.30
24" towel bar	BC@.280	Ea	23.40	10.10	33.50	57.00
Towel ring	BC@.280	Ea	19.40	10.10	29.50	50.20
Greenwich bath accessories, polished brass, Bath Unlimited.						
Robe hook	BC@.280	Ea	13.20	10.10	23.30	39.60
Soap dish	BC@.280	Ea	15.40	10.10	25.50	43.40
Toilet paper holder	BC@.280	Ea	20.40	10.10	30.50	51.90
Toothbrush/tumbler holder	BC@.280	Ea	15.70	10.10	25.80	43.90
Towel ring	BC@.280	Ea	19.40	10.10	29.50	50.20
18" towel bar	BC@.280	Ea	22.40	10.10	32.50	55.30
24" towel bar	BC@.280	Ea	23.40	10.10	33.50	57.00
Greenwich bath accessories, brass and chrome, Bath Unlimited.						
Robe hook	BC@.280	Ea	12.70	10.10	22.80	38.80
Tissue holder	BC@.280	Ea	20.10	10.10	30.20	51.30
Toothbrush/tumbler holder	BC@.280	Ea	15.20	10.10	25.30	43.00
18" Towel bar	BC@.280	Ea	21.20	10.10	31.30	53.20
24" Towel bar	BC@.280	Ea	23.30	10.10	33.40	56.80
Towel ring	BC@.280	Ea	19.00	10.10	29.10	49.50
Greenwich bath accessories, satin nickel, Bath Unlimited.						
Glass shelf	BC@.280	Ea	40.80	10.10	50.90	86.50
Soap dish	BC@.280	Ea	18.40	10.10	28.50	48.50
Toothbrush/tumbler holder	BC@.280	Ea	18.40	10.10	28.50	48.50
18" towel bar	BC@.280	Ea	24.50	10.10	34.60	58.80
Towel ring	BC@.280	Ea	20.40	10.10	30.50	51.90
Soap dish	BC@.280	Ea	12.30	10.10	22.40	38.10
Toothbrush/tumbler holder	BC@.280	Ea	12.30	10.10	22.40	38.10
24" towel bar	BC@.280	Ea	18.40	10.10	28.50	48.50
Towel ring	BC@.280	Ea	11.50	10.10	21.60	36.70
Soap dish	BC@.280	Ea	20.30	10.10	30.40	51.70
Toilet paper holder	BC@.280	Ea	26.20	10.10	36.30	61.70
Toothbrush/tumbler holder	BC@.280	Ea	20.70	10.10	30.80	52.40
18" towel bar	BC@.280	Ea	31.50	10.10	41.60	70.70
24" towel bar	BC@.280	Ea	33.60	10.10	43.70	74.30
Towel ring	BC@.280	Ea	22.10	10.10	32.20	54.70

	Craft@Hrs	Unit	Material	Labor	Total	Sell
Permanent shower rod. 72" long.						
Antique brass	P1@.250	Ea	16.50	8.79	25.29	43.00
Bone	P1@.250	Ea	13.40	8.79	22.19	37.70
Chrome	P1@.250	Ea	13.70	8.79	22.49	38.20
Decor rod with brass ends	P1@.250	Ea	17.00	8.79	25.79	43.80
Platinum	P1@.250	Ea	17.00	8.79	25.79	43.80
White	P1@.250	Ea	13.70	8.79	22.49	38.20

Bathroom Conversion Costs

Sump pumping systems for residential wastes. Including fiberglass sump basin, pump, 40' pipe run to septic tank or sewer line and automatic float switch. Add the cost of electrical work and excavation.

To 15' head	P1@8.00	Ea	1,700.00	281.00	1,981.00	3,370.00
To 25' head	P1@8.00	Ea	2,000.00	281.00	2,281.00	3,880.00
To 30' head	P1@8.00	Ea	3,000.00	281.00	3,281.00	5,580.00
Add for high water or pump failure alarm	P1@.500	Ea	500.00	17.60	517.60	880.00

Sheet metal area wall. Provides light well for basement windows. Galvanized corrugated steel. Labor includes the cost of setting in place only. Add the cost of excavation.

12" deep, 6" projection, 37" wide	SW@.410	Ea	31.40	16.60	48.00	81.60
18" deep, 6" projection, 37" wide	SW@.410	Ea	41.80	16.60	58.40	99.30
24" deep, 6" projection, 37" wide	SW@.410	Ea	54.40	16.60	71.00	121.00
30" deep, 6" projection, 37" wide	SW@.410	Ea	64.80	16.60	81.40	138.00
Add for steel grille cover	—	Ea	57.40	—	57.40	—

Extend plumbing vent through roof. Usually required when fixtures have been installed in a dormer addition. Add the cost of refinishing the wall interior and exterior, ceiling and roof.

Per roof vent	BC@5.50	Ea	47.00	198.00	245.00	417.00

Relocate sink vent stack. Usually required after a wall breakthrough. Add the cost of refinishing the wall interior and exterior, ceiling and roof.

Per vent stack	BC@9.50	Ea	52.00	342.00	394.00	670.00

Plumbing and HVAC

14

Not every older home needs an extra bathroom or a larger kitchen, but nearly every older home needs plumbing, heating and cooling systems brought up to modern standards. No home is truly comfortable without effective plumbing, heating and cooling. In a very old house, the entire mechanical system may have to be replaced.

Warm Air Heating

Many old houses still have gravity heating systems. The main difference between these and a modern forced-air system is that there's no fan to move the air. It depends on the physical property of heated air to rise, drawing in unheated air to fill the vacuum. This simple circulation system moves the warm air through the house. Gravity heating has its advantages: it's quiet, and there are no fan motors to repair. But they are very big units, with so many large ducts needed to funnel air to each room in the downstairs of the house that they almost fill the basement. They also tend to be heavy gas users. That may not have been a problem 50 years ago when gas was cheap, but fuel cost is something we have to consider today.

A new high-efficiency forced air furnace will recoup the replacement cost in just a few years. And since modern furnaces are small and use smaller ducts, eliminating the gravity system's octopus-like duct arms leaves most of the basement space available for conversion into living space. Forced-air heating has the added advantage of being able to bring the house up to a comfortable temperature in a matter of minutes, rather than the hour or two a gravity system takes.

The Btu capacity of a residential heating system depends on climate, window size and orientation, insulation and square footage to be heated. For cost estimating purposes, there's an easy way to calculate the Btu capacity of the furnace needed. Multiply the square feet of heated floor area by 53; then round up to the next larger furnace size. For example, to size a furnace for a 2,000 square foot home:

2,000 times 53 equals 106,000. The next larger furnace size is 125,000 Btu.

Altitude also affects the size of the furnace needed. Reduce the stated Btu rating of a furnace by 4 percent for each 1,000 feet above sea level. For example, the capacity of a 100,000 Btu furnace installed 5,000 feet above sea level would be 20 percent less, or 80,000 Btu.

When adding a room, enclosing a porch, expanding attic space or converting a garage, the existing furnace may not have capacity to serve the addition. Even if the existing furnace has enough capacity, a long duct run may reduce the volume of heat delivered at the register to below acceptable levels. The correct way to serve a remote room addition is to install a new trunk line directly to the furnace plenum. Talk to your heating subcontractor about installing a booster fan in the existing duct run to increase the flow of warm air.

If tapping into the existing HVAC system doesn't make sense, and there are no other practical options, install a floor or wall furnace in the new room addition. Direct-venting thru-wall furnaces are usually an acceptable alternative.

Hot Water Heating

One-pipe gravity steam heating systems are common in older homes. Radiators are fed by a single pipe on the entry (warm) side. The other side of the radiator has only an exhaust valve on the pipe. Once steam has cooled and condensed to water, it flows by gravity back to the boiler. This is an extremely simple system and, if properly installed and maintained, can be a reliable heat source. But again, one-pipe steam heating systems respond very slowly to requests for temperature change, and they offer far less control than modern hydronic systems. It may be possible to modernize a one-pipe system by replacing old radiators with baseboard heaters. Discuss it with your HVAC contractor.

There's a lot that can go wrong when modifying an old steam or hot water heating system. Results may not be what the owner expected. Include in your contract some language that limits both the scope of your work and your liability. For example: *Contractor makes no warranty on the condition, size or capacity of the existing boiler, radiators, or distribution piping.*

Radiant heating systems are less common, but tend to be trouble-free. Heated water flows through coils embedded in either a concrete slab or the plastered ceiling. Expert repair is required when a radiant heating system's piping develops leaks or becomes air locked. Breaks in ceiling coils can be repaired fairly easily. But repairing breaks in a floor is both difficult and expensive.

Electric Space Heaters

Assuming the home has sufficient capacity at the service panel, it's easy to add electric space heaters in a home if you need to extend heat to a room addition. However, electric heating is expensive and may not be practical in colder climates. Also, insulation standards are much higher for a home heated with electricity, so you must take that into consideration if you plan on this type of heat for your addition.

Water Supply

Low water pressure is usually due to an accumulation of lime or corrosion in supply lines, though it may also be that supply lines are too small. Main distribution pipes should be $^3/_4$" inside diameter. Branch lines serving a single fixture may be $^1/_2$" inside diameter.

If the house is served by a well on the premises, check the gauge on the pressure tank. A good operating pressure is 40 or 50 psi. If pressure is less than 20 psi, the pump isn't operating properly or the pressure setting is too low.

Water Hammer

Water hammer is a loud banging noise in supply lines caused when water is shut off abruptly. Over a period of time, water hammer can seriously damage valves and fixtures throughout the supply system. An air chamber placed on a supply line can absorb enough shock to prevent water hammer. If there's already an air chamber on the line, the air may have escaped. A waterlogged air chamber offers no protection against water hammer. To restore air to the chamber, turn off the water at the source. Open the air chamber to drain out any excess water, then reseal the chamber and restore the water pressure.

Plumbing Drain, Waste and Vent Lines

The drainage system consists of sewer laterals, drainage pipes and vents. It's common for pipes in older homes to break or become clogged. Most older homes have cast iron drain lines, which rust out after about 70 years. In some cases, the pipe diameter will be less than required by current codes. This is especially true for vent piping.

Sewer drain pipe in older homes may be bell-joint vitreous tile. Joints can break under pressure as the soil settles or heaves, and a broken pipe is an invitation for tree roots. Even if pipes aren't broken, tree roots can enter through the pipe joints and obstruct the flow of sewage. When this happens, remove the roots with a power-driven snake. If the system has to be snaked out every few months, you can assume there's a break or an open joint somewhere that needs repair. Tile or steel drainage pipe should last as long as the house. But these lines can accumulate sediment and need flushing out occasionally.

If you notice the smell of sewer gas, a vent stack is probably obstructed. The vent stacks you see running through the roof are designed to equalize air pressure in the drainage system. If a vent is obstructed, waste flowing through the system can create a vacuum that sucks water out of fixture traps, allowing sewer gas to enter the home. All fixtures should drain by gravity, not through suction. If you notice that water is being sucked out of the trap when a toilet is flushed, you can assume the roof vent is obstructed.

Checking Pipe Sizes

The nominal *(name)* size of pipe refers to the inside diameter.

Copper pipe

$^1/_2$" inside diameter pipe has $^5/_8$" outside diameter

$^3/_4$" inside diameter pipe has $^7/_8$" outside diameter

Galvanized pipe

$^1/_2$" inside diameter pipe has $^7/_8$" outside diameter

$^3/_4$" inside diameter pipe has $1^1/_8$" outside diameter

Water Heaters

Some older homes may still have domestic water coil systems in the furnace. That may provide enough hot water for cooking and bathing during the heating season. During summer months, when the furnace isn't running, a regular water heater will be required. A gas water heater for a three-bedroom home should have at least a 30-gallon capacity, though 40 is the current standard. An electric water heater for the same home should have a 50-gallon capacity.

Pricing Plumbing and HVAC Repairs

Home improvement contractors usually subcontract heating, air conditioning, and plumbing work to specialists. Plumbing and HVAC subcontractors usually quote the total installed prices only, without a breakdown of material and labor costs. However, to improve the data's versatility, this chapter provides a breakdown of the labor and material costs.

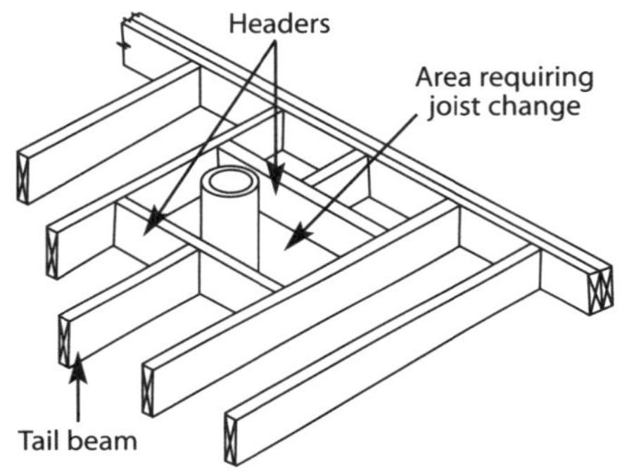

Figure 14-1

Header placement to accommodate plumbing lines

Framing for Plumbing and HVAC Improvements

Installing new heating duct, plumbing stacks, drains or water piping will usually require alteration of the framing. Plumbers and HVAC installers will arrive on the jobsite with all the tools needed to cut away structural wood framing. But a carpenter may not appreciate the result. Caution your plumbing and HVAC subs to avoid cutting if possible. When critical framing members must be cut, have a qualified carpenter add stiffeners or supports.

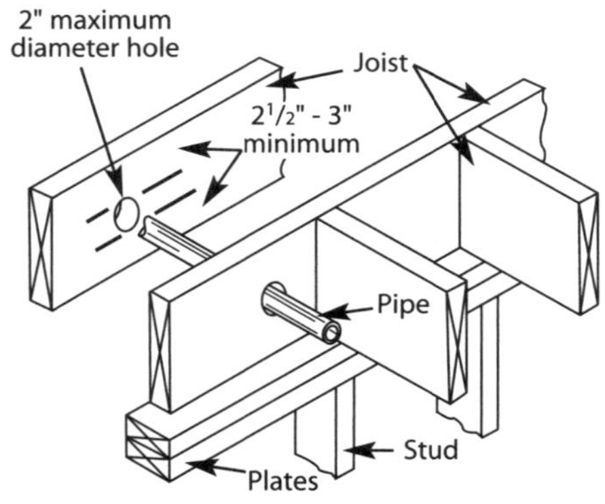

Figure 14-2

Dimensions required for passing pipe through joists

Cutting Floor Joists

When a floor joist has to make way for the passage of a stack or vent, install headers and tail beams (as in Figure 14-1) to reinforce the area around the joist that has been cut.

Figure 14-2 shows how to pass water pipe or conduit through a joist. The hole can't be any larger than 2" in diameter. Edges of the hole must be at least 2½" to 3" from the top and bottom of the joist. You can also cut a notch at the top or bottom edge of the joist. But the notch should be in the last quarter of the span and not more than 1/6th the depth of the joist. If a joist has to be cut and you can't comply with these rules, add an additional joist next to the cut joist or reinforce the cut joist by nailing scabs to each side.

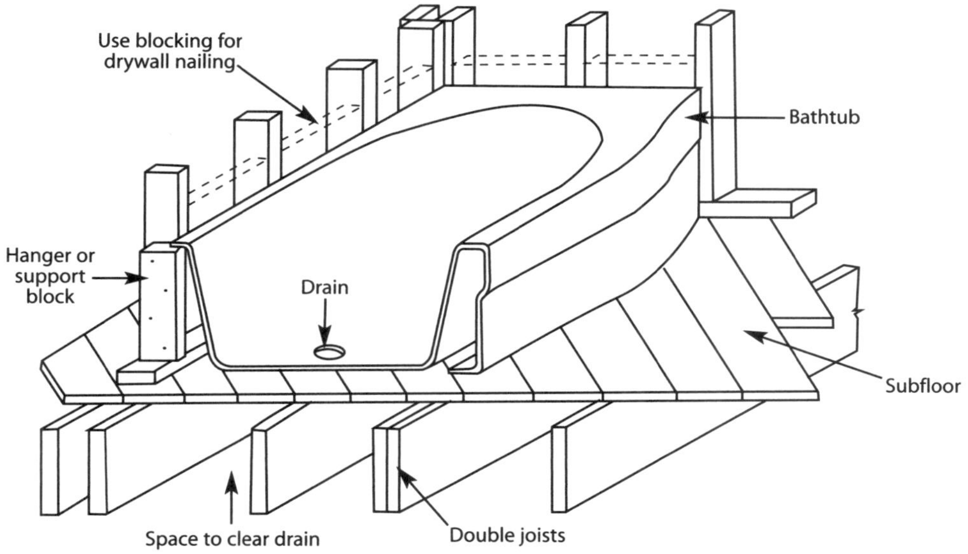

Figure 14-3
Framing for bathtub

Bathtub Framing

When adding a bathtub, more framing may be needed to support the heavy weight of a tub filled with water. Where joists are parallel to the length of the tub, use double joists under the outer edge of the tub. The inner edge is usually supported by wall framing. Use hangers or wood blocks as supports where the tub meets enclosing walls. See Figure 14-3.

Utility Walls

Walls that enclose new plumbing stacks or venting may require special framing. A 4" soil stack won't fit in a standard 2 x 4 stud wall. If a thicker wall is needed, frame it with 2 x 6 top and bottom plates and 2 x 4 studs placed flatwise at the edge of the plates. See Figure 14-4A. This leaves the center of the wall open for running both supply and drain pipes.

A 3" vent stack will fit in a 2 x 4 stud wall. But the hole for the vent will require cutting away most of the top plate. To repair the top plate, nail scabs cut from 2 x 4s on each side of the vent. See Figure 14-4B.

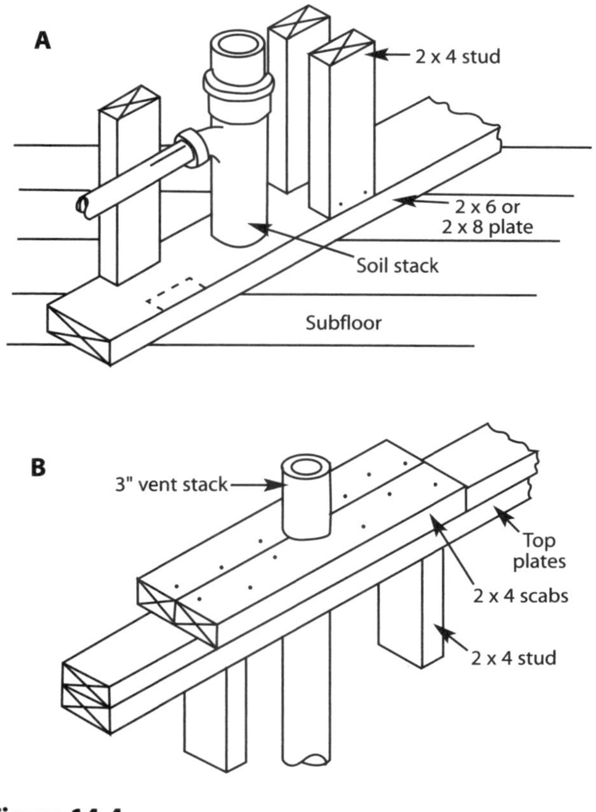

Figure 14-4
Framing around a stack

	Craft@Hrs	Unit	Material	Labor	Total	Sell

Heating System Renovations

Remove residential heating system components. Includes hauling components from the site. No salvage value assumed. Costs will be higher if components include detectable friable asbestos or if access to the site is limited. If the furnace or ductwork is wrapped in asbestos, you may be legally required to follow asbestos abatement procedures. This may require the services of a specialized asbestos abatement contractor.

	Craft@Hrs	Unit	Material	Labor	Total	Sell
Gravity feed furnace and plenum	P1@5.30	Ea	87.40	186.00	273.40	465.00
Forced air furnace and plenum	P1@2.50	Ea	53.60	87.90	141.50	241.00
Residential fuel oil storage tank	P1@2.50	Ea	53.60	87.90	141.50	241.00
Steam or hot water boiler	P1@2.50	Ea	59.00	87.90	146.90	250.00

Remove residential heating duct. Includes hauling components from the site. No salvage value assumed. Costs will be higher if components include detectable friable asbestos or if access to the site is limited. If the furnace or ductwork is wrapped in asbestos, you may be legally required to follow asbestos abatement procedures. This may require the services of a specialized asbestos abatement contractor. Add the cost of patching wall, floor or ceiling surfaces and the cost of closing openings after duct is removed. Large register openings in a gravity system can be converted to cold air returns in a forced air system. This avoids closing and patching large wall openings.

	Craft@Hrs	Unit	Material	Labor	Total	Sell
Duct runs to 30 feet, per run	P1@1.50	Ea	21.40	52.70	74.10	126.00
Duct runs 30 feet to 40 feet	P1@1.75	Ea	26.80	61.50	88.30	150.00
Duct runs over 40 feet	P1@2.00	Ea	32.10	70.30	102.40	174.00

Gas-fired forced air furnace — duct on interior walls. Includes a humidifier, filter, 5 to 8 galvanized sheet metal duct runs, registers and grilles, thermostat and electrical hookup, connection to gas line and gas piping, chimney vent and vent accessories, supply and return air plenums, system start-up and balancing. Installed in a one- or two-story home with conventional ducting. Add for wall, floor and ceiling patching as required.

	Craft@Hrs	Unit	Material	Labor	Total	Sell
80,000 Btu	P1@42.0	Ea	2,050.00	1,480.00	3,530.00	6,000.00
100,000 Btu	P1@43.5	Ea	2,220.00	1,530.00	3,750.00	6,380.00
125,000 Btu	P1@45.0	Ea	2,550.00	1,580.00	4,130.00	7,020.00
150,000 Btu	P1@47.0	Ea	2,750.00	1,650.00	4,400.00	7,480.00
175,000 Btu	P1@49.0	Ea	2,760.00	1,720.00	4,480.00	7,620.00
Add for each second floor duct run	P1@1.50	Ea	43.70	52.70	96.40	164.00
Add for trunk run to a room addition	P1@4.00	Ea	27.00	141.00	168.00	286.00

Gas-fired forced air furnace — duct on exterior walls (perimeter system). Includes a humidifier, filter, 5 to 8 galvanized sheet metal duct runs, registers and grilles, thermostat and electrical hookup, connection to gas line and gas piping, chimney vent and vent accessories, supply and return air plenums, system start-up and balancing. Installed in a one- or two-story home with perimeter ducting. Add for wall, floor and ceiling patching as required.

	Craft@Hrs	Unit	Material	Labor	Total	Sell
80,000 Btu	P1@46.0	Ea	2,170.00	1,620.00	3,790.00	6,440.00
100,000 Btu	P1@47.5	Ea	2,420.00	1,670.00	4,090.00	6,950.00
125,000 Btu	P1@49.0	Ea	2,590.00	1,720.00	4,310.00	7,330.00
150,000 Btu	P1@51.0	Ea	2,660.00	1,790.00	4,450.00	7,570.00
175,000 Btu	P1@53.0	Ea	2,670.00	1,860.00	4,530.00	7,700.00
Add for each second floor duct run	P1@2.00	Ea	43.70	70.30	114.00	194.00
Add for trunk run to a room addition	P1@4.00	Ea	431.00	141.00	572.00	972.00

Low-boy style gas-fired forced air furnace. When vertical space is limited, install a low-headroom gas furnace. Specs are the same as for a conventional gas furnace, but the furnace cost will be higher. Add the following to either conventional or perimeter gas furnace costs.

	Craft@Hrs	Unit	Material	Labor	Total	Sell
Add for low-boy 80,000 Btu furnace	P1@1.00	Ea	121.00	35.20	156.20	266.00
Add for low-boy 100,000 Btu furnace	P1@2.00	Ea	148.00	70.30	218.30	371.00
Add for low-boy 125,000 Btu furnace	P1@2.00	Ea	163.00	70.30	233.30	397.00
Add for low-boy 150,000 Btu furnace	P1@4.00	Ea	215.00	141.00	356.00	605.00
Add for low-boy 175,000 Btu furnace	P1@4.75	Ea	541.00	167.00	708.00	1,200.00

	Craft@Hrs	Unit	Material	Labor	Total	Sell
Counterflow gas-fired forced air furnace. When ducting is run under the floor and below the level of the furnace, a counterflow circulation furnace is required. Specs are the same as for a conventional gas furnace, but the furnace cost will be higher. Add the following to either conventional or perimeter gas furnace costs.						
Add for counterflow furnace to 125,000 Btu	P1@3.50	Ea	191.00	123.00	314.00	534.00
Add for counterflow furnace over 125,000 Btu	P1@6.00	Ea	570.00	211.00	781.00	1,330.00
Horizontal flow gas-fired forced air furnace. When space is restricted, a horizontal flow furnace may be the best option available. Specs are the same as for a conventional gas furnace, but the furnace cost will be higher. Add the following to either conventional or perimeter gas furnace costs.						
Add for horizontal furnace to 125,000 Btu	P1@2.50	Ea	149.00	87.90	236.90	403.00
Add for horizontal furnace over 125,000 Btu	P1@4.75	Ea	518.00	167.00	685.00	1,160.00
Oil-fired forced air furnace. Furnace and duct costs will be the same as for a gas-fired furnace. But add the following costs, which include storage tanks, gauge filter, vent and oil supply piping to the furnace. Add the cost of excavation for buried tanks or sight-screen fencing for above-ground tanks.						
275 gallon basement fuel tank	P1@3.75	Ea	963.00	132.00	1,095.00	1,860.00
Twin 275 gallon basement tanks	P1@5.25	LS	2,250.00	185.00	2,435.00	4,140.00
Concrete block foundation	P1@2.00	Ea	41.10	70.30	111.40	189.00
Exterior 550 gallon fuel tank	P1@8.00	Ea	1,650.00	281.00	1,931.00	3,280.00
Exterior 1,000 gallon fuel tank	P1@14.0	Ea	2,250.00	492.00	2,742.00	4,660.00
Remove and replace a gas-fired furnace — heating only. Using existing duct runs. Includes removal of the existing furnace and disposal, a new furnace with an A.G.A. certified ignition system, supply and return air plenums, filters and humidifier, connection to existing duct runs, new chimney pipe, and a new thermostat. Cost estimates for cooling assume installation with the gas furnace and include condenser, concrete pad, 20' of copper suction and liquid line refrigeration tubing, coil section and electrical connection.						
80,000 Btu	P1@22.0	Ea	2,080.00	774.00	2,854.00	4,850.00
100,000 Btu	P1@22.8	Ea	2,190.00	802.00	2,992.00	5,090.00
125,000 Btu	P1@24.0	Ea	2,380.00	844.00	3,224.00	5,480.00
150,000 Btu	P1@25.0	Ea	2,580.00	879.00	3,459.00	5,880.00
175,000 Btu	P1@25.8	Ea	2,950.00	907.00	3,857.00	6,560.00
Add when replacing a gravity furnace	P1@2.00	Ea	82.00	70.30	152.30	259.00
Add for 24,000 Btu cooling	P1@10.0	Ea	2,140.00	352.00	2,492.00	4,240.00
Add for 50,000 Btu cooling	P1@14.0	Ea	3,350.00	492.00	3,842.00	6,530.00
Add for 60,000 Btu cooling	P1@16.0	Ea	4,790.00	563.00	5,353.00	9,100.00
Add for a roof-mounted compressor	P1@6.00	Ea	1,290.00	211.00	1,501.00	2,550.00
Install duct runs. New work, such as for a room addition. Including diffusers and floor or wall registers. Duct suitable for either cold or warm air. Add the cost of wall, floor or ceiling refinishing, as required.						
Inside or outside wall, per run	P1@2.00	Ea	113.00	70.30	183.30	312.00
Cold air central return duct	P1@3.00	Ea	106.00	105.00	211.00	359.00
Add per run for second floor ducting	P1@1.50	Ea	38.80	52.70	91.50	156.00
Add for insulation, per SF of surface	P1@.030	SF	.74	1.05	1.79	3.04
Central humidifier. Added to an existing furnace or installed with a new furnace. Based on evaporation capacity. A 1,500-square-foot house (6 rooms) needs a 10-gallon-per-day humidifier. A 2,000-square-foot house (7 rooms) needs a 12-gallon-per-day humidifier. A 3,500-square-foot house (8 to 10 rooms) needs a 20-gallon-per-day humidifier.						
10 gallons per day, with new furnace	P1@3.75	Ea	206.00	132.00	338.00	575.00
10 gallons per day, existing furnace	P1@4.00	Ea	236.00	141.00	377.00	641.00
12 gallons per day, with new furnace	P1@4.00	Ea	234.00	141.00	375.00	638.00
12 gallons per day, existing furnace	P1@4.25	Ea	248.00	149.00	397.00	675.00
20 gallons per day, with new furnace	P1@4.50	Ea	297.00	158.00	455.00	774.00
20 gallons per day, existing furnace	P1@4.75	Ea	323.00	167.00	490.00	833.00

	Craft@Hrs	Unit	Material	Labor	Total	Sell
Central air conditioning. Added to an existing forced air furnace. Includes remote condenser and pad, coils and cabinet, refrigeration tubing, new thermostat, electric wiring and connection.						
2 ton, to 1,000 square feet	P1@16.0	Ea	2,520.00	563.00	3,083.00	5,240.00
3 ton, to 1,600 square feet	P1@18.0	Ea	3,590.00	633.00	4,223.00	7,180.00
5 ton, over 2,000 square feet	P1@20.0	Ea	4,760.00	703.00	5,463.00	9,290.00
Add or move steam heat or hot water radiator. Add a new cast iron radiator or change the location of an existing cast iron radiator. Based on exposed radiator piping. Add the cost of refinishing floor and wall surfaces if piping is to be concealed. First floor work. Based on average-size rooms.						
Add a new bathroom radiator	P1@7.50	Ea	314.00	264.00	578.00	983.00
Add a new bedroom radiator	P1@8.50	Ea	269.00	299.00	568.00	966.00
Add a new dining room radiator	P1@7.50	Ea	335.00	264.00	599.00	1,020.00
Add a new kitchen radiator	P1@8.50	Ea	432.00	299.00	731.00	1,240.00
Add a new living room radiator	P1@9.00	Ea	470.00	316.00	786.00	1,340.00
Relocate an existing radiator	P1@6.50	Ea	84.20	229.00	313.20	532.00
Add for second floor installations	—	%	20.0	—	—	—
Hydronic heating for room additions. Install up to 8 feet of baseboard convection unit with necessary supply and return lines. Based on an average-size single room.						
First floor work, per room	P1@9.00	Ea	804.00	316.00	1,120.00	1,900.00
Second floor work, per room	P1@11.0	Ea	927.00	387.00	1,314.00	2,230.00
Remove and replace a cast iron radiator. Replaced on the same wall using exposed piping. Add the cost of refinishing floor and wall surfaces if required.						
Replace with a baseboard radiator	P1@8.00	Ea	741.00	281.00	1,022.00	1,740.00
Replace with thin-line radiator	P1@8.00	Ea	733.00	281.00	1,014.00	1,720.00
Remove and replace a hydronic heating system. In an existing home. Remove and haul away the existing heating plant. Install a new gas- or propane-fired hydronic heating boiler with insulated steel jacket, new wall thermostat, electrical connections and wiring. Connect to existing gas, propane or oil supply lines and existing hydronic distribution piping. Includes electric or spark ignition, power vent controls, standard valves, gauges, and drain valves. Add the cost of radiators and refinishing floor, ceiling and wall surfaces as required.						
Gas-fired, 100,000 Btu	P1@26.0	Ea	3,160.00	914.00	4,074.00	6,930.00
Gas-fired, 125,000 Btu	P1@27.5	Ea	3,210.00	967.00	4,177.00	7,100.00
Oil-fired, 100,000 Btu	P1@27.0	Ea	3,530.00	949.00	4,479.00	7,610.00
Oil-fired, 125,000 Btu	P1@28.0	Ea	3,760.00	984.00	4,744.00	8,060.00
Add per bathroom radiator, copper pipe	P1@7.50	Ea	458.00	264.00	722.00	1,230.00
Add per bedroom radiator, copper pipe	P1@5.00	Ea	526.00	176.00	702.00	1,190.00
Add per dining room radiator, copper pipe	P1@8.50	Ea	555.00	299.00	854.00	1,450.00
Add per kitchen radiator, copper pipe	P1@6.00	Ea	604.00	211.00	815.00	1,390.00
Add per living room radiator, copper pipe	P1@7.50	Ea	497.00	264.00	761.00	1,290.00
Add per bathroom radiator, iron pipe	P1@5.00	Ea	399.00	176.00	575.00	978.00
Add per bedroom radiator, iron pipe	P1@8.50	Ea	463.00	299.00	762.00	1,300.00
Add per dining room radiator, iron pipe	P1@6.50	Ea	576.00	229.00	805.00	1,370.00
Add per kitchen radiator, iron pipe	P1@9.00	Ea	505.00	316.00	821.00	1,400.00
Add per living room radiator, iron pipe	P1@7.00	Ea	535.00	246.00	781.00	1,330.00
Add for second floor radiators	—	%	20.0	—	—	—
Add for circulating pump	P1@5.00	Ea	567.00	176.00	743.00	1,260.00
HEPA central air filtration retrofit. 240 CFM, 0.5 inches of water column pressure differential, with pre-filter and activated carbon filter. 2' x 8' duct collar, gasket and hanger assembly. Add the cost of control and electric wiring.						
HEPA filtration module, test and balance	SW@2.00	Ea	429.00	80.90	509.90	867.00
Cut hole in wall or roof	SW@1.00	Ea	37.50	40.40	77.90	132.00
Mount duct hangers	SW@0.25	Ea	3.75	10.10	13.85	23.50
Sheet metal duct, coated	SW@0.50	LF	5.78	20.20	25.98	44.20

	Craft@Hrs	Unit	Material	Labor	Total	Sell
Chimney liner (flue), aluminum. Prefabricated, 2-ply single-wall flexible aluminum gas or propane fuel chimney liner. Approved for installation in an existing masonry chimney. Maximum flue temperature of 270 degrees F. Check local codes for regulations regarding acceptable installation requirements.						
3" flexible chimney liner	P1@.085	LF	2.70	2.99	5.69	9.67
4" flexible chimney liner	P1@.090	LF	3.28	3.16	6.44	10.90
5" flexible chimney liner	P1@.095	LF	2.77	3.34	6.11	10.40
6" flexible chimney liner	P1@.100	LF	4.70	3.52	8.22	14.00
7" flexible chimney liner	P1@.115	LF	5.15	4.04	9.19	15.60
8" flexible chimney liner	P1@.120	LF	5.73	4.22	9.95	16.90
Chimney liner rain cap assembly, aluminum. Approved for installation in an existing masonry chimney. Maximum flue temperature of 270 degrees F. Check local codes for regulations regarding acceptable installation requirements.						
3" chimney liner cap assembly	P1@.450	Ea	53.80	15.80	69.60	118.00
4" chimney liner cap assembly	P1@.450	Ea	51.00	15.80	66.80	114.00
5" chimney liner cap assembly	P1@.450	Ea	58.90	15.80	74.70	127.00
6" chimney liner cap assembly	P1@.450	Ea	67.00	15.80	82.80	141.00
7" chimney liner cap assembly	P1@.500	Ea	75.60	17.60	93.20	158.00
8" chimney liner cap assembly	P1@.500	Ea	80.40	17.60	98.00	167.00
Chimney liner (flue), stainless steel. Prefabricated, 304/316 stainless steel single-wall flexible chimney liner. Approved for installation in an existing masonry chimney. Maximum flue temperature of 2,100 degrees F. Check local codes for regulations regarding acceptable installation requirements.						
3" flexible chimney liner	P1@.100	LF	15.60	3.52	19.12	32.50
4" flexible chimney liner	P1@.115	LF	16.50	4.04	20.54	34.90
5" flexible chimney liner	P1@.120	LF	20.30	4.22	24.52	41.70
6" flexible chimney liner	P1@.125	LF	22.80	4.40	27.20	46.20
7" flexible chimney liner	P1@.130	LF	24.80	4.57	29.37	49.90
8" flexible chimney liner	P1@.135	LF	33.80	4.75	38.55	65.50
Chimney (flue), C vent. Prefabricated single-wall 24-gauge galvanized steel snap lock gas or propane fuel chimney. Add for cutting new holes in ceiling and roof, and patching, if required. Check local codes for regulations regarding acceptable installation requirements.						
5" single wall	P1@.110	LF	1.53	3.87	5.40	9.18
6" single wall	P1@.115	LF	1.75	4.04	5.79	9.84
7" single wall	P1@.130	LF	1.97	4.57	6.54	11.10
8" single wall	P1@.135	LF	2.11	4.75	6.86	11.70
Chimney (flue), C vent, aluminum. Prefabricated .020" thickness, 24-gauge single-wall, aluminum snap lock gas or propane fuel chimney. Add for cutting new holes in ceiling and roof, and patching, if required. Check local codes for regulations regarding acceptable installation requirements.						
3" single wall	P1@.095	LF	.96	3.34	4.30	7.31
4" single wall	P1@.100	LF	1.14	3.52	4.66	7.92
5" single wall	P1@.110	LF	1.47	3.87	5.34	9.08
6" single wall	P1@.115	LF	1.72	4.04	5.76	9.79
7" single wall	P1@.130	LF	2.07	4.57	6.64	11.30
8" single wall	P1@.135	LF	2.41	4.75	7.16	12.20

	Craft@Hrs	Unit	Material	Labor	Total	Sell
Chimney (flue) base tee, C vent. Prefabricated single-wall galvanized steel gas or propane fuel chimney base tee, including cap. Add for cutting new holes in ceiling and roof, and patching, if required. Check local codes for regulations regarding acceptable installation requirements.						
3" single wall	P1@.150	Ea	22.70	5.27	27.97	47.50
4" single wall	P1@.155	Ea	23.20	5.45	28.65	48.70
5" single wall	P1@.165	Ea	28.40	5.80	34.20	58.10
6" single wall	P1@.180	Ea	32.10	6.33	38.43	65.30
7" single wall	P1@.195	Ea	37.90	6.86	44.76	76.10
8" single wall	P1@.220	Ea	52.40	7.74	60.14	102.00
Chimney (flue) base tee, C vent, aluminum. Prefabricated single-wall aluminum gas or propane fuel chimney base tee, including cap. Add for cutting new holes in ceiling and roof, and patching, if required. Check local codes for regulations regarding acceptable installation requirements.						
3" single wall	P1@.150	Ea	26.10	5.27	31.37	53.30
4" single wall	P1@.155	Ea	28.30	5.45	33.75	57.40
5" single wall	P1@.165	Ea	29.80	5.80	35.60	60.50
6" single wall	P1@.180	Ea	35.20	6.33	41.53	70.60
7" single wall	P1@.195	Ea	41.00	6.86	47.86	81.40
8" single wall	P1@.220	Ea	54.60	7.74	62.34	106.00
Chimney (flue) elbow, C vent. Prefabricated single-wall galvanized steel gas or propane fuel chimney adjustable elbow. Add for cutting new holes in ceiling and roof, and patching, if required. Check local codes for regulations regarding acceptable installation requirements.						
3" single wall, adjustable elbow	P1@.150	Ea	3.87	5.27	9.14	15.50
4" single wall, adjustable elbow	P1@.155	Ea	5.10	5.45	10.55	17.90
5" single wall, adjustable elbow	P1@.165	Ea	5.50	5.80	11.30	19.20
6" single wall, adjustable elbow	P1@.180	Ea	6.31	6.33	12.64	21.50
7" single wall, adjustable elbow	P1@.195	Ea	7.05	6.86	13.91	23.60
8" single wall, adjustable elbow	P1@.220	Ea	9.08	7.74	16.82	28.60
Chimney (flue) elbow, C vent, aluminum. Prefabricated single-wall aluminum gas or propane fuel chimney adjustable elbow. Add for cutting new holes in ceiling and roof, and patching, if required. Check local codes for regulations regarding acceptable installation requirements.						
3" single wall, adjustable elbow	P1@.150	Ea	4.56	5.27	9.83	16.70
4" single wall, adjustable elbow	P1@.155	Ea	5.57	5.45	11.02	18.70
5" single wall, adjustable elbow	P1@.165	Ea	6.32	5.80	12.12	20.60
6" single wall, adjustable elbow	P1@.180	Ea	8.03	6.33	14.36	24.40
7" single wall, adjustable elbow	P1@.195	Ea	11.80	6.86	18.66	31.70
8" single wall, adjustable elbow	P1@.220	Ea	16.40	7.74	24.14	41.00
Chimney (flue) storm collar, rain cap and roof flashing assembly, C vent. Prefabricated single-wall galvanized steel gas or propane fuel chimney termination assembly. Add for cutting new holes in ceiling and roof, and patching if required. Check local codes for regulations regarding acceptable installation requirements.						
3" single wall, termination assembly	P1@.600	Ea	37.70	21.10	58.80	100.00
4" single wall, termination assembly	P1@.650	Ea	40.50	22.90	63.40	108.00
5" single wall, termination assembly	P1@.700	Ea	43.30	24.60	67.90	115.00
6" single wall, termination assembly	P1@.750	Ea	51.20	26.40	77.60	132.00
7" single wall, termination assembly	P1@.900	Ea	60.90	31.60	92.50	157.00
8" single wall, termination assembly	P1@1.15	Ea	74.10	40.40	114.50	195.00

	Craft@Hrs	Unit	Material	Labor	Total	Sell
Chimney (flue), B vent. Prefabricated double-wall metal gas or propane fuel chimney. Aluminum inner liner with extruded aluminum outer jacket. Add for cutting new holes in ceiling and roof, and patching, if required. Check local codes for regulations regarding acceptable installation requirements.						
3" double wall	P1@.095	LF	6.02	3.34	9.36	15.90
4" double wall	P1@.100	LF	6.37	3.52	9.89	16.80
5" double wall	P1@.115	LF	7.24	4.04	11.28	19.20
6" double wall	P1@.120	LF	7.63	4.22	11.85	20.10
7" double wall	P1@.135	LF	9.86	4.75	14.61	24.80
8" double wall	P1@.140	LF	11.40	4.92	16.32	27.70
Chimney (flue) base tee, B vent. Prefabricated double-wall metal gas or propane fuel chimney base tee with cap. Aluminum inner liner with extruded aluminum outer jacket. Check local codes for regulations regarding acceptable installation requirements.						
3" double wall	P1@.150	Ea	43.70	5.27	48.97	83.20
4" double wall	P1@.155	Ea	44.90	5.45	50.35	85.60
5" double wall	P1@.165	Ea	51.30	5.80	57.10	97.10
6" double wall	P1@.180	Ea	55.00	6.33	61.33	104.00
7" double wall	P1@.195	Ea	64.10	6.86	70.96	121.00
8" double wall	P1@.220	Ea	74.10	7.74	81.84	139.00
Chimney (flue) elbow, B vent. Prefabricated double-wall metal gas or propane fuel chimney base tee with cap. Aluminum inner liner with extruded aluminum outer jacket. Check local codes for regulations regarding acceptable installation requirements.						
3" double wall, 45-degree elbow	P1@.150	Ea	20.60	5.27	25.87	44.00
3" double wall, 90-degree elbow	P1@.150	Ea	27.00	5.27	32.27	54.90
4" double wall, 45-degree elbow	P1@.155	Ea	21.10	5.45	26.55	45.10
4" double wall, 90-degree elbow	P1@.155	Ea	27.30	5.45	32.75	55.70
5" double wall, 45-degree elbow	P1@.165	Ea	21.40	5.80	27.20	46.20
5" double wall, 90-degree elbow	P1@.165	Ea	28.00	5.80	33.80	57.50
6" double wall, 45-degree elbow	P1@.180	Ea	24.60	6.33	30.93	52.60
6" double wall, 90-degree elbow	P1@.180	Ea	30.10	6.33	36.43	61.90
7" double wall, 45-degree elbow	P1@.195	Ea	29.80	6.86	36.66	62.30
7" double wall, 90-degree elbow	P1@.195	Ea	34.60	6.86	41.46	70.50
8" double wall, 45-degree elbow	P1@.220	Ea	36.20	7.74	43.94	74.70
8" double wall, 90-degree elbow	P1@.220	Ea	42.90	7.74	50.64	86.10
Chimney (flue) storm collar, rain cap and roof flashing assembly, B vent. Add for cutting new holes in ceiling and roof, and patching, if required. Check local codes for regulations regarding acceptable installation requirements.						
3" double wall, termination assembly	P1@.600	Ea	37.70	21.10	58.80	100.00
4" double wall, termination assembly	P1@.650	Ea	40.50	22.90	63.40	108.00
5" double wall, termination assembly	P1@.700	Ea	43.30	24.60	67.90	115.00
6" double wall, termination assembly	P1@.750	Ea	45.80	26.40	72.20	123.00
7" double wall, termination assembly	P1@.900	Ea	61.00	31.60	92.60	157.00
8" double wall, termination assembly	P1@1.15	Ea	74.10	40.40	114.50	195.00
Chimney (flue), L vent. Prefabricated double-wall metal gas or propane fuel chimney. Stainless steel inner liner with extruded aluminum outer jacket. High temperature and condensing boiler or furnace applications. Maximum flue temperature of 750 degrees F. Add for cutting new holes in ceiling and roof, and patching, if required. Check local codes for regulations regarding acceptable installation requirements.						
5" double wall	P1@.115	LF	22.20	4.04	26.24	44.60
6" double wall	P1@.120	LF	24.50	4.22	28.72	48.80
7" double wall	P1@.135	LF	28.70	4.75	33.45	56.90
8" double wall	P1@.140	LF	30.90	4.92	35.82	60.90

	Craft@Hrs	Unit	Material	Labor	Total	Sell
Chimney (flue) base tee, L vent. Prefabricated double-wall metal gas or propane fuel chimney base tee with cap. Stainless steel inner liner with extruded aluminum outer jacket. High temperature and condensing boiler or furnace applications. Maximum flue temperature of 750 degrees F. Check local codes for regulations regarding acceptable installation requirements.						
5" double wall	P1@.165	Ea	83.10	5.80	88.90	151.00
6" double wall	P1@.180	Ea	88.90	6.33	95.23	162.00
7" double wall	P1@.195	Ea	107.00	6.86	113.86	194.00
8" double wall	P1@.220	Ea	119.00	7.74	126.74	215.00
Chimney (flue) elbow, L vent. Prefabricated double-wall metal gas or propane fuel chimney. Stainless steel inner liner with extruded aluminum outer jacket. High temperature and condensing boiler or furnace applications. Maximum flue temperature of 750 degrees F. Check local codes for regulations regarding acceptable installation requirements.						
5" double wall, 45-degree elbow	P1@.165	Ea	65.80	5.80	71.60	122.00
5" double wall, 90-degree elbow	P1@.165	Ea	95.40	5.80	101.20	172.00
6" double wall, 45-degree elbow	P1@.180	Ea	70.70	6.33	77.03	131.00
6" double wall, 90-degree elbow	P1@.180	Ea	97.10	6.33	103.43	176.00
7" double wall, 45-degree elbow	P1@.195	Ea	99.00	6.86	105.86	180.00
7" double wall, 90-degree elbow	P1@.195	Ea	64.50	6.86	71.36	121.00
8" double wall, 45-degree elbow	P1@.220	Ea	79.00	7.74	86.74	147.00
8" double wall, 90-degree elbow	P1@.220	Ea	114.00	7.74	121.74	207.00
Chimney (flue) storm collar, rain cap and roof flashing assembly, L vent. Prefabricated double-wall metal gas or propane fuel chimney. Stainless steel inner liner with extruded aluminum outer jacket. High temperature and condensing boiler or furnace applications. Maximum flue temperature of 750 degrees F. Check local codes for regulations regarding acceptable installation requirements.						
5" double wall, termination assembly	P1@.700	Ea	98.60	24.60	123.20	209.00
6" double wall, termination assembly	P1@.750	Ea	115.00	26.40	141.40	240.00
7" double wall, termination assembly	P1@.900	Ea	127.00	31.60	158.60	270.00
8" double wall, termination assembly	P1@1.15	Ea	154.00	40.40	194.40	330.00
Chimney (flue), A vent. Prefabricated double-wall insulated stainless steel all-fuels chimney. Maximum flue temperature of 1,700 degrees F. Stainless steel inner liner and outer jacket. Twist-lock connection. Add for cutting new holes in ceiling and roof, and patching, if required. Check local codes for regulations regarding acceptable installation requirements.						
5" double wall	P1@.120	LF	43.00	4.22	47.22	80.30
6" double wall	P1@.125	LF	45.90	4.40	50.30	85.50
7" double wall	P1@.140	LF	48.00	4.92	52.92	90.00
8" double wall	P1@.145	LF	51.40	5.10	56.50	96.10
10" double wall	P1@.160	LF	83.00	5.63	88.63	151.00
Add for 2,100 degree maximum temp rating	—	%	45.0	—	—	—
Chimney (flue) base tee, A vent. Prefabricated double-wall insulated all-fuels chimney base tee including cap. Maximum flue temperature of 1,700 degrees F. Stainless steel inner liner and outer jacket. Twist-lock connection. Add for cutting new holes in ceiling and roof and patching, if required. Check local codes for regulations regarding acceptable installation requirements.						
5" double wall	P1@.170	Ea	187.00	5.98	192.98	328.00
6" double wall	P1@.185	Ea	201.00	6.50	207.50	353.00
7" double wall	P1@.200	Ea	212.00	7.03	219.03	372.00
8" double wall	P1@.225	Ea	222.00	7.91	229.91	391.00
10" double wall	P1@.260	Ea	265.00	9.14	274.14	466.00
Add for 2,100 degree F. maximum temp rating	—	%	25.0	—	—	—

	Craft@Hrs	Unit	Material	Labor	Total	Sell
Chimney (flue) elbow, A vent. Prefabricated double-wall insulated all-fuels chimney elbow. Maximum flue temperature of 1,700 degrees F. Stainless steel inner liner and outer jacket. Twist-lock connection. Add for cutting new holes in ceiling and roof and patching, if required. Check local codes for regulations regarding acceptable installation requirements.						
5" double wall, 15-degree elbow	P1@.170	Ea	69.60	5.98	75.58	128.00
5" double wall, 30-degree elbow	P1@.170	Ea	73.20	5.98	79.18	135.00
5" double wall, 45-degree elbow	P1@.170	Ea	105.00	5.98	110.98	189.00
6" double wall, 15-degree elbow	P1@.190	Ea	72.00	6.68	78.68	134.00
6" double wall, 30-degree elbow	P1@.190	Ea	75.90	6.68	82.58	140.00
6" double wall, 45-degree elbow	P1@.190	Ea	96.40	6.68	103.08	175.00
7" double wall, 15-degree elbow	P1@.225	Ea	82.50	7.91	90.41	154.00
7" double wall, 30-degree elbow	P1@.225	Ea	89.70	7.91	97.61	166.00
7" double wall, 45-degree elbow	P1@.225	Ea	115.00	7.91	122.91	209.00
8" double wall, 15-degree elbow	P1@.260	Ea	97.70	9.14	106.84	182.00
8" double wall, 30-degree elbow	P1@.260	Ea	107.00	9.14	116.14	197.00
8" double wall, 45-degree elbow	P1@.260	Ea	127.00	9.14	136.14	231.00
10" double wall, 15-degree elbow	P1@.330	Ea	118.00	11.60	129.60	220.00
10" double wall, 30-degree elbow	P1@.330	Ea	135.00	11.60	146.60	249.00
10" double wall, 45-degree elbow	P1@.330	Ea	164.00	11.60	175.60	299.00
Add for 2,100 degree maximum temp rating	—	%	25.0	—	—	—
Chimney (flue) storm collar, rain cap and roof flashing assembly, A vent. Prefabricated double-wall insulated all-fuels chimney. Maximum flue temperature of 1,700 degrees F. Twist-lock connection. Add for cutting new holes in ceiling and roof, and patching, if required. Check local codes for regulations regarding acceptable installation requirements.						
5" double wall, termination assembly	P1@.750	Ea	118.00	26.40	144.40	245.00
6" double wall, termination assembly	P1@.800	Ea	131.00	28.10	159.10	270.00
7" double wall, termination assembly	P1@.950	Ea	139.00	33.40	172.40	293.00
8" double wall, termination assembly	P1@1.20	Ea	154.00	42.20	196.20	334.00
10" double wall, termination assembly	P1@1.30	Ea	180.00	45.70	225.70	384.00
Chimney (flue) supports and brackets, A vent. Prefabricated double-wall insulated all-fuels chimney. Maximum flue temperature of 1,700 degrees F. Twist-lock connection. Add for cutting new holes in ceiling and roof, and patching, if required. Check local codes for regulations regarding acceptable installation requirements.						
5" double wall, anchor plate	P1@.200	Ea	40.10	7.03	47.13	80.10
5" double wall, adjustable wall support	P1@.300	Ea	64.60	10.50	75.10	128.00
5" double wall, wall or floor support	P1@.400	Ea	44.90	14.10	59.00	100.00
6" double wall, anchor plate	P1@.225	Ea	43.60	7.91	51.51	87.60
6" double wall, adjustable wall support	P1@.325	Ea	81.40	11.40	92.80	158.00
6" double wall, wall or floor support	P1@.425	Ea	67.40	14.90	82.30	140.00
7" double wall, anchor plate	P1@.230	Ea	46.30	8.09	54.39	92.50
7" double wall, adjustable wall support	P1@.330	Ea	91.30	11.60	102.90	175.00
7" double wall, wall or floor support	P1@.430	Ea	60.20	15.10	75.30	128.00
8" double wall, anchor plate	P1@.240	Ea	49.00	8.44	57.44	97.60
8" double wall, adjustable wall support	P1@.350	Ea	105.00	12.30	117.30	199.00
8" double wall, wall or floor support	P1@.450	Ea	71.60	15.80	87.40	149.00
10" double wall, anchor plate	P1@.260	Ea	82.20	9.14	91.34	155.00
10" double wall, adjustable wall support	P1@.400	Ea	124.00	14.10	138.10	235.00
10" double wall, wall or floor support	P1@.550	Ea	101.00	19.30	120.30	205.00

	Craft@Hrs	Unit	Material	Labor	Total	Sell
Electric baseboard heaters. Includes a remote thermostat, a separate circuit run to the point of use and connection. Each foot of electric baseboard heater delivers approximately 925 Btu of heating capacity.						
1,850 Btu, 540-watt, 2' heater	P1@4.00	Ea	168.00	141.00	309.00	525.00
2,775 Btu, 660-watt, 3' heater	P1@4.50	Ea	195.00	158.00	353.00	600.00
3,700 Btu, 1.0kW, 4' heater	P1@4.65	Ea	229.00	163.00	392.00	666.00
4,600 Btu, 1.35kW, 5' heater	P1@4.75	Ea	254.00	167.00	421.00	716.00
5,350 Btu, 1.55kW, 6' heater	P1@5.00	Ea	288.00	176.00	464.00	789.00
7,200 Btu, 2.0kW, 8' heater	P1@5.25	Ea	360.00	185.00	545.00	927.00
9,050 Btu, 2.65kW, 9' heater	P1@5.50	Ea	386.00	193.00	579.00	984.00
11,900 Btu, 3.5kW, 10' heater	P1@5.75	Ea	444.00	202.00	646.00	1,100.00
Central coal- or wood-burning furnace. Connected to existing ductwork and chimney flue. Triple flue heat exchanger, refractory lined chamber holds logs from 22" to 32" in diameter. Includes fan lift switch, blower system, cast iron grate, automatic controlled damper and removal of the existing furnace.						
Furnace	P1@25.8	Ea	2,690.00	907.00	3,597.00	6,110.00
Gas floor furnace. Includes thermostat, electrical connection, gas piping, flue and valve. Add the cost of interior and exterior patching.						
35,000 Btu, pilot ignition	P1@9.00	Ea	621.00	316.00	937.00	1,590.00
35,000 Btu, spark ignition	P1@9.00	Ea	738.00	316.00	1,054.00	1,790.00
50,000 Btu, pilot ignition	P1@9.50	Ea	691.00	334.00	1,025.00	1,740.00
50,000 Btu, spark ignition	P1@9.50	Ea	807.00	334.00	1,141.00	1,940.00
65,000 Btu, pilot ignition	P1@10.0	Ea	802.00	352.00	1,154.00	1,960.00
65,000 Btu, spark ignition	P1@10.0	Ea	935.00	352.00	1,287.00	2,190.00
Gas wall furnace. Includes thermostat, electrical connection, gas piping, valve, blower and debris removal. Add the cost of wall, ceiling and roof patching. Two room (rear grille) units have one thermostat and control the temperature in one room only. For a room under 300 square feet, a 25,000 Btu furnace usually has enough capacity if ceiling height is 8' or less.						
25,000 Btu, pilot ignition	P1@9.00	Ea	524.00	316.00	840.00	1,430.00
25,000 Btu, spark ignition	P1@9.00	Ea	524.00	316.00	840.00	1,430.00
40,000 Btu, pilot ignition	P1@9.50	Ea	651.00	334.00	985.00	1,670.00
40,000 Btu, spark ignition	P1@9.50	Ea	807.00	334.00	1,141.00	1,940.00
50,000 Btu, pilot ignition	P1@9.50	Ea	678.00	334.00	1,012.00	1,720.00
50,000 Btu, spark ignition	P1@9.50	Ea	833.00	334.00	1,167.00	1,980.00
60,000 Btu, pilot ignition	P1@10.0	Ea	738.00	352.00	1,090.00	1,850.00
60,000 Btu, spark ignition	P1@10.0	Ea	904.00	352.00	1,256.00	2,140.00
Add for dual wall unit (2 rooms)	P1@6.00	Ea	623.00	211.00	834.00	1,420.00
Thru-the-wall gas furnace. Includes thermostat, electrical connection, gas piping, valve, direct exterior wall vent and debris removal. Pilot ignition. Add the cost of interior and exterior wall patching if needed.						
100 square foot room, 14,000 Btu	P1@8.00	Ea	536.00	281.00	817.00	1,390.00
150 square foot room, 22,000 Btu	P1@8.50	Ea	568.00	299.00	867.00	1,470.00
200 square foot room, 25,000 Btu	P1@9.00	Ea	608.00	316.00	924.00	1,570.00
250 square foot room, 30,000 Btu	P1@9.00	Ea	640.00	316.00	956.00	1,630.00
300 square foot room, 40,000 Btu	P1@9.50	Ea	698.00	334.00	1,032.00	1,750.00
400 square foot room, 65,000 Btu	P1@10.0	Ea	852.00	352.00	1,204.00	2,050.00

	Craft@Hrs	Unit	Material	Labor	Total	Sell

Gas Water Heaters

30-gallon gas water heater. 6-year warranty. Includes temperature and pressure relief valve. Includes pipe connection labor. Add for flue, water and gas pipe, fitting materials and gas permit costs if required.

	Craft@Hrs	Unit	Material	Labor	Total	Sell
30 gallon	P1@2.50	Ea	350.00	87.90	437.90	744.00

40-gallon gas-fired water heater. 6-year warranty. Flammable Vapor Ignition Resistant (FVIR). Includes pipe connection labor. Add for flue, water and gas pipe, fitting materials and gas permit costs if required.

	Craft@Hrs	Unit	Material	Labor	Total	Sell
36,000 Btu	P1@2.75	Ea	363.00	96.70	459.70	781.00

50-gallon gas-fired water heater. 6-year warranty. Add for flue, water and gas pipe, fitting materials and gas permit costs if required.

	Craft@Hrs	Unit	Material	Labor	Total	Sell
40,000 Btu	P1@2.75	Ea	610.00	96.70	706.70	1,200.00

60-gallon gas water heater. 12-year warranty. Tall boy. Includes pipe connection labor. Add for flue, water and gas pipe, fitting materials and gas permit costs if required.]

	Craft@Hrs	Unit	Material	Labor	Total	Sell
50,000 Btu	P1@2.75	Ea	851.00	96.70	947.70	1,610.00

75-gallon gas water heater. 6-year warranty. Tall boy. Includes pipe connection labor. Add for flue, water and gas pipe, fitting materials and gas permit costs if required.

	Craft@Hrs	Unit	Material	Labor	Total	Sell
70,000 Btu, Low-NOx	P1@2.75	Ea	826.00	96.70	922.70	1,570.00

100-gallon commercial gas water heater. Includes pipe connection labor. Add for flue, water and gas pipe, fitting materials and gas permit costs if required.

	Craft@Hrs	Unit	Material	Labor	Total	Sell
77,000 Btu, 3-year warranty	P1@3.75	Ea	3,170.00	132.00	3,302.00	5,610.00
199,000 Btu, 3-year warranty	P1@3.75	Ea	6,520.00	132.00	6,652.00	11,308.00

50-gallon high-altitude gas water heater. Tall boy. Includes pipe connection labor. Add for flue, water and gas pipe, fitting materials and gas permit costs if required.

	Craft@Hrs	Unit	Material	Labor	Total	Sell
40,000 Btu	P1@2.75	Ea	662.00	96.70	758.70	1,290.00

75-gallon high-altitude gas water heater. 6-year warranty. For installations above 2,000 feet. Self-cleaning. Labor includes set in place only. Add for the cost of pipe and vent. Includes pipe connection labor. Add for flue, water and gas pipe, fitting materials and gas permit costs if required.

	Craft@Hrs	Unit	Material	Labor	Total	Sell
75,500 Btu	P1@3.50	Ea	850.00	123.00	973.00	1,650.00

Propane-fired tall boy water heater. 6-year warranty. Flammable Vapor Ignition Resistant (FVIR) water heater. Includes pipe connection labor. Add for flue, water and gas pipe, fitting materials and gas permit costs if required.

	Craft@Hrs	Unit	Material	Labor	Total	Sell
30 gallon, 27,000 Btu	P1@2.50	Ea	418.00	87.90	505.90	860.00
40 gallon, 36,000 Btu	P1@2.50	Ea	462.00	87.90	549.90	935.00
50 gallon, 36,000 Btu	P1@2.75	Ea	495.00	96.70	591.70	1,010.00

40-gallon power-vent propane water heater. 6-year warranty. Fan-assisted flue venting. Requires positive seal chimney vent. Power-assist permits combined horizontal and vertical venting. Air intake runs can be up to 80' (including one 90-degree elbow) using 3" Schedule 40 PVC, CPVC or ABS pipe. Includes pipe connection labor. Add for flue, water and gas pipe, fitting materials and gas permit costs if required.

	Craft@Hrs	Unit	Material	Labor	Total	Sell
42,000 Btu	P1@2.75	Ea	775.00	96.70	871.70	1,480.00

Oil-fired water heater. Labor includes setting and connecting only. Includes pipe connection labor. Add for flue, water and gas pipe, fitting materials and gas permit costs if required.

	Craft@Hrs	Unit	Material	Labor	Total	Sell
30 gallon, TF, tank only	P1@2.25	Ea	660.00	79.10	739.10	1,260.00
30 gallon, TF, with burner assembly	P1@2.25	Ea	1,050.00	79.10	1,129.10	1,920.00
50 gallon, TF, tank only	P1@2.75	Ea	1,250.00	96.70	1,346.70	2,290.00
50 gallon, TF, with burner assembly	P1@2.75	Ea	1,520.00	96.70	1,616.70	2,750.00

	Craft@Hrs	Unit	Material	Labor	Total	Sell

Electric Water Heaters

30-gallon electric water heater. 6-year warranty. Includes temperature and pressure relief valve. Includes pipe connection labor. Add for water pipe, fitting materials and electrical permit costs if required.

	Craft@Hrs	Unit	Material	Labor	Total	Sell
Two 3,800-watt elements	P1@2.00	Ea	346.00	70.30	416.30	708.00
Two 3,800-watt elements, low boy	P1@2.00	Ea	20.00	70.30	90.30	154.00

40-gallon electric water heater. 6-year warranty. Includes temperature and pressure relief valve. 42-gallon-per-hour recovery rate. 23" diameter, 32" height. $410 estimated annual operating costs. 0.90 energy factor. 11.5 R-factor on insulated jacket. Patented resistor-design heating elements prolong anode and tank life. Includes pipe connection labor. Add for water pipe, fitting materials and electrical permit costs if required.

	Craft@Hrs	Unit	Material	Labor	Total	Sell
Two 4,500-watt elements, low boy	P1@2.25	Ea	292.00	79.10	371.10	631.00

40-gallon electric water heater. 6-year warranty. Includes temperature and pressure relief valve. 49-gallon-per-hour recovery rate. 17" diameter, 59" height. $420 estimated annual operating costs. 0.88 energy factor. 11.5 R-factor on insulated jacket. Patented resistor-design heating elements prolong anode and tank life. Includes pipe connection labor. Add for water pipe, fitting materials and electrical permit costs if required.

	Craft@Hrs	Unit	Material	Labor	Total	Sell
Two 4,500-watt elements, tall boy	P1@2.25	Ea	304.00	79.10	383.10	651.00

47-gallon electric water heater. 6-year warranty. Includes temperature and pressure relief valve. 48-gallon-per-hour recovery rate. 27" diameter, 32" height. $425 estimated annual operating costs. 0.87 energy factor. 11.5 R-factor on insulated jacket. Patented resistor-design heating elements prolong anode and tank life. Includes pipe connection labor. Add for water pipe, fitting materials and electrical permit costs if required.

	Craft@Hrs	Unit	Material	Labor	Total	Sell
Two 4,500-watt elements, low boy	P1@2.50	Ea	405.00	87.90	492.90	838.00

50-gallon electric water heater. 6-year warranty. Includes temperature and pressure relief valve. Includes pipe connection labor. Add for water pipe, fitting materials and electrical permit costs if required.

	Craft@Hrs	Unit	Material	Labor	Total	Sell
Two 4,500-watt elements, low boy	P1@2.50	Ea	363.00	87.90	450.90	767.00
Two 4,500-watt elements, tall boy	P1@2.50	Ea	333.00	87.90	420.90	716.00

80-gallon electric water heater. 6-year warranty. Includes temperature and pressure relief valve. 23" diameter, 59" height. $425 estimated annual operating costs. 0.82 energy factor. 11.5 R-factor on insulated jacket. Includes pipe connection labor. Add for water pipe, fitting materials and electrical permit costs if required.

	Craft@Hrs	Unit	Material	Labor	Total	Sell
Two 4,500-watt elements	P1@3.25	Ea	550.00	114.00	664.00	1,130.00

Small electric water heater. 6-year warranty. Labor includes set in place only. Add for the cost of piping and electrical runs. Includes pipe connection labor. Add for water pipe, fitting materials and electrical permit costs if required.

	Craft@Hrs	Unit	Material	Labor	Total	Sell
4 gallons	P1@1.25	Ea	242.00	44.00	286.00	486.00

Thermal expansion tank. Control thermal expansion of water in domestic hot water systems. Absorbs the increased volume of water generated by the hot water heating sourcekeeping system pressure relief setting of the T&P Relief valve.

	Craft@Hrs	Unit	Material	Labor	Total	Sell
Thermal Expansion Tank 2.1 Gal.	P1@.300	Ea	41.70	10.50	52.20	88.70
Thermal Expansion Tank 4.5 Gal.	P1@.300	Ea	64.80	10.50	75.30	128.00

Retrofit installation increase labor to 1 man-hour per unit

Heat recovery ventilator central system retrofit. Fits onto an existing central air handler. Includes cabinet, twin ducts, duct collars, gaskets, mount plates, fasteners, air-to-air plate heat exchanger, temperature regulator sensor, damper, damper actuator and remote controller module. Add the cost of duct and wiring.

	Craft@Hrs	Unit	Material	Labor	Total	Sell
Central heat recovery module	SW@1.50	Ea	2,360.00	60.60	2,420.60	4,120.00
Install unit controls	BE@0.50	Ea	—	19.50	19.50	33.20
Air balance and test	SW@1.50	Ea	—	60.60	60.60	103.00

	Craft@Hrs	Unit	Material	Labor	Total	Sell

Tankless Water Heaters

Tankless gas water heater, Bosch AquaStar™. Gas tankless water heater with standing pilot. Hangs on the wall to save floor space. Provides endless hot water for one major application at a time. Made of high-quality plastic, brass, and copper. 18" wide x 30" high x 9" deep. 15-year warranty on heat exchanger. Replaces 40-gallon tank-style water heater. Natural gas or propane. Requires electric connection for ignition. Includes pipe connection labor. Add for water pipe, fitting materials and permit costs if required

	Craft@Hrs	Unit	Material	Labor	Total	Sell
117,000 Btu	P1@3.25	Ea	825.00	114.00	939.00	1,600.00

Tankless gas water heater power vent kit. For Bosch AquaStar.

	Craft@Hrs	Unit	Material	Labor	Total	Sell
Power vent kit	P1@2.50	Ea	387.00	87.90	474.90	807.00

Tankless electric water heater, PowerStar™. Designed to replace a 40-gallon electric storage tank water heater. Mounts on the wall. Electronic flow switch. Electronic thermostatic control. 10-year warranty. 99% efficiency rating. Glass reinforced plastic heat exchanger. 1/2" compression fitting connections. Filter screen on inlet. Thermal safety cut out. Add the cost of electrical source as required. Includes pipe connection labor. Add for pipe and fitting material costs if required.

	Craft@Hrs	Unit	Material	Labor	Total	Sell
240 volts, 17kW	P1@2.15	Ea	498.00	75.60	573.60	975.00
240 volts, 27kW	P1@2.15	Ea	712.00	75.60	787.60	1,340.00

Point-of-use electric water heater, Ariston GL4 Mini-Tank. 4-gallon capacity. 6-year warranty. Adjustable thermostatic control. Glass-lined tank. Includes temperature/pressure relief valve. Plugs into standard 110-volt outlet. Dimensions: 14" x 14" x 12". Installs independently or in-line with larger hot water source. Meets ASHRAE 90.1 Standard. UL listed. Includes pipe connection labor. Add for pipe and fitting material costs if required.

	Craft@Hrs	Unit	Material	Labor	Total	Sell
4 gallon	P1@2.15	Ea	242.00	75.60	317.60	540.00

Point-of-use electric water heater, PowerStream™. For under-sink applications. 5-year warranty. Solid copper heat exchanger. Mounts in any direction. Must be hardwired. Add the cost of electrical source as required. Includes pipe connection labor. Add for pipe and fitting material costs if required.

	Craft@Hrs	Unit	Material	Labor	Total	Sell
240 volts, 9.5kW	P1@2.15	Ea	216.00	75.60	291.60	496.00

Water Hammer Control

Water hammer arrester.

	Craft@Hrs	Unit	Material	Labor	Total	Sell
1/2" male threaded	P1@.250	Ea	9.15	8.79	17.94	30.50
3/4" male threaded	P1@.250	Ea	9.86	8.79	18.65	31.70

Water surge shock absorber. Water hammer arrestor. O-ring sealed air chamber maintains 60 PSI. Air charge absorbs high-pressure surge caused by quick closing of valves. Eliminates need for risers. Stops banging noises in pipes.

	Craft@Hrs	Unit	Material	Labor	Total	Sell
1/2" male threaded	P1@.250	Ea	14.80	8.79	23.59	40.10
For washing machine line	P1@.250	Ea	13.10	8.79	21.89	37.20

Plumbing

	Craft@Hrs	Unit	Material	Labor	Total	Sell
Remove and replace galvanized water pipe with copper pipe. Includes fittings and pipe clips. Based on Type M copper pipe with soft soldered joints. Includes disposal of scrap pipe and fittings. No salvage value assumed. Add for floor, wall or ceiling patching, if required.						
1/2" Type M copper pipe and fittings	P1@.145	LF	1.90	5.10	7.00	11.90
3/4" Type M copper pipe and fittings	P1@.160	LF	2.30	5.63	7.93	13.50
Prep and connect to existing 1/2" galvanized	P1@.550	Ea	9.46	19.30	28.76	48.90
Prep and connect to existing 3/4" galvanized	P1@.650	Ea	14.30	22.90	37.20	63.20
Remove and replace galvanized drain, waste and vent pipe (DWV) with copper pipe. Includes fittings and hangers. Based on type DWV copper pipe with soft soldered joints. Includes disposal of scrap pipe and fittings. No salvage value assumed. Add for floor, wall or ceiling patching, if required.						
1-1/2" DWV copper pipe and fittings	P1@.165	LF	6.73	5.80	12.53	21.30
2" DWV copper pipe and fittings	P1@.190	LF	8.44	6.68	15.12	25.70
Prep and connect to existing 1-1/2" galvanized	P1@.650	Ea	36.20	22.90	59.10	100.00
Prep and connect to existing 2" galvanized	P1@.850	Ea	41.00	29.90	70.90	121.00
Remove and replace cast iron drain, waste and vent pipe (DWV) with cast iron pipe. Includes fittings, couplings and hangers. Based on service weight hub-less cast iron pipe with mechanical joint couplings. Includes disposal of scrap pipe and fittings. No salvage value assumed. Add for floor, wall or ceiling patching, if required.						
2" cast iron MJ pipe and fittings	P1@.190	LF	7.37	6.68	14.05	23.90
3" cast iron pipe and fittings	P1@.210	LF	8.03	7.38	15.41	26.20
4" cast iron pipe and fittings	P1@.245	LF	9.56	8.61	18.17	30.90
Prep and connect to existing 2" cast iron	P1@.650	Ea	14.50	22.90	37.40	63.60
Prep and connect to existing 3" cast iron	P1@.700	Ea	19.60	24.60	44.20	75.10
Prep and connect to existing 4" cast iron	P1@.750	Ea	26.80	26.40	53.20	90.40
Remove and replace cast iron drain, waste and vent pipe (DWV) with ABS pipe. Includes fittings and hangers. Based on ABS plastic pipe with solvent weld joints. Includes disposal of scrap pipe and fittings. No salvage value assumed. Add for floor, wall or ceiling patching, if required.						
1-1/2" ABS pipe and fittings	P1@.110	LF	1.49	3.87	5.36	9.11
2" ABS pipe and fittings	P1@.120	LF	2.50	4.22	6.72	11.40
3" ABS pipe and fittings	P1@.145	LF	3.95	5.10	9.05	15.40
4" ABS pipe and fittings	P1@.160	LF	6.39	5.63	12.02	20.40
Prep and connect to existing 1-1/2" cast iron	P1@.550	Ea	29.20	19.30	48.50	82.50
Prep and connect to existing 2" cast iron	P1@.250	Ea	35.40	8.79	44.19	75.10
Prep and connect to existing 3" cast iron	P1@.750	Ea	45.10	26.40	71.50	122.00
Prep and connect to existing 4" cast iron	P1@.850	Ea	57.90	29.90	87.80	149.00
Remove and replace cast iron drain, waste and vent pipe (DWV) with PVC pipe. Includes fittings and hangers. Based on PVC DWV plastic pipe with solvent weld joints. Includes disposal of scrap pipe and fittings. No salvage value assumed. Add for floor, wall or ceiling patching, if required.						
1-1/2" PVC DWV pipe and fittings	P1@.115	LF	2.25	4.04	6.29	10.70
2" PVC DWV pipe and fittings	P1@.125	LF	3.94	4.40	8.34	14.20
3" PVC DWV pipe and fittings	P1@.150	LF	6.08	5.27	11.35	19.30
4" PVC DWV pipe and fittings	P1@.165	LF	9.74	5.80	15.54	26.40
Prep and connect to existing 1-1/2" cast iron	P1@.550	Ea	48.80	19.30	68.10	116.00
Prep and connect to existing 2" cast iron	P1@.250	Ea	57.60	8.79	66.39	113.00
Prep and connect to existing 3" cast iron	P1@.750	Ea	68.70	26.40	95.10	162.00
Prep and connect to existing 4" cast iron	P1@.850	Ea	85.20	29.90	115.10	196.00

Electrical

15

Most older homes have electrical systems that are in perfectly good working order — but completely inadequate by modern standards. They have too little power, too few circuits and far too few outlets. Even if the electrical service in the home has been upgraded in the last 25 years, more circuits may be needed to keep up with the demand created by today's multitude of electrical conveniences.

Few homes built before World War I were wired for electricity. Circuits were added later, usually gouged into plaster walls or run behind baseboards. In the 1920s and 1930s, most new homes were planned for 40 amps. The next jump was to four 15-amp circuits, or 60 amps total power. Many rooms had only a single duplex receptacle and a switched light fixture. By the 1950s, 100 amps was considered adequate power, unless the plan included an electric range or electric heat. Since the 1970s, 150 amps has been considered the minimum for a small home and 200 amps a better choice for most homes.

Enough Power?

The primary electrical shortcoming in older homes is too little amperage. Your first task is to determine how much power is available at the distribution panel (or fuse box). There are three ways to find out:

- ❖ The first limitation is the current-carrying capacity of the service entrance wire. The wire gauge may be marked on the insulation where the wire enters the weatherhead or where the cable emerges from underground conduit. If there's no marking, an electrician can measure the wire gauge and compute the maximum amperage. If you see only two wires instead of three running to the service head or emerging from underground conduit, there's no hope. Three wires are required for a modern 240-volt electrical system.
- ❖ Second, check the manufacturer's data plate or service disconnect. It will identify the maximum amperage of the main electrical panel.
- ❖ And third, the main circuit breaker will be marked with an amperage rating. It's always the same or less than the service drop and electrical panel.

Unfortunately, the amperage that can be delivered to loads in the house is determined by the *smallest* number among these three identifiers.

If the ampacity available isn't enough to carry the planned loads for your home improvement, you'll need a new service panel, and possibly a larger service drop from the electric company. If there's nothing wrong with the existing electrical system, you can just leave it in place. Then upgrade by adding more circuits and outlets and the new electrical service panel. Most building departments will allow non-electricians to do minor work, such as adding an outlet or switch, but a licensed electrician will be required for service upgrades. An electrician will be able to calculate what size panel you'll need to install. While doing that math, he can also figure the most efficient way to run circuits to new light fixtures, appliances and outlets. If you're not a licensed electrician, this is a job you'll have to sub out.

Licensing and Code Compliance

Electrical wiring can be dangerous, and a poorly-wired home is a fire hazard. Because of this, electricians have to be licensed and all their work must be done in compliance with the *National Electrical Code (NEC)*. Unfortunately, this isn't the last word on code compliance issues. Your electrical contractor will apply at the building department for a permit that covers the new work. That's easy enough. But the building department may require more. It's very unlikely that the existing electrical system in an older home complies with the current *NEC* standards. The question then becomes, how much of the old electrical system has to be brought up to code before the building inspector will sign off on the new work? If you're doing an extensive upgrade, the building department may require that you bring *all* the existing work up to meet current code requirements. This may involve more expense than the owner is willing, or able, to manage. And that's your best argument if you need to negotiate this issue with the building department. Too many requirements and nothing will get done — that benefits no one.

Resist the temptation to do major electrical work without a permit. You won't find a licensed electrician eager to bid on work like that, and a reputable home improvement contractor won't rely on an unlicensed electrician. A bootleg electrical job will make it hard for the owner to sell the home without first making the required code improvements. It's better for everyone to negotiate a fair compromise with the building official, use a professional electrician, and get the job signed off by a building inspector.

The Service Upgrade

The exterior portion of a service upgrade consists of replacing the meter, main service panel and the wire that connects the meter to the power grid. The lines up to the service drop belong to the utility company. The utility will probably have to do any work required on the service drop, but the homeowner will have to pay for the drop itself and the meter.

If the home has a basement, most of the service upgrade can be done there. Replace the old fuse box or boxes with a modern breaker panel. Then disconnect circuits running to the fuse box and reconnect them at the breaker box. Attach each circuit to a breaker with the correct capacity. The new breaker panel should have enough breaker spaces for existing circuits, plus a few extra for expansion. Most homes will need a 150- or 200-amp service panel.

Because the electrician's part of the work is done in the basement or outside, it can be done after all the carpentry and interior finishing is complete. However, any wiring that has to be run in the walls must be completed before drywall is installed. Otherwise, electrical wire has to be fished down between the studs.

Wiring Residences

Nearly all new homes today are wired with non-metallic sheathed cable, but that may not be what you find when you work on an older house. Depending on its age, you may have some surprises.

Knob and Tube Wiring

Homes built before about 1930 usually had knob and tube (K&T) wiring. Wires were strung between porcelain insulators driven into studs and joists. If wire had to pass through framing, a hollow porcelain tube was inserted in a hole drilled through the stud or joist. Conductors were usually single strands covered with cloth insulation. You won't find a ground wire on K&T. In those days, only lightning rods were grounded.

K&T that's given trouble-free service for nearly a century could probably do the same for another century, if no further demands were made on the system. But that's not likely. Because it isn't grounded, doesn't have enough capacity, and its insulation isn't worthy of the name, some insurance policies exclude coverage for homes with K&T wiring. For that reason alone, many of these older homes have already been upgraded.

If you find K&T wiring and the owner isn't willing to replace it, just bypass the K&T. Work around it with new circuits. *Don't extend it.* If possible, don't even touch it unless you find an actual or impending emergency.

Aluminum Wiring

Aluminum wire is another type that's no longer used for interior home electrical systems. Aluminum is a good and durable conductor and is usually less expensive than the more popular copper. However, late in the 1970s, electricians and code officials began to recognize a problem developing in homes with aluminum wiring. When aluminum wire carries current, it warms up and expands, just like copper. When it cools, the aluminum contracts, just like copper. But unlike copper, aluminum connections oxidize during the cool-down phase, creating resistance where conductors join. With time, the resistance grows into arcing — a spark that passes through the corrosion *(gap)* between the wire and the connector. Given the right conditions, that spark can ignite a fire.

The aluminum wire itself isn't the problem. It's the *wire connections* that are to blame. You don't need to rip out all the aluminum wire you find. But it's prudent to check connections in a home wired with aluminum. Electrical devices used with aluminum wire should be rated specifically for aluminum (usually stamped *CO/ALR* or *Al/Cu*). Look for signs of overheating, such as blackened connections or melted insulation. If you elect to extend an aluminum circuit using copper wire, your electrician will need to rent a special crimping tool made just for this purpose. When crimped, the wire connection must be covered with anti-oxidant grease.

Aluminum wire is still widely used for residential service entrance, though not for concealed wiring in walls and ceilings. Wires running between the house and the public utility grid are larger in diameter and require very few connections. That makes aluminum a good choice. For interior wiring, the price advantage of aluminum over copper usually isn't worth the risk or the extra trouble, though aluminum wire with a copper coating is used in some communities.

Non-Metallic Sheathed Cable

Non-metallic sheathed cable, called *Romex* or *"rope"* by electricians, is the most common wire type used in homes today. The *NEC* classifies it as *Type NM cable*. (Type NM-B cable is identical, but has a slightly-better temperature rating.) Romex has two or more insulated conductors and a ground wire, all covered in a plastic sheath. It's popular because it's inexpensive and easy to install. You can use NM cable in wall cavities where the wire is protected from physical damage and unlikely to get either wet or hot. When you run it through 2" x 4" stud walls, protect the cable at each stud with a metal plate to prevent damage from nails. Romex cable can be stapled to studs, rather than attached with nail-on hangers or supports. Most electrical codes permit the use of plastic (rather than metal) outlet boxes with NM cable.

You can't use Romex for exposed wiring on walls if it's within 5' of the floor. But most inspectors will approve *Type AC (armored cable)* for that purpose. AC is like Romex but includes a flexible aluminum cover that protects it from physical damage. If the inspector won't accept AC cable for exposed runs, you may have to install conduit. Unlike electric cable, conduit includes no wire. It's a protective tube through which wire is pulled. Conduit is used in most commercial buildings and occasionally in residences, such as in the service entrance mast where overhead wires terminate at the entrance cap. Flexible *(flex)*, EMT *(electric metallic tube)*, GRS *(galvanized rigid steel)* and IMC *(intermediate rigid conduit)* are the most common types of conduit. Flex is a hard metallic tube with enough flexibility to snake through studs. EMT is lightweight but not flexible. GRS conduit is heavier. IMC falls between EMT and GRS. Each of these types has specialized uses.

Electrical wire size is measured in American wire gauge (AWG) and usually abbreviated with the pound sign. For example, #14-3 indicates a 14-gauge wire with three conductors (and probably a separate bare ground wire). The smaller the gauge number, the bigger the wire and the greater its current-carrying capacity. Most circuits in a home are rated at 15 amps and use 14-gauge copper wire. Circuits for kitchen appliances should be rated at 20 amps and use 12-gauge copper wire. Circuits for an electric water heater, air conditioner or electric clothes dryer should be 30 amps and use 10-gauge copper wire. An electric range requires 6-gauge copper wire and a 50-amp breaker. All of these cables should include a copper ground wire.

Usually Type THHN wire is used with conduit because it has a heat-resistant thermoplastic cover. THHN is available in many colors to simplify the identification of conductors after the wire is pulled in the conduit. Small gauges are solid wire. Larger gauges are stranded wire, usually 19 strands per conductor. Stranded wire is a slightly better conductor than solid wire. It's also not as stiff, so it's easier to pull stranded wire in conduit.

Use Type UF cable for underground runs or in damp locations. It's not required by all codes, but common sense dictates that buried UF cable be trenched deep enough to make accidental damage unlikely. Either enclose the UF in PVC conduit or lay a warning tape in the trench over the wire before it's backfilled.

Protect cable with conduit wherever it's exposed, particularly where it exits the residence. For outdoor wiring, junction boxes and outlet boxes must be rated waterproof, and all receptacles must be GFCI protected.

Ground Wire

Electrical systems in old houses usually don't include a ground wire. If you're not sure about the house you're working on, just check any duplex receptacle. If it only has space for two contacts, you're dealing with an ungrounded electrical system. If you're not tearing into the walls, and you're working on a tight budget, it's best to simply leave this wiring as it is — assuming the building inspector doesn't intervene. An old electrical system that was properly installed is still safe and usable. But be aware that two-wire electrical cable can only support a 15-amp ungrounded circuit. That limits the number of receptacles and lighting fixtures on a circuit. If you're going to tear into the walls, plan to replace old two-wire circuits with 20-amp grounded circuits. If you disturb an old two-wire circuit, don't

Fishing Electrical Cable

In new construction, wire is threaded through studs and between joists before the framing cavities are enclosed with drywall. That's the easy way to string wire. In home improvement work, wall and ceiling cavities aren't always exposed, so the wire has to be *fished* through enclosed stud spaces using an electrician's fish tape. Fishing electrical cable through existing walls is slow work.

A fish tape is a reel of springy wire that retracts into a circular case. The wire is flexible enough to bend when needed but stiff enough to be self-supporting for at least a few feet. The leading end of the tape has a hook so electrical wire is easy to attach. If possible, you'll want to fish the cable vertically — down from the attic or up from the basement. From the attic, bore down through the wall's top plate and thread fish tape down between two studs. When tape is fished from beginning point to end point, attach electric cable to the leading end of the tape. Then reel in the tape, threading new wire through the stud and joist cavities as the tape retracts.

If you have no choice but to fish cable horizontally across studs and joists, the first step is to find the shortest and easiest route between where power is available and where power is needed. With the help of an electric stud finder, count the studs and joists between those two points. When you've settled on the path of least resistance, break into the wall at each stud or joist and cut out a small notch large enough for wiring to pass through. When all the framing has been notched, start threading fish tape through the framing from notch to notch. This is slow work and requires two people, one pushing fish tape and the other finding and guiding the tape end. When the wire ends have been connected and you're sure the electrical work is done, seal the holes in the drywall. Use fiberglass mesh tape and joint compound to patch and smooth the drywall surface.

count on it to be entirely reliable. Insulation on old wire can become brittle and crumble when moved, leaving dangerous bare wires that you'll have to replace.

Low Voltage Wiring

Phone lines, COAX TV cable, computer network cable and speaker wire don't present the same degree of risk that comes with 120-volt alternating current circuits. Because of this, the *NEC* has only a few simple rules for low voltage wiring. Primary among these is keeping low voltage wiring out of conduit, junction and outlet boxes that include 120-volt wiring. Low voltage wiring is dangerous if it's interconnected with regular AC circuits. You should also keep video, speaker and data cable at least a foot away from AC cable. That will minimize the effect of radio frequency interference (RFI) generated by alternating current circuits.

But don't let these rules impede progress. While adding or extending AC circuits, use the opportunity to also add new connections for modern low voltage conveniences. Most of the houses you'll be upgrading were built long before the need for phone jacks in every room, multiple cable TV outlets, and computer network cabling.

Adding Outlets and Switches

Most rooms in old houses have a single electrical outlet, and all the outlets in the house may be on the same 15-amp circuit. That's enough power for a few lamps and not much more. It certainly won't support a modern lifestyle. You may be able to turn on the lights and watch TV in the evening, but forget the microwave popcorn!

The *NEC* sets standards for residential electrical outlets. These standards may or may not be enforced in home improvement projects in your community. The degree that these regulations affect your project may be a matter for negotiation between the contractor and the building department. Select an electrical contractor with experience in negotiating with building inspectors. Remember, however, that the code exists for a reason and most of what it requires is simply good professional practice. Follow the code standards.

Spacing of Outlets — In most rooms, the code requires that no point along the floor line be more than 6' from a receptacle. That means you need an outlet at least every 12' along walls. Floor outlets don't help meet this requirement unless they're near the wall. Different standards apply to kitchens, bathrooms and laundry rooms. Spacing in hallways can be 20' and closets don't need any outlets at all.

Kitchens — Plan at least two 20-amp small appliance circuits to serve the kitchen, pantry and dining area. These circuits are in addition to circuits used by the refrigerator, dishwasher, oven, range, garbage disposer and lighting. Every kitchen counter wider than 12" needs at least one outlet. No point on a kitchen counter can be more than 24" from an outlet. That means you need an outlet at least every 4' over counters. Outlets have to be mounted on a wall, not face-up in the counter. The outlet next to the sink must be protected with a ground-fault circuit interrupter (GFCI). Plan on dedicated circuits for the range, dishwasher, garbage disposer, refrigerator, and microwave oven. A microwave oven will trip the breaker if it's on the same circuit with another large appliance, such as a refrigerator.

Bathrooms — Every bathroom needs at least one GFCI-protected outlet by the sink, even if the vanity lighting fixture includes an outlet. Keep the receptacle far enough away from the bathtub and shower to prevent the use of electric shavers or hairdryers while bathing.

Laundry — Provide at least one GFCI-protected duplex receptacle in the laundry area. The laundry receptacle must be on a dedicated 20-amp circuit.

Unfinished Basement or Attic — Provide at least one outlet.

Outdoors — Include a GFCI-protected duplex receptacle at the front and rear of the house.

Garage — Provide a GFCI-protected duplex receptacle for each parking space. Detached garages may not need any outlets.

Ground-Fault Circuit Interrupter — GFCI protection opens the circuit when a ground-fault is detected — such as when someone gets an electric shock. Outlets over the kitchen and bathroom sinks, in the laundry room, garage and outdoors have to be GFCI-protected. Installing GFCI outlets is easy, assuming the circuit includes a ground wire. Just remove the old outlet and replace it with a GFCI outlet. You don't have to buy a GFCI outlet for every receptacle that needs ground-fault protection. Several regular outlets can be wired to a single GFCI device. Circuitry in the GFCI outlet will protect all attached receptacles. That's the good news. The bad news is that GFCI outlets require a ground to work properly. If the old circuit has no ground wire, you'll have to run grounded cable to GFCI outlets.

Arc-Fault Interrupter — The *NEC* now requires these special breakers for circuits that serve sleeping rooms. Regular breakers open when there's an overload. Arc-fault breakers open any time the circuit is creating sparks, even if there's no overload.

Outlets per Circuit — Plan on six duplex outlets per 20-amp circuit. Your electrician may suggest ways to put 10 or even 12 outlets on a circuit and still meet *NEC* requirements. The code doesn't prohibit mixing light fixtures and outlets on the same *home run* (connection to the breaker panel).

Lights — Every room needs either a switch-operated overhead light or a switch-operated outlet. The code requires the switch to be located by the door at the room's entrance. You can run the wiring to any ceiling fixture that's still in good condition. If the fixture is worn, broken, or simply unattractive, replacing it is a simple task. But don't go overboard on wattage, such as replacing a 75-watt bathroom fixture with a 500-watt heat lamp. If your choice of replacement fixtures is a fluorescent, be sure there isn't a grounding problem. Many fluorescent fixtures require a ground wire. If you install a ceiling fan or a chandelier, plan to set a ceiling fixture box specifically rated for that purpose. Allow at least 3" between a recessed (non-IC) fixture and any insulation in the ceiling. Better yet, select a recessed fixture with an insulated case (IC) and thermal cutout specifically rated for installation touching insulation.

Smoke Detectors

Nearly all communities require some type of smoke detector. The most common are battery-powered ionization detectors that recognize products of combustion even before flame is visible. Other types of detectors recognize smoke or detect a rapid rise in temperature. If you have the opportunity, recommend an AC-powered detector rather than a battery-powered unit. Surveys show that a high percentage of battery-powered smoke detectors have a dead battery at any given time. For that reason, many building codes require AC detectors, some with a battery backup.

	Craft@Hrs	Unit	Material	Labor	Total	Sell

Labor Estimates for Wiring

Fish electrical cable through a stud wall. Add the cost of connecting wire at both ends.

	Craft@Hrs	Unit	Material	Labor	Total	Sell
Per job set-up	BE@.270	Ea	—	10.50	10.50	17.90
Per foot of cable fished	BE@.050	Ea	—	1.95	1.95	3.32
Per electric box set	BE@.761	Ea	—	29.70	29.70	50.50
Per stud notched and hole patched	BE@.400	Ea	—	15.60	15.60	26.50

Cable Snake™ flat wire fish tape, Gardner Bender. Perfect for fishing short runs of wire in existing structures. Ergonomic handle provides a sure grip for pulling. Aerodynamic tip reduces friction, requiring less force and decreased hang-ups. Reel handle allows for quick retrieval and locking for pulling.

	Craft@Hrs	Unit	Material	Labor	Total	Sell
20' x 1/4" x .031"	—	Ea	20.00	—	20.00	—

Connect new 220-volt appliance with plug. Does not include electrical rough-in. Includes a range, electric dryer or air conditioner using three-conductor cable with ground wire and connected plug. Add the cost of setting the appliance in place.

	Craft@Hrs	Unit	Material	Labor	Total	Sell
Per appliance	BE@.433	Ea	—	16.90	16.90	28.70

Connect new 220-volt appliance to an outlet box with cable. Includes a range, electric dryer or air conditioner to a service outlet box using three-conductor non-metallic cable with ground wire. Add the cost of setting the appliance in place.

	Craft@Hrs	Unit	Material	Labor	Total	Sell
Per appliance	BE@.530	Ea	—	20.70	20.70	35.20

Disconnect and reconnect an existing 220-volt appliance to an outlet box with cable. Includes a range, electric dryer or air conditioner to a service outlet box using three-conductor non-metallic cable with ground wire. Add the cost of setting the appliance in place.

	Craft@Hrs	Unit	Material	Labor	Total	Sell
Per appliance	BE@.510	Ea	—	19.90	19.90	33.80

Connect new 220-volt appliance using flex conduit. Includes a range, electric dryer or air conditioner connected to a service outlet box using wire pulled in flex conduit. Add the cost of setting the appliance in place.

	Craft@Hrs	Unit	Material	Labor	Total	Sell
Per appliance	BE@1.04	Ea	—	40.60	40.60	69.00

Remove and replace 220-volt cable from appliance. Includes a range, electric dryer or air conditioner. Add the cost of setting the appliance in place.

	Craft@Hrs	Unit	Material	Labor	Total	Sell
Per appliance	BE@.465	Ea	—	18.20	18.20	30.90

Connect or disconnect 110-volt appliance using three 8 gauge or smaller connectors. Includes connecting a dishwasher or garbage disposer to the supply box. Add the cost of setting or removing the appliance.

	Craft@Hrs	Unit	Material	Labor	Total	Sell
Per appliance	BE@.796	Ea	—	31.10	31.10	52.90

Type NM Non-Metallic Cable

14 gauge Type NM-B Romex® sheathed indoor cable. White. With full-size ground wire. Labor is based on installation in a wood-frame building, including boring out and pulling cable in exposed walls. Add the cost of fishing cable through enclosed walls, when required.

	Craft@Hrs	Unit	Material	Labor	Total	Sell
2 conductor, per foot	BE@.027	LF	..90	1.05	1.05	1.79
2 conductor, 25' coil	BE@.027	LF	.46	1.05	1.51	2.57
2 conductor, 50' coil	BE@.027	LF	.44	1.05	1.49	2.53
2 conductor, 100' coil	BE@.027	LF	.35	1.05	1.40	2.38
2 conductor, 250' coil	BE@.027	LF	.18	1.05	1.23	2.09
3 conductor, per foot	BE@.030	LF	1.08	1.17	2.25	3.83
3 conductor, 25' coil	BE@.030	LF	.69	1.17	1.86	3.16
3 conductor, 50' coil	BE@.030	LF	.66	1.17	1.83	3.11
3 conductor, 100' coil	BE@.030	LF	.55	1.17	1.72	2.92

	Craft@Hrs	Unit	Material	Labor	Total	Sell
3 conductor, 250' coil	BE@.030	LF	.29	1.17	1.46	2.48
14-2-2, per foot	BE@.032	LF	1.20	1.25	2.45	4.17
14-2-2, 250' coil	BE@.032	LF	.76	1.25	2.01	3.42

12 gauge Type NM-B Romex® sheathed indoor cable. Yellow. With full-size ground wire. Rated at 20 amps. Labor is based on installation in a wood-frame building, including boring out and pulling cable in exposed walls. Add the cost of fishing cable through enclosed walls, when required.

	Craft@Hrs	Unit	Material	Labor	Total	Sell
2 conductor, per foot	BE@.030	LF	1.10	1.17	2.27	3.86
2 conductor, 25' coil	BE@.030	LF	.70	1.17	1.87	3.18
2 conductor, 50' coil	BE@.030	LF	.67	1.17	1.84	3.13
2 conductor, 100' coil	BE@.030	LF	.48	1.17	1.65	2.81
2 conductor, 250' coil	BE@.030	LF	.27	1.17	1.44	2.45
3 conductor, per foot	BE@.031	LF	1.53	1.21	2.74	4.66
3 conductor, 25' coil	BE@.031	LF	1.15	1.21	2.36	4.01
3 conductor, 50' coil	BE@.031	LF	1.10	1.21	2.31	3.93
3 conductor, 100' coil	BE@.031	LF	.94	1.21	2.15	3.66
3 conductor, 250' coil	BE@.031	LF	.48	1.21	1.69	2.87
12-2-2, per foot	BE@.032	LF	1.39	1.25	2.64	4.49
12-2-2, 250' coil	BE@.032	LF	.90	1.25	2.15	3.66

10 gauge Type NM-B Romex® sheathed indoor cable. Orange. With ground. Rated at 30 amps. Labor is based on installation in a wood-frame building, including boring out and pulling cable in exposed walls. Add the cost of fishing cable through enclosed walls, when required.

	Craft@Hrs	Unit	Material	Labor	Total	Sell
2 conductor, per foot	BE@.031	LF	2.18	1.21	3.39	5.76
2 conductor, 25' coil	BE@.031	LF	1.29	1.21	2.50	4.25
2 conductor, 50' coil	BE@.031	LF	1.22	1.21	2.43	4.13
2 conductor, 100' coil	BE@.031	LF	1.07	1.21	2.28	3.88
2 conductor, 250' coil	BE@.031	LF	.63	1.21	1.84	3.13
3 conductor, per foot	BE@.035	LF	2.85	1.37	4.22	7.17
3 conductor, 25' coil	BE@.035	LF	1.74	1.37	3.11	5.29
3 conductor, 50' coil	BE@.035	LF	1.58	1.37	2.95	5.02
3 conductor, 100' coil	BE@.035	LF	1.20	1.37	2.57	4.37
3 conductor, 250' coil	BE@.035	LF	.79	1.37	2.16	3.67

8 gauge Type NM-B Romex® sheathed indoor cable. Black. With ground, except where noted. Rated at 40 amps. Labor is based on installation in a wood-frame building, including boring out and pulling cable in exposed walls. Add the cost of fishing cable through enclosed walls, when required.

	Craft@Hrs	Unit	Material	Labor	Total	Sell
2 conductor, per foot	BE@.033	LF	2.01	1.29	3.30	5.61
2 conductor, 125' coil	BE@.033	LF	.87	1.29	2.16	3.67
3 conductor, no ground, per foot	BE@.033	LF	2.69	1.29	3.98	6.77
3 conductor, per foot	BE@.037	LF	3.16	1.44	4.60	7.82
3 conductor, 125' coil	BE@.037	LF	1.38	1.44	2.82	4.79

6 gauge Type NM-B Romex® sheathed indoor cable. Black. With ground. Rated at 55 amps. Labor is based on installation in a wood-frame building, including boring out and pulling cable in exposed walls. Add the cost of fishing cable through enclosed walls when required.

	Craft@Hrs	Unit	Material	Labor	Total	Sell
2 conductor, per foot	BE@.038	LF	2.98	1.48	4.46	7.58
2 conductor, 125' coil	BE@.038	LF	1.21	1.48	2.69	4.57
3 conductor, per foot	BE@.044	LF	4.31	1.72	6.03	10.30
3 conductor, 125' reel	BE@.044	LF	1.77	1.72	3.49	5.93

	Craft@Hrs	Unit	Material	Labor	Total	Sell
Cable two-piece clamp connector. Zinc-plated steel. Butterfly style, does not require a locknut. For indoor use only in a dry location, Secures 14 to 10 gauge non-metallic sheathed cable to a steel outlet box or other metal enclosure. 3/8" size fits 1/2" knockout. UL listed.						
3/8"	BE@.050	Ea	.37	1.95	2.32	3.94
3/4" to 1"	BE@.050	Ea	.59	1.95	2.54	4.32
1-1/4"	BE@.050	Ea	1.21	1.95	3.16	5.37
1-1/2"	BE@.050	Ea	1.33	1.95	3.28	5.58
Stud nail plate. Steel. Installed at each stud to protect cable bored through framing, as defined in *NEC* Article 300-4.						
1-1/2" x 2-1/2", each	BE@.050	Ea	.28	1.95	2.23	3.79
1-1/2" x 2-1/2", pack of 50	BE@2.50	Ea	9.13	97.60	106.73	181.00
1-1/2" x 5", each	BE@.100	Ea	.36	3.91	4.27	7.26

Type UF Non-Metallic Cable

	Craft@Hrs	Unit	Material	Labor	Total	Sell
UF-B sheathed underground cable. Gray. With ground. Used to supply power to landscape lighting, pumps and other loads connected from the main building. Designed for burial in a trench. Add the cost of excavation and backfill.						
14 gauge, 2 conductor, per foot	BE@.010	LF	.93	.39	1.32	2.24
14 gauge, 2 conductor, 25' coil	BE@.010	LF	.63	.39	1.02	1.73
14 gauge, 2 conductor, 100' coil	BE@.010	LF	.52	.39	.91	1.55
14 gauge, 3 conductor, per foot	BE@.010	LF	1.17	.39	1.56	2.65
14 gauge, 3 conductor, 25' coil	BE@.010	LF	.93	.39	1.32	2.24
14 gauge, 3 conductor, 50' coil	BE@.010	LF	.82	.39	1.21	2.06
12 gauge, 2 conductor, per foot	BE@.010	LF	1.13	.39	1.52	2.58
12 gauge, 2 conductor, 25' coil	BE@.010	LF	.99	.39	1.38	2.35
12 gauge, 2 conductor, 50' coil	BE@.010	LF	.98	.39	1.37	2.33
12 gauge, 2 conductor, 100' coil	BE@.010	LF	.81	.39	1.20	2.04
12 gauge, 3 conductor, per foot	BE@.010	LF	1.90	.39	2.29	3.89
12 gauge, 3 conductor, 100' coil	BE@.010	LF	1.24	.39	1.63	2.77
10 gauge, 2 conductor, per foot	BE@.010	LF	1.26	.39	1.65	2.81
10 gauge, 3 conductor, per foot	BE@.010	LF	2.55	.39	2.94	5.00
UF non-metallic cable compression connector. Compression-type connector for underground feeder cable. Fits smaller UF cables as well as standard sizes. Box knockout size is the same as the nominal conduit size. UL listed.						
1/2", 14 to 12 gauge, 2 conductor	BE@.050	Ea	3.36	1.95	5.31	9.03
3/4", 14 to 12 gauge, 3 conductor	BE@.060	Ea	3.88	2.34	6.22	10.60
AC-90® steel armored flexible cable, American Flexible Conduit. Galvanized steel. 16 gauge integral bond wire-to-armor grounding path. Sometimes referred to as "BX" cable. Paper wrap conductor insulation covering. Maximum temperature rating is 90 degrees Celsius (dry). 600 volts. UL listed. By American Wire Gauge size and number of conductors. Add the cost of fishing cable through enclosed walls when required.						
14 gauge, 3 conductor, 25' coil	BE@.020	LF	1.23	.78	2.01	3.42
14 gauge, 4 conductor, per foot	BE@.021	LF	2.29	.82	3.11	5.29
14 gauge, 4 conductor, 250' coil	BE@.021	LF	.99	.82	1.81	3.08
12 gauge, 3 conductor, 50' coil	BE@.021	LF	1.09	.82	1.91	3.25
12 gauge, 3 conductor, 100' coil	BE@.021	LF	.97	.82	1.79	3.04
12 gauge, 4 conductor, per foot	BE@.023	LF	2.02	.90	2.92	4.96
12 gauge, 4 conductor, 250' reel	BE@.023	LF	1.06	.90	1.96	3.33
10 gauge, 3 conductor, 125' coil	BE@.027	LF	1.11	1.05	2.16	3.67

PVC Conduit and Fittings

PVC conduit, Schedule 40. For above- or below-ground applications subject to physical abuse. Also for use encased in concrete or for direct burial. Sunlight resistant. For use with conductors rated to 90 degrees Celsius. Sold in 10' lengths. Nominal sizes, by outside diameter (OD), inside diameter (ID) and wall thickness:

Nominal 1/2" measures 0.840" OD, 0.622" ID and 0.109" in wall thickness.
Nominal 3/4" measures 1.050" OD, 0.820" ID and 0.113" in wall thickness.
Nominal 1" measures 1.315" OD, 1.049" ID and 0.133" in wall thickness.
Nominal 1-1/4" measures 1.660" OD, 1.380" ID and 0.140" in wall thickness.
Nominal 1-1/2" measures 1.900" OD, 1.610" ID and 0.145" in wall thickness.
Nominal 2" measures 2.375" OD, 2.067" ID and 0.154" in wall thickness.
Nominal 2-1/2" measures 2.875" OD, 2.469" ID and 0.203" in wall thickness.

	Craft@Hrs	Unit	Material	Labor	Total	Sell
1/2" nominal diameter	BE@.031	LF	.13	1.21	1.34	2.28
3/4" nominal diameter	BE@.032	LF	.16	1.25	1.41	2.40
1" nominal diameter	BE@.033	LF	.27	1.29	1.56	2.65
1-1/4" nominal diameter	BE@.034	LF	.38	1.33	1.71	2.91
1-1/2" nominal diameter	BE@.035	LF	.45	1.37	1.82	3.09
2" nominal diameter	BE@.035	LF	.53	1.37	1.90	3.23
2-1/2" nominal diameter	BE@.039	LF	.99	1.52	2.51	4.27

PVC conduit, Schedule 80. For above- or below-ground applications subject to physical abuse. Also for use encased in concrete or for direct burial. Sunlight resistant. For use with conductors rated to 90 degrees Celsius. Meets specifications of Underwriters Laboratories 651 and the National Electrical Manufacturers Association TC 2.

	Craft@Hrs	Unit	Material	Labor	Total	Sell
1/2" nominal diameter	BE@.034	LF	.21	1.33	1.54	2.62
3/4" nominal diameter	BE@.035	LF	.24	1.37	1.61	2.74
1" nominal diameter	BE@.036	LF	.34	1.41	1.75	2.98
1-1/4" nominal diameter	BE@.034	LF	.56	1.33	1.89	3.21
1-1/2" nominal diameter	BE@.035	LF	.67	1.37	2.04	3.47
2" nominal diameter	BE@.040	LF	.97	1.56	2.53	4.30
2-1/2" nominal diameter	BE@.042	LF	1.59	1.64	3.23	5.49

Snap Strap PVC conduit clamp, Carlon®. For installation of polyvinyl chloride conduit. Can be used with rigid steel. Indoor use only. All plastic clamp allows conduit to expand and contract with temperature changes, eliminating bowing.

	Craft@Hrs	Unit	Material	Labor	Total	Sell
Single mount, 1/2"	BE@.050	Ea	.57	1.95	2.52	4.28
Single mount, 3/4"	BE@.050	Ea	.87	1.95	2.82	4.79
Single mount, 1"	BE@.050	Ea	.69	1.95	2.64	4.49
Double mount, 1-1/4"	BE@.050	Ea	.80	1.95	2.75	4.68
Double mount, 2"	BE@.050	Ea	1.00	1.95	2.95	5.02

Non-metallic reducer bushing, Carlon®. Rigid Schedule 40 PVC conduit reducer. Socket ends. Male x female. UL listed.

	Craft@Hrs	Unit	Material	Labor	Total	Sell
3/4" x 1/2"	BE@.030	Ea	.85	1.17	2.02	3.43
1" x 1/2"	BE@.030	Ea	1.13	1.17	2.30	3.91
1" x 3/4"	BE@.040	Ea	1.82	1.56	3.38	5.75
1-1/4" x 3/4"	BE@.050	Ea	1.77	1.95	3.72	6.32
1-1/4" x 1"	BE@.050	Ea	2.02	1.95	3.97	6.75
1-1/2" x 1"	BE@.050	Ea	1.93	1.95	3.88	6.60
1-1/2" x 1-1/4"	BE@.050	Ea	2.39	1.95	4.34	7.38
2" x 1-1/4"	BE@.060	Ea	2.65	2.34	4.99	8.48
2" x 1-1/2"	BE@.060	Ea	2.79	2.34	5.13	8.72
2-1/2" x 2"	BE@.060	Ea	6.99	2.34	9.33	15.90

	Craft@Hrs	Unit	Material	Labor	Total	Sell
PVC conduit box adapter, Carlon®. For non-metallic Schedule 40 or 80 conduit.						
1/2" diameter	BE@.050	Ea	.49	1.95	2.44	4.15
3/4" diameter	BE@.060	Ea	.53	2.34	2.87	4.88
1" diameter	BE@.080	Ea	.74	3.12	3.86	6.56
1-1/4" diameter	BE@.100	Ea	.90	3.91	4.81	8.18
1-1/2" diameter	BE@.100	Ea	.95	3.91	4.86	8.26
2" diameter	BE@.150	Ea	1.20	5.86	7.06	12.00
Non-metallic standard PVC coupling, Carlon®. For joining Schedule 40 PVC conduit.						
1/2" diameter	BE@.020	Ea	.24	.78	1.02	1.73
3/4" diameter	BE@.030	Ea	.30	1.17	1.47	2.50
1" diameter	BE@.050	Ea	.45	1.95	2.40	4.08
1-1/4" diameter	BE@.060	Ea	.63	2.34	2.97	5.05
1-1/2" diameter	BE@.060	Ea	.74	2.34	3.08	5.24
2" diameter	BE@.080	Ea	.94	3.12	4.06	6.90
2-1/2" diameter	BE@.090	Ea	1.36	3.51	4.87	8.28
Non-metallic 45-degree elbow, plain end, Carlon®. Standard radius. For Schedule 40 PVC conduit.						
1/2" diameter	BE@.050	Ea	.58	1.95	2.53	4.30
3/4" diameter	BE@.060	Ea	.75	2.34	3.09	5.25
1" diameter	BE@.080	Ea	1.82	3.12	4.94	8.40
1-1/4" diameter	BE@.080	Ea	1.49	3.12	4.61	7.84
1-1/2" diameter	BE@.100	Ea	1.61	3.91	5.52	9.38
2" diameter	BE@.150	Ea	2.61	5.86	8.47	14.40
2-1/2" diameter	BE@.150	Ea	3.75	5.86	9.61	16.30
Non-metallic 90-degree elbow with plain end, standard radius, Carlon®. For Schedule 40 PVC conduit.						
1/2" diameter	BE@.090	Ea	.61	3.51	4.12	7.00
3/4" diameter	BE@.090	Ea	.79	3.51	4.30	7.31
1" diameter	BE@.090	Ea	1.11	3.51	4.62	7.85
1-1/4" diameter	BE@.090	Ea	1.88	3.51	5.39	9.16
1-1/2" diameter	BE@.090	Ea	2.69	3.51	6.20	10.50
2" diameter	BE@.090	Ea	3.00	3.51	6.51	11.10
2-1/2" diameter	BE@.150	Ea	7.12	5.86	12.98	22.10
PVC conduit body, Type T, Carlon®. For easy access and pulling of wires through conduit runs. Will not rust or corrode. For use with Schedule 40 and 80 non-metallic conduit. Gasket included.						
1/2"	BE@.100	Ea	2.44	3.91	6.35	10.80
3/4"	BE@.150	Ea	3.04	5.86	8.90	15.10
1"	BE@.150	Ea	3.25	5.86	9.11	15.50
1-1/4"	BE@.150	Ea	4.06	5.86	9.92	16.90
1-1/2"	BE@.150	Ea	5.61	5.86	11.47	19.50
2"	BE@.150	Ea	9.66	5.86	15.52	26.40
PVC conduit body, Type LB, Carlon®. For easy access and pulling of wires through conduit run. For use with Schedule 40 and 80 non-metallic conduit. Gasket included. UL listed.						
1/2", 4 cubic inch	BE@.100	Ea	2.32	3.91	6.23	10.60
3/4", 12 cubic inch	BE@.100	Ea	3.03	3.91	6.94	11.80
1", 12 cubic inch	BE@.150	Ea	3.36	5.86	9.22	15.70
1-1/4", 32 cubic inch	BE@.150	Ea	4.56	5.86	10.42	17.70
1-1/2", 32 cubic inch	BE@.150	Ea	5.46	5.86	11.32	19.20
2", 63 cubic inch	BE@.200	Ea	8.64	7.81	16.45	28.00
2-1/2"	BE@.215	Ea	30.60	8.40	39.00	66.30

	Craft@Hrs	Unit	Material	Labor	Total	Sell
PVC service entrance cap. UL listed. Plus 40 and Plus 80.						
3/4"	BE@.150	Ea	3.04	5.86	8.90	15.10
1"	BE@.150	Ea	4.95	5.86	10.81	18.40
1-1/4"	BE@.300	Ea	6.50	11.70	18.20	30.90
1-1/2"	BE@.300	Ea	8.51	11.70	20.21	34.40
2"	BE@.500	Ea	10.40	19.50	29.90	50.80
Non-metallic two-gang conduit box, Super Blue, Carlon®. Two-gang hard shell box with captive nails. Depth 3-1/2", width 4-1/8", length 3-7/8". 8 integral clamps, 4 each side.						
35 cubic inch, wood studs	BE@.200	Ea	2.00	7.81	9.81	16.70
35 cubic inch, steel studs	BE@.150	Ea	2.49	5.86	8.35	14.20
Non-metallic three-gang conduit box, Super Blue, Carlon®. Three-gang hard shell box with captive nails. Depth 3-1/2", width 3-3/4", length 5-7/8". 12 integral clamps, 6 each side.						
53 cubic inch, wood studs	BE@.200	Ea	2.90	7.81	10.71	18.20
53 cubic inch, steel studs	BE@.150	Ea	2.96	5.86	8.82	15.00
Non-metallic ceiling fan box, Carlon®. 4", rated for fans up to 35 pounds. Rated for lighting fixtures up to 50 pounds. Includes mounting screws.						
1/2" deep, 8 cubic inch	BE@.300	Ea	3.47	11.70	15.17	25.80
2-1/4" deep, 20 cubic inch	BE@.300	Ea	4.59	11.70	16.29	27.70
Non-metallic old work switch box, Zip-Mount® retainers, Carlon®. For existing construction. Non-metallic cable clamps. UL listed.						
Single gang, 14 cubic inch	BE@.200	Ea	1.00	7.81	8.81	15.00
Single gang, 20 cubic inch	BE@.200	Ea	2.00	7.81	9.81	16.70
2 gang, 25 cubic inch	BE@.200	Ea	2.33	7.81	10.14	17.20
3 gang, 55 cubic inch	BE@.200	Ea	3.93	7.81	11.74	20.00
4 gang, 68 cubic inch	BE@.300	Ea	6.32	11.70	18.02	30.60
Non-metallic old work switch box, Zip-Mount® retainers, Carlon®. For existing construction. 4 integral non-metallic cable clamps. 3 Zip-Mount® retainers. 2-3/4" deep. High temperature resistance. UL listed.						
Round	BE@.300	Ea	6.32	11.70	18.02	30.60
Non-metallic low voltage device mounting bracket, Arlington. Low voltage device; Class 2 only. For both single- and double-gang installation on existing construction. Not to be used for AC circuits. 110 volt. 4.256" high. Non-conductive, smooth plastic construction. For communications, cable TV, computer wiring. CAT 5 listed. Adjusts to fit 1/4"- to 1"-thick wallboard, paneling, or drywall. Bracket is its own template for cutout. Specially designed wing flips up when mounting screw is tightened for a secure mount. UL listed.						
Single gang, 2.507" wide	BE@.050	Ea	1.41	1.95	3.36	5.71
Double gang, 4.185" wide	BE@.050	Ea	2.57	1.95	4.52	7.68

EMT Conduit and Fittings

	Craft@Hrs	Unit	Material	Labor	Total	Sell
Remove EMT conduit and wire. Removed in salvage condition.						
1/2" to 2" conduit diameter	BE@.030	LF	—	1.17	1.17	1.99
Electric metallic tube (EMT) conduit. Welded and galvanized. UL listed. *NEC* approved. Meets ANSI specifications. Sold in 10' lengths. Labor assumes installation in exposed frame walls.						
1/2" nominal diameter	BE@.033	LF	.20	1.29	1.49	2.53
3/4" nominal diameter	BE@.035	LF	.41	1.37	1.78	3.03
1" nominal diameter	BE@.040	LF	.70	1.56	2.26	3.84
1-1/4" nominal diameter	BE@.045	LF	1.05	1.76	2.81	4.78
1-1/2" nominal diameter	BE@.055	LF	1.27	2.15	3.42	5.81
2" nominal diameter	BE@.070	LF	1.60	2.73	4.33	7.36

	Craft@Hrs	Unit	Material	Labor	Total	Sell
EMT 45-degree elbow. Galvanized steel. Indoor use only. UL listed.						
1/2" nominal diameter	BE@.050	Ea	2.65	1.95	4.60	7.82
3/4" nominal diameter	BE@.060	Ea	2.75	2.34	5.09	8.65
1" nominal diameter	BE@.080	Ea	3.02	3.12	6.14	10.40
1-1/4" nominal diameter	BE@.100	Ea	4.28	3.91	8.19	13.90
1-1/2" nominal diameter	BE@.150	Ea	5.13	5.86	10.99	18.70
2" nominal diameter	BE@.200	Ea	8.57	7.81	16.38	27.80
EMT 90-degree elbow. Galvanized steel. Indoor use only. UL listed.						
1/2" nominal diameter	BE@.050	Ea	3.47	1.95	5.42	9.21
3/4" nominal diameter	BE@.060	Ea	3.90	2.34	6.24	10.60
1" nominal diameter	BE@.080	Ea	4.02	3.12	7.14	12.10
1-1/4" nominal diameter	BE@.100	Ea	5.41	3.91	9.32	15.80
1-1/2" nominal diameter	BE@.100	Ea	6.37	3.91	10.28	17.50
2" nominal diameter	BE@.150	Ea	10.90	5.86	16.76	28.50
EMT 90-degree pulling elbow, EMT to box. Die-cast zinc. Combination threaded/set screw. Includes cover and gasket. UL listed.						
1/2" nominal diameter	BE@.050	Ea	3.00	1.95	4.95	8.42
3/4" nominal diameter	BE@.060	Ea	3.99	2.34	6.33	10.80
1" nominal diameter	BE@.080	Ea	7.00	3.12	10.12	17.20
1-1/4" nominal diameter	BE@.100	Ea	9.30	3.91	13.21	22.50
EMT compression connector. Die-cast zinc. Indoor and outdoor use. Concrete-tight. Connects EMT conduit to steel outlet box, load center, or other metal enclosure. UL and CSA listed.						
1/2"	BE@.050	Ea	.56	1.95	2.51	4.27
3/4"	BE@.050	Ea	.93	1.95	2.88	4.90
1"	BE@.080	Ea	2.14	3.12	5.26	8.94
1-1/4"	BE@.100	Ea	2.91	3.91	6.82	11.60
1-1/2"	BE@.100	Ea	3.74	3.91	7.65	13.00
2"	BE@.150	Ea	5.07	5.86	10.93	18.60
EMT insulated throat compression connector. Die-cast zinc. Indoor and outdoor use. Concrete-tight. Connects EMT conduit to a steel outlet box, load center or other metal enclosure. UL listed.						
1/2"	BE@.050	Ea	.64	1.95	2.59	4.40
3/4"	BE@.060	Ea	.95	2.34	3.29	5.59
1"	BE@.080	Ea	2.12	3.12	5.24	8.91
1-1/4"	BE@.100	Ea	4.74	3.91	8.65	14.70
1-1/2"	BE@.100	Ea	5.24	3.91	9.15	15.60
2"	BE@.150	Ea	6.39	5.86	12.25	20.80
EMT set screw connector. Zinc plated steel. Concrete-tight when taped. Connects EMT conduit to a steel outlet box, load center or other metal enclosure.						
3/4"	BE@.050	Ea	.41	1.95	2.36	4.01
1"	BE@.080	Ea	1.29	3.12	4.41	7.50
1-1/4"	BE@.100	Ea	1.77	3.91	5.68	9.66
1-1/2"	BE@.100	Ea	2.64	3.91	6.55	11.10
2"	BE@.150	Ea	3.17	5.86	9.03	15.40
EMT offset screw connector. Die-cast zinc. Concrete-tight when taped. Connects EMT conduit to a metal outlet box or other enclosure where an offset is required. UL listed.						
1/2" conduit	BE@.100	Ea	2.04	3.91	5.95	10.10
3/4" conduit	BE@.100	Ea	2.91	3.91	6.82	11.60

	Craft@Hrs	Unit	Material	Labor	Total	Sell

EMT to box offset compression connector. Die-cast zinc. Indoor and outdoor use. Concrete-tight. Connects EMT conduit to a metal outlet box or other enclosure where an offset is required. UL listed.

	Craft@Hrs	Unit	Material	Labor	Total	Sell
1/2" conduit	BE@.100	Ea	3.54	3.91	7.45	12.70
3/4" conduit	BE@.100	Ea	4.87	3.91	8.78	14.90

EMT to flex combination coupling, compression to screw-in. Die-cast zinc. Indoor or outdoor use. Joins EMT conduit to flexible metal conduit. UL listed.

	Craft@Hrs	Unit	Material	Labor	Total	Sell
1/2"	BE@.050	Ea	2.32	1.95	4.27	7.26
3/4"	BE@.060	Ea	2.85	2.34	5.19	8.82
1"	BE@.150	Ea	3.48	5.86	9.34	15.90

EMT to non-metallic sheathed cable coupling.

	Craft@Hrs	Unit	Material	Labor	Total	Sell
1/2"	BE@.100	Ea	.63	3.91	4.54	7.72

EMT 1-hole strap. Zinc plated steel. Snap-on style. Indoor and outdoor use. Supports EMT conduit.

	Craft@Hrs	Unit	Material	Labor	Total	Sell
1/2"	BE@.050	Ea	.20	1.95	2.15	3.66
3/4"	BE@.050	Ea	.25	1.95	2.20	3.74
1"	BE@.080	Ea	.35	3.12	3.47	5.90
1-1/4"	BE@.100	Ea	.46	3.91	4.37	7.43
1-1/2"	BE@.100	Ea	.77	3.91	4.68	7.96
2"	BE@.100	Ea	.99	3.91	4.90	8.33

EMT 2-hole strap. Zinc plated steel. Indoor or outdoor use. Supports EMT conduit where greater load-bearing capacity is required. UL listed.

	Craft@Hrs	Unit	Material	Labor	Total	Sell
1/2"	BE@.030	Ea	.23	1.17	1.40	2.38
3/4"	BE@.040	Ea	.33	1.56	1.89	3.21
1-1/4"	BE@.050	Ea	.40	1.95	2.35	4.00
1-1/2"	BE@.050	Ea	.56	1.95	2.51	4.27
2"	BE@.100	Ea	.82	3.91	4.73	8.04

Metal Conduit Boxes

Install conduit box in confined area. Installed in an area such as an attic, crawlspace or inserted in a wall.

	Craft@Hrs	Unit	Material	Labor	Total	Sell
Junction, switch or outlet box	BE@.420	Ea	—	16.40	16.40	27.90

Drawn handy box. 4" x 2", 1-7/8" deep. 13-cubic-inch capacity. Used to mount switches or receptacles. Bracket for wood or metal studs. Drawn construction. UL listed. KO (knockout).

	Craft@Hrs	Unit	Material	Labor	Total	Sell
(3) 1/2" side KOs, (3) 1/2" bottom KOs	BE@.200	Ea	1.88	7.81	9.69	16.50
(3) 1/2" side KOs, (2) 1/2" end KOs, (3) 1/2" bottom KOs	BE@.200	Ea	2.47	7.81	10.28	17.50

Welded handy box, RACO®. 2-1/8" deep. 16.5-cubic-inch capacity. Used to mount switches or receptacles. Bracket for wood or metal studs. UL listed. KO (knockout).

	Craft@Hrs	Unit	Material	Labor	Total	Sell
4" x 2", (2) 1/2" end KOs, (3) 1/2" bottom KOs	BE@.200	Ea	2.09	7.81	9.90	16.80
4" x 2", (3) 1/2" side KOs, (2) 1/2" end KOs, (3) 1/2" bottom KOs	BE@.200	Ea	2.76	7.81	10.57	18.00

Handy box cover, RACO®. Covers also may be used as single gang wall plates. Includes captive screws.

	Craft@Hrs	Unit	Material	Labor	Total	Sell
20-amp receptacle	BE@.030	Ea	1.48	1.17	2.65	4.51
Duplex receptacle	BE@.030	Ea	.63	1.17	1.80	3.06
Toggle switch	BE@.030	Ea	.63	1.17	1.80	3.06

	Craft@Hrs	Unit	Material	Labor	Total	Sell
Welded square box, RACO®. Used to mount switches and receptacles. 42-cubic-inch wiring capacity. Welded construction. UL listed. KO (knockout).						
(2) 1/2" side KOs, (7) top KOs, (1) 1/2", (2) 3/4" bottom KOs	BE@.300	Ea	4.76	11.70	16.46	28.00
(2) 1/2" and (10) TKO side KOs, (2) 3/4" and (1) 1/2" TKO bottom KOs	BE@.300	Ea	4.45	11.70	16.15	27.50
(4) 3/4", (4) 1" side KOs, (3) 1/2", (2) 3/4" bottom KOs	BE@.300	Ea	4.35	11.70	16.05	27.30
Square surface box cover. 30- to 50-amp receptacle. Steel. Raised 1/2".						
2.156" diameter	BE@.060	Ea	5.85	2.34	8.19	13.90
Multi-gang switch box with conduit KOs, RACO®. Steel. 2-1/2" deep. Combination 1/2" and 3/4" KOs. UL listed. KO (knockout).						
3 gang, 5-19/32" wide, (6) top, (2) bottom, (3) back KOs	BE@.200	Ea	10.40	7.81	18.21	31.00
4 gang, 7-19/32" wide, (8) top, (2) bottom, (4) back KOs	BE@.200	Ea	12.80	7.81	20.61	35.00
Drawn 2-device switch box, RACO®. 4" x 4". Used to mount switches or receptacles. 30.3-cubic-inch wiring capacity. Bracket for wood or metal studs. Drawn construction. KO (knockout).						
1/2" side KOs, (5) 1/2" bottom KOs	BE@.200	Ea	5.84	7.81	13.65	23.20
(2) 1/2", (1) 3/4" side KOs, (3) 1/2", (2) 3/4" bottom KOs	BE@.200	Ea	5.47	7.81	13.28	22.60
(2) 3/4" side KOs, (3) 1/2", (2) 3/4" bottom KOs	BE@.200	Ea	6.40	7.81	14.21	24.20
Finished box cover. Baked Cadilite finished steel. 4" square. 1/2" deep. Fully depressed corners for easier mounting. UL listed.						
1 GFCI receptacle	BE@.050	Ea	1.38	1.95	3.33	5.66
1 single receptacle	BE@.050	Ea	1.49	1.95	3.44	5.85
1 duplex receptacle	BE@.050	Ea	1.22	1.95	3.17	5.39
2 duplex receptacles	BE@.050	Ea	1.22	1.95	3.17	5.39
1 toggle switch	BE@.050	Ea	1.48	1.95	3.43	5.83
2 toggle switches	BE@.050	Ea	1.48	1.95	3.43	5.83
Toggle and duplex	BE@.050	Ea	1.43	1.95	3.38	5.75
Ceiling fan box. Steel. UL listed for installation of ceiling fans up to 35 pounds. Fixture supports up to 50 pounds. Affixes to joist or cross brace. 4 side cable knockouts, (2) 1/2" side conduit knockouts, 1/2" bottom conduit knockout.						
4", includes fan-mounting screws with lock washers	BE@.200	Ea	5.08	7.81	12.89	21.90
Ceiling fan pancake box. 6-cubic-inch capacity. Steel. Includes fan-mounting screws with lock washers. UL listed for installation of ceiling fans up to 35 pounds. Fixture supports up to 50 pounds. Affixes to joist or cross brace. Includes non-metallic connector. KO (knockout).						
4", 1/2" deep, (4) 1/2" bottom KOs	BE@.200	Ea	3.98	7.81	11.79	20.00
R-3 drawn octagon box, RACO®. Used to support ceiling light fixture. Sides have (4) 1/2" knockouts, bottom has 1/2" knockout. 50-pound maximum light fixture support. Not designed for ceiling fan support. UL listed.						
11.8 cubic inch capacity	BE@.200	Ea	1.85	7.81	9.66	16.40

	Craft@Hrs	Unit	Material	Labor	Total	Sell
Steel octagon box cover, RACO®. Flat. Blank. UL listed.						
4" x 1/2"	BE@.050	Ea	.62	1.95	2.57	4.37

Flex Conduit and Fittings

	Craft@Hrs	Unit	Material	Labor	Total	Sell
Remove flexible metal conduit and wire. Includes removing one conduit box each 10'.						
1/2" to 2" conduit diameter	BE@.019	LF	—	.74	.74	1.26
Reduced wall flexible aluminum conduit. By nominal trade size.						
1/2", 25' coil	BE@.030	LF	.64	1.17	1.81	3.08
1/2", 100' coil	BE@.030	LF	.36	1.17	1.53	2.60
1/2", 500' coil	BE@.030	LF	.34	1.17	1.51	2.57
3/4", 100' coil	BE@.033	LF	.51	1.29	1.80	3.06
1", 50' coil	BE@.033	LF	1.26	1.29	2.55	4.34
Reduced wall flexible steel conduit. By nominal trade size.						
3/8", 100' coil	BE@.025	LF	.40	.98	1.38	2.35
3/8", 250' coil	BE@.025	LF	.35	.98	1.33	2.26
1/2", per foot	BE@.028	LF	.66	1.09	1.75	2.98
1/2", 25' coil	BE@.028	LF	.83	1.09	1.92	3.26
1/2", 100' coil	BE@.028	LF	.45	1.09	1.54	2.62
3/4", per foot	BE@.030	LF	1.20	1.17	2.37	4.03
3/4", 100' coil	BE@.030	LF	.64	1.17	1.81	3.08
1", 50' coil	BE@.030	LF	1.66	1.17	2.83	4.81
Flex screw-in connector. Die-cast zinc. Indoor use only. Connects flexible metallic conduit to steel outlet box, load center or other metal enclosure. 3/8" size fits 1/2" knockout. UL listed.						
3/8"	BE@.050	Ea	.71	1.95	2.66	4.52
1/2"	BE@.050	Ea	.47	1.95	2.42	4.11
3/4"	BE@.050	Ea	1.10	1.95	3.05	5.19
1"	BE@.060	Ea	1.66	2.34	4.00	6.80
Flex screw-in insulated throat connector. For flexible metal conduit. Zinc die-cast. Knockout size is the same as the conduit trade size.						
3/8" conduit	BE@.050	Ea	.93	1.95	2.88	4.90
1/2" conduit	BE@.050	Ea	.56	1.95	2.51	4.27
3/4" conduit	BE@.060	Ea	.56	2.34	2.90	4.93
1" conduit	BE@.080	Ea	.39	3.12	3.51	5.97
Flex 90-degree insulated throat connector.						
1/2" conduit	BE@.050	Ea	.98	1.95	2.93	4.98
3/4" conduit	BE@.050	Ea	1.42	1.95	3.37	5.73
Flex 90-degree connector. Die-cast zinc. Indoor use only. Use with flexible metal conduit, metal clad or armored cable. Use in a dry location to connect aluminum and steel flex to a steel outlet box or other metallic enclosure. 3/8" size fits 1/2" knockout. UL listed.						
3/8" conduit	BE@.050	Ea	1.06	1.95	3.01	5.12
1/2" conduit	BE@.050	Ea	1.29	1.95	3.24	5.51
3/4" conduit	BE@.060	Ea	2.20	2.34	4.54	7.72
1" conduit	BE@.080	Ea	5.37	3.12	8.49	14.40

	Craft@Hrs	Unit	Material	Labor	Total	Sell

Flex squeeze connector. Die-cast zinc. Indoor use only. In a dry location, use to connect flexible metallic conduit to a metal enclosure such as a steel outlet box or load center. 3/8" size fits 1/2" knockout. UL listed.

3/8" conduit	BE@.050	Ea	.66	1.95	2.61	4.44
1/2" conduit	BE@.050	Ea	.80	1.95	2.75	4.68
3/4" conduit	BE@.060	Ea	.95	2.34	3.29	5.59
1" conduit	BE@.080	Ea	1.48	3.12	4.60	7.82

Electrical non-metallic flex tubing (ENT) conduit, Carlon®. Flex-Plus® Blue™ ENT. No special tools required. Highly flexible. Cuts easily and cleanly. For use with 90-degree Celsius conductors. Corrugated design for easy wire pulling and pushing. NER-290 recognized by BOCA, ICC and SBCCI. UL listed. By nominal trade size.

1/2", per foot	BE@.021	LF	.40	.82	1.22	2.07
1/2", 10' coil	BE@.021	LF	.37	.82	1.19	2.02
1/2", 200' coil	BE@.021	LF	.24	.82	1.06	1.80
3/4", per foot	BE@.022	LF	.50	.86	1.36	2.31
3/4", 10' coil	BE@.022	LF	.45	.86	1.31	2.23
3/4", 100' coil	BE@.025	LF	.38	.98	1.36	2.31
1", 10' coil	BE@.040	LF	.47	1.56	2.03	3.45

Carflex® non-metallic liquid-tight flexible conduit, Carlon®. Non-conductive, non-corrosive, crush-, abrasion- and stain-resistant. Used to connect air conditioning and heating equipment for outdoor wiring and controls.

1/2", per foot	BE@.020	LF	.57	.78	1.35	2.30
1/2", 100' coil	BE@.020	LF	.35	.78	1.13	1.92
3/4", per foot	BE@.020	LF	.68	.78	1.46	2.48
3/4", 100' coil	BE@.020	LF	.53	.78	1.31	2.23
1", per foot	BE@.025	LF	1.06	.98	2.04	3.47
1", 100' coil	BE@.025	LF	.98	.98	1.96	3.33
2", per foot	BE@.030	LF	2.41	1.17	3.58	6.09

Carflex® conduit fitting, Carlon®. Requires no disassembly of components for installation. Locknut, O-rings and foam washer assembled onto fitting. Withstands temperatures up to 140 degrees F. UL listed.

1/2" 90-degree ell	BE@.100	Ea	2.69	3.91	6.60	11.20
3/4" 90-degree ell	BE@.100	Ea	3.03	3.91	6.94	11.80
1" 90-degree ell	BE@.150	Ea	4.00	5.86	9.86	16.80
1/2" straight coupling	BE@.100	Ea	2.01	3.91	5.92	10.10
3/4" straight coupling	BE@.100	Ea	2.28	3.91	6.19	10.50
1" straight coupling	BE@.150	Ea	2.67	5.86	8.53	14.50

Weatherproof Conduit Boxes and Fittings

1-gang rectangular steel weatherproof conduit box, Red Dot®. Includes mounting lugs, hole plugs and screws. UL listed. Silver.

1/2" hole, 3 holes	BE@.300	Ea	3.66	11.70	15.36	26.10
1/2" hole, 4 holes	BE@.300	Ea	4.76	11.70	16.46	28.00
1/2" hole, 5 holes	BE@.300	Ea	6.10	11.70	17.80	30.30
1/2" hole, 5 holes, 2 side holes	BE@.300	Ea	6.11	11.70	17.81	30.30
3/4" hole, 3 holes	BE@.300	Ea	5.55	11.70	17.25	29.30
3/4" hole, 4 holes	BE@.300	Ea	6.60	11.70	18.30	31.10
3/4" hole, 5 holes	BE@.300	Ea	7.03	11.70	18.73	31.80

2-gang steel weatherproof conduit box, Red Dot®. Includes mounting lugs, hole plugs and screws. UL listed. Silver.

1/2" hole, 3 holes	BE@.300	Ea	8.84	11.70	20.54	34.90
1/2" hole, 5 holes	BE@.300	Ea	10.15	11.70	21.85	37.10
1/2" hole, 7 holes	BE@.300	Ea	13.20	11.70	24.90	42.30
3/4" hole, 2 holes	BE@.300	Ea	9.86	11.70	21.56	36.70

	Craft@Hrs	Unit	Material	Labor	Total	Sell
3/4" hole, 5 holes	BE@.300	Ea	11.20	11.70	22.90	38.90
3/4" hole, 5 holes, round	BE@.300	Ea	9.31	11.70	21.01	35.70
3/4" hole, 7 holes	BE@.300	Ea	12.20	11.70	23.90	40.60
3/4" hole, 7 holes, 2 side holes	BE@.250	Ea	14.20	9.76	23.96	40.70
1" hole, 5 holes, deep	BE@.300	Ea	12.70	11.70	24.40	41.50

Non-metallic exposed weatherproof conduit box, Carlon®. With 1/2" threaded holes. Mounting feet included.

	Craft@Hrs	Unit	Material	Labor	Total	Sell
Rectangular, gray, 3 holes	BE@.300	Ea	5.91	11.70	17.61	29.90
Rectangular, white, 3 holes	BE@.300	Ea	5.91	11.70	17.61	29.90
Round, gray, 5 holes	BE@.300	Ea	7.95	11.70	19.65	33.40
Round, white, 5 holes	BE@.300	Ea	7.95	11.70	19.65	33.40

2-gang waterproof conduit box cover, Red Dot®. Includes gasket and screws. Silver.

	Craft@Hrs	Unit	Material	Labor	Total	Sell
Blank cover	BE@.150	Ea	2.40	5.86	8.26	14.00
Duplex receptacle cover	BE@.150	Ea	9.15	5.86	15.01	25.50
Switch cover	BE@.150	Ea	8.06	5.86	13.92	23.70

GFCI wet location conduit box receptacle cover, Red Dot®. Includes cover, gasket, screws and instructions. All-metal construction. Required by the *NEC*. For residential, commercial industrial or recreational use. UL listed.

	Craft@Hrs	Unit	Material	Labor	Total	Sell
2 gang	BE@.150	Ea	20.40	5.86	26.26	44.60

Horizontal duplex weatherproof conduit box receptacle cover, Red Dot®. Includes gasket and screws. Die-cast construction.

	Craft@Hrs	Unit	Material	Labor	Total	Sell
Silver	BE@.150	Ea	3.00	5.86	8.86	15.10
Silver, GFCI	BE@.150	Ea	4.82	5.86	10.68	18.20

Universal non-metallic wet location conduit box cover, Red Dot®, Carlon®. Thermoplastic cover with transparent finish. Lockable. Accommodates GFCI, single and duplex receptacles, toggle switches and single receptacle up to 1.59" diameter. Horizontal or vertical orientation. Mounts to a box or the device. All installation screws included. Device holes are keyed. Just back out the existing device screws and slip the cover over previously installed screws. Meets *NEC* requirements.

	Craft@Hrs	Unit	Material	Labor	Total	Sell
1 gang, clear	BE@.150	Ea	13.10	5.86	18.96	32.20
2 gang, clear	BE@.150	Ea	20.40	5.86	26.26	44.60

Non-metallic weatherproof single-gang conduit box cover. Rigid PVC. Non-conductive and non-corrosive. For industrial, commercial, and residential applications. UL listed. Gray.

	Craft@Hrs	Unit	Material	Labor	Total	Sell
15 amp	BE@.100	Ea	6.51	3.91	10.42	17.70
30 amp	BE@.100	Ea	5.74	3.91	9.65	16.40
50 amp	BE@.100	Ea	5.75	3.91	9.66	16.40
Blank	BE@.100	Ea	2.33	3.91	6.24	10.60
Single switch	BE@.100	Ea	5.70	3.91	9.61	16.30
Vertical duplex receptacle	BE@.100	Ea	6.55	3.91	10.46	17.80
Vertical GFCI receptacle	BE@.100	Ea	5.98	3.91	9.89	16.80

Weatherproof lampholder and cover, Red Dot®. Three-hole round cover, gaskets, and screws included.

	Craft@Hrs	Unit	Material	Labor	Total	Sell
Single lamp, rectangular	BE@.200	Ea	4.10	7.81	11.91	20.20
Dual lamps, round	BE@.800	Ea	11.10	31.20	42.30	71.90

Weatherproof switch and cover, Red Dot®. Gasket, screws, and spacers included. Switch is UL listed.

	Craft@Hrs	Unit	Material	Labor	Total	Sell
1 pole, 15 amp	BE@.250	Ea	3.03	9.76	12.79	21.70
3 way, 15 amp	BE@.250	Ea	3.40	9.76	13.16	22.40

Rigid and IMC Conduit and Fittings

	Craft@Hrs	Unit	Material	Labor	Total	Sell
Remove rigid steel conduit and wire. Includes conduit and wire removed in salvage condition.						
1/2" to 2" conduit diameter	BE@.030	LF	—	1.17	1.17	1.99
Install rigid steel conduit in confined area. Working in a confined area such as an attic, crawlspace or behind a wall.						
1/2" to 1" conduit diameter	BE@.076	LF	—	2.97	2.97	5.05
Galvanized rigid steel (GRS) conduit. *NEC* approved. Meets ANSI specifications. Made of high-strength strip steel. Provides radiation protection and magnetic shielding. Sold in 10' lengths.						
1/2" nominal diameter	BE@.040	LF	1.48	1.56	3.04	5.17
3/4" nominal diameter	BE@.045	LF	1.58	1.76	3.34	5.68
1" nominal diameter	BE@.050	LF	2.30	1.95	4.25	7.23
1-1/4" nominal diameter	BE@.070	LF	2.93	2.73	5.66	9.62
1-1/2" nominal diameter	BE@.080	LF	3.44	3.12	6.56	11.20
2" nominal diameter	BE@.110	LF	4.45	4.30	8.75	14.90
Intermediate metal conduit (IMC). UL listed. *NEC* approved. Meets ANSI specifications. Full-cut threads are galvanized after cutting and capped for protection. Interior coated with lubricating finish. Sold in 10' lengths.						
1/2" nominal diameter	BE@.038	LF	1.03	1.48	2.51	4.27
3/4" nominal diameter	BE@.040	LF	1.20	1.56	2.76	4.69
1" nominal diameter	BE@.045	LF	1.56	1.76	3.32	5.64
1-1/4" nominal diameter	BE@.065	LF	1.74	2.54	4.28	7.28
1-1/2" nominal diameter	BE@.065	LF	2.13	2.54	4.67	7.94
2" nominal diameter	BE@.090	LF	2.79	3.51	6.30	10.70
Rigid or IMC 45-degree conduit elbow. Galvanized steel. Indoor or outdoor use. UL listed.						
1/2" nominal diameter	BE@.100	Ea	2.96	3.91	6.87	11.70
3/4" nominal diameter	BE@.100	Ea	4.44	3.91	8.35	14.20
1" nominal diameter	BE@.120	Ea	5.29	4.69	9.98	17.00
1-1/4" nominal diameter	BE@.150	Ea	8.31	5.86	14.17	24.10
1-1/2" nominal diameter	BE@.150	Ea	9.84	5.86	15.70	26.70
2" nominal diameter	BE@.200	Ea	14.90	7.81	22.71	38.60
Rigid or IMC 90-degree conduit elbow. Galvanized steel. Indoor or outdoor use. UL listed.						
1/2" nominal diameter	BE@.100	Ea	4.18	3.91	8.09	13.80
3/4" nominal diameter	BE@.100	Ea	5.04	3.91	8.95	15.20
1" nominal diameter	BE@.120	Ea	7.10	4.69	11.79	20.00
1-1/4" nominal diameter	BE@.150	Ea	10.10	5.86	15.96	27.10
1-1/2" nominal diameter	BE@.150	Ea	11.90	5.86	17.76	30.20
2" nominal diameter	BE@.200	Ea	17.90	7.81	25.71	43.70
Rigid or IMC conduit coupling. Galvanized steel. Indoor and outdoor use. Joins threaded rigid or IMC conduit.						
1/2" conduit	BE@.050	Ea	1.48	1.95	3.43	5.83
3/4" conduit	BE@.050	Ea	1.89	1.95	3.84	6.53
1" conduit	BE@.080	Ea	2.81	3.12	5.93	10.10
1-1/4" conduit	BE@.100	Ea	3.11	3.91	7.02	11.90
1-1/2" conduit	BE@.100	Ea	3.57	3.91	7.48	12.70
2" conduit	BE@.150	Ea	4.23	5.86	10.09	17.20

	Craft@Hrs	Unit	Material	Labor	Total	Sell
Rigid or IMC compression conduit coupling. Steel. Concrete-tight. Indoor and outdoor use. Joins two lengths of threadless rigid or IMC conduit. UL listed.						
1/2" conduit	BE@.050	Ea	3.88	1.95	5.83	9.91
3/4" conduit	BE@.060	Ea	5.30	2.34	7.64	13.00
1" conduit	BE@.080	Ea	9.87	3.12	12.99	22.10
1-1/4" conduit	BE@.100	Ea	14.50	3.91	18.41	31.30
1-1/2" conduit	BE@.100	Ea	18.40	3.91	22.31	37.90
2" conduit	BE@.150	Ea	39.90	5.86	45.76	77.80
Rigid or IMC insulating bushing. High-impact thermoplastic. Flame-retardant. Indoor or outdoor. Use on threaded rigid or IMC conduit to protect wires. Required by *NEC*. UL listed.						
1/2"	BE@.020	Ea	.15	.78	.93	1.58
3/4"	BE@.020	Ea	.17	.78	.95	1.62
1"	BE@.030	Ea	.47	1.17	1.64	2.79
1-1/4"	BE@.040	Ea	.56	1.56	2.12	3.60
1-1/2"	BE@.040	Ea	.55	1.56	2.11	3.59
2"	BE@.050	Ea	.70	1.95	2.65	4.51
Rigid or IMC insulated metallic grounding bushing. Die-cast zinc. Lay-in lug. Indoor or outdoor use. Use with locknut to terminate service conduit to a cabinet. Lug provided for bonding jumper to a neutral bus bar. Thermoplastic liner rated at 150 degrees Celsius. UL listed.						
Trade size 1/2"	BE@.100	Ea	3.43	3.91	7.34	12.50
Trade size 3/4"	BE@.100	Ea	3.74	3.91	7.65	13.00
Trade size 1"	BE@.100	Ea	3.97	3.91	7.88	13.40
Trade size 1-1/4"	BE@.150	Ea	4.23	5.86	10.09	17.20
Trade size 1-1/2"	BE@.150	Ea	4.44	5.86	10.30	17.50
Trade size 2"	BE@.200	Ea	5.07	7.81	12.88	21.90
Rigid or IMC reducing bushing. Die-cast zinc. Dry locations only. Reduces the entry size of a run of threaded conduit. UL listed.						
3/4" x 1/2"	BE@.050	Ea	.95	1.95	2.90	4.93
1" x 1/2"	BE@.075	Ea	1.22	2.93	4.15	7.06
1" x 3/4"	BE@.100	Ea	1.35	3.91	5.26	8.94
1-1/2" x 1-1/4"	BE@.100	Ea	3.26	3.91	7.17	12.20
Rigid or IMC watertight conduit hub. With insulated throat. Zinc die-cast. UL listed.						
1/2" nominal diameter	BE@.200	Ea	4.02	7.81	11.83	20.10
3/4" nominal diameter	BE@.250	Ea	5.53	9.76	15.29	26.00
1" nominal diameter	BE@.300	Ea	6.95	11.70	18.65	31.70
1-1/4" nominal diameter	BE@.350	Ea	8.29	13.70	21.99	37.40
1-1/2" nominal diameter	BE@.375	Ea	11.10	14.60	25.70	43.70
2" nominal diameter	BE@.400	Ea	14.30	15.60	29.90	50.80
Steel sealing conduit locknut. With PVC molded seal. UL listed.						
1/2" nominal diameter	BE@.020	Ea	.79	.78	1.57	2.67
3/4" nominal diameter	BE@.020	Ea	1.51	.78	2.29	3.89
1" nominal diameter	BE@.020	Ea	1.58	.78	2.36	4.01
1-1/4" nominal diameter	BE@.020	Ea	1.97	.78	2.75	4.68
1-1/2" nominal diameter	BE@.020	Ea	2.25	.78	3.03	5.15
2" nominal diameter	BE@.020	Ea	3.11	.78	3.89	6.61

	Craft@Hrs	Unit	Material	Labor	Total	Sell
Rigid or IMC conduit nipple. Steel. Indoor or outdoor use. Connects two steel outlet boxes or enclosures. UL listed.						
1/2" x close trade size	BE@.050	Ea	1.12	1.95	3.07	5.22
1/2" x 4" trade size	BE@.050	Ea	1.54	1.95	3.49	5.93
3/4" x close trade size	BE@.060	Ea	1.38	2.34	3.72	6.32
1" x 2" trade size	BE@.080	Ea	1.32	3.12	4.44	7.55
1" x 2-1/2" trade size	BE@.080	Ea	1.58	3.12	4.70	7.99
1" x 3" trade size	BE@.080	Ea	2.05	3.12	5.17	8.79
1" x 5" trade size	BE@.080	Ea	2.77	3.12	5.89	10.00
1-1/4" x 2" trade size	BE@.100	Ea	1.82	3.91	5.73	9.74
1-1/4" x 3" trade size	BE@.100	Ea	2.78	3.91	6.69	11.40
1-1/2" x 2" trade size	BE@.100	Ea	2.35	3.91	6.26	10.60
1-1/2" x 3" trade size	BE@.100	Ea	3.81	3.91	7.72	13.10
2" x 3" trade size	BE@.150	Ea	3.45	5.86	9.31	15.80
2" x 5" trade size	BE@.150	Ea	5.49	5.86	11.35	19.30
2" x 7" trade size	BE@.150	Ea	6.90	5.86	12.76	21.70
2" x 8" trade size	BE@.150	Ea	7.32	5.86	13.18	22.40
Rigid or IMC conduit 1-hole strap. Zinc plated malleable iron. Dry locations. Secures rigid or IMC conduit to wood, masonry, or other surfaces. UL listed.						
1/2" conduit diameter	BE@.050	Ea	.39	1.95	2.34	3.98
3/4" conduit diameter	BE@.060	Ea	.49	2.34	2.83	4.81
1" conduit diameter	BE@.050	Ea	.59	1.95	2.54	4.32
1-1/4" conduit diameter	BE@.075	Ea	1.91	2.93	4.84	8.23
1-1/2" conduit diameter	BE@.075	Ea	2.45	2.93	5.38	9.15
2" conduit diameter	BE@.100	Ea	5.39	3.91	9.30	15.80
Rigid or IMC threaded aluminum conduit body. Copper-free die-cast aluminum. Indoor or outdoor use. Includes cover and gasket. Used to gain access to the interior of raceway for wire pulling, splicing and maintenance (as allowed by the *NEC*). UL listed.						
Type C, 1/2"	BE@.100	Ea	3.57	3.91	7.48	12.70
Type C, 3/4"	BE@.150	Ea	4.74	5.86	10.60	18.00
Type C, 1"	BE@.200	Ea	10.20	7.81	18.01	30.60
Type C, 1-1/4"	BE@.250	Ea	19.00	9.76	28.76	48.90
Type C, 1-1/2"	BE@.250	Ea	20.50	9.76	30.26	51.40
Type C, 2"	BE@.300	Ea	34.90	11.70	46.60	79.20
Type LB, 1/2"	BE@.100	Ea	3.57	3.91	7.48	12.70
Type LB, 3/4"	BE@.150	Ea	4.74	5.86	10.60	18.00
Type LB, 1"	BE@.150	Ea	10.60	5.86	16.46	28.00
Type LB, 1-1/4"	BE@.200	Ea	11.60	7.81	19.41	33.00
Type LB, 1-1/2"	BE@.200	Ea	15.80	7.81	23.61	40.10
Type LB, 2"	BE@.250	Ea	20.20	9.76	29.96	50.90
Type LL or LR, 1/2"	BE@.100	Ea	3.57	3.91	7.48	12.70
Type LL or LR, 3/4"	BE@.150	Ea	4.74	5.86	10.60	18.00
Type LL or LR, 1"	BE@.150	Ea	10.60	5.86	16.46	28.00
Type LL or LR, 1-1/4"	BE@.200	Ea	18.30	7.81	26.11	44.40
Type LL or LR, 1-1/2"	BE@.250	Ea	26.20	9.76	35.96	61.10
Type LL or LR, 2"	BE@.300	Ea	34.90	11.70	46.60	79.20
Type T, 1/2"	BE@.150	Ea	4.44	5.86	10.30	17.50
Type T, 3/4"	BE@.200	Ea	5.71	7.81	13.52	23.00
Type T, 1"	BE@.250	Ea	10.60	9.76	20.36	34.60
Type T, 1-1/4"	BE@.300	Ea	13.60	11.70	25.30	43.00
Type T, 1-1/2"	BE@.300	Ea	25.00	11.70	36.70	62.40
Type T, 2"	BE@.330	Ea	38.30	12.90	51.20	87.00

	Craft@Hrs	Unit	Material	Labor	Total	Sell
Steel service entrance cap. Copper-free die-cast aluminum. Clamp-on style. For overhead wiring service entrance. Mounts on top of EMT, rigid or IMC conduit. Serves as connecting point for service entrance wires. UL listed.						
1/2" conduit diameter	BE@.150	Ea	3.92	5.86	9.78	16.60
3/4" conduit diameter	BE@.200	Ea	4.41	7.81	12.22	20.80
1" conduit diameter	BE@.250	Ea	5.76	9.76	15.52	26.40
1-1/4" conduit diameter	BE@.300	Ea	4.28	11.70	15.98	27.20
1-1/2" conduit diameter	BE@.300	Ea	8.14	11.70	19.84	33.70
2" conduit diameter	BE@.500	Ea	10.70	19.50	30.20	51.30
2-1/2" conduit diameter	BE@.500	Ea	29.40	19.50	48.90	83.10
Rigid or IMC service entrance elbow. Copper-free die-cast aluminum. Indoor or outdoor use. Includes cover and gasket. Use to gain access to interior of raceway for wire pulling, inspection, and maintenance. UL listed.						
Trade size 1/2"	BE@.100	Ea	4.23	3.91	8.14	13.80
Trade size 3/4"	BE@.150	Ea	5.36	5.86	11.22	19.10
Trade size 1"	BE@.150	Ea	8.77	5.86	14.63	24.90
Trade size 1-1/4"	BE@.200	Ea	10.50	7.81	18.31	31.10
Trade size 1-1/2"	BE@.200	Ea	19.60	7.81	27.41	46.60
Trade size 2"	BE@.250	Ea	27.50	9.76	37.26	63.30
Service entrance roof flashing. Prevents leaks where conduit penetrates the roof surface. Slips over the conduit mast and mounts on the roof. Made of galvanized steel with neoprene seal.						
2" conduit diameter	BE@.250	Ea	7.10	9.76	16.86	28.70
2-1/2" conduit diameter	BE@.250	Ea	8.47	9.76	18.23	31.00
Service entrance conduit sill plate. Copper-free die-cast aluminum. Helps seal the hole where service entrance cable enters the home.						
1/2" to 3/4", 2 screws	BE@.100	Ea	1.70	3.91	5.61	9.54
1", 2 screws	BE@.100	Ea	2.27	3.91	6.18	10.50
1-1/4", 4 screws	BE@.200	Ea	3.02	7.81	10.83	18.40
1-1/2", 4 screws	BE@.200	Ea	3.69	7.81	11.50	19.60
2", 4 screws	BE@.200	Ea	5.69	7.81	13.50	23.00
Ground rod clamp. Bronze. Connects the grounding wire to the ground rod. Approved for direct burial. UL listed.						
1/2" rod	BE@.100	Ea	2.03	3.91	5.94	10.10
5/8" rod	BE@.100	Ea	2.13	3.91	6.04	10.30
3/4" rod	BE@.100	Ea	2.83	3.91	6.74	11.50
1-1/4" to 2", 8 to 4 gauge wire	BE@.100	Ea	3.65	3.91	7.56	12.90

Wire and Fittings

	Craft@Hrs	Unit	Material	Labor	Total	Sell
Type THHN copper building wire. Labor is based on wire pulled in conduit.						
14 gauge, solid, per foot	BE@.006	LF	.29	.23	.52	.88
14 gauge, solid, 50' coil	BE@.006	LF	.24	.23	.47	.80
14 gauge, solid, 100' coil	BE@.006	LF	.23	.23	.46	.78
14 gauge, solid, 500' coil	BE@.006	LF	.08	.23	.31	.53
14 gauge, stranded, per foot	BE@.006	LF	.38	.23	.61	1.04
14 gauge, stranded, 500' coil	BE@.006	LF	.09	.23	.32	.54
12 gauge, solid, per foot	BE@.007	LF	.50	.27	.77	1.31
12 gauge, solid, 50' coil	BE@.007	LF	.28	.27	.55	.94
12 gauge, solid, 100' coil	BE@.007	LF	.24	.27	.51	.87
12 gauge, solid, 500' coil	BE@.007	LF	.13	.27	.40	.68
12 gauge, stranded, per foot	BE@.007	LF	.57	.27	.84	1.43
12 gauge, stranded, 50' coil	BE@.007	LF	.33	.27	.60	1.02
12 gauge, stranded, 100' coil	BE@.007	LF	.31	.27	.58	.99
12 gauge, stranded, 500' coil	BE@.007	LF	.14	.27	.41	.70

	Craft@Hrs	Unit	Material	Labor	Total	Sell
10 gauge, stranded, per foot	BE@.008	LF	.74	.31	1.05	1.79
10 gauge, stranded, 50' coil	BE@.008	LF	.39	.31	.70	1.19
10 gauge, stranded, 100' coil	BE@.008	LF	.37	.31	.68	1.16
10 gauge, stranded, 500' coil	BE@.008	LF	.21	.31	.52	.88
8 gauge, stranded, per foot	BE@.009	LF	.77	.35	1.12	1.90
8 gauge, stranded, 500' coil	BE@.009	LF	.36	.35	.71	1.21
6 gauge, stranded, per foot	BE@.010	LF	.71	.39	1.10	1.87
6 gauge, stranded, 500' coil	BE@.010	LF	.52	.39	.91	1.55

20 gauge bell wire, Carol® Cable. For alarms, thermostat controls, touch plate systems, burglar alarms, doorbells and other low voltage installations.

2 conductor, 500' coil	BE@.006	LF	.08	.23	.31	.53

18 gauge thermostat cable, Carol® Cable. For thermostat control, heating and air conditioning, touch plate systems, burglar alarms, intercom systems, doorbells, remote control units, signal systems, and other low voltage installations. Annealed solid bare copper conductors. PVC insulation and white jacket. UL listed. Type CL-2.

2 conductor, 50' roll	BE@.006	LF	.18	.23	.41	.70
2 conductor, 100' roll	BE@.006	LF	.16	.23	.39	.66
5 conductor, 50' roll	BE@.006	LF	.38	.23	.61	1.04
5 conductor, 100' roll	BE@.006	LF	.28	.23	.51	.87

Speaker wire. Clear jacket. Low-noise copper wire.

16 gauge, 50' long	BE@.008	LF	.28	.31	.59	1.00
16 gauge, 100' long	BE@.008	LF	.23	.31	.54	.92
18 gauge, 50' long	BE@.008	LF	.18	.31	.49	.83
18 gauge, 100' long	BE@.008	LF	.16	.31	.47	.80

Coaxial cable. For RF signal transmission, MATV and CATV, drop cable, and FM broadcast. Solid bare copper-weld. AWG 18. Cellular polyethylene insulation. 100% aluminum/polyester tape shield with +55% aluminum braid. 0.308" outside diameter. Capacitance, pF/Ft. 17.3. Vel. propagation: 78% nominal impedance: 75 ohm. UL listed.

Black, per foot	BE@.007	LF	.25	.27	.52	.88
Black, 500' coil	BE@.007	LF	.09	.27	.36	.61

Coaxial F connector, RCA. For connecting to RG6 coaxial cable. Lifetime warranty.

Crimp-on	BE@.075	Ea	.51	2.93	3.44	5.85
Twist-on	BE@.075	Ea	.56	2.93	3.49	5.93

Coaxial cable wall plate. Fits standard electrical outlet box or flush-mount to drywall. Ivory or white.

75 ohm, single	BE@.150	Ea	4.56	5.86	10.42	17.70
75 ohm, duplex	BE@.200	Ea	5.49	7.81	13.30	22.60

Category 5E data communication cable, Carol® Cable. UTP; Spec. 4480; Type MPR/CMR. Suitable for high-speed data/LAN transmission up to 16 Mbps; 10 BASE-T, 4 Mbps token ring, 100 VG-Any LAN, 100 Mbps TP-PMD, 100 BASE-T, and 55/155 Mbps ATM. 24 AWG solid bare annealed copper construction. Flame retardant PVC insulation. Pairing: two twists per foot. Pull-Pac® cartons. Maximum DC resistance, 9.38 (omega)/100 m at 20°C (68°F). Mutual capacitance, 4.59pF/Ft. at 1 kHz. Characteristic impedance for 772 kHz, 87-117 (omega). For 1.0-100.0 MHz, 85-115 (omega). SRL for 1.0-20.0 MHz, 23 dB. For 20.0-100.0 MHz, 23-10 log. Skew 10 ns. NVP, 70% speed of light. UL listed. cPCCFT4 listed. Meets requirements: ANSI/TIA/EIA 568A, ANSI/ICEA S-90-661 and NEMA WC63.1. Ivory. 1" outside diameter. 24 gauge.

4 conductor, per foot	BE@.007	LF	.21	.27	.48	.82
4 conductor, 500' coil	BE@.007	LF	.14	.27	.41	.70

Decorator phone wall plate. Ivory or white.

Single line	BE@.270	Ea	5.50	10.50	16.00	27.20
Duplex	BE@.270	Ea	6.50	10.50	17.00	28.90

	Craft@Hrs	Unit	Material	Labor	Total	Sell
Switches and Receptacles						
Remove switch or receptacle. Includes removal of cover plate, disconnecting wires and taping ends.						
Single pole switch or receptacle	BE@.110	Ea	—	4.30	4.30	7.31
Double pole switch or receptacle	BE@.178	Ea	—	6.95	6.95	11.80
Residential-grade AC quiet toggle switch, Leviton. 120-volt, 15-amp, with grounding screw. Side-wired and quick-wired framed toggle push-in. Quickwire accepts conductors up to 12 gauge. Large head, triple-drive combination screws. Impact-resistant. Durable thermoplastic toggle and frame. UL listed.						
1 pole, almond	BE@.250	Ea	.75	9.76	10.51	17.90
1 pole, brown	BE@.250	Ea	.75	9.76	10.51	17.90
1 pole, ivory	BE@.250	Ea	.75	9.76	10.51	17.90
1 pole, white	BE@.250	Ea	.75	9.76	10.51	17.90
1 pole, clear illuminated	BE@.250	Ea	4.75	9.76	14.51	24.70
3 way, clear illuminated	BE@.250	Ea	6.62	9.76	16.38	27.80
Commercial-grade AC quiet toggle switch, Leviton. Side-wired 120/277-volt AC. 15-amp. Back-wiring clamps and large head. Double-drive terminal screws, backed out and staked. Accepts up to 10 gauge copper or copper-clad wire. Large silver and cadmium oxide contacts. Rust-resistant heavy-gauge steel mounting strap. Break-off plaster ears. One-piece brass alloy contact arm for reliable electrical performance. UL listed.						
1 pole, brown	BE@.250	Ea	3.05	9.76	12.81	21.80
1 pole, gray	BE@.250	Ea	3.05	9.76	12.81	21.80
1 pole, ivory	BE@.250	Ea	3.05	9.76	12.81	21.80
1 pole, white	BE@.250	Ea	3.05	9.76	12.81	21.80
3 way, ivory	BE@.300	Ea	6.62	11.70	18.32	31.10
3 way, white	BE@.300	Ea	6.62	11.70	18.32	31.10
4 way, framed, ivory	BE@.400	Ea	15.10	15.60	30.70	52.20
4 way, framed, white	BE@.400	Ea	15.10	15.60	30.70	52.20
4 way, grounded, ivory	BE@.400	Ea	15.10	15.60	30.70	52.20
4 way, grounded, white	BE@.400	Ea	15.10	15.60	30.70	52.20
1 pole, lighted, ivory	BE@.250	Ea	8.35	9.76	18.11	30.80
1 pole, lighted, white	BE@.250	Ea	8.35	9.76	18.11	30.80
3 way, lighted, ivory	BE@.300	Ea	10.80	11.70	22.50	38.30
3 way, lighted, white	BE@.300	Ea	10.80	11.70	22.50	38.30
Decora® AC quiet rocker switch, Leviton. 15-amp, 120/277-volt AC. Side wired with Quickwire push-in terminals accepting No.12 or No.14 copper or copper-clad wire. Switch frame shields against dust and fits in wallplate to prevent rocker binding. Sturdy construction for long service life. Full rated current capacity with tungsten, fluorescent and resistive loads. UL listed.						
1 pole	BE@.250	Ea	2.54	9.76	12.30	20.90
1 pole, illuminated	BE@.250	Ea	8.76	9.76	18.52	31.50
3 way	BE@.300	Ea	4.05	11.70	15.75	26.80
3 way, illuminated	BE@.300	Ea	10.70	11.70	22.40	38.10
4 way	BE@.350	Ea	13.20	13.70	26.90	45.70
Residential-grade duplex straight blade receptacle, Leviton. 15-amp, 125-volts. Quickwire push-in and side wired. Large head terminal screws accept up to 12 gauge copper or copper-clad wire. Quickwire push-in terminals accept up to 14 gauge copper or copper-clad wire. Double-wide power contacts. Plated steel mounting strap. Break-off plaster ears. Captive backed-out mounting screws. Break-off fins for 2-circuit conversion. (5-15R). UL listed.						
15 amps, almond	BE@.200	Ea	.60	7.81	8.41	14.30
15 amps, brown	BE@.200	Ea	.60	7.81	8.41	14.30
15 amps, ivory	BE@.200	Ea	.60	7.81	8.41	14.30
15 amps, white	BE@.200	Ea	.60	7.81	8.41	14.30
15 amps, self-grounding, ivory	BE@.200	Ea	1.25	7.81	9.06	15.40
15 amps, self-grounding, white	BE@.200	Ea	1.25	7.81	9.06	15.40

	Craft@Hrs	Unit	Material	Labor	Total	Sell
Commercial-grade duplex receptacle, Leviton. Side wired. 125 volts. Ground out. Large head triple-drive terminal screws, backed out and staked. Terminals and back-wiring clamps. Accepts 10 gauge copper or copper-clad wire. Rust-resistant mounting strap. Washer-type break-off plaster ears. Captive-mounting screws. 15-amp, NEMA 5-15R. 20-amp NEMA 2-20R. UL listed.						
15 amps	BE@.200	Ea	2.00	7.81	9.81	16.70
20 amps	BE@.200	Ea	3.00	7.81	10.81	18.40
CO/ALR duplex receptacle, Leviton. Residential grade. 125 volts. Side-wired grounding. Fits shallow receptacle. Large head terminal screws accept up to 12 gauge copper or copper-clad wire. Quickwire push-in terminals accept 14 gauge solid copper wire only. Rust-resistant steel mounting strap. Break-off plaster ears. Captive mounting screws. Break-off tabs allow easy two-circuit conversion. 5-15R. UL listed.						
15 amps, ivory	BE@.200	Ea	3.05	7.81	10.86	18.50
15 amps, white	BE@.200	Ea	3.05	7.81	10.86	18.50
GFCI hospital-grade receptacle.						
20 amps, ivory	BE@.300	Ea	14.70	11.70	26.40	44.90
20 amps, red	BE@.300	Ea	14.70	11.70	26.40	44.90
GFCI receptacle. 15-amp shock guard ground fault circuit interrupter. Side- and back-wired to accept 12 gauge wire. Shallow design allows for easy installation.						
15 amps, ivory	BE@.300	Ea	20.00	11.70	31.70	53.90
15 amps, white	BE@.300	Ea	20.00	11.70	31.70	53.90
Single receptacle, Leviton. Commercial grade. Large head triple-drive terminal screws, backed out and staked. Terminals and back-wiring clamps accept 10 gauge copper or copper-clad wire. Heavy-gauge, rust-resistant steel mounting strap. Convenient washer-type break-off plaster ears for best flush alignment. Triple-wide power contacts. Captive-mounting screws.						
15 amps, 125 volts	BE@.200	Ea	3.50	7.81	11.31	19.20
20 amps, 125 volts	BE@.200	Ea	4.50	7.81	12.31	20.90
20 amps, 250 volts	BE@.200	Ea	5.50	7.81	13.31	22.60
Appliance receptacle, flush mount, Pass & Seymour/Legrand. Straight blade receptacle. Industrial specification grade. 125/250-volt, 60 Hz, AC only. For copper or aluminum conductors. Easily identifiable color-coded terminals. Double-wipe copper alloy contacts. Terminals accept up to #4 conductors. Fits single- or 2-gang outlet boxes. Includes mounting hardware. UL listed.						
NEMA 14-30R, 30 amps, dryer	BE@.350	Ea	8.66	13.70	22.36	38.00
NEMA 14-50R, 50 amps, range	BE@.400	Ea	8.66	15.60	24.26	41.20
Single-gang plastic wall plate, Leviton. Residential grade. Standard size. Resists fading, discoloration, grease, oils, organic solvents and scratches. Round on edges to prevent injury and wall damage. Includes matching metal mounting screws. UL listed.						
1 duplex receptacle	BE@.050	Ea	.29	1.95	2.24	3.81
2 duplex receptacles	BE@.100	Ea	1.00	3.91	4.91	8.35
1 toggle switch	BE@.050	Ea	.29	1.95	2.24	3.81
1 toggle, 1 duplex receptacle	BE@.100	Ea	1.00	3.91	4.91	8.35
2 toggle switches	BE@.100	Ea	1.00	3.91	4.91	8.35
Blank	BE@.050	Ea	.75	1.95	2.70	4.59
Single-gang Decora® wall plate, Leviton. Smooth face and round edges that resist dust accumulation. Individual plastic wrapping to protect surfaces. Color-matched mounting screws included. UL listed.						
1 rocker switch	BE@.050	Ea	.60	1.95	2.55	4.34
2 rocker switches	BE@.100	Ea	1.06	3.91	4.97	8.45
1 rocker switch, 1 toggle	BE@.100	Ea	1.48	3.91	5.39	9.16

	Craft@Hrs	Unit	Material	Labor	Total	Sell
Incandescent slide dimmer. Commercial specification grade. 600 watt. Compact design. Single pole. Positive on/off switching with upper-end dimming bypass for maximum brightness. Solid-state circuitry with built-in radio/TV interference filter. UL listed.						
Ivory	BE@.250	Ea	10.20	9.76	19.96	33.90
White	BE@.250	Ea	10.20	9.76	19.96	33.90
3-way illuminated incandescent slide dimmer. Commercial specification grade. Separate heavy-duty on/off rocker for switching without disturbing preset brightness levels. 600 watt. Compact design. Solid-state circuitry with built-in radio/TV interference filter. Includes wall plate. UL listed. 120-volt, 60 hertz, AC.						
White, preset on/off	BE@.300	Ea	16.30	11.70	28.00	47.60
Spring-wound timer switch, Intermatic. Operation requires no electricity, automatically limiting length of time energy is used. Installs in most 2-1/2"-deep junction boxes. Compact design. Single pole, single throw.						
60 minute	BE@.200	Ea	15.30	7.81	23.11	39.30

Smoke Detectors

	Craft@Hrs	Unit	Material	Labor	Total	Sell
9-volt smoke detector, Kidde®. With test button. Flashing red light indicates unit is receiving power. Includes batteries. 3-year warranty.						
3-7/8", round	BE@.300	Ea	8.15	11.70	19.85	33.70
Dual-sensing smoke alarm, First Alert®. Dual-sensing ionization and photoelectric smoke-sensing technologies. Features remote-control test/silence. Unit can be silenced or tested from any location using most standard household remote controls. Microprocessor technology to reduce unwanted alarms. Low battery chirp, missing battery guard and blinking power indicator.						
10" diameter	BE@.300	Ea	15.30	11.70	27.00	45.90
Dual-ionization smoke detector. Battery and AC operation. Photoelectric sensor for detecting slow, smoldering fires. Dual-ionization sensor for detecting fast, flaming fires. Two test buttons to test each function, 30-day low battery signal.						
120-volt interconnectable	BE@.500	Ea	25.00	19.50	44.50	75.70
Interconnected Smoke and CO Alarm.						
120-volt interconnectable	BE@.500	Ea	48.90	19.50	68.40	116.00
Interconnected Smoke and CO Alarm.						
120-volt interconnectable	BE@.500	Ea	48.90	19.50	68.40	116.00
Battery Powered Carbon Monoxide Alarm						
Battery powered	BE@.300	Ea	48.90	11.70	60.60	103.00
Builder's door chime kit, Trine. Includes front and back door black and ivory push buttons, one chime, one transformer and wire. UL listed.						
Wired chime kit	BE@1.50	Ea	10.20	58.60	68.80	117.00
Wireless door bell, front and back, Trine. Weatherproof. Selectable front and back, push button and battery included. 128 selectable codes. 150-foot range. Westminster or ding-dong selection. Volume control and high-quality chime sound. Applications are 2 tones. Includes 12-volt remote control battery and mounting hardware. No wiring needed. Wall- or ceiling-mount. Coded transmission. No signal interference. Uses 3 standard D batteries. UL listed. White.						
Wireless chime	BE@.300	Ea	32.60	11.70	44.30	75.30

Electrical Distribution Panels

	Craft@Hrs	Unit	Material	Labor	Total	Sell
Disconnect and remove distribution panel. Add for disconnecting circuits. Single-phase breaker panel.						
50 to 100 amps, concrete wall	BE@.258	Ea	—	10.10	10.10	17.20
50 to 100 amps, frame wall	BE@.220	Ea	—	8.59	8.59	14.60
50 to 100 amps, steel column	BE@.416	Ea	—	16.20	16.20	27.50
200 amps, wood or concrete wall	BE@.300	Ea	—	11.70	11.70	19.90
200 amps, steel column	BE@.492	Ea	—	19.20	19.20	32.60
Add per circuit disconnected	BE@.063	Ea	—	2.46	2.46	4.18
HomeLine® indoor main lug load center, value pack, Square D®. Class 1170. Single-phase, 3-wire, 120/240-volt, 1-pole circuits. 10,000 RMS AIR. Factory-installed main lugs. Straight-in mains. Split-branch neutral. Rotates for top or bottom feed. 10-year limited warranty. Main wire size #6 to 250 AWG aluminum/copper. Main breaker frame size QOM2. Equipment ground bar. Combination surface/flush cover. Install Square D HOM type circuit breakers only. UL listed.						
200 amps, 20 spaces, 40 circuits, with (6) 15-amp and (6) 20-amp 1-pole and (1) 30-amp and (1) 50-amp 2-pole breakers	BE@1.50	Ea	80.60	58.60	139.20	237.00
HomeLine circuit breaker, Square D®. 1-pole, 120-volt. 10,000 AIC rated.						
15 amp	BE@.150	Ea	3.02	5.86	8.88	15.10
20 amp	BE@.150	Ea	3.02	5.86	8.88	15.10
30 amp	BE@.150	Ea	3.02	5.86	8.88	15.10
HomeLine arc-fault circuit breaker, Square D®. 1-pole, 120-volt arc fault interrupter. 10,000 AIC rated.						
15 amp	BE@.200	Ea	34.50	7.81	42.31	71.90
20 amp	BE@.200	Ea	34.50	7.81	42.31	71.90
HomeLine ground-fault interrupter circuit breaker, Square D®. 1-pole, 120-volt. 10,000 AIC rated. For use in Square D HomeLine load centers. UL listed.						
15 amp	BE@.200	Ea	34.50	7.81	42.31	71.90
20 amp	BE@.200	Ea	34.50	7.81	42.31	71.90
Combination service entrance all-in-one meter main load center, Square D®. Overhead or underground. Single-phase, 3-wire, 120/240-volt, 200-amp ring utility meter socket. 200-amp main circuit breaker and 30-space, 40-circuit, HOM load center with semi-flush outdoor enclosure. Meets EUSERC standards.						
Model CSEDVP1, 200 amps, 30 spaces, 40 circuits	BE@1.30	Ea	128.00	50.80	178.80	304.00
Combination ring-type meter socket and main load center, Square D®. Overhead or underground service entrance device. Dual main breaker with feed-through lugs. Single-phase, 3-wire, 120/240-volt AC. 10,000 amp short circuit current rating. Ring utility meter socket. Surface-mount convertible to semi-flush with SC200F flange kit. Flush-mount cover included with HOM load center. Enclosure is of rainproof construction and accepts A-type bolt-on hubs. NEMA 3R enclosure. Center mounted hub. Supplied with line side feed through lugs for #6 AWG-250 K circular mils (Al/Cu) conductors and provision for branch circuit breakers. Designed for single-family residential applications. Wire size #6 to 300 KCmil (Al/Cu). Box dimensions, 14-1/2" wide x 32-1/16" long x 6-15/16" deep. Meets EUSE standards. UL listed.						
No. SC816D200C, 200 amps, 8 spaces, 16 circuits	BE@1.30	Ea	127.00	50.80	177.80	302.00
Load center flange kit, Square D®. Converts 125- to 200-amp ring and ringless type surface mount load centers to semi-flush mount.						
Kit	BE@.050	Ea	4.06	1.95	6.01	10.20

	Craft@Hrs	Unit	Material	Labor	Total	Sell
Combination service entrance device, Square D®. All-in-one main circuit breaker panel. Surface-mount. Includes ring-type utility meter socket, service disconnect, and integral load center. HOM load center with surface outdoor enclosure. Single-phase, 3-wire. 120/240-volt AC, 10,000 amp short circuit current rating. NEMA Type 3R enclosure. Overhead or underground service feed. Left meter location. Hub style A. Suitable only for use as service equipment. UL listed. Meets EUSERC standards. 10-year limited warranty.						
No. SC1624M100S, 100 amp, 16 spaces, 24 circuits	BE@1.30	Ea	54.10	50.80	104.90	178.00
No. SC1624M125S2, 125 amp, 16 spaces, 24 circuits	BE@1.30	Ea	96.90	50.80	147.70	251.00
No. SC2040M200C, 200 amp, 20 spaces, 40 circuits	BE@1.30	Ea	123.00	50.80	173.80	295.00
Combination service entrance device, Square D®. All-in-one main circuit breaker panel. Surface mount. Includes ring-type utility meter socket, service disconnect, and integral load center. HOM load center with surface outdoor enclosure. Single-phase, 3-wire, 120/240-volt AC, 22,000 amp short circuit current rating. NEMA Type 3R enclosure. Overhead or underground service feed. Left meter location. Hub style A. Suitable only for use as service equipment. UL listed. Meets EUSERC standards. 10-year limited warranty.						
No. SC2040M200S, 200 amp, 20 spaces, 40 circuit	BE@1.30	Ea	123.00	50.80	173.80	295.00
No. SC3040M200SS, 200 amp, 30 spaces, 40 circuit	BE@1.30	Ea	138.00	50.80	188.80	321.00
No. SC40M200S, 200 amp, 40 spaces, 40 circuit	BE@1.30	Ea	143.00	50.80	193.80	329.00
Bolt-on A Type universal hub, Square D®. Bolt-on closing plate for A or A-L hub opening. Series A style. For Class 4131 meter socket. Class 4119 service entrance device. Aluminum. UL listed. Threaded for rigid conduit.						
1" conduit	BE@.050	Ea	7.47	1.95	9.42	16.00
1-1/4" conduit	BE@.050	Ea	7.47	1.95	9.42	16.00
1-1/2" conduit	BE@.050	Ea	7.47	1.95	9.42	16.00
2" conduit	BE@.050	Ea	7.47	1.95	9.42	16.00
Main circuit breaker, Class 1130 QO® main lugs, rainproof, Square D®. 2-pole, 22,000 AC rated. For convertible load centers only. Load center main rating. KCM is thousand circular mils.						
100 amp, AWG #4 to 2/0	BE@.150	Ea	51.20	5.86	57.06	97.00
125 amp, AWG #4 to 2/0	BE@.150	Ea	54.60	5.86	60.46	103.00
200 amp, AWG #4 to 300 KCM	BE@.150	Ea	77.00	5.86	82.86	141.00
QO® Indoor main lug load center, Square D®. Surface-mount. Class 736, 1130. Single-phase, 3-wire, 120/240-volt AC. Fixed mains. Factory-installed main lugs. 10,000 RMS amp short circuit current rating. Shielded copper bus. Straight-in mains. Rotates for top or bottom feed. Cover included. Ground bar kit sold separately. Uses QO, QOT, QO-EPD, QO-GFI or QO-PL branch circuit breakers separately. UL listed. 10-year limited warranty.						
No. QO2L30SCP, 30 amp, 2 spaces, 2 circuits, main wire #12-10 Al, #14-10 Cu	BE@.450	Ea	13.60	17.60	31.20	53.00
No. QO24L70SCP, 70 amp, 2 spaces, 4 circuits, main wire #12-3 Al, #14-4 Cu	BE@.500	Ea	17.10	19.50	36.60	62.20
No. QO612L100DS, 100 amp, 6 spaces, 12 circuits, main wire #8-1 Al and Cu	BE@.800	Ea	32.20	31.20	63.40	108.00

	Craft@Hrs	Unit	Material	Labor	Total	Sell
No. QO612L100SCP, 100 amp, 6 spaces, 12 circuits, main wire #8-1 Al and Cu	BE@.800	Ea	22.90	31.20	54.10	92.00
No. QO816L100DS, 100 amp, 8 spaces, 16 circuits, main wire #8-1 Al and Cu	BE@.900	Ea	50.60	35.10	85.70	146.00
No. QO816L100SCP, 100 amp, 8 spaces, 16 circuits main wire #8-1 Al and Cu	BE@.900	Ea	37.50	35.10	72.60	123.00

QO® outdoor main lug load center, QOM1 frame size, Square D®. Single-phase, 3-wire, 120/240-volt AC. Main wire #6-2/0 aluminum or copper. 65,000 RMS symmetrical amp short circuit current rating. Metallic enclosure. Convertible mains. Factory-installed main lug. QOM1 main frame size. Convertible to main circuit breaker. Straight-in mains. Includes rainproof cover. Order ground bar kit separately. Order QO, QOT, QO-GFI, QO-EPD, QO-AFI OR QO-APL branch circuit breakers separately. Install Square D QO type circuit breakers only. 10-year limited warranty. UL Listed.

	Craft@Hrs	Unit	Material	Labor	Total	Sell
No. QO11224L125GRB, 125 amp, 12 spaces, 24 circuits	BE@.800	Ea	94.90	31.20	126.10	214.00

QO® miniature circuit breaker, Square D®. Single pole. Square D QO miniature circuit breakers are plug-in products for use in QO load centers and NQOD panelboards. 120/240-volts AC. 10,000-amp interrupting rating (AIR) RMS symmetrical with VISI-TRIP® indicator. Class 730, 731 and 733. 1 space required. Use in Square D QO load centers only. UL listed.

	Craft@Hrs	Unit	Material	Labor	Total	Sell
15 amp, 1 pole	BE@.150	Ea	6.06	5.86	11.92	20.30
20 amp, 1 pole	BE@.150	Ea	6.06	5.86	11.92	20.30
25 amp, 1 pole	BE@.150	Ea	6.06	5.86	11.92	20.30
30 amp, 1 pole	BE@.150	Ea	6.06	5.86	11.92	20.30
40 amp, 1 pole	BE@.150	Ea	6.06	5.86	11.92	20.30
50 amp, 1 pole	BE@.150	Ea	6.06	5.86	11.92	20.30
15 amp, 2 pole	BE@.150	Ea	13.90	5.86	19.76	33.60
20 amp, 2 pole	BE@.150	Ea	13.90	5.86	19.76	33.60
25 amp, 2 pole	BE@.150	Ea	13.90	5.86	19.76	33.60
25 amp, 2 pole	BE@.150	Ea	13.90	5.86	19.76	33.60
30 amp, 2 pole	BE@.150	Ea	13.90	5.86	19.76	33.60
40 amp, 2 pole	BE@.150	Ea	13.90	5.86	19.76	33.60
50 amp, 2 pole	BE@.150	Ea	13.90	5.86	19.76	33.60

Replacement tandem circuit breaker, QO®, Square D®. 1 pole. 120-volts AC. 10,000 AIR. For use in Square D QO load centers manufactured before 1967. UL listed.

	Craft@Hrs	Unit	Material	Labor	Total	Sell
15 and 15 amp	BE@.150	Ea	34.10	5.86	39.96	67.90
20 and 20 amp	BE@.150	Ea	34.10	5.86	39.96	67.90

Tandem circuit breaker, Square D®. 1 pole. 120-volts AC. 10,000 AIR. HACR type. Visi-Trip® indicator. Requires handle tie No. QOTHT for common switching. For use in Square D QO load centers manufactured after 1967. UL listed as CTL.

	Craft@Hrs	Unit	Material	Labor	Total	Sell
15 and 15 amp	BE@.150	Ea	17.10	5.86	22.96	39.00
20 and 20 amp	BE@.150	Ea	17.10	5.86	22.96	39.00

	Craft@Hrs	Unit	Material	Labor	Total	Sell

Ground-fault interrupter plug-on circuit breaker, QO-GFI Qwik-Gard®, Square D®. QWIK-GARD® circuit breakers provide overload and short circuit protection, combined with Class A ground-fault protection. Class A denotes a ground-fault circuit interrupter that will trip when a ground-fault current is 6 milliamperes or greater, for people protection. Do not connect to more than 250' of load conductor for the total one-way run to prevent nuisance tripping. Class 685, 690, 730, 912 and 950. 10,000 AIR. HACR type. Visi-Trip® indicator. 1 poles are 120 volt and 2 poles are 120/240 volt. 2-pole breakers have common trip, 120-volts AC. 2 spaces required for 2-pole breakers. UL listed.

	Craft@Hrs	Unit	Material	Labor	Total	Sell
20 amp, 1 pole	BE@.150	Ea	74.00	5.86	79.86	136.00
20 amp, 2 pole	BE@.150	Ea	100.00	5.86	105.86	180.00
30 amp, 2 pole	BE@.150	Ea	105.00	5.86	110.86	188.00
50 amp, 2 pole	BE@.150	Ea	167.00	5.86	172.86	294.00
60 amp, 2 pole	BE@.150	Ea	150.00	5.86	155.86	265.00

QO® load center ground bar kit, Square D®. Sold separately from the QO® load centers. Indoor, single phase, main lug, class 736 and 1130. Aluminum. UL listed.

	Craft@Hrs	Unit	Material	Labor	Total	Sell
No. PK3GTA, 3 terminal	BE@.200	Ea	3.72	7.81	11.53	19.60
No. PK4GTA, 4 terminal	BE@.200	Ea	4.25	7.81	12.06	20.50
No. PK7GTA, 7 terminal	BE@.200	Ea	4.80	7.81	12.61	21.40
No. PK9GTA, 9 terminal	BE@.200	Ea	5.32	7.81	13.13	22.30
No. PK12GTACP, 12 terminal	BE@.200	Ea	5.32	7.81	13.13	22.30
No. PK15GTA, 15 terminal	BE@.200	Ea	6.40	7.81	14.21	24.20

Recessed Lighting

Recessed low-voltage baffle light kit, remodel, Commercial Electric. Not for direct contact with insulation. 4" black baffle trim. Uses (1) MR16 bulb, 50 watt (included). Minimizes glare in general, accent and task lighting. Pre-wired housing. Thermally protected. UL listed for damp locations. For suspended ceilings. Installation instructions and template enclosed. Compatible with standard wall dimmer.

	Craft@Hrs	Unit	Material	Labor	Total	Sell
4", with black baffle trim	BE@1.00	Ea	43.00	39.10	82.10	140.00

Recessed low-voltage slot aperture light kit, remodel, Commercial Electric. Not for direct contact with insulation. 4" white slot aperture trim. Uses (1) MR16 bulb, 50 watt (included). Pre-wired housing. Thermally protected. UL listed for damp locations. For suspended ceilings. Installation instructions and template enclosed. Compatible with standard wall dimmer.

	Craft@Hrs	Unit	Material	Labor	Total	Sell
4", white slot aperture trim	BE@1.00	Ea	48.00	39.10	87.10	148.00

Mini can recessed eyeball trim light kit, Commercial Electric. 5" mini non-IC housing. 5" white eyeball trim. For new or remodel construction. Not for direct contact with insulation. For suspended ceilings or joist support (hanger bars sold separately). Pre-wired housing. Thermally protected. UL listed for damp locations. Compatible with standard wall dimmer. 75-watt maximum, PAR-30 bulb (not included).

	Craft@Hrs	Unit	Material	Labor	Total	Sell
5", white	BE@1.00	Ea	13.40	39.10	52.50	89.30

Mini can recessed open light kit, Commercial Electric. 5" mini non-IC housing. 5" white open trim. For new or remodel construction. Not for direct contact with insulation. For suspended ceilings or joist support (hanger bars sold separately). Pre-wired housing. Thermally protected. UL listed for damp locations. Compatible with standard wall dimmer.

	Craft@Hrs	Unit	Material	Labor	Total	Sell
5", white	BE@1.00	Ea	11.70	39.10	50.80	86.40

Recessed light kit. For use in an insulated ceiling. Kit includes housing and trim. For new construction.

	Craft@Hrs	Unit	Material	Labor	Total	Sell
6", white trim	BE@.700	Ea	22.00	27.30	49.30	83.80

	Craft@Hrs	Unit	Material	Labor	Total	Sell
Lighting Fixtures						
Remove and replace ceiling light fixture. Disassemble and remove incandescent fixture and replace with incandescent fixture.						
Ceiling-mounted	BE@.787	Ea	—	30.70	30.70	52.20
Cambridge II residential fluorescent ceiling fixture, Lithonia Lighting. Electronic ballast. 32 watt, T8 tubes.						
1' x 4', 2 lamps, oak	BE@.800	Ea	105.00	31.20	136.20	232.00
1' x 4', 2 lamps, white	BE@.800	Ea	107.00	31.20	138.20	235.00
1.5' x 4', 4 lamps, oak	BE@.800	Ea	124.00	31.20	155.20	264.00
1.5' x 4', 4 lamps, dentil	BE@.800	Ea	129.00	31.20	160.20	272.00
1.5' x 4', 4 lamps, white	BE@.800	Ea	121.00	31.20	152.20	259.00
2' x 2', 2 U-lamps, oak	BE@.800	Ea	117.00	31.20	148.20	252.00
2' x 2', 2 U-lamps, white	BE@.800	Ea	117.00	31.20	148.20	252.00
Low-profile fluorescent diffuser fixture, Lithonia Lighting. Suitable for use in closets, laundry areas and pantries. Acrylic white diffuser. Wall- or ceiling-mount. UL listed.						
11", 22-watt lamp, round	BE@.900	Ea	30.60	35.10	65.70	112.00
12", 22-watt lamp, square	BE@.900	Ea	30.60	35.10	65.70	112.00
14", 2 lamps, 22/32 watts, round	BE@.900	Ea	40.80	35.10	75.90	129.00
15", 2 lamps, 22/32 watts, square	BE@.900	Ea	42.10	35.10	77.20	131.00
19", 2 lamps, 32/40 watts, round	BE@.900	Ea	49.90	35.10	85.00	145.00
20", 2 lamps, 32/40 watts, square	BE@.900	Ea	49.90	35.10	85.00	145.00
Wave bath bar light fixture, Hampton Bay™. Brushed nickel finish with etched marble glass. Uses 100-watt, medium base bulb(s), sold separately. 5" and 22" wide respectively by 7-1/4" high.						
1 light, brushed nickel	BE@.500	Ea	40.80	19.50	60.30	103.00
3 light, brushed nickel	BE@.500	Ea	82.50	19.50	102.00	173.00
Chrome bath bar light fixture, Hampton Bay™. Chrome finish with etched opal glass. Uses 100-watt, medium base bulb(s), not included. 9-5/8" high.						
1 light, chrome, 5-1/2" long	BE@.500	Ea	50.00	19.50	69.50	118.00
2 light, chrome, 13-1/4" long	BE@.500	Ea	63.00	19.50	82.50	140.00
3 light, chrome, 21-1/4" long	BE@.500	Ea	90.00	19.50	109.50	186.00
Beveled mirror bath bar, Hampton Bay™. Chrome finish. Beveled mirror, flat back. Uses 100-watt (maximum) globe bulbs, not included. Mounting hardware included. 4-1/2" high. Extends 4" from wall.						
24" wide, 4 light	BE@.800	Ea	32.30	31.20	63.50	108.00
36" wide, 6 light	BE@.800	Ea	46.00	31.20	77.20	131.00
48" wide, 8 light	BE@.800	Ea	58.00	31.20	89.20	152.00
Opal glass wall light, Hampton Bay™. Chrome finish. Hand-blown opal glass. Uses 100-watt bulbs. Chrome.						
1 light, 4-3/4" wide	BE@.800	Ea	30.60	31.20	61.80	105.00
2 lights, 14" wide	BE@.800	Ea	40.10	31.20	71.30	121.00
3 lights, 15" wide	BE@.800	Ea	45.90	31.20	77.10	131.00
4 lights, 26" wide	BE@.800	Ea	61.20	31.20	92.40	157.00
Two-light ceiling fixture. Flush-mount hidden pan. Frosted glass shade. Uses two frosted A-19 or torpedo 60-watt maximum lamps. 13" deep by 6-1/4" high.						
Chrome	BE@.800	Ea	45.90	31.20	77.10	131.00
Cobblestone finish	BE@.800	Ea	45.90	31.20	77.10	131.00
Mushroom ceiling light fixture drum, Westinghouse Lighting.						
6", white	BE@.800	Ea	5.50	31.20	36.70	62.40

	Craft@Hrs	Unit	Material	Labor	Total	Sell
Two-light incandescent flush-mount ceiling fixture, Sunburst. Screw-in bubble-style glass. Medium base.						
9" wide, white	BE@.800	Ea	12.20	31.20	43.40	73.80
Mushroom flush-mount light fixture, Hampton Bay™. 7-3/8" diameter x 4-1/2" high. Uses one medium base bulb, 60-watt maximum (not included).						
Satin nickel finish, frosted ribbed glass	BE@.800	Ea	11.20	31.20	42.40	72.10
Mushroom flush-mount light fixture, Hampton Bay™. 9-1/4" diameter x 5-5/8" high. Uses 2 medium base bulbs, 60-watt maximum (not included).						
Polished brass finish, clear ribbed glass	BE@.800	Ea	12.70	31.20	43.90	74.60
Satin nickel finish, frosted ribbed glass	BE@.800	Ea	14.60	31.20	45.80	77.90
Mushroom flush-mount ceiling fixture.						
8", polished brass	BE@.500	Ea	11.40	19.50	30.90	52.50
8", white on white, 75 watt	BE@.500	Ea	10.20	19.50	29.70	50.50
9-1/2", polished brass	BE@.500	Ea	12.50	19.50	32.00	54.40
10", white on white, 75 watt	BE@.500	Ea	12.20	19.50	31.70	53.90
Glass globe ceiling light fixture. 6" white dome fixture with polished brass fitter. Uses one 75-watt bulb, sold separately. UL listed. Flush mount.						
With switch, white on white	BE@.500	Ea	8.13	19.50	27.63	47.00
Without switch, polished brass	BE@.500	Ea	6.10	19.50	25.60	43.50
Without pull chain, white on white	BE@.500	Ea	6.10	19.50	25.60	43.50
Glass globe ceiling light fixture. 8" white dome fixture with polished brass fitter. Uses one 75-watt bulb, sold separately. UL listed.						
Flush mount	BE@.500	Ea	11.50	19.50	31.00	52.70
Swirl flush-mount dome light ceiling fixture. Polished brass finish. Clear swirled glass dome. Uses one 75-watt bulb, sold separately. Includes mounting hardware.						
11", clear	BE@.500	Ea	13.20	19.50	32.70	55.60
13", clear	BE@.500	Ea	20.40	19.50	39.90	67.80
6-light crystal chandelier.						
Antique copper finish	BE@1.10	Ea	475.00	43.00	518.00	881.00
Crystal chandelier	BE@1.10	Ea	425.00	43.00	468.00	796.00
White with shades	BE@1.10	Ea	525.00	43.00	568.00	966.00
With shades	BE@1.10	Ea	575.00	43.00	618.00	1,050.00
Flair collection glass chandelier, Hampton Bay™ Brushed nickel finish chandelier with etched marble glass shades with interchangeable blue and frosted white decorative rings. Uses 100-watt medium base bulbs, not included. Fixtures are 18-1/2" and 27-1/2" wide and 20" and 23" high respectively.						
4 light, brushed nickel	BE@1.10	Ea	155.00	43.00	198.00	337.00
6 light, brushed nickel	BE@1.10	Ea	165.00	43.00	208.00	354.00

	Craft@Hrs	Unit	Material	Labor	Total	Sell
Outdoor bulkhead oval light fixture, Hampton Bay™. 5" wide x 8-1/2" high. 4-1/2" extension from the wall. Cast aluminum construction. Frosted, ribbed glass lens. UL wet location listed. Uses 1 medium base bulb, 60-watt maximum (included).						
Black finish	BE@.800	Ea	13.20	31.20	44.40	75.50
White finish	BE@.800	Ea	13.20	31.20	44.40	75.50
Outdoor round bulkhead light fixture, Hampton Bay™. 8" wide x 8" high x 4-1/2" extension from the wall. Cast aluminum construction. Frosted, ribbed glass lens. UL wet location listed. Uses one medium base bulb, 60-watt maximum (included).						
White finish	BE@.800	Ea	15.30	31.20	46.50	79.10
Outdoor wall lantern, Hampton Bay™. 7" wide x 12-1/2" high. 7-3/8" extension from the wall. Solid brass construction. Clear bent beveled glass. Install with or without tail. Uses one 100-watt maximum medium base bulb (included).						
Antique brass finish	BE@.800	Ea	32.30	31.20	63.50	108.00
Polished brass finish	BE@.800	Ea	32.30	31.20	63.50	108.00
Curved glass wall-mount lantern with photo cell, Hampton Bay™. Cast aluminum. Six sides of clear bent beveled glass panels. Photo control turns light on at dusk and off at dawn. Weather-resistant finish. Uses one medium base bulb, 100-watt maximum (not included). 7" wide x 15" high.						
Black	BE@.800	Ea	35.70	31.20	66.90	114.00
White	BE@.800	Ea	35.70	31.20	66.90	114.00
Jelly jar fixture, Catalina Lighting. Extra-thick glass. Coated aluminum screw shell. 10-1/8" high by 6" wide and 5-3/4" diameter.						
Black	BE@.800	Ea	5.07	31.20	36.27	61.70
Polished brass	BE@.800	Ea	5.07	31.20	36.27	61.70
White	BE@.800	Ea	5.07	31.20	36.27	61.70
6-sided motion-sensing lantern, Hampton Bay™. Cast aluminum. Six sides of clear bent beveled glass panels. Motion sensor turns on the light when motion is detected. Adjustable to 30' away with a 180-degree angle of coverage for security and convenience. Weather-resistant finish. Uses one 100-watt maximum medium base bulb (not included).						
Black	BE@.800	Ea	40.80	31.20	72.00	122.00
White	BE@.800	Ea	40.50	31.20	71.70	122.00
Motion-sensor security light, Heath Zenith. Quartz. Daylight shut-off. Selectable walk test. Automatic power outage reset. Sensitivity adjustment. Selectable light timer. Manual override. Operating temperature • 22 to +120 degrees F. 150-degree angle lens. Instant-on operation. 70-foot sensor range. 2,400-square-foot sensor coverage area. Pulse count. Die-cast metal fixtures. Dual-Brite™ 2-level lighting. 3- and 5-hour dawn-to-dusk timer. Wall- or eave-mount. Two 150-watt quartz halogen bulbs included.						
Bronze	BE@.800	Ea	43.80	31.20	75.00	128.00
White	BE@.800	Ea	40.80	31.20	72.00	122.00
Heavy-duty motion-sensor security light, Heath Zenith. Metal fixtures. 150-degree angle lens. 70-foot illumination, 6,400-square-foot coverage. Indoor manual override wall switch. Weather-resistant die-cast aluminum construction. Bulbs not included. UL listed.						
Gray	BE@.800	Ea	23.40	31.20	54.60	92.80
White	BE@.800	Ea	23.40	31.20	54.60	92.80

Ceiling Fans

San Marino™ ceiling fan with light kit, Hampton Bay™. Stylish mesh spotlight kit and finish. Powerful 153mm x 15mm motor. Triple capacitor speed control. Accu-Arm blade arms and slide-on mounting bracket make for wobble-free installation. Tri-mount installation compatible. Five reversible rosewood and black blades.

	Craft@Hrs	Unit	Material	Labor	Total	Sell
36", brushed steel	BE@1.78	Ea	61.20	69.50	130.70	222.00

Indoor or outdoor ceiling fan, Hampton Bay™. Stainless steel hardware. UL rating for outdoor use. Lifetime warranty.

	Craft@Hrs	Unit	Material	Labor	Total	Sell
42", white	BE@1.50	Ea	80.00	58.60	138.60	236.00

Ceiling fan with light kit, indoor or outdoor, Hampton Bay™. Special galvanized finish. Five oak wood grain weather-resistant blades with PowerMax motor. Stepped dome light kit. Stainless steel hardware. UL rating for outdoor use. Accu-Arm for accurate, easy installation. Uses two 60-watt vibration resistant bulbs, sold separately. Separate (2) pull-chain switches for fan and light. Lifetime warranty.

	Craft@Hrs	Unit	Material	Labor	Total	Sell
52", copper patina	BE@1.78	Ea	125.00	69.50	194.50	331.00
52", verde, wood grain blades	BE@1.78	Ea	135.00	69.50	204.50	348.00
52", white, white blades	BE@1.78	Ea	125.00	69.50	194.50	331.00

Smart home electrical management system. Complies with U.S. Green Building Council standard for energy-efficient upgrades. Provides automated management of lights, appliances, audio/video and other IR-controlled devices. Includes sensors for occupancy, water, humidity and temperature and an emergency notification system (phone, flashing lights, email, etc.), remote control over locks for doors and gates, heating and cooling, integration with security alarm systems and sensors, plus energy savings through efficient control. Add the cost of consulting and programming (if required), permit, insurance inspection and an uninterruptible power supply (if required). Installed in an existing residence.

	Craft@Hrs	Unit	Material	Labor	Total	Sell
HVAC monitoring, thermostat and trending module	BE@4.00	Ea	850.00	156.00	1,006.00	1,710.00
Intruder motion-sensing and alarm module	BE@2.50	Ea	380.00	97.60	477.60	812.00
Remote PDA/iPhone-based communications monitoring and control module	BE@3.00	Ea	1,200.00	117.00	1,317.00	2,240.00
Main interior lighting and outdoor lighting control module	BE@3.00	Ea	650.00	117.00	767.00	1,300.00
Fire and emergency service direct interconnects	BE@2.00	Ea	400.00	78.10	478.10	813.00
Remote door lock actuator	BE@0.50	Ea	135.00	19.50	154.50	263.00
PC-recording CCD surveillance camera	BE@0.50	Ea	120.00	19.50	139.50	237.00
Mount, wire and test smart system controls	BE@0.25	LF	.35	9.76	10.11	17.20
RFI, lightning & ground fault wiring	BE@.184	LF	.35	7.19	7.54	12.80
Fire protection piping (if required)	P1@0.25	LF	8.00	8.79	16.79	28.50
Mount and wire automatic transfer switch to household main breaker box and electrical service	BE@.184	LF	.35	7.19	7.54	12.80
Commission and test	BE@2.00	Ea	—	78.10	78.10	133.00

LED lighting system retrofit. Light-emitting diodes use 10% to 20% of the electrical power required for equivalent incandescent or fluorescent lamps. LED life expectancy approximates 50,000 hours (25 years of normal use). Meets U.S. Green Building Council standards for energy efficient upgrades.

	Craft@Hrs	Unit	Material	Labor	Total	Sell
Fluorescent LED troffer substitute, interior	BE@0.50	Ea	230.00	19.50	249.50	424.00
Re-wire an existing troffer (remove ballast)	BE@0.75	Ea	125.00	29.30	154.30	262.00
Screw-in LED lamps, interior	BE@0.03	Ea	80.00	1.17	81.17	138.00
Track and spot LED lamps, interior	BE@0.03	Ea	60.00	1.17	61.17	104.00
External LED floodlights	BE@0.03	Ea	125.00	1.17	126.17	214.00
Wiring of electrical service (typical)	BE@.184	LF	.35	7.19	7.54	12.80
Test total system (per fixture)	BE@0.25	Ea	—	9.76	9.76	16.60

Porches and Decks 16

Adding a porch or a deck is a popular home improvement project. The advantage is obvious: Porches and decks add value at a lesser square foot cost than similar enclosed living area. What isn't so obvious is the disadvantage: Porches and decks are more exposed to the weather, and have a life expectancy considerably less than the home itself. That makes porch and deck work a common repair task. The result is a steady volume of work for porch and deck specialists. This chapter includes cost estimates for both the addition of new porches and decks, and repair work.

Inspecting for Decay

Exposure to sun, wind and rain tends to open wood grain. The result is a higher moisture content that supports the growth of wood-destroying organisms. Nearly all wood decays if it remains moist for long periods. Check for decay and insect damage with a screwdriver. Wood has lost nearly all its load-carrying capacity when you can push a common screwdriver blade into the grain. Deck flooring needs to be replaced when it sags under foot traffic.

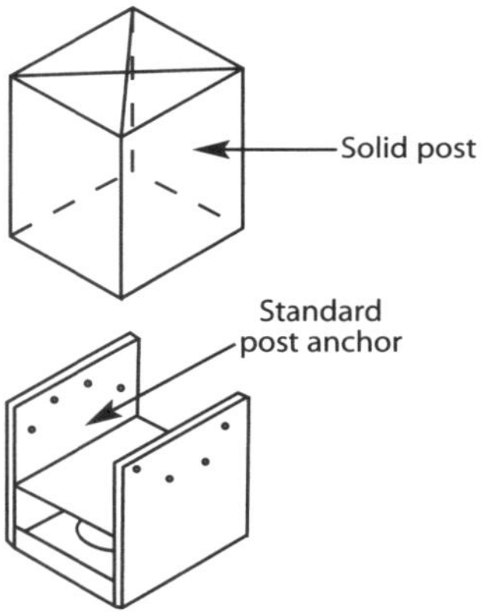

Figure 16-1

Base for post; standard post anchor for resistance to uplift

Check the crawlspace under the porch or deck for signs of dampness. Condensation forms on the underside of an elevated porch or deck when there isn't proper ventilation. Replace decayed framing with treated lumber.

Give particular attention to posts that are in contact with concrete. Posts should be replaced when heavily decayed. Support the new post slightly above the porch floor with a post anchor, as in Figure 16-1. Embed the post anchor in concrete to help protect the porch roof from uplift wind forces.

If a decayed post is strictly ornamental, consider cutting off the decayed base. Replace the decayed portion with a wood block secured to the concrete with a pin and a washer. Add base trim to conceal the patch (see Figure 16-2). Keep trim pieces slightly above the concrete so moisture can escape.

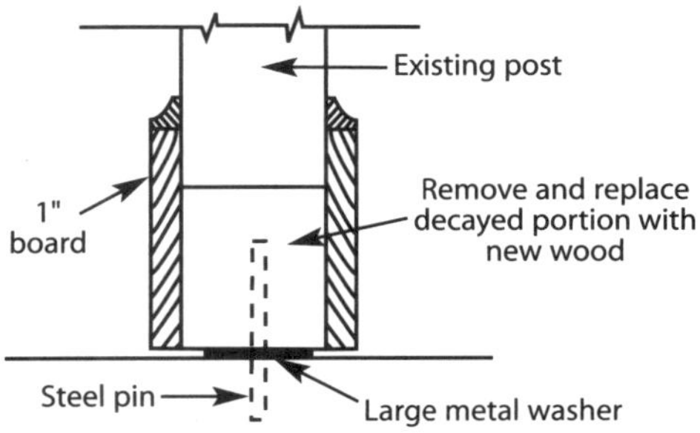

Figure 16-2

Replace end of decayed porch post

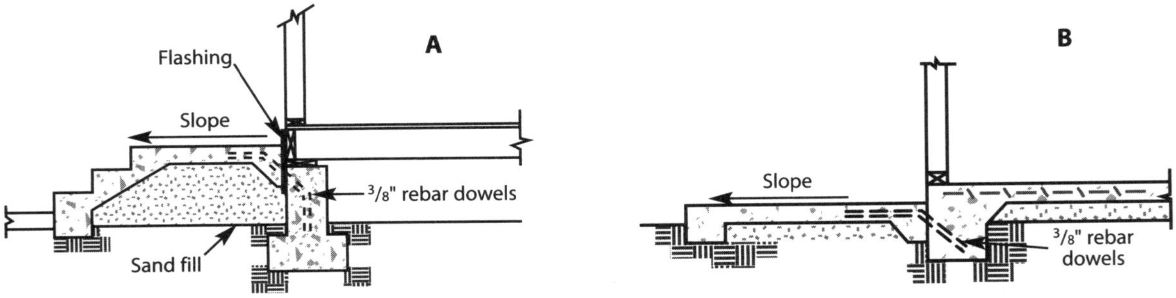

Figure 16-3
*Porch slab connection to house: **A**, concrete porch with wood-frame floor; **B**, porch with house slab*

Adding a Porch

Figure 16-3 shows how to tie a porch slab to a house with either a conventional crawlspace foundation (Figure 16-3A) or a concrete slab (Figure 16-3B). Tie the new concrete foundation, piers or slab to the existing foundation with anchor bolts. The new footing should go at least as deep as the footing on the existing building.

Figure 16-4 shows construction details for a porch addition built on a concrete slab. In colder climates and where the porch or deck can be built at least 18" above ground level, use conventional floor framing, as shown in Figure 16-5. Provide a polyethylene soil cover to control excessive moisture. Whether wood or concrete, slope the floor away from the house at least 1/8" per foot of width. That'll help keep water away from the existing foundation line.

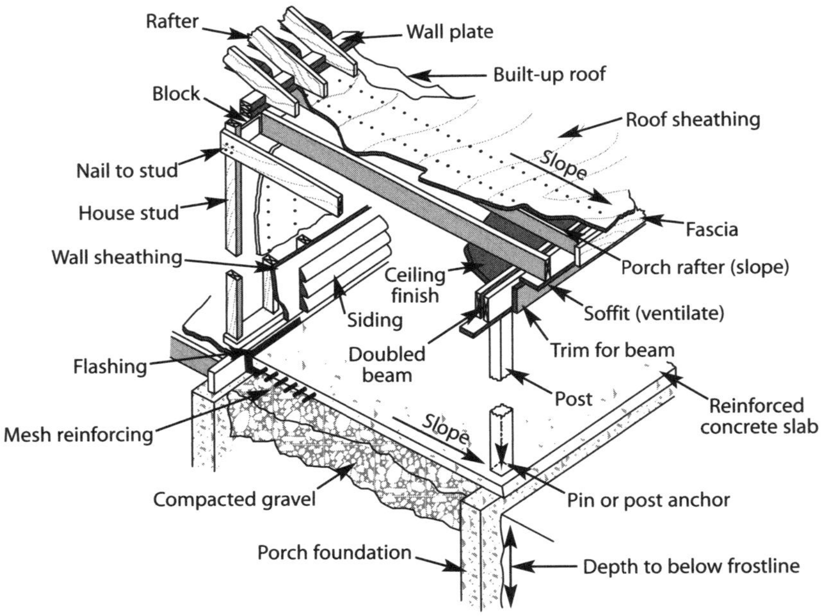

Figure 16-4
Details of porch construction for concrete slab

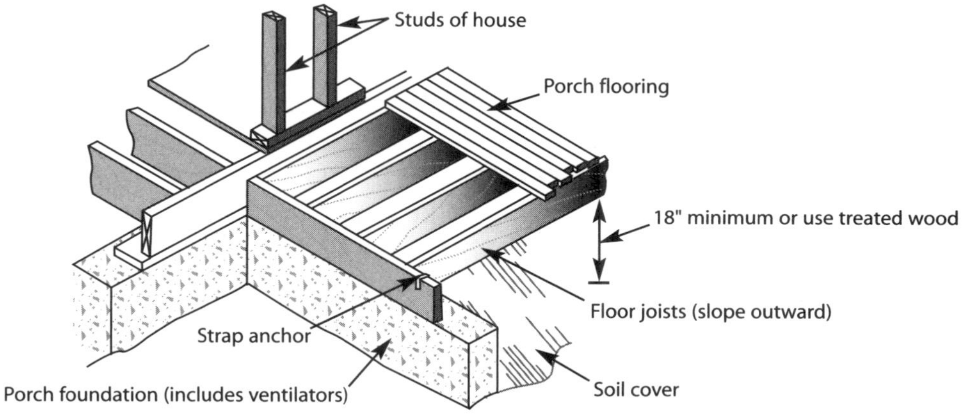

Figure 16-5
Porch floor with wood framing

Columns and Posts

Support the porch roof with either solid or built-up posts or columns. Figure 16-6A shows a column built from doubled 2 x 4s covered with 1 x 4" casing on two opposite sides and 1 x 6" finish casing on the other sides. Solid posts should be either 4 x 4" or 6 x 6" lumber. For a traditional-style home, use factory-made adjustable round columns.

Keep the post or column base out of any pocket or depression that will collect moisture and invite decay. If high winds aren't a design consideration, use a steel pin to join a one-piece post to the foundation. Set a large galvanized washer or spacer between the bottom of the post and the top of the floor or slab, as in Figure 16-6B. Apply wood preservative to cut ends at both the top and bottom of the post. For a cased post, install flashing under the base molding, as in Figure 16-6C. Build all deck posts and framing from pressure-treated lumber. Use 0.40 treated lumber for framing and 0.60 treated lumber for posts in contact with the ground.

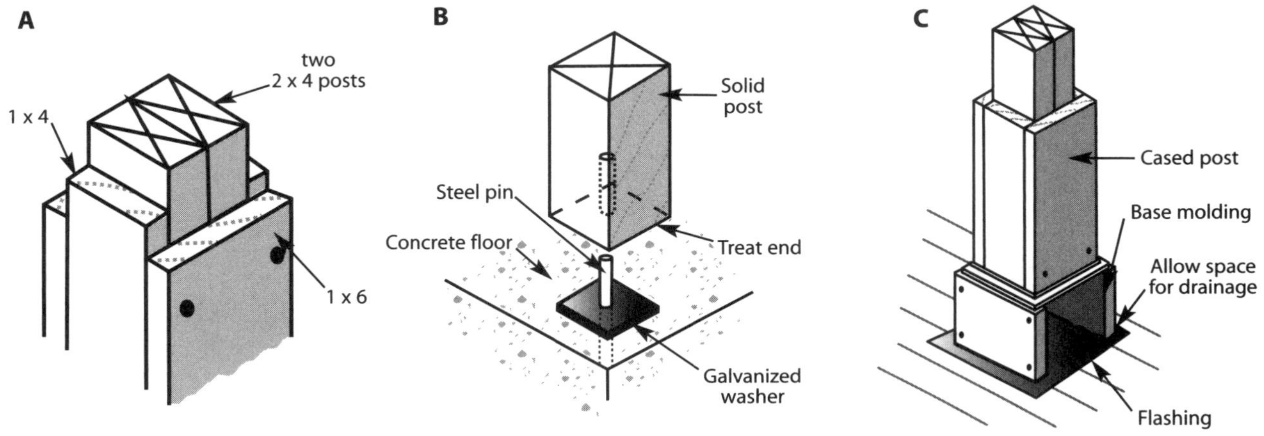

Figure 16-6
*Post details: **A**, cased post; **B**, pin anchor and space; **C**, flashing at base*

Balustrades

To exclude insects and flying debris, enclose the porch with either combination windows or window screen. See Figure 16-7A. An open balustrade adds safety as well as a decorative accent to a porch that isn't enclosed. See Figure 16-7B.

All wood railing exposed to weather should be shaped to shed water. Select railing that tapers slightly at the edges. Figure 16-8A shows a balustrade with a tapered top rail, a tapered bottom rail and balusters connecting the two. Design the joint between the post and rail to avoid trapping moisture. One method is to leave a small space between the post and the end of the railing, as shown in Figure 16-8B. Treat the cut rail end with water-repellent preservative. Wood posts, balusters and railings should be made from either the heartwood of a decay-resistant wood species or from treated wood. Place a small block under the bottom railing so it doesn't contact a concrete floor. Any wood in contact with concrete should be pressure-treated.

You must provide a railing around the perimeter of the porch if it's more than 30" above the existing grade. Generally the railing on a single-family residential porch or deck must be at least 36" high. Secure all hand rails and guard rails firmly to the deck. The space between balusters must be narrow enough so that nothing larger than a 4"-diameter sphere can pass through it.

Elevated Decks

An elevated wood deck is a good choice for a sloping lot where a concrete patio slab isn't practical. Use pressure-treated or decay-resistant lumber for support posts, beams and joists. Joist spacing can be 24", 16" or 12" on center, depending on the anticipated load and the type of deck material you use. The deck surface can be either wood that's naturally decay-resistant, pressure-treated, or a composite

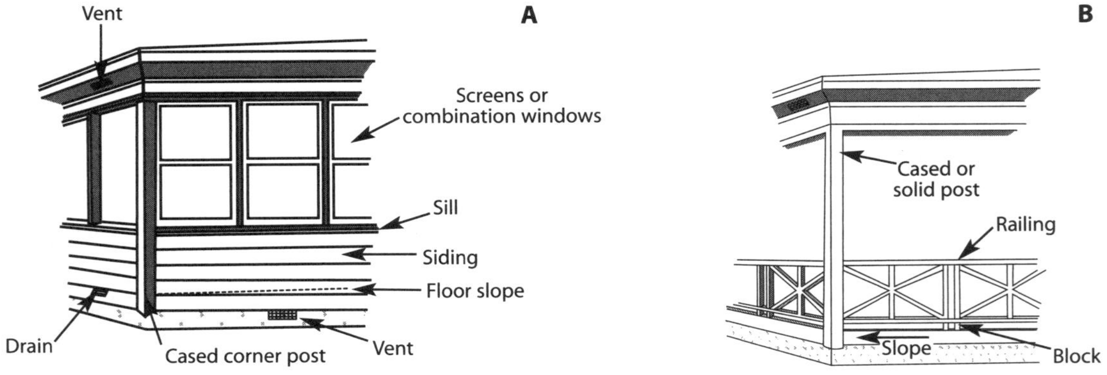

Figure 16-7
*Types of balustrades: **A**, closed balustrades; **B**, open balustrades*

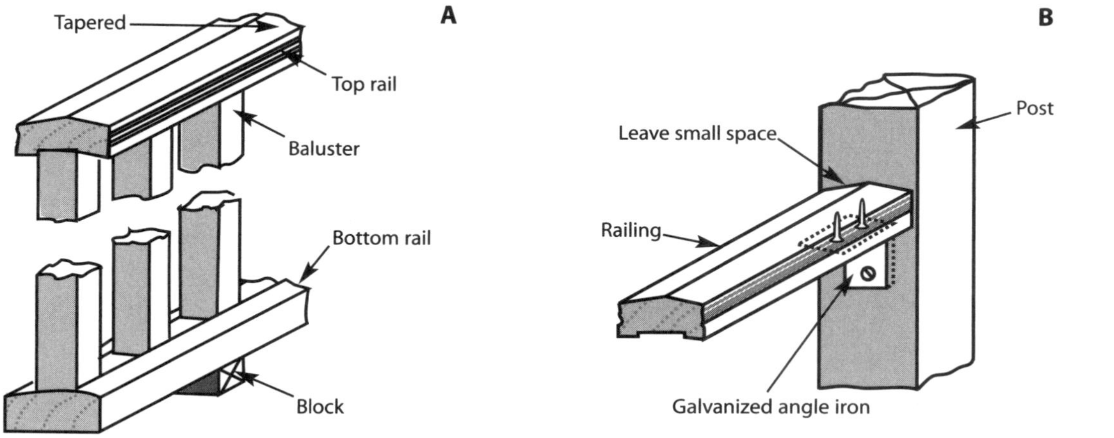

Figure 16-8

*Railing details: **A**, balustrade assembly; **B**, rail to post connection*

made from recycled plastics. Synthetic decking doesn't have the strength of wood decking and usually requires joists spaced 12" on center.

The bottom step of an elevated porch or deck stairway should never be in contact with the soil. Support the bottom step and stair carriage on a concrete pad or with a pressure-treated post. Figure 16-9 shows a porch stairway supported on treated posts. Posts like this should measure at least 5" in each dimension and should be embedded in the soil at least 3'. Nail and bolt a cross member between the posts. Then block the inner end of the horizontal support to the floor framing.

Where more than one step is required, use a 2 x 12 stringer at each end of the steps (see Figure 16-10). Bolt the lower end of each stringer to a treated post and attach the upper end of the stringer to the porch framing. You could also support the stairway on concrete or masonry piers.

Begin the construction of an elevated porch and deck by fastening a ledger securely to the existing sill with lag bolts (Figure 16-11). Then build a post and beam framework that'll support the joists and wood decking. Hang joists from the ledger using metal joist hangers. Joist spacing depends of the type of decking you select. As mentioned earlier, synthetic deck materials usually require joists no more than 12" on center. For treated $^5/_4$" fir and pine, you can usually space joists 16" on center. Install solid blocking between the joists at all supporting beams. Finally, fasten decking to the joists. You can face-nail, blind-nail or screw and plug wood decking in straight, diagonal, herringbone, or checkerboard patterns.

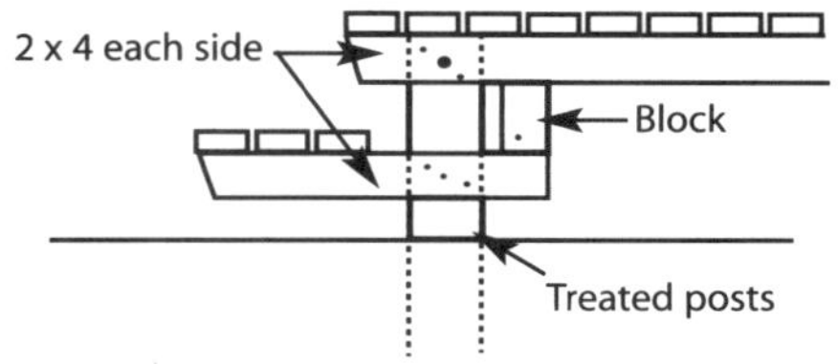

Figure 16-9

Single porch step supported on a treated post

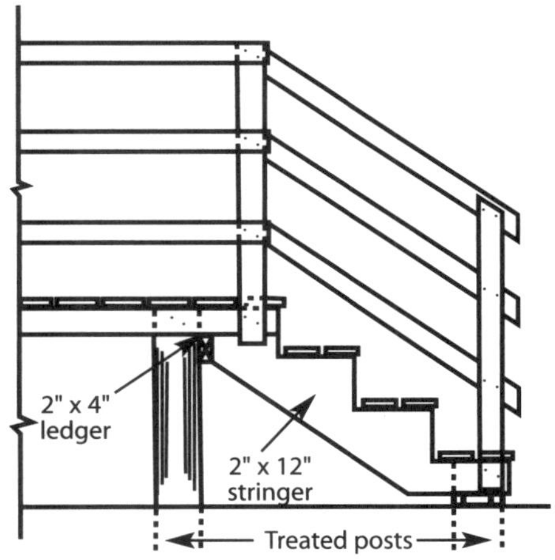

Figure 16-10

Step stringer supported by porch framing and posts

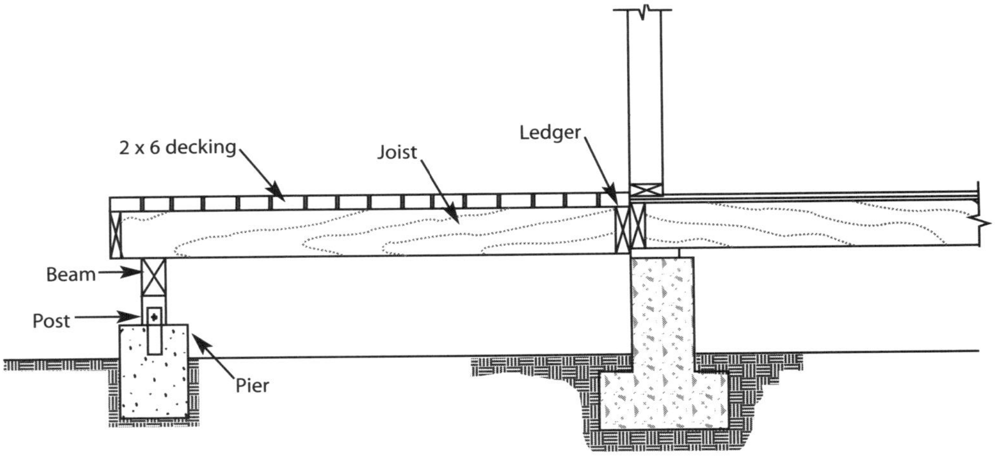

Figure 16-11
Wood deck attached to house

If you build or remodel a deck that is under a window, and the deck floor is less than 18" below the bottom edge of the window, that window may now be considered a hazardous location by the building code. If it is, the window must be special safety glass. Check for this when you estimate the job. That way if it applies, you can include the cost in your bid.

Choices in Porch and Deck Floor Plank

Western red cedar and redwood are naturally decay- and insect-resistant. The cost is higher than pressure-treated lumber, especially in areas remote from the west coast of the U.S. and Canada, where western red cedar and redwood are milled. For decks, use either construction common or construction heart redwood. "Merch" redwood is a lower grade ("merchantable") and has more knots and sapwood. Heart redwood is cinnamon-red and has more natural decay resistance. The sapwood is creamy yellow. The most common widths are 4" and 6". Lengths are usually 8', 12' and 16'. Both 5/4" and nominal 2" thickness are available. When left unfinished, both redwood and cedar turn a rustic gray after a few years. Apply water-repellent or stain to retard the color change. Varnishes and other film-forming coatings will crack and peel on a walking surface and aren't recommended. Both western red cedar and redwood are relatively soft and can be abraded under foot traffic.

Pressure-treated lumber makes sturdy and durable deck material. Most treated lumber sold for decking measures 5/4" (1 1/4") or nominal 2" thick and from 4" to 6" wide. Lengths are from 8' to 16'. Shorter lengths and narrower widths cost less per square foot of deck. Longer and wider pieces cost more but require less labor for installation. Pressure-treated wood decking costs less than other decking materials but should be fastened with galvanized nails or screws. All hangers and fasteners should be hot-dipped galvanized, such as Simpson Zmax. For 2" lumber, use

12d nails or 3" screws. For 5/4" decking, use 10d nails or 2 1/2" screws. Most pressure-treated lumber isn't pretty. It comes from the lumberyard tinted green from the residue of the ACQ (alkaline copper quaternary) or CA (copper azol) treatment. Soft areas in the wood absorb more stain than the harder portions. The result is a mottled appearance that won't win any prizes. Exposure to the sun and rain will fade the deck from green to honey to gray after a few years. Unless deck lumber has been coated already (Thompsonized), apply a water-repellent top coat immediately after installation.

All wood decks require occasional maintenance. Check for splits, warping, popped nails and splinters. Unless the owner likes the look of aged wood, he'll need to apply a coat of sealer every few years.

Synthetics include composites made from a combination of recycled wood fiber and recycled thermoplastics. These products (Fiberon is one name) look like real wood and have characteristics like real wood. For example, composites don't expand and contract as much as thermoplastics. But composites made from wood fiber will absorb moisture and tend to age like real wood. You can cut or drill composites with an ordinary hand drill or hand saw equipped with a carbide tip blade. Composites can be stained or painted and should never require maintenance. Deck planks made from rice hulls (Edeck®) are stronger than other composites and don't absorb moisture like deck lumber made from wood fiber. Edeck® requires installation with T-clips that permit expansion and contraction as the deck material heats and cools. Synthetics tend to be more slippery than real wood, especially when wet.

	Craft@Hrs	Unit	Material	Labor	Total	Sell
Porch and deck demolition. Includes demolition of framing, supports, flooring, posts, railing, wood stairs. Enclosed porch demolition also includes removal of roof framing, interior finish, gutters, downspouts and roofing. These figures assume use of hand tools (rather than excavation equipment). All debris piled on site. No salvage value assumed. Add the cost of debris hauling and dump fees. Add the cost of electric work, patching of exterior walls and patching of the roof, if required. Cost per square foot of floor demolished.						
Demolish enclosed wood-frame porch						
One story	BL@.250	SF	—	7.26	7.26	12.30
Two story	BL@.225	SF	—	6.54	6.54	11.10
Three story	BL@.196	SF	—	5.69	5.69	9.67
Demolish a screened porch built on a concrete slab. Includes demolition of one concrete or wood frame step to grade. Per square foot of floor area.						
Porch and one step (no slab demolition)	BL@.100	SF	—	2.91	2.91	4.95
Demolish a wood deck with railing or kneewall. Includes demolition of the wood deck and up to 7 steps to grade.						
Deck to 150 SF	BL@18.5	Ea	—	537.00	537.00	913.00
Add per SF for deck over 150 SF	BL@.100	SF	—	2.91	2.91	4.95
Porch and deck repair. Includes piling debris on site. No salvage value assumed.						
Remove and replace structural deck posts						
8' treated post, to 6" x 6"	B1@10.3	Ea	20.70	335.00	355.70	605.00
8' x 10" wood colonial column	B1@11.0	Ea	143.00	358.00	501.00	852.00
Built-up 8' to 10' post	B1@10.0	Ea	16.30	325.00	341.30	580.00
Cut post base and patch 6" x 6" post	B1@5.50	Ea	20.50	179.00	199.50	339.00
Deck and stair repairs						
Chip out decay in joist, reinforce with 2" x 6" to 2" x 10" sister joist per LF of joist	B1@.333	LF	1.77	10.80	11.98	20.40
Remove wood porch or deck floor and replace with 5/4" or 2" wood flooring per SF of flooring	B1@.166	SF	2.77	5.40	8.17	13.90
Remove and replace wood exterior open stairs, per SF of tread and landing	B1@.725	SF	19.00	23.60	42.60	72.40
Post and beam wood deck. Using pressure treated framing lumber with 4" x 6" posts set in concrete. Includes concrete, pressure treated posts, beams, joists 12" on center, deck plank as described, galvanized hardware, fasteners, 4' wide stairs from deck level to ground level, 36" high railing with balusters 4" on center and stair rail. Add for deck skirt and deck finish, if needed. Per square foot of floor area.						
Based on rectangular 12' x 10' deck with railing on four sides and stairway, 24" above ground level.						
With 5/4" pressure treated plank	B1@.260	SF	10.90	8.46	19.36	32.90
With 2" Thompsonized plank	B1@.260	SF	11.30	8.46	19.76	33.60
With 5/4" western red cedar plank	B1@.260	SF	11.30	8.46	19.76	33.60
With 5/4" Fiberon deck plank	B1@.260	SF	15.00	8.46	23.46	39.90
With 5/4" Edeck® plank	B1@.260	SF	20.10	8.46	28.56	48.60
Based on L-shaped 12' x 12' deck with railing on three sides and stairway, 120" above ground level.						
With 5/4" pressure treated plank	B1@.330	SF	13.30	10.70	24.00	40.80
With 2" Thompsonized plank	B1@.330	SF	13.90	10.70	24.60	41.80
With 5/4" western red cedar plank	B1@.330	SF	13.90	10.70	24.60	41.80
With 5/4" Fiberon deck plank	B1@.330	SF	18.40	10.70	29.10	49.50
With 5/4" Edeck® plank	B1@.330	SF	22.29	10.70	32.99	56.10
Add for framing to support a spa	B1@8.50	LS	114.00	277.00	391.00	665.00
Add for split level deck, 6" difference	B1@6.00	Ea	119.00	195.00	314.00	534.00
Add for stain and sealer finish	B1@.006	SF	.12	.20	.32	.54
Pressure treated construction select deck framing lumber. For use above ground. .40 treatment. Nominal sizes.						
2" x 4" x 8'	—	Ea	5.07	—	5.07	—
2" x 4" x 10'	—	Ea	7.11	—	7.11	—
2" x 4" x 16'	—	Ea	12.20	—	12.20	—

	Craft@Hrs	Unit	Material	Labor	Total	Sell
2" x 6" x 8'	—	Ea	9.15	—	9.15	—
2" x 6" x 10'	—	Ea	11.20	—	11.20	—
2" x 6" x 16'	—	Ea	18.30	—	18.30	—
2" x 8" x 8'	—	Ea	13.20	—	13.20	—
2" x 8" x 10'	—	Ea	16.30	—	16.30	—
2" x 8" x 12'	—	Ea	19.40	—	19.40	—
2" x 8" x 16'	—	Ea	25.50	—	25.50	—
2" x 10" x 8'	—	Ea	16.90	—	16.90	—
2" x 10" x 10'	—	Ea	19.97	—	19.97	—
2" x 10" x 12'	—	Ea	25.50	—	25.50	—
2" x 10" x 16'	—	Ea	34.24	—	34.24	—
2" x 12" x 8'	—	Ea	20.40	—	20.40	—
2" x 12" x 10'	—	Ea	25.50	—	25.50	—
2" x 12" x 12'	—	Ea	29.60	—	29.60	—
2" x 12" x 16'	—	Ea	42.80	—	42.80	—
4" x 4" x 6'	—	Ea	28.50	—	28.50	—
4" x 4" x 10'	—	Ea	16.30	—	16.30	—
4" x 4" x 12'	—	Ea	21.40	—	21.40	—
4" x 6" x 8'	—	Ea	21.40	—	21.40	—
4" x 6" x 10'	—	Ea	26.50	—	26.50	—
4" x 6" x 12'	—	Ea	31.60	—	31.60	—
4" x 6" x 16'	—	Ea	41.40	—	41.40	—
4" x 8" x 10'	—	Ea	33.60	—	33.60	—
4" x 8" x 12'	—	Ea	40.80	—	40.80	—
4" x 8" x 16'	—	Ea	57.30	—	57.30	—
4" x 8" x 20'	—	Ea	68.20	—	68.20	—
4" x 10" x 16'	—	Ea	71.20	—	71.20	—
4" x 12" x 12'	—	Ea	61.20	—	61.20	—
4" x 12" x 16'	—	Ea	83.60	—	83.60	—
4" x 12" x 20'	—	Ea	109.00	—	109.00	—
6" x 6" x 8'	—	Ea	28.50	—	28.50	—
6" x 6" x 10'	—	Ea	51.70	—	51.70	—

Pressure treated #2 grade deck posts. For use in contact with the ground. .60 treatment.

	Craft@Hrs	Unit	Material	Labor	Total	Sell
4" x 6" x 8'	—	Ea	17.30	—	17.30	—
4" x 6" x 10'	—	Ea	15.30	—	15.30	—
4" x 6" x 12'	—	Ea	30.60	—	30.60	—
4" x 6" x 14'	—	Ea	31.60	—	31.60	—
4" x 6" x 16'	—	Ea	41.40	—	41.40	—
6" x 6" x 8'	—	Ea	28.50	—	28.50	—
6" x 6" x 10'	—	Ea	23.40	—	23.40	—
6" x 6" x 12'	—	Ea	36.70	—	36.70	—
6" x 6" x 16'	—	Ea	45.90	—	45.90	—

Western red cedar deck lumber. Round edge, surfaced four sides.

	Craft@Hrs	Unit	Material	Labor	Total	Sell
5/4" x 4" x 8'	—	Ea	3.49	—	3.49	—
5/4" x 4" x 10'	—	Ea	4.35	—	4.35	—
5/4" x 4" x 12'	—	Ea	5.22	—	5.22	—
5/4" x 4" x 16'	—	Ea	6.97	—	7.97	—
5/4" x 6" x 8'	—	Ea	8.00	—	8.00	—
5/4" x 6" x 10'	—	Ea	10.00	—	10.00	—
5/4" x 6" x 12'	—	Ea	12.00	—	12.00	—
5/4" x 6" x 16'	—	Ea	16.00	—	16.00	—
2" x 6" x 8', Select	—	Ea	11.80	—	11.80	—

	Craft@Hrs	Unit	Material	Labor	Total	Sell
Pressure treated deck lumber. .40 treatment. Nominal sizes.						
5/4" x 4" x 8', Premium	—	Ea	4.05	—	4.05	—
5/4" x 4" x 10', Premium	—	Ea	5.38	—	5.38	—
5/4" x 4" x 12', Premium	—	Ea	7.11	—	7.11	—
5/4" x 4" x 16', Premium	—	Ea	9.50	—	9.50	—
5/4" x 6" x 8', Standard and Better	—	Ea	3.85	—	3.85	—
5/4" x 6" x 8', Premium	—	Ea	5.68	—	5.68	—
5/4" x 6" x 10', Standard and Better	—	Ea	4.87	—	4.87	—
5/4" x 6" x 10', Premium	—	Ea	7.72	—	7.72	—
5/4" x 6" x 12', Standard and Better	—	Ea	5.48	—	5.48	—
5/4" x 6" x 12', Premium	—	Ea	8.95	—	8.95	—
5/4" x 6" x 14', Standard and Better	—	Ea	6.60	—	6.60	—
5/4" x 6" x 14', Premium	—	Ea	12.74	—	12.74	—
5/4" x 6" x 16', Standard and Better	—	Ea	8.13	—	8.13	—
5/4" x 6" x 16', Premium	—	Ea	15.90	—	15.90	—
Thompsonized outdoor decking. Nominal sizes. Actual size is about 1/2" less when dry. Pressure treated and water repellent.						
2" x 4" x 8'	—	Ea	3.68	—	3.68	—
2" x 4" x 10'	—	Ea	4.92	—	4.92	—
2" x 4" x 12'	—	Ea	6.15	—	6.15	—
2" x 4" x 16'	—	Ea	8.63	—	8.63	—
2" x 6" x 8'	—	Ea	6.15	—	6.15	—
2" x 6" x 10'	—	Ea	8.07	—	8.07	—
2" x 6" x 12'	—	Ea	9.30	—	9.30	—
2" x 6" x 16'	—	Ea	12.40	—	12.40	—
Deck post surface-mount kit	—	Ea	10.90	—	10.90	—
Deck post, 4" x 4" x 48"	—	Ea	17.60	—	17.60	—
End cut preservative, quart	—	Ea	14.20	—	14.20	—
Composite decking. Composite deck made from recycled fiber and recycled polyethylene plastic. Will not split, warp, rot, cup, or crack. Slip-resistant when wet. Installs with conventional woodworking tools. Use stainless steel nails or screws for best results.						
5/4" x 6" x 8'	—	Ea	18.30	—	18.30	—
5/4" x 6" x 12'	—	Ea	27.50	—	27.50	—
5/4" x 6" x 16'	—	Ea	36.70	—	36.70	—
4" x 4" x 48" deck post	—	Ea	24.50	—	24.50	—
Post cap	—	Ea	6.48	—	6.48	—
Post bracket, joist mount	—	Ea	18.90	—	18.90	—
Post hardware for concrete	—	Ea	58.20	—	58.20	—
Post jacket top	—	Ea	5.04	—	5.04	—
Post sleeve, 4-1/2" x 4-1/2" x 42"	—	Ea	18.90	—	18.90	—
Bevel post cap	—	Ea	2.51	—	2.51	—
Top or bottom rail, 8' long	—	Ea	42.60	—	42.60	—
Square baluster, 32" long	—	Ea	3.91	—	3.91	—
Line rail hardware kit	—	Ea	16.90	—	16.90	—
Stair rail hardware kit	—	Ea	26.20	—	26.20	—
Edeck® composite deck. Composite deck made from rice hulls. Doesn't absorb moisture.						
5/4" x 6" x 12' Edeck® plank	—	Ea	22.20	—	22.20	—
5/4" x 6" x 16' Edeck® plank	—	Ea	33.00	—	33.00	—
5/4" x 6" x 20' Edeck® plank	—	Ea	33.40	—	33.40	—
2" x 6" x 10' Edeck® plank	—	Ea	24.10	—	24.10	—
Edeck® #8 x 3/4" screws, per box	—	Ea	7.28	—	7.28	—
Edeck® screw hole plug	—	Ea	10.60	—	10.60	—

	Craft@Hrs	Unit	Material	Labor	Total	Sell
Edeck® rail bracket	—	Ea	7.10	—	7.10	—
5" x 5" x 53" Edeck® rail post	—	Ea	26.60	—	26.60	—
5" x 5" peaked Edeck® post cap	—	Ea	2.58	—	2.58	—
E-Clips kit, pack of 100	—	Ea	34.00	—	34.00	—
1.5" x 1.5" x 2.5' Edeck® steel bar	—	Ea	8.22	—	8.22	—
Turned colonial-style wood porch post.						
4" x 4" x 8'	B1@1.18	Ea	49.00	38.40	87.40	149.00
4-1/4" x 4-1/4" x 8'	B1@1.18	Ea	67.30	38.40	105.70	180.00
6" x 6" x 8'	B1@1.18	Ea	80.00	38.40	118.40	201.00
8" x 8" x 8'	B1@1.18	Ea	100.00	38.40	138.40	235.00
Plinth for turned wood porch post.						
4" plinth	B1@.250	Ea	5.08	8.13	13.21	22.50
5" plinth	B1@.250	Ea	5.73	8.13	13.86	23.60
6" plinth	B1@.250	Ea	6.10	8.13	14.23	24.20

Structural aluminum porch column. White decorative column. Includes column, cap and base. Integral interlocking device permanently holds column together. Electrostatically painted. 6" load capacity is 16,000 pounds. 8" load capacity is 20,000 pounds.

	Craft@Hrs	Unit	Material	Labor	Total	Sell
6" x 8' column, round	B1@2.00	Ea	89.30	65.10	154.40	262.00
8" x 8' column, round	B1@2.00	Ea	120.00	65.10	185.10	315.00
6" x 6" x 8' column, square	B1@2.00	Ea	117.00	65.10	182.10	310.00

Round repair column. Colonial column can wrap around existing posts or be used as load-bearing column for new construction. Fluted design. Maintenance-free baked-on enamel finish.

	Craft@Hrs	Unit	Material	Labor	Total	Sell
6" x 8' round column	B1@2.00	Ea	98.00	65.10	163.10	277.00
8" x 8' round column	B1@2.00	Ea	123.00	65.10	188.10	320.00

Porch roof framing. Conventionally framed beams and joists. Per square foot of area covered. Figures in parentheses indicate board feet per square foot of ceiling including end joists, header joists and 5% waste. Based on a 200 SF job. Add the cost of support posts and finish roof.

	Craft@Hrs	Unit	Material	Labor	Total	Sell
2" x 4", Standard and Better grade						
16" centers (.59 BF per SF)	B1@.020	SF	.34	.65	.99	1.68
24" centers (.42 BF per SF)	B1@.014	SF	.24	.46	.70	1.19
2" x 6", Standard and Better grade						
16" centers (.88 BF per SF)	B1@.020	SF	.51	.65	1.16	1.97
24" centers (.63 BF per SF)	B1@.015	SF	.37	.49	.86	1.46

Add for porch support posts. Per post, including one sack of concrete per post. Add the cost of excavation and post anchors.

	Craft@Hrs	Unit	Material	Labor	Total	Sell
3" x 3" yellow pine, pressure treated, 10' long	B1@.500	Ea	10.50	16.30	26.80	45.60
4" x 4" pressure treated, 10' long	B1@.750	Ea	9.82	24.40	34.22	58.20
6" x 6" pressure treated, 10' long	B1@.750	Ea	25.80	24.40	50.20	85.30

Porch roofing. Cover over conventionally framed wood ceiling joists. Flat roof with built-up roofing. Using 1/2" CDX roof sheathing with 3 ply asphalt consisting of 2 plies of 15 lb. felt with 90 lb. mineral surface cap sheet and 2 hot mop coats of asphalt. Add the cost of roof framing.

	Craft@Hrs	Unit	Material	Labor	Total	Sell
Total flat roof assembly cost per SF	B1@.029	SF	1.43	.94	2.37	4.03

Deck railing. Typical 42"-high deck rail and baluster consisting of top and bottom rail and 2" x 2" baluster 5" OC with posts lag bolted to the edge of the deck. See detailed railing material costs below. Cost per linear foot of deck railing.

	Craft@Hrs	Unit	Material	Labor	Total	Sell
Pine, pressure treated	B1@.333	LF	3.27	10.80	14.07	23.90
Redwood, select heart	B1@.333	LF	8.07	10.80	18.87	32.10
Recycled plastic lumber	B1@.333	LF	19.50	10.80	30.30	51.50

	Craft@Hrs	Unit	Material	Labor	Total	Sell
Pressure treated deck posts. 4" x 4" x 4" high.						
Chamfered dado post	—	Ea	14.90	—	14.90	—
Treated cedar post	—	Ea	18.20	—	18.20	—
Laminated finial-ready post	—	Ea	14.90	—	14.90	—
Notched cedar post	—	Ea	20.30	—	20.30	—
V-groove cedar post	—	Ea	21.80	—	21.80	—
Turned deck post	—	Ea	20.40	—	20.40	—
Premium ball-top post	—	Ea	21.70	—	21.70	—
Pressure treated deck post cap.						
3-1/2" x 3-1/2" square cap	—	Ea	1.24	—	1.24	—
4" x 4" plain ball cap	—	Ea	3.34	—	3.34	—
4" x 4" royal coachman cap	—	Ea	3.41	—	3.41	—
4" x 4" Hampton flat cap	—	Ea	4.55	—	4.55	—
4" x 4" cannon ball post cap	—	Ea	7.29	—	7.29	—
4" x 4" Newport pyramid cap	—	Ea	7.11	—	7.11	—
4" x 4" Verona copper cap	—	Ea	12.20	—	12.20	—
4" x 4" Victoria copper cap	—	Ea	14.20	—	14.20	—
6" x 6" Verona copper cap	—	Ea	17.30	—	17.30	—
6" x 6" Victoria copper cap	—	Ea	20.80	—	20.80	—
Pressure treated deck rail.						
2" x 3" x 8' pine rail insert	—	Ea	6.09	—	6.09	—
2" x 4" x 8' molded rail	—	Ea	7.17	—	7.17	—
2" x 4" x 8' combo rail	—	Ea	12.50	—	12.50	—
Contemporary finial	—	Ea	3.54	—	3.54	—
Pressure treated deck rail baluster.						
2" x 2" x 32", square ends	—	Ea	.82	—	.82	—
2" x 2" x 36", square ends	—	Ea	.91	—	.91	—
2" x 2" x 42", beveled one end	—	Ea	.99	—	.99	—
2" x 2" x 48", beveled one end	—	Ea	1.52	—	1.52	—
2" x 2" x 42", beveled two ends	—	Ea	.99	—	.99	—
Pressure treated deck rail turned spindle.						
2" x 2" x 32" classic spindle	—	Ea	1.74	—	1.74	—
2" x 2" x 36" classic spindle	—	Ea	2.01	—	2.01	—
2" x 3" x 36" classic spindle	—	Ea	2.85	—	2.85	—
2" x 2" x 36" colonial spindle	—	Ea	4.18	—	4.18	—
3' x 3" x 36" jumbo spindle	—	Ea	6.09	—	6.09	—
Outdoor deck rail.						
2" x 4" x 8' redwood dado rail	—	Ea	15.60	—	15.60	—
2" x 5" x 8' primed handrail	—	Ea	12.00	—	12.00	—
2" x 4" x 8' deluxe cedar rail	—	Ea	16.60	—	16.60	—
2" x 6" x 8' western red cedar rail	—	Ea	18.60	—	18.60	—
2" x 4" x 8' western red cedar rail	—	Ea	18.90	—	18.90	—
Rail bracket kit, 4-piece	—	Ea	10.20	—	10.20	—
Rail angle bracket kit, 4-piece	—	Ea	20.40	—	20.40	—
Outdoor deck rail baluster.						
2" x 2" x 24" Douglas fir baluster	—	Ea	1.30	—	1.30	—
2" x 2" x 30" Douglas fir baluster	—	Ea	1.83	—	1.83	—
2" x 2" x 36" Douglas fir baluster	—	Ea	2.03	—	2.03	—
2" x 2" x 36", beveled redwood	—	Ea	2.01	—	2.01	—

	Craft@Hrs	Unit	Material	Labor	Total	Sell
2" x 2" x 42", one end beveled cedar	—	Ea	2.33	—	2.33	—
1-3/8" x 1-3/8" x 32", beveled redwood	—	Ea	2.97	—	2.97	—
1-3/8" x 1-3/8" x 48", redwood	—	Ea	3.56	—	3.56	—
8' Western red cedar fillet strip	—	Ea	2.60	—	2.60	—
Outdoor deck post cap.						
5" cedar post cap	—	Ea	3.21	—	3.21	—
4" x 4" suburban pyramid	—	Ea	6.09	—	6.09	—
Cedar ball top	—	Ea	5.41	—	5.41	—
4" x 4" cascade peak cap	—	Ea	7.07	—	7.07	—
Gothic post cap	—	Ea	7.11	—	7.11	—
Newport classic cap	—	Ea	9.15	—	9.15	—
4" x 4" Newport pyramid, cedar	—	Ea	11.20	—	11.20	—
4" x 4" copper high point cap	—	Ea	12.20	—	12.20	—
4" x 4" redwood ball cap	—	Ea	14.40	—	14.40	—
4" x 4" cedar ball cap	—	Ea	14.30	—	14.30	—

Porch or deck stairs. All stairs assume 12" tread (step) composed of two 2" x 6" lumber and riser. Add the cost of a concrete footing.

One or two steps high, box construction. Prices are per LF of step. A 4' wide stairway with 2 steps has 8 LF of step. Use 8 LF as the quantity.

	Craft@Hrs	Unit	Material	Labor	Total	Sell
Pine treads, open risers	B1@.270	LF	4.34	8.78	13.12	22.30
Pine treads, with pine risers	B1@.350	LF	7.79	11.40	19.19	32.60
Redwood treads, open risers	B1@.270	LF	4.14	8.78	12.92	22.00
Redwood treads, with redwood risers	B1@.350	LF	7.18	11.40	18.58	31.60
5/4" recycled plastic lumber treads, open risers	B1@.270	LF	9.25	8.78	18.03	30.70
5/4" recycled plastic lumber treads, with recycled plastic lumber risers	B1@.350	LF	20.10	11.40	31.50	53.60
2" x 6" recycled plastic lumber treads, open risers	B1@.270	LF	9.79	8.78	18.57	31.60
2" x 6" recycled plastic lumber treads, with recycled plastic lumber risers	B1@.350	LF	21.70	11.40	33.10	56.30

Three or more steps high, 36" wide. 3 stringers cut from 2" x 12" pressure treated lumber.

Cost per riser.

	Craft@Hrs	Unit	Material	Labor	Total	Sell
Pine treads, open risers	B1@.530	Ea	15.70	17.20	32.90	55.90
Pine treads, with pine risers	B1@.620	Ea	26.00	20.20	46.20	78.50
Redwood treads, open risers	B1@.530	Ea	15.00	17.20	32.20	54.70
Redwood treads, with redwood risers	B1@.620	Ea	24.20	20.20	44.40	75.50
5/4" recycled plastic lumber treads, open risers	B1@.810	Ea	31.60	26.30	57.90	98.40
5/4" recycled plastic lumber treads, with recycled plastic lumber risers	B1@1.05	Ea	64.10	34.20	98.30	167.00
2" x 6" recycled plastic lumber treads, open risers	B1@.810	Ea	33.20	26.30	59.50	101.00
2" x 6" recycled plastic lumber treads, with recycled plastic lumber risers	B1@1.05	Ea	68.90	34.20	103.10	175.00

	Craft@Hrs	Unit	Material	Labor	Total	Sell
Precut outdoor deck stair stringers. Cost per stringer.						
2" x 12" x 3 step, redwood	—	Ea	17.30	—	17.30	—
2" x 12" x 4 step, redwood	—	Ea	22.40	—	22.40	—
2" x 12" x 5 step, redwood	—	Ea	27.50	—	27.50	—
2 step, treated lumber	—	Ea	13.20	—	13.20	—
3 step, treated lumber	—	Ea	8.13	—	8.13	—
4 step, treated lumber	—	Ea	10.20	—	10.20	—
5 step, treated lumber	—	Ea	25.50	—	25.50	—
2 step, steel	—	Ea	14.10	—	14.10	—
3 step, steel	—	Ea	21.10	—	21.10	—
4 step, steel	—	Ea	26.70	—	26.70	—
EZ stair building hardware kit	—	Ea	9.18	—	9.18	—
Outdoor deck stair stepping. Pressure treated lumber.						
2" x12" x 48" tread	—	Ea	13.20	—	13.20	—

Porch and patio misting (fog) system. Installed 8' high on a porch exterior. Based on .008" stainless steel and brass nozzles installed each 18" along 3/8" Type L copper refrigeration tubing. Includes one electric operated 1 HP, 800 PSI, 1 GPM water pump, supply valve, auto drain valve, a wireless remote control switch, 50' of underground water connection line tapped to the existing service and trenching. Add the cost of 20 amp electrical connection.

	Craft@Hrs	Unit	Material	Labor	Total	Sell
Set up water pump, tap to existing water line	P1@5.00	Ea	1,600.00	176.00	1,776.00	3,020.00
Add per nozzle (each 18" of tube)	P1@.160	Ea	7.00	5.63	12.63	21.50

Porch enclosure screening. Includes medium quality aluminum frame stock, screening and insertion of vinyl screen spline in the frame. Cost per square foot of screen installed, either from a kneewall to the header or from the deck to the header.

	Craft@Hrs	Unit	Material	Labor	Total	Sell
With fiberglass screen wire	B1@.040	SF	1.18	1.30	2.48	4.22
With aluminum screen wire	B1@.040	SF	1.35	1.30	2.65	4.51
With solar bronze screen wire	B1@.040	SF	1.75	1.30	3.05	5.19
Porch enclosure screen wire. Per roll.						
Aluminum, 36" x 25'	—	Ea	20.40	—	20.40	—
Aluminum, 48" x 25'	—	Ea	26.50	—	26.50	—
Fiberglass, 72" x 100'	—	Ea	123.00	—	123.00	—
Solar screen, 36" x 25'	—	Ea	53.00	—	53.00	—
Solar screen, 48" x 25'	—	Ea	64.30	—	64.30	—

Aluminum porch screen frame stock. Framing for screen porch enclosure. With integrated groove ready to accept screen and flat spline when framework assembly is complete. Supports the top of a kickplate coil when attached between uprights. Bronze or white finish.

	Craft@Hrs	Unit	Material	Labor	Total	Sell
1" x 2", 8'	—	Ea	9.19	—	9.19	—
1" x 2", 10'	—	Ea	11.20	—	11.20	—
2" x 2", 8'	—	Ea	15.30	—	15.30	—
2" x 2", 10'	—	Ea	20.50	—	20.50	—
2" x 3", 8'	—	Ea	26.60	—	26.60	—

Angle anchor clip. Aluminum. Use to attach 2" x 3" uprights to 1" x 2" and concrete. Bronze or white finish.

	Craft@Hrs	Unit	Material	Labor	Total	Sell
2" x 2" x 1/8", bag of 25	—	Ea	9.93	—	9.93	—

Capri clip angle fastener. Aluminum. Attaches top of uprights to framework header with self-drilling screws. Bronze or white finish.

	Craft@Hrs	Unit	Material	Labor	Total	Sell
Clip with Fasteners	—	Ea	1.14	—	1.14	—

	Craft@Hrs	Unit	Material	Labor	Total	Sell
Castle clip U-fastener. Aluminum. Attaches kickplate rails to uprights with self-drilling screws. Bronze or white finish.						
Fastener	—	Ea	.81	—	.81	—
F-channel. Aluminum. Attaches to soffit or fascia. Brown or white finish.						
12' long	—	Ea	8.89	—	8.89	—
Kickplate channel. Aluminum. Trims edges of kickplate coil. Slips over edges of kickplate coil and is screwed to coil and framework. Bronze or white finish.						
10' long	—	Ea	4.66	—	4.66	—
Kickplate coil. Aluminum. Covers area under kickplate rails. Helps keep dirt from blowing and rain from splashing into the room. Bronze or white finish.						
16" x 16'	—	Ea	36.80	—	36.80	—
Screen enclosure screws. Weatherproof coating provides corrosion resistance. Use 2" screws to 1" x 2" of side wall to front wall corner upright. Use 3" screws to attach header to wood framing.						
SKU 457960,Size 8 self-drilling hex washer head, 9/16", bag of 100	—	Ea	4.66	—	4.66	—
SKU 457764,Size 10 hex head self-drilling hex washer head,						
2", bag of 25	—	Ea	2.31	—	2.31	—
3", bag of 25	—	Ea	2.66	—	2.66	—
Screen spline. Vinyl. Rolls into spline grooves of framework to hold screen in place.						
100' roll, flat black	—	Ea	6.10	—	6.10	—
Deck screws. Includes drive bit.						
Coated composite, 2-1/2", 5 pounds	—	Ea	30.60	—	30.60	—
Stainless steel, 3", 5 pounds	—	Ea	43.30	—	43.30	—
Stainless trim screw, 2-1/4", 1 pound	—	Ea	15.20	—	15.20	—

Aluminum Patio Covers

Standard aluminum patio cover. 12" wide V aluminum pans supported by scroll aluminum posts bolted to an existing concrete slab or wood deck. One side secured to an existing wall, roof or fascia. With rain gutter and downspout. Designed for 10 psf live load and 90 mph wind speed to comply with most building codes. To 10' high. Cost per square foot of cover (length times width) based on projection measured from the existing wall, roof or fascia.

	Craft@Hrs	Unit	Material	Labor	Total	Sell
10' projection	B1@.050	SF	6.43	1.63	8.06	13.70
12' projection	B1@.050	SF	6.17	1.63	7.80	13.30
14' projection	B1@.050	SF	6.05	1.63	7.68	13.10
16' projection	B1@.050	SF	5.79	1.63	7.42	12.60

Embossed aluminum patio cover. 12" wide W .018 aluminum pans supported by 3" x 8" aluminum header beam and 2" x 6" aluminum rafters on 3" square aluminum posts bolted to an existing concrete slab or wood deck. One side attached to an existing wall, roof or fascia. With rain gutter and downspout. Designed for 10 psf live load and 90 mph wind speed to comply with most building codes. To 12' high. Cost per square foot of cover (length times width) based on projection measured from the existing wall, roof or fascia.

	Craft@Hrs	Unit	Material	Labor	Total	Sell
10' projection	B1@.050	SF	10.30	1.63	11.93	20.30
12' projection	B1@.050	SF	9.65	1.63	11.28	19.20
14' projection	B1@.050	SF	9.00	1.63	10.63	18.10
16' to 20' projection	B1@.050	SF	8.36	1.63	9.99	17.00
Add for 2" square lattice in place of W pans	—	SF	1.93	—	1.93	—

	Craft@Hrs	Unit	Material	Labor	Total	Sell
Add for decorative 8" diameter Roman column	—	Ea	354.00	—	354.00	—
Add for 30 psf snow load	—	%	20.0	—	—	—
Add for free-standing patio cover including footings	—	%	50.0	—	—	—

Pool and Spa Replastering

Pool replaster preparation. Drain pool, remove pool fixtures for reinstallation, chip out existing deteriorated plaster with pneumatic tools and dispose of debris. Add the cost of a compressor, hose and chipping hammers. Per linear foot of pool perimeter, based on an average depth of 4 feet.

	Craft@Hrs	Unit	Material	Labor	Total	Sell
Chip out existing plaster surface	B8@.333	LF	.36	11.60	11.96	20.30
Add per LF of cut around tile or obstructions	B8@.200	LF	—	6.98	6.98	11.90

Pool replastering and resetting pool fixtures. Includes setting new ceramic trim tile on steps, repairing rebar, if needed, and replacing the main drain cover and rope anchors. Add the cost of replacing or regrouting water line tile, if required. Per linear foot of pool perimeter, based on an average depth of 4 feet, and using Colorquartz or similar plaster aggregate.

	Craft@Hrs	Unit	Material	Labor	Total	Sell
White plaster	B6@.500	LF	14.30	15.70	30.00	51.00
Pastel plaster aggregate	B6@.500	LF	20.70	15.70	36.40	61.90
Deep color plaster aggregate	B6@.500	LF	33.70	15.70	49.40	84.00

Replace pop-up cleaning heads. Cost is per head replaced. Includes chipping away plaster collar, cutting off old head with a saber saw, installing new riser and temporary cap, and performing system water pressure test. Based on a job with 15 quick-clean pop-up heads.

	Craft@Hrs	Unit	Material	Labor	Total	Sell
Per head installed	P1@.200	Ea	75.00	7.03	82.03	139.00

Spa replaster preparation. Drain spa, remove pool fittings for reinstallation, chip out the existing deteriorated plaster with pneumatic tools and dispose of debris. Add the cost of a compressor, hose and chipping hammers. Cost per linear foot of the maximum spa diameter, based on an average depth of 3 feet. Deduct 15% when a pool and spa are refinished as part of the same job.

	Craft@Hrs	Unit	Material	Labor	Total	Sell
Chip out existing plaster surface	B8@.430	LF	.72	15.00	15.72	26.70
Add per LF of cut around tile or obstructions	B8@.200	LF	—	6.98	6.98	11.90

Spa replastering and resetting spa fixtures. Includes setting new ceramic trim tile on steps, repairing rebar if needed, and replacing the main drain cover. Add the cost of replacing or regrouting water line tile, if required. Per linear foot of maximum spa diameter, based on an average depth of 3 feet, and using Colorquartz or similar plaster aggregate. Deduct 15% when a pool and spa are refinished as part of the same job.

	Craft@Hrs	Unit	Material	Labor	Total	Sell
White plaster	B6@4.30	LF	30.30	135.00	165.30	281.00
Pastel plaster aggregate	B6@4.30	LF	41.30	135.00	176.30	300.00
Deep color plaster aggregate	B6@4.30	LF	48.30	135.00	183.30	312.00

Painting and Finishing

17

Nothing improves the looks of an old house like a new coat of paint. But before deciding to repaint, consider washing the surface or spot painting. Excessive paint buildup increases the chance of chipping and peeling from incompatible paint layers. Paint film that's too thick is likely to crack. The only remedy for cross-grain cracking or intercoat peeling may be to strip the surface down to bare wood. Areas sheltered from the sun, such as eaves and porch ceilings, don't need to be painted as often as exposed areas. Instead of repainting the entire surface, spot paint only in areas showing the most wear. Matching colors is easy with modern optical paint-matching equipment. Most paint suppliers can match nearly any sample you bring in.

Guidelines for Recoating

Nearly all surfaces in a home can be recoated successfully with the right surface preparation. The challenge is to identify what's on the surface so you know what treatment to use, and then to select the appropriate coating.

- *Kalsomine* (also called *calcimine*) is a whitewash made from zinc oxide and glue. Kalsomine turns to powder as it ages, which makes it an unsuitable surface for any type of coating. Either remove the existing surface by sandblasting or cover the surface, such as with new siding.

- *Grease* falls in the same category as kalsomine. No paint will adhere to walls or ceilings coated with grease. This is a common problem in kitchens. Remove all grease before recoating. If paint has been applied over grease, remove all coatings down to the bare surface.

- *Non-compatible paints* can result in intercoat peeling. Adding another coat on top of peeling paint will only compound the problem. You'll generally need to remove the existing paint. If the surface you're to paint is incompatible with the paint you're going to use, apply a coat of primer between the non-compatible paints. That will usually solve the problem.

- *Stains* caused by water damage, smoke or a foreign substance need special attention. Be sure the surface is dry and clean, then coat with a primer-sealer such as Kilz®.
- *Plaster cracks* should be repaired with fiberglass tape and joint compound before recoating. Don't think that paint will fill the cracks for you.
- *Varnish* should be cleaned with a strong tri-sodium phosphate solution and then painted soon after drying. Heavily-alligatored varnish must be removed before coating.
- *Metals* should be wire brushed to remove loose material and then coated with a metal primer.
- *Hardwood floors* of oak, birch, beech, and maple are usually finished by applying two coats of sealer. Sand lightly or buff with steel wool between coats. Then apply a coat of paste wax and buff. Maintain the finish by rewaxing. For a high-gloss finish, instead of applying two coats of sealer, make the second coat varnish.
- *Wood trim and paneling* are usually finished with a clear wood sealer or a stain-sealer combination and then top-coated with at least one additional coat of sealer or varnish. Sand lightly between sealing and top-coating. For a deeper finish, apply one coat of high-gloss varnish followed by a coat of semi-gloss varnish.
- *Painted wood trim* requires a primer or undercoat and then a coat of acrylic latex, either flat or semi-gloss. Paint with at least some gloss resists fingerprints better and can be cleaned with soap and water.
- *Kitchen and bathroom* walls need a coat of semi-gloss enamel. This type of finish wears well, is easy to clean, and resists moisture.
- *Drywall*, in rooms other than the kitchen and bathroom, needs a coat of flat latex. New drywall, since it's highly porous, should have a coat of primer before the final coat of latex paint.

Paint Color and Gloss

Paint comes in an infinite range of colors. Your paint store probably stocks a few of the basic colors, such as white and black, but can mix any color you want. You buy the tinting base and the paint technician adds just the right amount of pigment to yield the color needed. Any color you select will be the same price. However, you'll pay more for high-gloss paints and less for low-gloss *(flat)* paints. High-gloss paint resists fingerprints and holds up better when scrubbed. Use high-gloss in kitchens and bathrooms where you need a washable surface. High-gloss paint tends to reveal surface imperfections and any application defects. Flat paints cost a little less than high-gloss and hide flaws better, but don't stand up as well to scrubbing. Use a flat paint in bedrooms, halls and living rooms. Eggshell paint has slightly more gloss than flat paint. Semi-gloss and satin finishes have more gloss than eggshell, but are still not as glossy as high-gloss.

Paint is also classified by its reflectance value — how much light is reflected off the surface. Light colors have a light reflectance value (LRV) of 50 or more and are a good choice for interiors. Darker colors are better for exteriors.

Paint selection isn't the place to try and save a few bucks on materials. Top quality paints made by well-known and respected brands will yield better results. Lesser quality paints may require additional coats, which will end up costing more in the end.

Exterior Trim

Expect wood trim on the exterior of an older home to be in poor condtion. If the finish is worn but the surface is smooth, another coat of paint should be enough. Trim that's damaged or badly chipped may need to be replaced. Matching old trim isn't easy. Patterns popular 50 or 100 years ago may no longer be available at your local lumberyard. You can have trim pieces custom made to order, but the cost may greater than replacing all similar trim.

Wood Preservatives

Remove blotchy discoloration and signs of rain spatter on wood by wire brushing. Brush *with* the grain. Then treat the surface with a water-repellent preservative. Applying a coat of water-repellent preservative every few years will prevent nearly all decay.

Wood treated with water-soluble preservatives can be painted the same as untreated wood. The coating may not last as long as with untreated wood, but there's no major difference, especially when the treated wood has weathered for several months.

Penetrating Stains

If the surface hasn't been coated previously, consider a penetrating stain. These stains make a good finish for most wood surfaces, especially rough-sawn, textured wood and knotty wood that would be difficult to paint. Penetrating stain doesn't form a film on the surface. Because the surface isn't sealed, the coating won't crack, peel or blister. Use pigmented penetrating stains to add shades of brown, green, red, or gray to the wood. To avoid lap marks, coat full lengths of siding or trim without stopping. On smooth surfaces, apply a single coat. Two or three years later, when the surface has weathered, apply another coat of penetrating stain. This time the surface will be rougher and will absorb more stain. That will protect the wood longer — up to eight years.

On rough-sawn or weathered surfaces, apply two coats of penetrating stain. Both coats should be applied within a few hours. Don't let the first coat dry before applying the second. A dry first coat will keep the second from penetrating into the wood. Rough wood that's been double-coated with penetrating stain has a finish life of ten years or more.

Refinish a stained wood surface when the color has faded and bare wood is beginning to show through. Prepare the surface by brushing with steel wool. Brush in the direction of the grain. Then hose down with water. When the surface is thoroughly dry, apply another coat of stain.

Paint Application Methods

The three most common paint application techniques are brush, roller and spray. Brushwork takes the most time per square foot coated. Productivity is highest when a spray rig can be put to good use, but not all paints are suitable for spraying. Sprayers generally require the paint to be thinned, and some paints are designated "Do not thin." Applying paint by roller takes less time per square foot than brushing, but more time than spraying. The cost of a painting task will differ depending on the application method used. That's why labor estimates and material costs are separated in this chapter. Labor estimates come first, followed by material costs. The labor estimates include figures for each of the three application methods.

The best tool for applying paint depends on what's being coated and what's not supposed to get coated. Spray painting isn't as precise as painting with a brush or even a roller. Spray painting requires much more attention to masking and covering adjacent surfaces, such as the floor, doors, windows, hardware and trim. When painting a wall or ceiling with brush or roller, it may be enough to lay down several drop cloths. That takes only a minute or two. Coating the same surface with a spray rig will require masking off everything in the room that's not supposed to get coated — and then closing off openings to adjacent rooms. When spray painting,

Brush, Roller or Spray?

- ❖ Spray application is usually impractical (takes longer than brushing or rolling) in small rooms with multiple openings and installed hardware or fixtures.
- ❖ Bedrooms and bathrooms are seldom candidates for spray painting, especially if the home is occupied.
- ❖ Even in larger rooms, such as a kitchen or living room, spray painting isn't usually practical in an occupied home.
- ❖ Building exteriors are good candidates for spray painting, especially stucco, masonry and panel exteriors where less masking is required.
- ❖ Interiors of most occupied homes are usually painted with a brush (hand cut in) and roller. The labor tables that follow identify this work as "roll and brush."
- ❖ It's safe to assume that recoating railings, trim, doors, windows, metals and cabinets will be done with a brush.

as much as 80 percent of the job may be masking and protecting adjacent surfaces. That's especially true in an occupied building where it's impractical to remove every ornament and piece of hardware — and where overspray poses a major risk to the dwelling contents. If spray painting a large room can be expected to take an hour, expect painters will spend two to four hours on masking and protection from overspray.

Brush, roller and spray don't necessarily produce equivalent results. For example, paint brushed on a metal surface makes better contact with the metal than paint sprayed on the same surface. Contact and adhesion affect the life of the coating, especially on metal. So most metal surfaces are coated with brush or roller, at least in home improvement work. Likewise, when applying semi-gloss to doors, most painters prefer to brush coat four sides (front, back, hinge edge and latch edge) while the door is hung in the opening. That results in a much smoother finish. Even when paint is sprayed on, some jobs are *back-rolled*. The spray coat is smoothed out with a roller immediately after application.

When a spray job is complete, good practice requires that someone inspect the surface carefully, looking for voids, imperfections and painter's holidays (areas missed). Assume that about 10 percent of the surface will require brush or roller touch-up after a spray application is complete.

When working with a roller, most jobs require hand *cut in*. Paint is applied by brush at corners, where different materials intersect and around obstructions such as hardware, vents, cases, trim and fixtures. When hand cut-in is complete, the remainder of the surface is rolled.

Additional Costs

None of the figures in this chapter includes moving furniture, fixtures, wall hangings, objects of art, etc. All assume that loose contents are moved to the center of the room or removed entirely before painting begins. Reduce productivity by 30 to 50 percent if painters will be moving furniture, removing and rehanging pictures, and removing and replacing drapery valances.

Straight, uninterrupted walls with square corners are quick and easy paint jobs. Painting cut-up surfaces with alcoves, soffits, light wells and offsets takes longer. For a cut-up job, allow 20 to 40 percent more time than indicated in the following tables.

These estimates assume landscaping materials (shrubs and trees) are cut back or pulled away from exterior surfaces before exterior painting begins.

	Craft@Hrs	Unit	Material	Labor	Total	Sell

Labor Estimates for Painting

Cost to repaint a home interior. Typical costs per 100 square feet of floor for roller and brush application of a single coat to walls and ceilings in all rooms of an occupied dwelling, including bathrooms and closets. Add the cost of coating cabinets, trim, doors and window trim. Includes minimum surface preparation, spackle of minor defects in wallboard or plaster, masking of adjacent surfaces and priming of stained or discolored surfaces. Use these figures for preliminary estimates and to check completed bids. These estimates equate to 3 painter-hours and 1 gallon of paint per 100 square feet of floor.

	Craft@Hrs	Unit	Material	Labor	Total	Sell
Repaint a home interior, per 100 SF of floor	PT@3.00	CSF	19.70	110.00	129.70	220.00

Cost to repaint a home exterior. Typical costs per 100 square feet of floor for spray painting a single coat on exterior walls and soffits of an occupied dwelling. Add the cost of painting trim, doors and windows. Includes a waterblast of the exterior, repair of minor surface defects in siding and trim, and priming of stained, cracked or peeling surfaces. Use these figures for preliminary estimates and to check completed bids. These figures equate to 2 painter-hours and 1 gallon of paint per 100 square feet of floor.

	Craft@Hrs	Unit	Material	Labor	Total	Sell
Repaint a home exterior, per 100 SF of floor	PT@2.00	CSF	20.80	73.50	94.30	160.00

High time difficulty factors for surface preparation and painting. Painting takes longer when heights exceed 8' above the floor. Productivity is lower when application requires a roller pole or wand on a spray gun or when work is done from a ladder or scaffold. When painting above 8', apply the following factors.

Add 30% to the area for heights from 8' to 13' (multiply by 1.3)
Add 60% to the area for heights from 13' to 17' (multiply by 1.6)
Add 90% to the area for heights from 17' to 19' (multiply by 1.9)
Add 120% to the area for heights from 19' to 21' (multiply by 2.2)

Labor for acid wash of gutters and downspouts. For heights above one story, use the high time difficulty factors above.

	Craft@Hrs	Unit	Material	Labor	Total	Sell
Per 100 linear feet	PT@1.05	CLF	—	38.60	38.60	65.60

Labor to burn off paint. Using a heat gun or torch. For heights above 8', use the high time difficulty factors above.

	Craft@Hrs	Unit	Material	Labor	Total	Sell
Beveled wood siding, per 100 square feet	PT@3.33	CSF	—	122.00	122.00	207.00
Exterior plain surfaces, per 100 square feet	PT@2.50	CSF	—	91.90	91.90	156.00
Exterior trim, per 100 square feet	PT@5.00	CSF	—	184.00	184.00	313.00
Interior plain surfaces, per 100 square feet	PT@4.00	CSF	—	147.00	147.00	250.00
Interior trim, per 100 square feet	PT@6.67	CSF	—	245.00	245.00	417.00

Labor for caulking. Add extra time for digging out and removing old caulk and residue, protecting adjacent surfaces and masking.

	Craft@Hrs	Unit	Material	Labor	Total	Sell
1/8" gap, per 100 linear feet	PT@1.54	CLF	—	56.60	56.60	96.20
1/4" gap, per 100 linear feet	PT@1.82	CLF	—	66.90	66.90	114.00
3/8" gap, per 100 linear feet	PT@2.22	CLF	—	81.60	81.60	139.00
1/2" gap, per 100 linear feet	PT@2.63	CLF	—	96.70	96.70	164.00

Labor for masking.

	Craft@Hrs	Unit	Material	Labor	Total	Sell
Masking per 100 linear feet	PT@2.50	CLF	—	91.90	91.90	156.00

Labor for cutting-in.

	Craft@Hrs	Unit	Material	Labor	Total	Sell
Cutting-in per 100 linear feet	PT@2.75	CLF	—	101.00	101.00	172.00

Labor for interior sanding. For heights above 8', use the high time difficulty factors above.

	Craft@Hrs	Unit	Material	Labor	Total	Sell
Sand flat wall areas before first coat, per 100 square feet	PT@.333	CSF	—	12.20	12.20	20.70
Sand and putty before second flat coat, per 100 square feet	PT@.500	CSF	—	18.40	18.40	31.30
Sand and putty before second enamel coat, per 100 square feet	PT@.800	CSF	—	29.40	29.40	50.00

	Craft@Hrs	Unit	Material	Labor	Total	Sell
Sand and putty cabinets before second coat, per 100 square feet	PT@.667	CSF	—	24.50	24.50	41.70
Sand and putty bookshelves before second coat, per 100 square feet	PT@.800	CSF	—	29.40	29.40	50.00
Sand bookshelves with light grit before third coat, per 100 square feet	PT@.444	CSF	—	16.30	16.30	27.70
Sand with light grit before third enamel coat, per 100 square feet	PT@.714	CSF	—	26.30	26.30	44.70

Labor for exterior sanding. For heights above 8', use the high time difficulty factors above.

	Craft@Hrs	Unit	Material	Labor	Total	Sell
Sand and putty siding and trim, before second coat, per 100 square feet	PT@.500	CSF	—	18.40	18.40	31.30
Sand and putty trim only, before second coat, per 100 square feet	PT@.909	CSF	—	33.40	33.40	56.80
Sand trim only with light grit, before third coat, per 100 square feet	PT@.571	CSF	—	21.00	21.00	35.70

Labor for sandblasting. Sandblast only. Add the cost of equipment, protecting adjacent surfaces, mobilization and cleanup. For heights above 8', use the high time difficulty factors above.

	Craft@Hrs	Unit	Material	Labor	Total	Sell
Brush-off sandblast to remove cement base paint, per 100 square feet	PT@.571	CSF	—	21.00	21.00	35.70
Brush-off sandblast to remove oil or latex paint, per 100 square feet	PT@.800	CSF	—	29.40	29.40	50.00

Labor for stripping paint. For heights above 8', use the high time difficulty factors above.

	Craft@Hrs	Unit	Material	Labor	Total	Sell
Strip, remove, or bleach, per 100 square feet	PT@1.43	CSF	—	52.60	52.60	89.40
Strip varnish with light-duty liquid remover, per 100 square feet	PT@2.86	CSF	—	105.00	105.00	179.00
Strip varnish with heavy-duty liquid remover, per 100 square feet	PT@3.33	CSF	—	122.00	122.00	207.00
Paint removal with light-duty liquid remover, per 100 square feet	PT@3.33	CSF	—	122.00	122.00	207.00
Paint removal with heavy-duty liquid remover, per 100 square feet	PT@4.00	CSF	—	147.00	147.00	250.00

Labor for wash down. For heights above 8', use the high time difficulty factors above.

	Craft@Hrs	Unit	Material	Labor	Total	Sell
Interior smooth flat wall, per 100 square feet	PT@.500	CSF	—	18.40	18.40	31.30
Wash and touch up a smooth flat interior wall, per 100 square feet	PT@.625	CSF	—	23.00	23.00	39.10
Wash down a rough surface flat interior wall, per 100 square feet	PT@.667	CSF	—	24.50	24.50	41.70
Wash and touch up a rough surface flat interior wall, per 100 square feet	PT@1.05	CSF	—	38.60	38.60	65.60
Wash down an interior enamel wall, per 100 square feet	PT@.465	CSF	—	17.10	17.10	29.10
Wash down and touch up an interior enamel wall, per 100 square feet	PT@.885	CSF	—	32.50	32.50	55.30
Wash down interior enamel trim, per 100 square feet	PT@.667	CSF	—	24.50	24.50	41.70

	Craft@Hrs	Unit	Material	Labor	Total	Sell
Wash and touch up interior enamel trim, per 100 square feet	PT@.833	CSF	—	30.60	30.60	52.00
Wash interior varnish trim, per 100 square feet	PT@.513	CSF	—	18.90	18.90	32.10
Wash and touch up interior varnish trim, per 100 square feet	PT@.714	CSF	—	26.30	26.30	44.70
Wash interior varnished floors, per 100 square feet	PT@.476	CSF	—	17.50	17.50	29.80
Wash and touch up interior varnish floors, per 100 square feet	PT@.654	CSF	—	24.00	24.00	40.80
Wash interior smooth plaster, per 100 square feet	PT@.571	CSF	—	21.00	21.00	35.70
Wash and touch up interior smooth plaster, per 100 square feet	PT@.714	CSF	—	26.30	26.30	44.70
Wash interior sand finish plaster, per 100 square feet	PT@.741	CSF	—	27.20	27.20	46.20
Wash and touch up interior sand finish plaster, per 100 square feet	PT@.909	CSF	—	33.40	33.40	56.80

Labor for waterblasting (pressure washing). Waterblast to remove deteriorated, cracked or flaking paint from wood, concrete, brick, block, plaster or stucco surfaces. Add the cost of waterblast equipment, protecting adjacent surfaces and mobilization. For heights above 8', use the high time difficulty factors. These production rates will apply on larger jobs, such as an entire home. For smaller jobs, use a minimum charge of 3 or 4 hours to move on, set up, waterblast, clean up and move off. Production rates are based on medium-pressure equipment (2,500 PSI). High-pressure equipment (3,000 PSI) will increase productivity 10% to 50%. Low-pressure equipment (1,700 PSI) will reduce productivity by 10% to 50%. Rates assume a 1/4"-diameter nozzle. Productivity will be much higher when the surface is not deteriorated or contaminated with grease or grime. Note that the surface must dry thoroughly before coating begins. Per 100 square feet of surface cleaned.

	Craft@Hrs	Unit	Material	Labor	Total	Sell
Light blast (250 SF per hour)	PT@.011	SF	—	.40	.40	.68
Most jobs (150 SF per hour)	PT@.018	SF	—	.66	.66	1.12
Difficult jobs (75 SF per hour)	PT@.037	SF	—	1.36	1.36	2.31

Labor for painting baseboard. Paint application only. Add the cost of surface preparation, protecting adjacent surfaces, mobilization, cleanup and callbacks.

	Craft@Hrs	Unit	Material	Labor	Total	Sell
Brush 1 coat, 700 linear feet per gallon, per 100 linear feet	PT@.833	CLF	—	30.60	30.60	52.00
Roll 1 coat, 750 linear feet per gallon, per 100 linear feet	PT@.083	CLF	—	3.05	3.05	5.19

Labor for painting bookcases and shelves. Paint application only. Add the cost of surface preparation, protecting adjacent surfaces, mobilization, cleanup and callbacks.

	Craft@Hrs	Unit	Material	Labor	Total	Sell
Roll and brush undercoat, 280 square feet per gallon, per 100 square feet	PT@3.33	CSF	—	122.00	122.00	207.00
Roll and brush finish coat, 340 square feet per gallon, per 100 square feet	PT@2.50	CSF	—	91.90	91.90	156.00
Spray undercoat, 133 square feet per gallon, per 100 square feet	PT@.606	CSF	—	22.30	22.30	37.90
Spray finish coat, 158 square feet per gallon, per 100 square feet	PT@.400	CSF	—	14.70	14.70	25.00

	Craft@Hrs	Unit	Material	Labor	Total	Sell
Labor for painting cabinets. Paint application only. Add the cost of surface preparation, protecting adjacent surfaces, mobilization, cleanup and callbacks.						
Brush cabinet back, 275 square feet per gallon, per 100 square feet	PT@.769	CSF	—	28.30	28.30	48.10
Roll and brush cabinet face, 250 square feet per gallon, per 100 square feet of face	PT@1.08	CSF	—	39.70	39.70	67.50
Spray cabinet face, 113 square feet per gallon, per 100 square feet of face	PT@.714	CSF	—	26.30	26.30	44.70
Labor for applying acoustic ceiling texture. Application only. Add the cost of surface preparation, protecting adjacent surfaces, mobilization, cleanup and callbacks. For work more than 8' above floor level, use the high time difficulty factors.						
Spray prime coat for acoustic textured ceiling, 90 square feet per gallon, per 100 square feet	PT@.333	CSF	—	12.20	12.20	20.70
Spray finish coat for acoustic textured ceiling, 170 square feet per gallon, per 100 square feet	PT@.222	CSF	—	8.16	8.16	13.90
Spray stipple finish texture paint, light coverage, dry powder mix, 7.5 square feet per gallon, per 100 square feet	PT@.400	CSF	—	14.70	14.70	25.00
Labor for painting tongue-and-groove ceiling — brush. Paint application only. Add the cost of surface preparation, protecting adjacent surfaces, mobilization, cleanup and callbacks. For heights above 8', use the high time difficulty factors.						
Brush 1st coat of water-base flat latex, 288 square feet per gallon, per 100 square feet	PT@1.54	CSF	—	56.60	56.60	96.20
Brush additional coats of water-base flat latex, 338 square feet per gallon, per 100 square feet	PT@1.33	CSF	—	48.90	48.90	83.10
Brush prime coat of water-base sealer, 288 square feet per gallon, per 100 square feet	PT@1.43	CSF	—	52.60	52.60	89.40
Brush 1st finish coat of water-base enamel, 288 square feet per gallon, per 100 square feet	PT@1.33	CSF	—	48.90	48.90	83.10
Brush additional coats of water-base enamel, 363 square feet per gallon, per 100 square feet	PT@1.11	CSF	—	40.80	40.80	69.40
Labor for painting tongue-and-groove ceiling — roll. Paint application only. Add the cost of surface preparation, protecting adjacent surfaces, mobilization, cleanup and callbacks. For heights above 8', use the high time difficulty factors.						
Roll 1st coat of water-base flat latex, 263 square feet per gallon, per 100 square feet	PT@.769	CSF	—	28.30	28.30	48.10
Roll additional coats of water-base flat latex, 313 square feet per gallon, per 100 square feet	PT@.645	CSF	—	23.70	23.70	40.30

	Craft@Hrs	Unit	Material	Labor	Total	Sell
Roll prime coat of water-base sealer, 263 square feet per gallon, per 100 square feet	PT@.833	CSF	—	30.60	30.60	52.00
Roll 1st finish coat of water-base enamel, 313 square feet per gallon, per 100 square feet	PT@.690	CSF	—	25.40	25.40	43.20
Roll additional coats of water-base enamel, 338 square feet per gallon, per 100 square feet	PT@.526	CSF	—	19.30	19.30	32.80

Labor for painting tongue-and-groove ceiling — spray. Paint application only. Add the cost of surface preparation, protecting adjacent surfaces, mobilization, cleanup and callbacks. For heights above 8', use the high time difficulty factors.

	Craft@Hrs	Unit	Material	Labor	Total	Sell
Spray 1st coat of water-base flat latex, 155 square feet per gallon, per 100 square feet	PT@.278	CSF	—	10.20	10.20	17.30
Spray additional coats of water-base flat latex, 225 square feet per gallon, per 100 square feet	PT@.213	CSF	—	7.83	7.83	13.30
Spray prime coat of water-base sealer, 155 square feet per gallon, per 100 square feet	PT@.263	CSF	—	9.67	9.67	16.40
Spray 1st finish coat of water-base enamel, 225 square feet per gallon, per 100 square feet	PT@.222	CSF	—	8.16	8.16	13.90
Spray additional coats of water-base enamel, 300 square feet per gallon, per 100 square feet	PT@.182	CSF	—	6.69	6.69	11.40

Labor for staining tongue-and-groove ceiling. Stain application only. Add the cost of surface preparation, protecting adjacent surfaces, mobilization, cleanup and callbacks. For heights above 8', use the high time difficulty factors.

	Craft@Hrs	Unit	Material	Labor	Total	Sell
Roll and brush water-base semi-transparent stain, 275 square feet per gallon, per 100 square feet	PT@.417	CSF	—	15.30	15.30	26.00

Labor for painting closet shelf and pole. Paint application only. Add the cost of surface preparation, protecting adjacent surfaces, mobilization, cleanup and callbacks.

	Craft@Hrs	Unit	Material	Labor	Total	Sell
Brush 1 coat of water-base undercoat, 70 linear feet per gallon, per 100 linear feet	PT@4.55	CLF	—	167.00	167.00	284.00
Brush split coat of water-base undercoat and enamel, 70 linear feet per gallon, per 100 linear feet	PT@4.76	CLF	—	175.00	175.00	298.00
Brush finish coat of water-base enamel, 70 linear feet per gallon, per 100 linear feet	PT@5.00	CLF	—	184.00	184.00	313.00

	Craft@Hrs	Unit	Material	Labor	Total	Sell
Labor for painting door frame and trim only. Paint application only. Opening refers to both sides of a doorway. Add the cost of surface preparation, protecting adjacent surfaces, mobilization, cleanup and callbacks.						
Brush 1 coat of water-base undercoat, 28 openings per gallon, per opening	PT@.063	Ea	—	2.32	2.32	3.94
Brush split coat of water-base undercoat and enamel, 31 openings per gallon, per opening	PT@.071	Ea	—	2.61	2.61	4.44
Brush water-base enamel, 31 openings per gallon, per opening	PT@.077	Ea	—	2.83	2.83	4.81
Labor for painting flush exterior door, frame and trim. Paint application only. Add the cost of surface preparation, protecting adjacent surfaces, mobilization, cleanup and callbacks.						
Roll and brush 2 coats of water-base exterior enamel, 6 doors per gallon, per door	PT@.400	Ea	—	14.70	14.70	25.00
Roll and brush 2 coats of oil-base exterior enamel, 6 doors per gallon, per door	PT@.400	Ea	—	14.70	14.70	25.00
Brush 2 coats of polyurethane, 4.5 doors per gallon, per door	PT@.600	Ea	—	22.10	22.10	37.60
Labor for painting exterior French door, frame and trim. Paint application only. Add preparation time for sanding, puttying, protecting adjacent surfaces and masking or waxing glass panes.						
Roll and brush 2 coats of water-base exterior enamel, 10 doors per gallon, per door	PT@.800	Ea	—	29.40	29.40	50.00
Roll and brush 2 coats of oil-base exterior enamel, 12 doors per gallon, per door	PT@.800	Ea	—	29.40	29.40	50.00
Brush 2 coats of polyurethane, 7.5 doors per gallon, per door	PT@1.30	Ea	—	47.80	47.80	81.30
Labor for painting exterior louver door, frame and trim. Paint application only. Add the cost of surface preparation, protecting adjacent surfaces, mobilization, cleanup and callbacks.						
Roll and brush 2 coats of water-base exterior enamel, 6 doors per gallon, per door	PT@1.10	Ea	—	40.40	40.40	68.70
Brush 2 coats of polyurethane, 4.5 doors per gallon, per door	PT@1.50	Ea	—	55.20	55.20	93.80
Labor for painting exterior panel (entry) door, frame and trim. Paint application only. Add the cost of surface preparation, protecting adjacent surfaces, mobilization, cleanup and callbacks.						
Roll and brush 2 coats of water-base exterior enamel, 3 doors per gallon, per door	PT@1.10	Ea	—	40.40	40.40	68.70
Brush 2 coats of polyurethane, 3 doors per gallon, per door	PT@1.50	Ea	—	55.20	55.20	93.80
Labor for painting interior flush door, frame and trim — roll and brush. Paint application only. Add the cost of surface preparation, protecting adjacent surfaces, mobilization, cleanup and callbacks.						
Roll and brush 1 coat of water-base undercoat, 13 doors per gallon, per door	PT@.300	Ea	—	11.00	11.00	18.70
Roll and brush 1st finish coat of water-base enamel, 12.5 doors per gallon, per door	PT@.250	Ea	—	9.19	9.19	15.60
Roll and brush additional finish coats of water-base enamel, 13.5 doors per gallon, per door	PT@.200	Ea	—	7.35	7.35	12.50

	Craft@Hrs	Unit	Material	Labor	Total	Sell
Labor for painting interior flush door, frame and trim — spray. New installations only. Includes paint application only. Add the cost of surface preparation, protecting adjacent surfaces, mobilization, cleanup and callbacks.						
Spray 1 coat of water-base undercoat, 16 doors per gallon, per door	PT@.090	Ea	—	3.31	3.31	5.63
Spray 1 coat of oil-base undercoat, 16 doors per gallon, per door	PT@.090	Ea	—	3.31	3.31	5.63
Spray 1st finish coat of water-base enamel, 17 doors per gallon, per door	PT@.080	Ea	—	2.94	2.94	5.00
Spray additional coats of water-base enamel, 18 doors per gallon, per door	PT@.070	Ea	—	2.57	2.57	4.37
Labor for painting interior French door, frame and trim. Paint application only. Add preparation time for sanding, puttying, protecting adjacent surfaces and masking or waxing glass panes.						
Roll and brush 1 coat of water-base undercoat, 13 doors per gallon, per door	PT@.380	Ea	—	14.00	14.00	23.80
Roll and brush 1st finish coat of water-base enamel, 14 doors per gallon, per door	PT@.350	Ea	—	12.90	12.90	21.90
Roll and brush additional finish coats of water-base enamel, 15 doors per gallon, per door	PT@.330	Ea	—	12.10	12.10	20.60
Labor for painting interior louver door, frame and trim — roll and brush. Paint application only. Add the cost of surface preparation, protecting adjacent surfaces, mobilization, cleanup and callbacks.						
Roll and brush 1 coat of water-base undercoat, 7 doors per gallon, per door	PT@.540	Ea	—	19.90	19.90	33.80
Roll and brush 1st finish coat of water-base enamel, 8 doors per gallon, per door	PT@.420	Ea	—	15.40	15.40	26.20
Roll and brush additional finish coats of water-base enamel, 9 doors per gallon, per door	PT@.300	Ea	—	11.00	11.00	18.70
Labor for painting interior louver door, frame and trim — spray. New installations only. Includes paint application only. Add the cost of surface preparation, protecting adjacent surfaces, mobilization, cleanup and callbacks.						
Spray 1 coat of water-base undercoat, 11 doors per gallon, per door	PT@.140	Ea	—	5.15	5.15	8.76
Spray 1st finish coat of water-base enamel, 12 doors per gallon, per door	PT@.110	Ea	—	4.04	4.04	6.87
Spray additional finish coats of water-base enamel, 13 doors per gallon, per door	PT@.090	Ea	—	3.31	3.31	5.63
Labor for painting interior panel door, frame and trim — roll and brush. Paint application only. Add the cost of surface preparation, protecting adjacent surfaces, mobilization, cleanup and callbacks.						
Roll and brush 1 coat of water-base undercoat, 7 doors per gallon, per door	PT@.330	Ea	—	12.10	12.10	20.60
Roll and brush 1st finish coat of water-base enamel, 10 doors per gallon, per door	PT@.300	Ea	—	11.00	11.00	18.70
Roll and brush additional finish coats of water-base enamel, 11 doors per gallon, per door	PT@.250	Ea	—	9.19	9.19	15.60

	Craft@Hrs	Unit	Material	Labor	Total	Sell
Labor for painting interior panel door, frame and trim — spray. New installations only. Includes paint application only. Add the cost of surface preparation, protecting adjacent surfaces, mobilization, cleanup and callbacks.						
Spray 1 coat of water-base undercoat, 14 doors per gallon, per door	PT@.110	Ea	—	4.04	4.04	6.87
Spray 1st finish coat of water-base enamel, 15 doors per gallon, per door	PT@.080	Ea	—	2.94	2.94	5.00
Spray additional finish coats of water-base enamel, 16 doors per gallon, per door	PT@.080	Ea	—	2.94	2.94	5.00
Labor for staining 2" x 4" fascia. Front side and bottom edge only. Stain application only. Add the cost of surface preparation, protecting adjacent surfaces, mobilization, cleanup and callbacks. For heights above 8', use the high time difficulty factors.						
Roll and brush each coat of solid-body stain, 130 linear feet per gallon, per 100 linear feet	PT@.488	CLF	—	17.90	17.90	30.40
Roll and brush each coat of semi-transparent stain, 150 linear feet per gallon, per 100 linear feet	PT@.444	CLF	—	16.30	16.30	27.70
Spray each coat of solid-body stain, 100 linear feet per gallon, per 100 linear feet	PT@.308	CLF	—	11.30	11.30	19.20
Spray each coat of semi-transparent stain, 115 linear feet per gallon, per 100 linear feet	PT@.286	CLF	—	10.50	10.50	17.90
Labor for staining 2" x 6" or 2" x 10" fascia. Front side and bottom edge only. Stain application only. Add the cost of surface preparation, protecting adjacent surfaces, mobilization, cleanup and callbacks. For heights above 8', use the high time difficulty factors.						
Roll and brush each coat of solid-body stain, 110 linear feet per gallon, per 100 linear feet	PT@.571	CLF	—	21.00	21.00	35.70
Roll and brush each coat of semi-transparent stain, 130 linear feet per gallon, per 100 linear feet	PT@.513	CLF	—	18.90	18.90	32.10
Spray each coat of solid-body stain, 80 linear feet per gallon, per 100 linear feet	PT@.333	CLF	—	12.20	12.20	20.70
Spray each coat of semi-transparent stain, 95 linear feet per gallon, per 100 linear feet	PT@.319	CLF	—	11.70	11.70	19.90
Labor for staining 2" x 12" fascia. Front side and bottom edge only. Stain application only. Add the cost of surface preparation, protecting adjacent surfaces, mobilization, cleanup and callbacks. For heights above 8', use the high time difficulty factors.						
Roll and brush each coat of solid-body stain, 75 linear feet per gallon, per 100 linear feet	PT@.769	CLF	—	28.30	28.30	48.10
Roll and brush each coat of semi-transparent stain, 95 linear feet per gallon, per 100 linear feet	PT@.667	CLF	—	24.50	24.50	41.70

	Craft@Hrs	Unit	Material	Labor	Total	Sell
Spray each coat of solid-body stain, 50 linear feet per gallon, per 100 linear feet	PT@.380	CLF	—	14.00	14.00	23.80
Spray each coat of semi-transparent stain, 65 linear feet per gallon, per 100 linear feet	PT@.347	CLF	—	12.80	12.80	21.80

Labor for staining picket fence. Area is the fence length times the fence height times two. Stain application only. Add the cost of surface preparation, protecting adjacent surfaces, mobilization, cleanup and callbacks.

	Craft@Hrs	Unit	Material	Labor	Total	Sell
Roll and brush 1st coat of solid-body or semi-transparent stain, 343 square feet per gallon, per 100 square feet	PT@.690	CSF	—	25.40	25.40	43.20
Roll and brush additional coats of solid-body or semi-transparent stain, 388 square feet per gallon, per 100 square feet	PT@.444	CSF	—	16.30	16.30	27.70
Spray 1st coat of solid-body or semi-transparent stain, 275 square feet per gallon, per 100 square feet	PT@.200	CSF	—	7.35	7.35	12.50
Spray additional coats of solid-body or semi-transparent stain, 325 square feet per gallon, per 100 square feet	PT@.167	CSF	—	6.14	6.14	10.40

Labor for painting concrete floor — roll. Paint application only. Add the cost of surface preparation, protecting adjacent surfaces, mobilization, cleanup and callbacks.

	Craft@Hrs	Unit	Material	Labor	Total	Sell
Roll and brush 1st coat of water-base masonry paint, 263 square feet per gallon, per 100 square feet	PT@.459	CSF	—	16.90	16.90	28.70
Roll and brush 2nd coat of water-base masonry paint, 325 square feet per gallon, per 100 square feet	PT@.373	CSF	—	13.70	13.70	23.30
Roll and brush 3rd or additional coats of water-base masonry paint, 350 square feet per gallon, per 100 square feet	PT@.333	CSF	—	12.20	12.20	20.70
Roll and brush each coat of 1-part water-base epoxy, 488 square feet per gallon, per 100 square feet	PT@.444	CSF	—	16.30	16.30	27.70
Roll and brush each coat of 2-part water-base epoxy, 488 square feet per gallon, per 100 square feet	PT@.481	CSF	—	17.70	17.70	30.10
Roll and brush 1st coat of penetrating oil stain, 425 square feet per gallon, per 100 square feet	PT@.400	CSF	—	14.70	14.70	25.00
Roll and brush 2nd coat of penetrating oil stain, 475 square feet per gallon, per 100 square feet	PT@.290	CSF	—	10.70	10.70	18.20

	Craft@Hrs	Unit	Material	Labor	Total	Sell
Roll and brush 3rd or additional coats of penetrating oil stain, 500 square feet per gallon, per 100 square feet	PT@.261	CSF	—	9.60	9.60	16.30

Labor for painting concrete floor — spray. Paint application only. Add the cost of surface preparation, protecting adjacent surfaces, mobilization, cleanup and callbacks.

	Craft@Hrs	Unit	Material	Labor	Total	Sell
Spray 1st coat of water-base masonry paint, 163 square feet per gallon, per 100 square feet	PT@.111	CSF	—	4.08	4.08	6.94
Spray 2nd coat of water-base masonry paint, 263 square feet per gallon, per 100 square feet	PT@.100	CSF	—	3.68	3.68	6.26
Spray 3rd or additional coats of water-base masonry paint, 313 square feet per gallon, per 100 square feet	PT@.091	CSF	—	3.35	3.35	5.70

Labor for painting wood deck or floor — roll and brush. Paint application only. Add the cost of surface preparation, protecting adjacent surfaces, mobilization, cleanup and callbacks.

	Craft@Hrs	Unit	Material	Labor	Total	Sell
Roll and brush water-base prime undercoat, 425 square feet per gallon, per 100 square feet	PT@.228	CSF	—	8.38	8.38	14.20
Roll and brush porch and deck enamel, 450 square feet per gallon, per 100 square feet	PT@.216	CSF	—	7.94	7.94	13.50
Brush 1-part water-base epoxy, 400 square feet per gallon, per 100 square feet	PT@.400	CSF	—	14.70	14.70	25.00
Roll 2-part epoxy, 225 square feet per gallon, per 100 square feet	PT@.444	CSF	—	16.30	16.30	27.70

Labor for staining wood deck or floor — brush. Stain application only. Add the cost of surface preparation, protecting adjacent surfaces, mobilization, cleanup and callbacks.

	Craft@Hrs	Unit	Material	Labor	Total	Sell
Brush 1st coat oil-base wiping stain, wipe and fill wood floor, 475 square feet per gallon, per 100 square feet	PT@.400	CSF	—	14.70	14.70	25.00
Brush 2nd coat oil-base stain, wipe and fill wood floor, 500 square feet per gallon, per 100 square feet	PT@.235	CSF	—	8.64	8.64	14.70
Brush 3rd or additional coats of oil-base stain, wipe and fill wood floor, 525 square feet per gallon, per 100 square feet	PT@.222	CSF	—	8.16	8.16	13.90
Brush 1st coat of sanding sealer on maple or pine floor, 450 square feet per gallon, per 100 square feet	PT@.250	CSF	—	9.19	9.19	15.60
Brush additional coats of sanding sealer on maple or pine floor, 525 square feet per gallon, per 100 square feet	PT@.222	CSF	—	8.16	8.16	13.90

	Craft@Hrs	Unit	Material	Labor	Total	Sell
Brush 1st coat of sanding sealer on oak floor, 500 square feet per gallon, per 100 square feet	PT@.235	CSF	—	8.64	8.64	14.70
Brush additional coats of sanding sealer on oak floor, 600 square feet per gallon, per 100 square feet	PT@.190	CSF	—	6.99	6.99	11.90
Brush 1st coat of penetrating stain, wax and wipe wood floor, 525 square feet per gallon, per 100 square feet	PT@.400	CSF	—	14.70	14.70	25.00
Brush additional coats of penetrating stain, wax and wipe wood floor, 525 square feet per gallon, per 100 square feet	PT@.333	CSF	—	12.20	12.20	20.70
Wax and polish wood floors by hand, per 100 square feet	PT@.500	CSF	—	18.40	18.40	31.30
Wax and polish wood floors by machine, per 100 square feet	PT@.222	CSF	—	8.16	8.16	13.90

Labor for painting gutters and downspouts. Paint application only. Add the cost of surface preparation, protecting adjacent surfaces, mobilization, cleanup and callbacks. For heights above 8', use the high time difficulty factors.

	Craft@Hrs	Unit	Material	Labor	Total	Sell
Brush prime coat rust inhibitor on metal gutters, 375 linear feet per gallon, per 100 linear feet	PT@1.11	CLF	—	40.80	40.80	69.40
Brush 1st finish coat on metal gutters, 400 linear feet per gallon, per 100 linear feet	PT@.909	CLF	—	33.40	33.40	56.80
Brush additional finish coats on metal gutters, 425 linear feet per gallon, per 100 linear feet	PT@.769	CLF	—	28.30	28.30	48.10
Brush prime coat rust inhibitor on metal downspouts, 225 linear feet per gallon, per 100 linear feet	PT@2.86	CLF	—	105.00	105.00	179.00

Labor for applying masonry block filler on brick masonry. Filler application only. Add the cost of surface preparation, protecting adjacent surfaces, mobilization, cleanup and callbacks. For heights above 8', use the high time difficulty factors.

	Craft@Hrs	Unit	Material	Labor	Total	Sell
Brush one coat of masonry block filler, 65 square feet per gallon, per 100 square feet	PT@.800	CSF	—	29.40	29.40	50.00
Roll one coat of masonry block filler, 60 square feet per gallon, per 100 square feet	PT@.465	CSF	—	17.10	17.10	29.10
Spray one coat of masonry block filler, 55 square feet per gallon, per 100 square feet	PT@.190	CSF	—	6.99	6.99	11.90

	Craft@Hrs	Unit	Material	Labor	Total	Sell
Labor for painting smooth brick masonry — brush. Paint application only. Add the cost of surface preparation, protecting adjacent surfaces, mobilization, cleanup and callbacks. For heights above 8', use the high time difficulty factors.						
Brush 1st coat of water-base masonry paint, 275 square feet per gallon, per 100 square feet	PT@.444	CSF	—	16.30	16.30	27.70
Brush additional coats of water-base masonry paint, 300 square feet per gallon, per 100 square feet	PT@.364	CSF	—	13.40	13.40	22.80
Labor for painting smooth brick masonry — roll. Paint application only. Add the cost of surface preparation, protecting adjacent surfaces, mobilization, cleanup and callbacks. For heights above 8', use the high time difficulty factors.						
Roll 1st coat of water-base masonry paint, 213 square feet per gallon, per 100 square feet	PT@.286	CSF	—	10.50	10.50	17.90
Roll additional coats of water-base masonry paint, 250 square feet per gallon, per 100 square feet	PT@.250	CSF	—	9.19	9.19	15.60
Roll 1st coat of clear hydro sealer, 150 square feet per gallon, per 100 square feet	PT@.444	CSF	—	16.30	16.30	27.70
Roll additional coats of clear hydro sealer, 190 square feet per gallon, per 100 square feet	PT@.400	CSF	—	14.70	14.70	25.00
Labor for painting smooth brick masonry — spray. Paint application only. Add the cost of surface preparation, protecting adjacent surfaces, mobilization, cleanup and callbacks. For heights above 8', use the high time difficulty factors.						
Spray 1st coat of water-base masonry paint, 225 square feet per gallon, per 100 square feet	PT@.133	CSF	—	4.89	4.89	8.31
Spray additional coats of water-base masonry paint, 238 square feet per gallon, per 100 square feet	PT@.121	CSF	—	4.45	4.45	7.57
Spray 1st coat of clear hydro sealer, 100 square feet per gallon, per 100 square feet	PT@.125	CSF	—	4.60	4.60	7.82
Spray additional coats of clear hydro sealer, 138 square feet per gallon, per 100 square feet	PT@.111	CSF	—	4.08	4.08	6.94
Labor for painting concrete masonry units (CMU) — brush. Paint application only. Add the cost of surface preparation, protecting adjacent surfaces, mobilization, cleanup and callbacks. For heights above 8', use the high time difficulty factors.						
Brush 1st coat of water-base masonry paint, 88 square feet per gallon, per 100 square feet	PT@.769	CSF	—	28.30	28.30	48.10
Brush additional coats of water-base masonry paint, 168 square feet per gallon, per 100 square feet	PT@.476	CSF	—	17.50	17.50	29.80

	Craft@Hrs	Unit	Material	Labor	Total	Sell
Brush 1st coat of 2-part clear epoxy, 98 square feet per gallon, per 100 square feet	PT@.870	CSF	—	32.00	32.00	54.40
Brush additional coats of 2-part clear epoxy, 188 square feet per gallon, per 100 square feet	PT@.526	CSF	—	19.30	19.30	32.80
Brush 1st coat of clear hydro sealer, 80 square feet per gallon, per 100 square feet	PT@.667	CSF	—	24.50	24.50	41.70
Brush additional coats of clear hydro sealer, 110 square feet per gallon, per 100 square feet	PT@.364	CSF	—	13.40	13.40	22.80

Labor for painting concrete masonry units (CMU) — roll. Paint application only. Add the cost of surface preparation, protecting adjacent surfaces, mobilization, cleanup and callbacks. For heights above 8', use the high time difficulty factors.

	Craft@Hrs	Unit	Material	Labor	Total	Sell
Roll 1st coat of water-base masonry paint, 78 square feet per gallon, per 100 square feet	PT@.333	CSF	—	12.20	12.20	20.70
Roll additional coats of water-base masonry paint, 143 square feet per gallon, per 100 square feet	PT@.308	CSF	—	11.30	11.30	19.20
Roll 1st coat 2-part epoxy coating, 88 square feet per gallon, per 100 square feet	PT@.364	CSF	—	13.40	13.40	22.80
Roll additional costs of 2-part epoxy coating, 160 square feet per gallon, per 100 square feet	PT@.333	CSF	—	12.20	12.20	20.70

Labor for painting concrete masonry units (CMU) — spray. Paint application only. Add the cost of surface preparation, protecting adjacent surfaces, mobilization, cleanup and callbacks. For heights above 8', use the high time difficulty factors.

	Craft@Hrs	Unit	Material	Labor	Total	Sell
Spray 1st coat of water-base masonry paint, 78 square feet per gallon, per 100 square feet	PT@.143	CSF	—	5.26	5.26	8.94
Spray additional coats of water-base masonry paint, 133 square feet per gallon, per 100 square feet	PT@.125	CSF	—	4.60	4.60	7.82
Spray 1st coat of 2-part clear epoxy, 68 square feet per gallon, per 100 square feet	PT@.167	CSF	—	6.14	6.14	10.40
Spray additional coats of 2-part clear epoxy, 130 square feet per gallon, per 100 square feet	PT@.143	CSF	—	5.26	5.26	8.94
Spray 1st coat of clear hydro sealer, 50 square feet per gallon, per 100 square feet	PT@.143	CSF	—	5.26	5.26	8.94
Spray additional coats of clear hydro sealer, 75 square feet per gallon, per 100 square feet	PT@.125	CSF	—	4.60	4.60	7.82

	Craft@Hrs	Unit	Material	Labor	Total	Sell

Labor for painting molding or trim. Paint application only. Add the cost of surface preparation, protecting adjacent surfaces, mobilization, cleanup and callbacks. For heights above 8', use the high time difficulty factors.

	Craft@Hrs	Unit	Material	Labor	Total	Sell
Brush prime coat, 600 linear feet per gallon, per 100 linear feet	PT@.488	CLF	—	17.90	17.90	30.40
Brush split coat of 1/2 undercoat and 1/2 enamel, 675 linear feet per gallon, per 100 linear feet	PT@.741	CLF	—	27.20	27.20	46.20
Brush 1st finish coat of enamel, 675 linear feet per gallon, per 100 linear feet	PT@.625	CLF	—	23.00	23.00	39.10
Brush 2nd or additional coats of enamel, 675 linear feet per gallon, per 100 linear feet	PT@.667	CLF	—	24.50	24.50	41.70
Brush stipple finish enamel on smooth exterior molding, per 100 linear feet	PT@1.11	CLF	—	40.80	40.80	69.40
Brush glazing or mottling over enamel on smooth molding, 900 linear feet per gallon, per 100 linear feet	PT@1.54	CLF	—	56.60	56.60	96.20

Labor for painting roof overhang to 30" wide — roll and brush. Paint application only. Add the cost of surface preparation, protecting adjacent surfaces, mobilization, cleanup and callbacks. See the overhang difficulty factors that follow.

	Craft@Hrs	Unit	Material	Labor	Total	Sell
Roll and brush 1st coat of solid-body stain, 185 square feet per gallon, per 100 square feet	PT@.800	CSF	—	29.40	29.40	50.00
Roll and brush 2nd coat of solid-body stain, 240 square feet per gallon, per 100 square feet	PT@.541	CSF	—	19.90	19.90	33.80
Roll and brush additional coats of solid body stain, 270 square feet per gallon, per 100 square feet	PT@.444	CSF	—	16.30	16.30	27.70

Labor for painting roof overhang to 30" wide — spray. Paint application only. Add the cost of surface preparation, protecting adjacent surfaces, mobilization, cleanup and callbacks. See the overhang difficulty factors that follow.

	Craft@Hrs	Unit	Material	Labor	Total	Sell
Spray 1st coat of semi-transparent stain, 125 square feet per gallon, per 100 square feet	PT@.286	CSF	—	10.50	10.50	17.90
Spray 2nd coat of semi-transparent stain, 163 square feet per gallon, per 100 square feet	PT@.222	CSF	—	8.16	8.16	13.90
Spray 3rd or additional coats of semi-transparent stain, 213 square feet per gallon, per 100 square feet	PT@.190	CSF	—	6.99	6.99	11.90

	Craft@Hrs	Unit	Material	Labor	Total	Sell
Labor for painting large roof overhang over 30" wide — roll and brush. Paint application only. Add the cost of surface preparation, protecting adjacent surfaces, mobilization, cleanup and callbacks. See the overhang difficulty factors below.						
Roll and brush 1st coat of solid-body stain, 210 square feet per gallon, per 100 square feet	PT@.556	CSF	—	20.40	20.40	34.70
Roll and brush 2nd coat of solid-body stain, 275 square feet per gallon, per 100 square feet	PT@.444	CSF	—	16.30	16.30	27.70
Roll and brush 3rd coat of solid-body stain, 280 square feet per gallon, per 100 square feet	PT@.357	CSF	—	13.10	13.10	22.30
Labor for painting large roof overhang over 30" wide — spray. Paint application only. Add the cost of surface preparation, protecting adjacent surfaces, mobilization, cleanup and callbacks. See the overhang difficulty factors below.						
Spray 1st coat of solid-body or semi-transparent stain, 90 square feet per gallon, per 100 square feet	PT@.167	CSF	—	6.14	6.14	10.40
Spray 2nd coat of solid-body or semi-transparent stain, 163 square feet per gallon, per 100 square feet	PT@.154	CSF	—	5.66	5.66	9.62
Spray 3rd or additional coats of solid-body or semi-transparent stain, 213 square feet per gallon, per 100 square feet	PT@.143	CSF	—	5.26	5.26	8.94

Overhang difficulty factors for eaves and cornices. The labor estimates above are for roof overhang with boxed eaves (no exposed rafter tails) on new single-story buildings.

Add 50% for repainting overhang or painting overhang with exposed rafter tails or second color (multiply the area by 1.5).

Add 50% for painting second-story overhang from scaffolding (multiply the area by 1.5).

Add 100% for second story with no scaffolding or exposed rafter tails (multiply the area by 2.0)

	Craft@Hrs	Unit	Material	Labor	Total	Sell
Labor for painting exterior plaster or stucco — brush. Paint application only. Add the cost of surface preparation, protecting adjacent surfaces, mobilization, cleanup and callbacks. For heights above 8', use the high time difficulty factors.						
Brush water-base masonry paint, 213 square feet per gallon, per 100 square feet	PT@.613	CSF	—	22.50	22.50	38.30
Brush 2nd coat water-base masonry paint, 230 square feet per gallon, per 100 square feet	PT@.613	CSF	—	22.50	22.50	38.30
Brush 3rd or additional coat water-base masonry paint, 245 square feet per gallon, per 100 square feet	PT@.578	CSF	—	21.30	21.30	36.20
Brush 1st coat of clear hydro sealer, 163 square feet per gallon, per 100 square feet	PT@.667	CSF	—	24.50	24.50	41.70
Brush 2nd coat of clear hydro sealer, 188 square feet per gallon, per 100 square feet	PT@.500	CSF	—	18.40	18.40	31.30

	Craft@Hrs	Unit	Material	Labor	Total	Sell
Labor for painting exterior plaster or stucco — roll. Paint application only. Add the cost of surface preparation, protecting adjacent surfaces, mobilization, cleanup and callbacks. For heights above 8', use the high time difficulty factors.						
Roll 1st water-base masonry prime coat, 175 square feet per gallon, per 100 square feet	PT@.366	CSF	—	13.50	13.50	23.00
Roll 2nd water-base masonry coat, 200 square feet per gallon, per 100 square feet	PT@.313	CSF	—	11.50	11.50	19.60
Roll 3rd or additional water-base masonry coats, 225 square feet per gallon, per 100 square feet	PT@.294	CSF	—	10.80	10.80	18.40
Roll 1st coat of clear hydro sealer, 138 square feet per gallon, per 100 square feet	PT@.275	CSF	—	10.10	10.10	17.20
Roll 2nd or additional coats of clear hydro sealer, 163 square feet per gallon, per 100 square feet	PT@.235	CSF	—	8.64	8.64	14.70
Labor for painting exterior plaster or stucco — spray. Paint application only. Add the cost of surface preparation, protecting adjacent surfaces, mobilization, cleanup and callbacks. For heights above 8', use the high time difficulty factors.						
Spray 1st water-base masonry prime coat, 120 square feet per gallon, per 100 square feet	PT@.148	CSF	—	5.44	5.44	9.25
Spray 2nd water-base masonry coat, 150 square feet per gallon, per 100 square feet	PT@.125	CSF	—	4.60	4.60	7.82
Spray 3rd or additional water-base masonry coat, 165 square feet per gallon, per 100 square feet	PT@.118	CSF	—	4.34	4.34	7.38
Spray 1st coat of clear hydro sealer, 113 square feet per gallon, per 100 square feet	PT@.143	CSF	—	5.26	5.26	8.94
Spray 2nd or additional coats of clear hydro sealer, 138 square feet per gallon, per 100 square feet	PT@.121	CSF	—	4.45	4.45	7.57
Labor for painting decorative wood handrail — brush. Paint application only. Add the cost of surface preparation, protecting adjacent surfaces, mobilization, cleanup and callbacks.						
Brush undercoat or enamel, 110 linear feet per gallon, per 100 linear feet	PT@2.86	CLF	—	105.00	105.00	179.00
Labor for painting 2' x 4' shutters or blinds — brush. Paint application only. Add the cost of surface preparation, protecting adjacent surfaces, mobilization, cleanup and callbacks. For heights above 8', use the high time difficulty factors.						
Brush undercoat, 11 shutters per gallon, per shutter	PT@.333	Ea	—	12.20	12.20	20.70
Brush split coat (1/2 undercoat + 1/2 enamel), 14 shutters per gallon, per shutter	PT@.286	Ea	—	10.50	10.50	17.90
Brush exterior enamel, 14 shutters per gallon, per shutter	PT@.400	Ea	—	14.70	14.70	25.00

	Craft@Hrs	Unit	Material	Labor	Total	Sell

Labor for painting 2' x 4' shutters or blinds — spray. Paint application only. Add the cost of surface preparation, protecting adjacent surfaces, mobilization, cleanup and callbacks. For heights above 8', use the high time difficulty factors.

	Craft@Hrs	Unit	Material	Labor	Total	Sell
Spray undercoat, 9 shutters per gallon, per shutter	PT@.111	Ea	—	4.08	4.08	6.94
Spray split coat (1/2 undercoat + 1/2 enamel), 11 shutters per gallon, per shutter	PT@.100	Ea	—	3.68	3.68	6.26
Spray exterior enamel, 11 shutters per gallon, per shutter	PT@.125	Ea	—	4.60	4.60	7.82

Labor for painting aluminum siding — brush. Paint application only. Add the cost of surface preparation, protecting adjacent surfaces, mobilization, cleanup and callbacks. For heights above 8', use the high time difficulty factors.

	Craft@Hrs	Unit	Material	Labor	Total	Sell
Brush metal primer, 420 square feet per gallon, per 100 square feet	PT@.426	CSF	—	15.70	15.70	26.70
Brush 1st or additional finish coats on aluminum siding, 465 square feet per gallon, per 100 square feet	PT@.351	CSF	—	12.90	12.90	21.90

Labor for staining rough wood siding — roll and brush. Stain application only. Add the cost of surface preparation, protecting adjacent surfaces, mobilization, cleanup and callbacks. For heights above 8', use the high time difficulty factors.

	Craft@Hrs	Unit	Material	Labor	Total	Sell
Roll and brush 1st coat of water-base stain, 213 square feet per gallon, per 100 square feet	PT@.500	CSF	—	18.40	18.40	31.30
Roll and brush 2nd coat of water-base stain, 275 square feet per gallon, per 100 square feet	PT@.400	CSF	—	14.70	14.70	25.00
Roll and brush 3rd or additional coats of water-base stain, 325 square feet per gallon, per 100 square feet	PT@.347	CSF	—	12.80	12.80	21.80
Roll and brush 1st coat of penetrating oil stain, 125 square feet per gallon, per 100 square feet	PT@.444	CSF	—	16.30	16.30	27.70
Roll and brush 2nd coat of penetrating oil stain, 240 square feet per gallon, per 100 square feet	PT@.364	CSF	—	13.40	13.40	22.80
Roll and brush 3rd or additional coats of penetrating oil stain, 330 square feet per gallon, per 100 square feet	PT@.299	CSF	—	11.00	11.00	18.70

Labor for staining rough wood siding — spray. Stain application only. Add the cost of surface preparation, protecting adjacent surfaces, mobilization, cleanup and callbacks. For heights above 8', use the high time difficulty factors.

	Craft@Hrs	Unit	Material	Labor	Total	Sell
Spray 1st coat of water-base stain, 115 square feet per gallon, per 100 square feet	PT@.200	CSF	—	7.35	7.35	12.50
Spray 2nd coat of water-base stain, 125 square feet per gallon, per 100 square feet	PT@.167	CSF	—	6.14	6.14	10.40

	Craft@Hrs	Unit	Material	Labor	Total	Sell
Spray 3rd or additional coats of water-base stain, 170 square feet per gallon, per 100 square feet	PT@.154	CSF	—	5.66	5.66	9.62
Spray 1st coat penetrating oil stain, 180 square feet per gallon, per 100 square feet	PT@.200	CSF	—	7.35	7.35	12.50
Spray 2nd coat of penetrating oil stain, 245 square feet per gallon, per 100 square feet	PT@.182	CSF	—	6.69	6.69	11.40
Spray 3rd or additional coats of penetrating oil stain, 360 square feet per gallon, per 100 square feet	PT@.154	CSF	—	5.66	5.66	9.62
Spray 1st coat of clear hydro sealer, 113 square feet per gallon, per 100 square feet	PT@.174	CSF	—	6.40	6.40	10.90
Spray 2nd coat of clear hydro sealer, 150 square feet per gallon, per 100 square feet	PT@.148	CSF	—	5.44	5.44	9.25
Spray 3rd or additional coats of clear hydro sealer, 175 square feet per gallon, per 100 square feet	PT@.133	CSF	—	4.89	4.89	8.31

Labor for painting gypsum drywall walls — roll and brush. Paint application only. Add the cost of surface preparation, protecting adjacent surfaces, mobilization, cleanup and callbacks. For heights above 8', use the high time difficulty factors.

	Craft@Hrs	Unit	Material	Labor	Total	Sell
Roll and brush 1st coat of water-base flat latex, 275 square feet per gallon, per 100 square feet	PT@.186	CSF	—	6.84	6.84	11.60
Roll and brush 2nd coat of water-base flat latex, 313 square feet per gallon, per 100 square feet	PT@.167	CSF	—	6.14	6.14	10.40
Roll and brush additional coats of water-base flat latex, 338 square feet per gallon, per 100 square feet	PT@.154	CSF	—	5.66	5.66	9.62
Roll and brush prime coat of water-base sealer, 263 square feet per gallon, per 100 square feet	PT@.200	CSF	—	7.35	7.35	12.50
Roll and brush 1st finish coat of water-base enamel, 263 square feet per gallon, per 100 square feet	PT@.222	CSF	—	8.16	8.16	13.90
Roll and brush 2nd finish coat of water-base enamel, 288 square feet per gallon, per 100 square feet	PT@.211	CSF	—	7.76	7.76	13.20

Labor for painting gypsum drywall walls — spray. Paint application only. Add the cost of surface preparation, protecting adjacent surfaces, mobilization, cleanup and callbacks. For heights above 8', use the high time difficulty factors.

	Craft@Hrs	Unit	Material	Labor	Total	Sell
Spray 1st coat of water-base flat latex, 225 square feet per gallon, per 100 square feet	PT@.125	CSF	—	4.60	4.60	7.82
Spray 2nd coat of water-base flat latex, 275 square feet per gallon, per 100 square feet	PT@.111	CSF	—	4.08	4.08	6.94

	Craft@Hrs	Unit	Material	Labor	Total	Sell
Spray additional coats of water-base flat latex, 300 square feet per gallon, per 100 square feet	PT@.105	CSF	—	3.86	3.86	6.56
Spray prime coat of water-base sealer, 225 square feet per gallon, per 100 square feet	PT@.136	CSF	—	5.00	5.00	8.50
Spray 1st finish coat of water-base enamel, 238 square feet per gallon, per 100 square feet	PT@.148	CSF	—	5.44	5.44	9.25
Spray 2nd finish coat of water-base enamel, 263 square feet per gallon, per 100 square feet	PT@.143	CSF	—	5.26	5.26	8.94
Spray additional finish coats of water-base enamel, 275 square feet per gallon, per 100 square feet	PT@.129	CSF	—	4.74	4.74	8.06

Labor for painting gypsum drywall ceiling — roll and brush. Paint application only. Add the cost of surface preparation, protecting adjacent surfaces, mobilization, cleanup and callbacks. For heights above 8', use the high time difficulty factors.

	Craft@Hrs	Unit	Material	Labor	Total	Sell
Roll and brush 1st coat of flat latex, 275 square feet per gallon, per 100 square feet	PT@.286	CSF	—	10.50	10.50	17.90
Roll and brush 2nd coat of flat latex, 313 square feet per gallon, per 100 square feet	PT@.267	CSF	—	9.82	9.82	16.70
Roll and brush 3rd coat of flat latex, 338 square feet per gallon, per 100 square feet	PT@.235	CSF	—	8.64	8.64	14.70
Roll and brush prime coat of water-base sealer, 275 square feet per gallon, per 100 square feet	PT@.267	CSF	—	9.82	9.82	16.70
Roll and brush 1st finish coat of water-base enamel, 313 square feet per gallon, per 100 square feet	PT@.286	CSF	—	10.50	10.50	17.90
Roll and brush additional coats of water-base enamel, 338 square feet per gallon, per 100 square feet	PT@.250	CSF	—	9.19	9.19	15.60
Roll and brush 1st coat of white epoxy coating, 288 square feet per gallon, per 100 square feet	PT@.308	CSF	—	11.30	11.30	19.20
Roll and brush additional coats of white epoxy coating, 288 square feet per gallon, per 100 square feet	PT@.267	CSF	—	9.82	9.82	16.70

Labor for painting gypsum drywall ceiling — spray. Paint application only. Add the cost of surface preparation, protecting adjacent surfaces, mobilization, cleanup and callbacks. For heights above 8', use the high time difficulty factors.

	Craft@Hrs	Unit	Material	Labor	Total	Sell
Spray 1st coat of flat latex, 200 square feet per gallon, per 100 square feet	PT@.133	CSF	—	4.89	4.89	8.31
Spray 2nd coat of flat latex, 225 square feet per gallon, per 100 square feet	PT@.114	CSF	—	4.19	4.19	7.12

	Craft@Hrs	Unit	Material	Labor	Total	Sell
Spray additional coats of flat latex, 250 square feet per gallon, per 100 square feet	PT@.108	CSF	—	3.97	3.97	6.75
Spray prime coat of water-base sealer, 200 square feet per gallon, per 100 square feet	PT@.125	CSF	—	4.60	4.60	7.82
Spray 1st finish coat of water-base enamel, 225 square feet per gallon, per 100 square feet	PT@.121	CSF	—	4.45	4.45	7.57
Spray additional coats of water-base enamel, 250 square feet per gallon, per 100 square feet	PT@.114	CSF	—	4.19	4.19	7.12

Labor for painting wood paneled walls — roll and brush. Paint application only. Add the cost of surface preparation, protecting adjacent surfaces, mobilization, cleanup and callbacks. For heights above 8', use the high time difficulty factors.

	Craft@Hrs	Unit	Material	Labor	Total	Sell
Roll and brush water-base undercoat, 263 square feet per gallon, per 100 square feet	PT@.333	CSF	—	12.20	12.20	20.70
Roll and brush water-base split coat (1/2 undercoat + 1/2 enamel), 263 square feet per gallon, per 100 square feet	PT@.364	CSF	—	13.40	13.40	22.80
Roll and brush 1st finish coat of water-base enamel, 313 square feet per gallon, per 100 square feet	PT@.235	CSF	—	8.64	8.64	14.70

Labor for painting wood paneled walls — spray. Paint application only. Add the cost of surface preparation, protecting adjacent surfaces, mobilization, cleanup and callbacks. For heights above 8', use the high time difficulty factors.

	Craft@Hrs	Unit	Material	Labor	Total	Sell
Spray water-base undercoat, 150 square feet per gallon, per 100 square feet	PT@.235	CSF	—	8.64	8.64	14.70
Spray water-base split coat (1/2 undercoat + 1/2 enamel), 150 square feet per gallon, per 100 square feet	PT@.250	CSF	—	9.19	9.19	15.60
Spray 1st finish coat of water-base enamel, 225 square feet per gallon, per 100 square feet	PT@.182	CSF	—	6.69	6.69	11.40
Spray additional finish coats of water-base enamel, 325 square feet per gallon, per 100 square feet	PT@.154	CSF	—	5.66	5.66	9.62

Labor for staining wood paneled walls. Stain application only. Add the cost of surface preparation, protecting adjacent surfaces, mobilization, cleanup and callbacks. For heights above 8', use the high time difficulty factors.

	Craft@Hrs	Unit	Material	Labor	Total	Sell
Roll and brush each coat of solid or semi-transparent stain, 450 square feet per gallon, per 100 square feet	PT@.380	CSF	—	14.00	14.00	23.80
Spray solid or semi-transparent stain, 250 square feet per gallon, per 100 square feet	PT@.250	CSF	—	9.19	9.19	15.60

	Craft@Hrs	Unit	Material	Labor	Total	Sell
Labor for painting windows to 15 square feet, 1, 2 or 3 panes, per side painted. Paint application only. Add preparation time for sanding, puttying, protecting adjacent surfaces and masking or waxing window panes. For heights above 8', use the high time difficulty factors.						
Brush undercoat, 14.5 windows per gallon, per window	PT@.200	Ea	—	7.35	7.35	12.50
Brush split coat (1/2 undercoat + 1/2 enamel), 16.5 windows per gallon, per window	PT@.250	Ea	—	9.19	9.19	15.60
Brush finish coat of enamel, 15.5 windows per gallon, per window	PT@.300	Ea	—	11.00	11.00	18.70
Labor for painting windows to 15 square feet, 4, 5 or 6 panes, per side painted. Paint application only. Add preparation time for sanding, puttying, protecting adjacent surfaces and masking or waxing window panes. For heights above 8', use the high time difficulty factors.						
Brush undercoat, 13.5 windows per gallon, per window	PT@.300	Ea	—	11.00	11.00	18.70
Brush split coat (1/2 undercoat + 1/2 enamel), 15.5 windows per gallon, per window	PT@.350	Ea	—	12.90	12.90	21.90
Brush finish coat of enamel, 14.5 windows per gallon, per window	PT@.450	Ea	—	16.50	16.50	28.10
Labor for painting windows to 15 square feet, 7 or 8 panes, per side painted. Paint application only. Add preparation time for sanding, puttying, protecting adjacent surfaces and masking or waxing window panes. For heights above 8', use the high time difficulty factors.						
Brush undercoat, 13 windows per gallon, per window	PT@.400	Ea	—	14.70	14.70	25.00
Brush split coat (1/2 undercoat + 1/2 enamel), 15 windows per gallon, per window	PT@.500	Ea	—	18.40	18.40	31.30
Brush finish coat of enamel, 14 windows per gallon, per window	PT@.620	Ea	—	22.80	22.80	38.80
Labor for painting windows to 15 square feet, 9, 10 or 11 panes, per side painted. Paint application only. Add preparation time for sanding, puttying, protecting adjacent surfaces and masking or waxing window panes. For heights above 8', use the high time difficulty factors.						
Brush undercoat, 12 windows per gallon, per window	PT@.500	Ea	—	18.40	18.40	31.30
Brush split coat (1/2 undercoat + 1/2 enamel), 14 windows per gallon, per window	PT@.620	Ea	—	22.80	22.80	38.80
Brush finish coat of enamel, 13 windows per gallon, per window	PT@.730	Ea	—	26.80	26.80	45.60
Labor for painting windows to 15 square feet, 12 panes, per side painted. Paint application only. Add preparation time for sanding, puttying, protecting adjacent surfaces and masking or waxing window panes. For heights above 8', use the high time difficulty factors.						
Brush undercoat, 11 windows per gallon, per window	PT@.620	Ea	—	22.80	22.80	38.80
Brush split coat (1/2 undercoat + 1/2 enamel), 13 windows per gallon, per window	PT@.670	Ea	—	24.60	24.60	41.80
Brush finish coat of enamel, 12 windows per gallon, per window	PT@.800	Ea	—	29.40	29.40	50.00

	Craft@Hrs	Unit	Material	Labor	Total	Sell
Labor for painting windows, per square foot of window, per side painted. Use figures in the following table to estimate labor for windows larger than 15 square feet (length times width). To calculate the window area, add 1 foot to each dimension (top, bottom, left and right side). Then multiply the width times length. For example, a window measuring 4' x 4' with 1 foot added to the top, bottom, right side and left side would be 6' x 6' or 36 square feet. Then, add 2 square feet for each window pane to allow time for painting the mullions, muntins and sash. The calculation for a 4' x 4' double-hung window with two panes would be 6' x 6' + 4', or 40 square feet. Add preparation time for sanding, puttying, protecting adjacent surfaces and masking or waxing window panes. For heights above 8', use the high time difficulty factors.						
Brush undercoat,						
450 square feet per gallon, per 100 square feet	PT@.606	CSF	—	22.30	22.30	37.90
Brush split coat (1/2 undercoat + 1/2 enamel),						
500 square feet per gallon, per 100 square feet	PT@.741	CSF	—	27.20	27.20	46.20
Brush finish coat of enamel,						
480 square feet per gallon, per 100 square feet	PT@.885	CSF	—	32.50	32.50	55.30

Exterior Solid Stain

Deck Plus® solid color deck and siding stain, Behr. 100% acrylic latex. Opaque. For decks (including pressure-treated), siding, fences and patio furniture. Water-repellent and mildew-resistant. Water cleanup. Resists scuffing, cracking and peeling. Covers 400 square feet per gallon. Dries to touch in 2 to 3 hours; to recoat in 24 hours. Full cure in 30 days.

	Craft@Hrs	Unit	Material	Labor	Total	Sell
Gallon	—	Ea	29.60	—	29.60	—
5 gallons	—	Ea	109.00	—	109.00	—

Premium Plus® solid color exterior stain, Behr. Oil and latex formula. Water cleanup. Protects against fading, weathering and chalking for up to 10 years on vertical surfaces. Prevents cracking, peeling and blistering. Opaque finish covers wood grain and allows wood texture to remain. For exterior shakes, shingles, wood siding, fencing, trim, stucco and masonry surfaces.

	Craft@Hrs	Unit	Material	Labor	Total	Sell
Gallon	—	Ea	28.70	—	28.70	—
5 gallons	—	Ea	133.00	—	133.00	—

Acrylic exterior and interior stain, Behr.

	Craft@Hrs	Unit	Material	Labor	Total	Sell
Gallon	—	Ea	28.50	—	28.50	—
5 gallons	—	Ea	117.00	—	117.00	—

Preserva-Wood penetrating finish, Preservaproducts. Will not crack or peel. Tinted natural finishes to enhance the natural beauty and color of various types of wood. Ultraviolet protection, environmentally responsible. Clear or tinted.

	Craft@Hrs	Unit	Material	Labor	Total	Sell
Quart	—	Ea	10.60	—	10.60	—
Gallon	—	Ea	26.50	—	26.50	—
5 gallons	—	Ea	118.00	—	118.00	—

Wood bleach kit, Klean-Strip. 2-part solution. Lightens stripped or unfinished wood. Prepares bare wood for staining. Pint covers 15 to 20 square feet.

	Craft@Hrs	Unit	Material	Labor	Total	Sell
2 Gallon A+B	—	Ea	86.70	—	86.70	—

Interior Oil Wiping Stain

Wood finish wiping stain, Minwax®. Use on interior bare or stripped wood surface. Penetrates into wood fibers to highlight grain. Available in 20 wood-tone colors.

	Craft@Hrs	Unit	Material	Labor	Total	Sell
Quart	—	Ea	7.93	—	7.93	—
Gallon	—	Ea	27.30	—	27.30	—

	Craft@Hrs	Unit	Material	Labor	Total	Sell
Watco™ Danish Oil™, Rust-Oleum®. Natural, hand-rubbed appearance. Penetrating stain soaks into pores of wood. Won't crack, chip or peel. Walnut, fruitwood or natural tint.						
Quart	—	Ea	11.10	—	11.10	—
Gallon	—	Ea	29.20	—	29.20	—

Waterborne Stain

	Craft@Hrs	Unit	Material	Labor	Total	Sell
Water-based pickling stain, Minwax®. Leaves a unique, subtle shade of white while allowing the natural wood grain to show through.						
Quart	—	Ea	11.00	—	11.00	—
Water-based wood stain, Minwax®. Provides rich, even stain penetration. Fast drying and easily cleans up with soap and water.						
Half pint	—	Ea	7.56	—	7.56	—
Quart	—	Ea	11.00	—	11.00	—

Gel Stain

	Craft@Hrs	Unit	Material	Labor	Total	Sell
Gel stain, Minwax®. Oil-based, gelled stain for finishing vertical surfaces and non-wood surfaces like fiberglass, metal, veneer and fiberboard. Non-drip formula. Suitable for fine woods. Controlled penetration for uniform color.						
Quart	—	Ea	16.10	—	16.10	—
Polyurethane stain, Minwax®. Stain color and tough polyurethane protection in one easy step. Available in nine colors.						
Half pint	—	Ea	8.14	—	8.14	—
Quart	—	Ea	13.10	—	13.10	—

Exterior Clear Wood Coating

	Craft@Hrs	Unit	Material	Labor	Total	Sell
Spar varnish, Minwax®. For outdoor furniture, doors, entryways, finish, and trim. Resists water, salt, and sun. Low volatile organic compounds (VOCs).						
Gloss, quart	—	Ea	16.90	—	16.90	—
Gloss, gallon	—	Ea	48.30	—	48.30	—
Satin, quart	—	Ea	16.90	—	16.90	—
Satin, gallon	—	Ea	48.30	—	48.30	—
Man-O-War marine spar varnish, Valspar. Protection against weathering and salt water. Resists cracking and peeling. Better than polyurethane. Expands and contracts with weather conditions. For fences and outdoor furniture. Gloss or satin finish.						
Gloss, quart	—	Ea	25.50	—	25.50	—
Gloss, gallon	—	Ea	71.30	—	71.30	—
Helmsman® spar urethane, Minwax. Clear finish protects wood exposed to sunlight, water or temperature changes. Ultraviolet absorbers help protect wood from graying and fading in the sun. Contains oils that help the finish expand and contract with wood as the seasons and temperature change. 24-hour drying time. 400 square feet per gallon. 2-3 coat application. Clean up with mineral spirits. High gloss, semi-gloss or satin finish.						
Quart	—	Ea	16.90	—	16.90	—
Gallon	—	Ea	40.80	—	40.80	—
Clearshield™ weather-resistant wood coating, Minwax®. Clear protective topcoat with ultraviolet absorbers to help protect wood from weather damage. Satin or semi-gloss.						
Quart	—	Ea	19.00	—	19.00	—

	Craft@Hrs	Unit	Material	Labor	Total	Sell

Varathane® exterior classic clear finish, water-based, Rust-Oleum®. For outdoor patio furniture, garage doors and front doors. Excellent scratch and impact resistance. Transparent, non-yellowing, and low odor. Fast drying and easy water clean up. Gloss, semi-gloss or satin finish.

	Craft@Hrs	Unit	Material	Labor	Total	Sell
12-ounce aerosol	—	Ea	7.12	—	7.12	—
Quart	—	Ea	16.30	—	16.30	—
Gallon	—	Ea	51.00	—	51.00	—

Varathane® exterior classic clear finish, oil-based, Rust-Oleum®. For outdoor furniture, garage doors and front doors. Excellent scratch and impact resistance. Expands and contracts with weather conditions. Fast drying and self-leveling. Gloss or satin finish.

	Craft@Hrs	Unit	Material	Labor	Total	Sell
Quart	—	Ea	13.00	—	13.00	—
Gallon	—	Ea	37.20	—	37.20	—

Log home gloss finish. For log and timber frame homes, siding, fencing and railings. Contains mildewcide and offers UV protection. Water clean-up. Dries to the touch in 8 to 12 hours. Recoat after 24 hours. 3 to 4 coats provide maximum protection. Clear or cedar finish.

	Craft@Hrs	Unit	Material	Labor	Total	Sell
Gallon	—	Ea	35.20	—	35.20	—
5 gallons	—	Ea	158.00	—	158.00	—

Oil-based Polyurethane Interior Clear Finish

Fast-drying clear polyurethane, Minwax®. A clear, oil-based, durable protective finish for wood. Provides long-lasting protection and beauty to interior wood surfaces such as furniture, doors, cabinets and floors. Gloss, semi-gloss or satin finish.

	Craft@Hrs	Unit	Material	Labor	Total	Sell
Half pint	—	Ea	6.61	—	6.61	—
Quart	—	Ea	12.90	—	12.90	—
Gallon	—	Ea	33.10	—	33.10	—
2-1/2 gallons	—	Ea	71.40	—	71.60	—

Varathane® oil-based floor finish, Rust-Oleum®. High impact resistance. Self-leveling to eliminate brush marks. Scratch and mar resistance. High solids content. Gloss, semi-gloss or satin.

	Craft@Hrs	Unit	Material	Labor	Total	Sell
Gallon	—	Ea	40.80	—	40.80	—

Varathane® interior classic clear finish, oil-based, Rust-Oleum®. Extra hard finish resists scuffing, chipping, and abrasion. Self-leveling. Outlasts varnish. Gloss, semi-gloss or satin.

	Craft@Hrs	Unit	Material	Labor	Total	Sell
12-ounce aerosol	—	Ea	6.47	—	6.47	—
Quart	—	Ea	11.70	—	11.70	—
Gallon	—	Ea	40.80	—	40.80	—

Waterborne Clear Finish

Polycrylic® protective finish, Minwax®. A hard, clear, ultra fast-drying protective finish. Non-flammable and very little odor. Cleans up with water. For use over light woods, pastel stained woods, painted surfaces, both latex- and oil-based and well-bonded wall coverings. Available in clear gloss, clear semi-gloss and clear satin. Covers 125 square feet per quart. Dries clear in 2 hours.

	Craft@Hrs	Unit	Material	Labor	Total	Sell
12-ounce aerosol	—	Ea	12.20	—	12.20	—
Half pint	—	Ea	9.16	—	9.16	—
Quart	—	Ea	18.30	—	18.30	—
Gallon	—	Ea	48.90	—	48.90	—

Parks PRO Finisher waterborne sanding sealer, Parks Corporation. A fast-drying water-based sanding sealer to be used on unfinished wood before applying a waterborne top coat. Thoroughly sand sealer to retard grain raise.

	Craft@Hrs	Unit	Material	Labor	Total	Sell
Gallon	—	Ea	40.80	—	40.80	—

	Craft@Hrs	Unit	Material	Labor	Total	Sell

Shellac

Bulls Eye shellac, Zinsser. Protects cabinetry, trim work, and furniture. Seals in stain to protect from water leaks, grease marks, smoke, and soot. Adheres to glossy surfaces without sanding. Seals knots and tannins. May be tinted to produce fast-drying deep color primer. Interior use only. Clear or amber color.

	Craft@Hrs	Unit	Material	Labor	Total	Sell
Amber, quart	—	Ea	14.10	—	14.10	—
Amber, gallon	—	Ea	43.20	—	43.20	—
Clear, quart	—	Ea	14.10	—	14.10	—
Clear, gallon	—	Ea	44.90	—	44.90	—

Water Seal for Wood

WaterSeal® clear wood protector, Thompson's. Enhance and maintain wood's natural color. Protects against damage from water and sun and provides a mildew-resistant coating. Oil-enriched formula provides long-lasting protection for all types of wood.

	Craft@Hrs	Unit	Material	Labor	Total	Sell
Gallon	—	Ea	15.00	—	15.00	—
5 gallons	—	Ea	66.30	—	66.30	—

WaterSeal® clear multi-surface waterproofer, Thompson's. For wood, brick, concrete, stucco and masonry. Can be applied to ACQ pressure-treated wood. Protects against warping, cracking, swelling, and water damage. Paintable. One-coat coverage. Withstands abrasion. Allows wood to weather naturally. Prevents moisture from penetrating by setting up barrier beneath. Goes on milky white, dries clear.

	Craft@Hrs	Unit	Material	Labor	Total	Sell
12-ounce aerosol	—	Ea	5.08	—	5.08	—
Gallon	—	Ea	11.40	—	11.40	—
5 gallons	—	Ea	51.00	—	51.00	—

Waterproofing wood protector, Behr. For decks, fences, siding and furniture. Lasts up to one year on decks and up to two years on fences and siding.

	Craft@Hrs	Unit	Material	Labor	Total	Sell
Gallon	—	Ea	16.30	—	16.30	—
5 gallons	—	Ea	72.40	—	72.40	—

Advanced wood protector, Sherwin-Williams. Clear or tinted.

	Craft@Hrs	Unit	Material	Labor	Total	Sell
Gallon	—	Ea	19.60	—	19.60	—
5 gallons	—	Ea	98.00	—	98.00	—

Wood Preservative

Dock and fence post preservative, No. 90, Behr. Exterior, green, water-repellent. For marine or below-ground use. Deep-penetrating formula. For wood in contact with soil and water. Protects against wood rot, warping, swelling, termite damage, mildew and moisture. Water clean up.

	Craft@Hrs	Unit	Material	Labor	Total	Sell
Gallon	—	Ea	23.30	—	23.30	—

Seasonite® new wood treatment, The Flood Company. Protects pressure-treated and untreated wood against moisture. Provides a gradual seasoning process to reduce warping, cupping, splitting, and checking. Readies surface for painting, staining, or finishing. Ideal for new deck and patio lumber. Mildew resistant. Soap and water clean up.

	Craft@Hrs	Unit	Material	Labor	Total	Sell
Gallon	—	Ea	19.40	—	19.40	—

Termin-8 wood preservative, green, Jasco. Oil-based. Preserves and protects wood from carpenter ants, termites, and powder post beetles. Protects against rot, warping, mildew, swelling, fungus, and moss. Slow drying for deep absorption. Exterior or marine use. Gallon covers 100 to 300 square feet. Paintable after 2 to 7 days

	Craft@Hrs	Unit	Material	Labor	Total	Sell
Quart	—	Ea	10.70	—	10.70	—
Gallon	—	Ea	23.80	—	23.80	—
5 gallons	—	Ea	100.00	—	100.00	—

	Craft@Hrs	Unit	Material	Labor	Total	Sell
Copper wood preservative, Jasco.						
Brown, quart	—	Ea	12.60	—	12.60	—
Brown, gallon	—	Ea	31.60	—	31.60	—
Clear, quart	—	Ea	12.60	—	12.60	—
Clear, gallon	—	Ea	31.60	—	31.60	—

ZPW clear wood preservative, Jasco. Kills termites, carpenter ants, protects decks and fences. Use inside or outside house above or below ground. Paintable and tintable. Controls warping, swelling and cracking.

Quart	—	Ea	11.30	—	11.30	—
Gallon	—	Ea	22.80	—	22.80	—

Wood Conditioners

Tung oil.

Quart	—	Ea	20.30	—	20.30	—

Boston polish, Butcher's. Like White Diamond bowling alley wax except for color. Amber tint hides kinks and scratches on wood.

Amber, pound	—	Ea	19.40	—	19.40	—

White Diamond bowling alley wax, Butcher's. Clear wax in turpentine and other select solvents. Cleans as it waxes. Polishes easily to a deep luster. Also recommended for shoes, boots, leather, copper, brass, fiberglass, marble, sealed brick and flagstone, linoleum and cork floors.

Clear, pcund	—	Ea	15.30	—	15.30	—

Teak oil, Valspar.

Quart	—	Ea	11.20	—	11.20	—

Finishing paste wax, Minwax®. Protects and adds luster to any stained or finished wood surface.

1 pound	—	Ea	10.20	—	10.20	—

Wood conditioner, Minwax®. Ensures even stain penetration on soft or porous woods such as pine, fir, alder and maple. When working with soft or porous woods, wood conditioner helps prevent stains from blotching and streaking.

Pint	—	Ea	6.10	—	6.10	—
Quart	—	Ea	12.20	—	12.20	—
Gallon	—	Ea	28.90	—	28.90	—

Interior Primers

Interior enamel undercoater. Improves sheen, hide, adhesion, flow and leveling of enamel topcoats. Blocks stains. Conceals surface blemishes. For use with wood, drywall, plaster, stucco, masonry, brick and metal. Use prior to top-coating with satin, semi-gloss and high-gloss topcoats.

Quart	—	Ea	9.16	—	9.16	—
Gallon	—	Ea	19.33	—	19.33	—
5 gallons	—	Ea	87.00	—	87.00	—

Interior polyvinyl acrylic latex primer, Speed-Wall®. General purpose. Easy application, fast-drying and high-hiding. Covers 400 square feet per gallon.

Gallon	—	Ea	10.20	—	10.20	—
5 gallons	—	Ea	40.80	—	40.80	—

	Craft@Hrs	Unit	Material	Labor	Total	Sell
KILZ® Premium sealer-primer-stainblocker, Masterchem. Interior/exterior, water-base formula. Superior stainblocker. Great for mildew-prone areas. Excellent enamel undercoater. High hide and durability.						
Quart	—	Ea	9.67	—	9.67	—
Gallon	—	Ea	21.40	—	21.40	—
5 gallons	—	Ea	96.30	—	96.30	—

Interior Paint

	Craft@Hrs	Unit	Material	Labor	Total	Sell
Interior latex ceiling paint. Minimal splatter and excellent coverage. Hides minor surface imperfections and dries quickly to a smooth, even finish. Clean up with water. Covers 300 square feet per gallon.						
Gallon	—	Ea	24.50	—	24.50	—
5 gallons	—	Ea	105.00	—	105.00	—
Interior acrylic latex flat wall and trim paint. Dries to a smooth, even finish. Hides minor surface imperfections. Easy to clean up and dries quickly. Covers 400 square feet per gallon.						
Gallon	—	Ea	12.20	—	12.20	—
5 gallons	—	Ea	55.00	—	55.00	—
Ultra-Hide® interior latex flat finish, Glidden. For use on interior walls, ceilings, and low-traffic areas. High-hiding with excellent touch-up ability when applied by spray, brush, or roller. Quick drying and recoat. Covers 400 square feet per gallon.						
Gallon	—	Ea	18.30	—	18.30	—
5 gallons	—	Ea	91.70	—	91.70	—
Premium Plus® interior flat paint, Behr. 100% acrylic latex. Hides minor surface imperfections. For walls, woodwork and ceilings. Lifetime guarantee. Matte, flat finish. Easy water clean up.						
Pastel base, gallon	—	Ea	25.20	—	25.20	—
Pastel base, 5 gallons	—	Ea	112.00	—	112.00	—
Ultra white, quart	—	Ea	11.40	—	11.40	—
Ultra white, gallon	—	Ea	23.40	—	23.40	—
Ultra white, 5 gallons	—	Ea	114.00	—	114.00	—
Latex eggshell wall and trim paint. Acrylic latex. 15-year durability.						
Quart	—	Ea	11.90	—	11.90	—
Gallon	—	Ea	26.50	—	26.50	—
5 gallons	—	Ea	110.00	—	110.00	—
Ultra-Hide® interior latex eggshell, Glidden.						
Gallon	—	Ea	18.30	—	18.30	—
5 gallons	—	Ea	82.50	—	82.50	—
Latex satin wall and trim enamel, Behr. 100% acrylic latex. For kitchens and bathrooms. Exceptional washability, scrubbability, stain and mildew-resistance. Pearl-like finish. Hides minor surface imperfections. Superior adhesion. Lifetime guarantee.						
Quart	—	Ea	13.70	—	13.70	—
Gallon	—	Ea	28.60	—	28.60	—
5 gallons	—	Ea	128.00	—	128.00	—
Ultra-Hide® interior latex semi-gloss wall and trim enamel, Glidden. High-hiding. Good adhesion and moisture resistance. Easy application. Fast-drying. Covers 400 square feet per gallon.						
Gallon	—	Ea	20.40	—	20.40	—
5 gallons	—	Ea	91.70	—	91.70	—

	Craft@Hrs	Unit	Material	Labor	Total	Sell
Speed-Wall® interior latex semi-gloss wall and trim enamel, Glidden. Good dry hide. Hard and durable. Easy application. Fast-drying. Covers 400 to 450 square feet per gallon.						
Gallon	—	Ea	15.30	—	15.30	—
5 gallons	—	Ea	68.70	—	68.70	—
Enamel semi-gloss interior latex paint, Behr. For walls, ceilings, trim, previously painted or primed masonry, previously painted or primed metal. Scrubbable finish. Resists scuffs and stains. Quick-drying.						
Quart	—	Ea	18.30	—	18.30	—
Gallon	—	Ea	40.80	—	40.80	—
5 gallons	—	Ea	184.00	—	184.00	—
Premium Plus® interior and exterior high-gloss enamel, Behr. 100% acrylic latex. Mildew-proof. Glass-like finish. For kitchen, bath, railings, trim, doors, cabinets, and woodwork. Exceptional washability and stain removal.						
Quart	—	Ea	13.80	—	13.80	—
Gallon	—	Ea	29.40	—	29.40	—
5 gallons	—	Ea	138.00	—	138.00	—
Ultra-Hide® interior and exterior alkyd gloss enamel, Glidden. Durable high-gloss finish. High solids and low VOC. Good abrasion resistance. Excellent resistance to grease, oil and water. Covers 350 to 400 square feet per gallon.						
Quart	—	Ea	13.50	—	13.50	—
Gallon	—	Ea	28.10	—	28.10	—
5 gallons	—	Ea	122.00	—	122.00	—
Interior texture paint, Behr. Decorative finish for walls and ceilings. Alternative to wallpaper, borders, and stenciling. Reinforced with interwoven fibers to easily repair old textured surfaces. Usable with topcoat. Brightest white, ready to use. 2-gallon container.						
Trowel-on Mediterranean ceiling	—	Ea	28.50	—	28.50	—
Sand stucco knockdown	—	Ea	28.50	—	28.50	—
Roll-on smooth acoustic	—	Ea	28.50	—	28.50	—
Roll-on stucco texture, Litex. Interior and exterior.						
3.5 gallons	—	Ea	42.40	—	42.40	—
Sand-Texture™, Homax®. A classic wall texture for a sandstone-like finish. Mixes with any paint, won't change the color of the paint. Use one package per gallon.						
6 ounces	—	Ea	5.07	—	5.07	—
Perlite texturing paint additive, Litex. Add to any paint to create texture and sand finishes.						
Fine, quart	—	Ea	4.11	—	4.11	—
Medium, quart	—	Ea	4.11	—	4.11	—
Coarse, quart	—	Ea	4.11	—	4.11	—

Exterior Primer

	Craft@Hrs	Unit	Material	Labor	Total	Sell
Premium Plus® oil-based exterior primer sealer, Behr. Universal undercoat for use on wood, hardboard, aluminum and wrought iron. Prevents extractive bleeding.						
Quart	—	Ea	11.70	—	11.70	—
Gallon	—	Ea	25.50	—	25.50	—
Water-based concrete bonding primer, No. 880, Behr. Clear epoxy treatment for porous concrete, garage floors, driveways, walkways, patios and stucco. Penetrates and bonds. Resistant to water, alkali, and efflorescence. Creates a sound surface for painting or staining. Promotes uniform finish of topcoats. Dries to touch in 2 to 4 hours. Approximately 350 to 400 square feet per gallon coverage.						
Clear, gallon	—	Ea	18.30	—	18.30	—

	Craft@Hrs	Unit	Material	Labor	Total	Sell
Premium Plus® exterior water-base primer and sealer, Behr. 100% acrylic latex. Multi-purpose stain blocking. Use on exterior wood, hardboard, masonry, stucco, brick, galvanized metal, and aluminum.						
Quart	—	Ea	10.70	—	10.70	—
Gallon	—	Ea	21.40	—	21.40	—
5 gallons	—	Ea	95.90	—	95.90	—
KILZ® exterior sealer primer stainblocker, Masterchem. Exterior oil-base formula. Resists cracking and peeling. Mildew-resistant coating. Blocks tannin bleed on redwood and cedar.						
Gallon	—	Ea	33.60	—	33.60	—
5 gallons	—	Ea	153.00	—	153.00	—

Exterior Paint

	Craft@Hrs	Unit	Material	Labor	Total	Sell
Glidden Premium® exterior latex house and trim paint, 20 year. 100% acrylic to protect against UV rays, color fading and severe weather conditions. Hides minor surface imperfections, gives a mildew-resistant coating. Non-chalking. Good weather resistance against dirt, blistering and peeling. For siding, wood, stucco, brick and masonry surfaces. Use the satin finish on windows and doors. Covers 300 to 400 square feet per gallon.						
Flat, quart	—	Ea	10.30	—	10.30	—
Flat, gallon	—	Ea	20.40	—	20.40	—
Flat, 5 gallon	—	Ea	102.00	—	102.00	—
Satin, quart	—	Ea	11.50	—	11.50	—
Satin, gallon	—	Ea	25.50	—	25.50	—
Satin, 5 gallons	—	Ea	114.00	—	114.00	—
Semi-gloss, quart	—	Ea	12.60	—	12.60	—
Semi-gloss, gallon	—	Ea	27.50	—	27.50	—
Semi-gloss, 5 gallons	—	Ea	123.00	—	123.00	—
Gloss, quart	—	Ea	14.10	—	14.10	—
Gloss, gallon	—	Ea	29.60	—	29.60	—
Premium Plus® exterior paint, Behr. 100% acrylic latex. Mildew and fade resistant. For use on wood, masonry, brick, stucco, vinyl and aluminum siding, and metal.						
Flat, quart	—	Ea	12.50	—	12.50	—
Flat, gallon	—	Ea	26.40	—	26.40	—
Flat, 5 gallons	—	Ea	118.00	—	118.00	—
Satin, gallon	—	Ea	28.40	—	28.40	—
Satin, 5 gallons	—	Ea	128.00	—	128.00	—
Semi-gloss, quart	—	Ea	14.60	—	14.60	—
Semi-gloss, gallon	—	Ea	29.50	—	29.50	—
Semi-gloss, 5 gallons	—	Ea	132.00	—	132.00	—

Masonry and Stucco Paint

	Craft@Hrs	Unit	Material	Labor	Total	Sell
Plus 10® latex flat masonry and stucco paint, Behr. For interior or exterior use. Acrylic latex paint for smooth, rough and textured masonry surfaces. Weather- and temperature-resistant.						
Gallon	—	Ea	20.40	—	20.40	—
5 gallons	—	Ea	90.80	—	90.80	—
Plus 10® elastomeric waterproofing paint, Behr. Acrylic latex for masonry, stucco, and concrete. Exterior, flexible, high-build coating. Expands and contracts, bridging hairline cracks. Stretches up to 600%. Resists mildew and dirt. Retains color. Passes Federal Specification TT-C-555B sec. 3.3.3.						
Gallon	—	Ea	28.50	—	28.50	—
5 gallons	—	Ea	129.00	—	129.00	—

	Craft@Hrs	Unit	Material	Labor	Total	Sell

Concrete Coatings

Waterproofing sealer, Seal-Krete. Waterborne, high-binding acrylic base. Penetrating sealer formulated for use on all vertical concrete, cement, masonry and stucco surfaces. Non-toxic.

	Craft@Hrs	Unit	Material	Labor	Total	Sell
Gallon	—	Ea	13.10	—	13.10	—
5 gallons	—	Ea	60.10	—	60.10	—

Exposed aggregate sealer.

	Craft@Hrs	Unit	Material	Labor	Total	Sell
Brown or clear, gallon	—	Ea	20.40	—	20.40	—
Brown or clear, 5 gallons	—	Ea	79.50	—	79.50	—

Concrete and masonry waterproofer, Behr. Penetrating water-based silicone formula reduces water damage and deterioration for up to 10 years. Forms a breathable barrier that reduces corrosion, freeze-thaw damage, mildew and algae staining.

	Craft@Hrs	Unit	Material	Labor	Total	Sell
Gallon	—	Ea	22.40	—	22.40	—
5 gallons	—	Ea	100.00	—	100.00	—

Thoroseal® protective coating, Thoro. For walls, pool floors, cisterns, tanks, and masonry reservoirs. Finishes masonry of all types. Seals water and dampness out of walls, above or below grade, inside or out. Leakproofs water containers. Under ordinary water conditions, 2 pounds cover 1 square yard as a base coat; 1 pound covers 1 square yard as a second coat. Under extreme water pressures, 1 brush coat at 2 pounds per square yard and 1 trowel coat. White or gray.

	Craft@Hrs	Unit	Material	Labor	Total	Sell
50 pound bag	—	Ea	26.50	—	26.50	—

Epoxy floor coating, 2 part. Prevents hot tire pickup. Resists gasoline, oil, grease, acid, transmission fluid, saltwater, bleach, and alkali.

	Craft@Hrs	Unit	Material	Labor	Total	Sell
Clear, 2 gallon	—	Ea	109.00	—	109.00	—
Gray, 2 gallon	—	Ea	109.00	—	109.00	—

Cure-Seal wet-look concrete sealer, Jasco. Protects exposed aggregate, brick, flagstone, and concrete from oil, rust, and water staining. Dries clear. Covers 150 to 400 square feet per gallon. Ready for traffic in eight to 24 hours.

	Craft@Hrs	Unit	Material	Labor	Total	Sell
Gallon	—	Ea	19.40	—	19.40	—
5 gallons	—	Ea	79.30	—	79.30	—

Plus 10® concrete stain. Durable, water-repellent. Interior and exterior. Forms tough, longwearing film on all properly-prepared surfaces such as patios, basements, pool decks, pillars, bricks and sidewalks. Natural slate color. 100% acrylic resins.

	Craft@Hrs	Unit	Material	Labor	Total	Sell
Gallon	—	Ea	28.50	—	28.50	—
5 gallons	—	Ea	126.00	—	126.00	—

Premium Plus® gloss porch and floor enamel, Behr. 100% acrylic. Mildew-resistant. Dries to touch in 2 hours. Recoat in 24 to 48 hours. Full cure in 30 days.

	Craft@Hrs	Unit	Material	Labor	Total	Sell
Gloss, quart	—	Ea	13.20	—	13.20	—
Gloss, gallon	—	Ea	30.50	—	30.50	—
Slate gray, low luster, gallon	—	Ea	28.50	—	28.50	—
Slate gray, low luster, 5 gallons	—	Ea	127.00	—	127.00	—

Porch and floor polyurethane oil gloss, Glidden. For interior or exterior wood and concrete floors, stairs, porches, railings and steps. Covers 400 to 500 square feet per gallon.

	Craft@Hrs	Unit	Material	Labor	Total	Sell
Gallon	—	Ea	25.50	—	25.50	—
5 gallons	—	Ea	102.00	—	102.00	—

	Craft@Hrs	Unit	Material	Labor	Total	Sell

Rust Preventative

Clean metal primer. White. For use on bare, lightly rusted or previously-painted surfaces. Creates a strong surface for better topcoat adhesion and weather resistance.

	Craft@Hrs	Unit	Material	Labor	Total	Sell
Quart	—	Ea	9.16	—	9.16	—
Quart, low VOC	—	Ea	9.07	—	9.07	—

Rust converting metal primer, Rust Destroyer®. Non-toxic, USDA and FDA approved. Heat-resistant up to 800 degrees F. Five-year guarantee. Works on clean or rusty steel, zinc galvanized metal, core tan steel, aluminum, tin, previously painted surfaces and barbeques. When applied over new metal, will prevent rusting.

	Craft@Hrs	Unit	Material	Labor	Total	Sell
Quart	—	Ea	15.00	—	15.00	—
Gallon	—	Ea	50.00	—	50.00	—

Professional oil-based enamel, Rust-Oleum®. Tough industrial enamel. Prevents rust, resists cracking, peeling, chipping, and fading. Scuff-, abrasion- and moisture-resistant. Lead-free. Not low-volatile-organic-compound compliant.

	Craft@Hrs	Unit	Material	Labor	Total	Sell
Gallon	—	Ea	28.00	—	28.00	—
5 gallons	—	Ea	104.00	—	104.00	—

Stops Rust, Rust-Oleum®. Resists moisture and corrosion. For metal, wood, concrete and masonry.

	Craft@Hrs	Unit	Material	Labor	Total	Sell
Half pint	—	Ea	4.66	—	4.66	—
Quart	—	Ea	9.67	—	9.67	—

Rusty metal primer. Fish oil based formula penetrates rust to bare metal. Drives out corrosive air and moisture. Bonds tightly to rust to form a surface that topcoats can adhere to. Volatile organic compound (VOC) compliant.

	Craft@Hrs	Unit	Material	Labor	Total	Sell
Half pint	—	Ea	5.57	—	5.57	—
Quart	—	Ea	9.67	—	9.67	—

Specialty Coatings

Chlorinated rubber swimming pool paint, Insl-X. For new or existing pools, indoor or outdoor. Semi-gloss, self-priming for gunite and marcite pools. Durable in salt or fresh water. Resists fading, fungus, algae, abrasion and alkalis.

	Craft@Hrs	Unit	Material	Labor	Total	Sell
Gallon	—	Ea	50.00	—	50.00	—

Barn and fence paint.

	Craft@Hrs	Unit	Material	Labor	Total	Sell
Gallon	—	Ea	14.20	—	14.20	—
5 gallons	—	Ea	68.30	—	68.30	—

Tru-Glaze 4508™ epoxy gloss coating, Devoe. Waterborne. Hard, durable high-performance architectural coating for use on interior concrete, concrete block, drywall, metal and wood.

	Craft@Hrs	Unit	Material	Labor	Total	Sell
5 Gallons	—	Ea	418.00	—	418.00	—

Tru-Glaze 4160™ epoxy primer, Devoe. High-performance waterborne epoxy rust-inhibitive primer for steel, aluminum and certain galvanized metal surfaces. Can also be used over concrete, masonry, glazed brick or ceramic tile. Excellent adhesion, chemical resistance, and good resistance to splash, spillage or fumes for a wide range of chemicals. May be top-coated with epoxy, urethane, alkyd or latex coatings.

	Craft@Hrs	Unit	Material	Labor	Total	Sell
Gallon	—	Ea	30.23	—	30.23	—

Earthquake Damage

18

Most homes built more than 100 years ago were constructed with little or no regard for seismic safety. Earthquakes were considered random acts of nature that couldn't be forecast or prevented. Damage was inevitable — but only for the unfortunate few in the area affected. And there was no way to anticipate who would be affected. So there was little reason, and no practical way, to build for earthquake safety.

Since the 1930s, our perspective has changed, largely due to analysis of damage and loss of life from earthquakes in California and elsewhere. Earthquakes remain unpredictable and inevitable. But damage and loss of life is not. Materials have improved, codes have improved, our understanding of how earthquakes affect structures has improved. Homes and commercial buildings constructed under modern codes, such as the *International Residential Code (IRC)* and the *International Building Code (IBC)*, have to meet the seismic safety standards appropriate for the community where built.

But for many years to come, older buildings along the Pacific rim will be susceptible to damage or destruction from earthquake. Buildings along the central Mississippi valley, the eastern seaboard and in other areas may also be damaged by seismic activity. In the 20 years between 1975 and 1995, there were only four states that did not have at least one earthquake: Florida, Iowa, North Dakota, and Wisconsin.

Types of Damage

Damage from earthquakes can be either structural or cosmetic. Any compromise in the ability of a building to resist expected loads is considered structural damage. Some structural damage can be repaired. For example, it's common for wood-frame walls, window openings and doorways to be twisted out of alignment by an earthquake. If the framing is otherwise sound, the wall can be braced back into alignment and secured in place, and may last for many years.

Other structural damage is more difficult to repair. For example, the massive structural failure in Figure 18-1 makes this home a candidate for demolition rather than improvement. Expect local authorities to prohibit entry into this home, at least until a licensed structural engineer makes an evaluation of whether it's safe to re-enter the building. The *International Property Maintenance Code (IPMC)*, where

(Photo courtesy of FEMA)

Figure 18-1
A candidate for demolition

adopted, gives the building department authority to deal with public hazards like this. A building official will usually notify the property owner of the action required. If the owner doesn't comply, the building official can expel the occupants, secure the building or even demolish the structure.

Nearly all cosmetic damage can be repaired. Stiff, brittle materials, such as wallboard, brick masonry and glass tend to fracture under earthquake loads. When placed under stress by sudden ground movement, well-anchored wood framing will deflect and then rebound. But finish materials (such as plaster, wallboard or stucco) will show evidence of stress by developing cracks at joints and corners. Most repair of earthquake damage involves patching and matching. These cosmetic repairs won't prevent similar damage in the future. But the next earthquake may be years or decades into the future, perhaps well beyond the useful life of the building.

Structural Improvements

When making earthquake repairs, consider making upgrades that can prevent damage in the future, especially serious structural damage. The Federal Emergency Management Agency (FEMA) offers a list of cost-effective structural improvements in *Home Builder's Guide to Seismic Resistant Construction*. These seismic improvements are especially cost-effective when made at the same time earthquake damage is repaired. What FEMA recommends won't bring an older building up to the standard set by modern building codes. But FEMA recommendations address the most common failures found when older buildings are damaged by earthquake.

Structural home improvements require a permit nearly everywhere in the U.S. All construction contractors understand the risk of working without a permit. But you've probably met owners who don't want an inspector on the property. Their concern is that an inspector will require the entire building to be brought up to code. That's possible. But it isn't likely. And it isn't required by the code. *International Residential Code* Section R102.7.1 requires only that new work conform to the *IRC*. Existing construction is grandfathered in unless it's inherently unsafe (such as an unreinforced masonry chimney) or reduces performance of the house. What that means is a matter for negotiation. But in any negotiation over improvements to an older building, both the contractor and the owner have an advantage. What you plan to do will increase the durability, safety and the value of the structure. That's a benefit to everyone. If the building department makes demands the owner can't meet, nothing is going to happen — neither the improvements you plan nor what the building department requests. On that basis, the building department will usually be quite reasonable. But it's always good practice to discuss the issue of code upgrades when applying for a permit to work on an existing building.

In any case, professional home improvement contractors who value their reputation (and their license) are well-advised to resist the temptation to work without a permit.

What FEMA Recommends

FEMA has studied hundreds of structural failures caused by earthquakes. Many of these failures could have been limited or eliminated entirely with a few straps, anchors or braces. Consider making the changes recommended in this chapter any time you repair earthquake damage. Most of these changes can be made at a very reasonable cost.

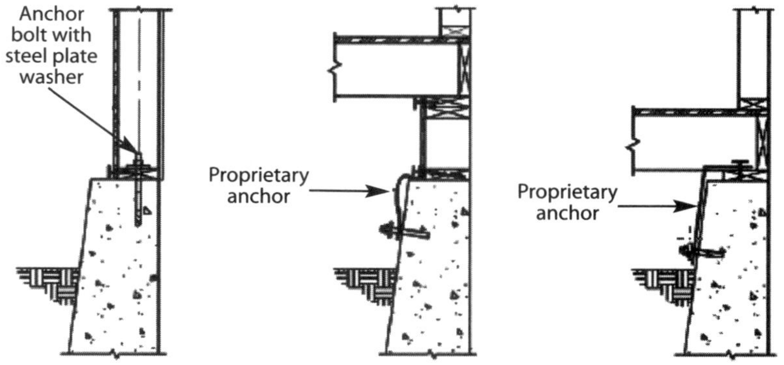

(*Courtesy of FEMA*)

Figure 18-2
Three types of foundation anchors

Here are the defects FEMA has cited as most common in older homes:

- missing or inadequate bolting to the foundation
- inadequate cripple wall bracing
- damage to bracing and finish materials
- excessive drift at garage fronts
- partial or complete collapse of hillside houses
- separation and loss of vertical support at "split-level" floor offsets
- damage and collapse of masonry chimneys

Nearly every older home in a seismic zone is a candidate for one or more of these improvements.

Foundation Bolting

Many houses constructed in California before 1950, and even later in other parts of the United States, do not have anchor bolts that attach the wall bottom plate or foundation sill plate securely to the foundation. During an earthquake, wood framing can slide off the foundation. If you're doing cosmetic earthquake repairs in an older building, note the presence or absence of anchor bolts. *IRC* code section 403.1.6 requires a ½" anchor bolt embedded 7 inches into concrete every 6 feet.

If you don't find anchor bolts where expected, adding additional anchor bolts isn't particularly difficult or expensive. Figure 18-2 suggests three good options. Drill expansion-type anchors into a reinforced concrete or masonry foundation. Adhesive anchors are a good choice for use in unreinforced masonry or weaker concrete foundations, but can be used with any type of foundation. Be sure to place a steel plate washer under each anchor head.

If there isn't enough working area to drill an anchor bolt hole through the bottom plate and into the foundation, use one of the other type of anchors shown in Figure 18-2. All of these anchors will transfer horizontal earthquake loads from framing to the foundation.

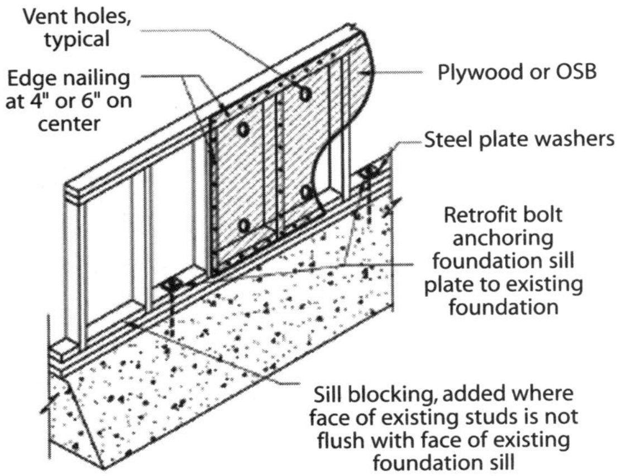

(Courtesy of FEMA)

Figure 18-3
Cripple wall bracing

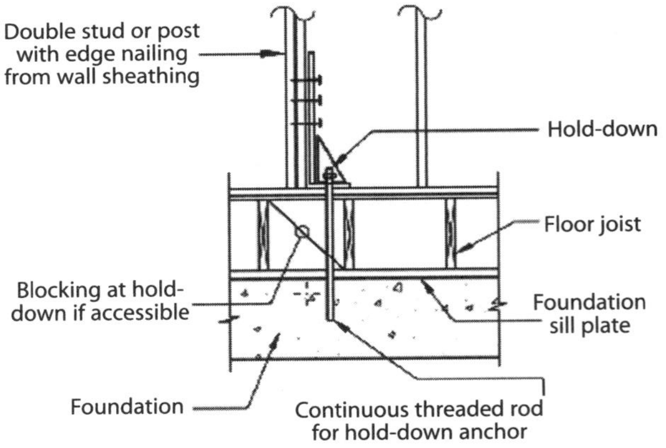

(Courtesy of FEMA)

Figure 18-4
Adding a hold-down anchor

Cripple Wall Bracing

Any framed wall above the foundation and below the first framed floor is a candidate for extra bracing. These *cripple walls* are highly susceptible to damage during an earthquake. Cripple walls in houses built prior to about 1960 seldom have adequate bracing. The sheathing won't be structural grade panels and may not be attached securely at both the bottom plate and the first floor. It's easy to upgrade cripple walls for earthquake resistance.

Before starting the upgrade, check existing cripple wall framing for signs of decay. Replace any framing lumber that appears to be decayed. Use treated lumber. Check to be sure there are enough foundation anchors. Then apply structural panel sheathing to either the exterior or interior face of the cripple wall. Sheathing and connections should meet *IRC* requirements for new construction. Figure 18-3 shows cripple wall bracing details. The Association of Bay Area Governments (San Francisco) offers a detailed plan sheet for bracing of cripple walls at www.abag.ca.gov

Apply $^{3}/_{8}$" plywood or OSB panels to the interior face of cripple walls if there's enough room to work in the crawlspace. If not, install panels on the exterior with nails spaced no more than 4 inches apart. Reinforcing a poorly-braced cripple wall is probably the single most cost-effective improvement you can make to protect a pre-1960 structure from earthquake damage.

Weak- and Soft-Story Bracing

Ground floors of older multi-story buildings seldom have adequate bracing. To reduce the risk of earthquake damage, consider adding hold-down anchors at corners any time you remove interior finish materials. Attach these anchors to the end studs (or other studs that have sheathing edge nailing). Sink a vertical rod or bolt into the foundation below. See Figure 18-4. Removing a 16"-wide by 4'-high section of drywall at each corner should provide enough working room.

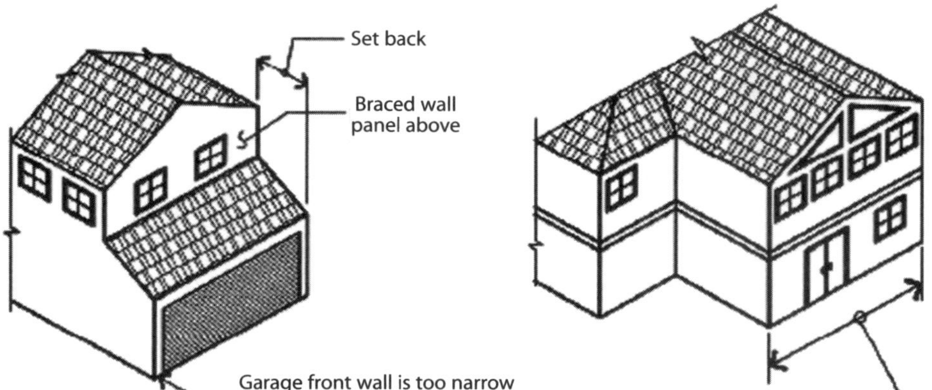

(Courtesy of FEMA)

Figure 18-5
Open fronts in need of extra bracing

This is an effective upgrade anywhere you can tie into a reinforced continuous concrete or masonry foundation. Anchors set in isolated footings or unreinforced masonry footings will be much less effective.

Open-Front Bracing

Figure 18-5 shows two buildings with what could be called an *open-front*. The double garage door in the building at the left occupies most of the building wall of the lower story. The building on the right has windows nearly all the way across one wall of the second story. In both cases, wall openings leave little room for adequate wall bracing. Open-front garages (at the left in Figure 18-5) are especially vulnerable when a bedroom is located above the garage.

In older homes without engineered steel bracing, every wall should have at least one solid wall panel at least 32 inches wide. Walls over 25 feet long should have at least one solid wall panel for each 25 feet of length. In newer homes, narrow wall segments at each side of a garage door will usually conceal engineered steel bracing or a fabricated steel panel. You'll probably see steel hold-down connectors or straps embedded in the foundation. If the home doesn't meet one of these standards, consider installing either steel wall braces or structural panel sheathing. Tie the braces or sheathing to the foundation as required by the *IRC*. Figure 18-6 shows a typical application.

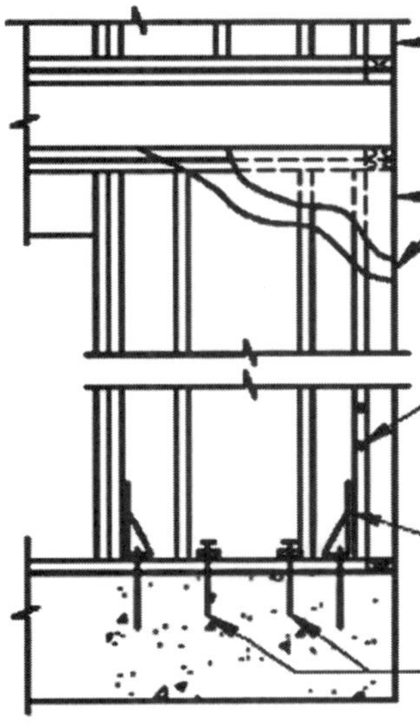

(Courtesy of FEMA)

Figure 18-6
Bracing a narrow wall segment

(Photo courtesy of FEMA)

Figure 18-7

Poorly-anchored hillside home

Extra hold-downs will add earthquake resistance if the existing foundation is continuous reinforced concrete or masonry. If the foundation isn't reinforced concrete or masonry, a structural engineer may recommend a steel moment frame that can transfer loads to other parts of the building.

Bracing Hillside Homes

Beginning in the 1950s, it was common to see homes built over steep hillsides. Usually the main floor of such a house is at street level. A system of stilts or tall wood-frame walls support the portion of the home built over the lower part of the lot. In an earthquake, the floor framing can pull away from the uphill foundation wall. Figure 18-7 shows a common result. An earthquake moved the home a step or two down the slope. Today we know better ways to anchor a home securely to the uphill foundation. But older hillside homes need extra protection from earthquakes.

The *IRC* doesn't include any approved design for reducing the risk of collapse in a hillside home. But an engineer will recommend upgrades that include:

- additional hold-down connectors or adhesive anchors to tie each floor joist securely to the upper level foundation, as in Figure 18-8.
- attaching bottom plates of stepped (or sloped) downhill cripple walls to the upper level foundation. This is particularly important when the top of the side foundation is sloped rather than stepped.

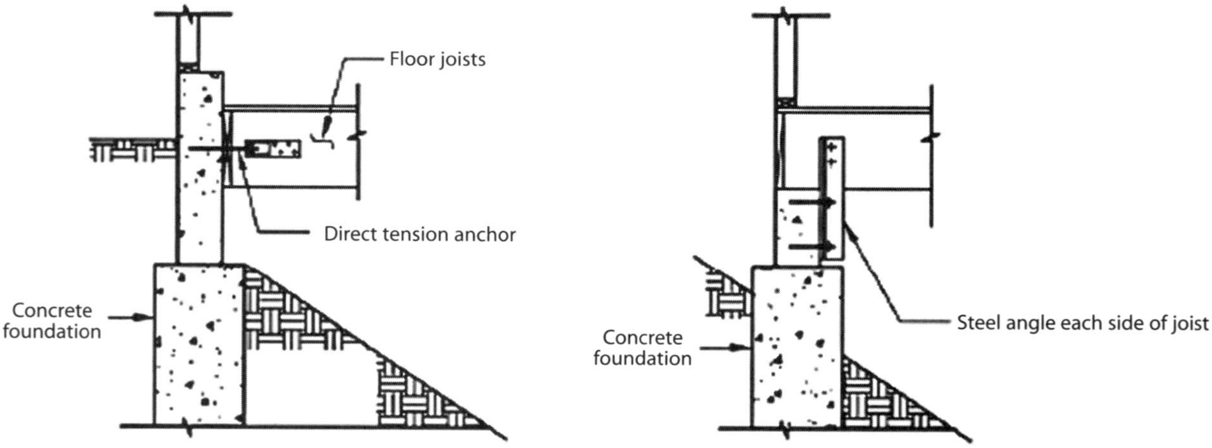

(Courtesy of FEMA)

Figure 18-8

Anchors for a hillside home

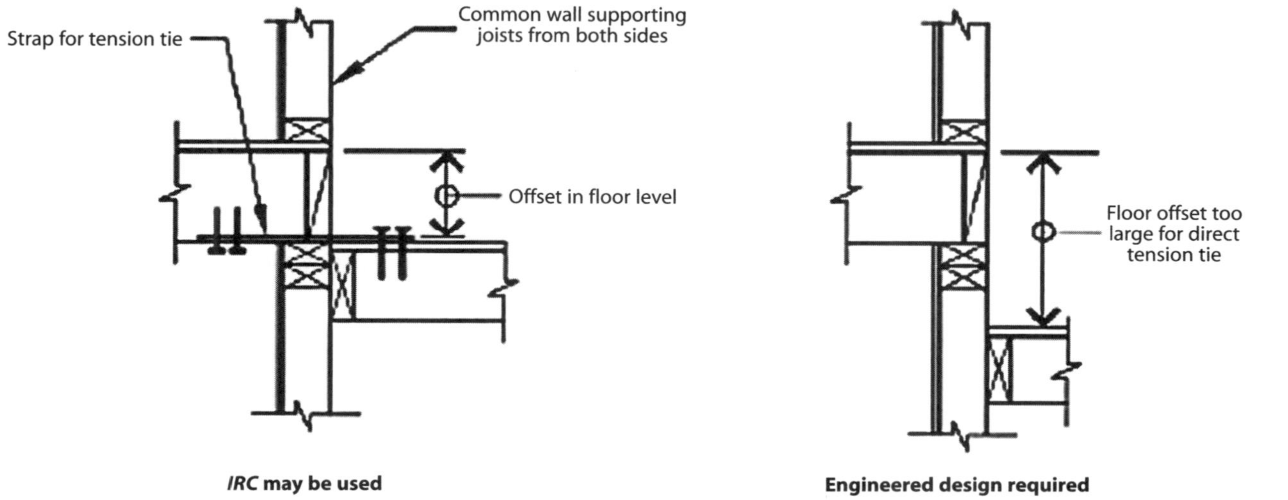

(Courtesy of FEMA)

Figure 18-9
Ties for a split-level home

- joining stepped or sloped side cripple wall top plates at each splice joint with strap connectors. Heavy steel straps should connect securely on both sides of the splice. Make the top plate of a stepped cripple wall act as though it were one piece along the entire length of the wall.
- structural panel sheathing applied over the stepped wall and on the downhill wall.

Split-Level Homes

The weak point in a split-level home is where vertical offsets join a common wall or other support at different levels. See Figure 18-9. In an earthquake, floor and roof framing can pull away from the common wall.

To reduce the risk of damage due to earth movement, add ties to the floor framing on either side of the common wall. Where the offset in floor elevations is no more than the depth of floor joists, install steel straps every 8 feet along the common wall. Where the offset is more than the depth of a floor joist, the *IRC* doesn't offer a solution. You'll need help from a structural engineer.

Anchor Masonry Chimneys

Fireplace chimneys are heavy, rigid, brittle and very susceptible to earthquake damage. Chimneys in older homes may not have reinforcing steel and usually aren't anchored securely to the house. In an earthquake, large sections of the chimney can fall away from the house — even into the house. The *IRC* now requires

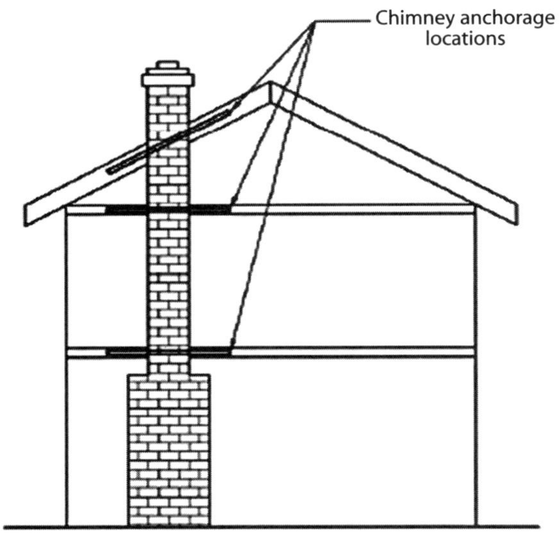

(Courtesy of FEMA)

Figure 18-10

Anchoring a chimney to the building frame

both horizontal and vertical reinforcing and chimney anchors that tie into the floor, ceiling and roof.

If the chimney is outside the frame walls, consider wrapping steel straps around the outside of the chimney. Figure 18-10 shows straps at the floor, ceiling and roof level. These straps will be exposed to the elements, so use straps with corrosion protection. Straps aren't likely to reduce earthquake damage to the chimney itself but will reduce the chance of injury due to falling debris.

Where the chimney extends well above the roof line, add a brace to connect the top of the chimney to the roof surface.

A better choice is to replace a masonry fireplace and chimney with a factory-built fireplace and flue surrounded by light-frame walls. Factory-built fireplaces and chimneys withstand earthquakes much better than conventional masonry fireplaces and chimneys.

When Making Alterations

Consider earthquake risk any time you change or remove bracing from a wall, floor or roof. Even interior remodeling projects can increase the risk of earthquake damage. Many interior walls actually provide protection against earthquake and wind loads.

Figure 18-11 shows two popular remodeling projects that can add risk of earthquake damage. At the left in Figure 18-11, dormer windows have been added to an existing roof. Any time you add an opening to an existing roof, check the opening size against permitted maximum sizes in the *IRC*. Doubled rafters and headers may be required to meet gravity load requirements.

At the right in Figure 18-11, a sun room has been added to the rear of a home. If you remove wall bracing when making an alteration such as this, check the remaining wall braces to be sure location, length and type of bracing comply with the *IRC*. More bracing will usually be required in walls on either side of the alteration.

When Adding a Room

Additions to a home are either horizontal or vertical, or both. Figure 18-12 shows a horizontal addition (upper drawing) and a vertical addition (lower drawing). Many horizontal additions built at the side of an existing house will require reframing the roof, as

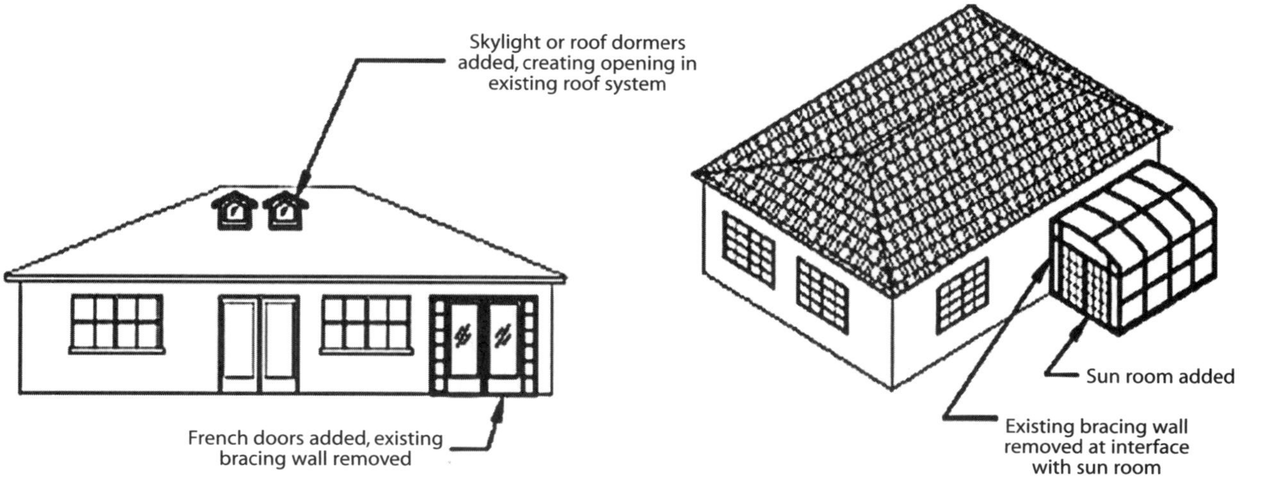

(Courtesy of FEMA)

Figure 18-11

Remodeling projects can increase the risk of earthquake damage

shown in the upper drawing in Figure 18-12. Vertical additions always require a change in the roof line.

Think about earthquake resistance when you add living space to a home. Removing any wall bracing will increase the risk of earthquake damage. Any increase in the living area, the roof area or the floor area will add extra weight and thus increase earthquake loading.

Horizontal additions will require extra bracing around new wall penetrations — and possibly extra bracing of interior walls. Check to be sure that bracing and anchors in the completed wall comply with the *IRC*.

Tie the addition to the existing house with steel straps. Ideally, the joint between old and new should be as secure as though both sections were built at the same time. You can't join the old and new foundations as though both were poured at the same time. But you *can* strap the new top plates and sill plates securely to existing top plates and sill plates. If possible, connect structural sheathing panels across the joint between old and new construction.

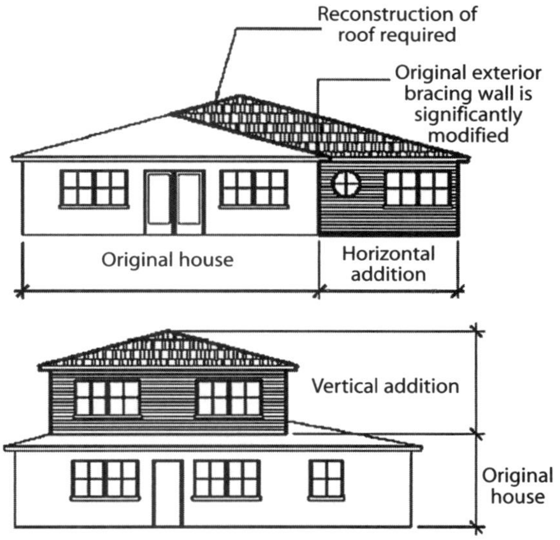

(Courtesy of FEMA)

Figure 18-12

Horizontal addition above, and a vertical addition below

A vertical addition will more than double the gravity and lateral loads on walls of the floor below. That usually means you'll have to bring the entire home into compliance with the current code. In theory, you could probably meet *IRC* requirements when adding a second story. But it's usually more practical to prepare a plan that meets design requirements of the *International Building Code*.

Checking for Earthquake Damage

Serious structural damage to a building deserves evaluation by a structural engineer. The decision to get advice from a licensed structural engineer rests with the building owner (or the building department), not with the contractor called on to make repairs. If you sense that a structural evaluation is required and if none has been performed, raise that issue with the owner. For the purposes of this chapter, I'm going to assume that engineering issues have been resolved before you step on the site.

Most earthquake damage is obvious from the exterior — walls and walks will show signs of stress or collapse. Walk around the building and through the building interior and you'll see most of what has to be repaired or replaced. If you don't see damage at the foundation level, crawling under the building won't provide much additional information. If you see obvious damage to a cripple wall, investigate from the interior. Decide if it's practical to add bracing to the wall interior.

What applies to crawlspaces also applies to attics. True, an unfinished attic is the best place to inspect framing members without removing finish materials. But any earthquake damage to ceiling joists or rafters is likely to be obvious from the condition of walls and ceilings below the attic. Crawling around in the attic isn't likely to add much to your damage assessment.

Check joints where a slab meets the foundation wall. When a home is built partially on filled ground and partially on cut ground, it's common to see gaps open between the slab and the foundation. Check the slab perimeter for gaps anywhere the slab is exposed, such as in the garage or in closets. Roll back the carpets at the room perimeter. Where finish floors are hard tile, stone or wood, you'll see loose flooring if the slab has separated from the foundation stem wall.

A minor gap between a footing and a floor slab isn't a structural risk. Caulk the opening to help keep insects and moisture out of the building interior. If a crack is wider than ¼", clean the opening of any foreign material and then fill the void with a masonry crack filler. If either side of the crack has been displaced up or down, finish the job with a coat of floor leveling compound.

Before Making Repairs

Before removing any finishes, get an estimate of the year when the building was constructed. The Environmental Protection Agency's *Renovation, Repair and Painting (RRP)* rule sets standards for repair work done on homes built in 1978 or earlier. That was the last year lead-based paints were commonly used in residences. Full information on the rule can be found on the EPA site: www.epa.gov/lead/pubs/renovation.htm. If in doubt about the year of construction, test the paint for lead content. Many building supply stores sell inexpensive lead test kits that meet requirements of the RRP rule.

Cosmetic repair of earthquake damage such as surface cracks, nail pops and minor misalignment doesn't require a building permit in most communities. But some building departments require a permit if the cost of cosmetic repairs exceeds a specific percentage (such as 10 percent) of the cost of a new wall or ceiling.

Repair of Foundations and Floor Slabs

Cracks up to 1/32" wide (less than the thickness of a dime) in concrete and masonry are usually the result of shrinkage and are more a cosmetic issue than a structural risk. No repair is required. Wider cracks in patios and walkways can be repaired by grinding away raised edges and filling gaps with mortar mix. Where load-bearing capacity is important, such as in a driveway, floor slab or foundation, structural repair may be required.

Cracks in concrete can expose steel reinforcing embedded in the concrete. Steel exposed to the air will corrode. Badly corroded steel reinforcing or foundation bolts can result in structural failure. Corroded concrete reinforcing may be a more serious structural problem than fractures in the concrete itself. Fortunately, there's a good way to prevent corrosion and restore the load-carrying capacity of most concrete foundations and slabs.

After the Northridge, California earthquake on January 17, 1994, several insurance carriers agreed to pay for concrete crack repair done with injected epoxy. At the time, these repairs were controversial. In 1994, no structural testing had been done on reinforced concrete repaired with injected epoxy. Today, we know that injected epoxy can restore the original load-carrying capacity of concrete foundations and slabs with cracks up to about 1/2" in width. Cracks wider than ¼" may require pre-packing with sand or aggregate.

Epoxy Injection

Before starting work, inquire at your building department about any special requirements. Your building department may require a building permit or other documentation for structural crack repair with epoxy.

Repair starts by drilling injection ports every 4" to 8" along the crack. Port depth depends on thickness of the concrete. Most slabs are 4" thick. Foundation stem walls are usually 8" thick. Spacing and diameter of the ports depend on the width of the crack and the thickness of the concrete. Tighter cracks and thicker concrete require more ports. The drill should have an attached vacuum system to keep concrete dust out of the repair area. When drilling is done, flush the crack with a stream of water or an air blast. Place a plastic injection cap over each drilled port. Then seal the entire surface of the crack with epoxy gel. When the gel has set enough to seal the crack, begin injecting epoxy resin through the port caps using a binary caulking gun. High pressure isn't required. A thin mix of epoxy will penetrate very small cracks even in cold weather.

Start filling ports at the lowest point. Inject epoxy until material begins to flow out of an adjacent port. Then move on to the next port.

One word of caution: A concrete wall or floor slab that doesn't have proper expansion joints can be put at risk by too much crack repair. Generally, walls and slabs need an expansion joint every 8 to 10 feet. Without joints, concrete walls and slabs will crack due to changes in temperature.

Epoxy injection can also be used to seal cracks that leak water into a basement. The application procedure is the same except that full penetration of epoxy into the cracks isn't essential.

Floors Out of Level

Earth movement can cause uneven settling of a foundation even when the foundation remains intact and structurally sound. No building is completely level. But if the floor above a conventional crawlspace foundation is out of level by more than 1" in 20 feet, it may be possible to re-level the home.

Before making repairs, satisfy yourself that future soil movement is unlikely. A soils engineer may recommend either pressure-injected grout or pipe piles. Both are expensive choices and may be impractical for most homes. The *National Construction Estimator* includes cost estimates for both pressure grouting and steel pipe piles.

Inspect the crawlspace under a home that's out of level. You'll probably find tilted support posts, floor beams that lack adequate support and fractured or slipping mudsills. If you plan to re-level a home, make repairs to posts, beams and sills part of the plan. Otherwise, the home isn't likely to remain level for long.

To re-level a home with a conventional crawlspace foundation, jack the floors back to level using bottle jacks and needle beams. Chapter 4 has details and cost estimates. Start by backing off the foundation bolts. Then set needle beams under floor joists and begin raising the floor. This should be a gradual process, no more than ¼" per day. When the floor is back to level, inject grout under the mudsills and shim other supports. Some cosmetic damage to interior finishes at wall corners and around wall openings is almost inevitable. Exterior walls will show cracks at the sill line.

Framing Repairs

It's common for large beams to split and twist along the long axis as the wood dries. Frame walls tend to creep at least slightly out of square and plumb as lumber dries. Normal splitting and twisting of lumber is considered a natural event, not a structural problem. Earthquake damage to framing will be more serious — such as walls racked so far out of alignment that windows or doors no longer operate smoothly.

The Cape Mendocino earthquake of April 25, 1992 left the first floor of the home in Figure 18-13 displaced about 6 inches to the left. Notice the shape of the garage door opening.

(Photo courtesy of FEMA)

Figure 18-13
Racked first floor wall

You'll seldom see damage this obvious. Use a spirit level to check walls for plumb. Earthquake damage will usually leave most walls out of plumb in the same direction. Use a framing square to check openings for square. If windows or doors no longer fit the frame or bind when opened and closed, the wall probably needs to be straightened. In general, any 8' wall more than ½" out of vertical alignment needs repair.

To straighten a wood-frame wall, remove the finish on one side. Set wedge bracing or jackposts and jack the wall back into alignment. (Notice the wedge brace at the left in Figure 18-13.) When re-aligned and when windows and doors operate normally, re-nail the framing connections and install either diagonal blocking, let-in wall bracing or structural grade shear panels. Shear panels nailed every 4" at the base and top of each panel should prevent similar damage from future earthquakes. Then re-install the wall finish.

Cracks in Plaster and Stucco

Most movement of ground during an earthquake is lateral — left and then right. Lateral movement in a frame building creates horizontal cracks starting at the corners of framed openings such as doors and windows. On interior walls, you'll see a pattern of similar cracks on both sides of the wall. If you see cracks running at random or mostly vertical on a plaster or stucco wall, expect that the cause is settling or shrinkage of framing members rather than an earthquake.

Don't bother patching hairline cracks in stucco or plaster. A good coat of paint will fill and cover cracks no wider than the stroke from a ballpoint pen ($^{1}/_{64}$"). Any crack wide enough to insert the blade of a table knife is a candidate for repair.

Scribe out the crack with the tip of a screwdriver, putty knife or a linoleum blade. Cut deep enough to get down to the brown coat. On exteriors, fill the crack with premixed stucco patch. On interior surfaces, use premixed spackling compound. Several thin coats are better than a single thick coat. Sand when dry and then repaint.

According to *The General Guidelines for the Assessment and Repair of Earthquake Damage* published by CUREE (The Consortium of Universities for Research in Earthquake Engineering), building paper under a stucco finish is far more flexible than the stucco itself. Even where the stucco is badly fractured, tears in the building paper are unlikely. If the stucco can be repaired without being

(Photo courtesy of FEMA)

Figure 18-14

Factory-built firebox enclosed in frame walls

removed, don't worry about vapor barrier under the stucco. No inspection of the building paper is required.

Gypsum Drywall

Minor cracks (to 1/8" wide) usually show up at taped panel joints, at the corners of wall openings, at interior wall corners and where panels join corner bead or trim. Cut away any existing joint tape. Scrape off loose joint compound. Then re-tape, re-texture, sand smooth, and re-paint.

Nail head pops can happen anywhere in a panel. Pull the nail. Set a drywall screw about 1" from the original nail. Drive the replacement screw just deep enough to dimple but not tear the paper cover. Then re-texture, sand and re-paint.

When shaking is severe, drywall panels can buckle or work loose from the framing. Heavily stressed drywall panels develop an "X" pattern of ripples that extends diagonally from one corner to the opposite corner across the face of the board. If the surface of drywall has ripples or if cracks extend through the board, it's best to remove the damaged section. If a large section is damaged, remove the entire panel. If damage is limited to a small area, strip off a section 48" high and wide enough to expose the middle of the stud or joist on each side. Cut a section of board to fit the opening. Apply the replacement section with drywall screws. Tape, texture, sand and finish.

Fireplaces and Chimneys

Masonry fireplaces and chimneys are usually built on a separate foundation and are intended to be self-supporting. Most masonry chimneys built before about 1930 have mortar made from lime rather than Portland cement. Lime mortar is weak. Any chimney with lime mortar is a good candidate for replacement. To check for lime mortar, scratch the surface of a joint with a screwdriver. If you cut into the joint, it's made with lime mortar.

Today, most fireplaces in seismic zones are prefabricated metal. An insulated sheet metal flue is enclosed with wood-frame walls. The finish can be stone or masonry veneer or stucco. A prefabricated firebox and chimney, such as in Figure 18-14, has little risk of earthquake damage.

Chimneys strapped to the main building, as in Figure 18-10, aren't likely to topple over. But a strapped chimney that isn't adequately reinforced can still fracture. If you see cracks in a chimney, shine a flashlight down the chimney to check the flue lining for visible damage. Broken flue tile will be a fire hazard the next time there's a fire in the fireplace.

A masonry firebox at the base of a chimney can develop minor cracks under stress from an earthquake. These cracks don't present a fire hazard. But stress from earth movement can snap masonry at the point where the fire box joins the chimney. If you see cracks at this point or suspect damage to the chimney flue, build a smoky fire in the fireplace and check for smoke leaking from the chimney.

Masonry and concrete chimneys usually have a flashed, counter-flashed and caulked construction joint where the chimney joins the exterior finish. This joint allows the chimney to move independently of the remainder of the home during an earthquake. Even light ground-shaking can cause this joint to open. If the flashing is intact, running a new bead of caulk down the open joint will restore weather-resistance.

Repair of damaged masonry chimneys is usually prohibited in seismic zones. The chimney has to be removed completely and re-built to current code standards. Although the code permits adding a new metal chimney to an existing firebox, it's usually more practical to replace both the firebox and chimney with a new prefabricated unit.

Mechanical, Electrical, and Plumbing Systems

The most likely damage to plumbing will be a water heater that's tumbled over or a water heater vent that's become dislodged. If the water heater remains upright, check the gas and water lines for evidence of a leak — especially if the connection is made with rigid metal pipe or tube. Proper restraints on the water heater and flex connections can prevent most earthquake damage.

Duct, flues, piping and electrical lines can rupture when framing is flexed by ground movement. But there's seldom a need to look for trouble. The problem will usually be obvious: A vent or duct that's blocked; an electrical circuit that's out; moisture leaking out of a concealed plumbing line; a suspended light fixture that's hanging by the electrical wire conductors; an appliance (such as a furnace) that's tumbled over and ruptured the connections. Unless displacement or shifting is obvious, it's usually safe to assume that plumbing and electrical lines, vents and duct have survived an earthquake intact.

There is one exception. Ground shaking can dislodge deposits that accumulate in water supply lines and water softeners. If there's a noticeable drop in pressure at a fixture or a water tap, suspect blockage in the line. Remove the screen filter or valve at the fixture. Then use water pressure to blow out the line.

	Craft@Hrs	Unit	Material	Labor	Total	Sell
Foundation anchors Drilled through the blocking and mudsill of a frame wall and 5" into an existing reinforced concrete or masonry foundation. Attached with a 5/8" diameter and a steel 3" x 3" plate washer. Add the cost of removing and replacing wall finish materials, if required.						
Drill and set 5/8" wedge anchor bolt 5" in concrete	BC@.667	Ea	4.62	24.00	28.62	48.70
Drill and set 5/8" adhesive anchor bolt 5" in concrete	BC@.800	Ea	7.62	28.80	36.42	61.90
Equipment rental (rotary hammer drill and bits)	—	Day	50.00	—	50.00	—
Foundation brackets (UFP 10) Installed to secure the bottom plate of an existing frame wall to an existing reinforced concrete or masonry foundation. 7" x 10" bracket with five 3" wood screws. Each bracket requires two 5" drilled into the foundation						
Drill and set foundation bracket on a bottom plate	BC@.333	Ea	19.30	12.00	31.30	53.20
Drill and set two 5" wedge anchors, per bracket	BC@.667	Pair	2.85	24.00	26.85	45.60
Equipment rental (rotary hammer drill and bits)	—	Day	50.00	—	50.00	—
Cripple wall bracing Structural panels secured with 8d nails each 4" at panel edges and each 12" at intermediate supports. Add the cost of removing and replacing the existing wall cover, if required. Add the cost of hold-down straps, posts, shear blocking or extra studs, if required.						
3/8" exterior plywood panel, in crawl spaces	BC@.038	SF	.71	1.37	2.08	3.54
3/8" exterior plywood panel, on exterior walls	BC@.025	SF	.71	.90	1.61	2.74
1/2" exterior plywood panel, in crawl spaces	BC@.038	SF	.84	1.37	2.21	3.76
1/2" exterior plywood panel, on exterior walls	BC@.025	SF	.84	.90	1.74	2.96
1/2" structural OSB panel, in crawl spaces	BC@.038	SF	.45	1.37	1.82	3.09
1/2" structural OSB panel, on exterior walls	BC@.025	SF	.45	.90	1.35	2.30
Hold-down anchors (HTT5) 2" x 16" high. Set with an anchor bolt drilled 5" into an existing reinforced concrete or masonry foundation and nailed to a corner stud. Add the cost of removing and replacing wall finish materials, if required.						
Drill and set wedge anchor bolt	BC@.667	Ea	3.17	24.00	27.17	46.20
Install and secure hold-down	BC@.400	Ea	21.80	14.40	36.20	61.50
Equipment rental (rotary hammer drill and bits)	—	Day	50.00	—	50.00	—
Brace a water heater Using a galvanized water heater strap kit: two straps on a water heater under 75 gallons, three straps on a 75 gallon water heater and four straps on a 100 gallon water heater. Kit includes strapping, lag screws, washers, spacers and tension bolts. Cost per water heater braced.						
Water heater under 75 gallons (2 straps)	PM@.500	EA	38.70	20.60	59.30	101.00
75 gallon water heater (3 straps)	PM@.750	LF	58.10	31.00	89.10	151.00
100 gallon water heater (4 straps)	PM@1.00	LF	77.40	41.30	118.70	202.00
Replace and gas lines with flex connections	PM@.750	LF	26.70	31.00	57.70	98.10
Framing straps and braces Nailed to exposed wood framing.						
L-straps, 12 gauge (L)						
5" x 5", 1" wide (55L)	BC@.132	Ea	2.07	4.75	6.82	11.60
6" x 6", 1-1/2" wide (66L)	BC@.132	Ea	3.50	4.75	8.25	14.00
8" x 8", 2" wide (88L)	BC@.132	Ea	5.40	4.75	10.15	17.30
12" x 12", 2" wide (1212L)	BC@.165	Ea	6.63	5.94	12.57	21.40
Add for heavy duty L-straps, 7 gauge	—	Ea	5.80	—	5.80	—
Long straps (MST), 12 gauge						
27" (MST27)	BC@.165	Ea	6.10	5.94	12.04	20.50
37" (MST37)	BC@.165	Ea	7.65	5.94	13.59	23.10
48" (MST48)	BC@.265	Ea	10.20	9.54	19.74	33.60
60" (MST60)	BC@.265	Ea	16.40	9.54	25.94	44.10
72" (MST72)	BC@.285	Ea	30.60	10.30	40.90	69.50

	Craft@Hrs	Unit	Material	Labor	Total	Sell
Plate and stud straps (FHA), galvanized, 1-1/2" wide, 12 gauge						
6" (FHA6)	BC@.100	Ea	2.72	3.60	6.32	10.70
9" (FHA9)	BC@.132	Ea	2.73	4.75	7.48	12.70
12" (FHA12)	BC@.132	Ea	3.04	4.75	7.79	13.20
18" (FHA18)	BC@.132	Ea	4.05	4.75	8.80	15.00
24" (FHA24)	BC@.165	Ea	5.28	5.94	11.22	19.10
30" (FHA30)	BC@.165	Ea	8.77	5.94	14.71	25.00
Tie straps (ST)						
2-1/16" x 9-5/16" 20 ga. (ST292)	BC@.132	Ea	1.70	4.75	6.45	11.00
3/4" x 16-5/16" 20 ga. (ST2115)	BC@.046	Ea	1.61	1.66	3.27	5.56
2-1/16" x 16-5/16" 20 ga. (ST2215)	BC@.132	Ea	2.09	4.75	6.84	11.60
2-1/16" x 23-5/16" 16 ga. (ST6224)	BC@.165	Ea	4.66	5.94	10.60	18.00
2-1/16" x 33-13/16" 14 ga. (ST6236)	BC@.165	Ea	5.63	5.94	11.57	19.70
T-straps, 14 gauge (T)						
6" x 6", 1-1/2" wide (66T)	BC@.132	Ea	4.16	4.75	8.91	15.10
12" x 8", 2" wide (128T)	BC@.132	Ea	9.45	4.75	14.20	24.10
12" x 12", 2" wide (1212T)	BC@.132	Ea	14.10	4.75	18.85	32.00
Add for heavy-duty T-straps, 7 gauge	—	Ea	5.80	—	5.80	—
Wall braces (WB), diagonal, 1-1/4" wide, 16 gauge						
10' long, corrugated or flat (WB106C or WB106)	BC@.101	Ea	12.90	3.64	16.54	28.10
12' long, flat (WB126)	BC@.101	Ea	14.40	3.64	18.04	30.70
14' long, flat (WB143C)	BC@.101	Ea	27.40	3.64	31.04	52.80
Wall braces (CWB), diagonal, 18 gauge, 1" wide						
10' long, L-shaped (CWB106)	BC@.125	Ea	13.00	4.50	17.50	29.80
12' long, L-shaped (CWB126)	BC@.125	Ea	15.20	4.50	19.70	33.50

Repair concrete cracks Structural repair using injected epoxy. Drilling an injection port every 4" to 8" along a crack and injecting low viscosity epoxy resin with a minimum bond strength of 7,000 P.S.I. Add the cost of excavation or removal of surface materials if required to expose the cracked concrete. Equipment is an electric drill with a swivel chuck, hollow drill bits, oil-free compressed air and a two component epoxy injection gun. Add the cost of concrete coring if core samples are required to confirm that all voids are filled.

	Craft@Hrs	Unit	Material	Labor	Total	Sell
Remove foreign material from within 2" of the crack	CF@.060	LF	—	2.03	2.03	3.45
Drill 4" slab and set an injection port, per port	CF@.330	Ea	.75	11.20	11.95	20.30
Drill 8" foundation and set an injection port, per port	CF@.500	Ea	.75	16.90	17.65	30.00
Flush out debris, air blast, per port	CF@.040	Ea	—	1.35	1.35	2.30
Mix and apply 1/8" thick epoxy gel surface sealer	CF@.050	LF	1.00	1.69	2.69	4.57
Inject epoxy into port, per port	CF@.100	Ea	3.50	3.39	6.89	11.70
Remove ports and excess material	CF@.030	LF	—	1.02	1.02	1.73
Add for equipment rental (drill, bits, air and gun)	—	Day	50.00	—	50.00	—

Grind off a raised concrete seam Taper a raised joint or crack in a concrete walk or slab to not more than a 7-1/2 degree incline (1" in 12") using an 8" gas-powered walk-behind concrete planer. Per linear foot of seam tapered.

	Craft@Hrs	Unit	Material	Labor	Total	Sell
Taper 1/4" raised seam, 3" wide	OE@.050	LF	—	2.04	2.04	3.47
Taper 1/2" raised seam, 6" wide	OE@.083	LF	—	3.38	3.38	5.75
Taper 3/4" raised seam, 9" wide	OE@.125	LF	—	5.09	5.09	8.65
Taper 1" raised seam, 12" wide	OE@.167	LF	—	6.80	6.80	11.60
Add for equipment rental (planer and fuel)	—	Day	315.00	—	315.00	—

Level a concrete slab Apply concrete patching and leveling compound with a trowel. 25 pound bag ($17.80) covers 25 square feet at 1/4" depth.

	Craft@Hrs	Unit	Material	Labor	Total	Sell
Per SF of surface covered at 1/4" depth	BF@.016	LF	.73	.52	1.25	2.13

	Craft@Hrs	Unit	Material	Labor	Total	Sell
Straighten a frame wall Using wedge braces and jack posts to restore a wall to vertical alignment. Add the cost of removing and reinstalling wall finish materials, if required. Vertical jacks may be required to relieve the weight on a load-bearing wall. Add the cost of installing wall straps and shear panels once the wall is plumb.						
Set and remove diagonal brace, per brace	B3@.350	Ea	—	11.00	11.00	18.70
Set and remove vertical jack, per jack	B3@.350	Ea	—	11.00	11.00	18.70
Wedge the wall into vertical alignment, per LF of wall	B3@.200	LF	—	6.27	6.27	10.70
Straighten one stud, cut saw kerf and drive shim or screw into the kerf, reinforce with a plywood scab	BC@.300	Ea	—	10.80	10.80	18.40
Remove and replace a badly warped stud	BC@.300	Ea	2.58	10.80	13.38	22.70
Adjustable jack post, 20,000 lb., purchase	—	Ea	50.00	—	50.00	—
Wedge braces, 3 uses, #2 and better lumber						
2" x 4" x 12'	B3@.012	Ea	1.33	.38	1.71	2.91
2" x 6" x 12'	B3@.018	Ea	2.26	.56	2.82	4.79
Straighten a window opening Remove a damaged window from a wood-frame wall opening and straighten the framed opening to accept a replacement window. Includes prying off window trim, removing the stops and sash, removing exterior finish to expose the jamb, pulling or cutting nails securing the window frame, lifting out the damaged window and restoring the opening to square with braces, extended bar clamps or corner clamps. Add the cost of straps, braces, the replacement window and patching the exterior wall finish.						
Remove an existing damaged window	BC@1.50	Ea	—	54.00	54.00	91.80
Re-square the window frame and sill	BC@.500	Ea	—	18.00	18.00	30.60
Plaster and stucco repair Using hand tools and working from a ladder.						
Scribe cracked plaster or stucco to 1/4" depth and fill the crack with:						
Elastomeric caulk, per linear foot filled	CF@.033	LF	—	1.12	1.12	1.90
Elastomeric patch and caulk, 10 oz. tube	—	Ea	3.85	—	3.85	—
Patch hole in exterior stucco. Chip and wire brush loose stucco, moisten surface, apply stucco patch with a trowel, per 4" x 4" patch	CF@.250	Ea	0.45	8.47	8.92	15.20
Premixed stucco patch, quart	—	Ea	7.63	—	7.63	—
Touch-up painting to match	PT@.333	Ea	4.30	12.20	16.50	28.10

Gypsum drywall repair Cost estimates for removing, repairing and replacing interior finishes appear in chapter 10 beginning on page 242.

Fireplaces and chimneys Cost estimates for removing, repairing and replacing fireplaces and chimneys appear in chapter 6 beginning on page 115.

Water Damage and Mold Remediation

19

The first part of this chapter focuses on cost estimates for dryout (also known as *restorative drying*). The second part covers mold remediation. The two topics (dryout and mold remediation) are often complementary. Prompt dryout avoids mold. Even when mold isn't the direct result of a flood or a leak, much of the work required for mold remediation is similar to what's needed when a building is flooded. The estimates in this chapter cover both dryout and mold remediation from beginning to end.

Three notes of caution:

- *First,* some states license either dryout contractors or mold remediation contractors, or both. Be sure you have the license or certification required in your community before taking on a dryout or mold remediation job.
- *Second,* mold may be a problem on some dryout jobs. If you see evidence of mold when handling a dryout job, address the mold issue before starting to move air (and mold spores) around the premises.
- *Third,* most dryout and mold remediation won't be covered by insurance. For example, melting snow, leaks from a wornout roof, dry rot, and rising water aren't usually covered. The owner probably has coverage for sudden and accidental losses like a broken pipe, wind-driven rain or an overflowing toilet.

Building Dryout

Flooding affects a home in at least three ways:

1. Water will damage most construction materials. For example, wallboard disintegrates when wet for too long. Wood can swell, warp and rot. Electrical parts can short out and corrode. Wet fiber insulation loses nearly all insulating value.
2. Mud, silt and waterborne contaminants create a health hazard.
3. Dampness promotes growth of mildew, mold and fungus.

A wet building and the contents should be dried out as soon as possible, whether moisture is from flooding (rising water) or casual water (such as from a broken pipe or roof leak), or the result of fighting a fire.

We're going to assume that damage was caused by clean water. If floodwater is contaminated with sewage, chemical or biological pollutants, Occupational Safety and Health Administration (OSHA) rules require personal protective gear and containment. Even clean water should be treated as contaminated after 72 hours. You may need the advice of an industrial hygienist, as explained the section that follows on mold remediation.

Start work when floodwater has receded or the leak has stopped. Begin at the exterior. Check the building perimeter for a gas leak, toxics (such as a sewage spill) or a downed power line. Check for obvious structural damage: a cracked foundation, supports or framing that have failed, buckling, or subsidence. Some communities require a sign-off by the building inspector before repairs can begin on a flooded building. When flood damage exceeds 50 percent of the market value of a home, the building department may require demolition and rebuilding above flood level. The code in some flood-prone areas may not permit rebuilding at all.

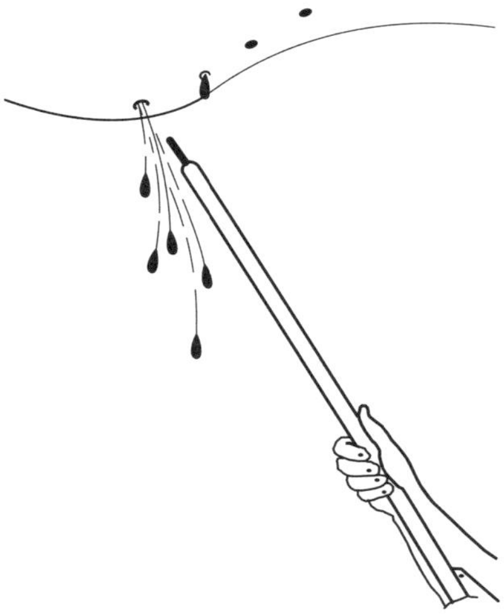

Figure 19-1
Release trapped moisture

If flooding is extensive, turn the power off at the main breaker, even if power is already off in the neighborhood. You don't want power to come back on without warning while work is under way. Turn off gas at the meter. Close the valve to any fuel oil or propane tank. Be alert for leaking water pipes. If necessary, shut off the water valve at the main.

Begin Interior Work

Start with the ceiling. Look for anything that seems likely to collapse. Wet plaster or wallboard is dangerous when it falls. If the ceiling is sagging, drive a nail into the end of a broomstick and poke holes in the plaster or wallboard. See Figure 19-1. Use hand tools to knock down a sagging ceiling. Check for loose flooring, cabinets or tall furniture that might be ready to fall. Remove mirrors and heavy pictures from wet walls. They won't stay up for long on saturated wallboard.

When it's safe to work inside, unplug all appliances and lamps. Remove the cover plates on wall switches and outlets that got wet. Either disconnect the wiring or leave the wires connected and pull the fixture out of the box, as in Figure 19-2.

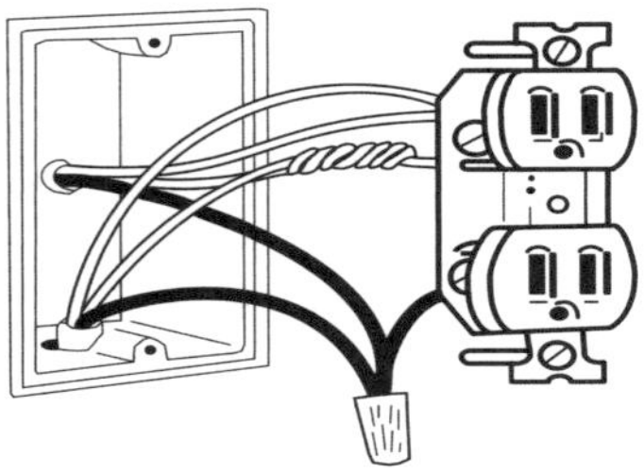

Figure 19-2
Pull the fixture out of the box

Secure the valuables. If a restoration specialist isn't available, wash and blow dry high value items (securities and jewelry) and anything irreplaceable (photographs, art, insurance papers). Then pack in cardboard cartons. You'll need several box sizes, including wardrobe boxes, bubble wrap and sealing tape. Label or number each box. Create a written inventory of everything boxed for pack-out. That saves time, both when the owner needs to find some item in storage and when moving the contents back into the dried-out building. For insurance proof of loss purposes, note the item name, room where located at the time of loss and condition after the loss.

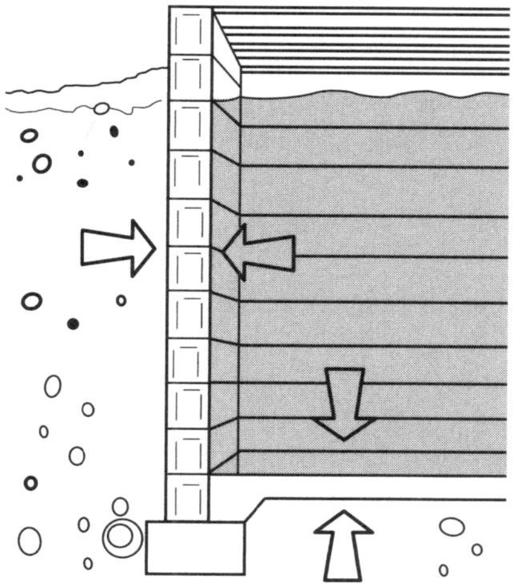

Figure 19-3

Flooded basement with equal pressure

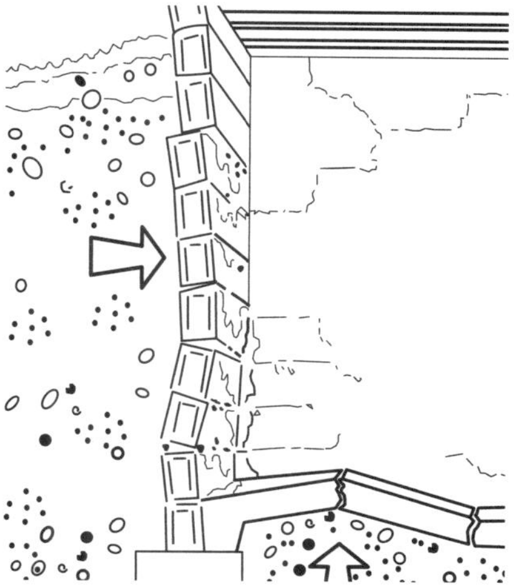

Figure 19-4

Dry basement with unequal pressure

Relocate heavy items. The best choice for relocating building contents is another room on site, a second floor or a storage container placed temporarily in the driveway. If there's going to be a delay in moving some heavy items, set foam blocks or plastic sheeting under legs or supports to eliminate contact with anything that's wet. Moisture will "wick" up wood furniture legs, discoloring the wood as it goes. If appliances and plumbing fixtures have to be moved, cap waste and supply lines (water, electric and gas). If a water heater has to be moved, start draining the tank right away.

Get the mud and debris out. Clear out the mud before it dries. Shovel or squeegee as much mud as possible. If you have water pressure, hose out the interior. Clean the mud out of electrical outlets, switch boxes and light sockets.

Remove the vents and registers from a flooded HVAC system. Duct in a flooded basement or crawl space will be contaminated. Remove a section of duct to get access to all areas. Then thoroughly wash out all flooded duct. When the mud is out, clean the duct with a disinfectant or sanitizer. Don't restart the HVAC system until you're sure it's free of contaminants.

Drain the basement carefully. Don't be in a hurry to pump out a flooded basement. Here's why. Groundwater around the exterior puts pressure on basement walls. As long as there's water in the basement, pressure will be approximately equal on both sides of the wall. See Figure 19-3.

If you drain the basement too quickly, the pressure won't be equal. See Figure 19-4. Basement walls or the floor can crack or deflect. To avoid problems, follow these steps:

- ❖ Delay pumping until soil around the exterior of the basement is free of standing water.
- ❖ Then pump the water level down 2 or 3 feet. Mark the level and wait overnight.

- ❖ If water rises to the original level by the next day, it's still too early to drain the basement.
- ❖ Wait a day or two. Then pump basement water down several feet again.
- ❖ When water stops rising in the basement, pump the water down another 2 feet.
- ❖ Wait overnight. Then pump another 2 feet each day until the basement is dry.

The water in a flooded basement is sure to be dirty — not suitable for a clean water pump. You'll need a trash pump with a screened inlet.

Start the drying. There are six ways. You'll probably need to use all six:

1. Use air movers to get the air circulating. Don't use central air conditioning or the furnace blower until you're sure ducting is free of contaminants. Carpet saturated with clean water can usually be salvaged by "floating" the carpet and pad on a stream of dry air. Raise the carpet on blocks in several places. Then use an air mover to force air under the carpet.
2. Use dehumidifiers. A high-capacity dehumidifier can reduce excess moisture very quickly.
3. Open closet and cabinet doors. Remove drawers to promote air circulation. Wood drawers swell when wet. Some may be stuck shut. You should be able to open stuck drawers once drying has started.
4. Begin drying ceiling and wall cavities. We'll cover structural drying later in this chapter.
5. Use desiccants. Materials that absorb moisture are particularly useful when drying an enclosed area with limited air circulation. Desiccants include chemicals, cat litter made of clay, and calcium chloride pellets. Hang the desiccant in a pillowcase, nylon stocking or other porous bag. Put a bucket under the bag to catch any dripping water. Then close the closet or cabinet being dried.
6. If the humidity is lower than indoors, and if weather and security conditions permit, consider opening up the house. Use a hygrometer to check indoor and outdoor humidity. Opening all the doors and windows can exchange moist indoor air for drier outdoor air. Close up the house at night or any time humidity is higher outdoors.

Be patient. Even with plenty of equipment, drying a home will take several days. Until the home is reasonably dry, damage caused by mildew and decay will continue. The musty odor will remain until the home is thoroughly dry.

Dryout by type of material

How you restore saturated floors, ceilings and walls depends on the materials involved.

Wallboard. Ceilings and walls in most homes built after about 1950 are finished with gypsum wallboard. "Gypboard" acts like a sponge, drawing moisture up the wall above flood level. Any mud or contaminants in the floodwater will bond to the gypsum, making the wallboard a permanent health hazard. Wallboard soaked in a flood usually has to be removed and discarded. If flooding hasn't soaked more than the lower 2 or 3 feet of board, consider removing just the lower 4 feet of each panel. Toenail 2" x 4" wood blocks horizontally between the studs 4 feet above the floor. Then hang 4' x 8' sheets of gypboard with the long edge horizontal, as in Figure 19-5.

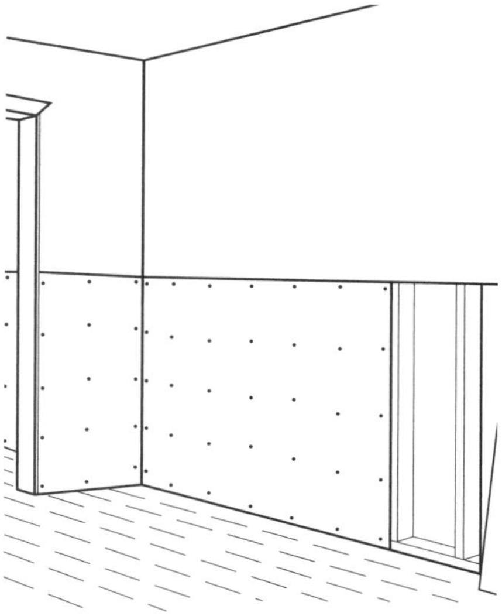

Figure 19-5
Drywall hung horizontally

Wallboard on interior walls that's been soaked with clean water will be worth saving if drying starts within the first 48 hours. After that, gypboard deflects, gets fragile and will usually need to be replaced.

Stucco and plaster will survive a flood better than gypsum wallboard. You'll have to remove stucco or plaster wall finish only if finish coats have separated from the lath below or if saturated insulation in the wall has to be removed. Vapor barrier in the wall won't be damaged by flooding. If a plaster or stucco wall is in good shape and doesn't include saturated insulation, follow the following instructions for structural drying of stud cavities. Drying stucco or plaster is slow work, even with plenty of air drying equipment.

Insulation. Flooded foam insulation board usually requires only a rinse down. Wet fiberglass batts should be removed and discarded so the rest of the wall can dry. Once the wall is dry, install new batts. Cellulose insulation will hold moisture for weeks. Even when dry, most of the antifungal and fire retardant rating will be gone. Plan to replace saturated cellulose insulation.

Wood. Wood framing lumber in an existing home usually has a water content of about 10 percent. Lumber that's been submerged can have a water content of 30 percent or more. The extra moisture causes swelling which warps lumber out of shape. If allowed to dry, most wood framing will return to the original condition. Wet studs and sills that swell out of shape won't need to be torn out. But wood framing soaked in contaminated water will need to be treated before being covered with finish materials.

A moisture meter will be useful when drying wood, especially wood flooring. Most moisture meters have two thin probes. Hold the probes against the material being tested to find the moisture content. Moisture meters work on carpet, wallboard, wood, brick, and concrete.

Figure 19-6

Wedge paneling away from the wall

Most hollow wood doors have cardboard spacers between the two veneers. Even if the veneer surfaces survive a flood intact, the cardboard core will expand, causing the door to deflect. Saturated hollow-core doors usually have to be replaced.

Laminated wood (such as plywood) tends to delaminate when the layers dry at different rates. Composition sheathing and underlayment (such as oriented strand board) start to disintegrate when soaked. Plywood or OSB that has lost nail-holding ability should be replaced.

Concrete block. The cavities in a concrete block wall will drain on their own. Water held temporarily in concrete wall cavities won't do any damage.

Wallcovering. Vinyl wallcover seals the wall and prevents drying. Wallpaper paste supports growth of mold and mildew. In most cases, you'll want to remove and discard wallcover that gets wet. But you may be able to save vinyl wallcover that's only loose at the base of the wall. Peel vinyl off the wall up to flood level. When the wall is dry and clean, reapply the vinyl.

Paneling. Carefully pry the bottom of each panel away from the wall. See Figure 19-6. Then wedge a block between the panel and the framing so air can circulate into the wall cavity. Once the wall is dry, re-nail the panels. On an exterior wall filled with wet insulation, you'll have to remove paneling to get access to the insulation.

Floor cover. Small throw rugs and loose-lay indoor/outdoor carpet can be hosed off and hung up to dry. Persian carpet may be worth professional cleaning. Wall-to-wall carpet submerged under rising water should be discarded. Cut the carpet and pad into strips that are easy to carry. Also remove and discard tack strips laid around the room perimeter. You'll need solvent and a spudding tool to remove the residue from foam-back carpet laid in adhesive.

Wall-to-wall carpet soaked by clean water can be float-dried in place. But note that saturated carpet may shrink when it dries. You'll need a professional carpet kicker if you plan to re-install float-dried carpet. Resilient flooring (tile, linoleum or seamless vinyl) doesn't have to be removed unless loose from the floor or unless the subfloor or underlayment has to be replaced.

Prices for floor cover vary widely. If you're discarding expensive carpet, be sure to save a couple of square feet for a sample the owner can give the adjuster.

Floors. If a crawl space is flooded, pump it out. Remove any vapor barrier or insulation from below the floor. Both will have to be replaced when the floor system is completely dry. Then begin circulating air both above and below the floor.

Basement. If a basement ceiling was flooded, you'll have to remove the ceiling finish to get access to insulation between the floor joists. If wet insulation isn't a problem, drill a hole through the ceiling finish into each framing cavity. Then dry joist cavities as explained in the next section. Avoid cutting or drilling near electric lines or pipes.

Gas and oil appliances. If a furnace, water heater, stove or other gas or oil appliances was flooded to the level of the burner, the appliance may have a cracked heating element. The appliance should be cleaned and checked by an expert. If you decide to restart a gas appliance, first be sure the room is well-ventilated and that there is no open flame. Turn on the gas valve and let the gas flow for a few seconds. That should clear any air or impurities out of the pipe. Then shut the valve for a minute so gas in the air can escape. Finally, try lighting the appliance.

Oil appliances have an oil pump. If the pump was flooded, it should be cleaned and checked by a professional. Look for any sign that supply piping or the oil tank moved during the flood. Oil tanks, even buried oil tanks, will float when flooded. When the electricity is back on, open the main oil valve and turn the pump on.

The electric system. Most building codes require that electrical work be done by a licensed electrician. For example, a service entrance box that was flooded should be cleaned and checked by an electrician. But an electrician isn't required to clean flooded circuits.

1. Be sure the power is still off. Switch all breakers off or remove the fuses.
2. Wash out any mud or dirt you find in switch boxes, outlets boxes and light fixtures.
3. Clean or replace any contaminated or corroded switch, outlet or light fixture. You can clean most switches, outlets and light fixtures in a pail of tap water. Allow 24 hours for drying.
4. Plastic-covered copper wire survives submersion, even for long periods. But replace aluminum wiring that's been flooded by salt water. Replace any fabric-covered wire that has been submerged.
5. When drying is complete, re-install the switches, outlets and light fixtures.
6. Test the circuits one at a time. Turn all circuits off. Then turn on the main. Energize the circuits one at a time. If a breaker trips or a fuse blows on any circuit, there's a short somewhere. Walk through the building searching for signs of sparking or the smell of charring. Get help from an electrician if the source of the short isn't obvious.
7. Bathroom, kitchen and outdoor circuits are usually protected by a GFCI (ground fault circuit interrupter), either at the breaker box or at a wall outlet. If a circuit doesn't come back on line, the GFCI is probably tripped. Look for a popped GFCI button somewhere on the circuit. A single GFCI can serve several outlets in several rooms.

Dry electric motors, switches and receptacles with a blow dryer. Then spray on a moisture displacement such as WD-40 to stop corrosion. Appliance motors may have to be disassembled for a thorough cleaning.

Structural Drying

The quickest way to dry wall and ceiling cavities is to remove the finish material — wallboard, plaster, paneling, sheathing, etc. — and the insulation. But that's expensive and may not be necessary, especially on interior walls and ceilings that don't have sheathing and aren't insulated. If the finish material is in good condition, there's a better choice. If it's an insurance job, don't do any demolition without clearance by the adjuster.

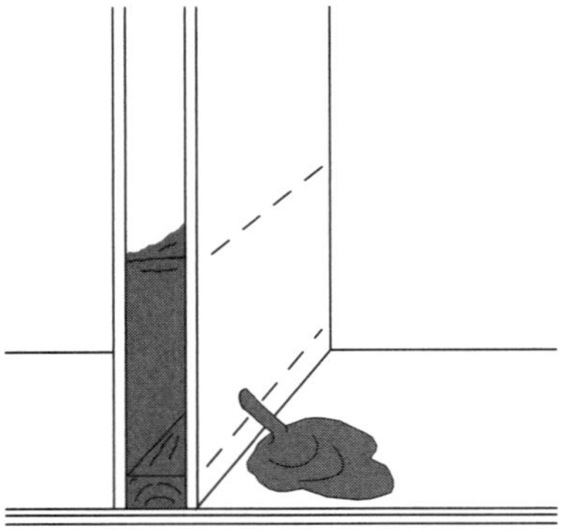

Figure 19-7
Drain wall cavities

Ceilings. Plan to replace a wallboard ceiling that's been covered by floodwater. A plaster ceiling won't have to be replaced unless there are too many cracks or sags. If the ceiling was flooded by a roof leak or water from an upper floor, punch or drill holes in the ceiling to drain away puddled water. If you have access to attic crawl space over the ceiling, gather up any wet insulation. Then install air-moving equipment in the crawl space to exhaust excess moisture. If you don't have access above the ceiling, dry the ceiling cavity as explained next for stud walls.

Walls. Start by removing the baseboard. Drill or puncture a hole in the wallboard about 2 inches above the floor (just above the sill plate). See Figure 19-7. The hole should be large enough to drain standing water and accommodate the air tube you'll insert into each stud cavity. When drilling, be careful to avoid wiring. Most wiring will be at about the same wall height as the electrical outlets.

Figure 19-8
Air drying manifold

If walls are plaster, use a drill to create these holes. A chisel would shatter the plaster. If a flooded home or office has steel framing, the bottom sill will collect water like a trough. Drill a hole at floor level to drain water from the bottom channel.

The size of the hole depends on the type of air manifold you plan to use. Some equipment distributes air through 1/8" spaghetti tube. Other systems, such as in Figure 19-8, use hose up to 5/8" in diameter. Each will be effective if used according to the manufacturer's instructions. Some air injection systems come with an air blower. Others attach to a standard air mover, as in Figure 19-8.

Figure 19-9
Air mover

The most basic dryout tool is an air mover. These units can dry any wet surface, including carpets, floors, walls, drapes and upholstery. Most can be installed in several positions, operate at several speeds and are designed for continuous service over long periods. See Figure 19-9. You'll need at least one air mover for each room or flooded area. A typical job requires one air mover for each 150 square feet of floor. A one horsepower air mover rated at 3,500 cubic feet per minute draws about 10 amps of power when running at full speed. When run at full speed, you won't be able to connect more than one of these units on a 15-amp circuit. Kitchens and laundry rooms usually have 20-amp circuits. A temporary power distribution panel (*spider box*) can tap into the 220-volt outlet

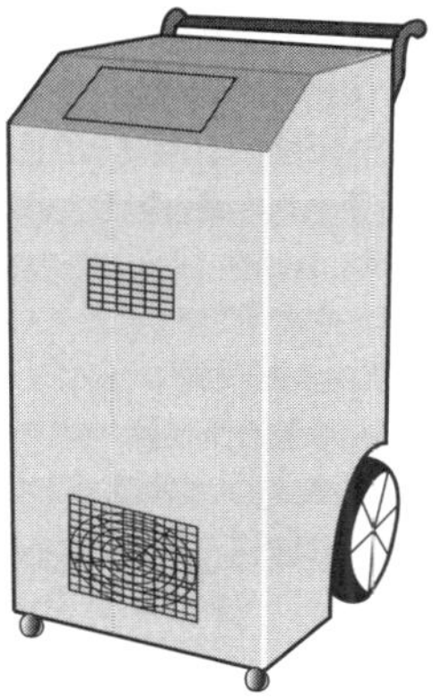

Figure 19-10
Portable Dehumidifier

Figure 19-11
Negative Air Scrubber

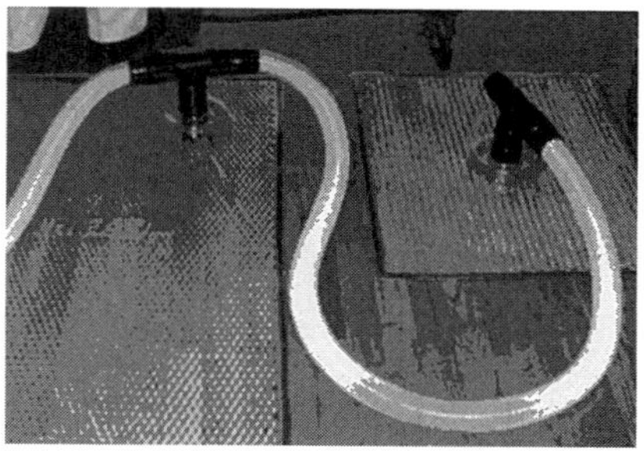

Figure 19-12
Mat drying a hardwood floor

behind an electric range or clothes dryer and deliver 110-volt GFCI-protected power to several air movers.

Air movers remove water from wet materials through evaporation. Air moving across the wet surface collects available moisture in the form of vapor. Dehumidifiers wring water out of the vapor by passing moist air over cooling coils or desiccant. Refrigerant coils in a dehumidifier collect moisture like a cold glass collects water droplets on a humid day. Desiccant dehumidifiers cost more but are more effective in cold temperatures. The portable dehumidifier in Figure 19-10 weighs about 125 pounds and will extract about 100 pints a day when the air temperature is 80 degrees and relative humidity is 60 percent.

A negative air scrubber will be needed when working with toxics in a containment structure. (See Figure 19-11) The scrubber draws air out of a room or containment structure so air flows continually from outside the room or structure to the interior. That prevents escape of airborne toxics from the area where work is being done. The scrubber has to be placed inside the containment area. A HEPA filter in the scrubber traps nearly all airborne particles before exhausting air through 12" or 14" flex duct, either to the exterior or outside the containment structure. Some negative air scrubbers also include an odor absorption filter. Other units can be reversed to provide clean positive air to a room that has to be kept free of airborne particles. A single 1,400-CFM scrubber will provide one clean air change a minute in a 12- x 14-foot room.

Scrubbers usually have two or three filters. The primary filter should be replaced daily. The secondary filter, if there is one, should be replaced every three to five days. The HEPA filter will require replacement after 500 to 1,000 hours, depending on conditions.

Hardwood flooring presents a special problem. Hardwood swells and cups at the edges when wet for more than a few hours. Drying the floor promptly will prevent most damage. It's easy to remove water standing on a hardwood floor. Any wet-vac or sponge mop will work fine. But it's harder to remove water that's soaked through wood seams into the subfloor. Mat drying, as in Figure 19-12, is one good solution. A vacuum extracts moist air from under mats laid over wet portions of the floor. A hardwood floor with 30 percent moisture should dry to a normal moisture level (under 10 percent) in four or five days. Cupping and swelling will disappear as the floor dries, leaving a floor that may need refinishing but won't have to be torn out.

Most engineered wood floors, such as parquet and laminate, include synthetic materials that absorb water selectively, causing the material to swell and distort unevenly. Drying is difficult and may not be successful. It's usually better to tear out and replace a wet engineered wood floor.

For good coverage of the equipment needed for dryout work and how to use that equipment to best advantage, see *Insurance Restoration Contracting: Startup to Success*. An order form is at the back of this book. Or go to http://Craftsman-Book.com on the Web.

Cleanup

The last step is cleaning all affected surfaces. You'll need: brooms, mops, brushes, sponges, buckets, hose, rubber gloves, rags, cleaner, disinfectants, lubricating oil, trash bags and a hair dryer. The best cleaner for dryout is a non-sudsing household cleaner. A good second choice is laundry soap or detergent. Many commercial disinfectants are available. Look on the ingredients list for quaternary, phenolic, or pine oil. Two ounces of laundry bleach in a gallon of water makes a good disinfectant. To inhibit mold, apply a mildew remover or mildewcide. Milban 1-2-3 is a disinfectant, cleaner, mildewstat, sanitizer, virucide and deodorizer. A second choice for removing mold is five tablespoons of tri-sodium phosphate in a gallon of water.

Tackle one room at a time. A two-bucket approach is most efficient. One bucket has rinse water and the other has cleaner. Rinse out your sponge, mop, or cleaning cloth in the rinse bucket. Wring it as dry as possible. Keep it rolled up tight until submerged in the cleaner bucket. Then let it unroll and absorb the cleaner. Using two buckets keeps most of the dirty rinse water out of your cleaning solution. Replace the rinse water frequently.

Walls. Start cleaning walls at the bottom or where damage is the worst. If you had to remove the wallboard or plaster, wash and disinfect the framing. If you didn't remove the wall finish, wash the wall to remove flood residue and discoloration. Then apply a disinfectant.

Furniture. Solid wood furniture usually requires only a cleaning. Wood alcohol or turpentine applied with a cotton ball may remove white mildew spots on wood. Use a cream wood restorer with lanolin to restore the finish. Upholstered furniture that has soaked up contaminants should be either discarded or cleaned by a professional.

Paper and books. Valuable papers such as photographs or a stamp collection can be salvaged with a great deal of effort. The owner may want to put valuable papers in a frost-free freezer or commercial meat locker until time is available to do page-by-page drying with a blow dryer. Don't try to force wet pages of paper apart. Just keep drying. Consider making copies of valuable papers. Contaminants in floodwater can turn to dark stains as the paper dries.

If a computer hard drive, cellphone, hand-held computer or memory card has been flooded, rinse in clear water and put it in a plastic bag in the refrigerator. Most information can be salvaged by a data recovery specialist. Search on the Web for a computer data recovery service. Consider using a countertop ultrasonic cleaning unit to remove contaminants from dishes, jewelry and art objects.

Mold Remediation

Mold thrives everywhere there is air, moisture and something organic to feed on. Mold spreads from microscopic spores. What's obvious to the eye will be discoloration on framing, sheathing, flooring, drywall, carpet, insulation or in HVAC duct. It's impossible to eliminate all mold and mold spores from a building. The goal is to reduce mold levels to about what exists outside the building. Once high concentrations of mold are gone, eliminating excess moisture will prevent regrowth.

Cleaning away mold is easy. The hard part is doing that cleanup without spreading mold spores throughout the building. Spores float anywhere air can penetrate, including into unprotected lungs and eyes. There are thousands of types of mold. Only a few are known to create toxins. But some people are more sensitive to mold spores than others. High concentrations of mold spores can make anyone sick.

Any home improvement project can turn into a mold remediation job. Mold may not be obvious until you start removing drywall, wallpaper, paneling, ceiling tile, sheathing or carpet. Pipe chases, walls behind cabinets and plumbing fixtures, condensate drain pans, porous duct liners and roof sheathing are favorite habitats for mold. Vinyl wall cover creates a vapor barrier, trapping moisture that supports growth of mold in the wall. Stripping vinyl cover from a wall can release a cloud of spores into the building.

If you discover a mold problem, you have an obligation to inform the owner. After that, you have several choices:

- Pack up your tools and walk off the job, advising the owner to call when the mold problem is solved.
- Develop a plan for dealing with the mold and getting on with the job.
- Offer to get quotes from remediation subcontractors.

The owner, of course, has to make the decision. But most owners will be eager to get your advice on:

- The extent of the problem, including a rough estimate of what's required and the cost.
- Insurance coverage. Generally, mold remediation isn't an insured loss. Still, the owner should check for coverage.

- Licensing. If you don't have a license that's required for mold removal, make that point clear.

Mold Remediation by Job Size

The Environmental Protection Agency hasn't set standards for mold contamination. But the EPA does suggest a classification system for mold remediation work based on the square feet affected.

- Small — less than 10 square feet.
- Medium — More than 10 square feet but less than 100 square feet.
- Large — More than 100 square feet or where there is greater risk of spreading mold during remediation.

The cleaning method recommended by the EPA is essentially the same for all three classes:

- Wet vacuum to remove excess moisture.
- Hand wash.
- HEPA vacuum when the surface is dry.

The chief difference among the three job sizes is in the containment structure and the personal protective equipment (PPE) required. Larger jobs need more containment and more PPE.

The EPA classification system makes sense. Larger jobs usually have a greater mold density and take longer to remediate. But the EPA also recommends using professional judgment when deciding how to handle a remediation job. For example:

- Access to the mold may be easy or difficult. A 20-foot-long wall finished in vinyl and concealing concentrations of mold would qualify as a "large" job by EPA standards. But the vinyl can be stripped a little at a time and cleaned immediately. Access is easy. The job shouldn't take more than a few hours for a crew of two. Building a full containment structure would be overkill.
- The job described above might warrant full containment if mold also contaminates the framing, especially if an occupant in the home is elderly or has respiratory problems.

Obviously, your good judgment is required. If in doubt, get the advice of an industrial hygienist, as described later in this chapter.

Here's what the EPA recommends for each of the three job classifications:

- Small jobs don't require containment. The only PPE needed will be an N-95 respirator, gloves and goggles.
- Medium jobs require polyethylene sheeting from ceiling to floor around the affected area. Provide a slit entry and a covering

flap. Use double-sided tape to secure poly sheeting to the floors, walls and ceiling. Seal off cracks and openings, HVAC vents and registers. Maintain negative air pressure with a HEPA scrubber. PPE should include gloves, an N-95 respirator or half-face respirator with HEPA filter, disposable overalls and goggles.

- Large jobs require containment made from two layers of fire-retardant poly supported on a frame made of framing lumber or PVC pipe. The containment structure should include an egress pathway from where work is done all the way to the building exterior. Seal off cracks and openings, HVAC vents and registers. Maintain negative air pressure with a HEPA scrubber that vents to the exterior. Those working in the containment structure will need full PPE: gloves, disposable full-body clothing, head cover, foot cover, and a full-face respirator with HEPA filter. Provide an airlock chamber large enough for changing into and out of PPE. All contaminated PPE, except respirators, should be placed in a sealed bag when leaving the airlock. Your crew should wear respirators until outside the airlock chamber.

Sampling and Your Protocol

Sampling isn't necessary on most jobs. But recommend sampling to the owner if:

- Litigation is expected,
- The source of contamination isn't clear,
- Health concerns are a problem,
- Cleanup is going to take more than a few days.

An industrial hygienist can collect samples and arrange for testing. The lab report will show spore count (by fungal type) at each of the locations tested. When cleanup is done, a second sample should confirm that spore count inside and outside the building are approximately the same. From a contractor's perspective, the second test is proof positive that work was done correctly.

All samples should be tested under standards published by the American Industrial Hygiene Association (AIHA) or the American Conference of Governmental Industrial Hygienists (ACGIH).

An industrial hygienist will also draft a *protocol*, the plan for remediation work. Even if the owner elects to skip sampling, it's good practice to have a protocol for any but the most routine mold remediation jobs. The protocol will describe the containment required, the PPE recommended, what has to be removed, the cleaning procedure, the chemicals, and the equipment to use. In short, the protocol is to mold remediation what plans and specs are to a home improvement job. Most construction professionals feel comfortable following plans. You'll feel better about the job with a good written protocol.

Cleaning Methods

Non-porous materials (such as metals, glass and hard plastics) are easy to clean. Semi-porous materials (such as wood, concrete, and masonry) should be cleaned if structurally sound. It's usually better to discard more porous materials (such as ceiling tile, insulation and drywall). If you try to clean drywall, keep cleaning at least 6 inches beyond where the board is discolored by mold. Before you decide to clean the face of drywall, check the back side. The back may have a higher concentration of mold than the front side.

Start breakout of what's to be discarded by misting the surface to minimize release of spores. Keep using a HEPA vacuum to collect spores as work continues. Seal everything you remove in 6-mil plastic bags while inside the containment structure. Wipe the exterior of each bag before removing it from containment. Bag disposal doesn't require any special handling once bags are out of the building.

The best liquid for mold cleaning is a gallon of warm water mixed with two to four ounces of a disinfectant-sanitizer. Again, mist the surface to minimize release of spores. Use a mister with a coarse spray tip. Scrape any gross filth or heavy soil off the surface. Keep the surface moist for 10 minutes so the disinfectant-sanitizer has time to work. Then wipe the surface clean using two buckets, the first with plain water for rinsing and the second with the diluted disinfectant-sanitizer. Gentle cleaning helps minimize release of spores. Clean beyond the area where you see signs of mold. When fully dry, go over the area cleaned with a HEPA vacuum. Change filters as recommended by the manufacturer. Do filter changes *inside* the containment area. Dispose of filters in sealed plastic bags.

Abrading (sanding and scraping) may be required on semi-porous surfaces. Where access is difficult, such as in a crawl space or attic, blasting is usually more practical than sanding and scraping by hand. It might take 120 manhours to sand and scrape a mold-infested attic. Blasting could reduce the labor by half. Blasting done with sand, glass beads or bicarbonate of soda leaves a residue which has to be cleaned up with a HEPA filter vacuum. You'll need both an air source and a scrubber when blasting inside a containment structure.

Blasting with dry ice leaves much less residue. Dry-ice pellets are soft. They don't scratch the surface as much as other blasting materials. Instead, sudden cooling makes the mold easy to flake off. The dry ice simply evaporates, leaving no chemical residue.

Regardless of the cleaning method, some mold residue will remain on semi-porous surfaces such as concrete, masonry or framing. Once the surface is clean and dry, encapsulate any mold that remains on the surface with an anti-microbial coating, either by brush, roller or spray.

When done with the containment structure, run a damp cloth or sponge mop over the entire interior, including the airlock. When dry, vacuum the interior with a HEPA vacuum. Dispose of the containment structure's poly sheeting and air lock in 6-mil bags. The entire area should be left dry and visibly free from mold, dust and debris.

	Craft@Hrs	Unit	Material	Labor	Total	Sell

Water Extraction. ("dry-out") Removing water from the interior of an existing building in compliance with The Clean Trust (IICRC) standards. Category 3 "black" water extraction (toxics, sewage) requires protective clothing. Some Cat 2 jobs require a respirator and gloves. See the sections that follow. Work done outside regular business hours (7 AM to 6 PM) may cost up to 50 percent more. These are costs to the dry-out contractor. Add overhead and profit.

Emergency dry-out service call. Includes dispatch of a van equipped for water extraction work, a crew of one certified dry-out technician and one helper, equipment setup, two hours of work and demobilization when work is complete. These are minimum charges for job requiring extraction of category 1 clean water (roof leak or broken water pipe) or category 2 (gray) water such as from a flood. For category 3 (black) water, add the cost of personal protection equipment (PPE) from the section that follows. For larger jobs, add the cost of pre-drying, structural water extraction (walls and ceilings), demolition, repairs, disposal of debris, consumables and equipment rental from the sections that follow.

	Craft@Hrs	Unit	Material	Labor	Total	Sell
Minimum charge, 100 SF job	—	Ea	—	—	250.00	250.00
Charge per SF beyond 100 SF	B8@.016	Ea	—	.56	.56	.95
Charge per crew hour beyond 2 hours	—	Hour	—	—	100.00	100.00
Delivery of dry-out equipment, per load	B8@1.00	Ea	—	34.90	34.90	59.30

Structural water extraction. Category 1 or category 2 water extraction. Based on a 1,000 SF water map with 200 LF of wall saturated to 6" maximum water height. Job duration is four to six days using two dehumidifiers, 10 air movers and 200 LF of wall manifold. Crew is one dry-out technician and one helper. Add the cost of pre-drying, demolition, cleanup after extraction, consumables, personal protection gear and air scrubbers, if required. Larger jobs and jobs with hardwood floors, flooding deeper than 6" or black (category 3) water will cost more. No equipment decontamination is required on Cat 1 jobs.

	Craft@Hrs	Unit	Material	Labor	Total	Sell
Equipment, deliver, install and startup	B8@4.00	Ea	—	140.00	140.00	238.00
Equipment monitoring (1 hour per day for 5 days)	B8@5.00	Ea	—	175.00	175.00	298.00
Equipment rental, 6 days	—	Ea	—	—	1,900.00	3,230.00
Equipment demobilization	B8@2.00	Ea	—	69.80	69.80	119.00
Equipment decontamination, per job	B8@2.00	Ea	—	69.80	69.80	119.00
Total cost per 1,000 SF of floor	—	LS	—	454.60	2,354.60	4,004.00
Total cost per SF of floor	—	LS	—	.45	2.35	4.00

Pre-drying. Preparing a building interior for water extraction.

Block and pad furniture in preparation for moving. Add the cost of tape and furniture pads from the section Consumables for water extraction and mold remediation.

	Craft@Hrs	Unit	Material	Labor	Total	Sell
Small room	B8@.268	Ea	—	9.35	9.35	15.90
Bedroom	B8@.448	Ea	—	15.60	15.60	26.50
Larger room	B8@.627	Ea	—	21.90	21.90	37.20

Move room contents to a new location on the same site. Add the cost of boxes and bubble wrap from the section Consumables for water extraction and mold remediation.

Box and hand-carry small items

	Craft@Hrs	Unit	Material	Labor	Total	Sell
Small room	B8@.322	Ea	—	11.20	11.20	19.00
Bedroom	B8@.483	Ea	—	16.90	16.90	28.70
Larger room	B8@.725	Ea	—	25.30	25.30	43.00
Large items (sofa, table, etc.), per item	B8@.145	Ea	—	5.06	5.06	8.60
Add for replacing contents after dry-out	—	%	—	100.0	—	—
Deduct when room contents can be shifted to another side of the same room	—	%	—	−50.0	—	—
Pack out inventory of items removed, by description, condition and packing box number, with report, per item	B8@.012	Ea	—	.42	.42	.71

Remove accumulated debris, mud or silt. Load in a 5 C.F. wheelbarrow with a shovel, move 100' and dump.

	Craft@Hrs	Unit	Material	Labor	Total	Sell
Per cubic foot removed	BL@.056	CF	—	1.63	1.63	2.77

	Craft@Hrs	Unit	Material	Labor	Total	Sell
Pressure wash building interior. Add the cost of equipment rental.						
Floors and walls, per SF washed	B8@.002	SF	—	.07	.07	.12
HVAC vent or register	B8@.033	Ea	—	1.15	1.15	1.96
Deploy and remove a 6" high re-usable plastic berm at the building perimeter	B8@.010	LF	.91	.35	1.26	2.14
Ultrasonic cleaning of building contents, dishes, jewelry, art objects, etc. Using a 10 gallon countertop ultrasonic cleaner. 1.3 CF basket. Add the cost of equipment rental.						
Per load, 5 minute cleaning cycle	B8@.167	Ea	—	5.83	5.83	9.91
Cleaning solution, per diluted gallon	—	Gal	2.00	—	2.00	

Remove appliances. Disconnect, wrap unit in a pad, mount on a hand dolly and move to a new location on the same site for cleaning or disposal. Removal of gas and oil appliances includes closing and locking the gas or oil line. Per appliance.

	Craft@Hrs	Unit	Material	Labor	Total	Sell
Most electric appliances	B8@.485	Ea	—	16.90	16.90	28.70
Most gas and oil appliances	B8@.667	Ea	—	23.30	23.30	39.60
Water heater, drain and remove	B8@.667	Ea	—	23.30	23.30	39.60
Undercounter dishwasher	B8@.900	Ea	—	31.40	31.40	53.40
Undercounter trash compactor	B8@.250	Ea	—	8.73	8.73	14.80
Undercounter refrigerator	B8@.250	Ea	—	8.73	8.73	14.80
Wall oven or microwave	B8@.450	Ea	—	15.70	15.70	26.70
Range hood	B8@.500	Ea	—	17.50	17.50	29.80
Garbage disposer	B8@.600	Ea	—	20.90	20.90	35.50
Add for re-installing appliances	—	%	—	90.0	—	—

Remove fixtures. Includes turning off the water, disconnecting the drain, capping supply and waste lines, disconnecting faucets and fittings and stack on site.

	Craft@Hrs	Unit	Material	Labor	Total	Sell
Self-rimming sink	B8@1.00	Ea	—	34.90	34.90	59.30
Ledge-type or hudee ring sink	B8@1.65	Ea	—	57.60	57.60	97.90
Self-rimming sink with disposer	B8@1.50	Ea	—	52.40	52.40	89.10
Ledge-type or hudee ring sink with disposer	B8@2.00	Ea	—	69.80	69.80	119.00
Bathtub	B8@2.00	Ea	—	69.80	69.80	119.00
Toilet or bidet	B8@.550	Ea	—	19.20	19.20	32.60
Vanity cabinet with sink	B8@1.25	Ea	—	43.60	43.60	74.10
Electric switch or receptacle and plate	B8@.110	Ea	—	3.84	3.84	6.53
Pendent, wall or recessed light fixture	B8@.295	Ea	—	10.30	10.30	17.50
Add for re-installing fixtures	—	%	—	90.0	—	—

HVAC system cleaning

	Craft@Hrs	Unit	Material	Labor	Total	Sell
Heat pump, disconnect, unbolt, remove	B8@2.33	Ea	—	81.30	81.30	138.00
Air handler, clean and dry out interior, replace filters lubricate fan bearing, perform test	B8@.450	Ea	3.00	15.70	18.70	31.80
Air handler, disconnect, unbolt, remove	B8@.750	Ea	—	26.20	26.20	44.50
Register or drop, clean	B8@.240	Ea	—	8.38	8.38	14.20
HVAC duct, clean per NADCA standards. Per LF of duct. Typical home has 150 LF of duct.	B8@.050	LF	—	1.75	1.75	2.98
Environmental control, remove, dry, clean, inspect, replace and test	B8@.500	Ea	—	17.50	17.50	29.80

	Craft@Hrs	Unit	Material	Labor	Total	Sell
Carpet removal. Remove saturated carpet or pad. Cut and roll into manageable sizes and remove to the exterior for disposal. (35 SY per CY)						
Carpet, loose lay or on 2-sided tape	B8@.040	SY	—	1.40	1.40	2.38
Loose lay pad	B8@.035	SY	—	1.22	1.22	2.07
Carpet laid on tack strips	B8@.056	SY	—	1.95	1.95	3.32
Remove carpet tack ("tackless") strip	B8@.011	LF	—	.38	.38	.65
Replace pad, 6 lb., rebond urethane	B8@.025	SY	4.40	.87	5.27	8.96
Replace carpet, 40 oz., standard grade	B8@.120	SY	25.00	4.19	29.19	49.60
Remove saturated foam-backed carpet laid in adhesive. Move debris to the exterior for disposal. Add the cost per square foot to the cost per linear foot of room perimeter. (35 SY per CY)						
Per square foot of carpet	B8@.010	SF	—	.35	.35	.60
Per linear foot of room perimeter	B8@.004	LF	—	.14	.14	.24
Add for removing adhesive residue using hot water	B8@.012	SF	—	.42	.42	.71
Water extraction. When required by toxic conditions, add the cost of personal protection equipment (jumpsuit, protective gloves, waterproof boots with steel toes, eye protection goggles and respirators). Add the cost of personal protection equipment (PPE) from the section that follows.						
Hard surface floors (concrete, tile, wood, sheet goods). Remove debris and standing water. Per 1,000 SF. Add the cost of equipment, if required. For equipment costs, see the sections that follow.						
Light (squeegee and mop only)	B8@1.56	MSF	—	54.40	54.40	92.50
Medium (squeegee, rinse and wet-vac)	B8@2.50	MSF	—	87.30	87.30	148.00
Heavy (remove silt, rinse and wet-vac)	B8@3.50	MSF	—	122.00	122.00	207.00
Shovel mud, rinse and wet-vac	B8@4.50	MSF	—	157.00	157.00	267.00
Carpeted floors with no pad. Remove standing water and surface moisture with a truck-mounted vacuum. For equipment costs, see the sections that follow.						
800 SF per hour	B8@1.25	MSF	—	43.60	43.60	74.10
Carpeted floors with pad. Remove standing water and sub-surface water. For equipment costs, see the sections that follow.						
Remove standing water, 13 CF per hour	B8@.182	CF	—	6.35	6.35	10.80
Light work, moist (155 SF per hour)	B8@6.45	MSF	—	225.00	225.00	383.00
Medium work, saturated (105 SF per hour)	B8@9.52	MSF	—	332.00	332.00	564.00
Heavy work, puddled (78 SF per hour)	B8@12.8	MSF	—	447.00	447.00	760.00
Small Water Claw (200 SF per hour)	B8@5.00	MSF	—	175.00	175.00	298.00
Medium Water Claw (333 SF per hour)	B8@3.00	MSF	—	105.00	105.00	179.00
Large Water Claw (500 SF per hour)	B8@2.00	MSF	—	69.80	69.80	119.00
Hydro Xtreme water extraction. Self-propelled. With Hydro-X Vacuum pack. No carpet removal required. For equipment costs, see the sections that follow.						
Equipment setup, per room	B8@.250	Ea	—	8.73	8.73	14.80
Extracting at 500 SF per hour	B8@2.00	MSF	—	69.80	69.80	119.00
Float carpet. Lift and block carpet to allow air circulation. Add the cost of air moving equipment and monitoring. For equipment costs, see the sections that follow. Cost per room.						
Small room, under 150 SF, per room	B8@.655	Ea	—	22.90	22.90	38.90
Most rooms, 150 SF to 300 SF, per room	B8@1.05	Ea	—	36.60	36.60	62.20
Room over 300 to 600 SF, per room	B8@1.57	Ea	—	54.80	54.80	93.20
Room over 600 SF, per 1,000 SF	B8@2.33	MSF	—	81.30	81.30	138.00
Add for re-installation (carpet stretching)	—	%	—	90.0	—	—
Hang dry Oriental rug at remote carpet dry cleaning plant. Including transportation to and from the plant site. Add the cost of carpet removal and reinstallation.						
Minimum charge	—	Ea	—	—	200.00	200.00
Per square yard of carpet or rug	—	SY	—	—	4.50	
Anti-mold, anti mildew sanitizer, Milban 1-2-3 sprayed on wet areas ($25 gallon covers 1,600 SF)	B8@.500	MSF	15.60	17.50	33.10	56.30

	Craft@Hrs	Unit	Material	Labor	Total	Sell
Clean and deodorize carpet, no water extraction, no furniture moving, subcontract	—	MSF	—	—	210.00	357.00
Scotchguard post-extraction cleaner concentrate ($20 gallon covers 3,200 SF)	—	MSF	6.25	—	6.25	10.60

Drywall repairs. Remove and replace drywall on ceilings and walls. Remove the damaged drywall with hand tools at heights to 9' and handle debris to a trash bin on site. Add the cost of debris disposal. 250 SF of demolished drywall debris weighs about 600 pounds and has a loose volume of about one cubic yard.

	Craft@Hrs	Unit	Material	Labor	Total	Sell
Remove saturated drywall panels. Includes the cost of pulling or driving the old fasteners.						
On ceilings	B8@.014	SF	—	.49	.49	.83
On walls	B8@.012	SF	—	.42	.42	.71
Replace full drywall panels. Includes joint tape, three coats of joint compound and sanding.						
1/2" board on walls, taped and finished	B8@.018	SF	0.38	.63	1.01	1.72
1/2" board on ceilings, taped and finished	B8@.024	SF	0.38	.84	1.22	2.07
Add for job setup, per room	B8@.200	Ea	—	6.98	6.98	11.90
Repair damaged drywall section. Cost per replacement section to 4' x 4'. Includes drywall, joint tape, three coats of joint compound and sanding.						
Cut out damaged section	B8@.432	Ea	—	15.10	15.10	25.70
Add 2" x 4" blocking at section edge	B8@.350	Ea	1.41	12.20	13.61	23.10
Cut, fix and secure replacement section	B8@.490	Ea	6.08	17.10	23.18	39.40
Tape and finish joints	B8@.160	Ea	0.08	5.58	5.66	9.62
Total cost per section to 4' x 4'	B8@1.29	Ea	7.57	45.00	52.57	89.40
Total cost per linear foot of repair	B8@.323	Ea	1.89	11.30	13.19	22.40
Patch hole to 6" diameter using repair clips	B8@.750	Ea	2.00	26.20	28.20	47.90
Add for work done in containment and PPE	—	%	—	50.0	—	—

Demolition. Remove damaged, contaminated or saturated building components as part of the water extraction process. Includes breaking materials into manageable pieces with hand tools and handling debris to a trash bin on site. Figures in parentheses show the approximate "loose" volume and weight of the materials after demolition. No salvage of materials except as noted. Add the cost of disposal of debris. Add 50% for work done in personal protective equipment.

Wall Finish Demolition. See also Drywall repairs above.

	Craft@Hrs	Unit	Material	Labor	Total	Sell
Ceramic tile, using hand tools, (25 SY per CY and 34 lbs. per SY)	BL@.029	SF	—	.84	.84	1.43
Ceramic tile and backer, using hand tools, (12 SY per CY and 70 lbs. per SY)	BL@.058	SF	—	1.68	1.68	2.86
Plywood sheathing to 1" thick, (200 SF per CY and 2 lbs. per SF)	BL@.017	SF	—	.49	.49	.83
Wood board sheathing, (250 SF per CY and 2 lbs. per SF)	BL@.030	SF	—	.87	.87	1.48
Metal siding, (200 SF per CY and 2 lbs. per SF)	BL@.027	SF	—	.78	.78	1.33
Stucco, (150 SF per CY and 8 lbs. per SF)	BL@.036	SF	—	1.05	1.05	1.79

	Craft@Hrs	Unit	Material	Labor	Total	Sell
Plywood or insulation board,						
(200 SF per CY and 2 lbs. per SF)	BL@.018	SF	—	.52	.52	.88
Plywood or plank paneling,						
(200 SF per CY and 2 lbs. per SF)	BL@.017	SF	—	.49	.49	.83
Insulation, fiberglass batts or rolls, per square foot of coverage						
(500 SF per CY and .3 lb per SF)	BL@.003	SF	—	.09	.09	.15
Plaster and lath						
(150 SF per CY and 8 lbs. per SF)	BL@.023	SF	—	.67	.67	1.14
Plaster, lath and furring						
(130 SF per CY and 10 lbs. per SF)	BL@.032	SF	—	.93	.93	1.58
Wall mirror, remove in salvage condition	BL@.025	SF	—	.73	.73	1.24
Wallpaper, scrape off up to 2 layers						
(1,000 SF per CY and .12 lbs. per SF)	BL@.015	SF	—	.44	.44	.75
Baseboard molding						
Single member base, no salvage	BL@.022	LF	—	.64	.64	1.09
Base and base shoe, no salvage	BL@.038	LF	—	1.10	1.10	1.87
Remove and salvage single member base	BL@.035	LF	—	1.02	1.02	1.73
Remove and salvage base and base shoe	BL@.050	LF	—	1.45	1.45	2.47
Drill injection port at wall base	B8@.045	Ea	—	1.57	1.57	2.67
Add for work done in containment and PPE	—	%	—	50.0	—	—

Ceiling demolition. Heights to 9'. Add 25 percent to the labor cost for work from ladders. See also Drywall repairs above.

	Craft@Hrs	Unit	Material	Labor	Total	Sell
Plaster ceiling (typically 175 to 200 SF per CY). Includes the cost of protecting the floor from falling debris.						
Including lath and furring	BL@.025	SF	—	.73	.73	1.24
Plaster suspension grid only	BL@.020	SF	—	.58	.58	.99
Acoustic tile ceiling (typically 200 to 250 SF per CY).						
Including suspended grid	BL@.010	SF	—	.29	.29	.49
Including grid in salvage condition	BL@.019	SF	—	.55	.55	.94
Tile glued or stapled to ceiling	BL@.015	SF	—	.44	.44	.75
Tile on strip furring, including furring	BL@.025	SF	—	.73	.73	1.24
Insulation, loose fill, fiberglass batts or rolls in attic, per square foot of coverage.						
(500 SF per CY and .3 lb per SF)	BL@.005	SF	—	.15	.15	.26

Flooring demolition. Per square yard of flooring removed. (1 square yard = 9 square feet).

	Craft@Hrs	Unit	Material	Labor	Total	Sell
Ceramic tile, using hand tools,						
(25 SY per CY and 34 lbs. per SY)	BL@.263	SY	—	7.64	7.64	13.00
Ceramic tile and backer, using hand tools,						
(12 SY per CY and 70 lbs. per SY)	BL@.522	SY	—	15.20	15.20	25.80
Hardwood, nailed,						
(25 SY per CY and 18 lbs. per SY)	BL@.288	SY	—	8.37	8.37	14.20
Hardwood, glued,						
(25 SY per CY and 18 lbs. per SY)	BL@.503	SY	—	14.60	14.60	24.80
Floating laminated hardwood flooring,						
(25 SY per CY and 15 lbs. per SY)	BL@.025	SF	—	.73	.73	1.24
Sheet vinyl or linoleum,						
(30 SY per CY and 3 lbs. per SY)	BL@.056	SY	—	1.63	1.63	2.77
Resilient tile,						
(30 SY per CY and 3 lbs. per SY)	BL@.096	SY	—	2.79	2.79	4.74

	Craft@Hrs	Unit	Material	Labor	Total	Sell
Loose lay sheet flooring, (30 SY per CY and 3 lbs. per SY)	BL@.012	SY	—	.35	.35	.60
Vinyl cove base, for salvage	BL@.025	LF	—	.73	.73	1.24
Flooring underlay, (200 SY per CY and 3 lbs. per SF)	BL@.017	SF	—	.49	.49	.83
Subfloor, (120 SY per CY and 4 lbs. per SF)	BL@.023	SF	—	.67	.67	1.14
Interior demolition.						
Remove, dry and re-hang wood door to 4' x 8'	BL@.300	Ea	—	8.72	8.72	14.80
Door trim only, remove two sides, per opening	BL@.450	Ea	—	13.10	13.10	22.30
Remove hollow metal door and door frame in a masonry wall. (2 doors per CY)						
Single door to 4' x 8'	BL@1.00	Ea	—	29.10	29.10	49.50
Two doors, per opening to 8' x 8'	BL@1.50	Ea	—	43.60	43.60	74.10
Remove wood door and door frame in a wood frame wall. (2 doors per CY)						
Single door to 4' x 8'	BL@.500	Ea	—	14.50	14.50	24.70
Two doors, per opening to 8' x 8'	BL@.750	Ea	—	21.80	21.80	37.10
Remove window, frame and hardware, per SF of window.						
Metal or vinyl (36 SF per CY)	BL@.058	SF	—	1.68	1.68	2.86
Wood (36 SF per CY)	BL@.063	SF	—	1.83	1.83	3.11

Cabinet demolition. Per linear foot of cabinet face or back (whichever is longer). The cost of removing base cabinets includes the cost of removing the countertop.

	Craft@Hrs	Unit	Material	Labor	Total	Sell
Wall cabinets, wood	BL@.250	LF	—	7.26	7.26	12.30
Wall cabinets, metal	BL@.350	LF	—	10.20	10.20	17.30
Base cabinets, wood	BL@.400	LF	—	11.60	11.60	19.70
Base cabinets, metal	BL@.550	LF	—	16.00	16.00	27.20
Cabinet door only	BL@.250	Ea	—	7.26	7.26	12.30
Remove and reinstall cabinet door	BL@.580	Ea	—	16.80	16.80	28.60
Remove countertop only	BL@.080	SF	—	2.32	2.32	3.94
Built-in shelving, per SF of face	BL@.070	SF	—	2.03	2.03	3.45

Cleanup after water extraction. Removing water residue using a sponge or mop and commercial cleaning products.

	Craft@Hrs	Unit	Material	Labor	Total	Sell
Cabinets, per linear foot of cabinet face. Wall or base cabinet, bathroom, kitchen, pantry or linen.						
Clean inside and outside	BL@.048	LF	0.08	1.39	1.47	2.50
Add for polish, exposed surfaces only	BL@.033	LF	0.06	.96	1.02	1.73
Countertops, per linear foot of front, including the sink.						
Marble, granite, laminated or composition	BL@.012	LF	0.06	.35	.41	.70
Tile, including grout joints	BL@.020	LF	0.12	.58	.70	1.19
Painted surfaces, wipe down by hand.						
Baseboard, trim or molding	BL@.003	LF	0.04	.09	.13	.22
Cased openings	BL@.060	Ea	0.03	1.74	1.77	3.01
Ceilings, painted	BL@2.00	MSF	6.00	58.10	64.10	109.00
Walls, painted or papered	BL@1.50	MSF	6.00	43.60	49.60	84.30
Window sills	BL@.004	LF	0.02	.12	.14	.24
Ceramic tile, including grout cleaning.						
Walls to 6'6" high	BL@.004	SF	0.06	.12	.18	.31
Floors	BL@.003	SF	0.06	.09	.15	.26
Doors, painted or stained.						
Cleaned two sides, including casing	BL@.150	Ea	0.10	4.36	4.46	7.58

	Craft@Hrs	Unit	Material	Labor	Total	Sell
Floors, per 1,000 SF.						
Vacuum carpet or sweep floor	BL@.650	MSF	—	18.90	18.90	32.10
Mop floor	BL@1.25	MSF	.50	36.30	36.80	62.60
Shampoo carpet	BL@3.30	MSF	4.50	95.90	100.40	171.00
Hardwood floor, clean, wax and polish	BL@5.50	MSF	7.50	160.00	167.50	285.00
Plumbing fixtures.						
Bar sink, including faucet	BL@.285	Ea	0.16	8.28	8.44	14.30
Bathtub, including tub surround	BL@.560	Ea	0.32	16.30	16.62	28.30
Shower stall, including door	BL@.560	Ea	0.32	16.30	16.62	28.30
Kitchen sink, including faucets	BL@.375	Ea	0.16	10.90	11.06	18.80
Lavatories, including faucets	BL@.375	Ea	0.16	10.90	11.06	18.80
Toilets, including seat and tank	BL@.450	Ea	0.16	13.10	13.26	22.50
Sliding glass doors, clean both sides including track and screen.						
Up to 8' wide	BL@.650	Ea	0.08	18.90	18.98	32.30
Over 8' wide	BL@.750	Ea	0.13	21.80	21.93	37.30
Venetian blinds, horizontal or vertical slats, remove, pressure wash, dry and re-install.						
Window blinds to 20 SF, per window	BL@.500	Ea	—	14.50	14.50	24.70
Larger blinds, per SF blind	BL@.025	Ea	—	.73	.73	1.24
Windows, clean inside or outside, first or second floors, per SF cleaned.						
Sliding, casement or picture windows	BL@.003	SF	0.01	.09	.10	.17
Jalousie type	BL@.004	SF	0.01	.12	.13	.22
Remove screen, hose wash, replace	BL@.100	Ea	—	2.91	2.91	4.95
Mirrors, per SF of face	BL@.003	SF	.01	.09	.10	.17
Add for work done in containment and PPE	—	%	25.0	50.0	—	—

Mold inspection. Work done by a certified mold remediation professional or an industrial hygienist. Includes an inspection for microbial activity, water stains, moisture intrusion and musty odors. May include testing with a moisture meter or thermal imaging. If the full extent of mold infestation can not be identified by inspection, a battery of samples will be collected from the air and surfaces (by swab or tape). The lab report will identify the type of mold, levels of mold and probable exposure symptoms.

	Craft@Hrs	Unit	Material	Labor	Total	Sell
Visual inspection	—	LS	—	—	195.00	—
Mold inspection, visual and indoor/outdoor sampling	—	LS	—	—	350.00	—
Air sampling, per sample	—	Ea	—	—	125.00	—
Air sampling in wall cavities, per sample	—	Ea	—	—	150.00	—
Surface sampling, swab or tape, per sample	—	Ea	—	—	125.00	—
Deduct additional samples of the same type	—	Ea	—	—	-50.00	—
Add for post-remediation testing	—	%	—	—	100.0	—

Mold remediation protocol. Prepared by a licensed mold assessment consultant. Specifies the materials to be remediated, the recommended remediation method, the PPE and containment required and the criteria for clearance. Larger and more complex jobs will cost more.

	Craft@Hrs	Unit	Material	Labor	Total	Sell
Remediation protocol, small job	—	LS	—	—	400.00	—
Remediation protocol, larger job	—	LS	—	—	650.00	—
Hourly rate for industrial hygienist	—	Hr	—	—	115.00	—

	Craft@Hrs	Unit	Material	Labor	Total	Sell
Containment structure. Polyethylene sheeting to enclose the contaminated area. For a large job, containment should extend from the affected area to the building exterior. Includes fabrication, cleaning, removal and typical loss due to waste. Cost of the support frame assumes five uses of materials.						
Typical medium job, 600 CF enclosure, 15 feet x 5 feet x 8 feet 4 mil poly taped to walls, ceiling and floor, per cubic foot enclosed	B8@.006	CF	.07	.21	.28	.48
Sealing surface cracks.						
To 1/4" wide	B8@.025	LF	.01	.87	.88	1.50
To 1/2" wide	B8@.033	LF	.03	1.15	1.18	2.01
Poly sheeting, clear.						
4 mil, single layer	B8@.003	SF	.07	.10	.17	.29
6 mil, single layer	B8@.003	SF	.08	.10	.18	.31
Flame retardant, 6 mil, double layer	B8@.006	SF	.16	.21	.37	.63
Double-sided tape.						
1" wide, 60 yard roll	B8@.003	LF	.04	.10	.14	.24
2" wide, 60 yard roll	B8@.003	LF	.07	.10	.17	.29
Containment frame, 2" x 4" lumber, assumes five uses of material.						
Top and bottom wall plates, per LF of wall	B8@.024	LF	.18	.84	1.02	1.73
Wall studs 48" O.C., per SF of wall	B8@.008	SF	.03	.28	.31	.53
Containment frame, 1" PVC pipe, assumes five uses of materials.						
Per LF of 1" PVC pipe	B8@.006	LF	.06	.21	.27	.46
Per 1" PVC pipe fitting	B8@.020	Ea	.13	.70	.83	1.41
Poly adhesive zipper, ZipWall	B8@.250	Ea	10.00	8.73	18.73	31.80
Doorway kit, 3' x 7' opening	B8@.250	Ea	31.50	8.73	40.23	68.40
Adjustable support rods, 63" to 113" high	B8@.083	Ea	29.00	2.90	31.90	54.20
Ceiling support rail, 63" long, with clips	B8@.167	Pair	46.00	5.83	51.83	88.10
Antimicrobial cleaning. Using a sprayer, sponge and sponge mop on non-porous or semi-porous surfaces. Cost per 100 square feet (CSF) of surface sprayed or cleaned. An open wall or ceiling cavity has about 1.5 SF of surface area for each square foot of wall or ceiling. Spraying assumes the surface is kept moist for 10 minutes. Add the cost of an EPA-registered disinfectant-sanitizer-cleaner (bactericidal, virucidal, tuberculocidal, fungicidal). Work done in containment and personal protective equipment (PPE) will take about 50 percent longer.						
Spray surfaces only.						
Using a hand sprayer	B8@.063	CSF	—	2.20	2.20	3.74
Using a garden sprayer	B8@.038	CSF	—	1.33	1.33	2.26
Clean floors.						
By sponge mop	B8@.083	CSF	—	2.90	2.90	4.93
By hand with a sponge	B8@.150	CSF	—	5.24	5.24	8.91
Clean walls or wall cavities.						
By sponge mop	B8@.104	CSF	—	3.63	3.63	6.17
By hand with a sponge	B8@.168	CSF	—	5.86	5.86	9.96
Clean ceiling or ceiling cavity.						
By sponge mop	B8@.113	CSF	—	3.94	3.94	6.70
By hand with a sponge	B8@.250	CSF	—	8.73	8.73	14.80
Disinfectant-sanitizer-cleaner, 1,600 SF per gallon.						
ShockWave, $45 per gallon	—	CSF	2.81	—	2.81	4.78
Milban 1-2-3, $25 per gallon	—	CSF	1.56	—	1.56	2.65
Add for work done in containment and PPE	—	%	—	50.0	—	—

	Craft@Hrs	Unit	Material	Labor	Total	Sell
Remove mold by abrading. Manually remove mold spores from semi-porous surfaces by sanding, brushing or scraping. Includes removal of abraded residue with a HEPA filter vacuum but not the cost of a HEPA vacuum. Work done in containment and personal protective equipment (PPE) will take about 50 percent longer.						
Light contamination (33 SF per manhour)	BL@.030	SF	—	.87	.87	1.48
Medium contamination (29 SF per manhour)	BL@.035	SF	—	1.02	1.02	1.73
Heavy contamination (25 SF per manhour)	BL@.040	SF	—	1.16	1.16	1.97
Add for work done in containment and PPE	—	%	—	50.0	—	—
Dry ice blasting. Per square foot of surface blasted. Assumes work is done on framing materials in a confined area (attic or crawl space). Add the cost of a dry ice blaster and dry ice. Includes removal of the blasted residue with a HEPA filter vacuum but not the cost of the vacuum.						
Smooth face (50 SF per manhour)	B8@.020	SF	—	.70	.70	1.19
Rough face (40 SF per manhour)	B8@.025	SF	—	.87	.87	1.48
Deduct for work done in open areas	—	%	—	50.0	—	—

	Day	Week	Month
Dry ice blaster rental. Blaster with a 40 lb. hopper, hoses and nozzle. Feed rate in pounds per minute (ppm) of dry ice.			
Aero 40, 20 to 140 psi, 0 to 4 ppm	400.00	1,000.00	3,000.00
Aero 40, 20 to 250 psi, 0-4 ppm	440.00	1,100.00	3,300.00
Aero 80, 30 to 140 psi, 0-7 ppm	640.00	1,600.00	4,800.00
Aero 100, 20 to 140 psi, 1-7 ppm	480.00	1,200.00	3,600.00

	Craft@Hrs	Unit	Material	Labor	Total	Sell
Dry ice pellets. 1/8" (2.9mm) high density dry ice pellets. Usage can vary from 1 pound per minute to 7 pounds per minute. Estimate 3 pounds per minute on most jobs. As a rule of thumb, allow 1,000 pounds of blasting ice per 8 hour shift. At 3 pounds per minute, 1,000 pounds of blasting ice will provide 5-1/2 hours of uninterrupted blasting.						
1/8" (2.9 mm) high density dry ice pellets	—	Lb	1.00	—	1.00	1.70
Encapsulation. Encapsulation of pre-cleaned surfaces to seal mold residue and inhibit future growth.						
Fungicide, Foster Full Defense 40-25, per 100 square feet coated (CSF).						
Purchase, per gallon	—	Gallon	51.00	—	51.00	—
Semi-porous surface, 200 SF per gallon	—	CSF	25.50	—	25.50	—
Non-porous or surface, 250 SF per gallon	—	CSF	20.40	—	20.40	—
Anti-microbial, Caliwel™ with BNA, per 100 square feet coated (CSF).						
Purchase, per gallon	—	Ea	75.00	—	75.00	—
Semi-porous surface, 200 SF per gallon	—	CSF	34.50	—	34.50	—
Non-porous surface, 250 SF per gallon	—	CSF	28.40	—	28.40	—
Apply mold sealer on semi-porous (unprimed) surfaces.						
Brush, 140 SF per hour	B8@.717	CSF	—	25.00	25.00	42.50
Roller, 220 SF per hour	B8@.450	CSF	—	15.70	15.70	26.70
Spray, 575 SF per hour	B8@.173	CSF	—	6.04	6.04	10.30
Apply mold sealer on non-porous (or primed) surfaces.						
Brush, 175 SF per hour	B8@.567	CSF	—	19.80	19.80	33.70
Roller, 275 SF per hour	B8@.363	CSF	—	12.70	12.70	21.60
Spray, 700 SF per hour	B8@.143	CSF	—	4.99	4.99	8.48

	Craft@Hrs	Unit	Material	Labor	Total	Sell
Consumables for water extraction and mold remediation. Purchase.						
Adhesive spray, 10 ounce aerosol can	—	Ea	8.00	—	8.00	—
Duct tape, 2" wide roll, 60 yards	—	Ea	3.50	—	3.50	—
Dumpster, 3 CY, rented and emptied once	—	Ea	—	—	100.00	—
Dumpster, 14 CY low boy, per load	—	Ea	—	—	250.00	—
Furniture pad, 72" x 80", purchase	—	Ea	15.00	—	15.00	—
Stretch wrap, 5" x 1,000' roll	—	Ea	9.00	—	9.00	—
Bubble wrap, 12' x 80' roll	—	Ea	35.00	—	35.00	—
Poly furniture bag, 2-mil, 72" x 48"	—	Ea	2.40	—	2.40	—
Poly trash bag, 6-mil, 38" x 64"	—	Ea	3.40	—	3.40	—
Packing boxes. The average room requires eight small boxes, eight medium boxes and four large boxes. No salvage value assumed.						
Small box, 1 cubic foot	—	Ea	1.50	—	1.50	—
Medium box, 2 cubic feet	—	Ea	2.25	—	2.25	—
Large box, 3.5 cubic feet	—	Ea	3.50	—	3.50	—
Wardrobe box, metal rod	—	Ea	15.00	—	15.00	—
Box sealing tape, 660 yard roll	—	Ea	2.50	—	2.50	—
Typical cost per room for packing boxes	—	Ea	44.00	—	44.00	—
Storage area, climate controlled, monthly cost per square foot.						
10' x 10'	—	Mo	—	—	1.20	—
10' x 20'	—	Mo	—	—	.80	—
20' x 20'	—	Mo	—	—	.60	—
Personal protection equipment. Purchase.						
Whole body suit, disposable, Tyvek™	—	Ea	5.75	—	5.75	—
Gloves, hazmat, per pair	—	Ea	18.80	—	18.80	—
Eye protection	—	Ea	2.50	—	2.50	—
Respirator, half-face	—	Ea	11.00	—	11.00	—
Respirator, full-face	—	Ea	138.00	—	138.00	—
HEPA P100 filter for respirator, per pair	—	Ea	8.00	—	8.00	—
Respirator, N-95	—	Ea	2.00	—	2.00	—

	Purchase	Rental Day	Rental Week
Dehumidifiers. Rating in pints per day by Association of Home Appliance Manufacturers standard.			
Atlantic, 140 pints, low grain	2,400.00	100.00	—
DrizAir 110, 58 pints, conventional	1,000.00	39.00	—
DrizAir 1200, 64 pints, conventional	1,300.00	50.00	207.00
DrizAir 2000, 110 pints, low grain	2,300.00	85.00	—
DrizAir 2400, 140 pints, low grain	2,600.00	125.00	—
DrizAir 80DX, 40 pints, conventional	870.00	25.00	150.00
DrizAir Evolution, 70 pints, low grain	1,710.00	40.00	200.00
Dry Pro 5000, 54 pints, conventional	900.00	28.00	116.00
Dry Pro 7000, 70 pints, conventional	1,800.00	60.00	240.00
Ebac Triton, 51 pints, conventional	920.00	20.00	—
Phoenix 200 HT, 140 pints, low grain	1,990.00	50.00	250.00
Phoenix 200 Max, 133 pints	2,100.00	53.00	310.00
Phoenix 300, 176 pints	3,300.00	175.00	—
Phoenix R175, 92 pints, low grain	1,450.00	40.00	266.00
Ultimate 340, 145 pints, low grain	2,200.00	65.00	280.00

	Purchase	Rental Day	Rental Week
Air movers. By rated cubic feet per minute.			
B-air Cub, 375 CFM	145.00	12.50	56.00
Dri-Eaz Ace, 1,881 CFM	429.00	28.00	120.00
Dri-Eaz Vortex, 2,041 CFM	515.00	35.00	190.00
Dri-Eaz Jet CXV, 3,200 CFM	495.00	24.00	—
Dri-Eaz Sahara HD, 3,500 CFM	445.00	20.00	64.00
Dry Air Force 9	380.00	35.00	125.00
Dry Air Gale Force, 2,260 CFM	228.00	30.00	—
Dry Air Tempest, 230V	174.00	10.00	50.00
Dry Air Twister, 110V	174.00	10.00	50.00
Dry Air Typhoon	205.00	20.00	70.00
Phoenix CAM Pro, 2,700 CFM	309.00	10.00	50.00
Viking 2200EX, 2,200 CFM	183.00	10.00	50.00
Viking AX3000, 3,000 CFM	335.00	12.00	70.00
Viking Windjammer 2,300 CFM	191.00	10.00	50.00
Wall Cavity Dryer. No air movers included except as noted.			
Adaptidry 9-port, 4" manifold, with hose	800.00	30.00	—
Adaptidry 5-port, 4" manifold, with hose	550.00	30.00	—
Dri-Eaz Turbo Vent 48" wall cavity dryer	205.00	6.00	30.00
Dri Force Drying System, 90 LF of wall, with blower	2,050.00	135.00	—
Injectidry HP60i wall drying system, with blower/vacuum	3,000.00	150.00	525.00
Injectidry floor or wall panel, 30" x 46"	83.00	14.00	—
Octi-Dry 8' wall manifold, add for a blower	400.00	45.00	210.00
Floor Drying System. No air movers included except as noted.			
Dri-Eaz AirWolf floor dryer, with duct and clamps	520.00	95.00	380.00
Dri-Eaz Rescue Mat System, 15 SF of mat, per set	570.00	95.00	380.00
Injectidry water separator	120.00	—	—
Injectidry HEPA inline filter	400.00	—	—

Negative Air/Scrubbers. Prevents escape of contaminated air during water extraction. Exhausts air through a 99.97 percent HEPA filter to maintain negative (or positive) air pressure. By maximum rated cubic feet per minute of air filtration. A 900 CFM scrubber will provide six air changes an hour in an 8,000 cubic foot (25' x 40' x 8') space.

	Purchase	Rental Day	Rental Week
Abatement Tech PAS2400, 1,950 CFM	3,500.00	119.00	—
ACSI Force Air 700 (used), 700 CFM	596.00	36.00	—
ACSI Force Air 2000, 1,975 CFM	967.00	99.00	—
Dri-Eaz 500, 500 CFM	860.00	65.00	280.00
Novair 2000, 2,000 CFM	910.00	15.00	250.00
Phoenix Guardian HEPA System, 1,400 CFM	1,775.00	125.00	350.00

	Craft@Hrs	Unit	Material	Labor	Total	Sell
Consumables.						
HEPA filter replacement for negative air units						
Typical pre-filter	—	Ea	1.50	—	1.50	—
Typical second stage filter	—	Ea	150.00	—	150.00	—
Activated carbon filter (odor control)	—	Ea	165.00	—	165.00	—

	Purchase	Rental Day	Rental Week
Other Dryout and Mold Equipment.			
Temporary power distribution (spider) box, 100 amps	850.00	55.00	280.00
Environmental control unit, cool/heat	670.00	33.50	134.00
Submersible trash pump, 1/2 HP, 1-1/2" to 2" outlet	300.00	44.00	176.00
Wheelbarrow, 2-wheel, 5 cubic foot capacity	65.00	8.00	44.00
Ultrasonic cleaner, 10 gallon, 1.3 CF	1,800.00	85.00	340.00
HEPA filter vacuum, 100 CFM	450.00	40.00	180.00
Power generators. Drying equipment draws approximately 1.2 KW for each 100 SF of water extraction map.			
5 KW, gasoline	770.00	60.00	235.00
15 KW, diesel	2,400.00	115.00	385.00
60 KW, diesel, towable	36,000.00	250.00	760.00
Water Claw extractor			
Small, 8" x 14"	300.00	—	—
Medium, 10" x 17" (650 SF per hour)	370.00	—	—
Large, 12" x 21" (1,000 SF per hour)	525.00	—	—
Van with truck-mounted blower-vacuum package, add $1 per mile for travel			
23 HP, 288 CFM, 150' hose	36,000.00	40.00	225.00
26 HP, 330 CFM, 250' hose	38,000.00	45.00	255.00
29 HP, 495 CFM, 300' hose	45,000.00	50.00	290.00
Pressure washer			
2,000 PSI, electric	200.00	20.00	50.00
3,000 PSI, gas, hot water	2,650.00	130.00	660.00

Index

C

D

G

H

L

Q

R

S

T

W

X – Y – Z

Practical References for Builders

Construction Contract Writer

Relying on a "one-size-fits-all" boilerplate construction contract to fit your jobs can be dangerous — almost as dangerous as a handshake agreement. *Construction Contract Writer* lets you draft a contract in minutes that precisely fits your needs and the particular job, and meets both state and federal requirements. You just answer a series of questions — like an interview — to construct a legal contract for each project you take on. Anticipate where disputes could arise and settle them in the contract before they happen. Include the warranty protection you intend, the payment schedule, and create subcontracts from the prime contract by just clicking a box. Includes a feedback button to an attorney on the Craftsman staff to help should you get stumped — *No extra charge.* **$99.95.** Download the *Construction Contract Writer* at: http://www.constructioncontractwriter.com

Construction Forms for Contractors

This practical guide contains 78 useful forms, letters and checklists, guaranteed to help you streamline your office, manage your jobsites, gather and organize records and documents, keep a handle on your subs, reduce estimating errors, administer change orders and lien issues, monitor crew productivity, track your equipment use, and more. Includes accounting forms, change order forms, forms for customers, estimating forms, field work forms, HR forms, lien forms, office forms, bids and proposals, subcontracts, and more. All are also on the CD-ROM included, in *Excel* spreadsheets, as formatted Rich Text that you can fill out on your computer, and as PDFs. **360 pages, 8½ x 11, $48.50**

CD Estimator

If your computer has *Windows*™ and a CD-ROM drive, CD Estimator puts at your fingertips over 150,000 construction costs for new construction, remodeling, renovation & insurance repair, home improvement, framing & finish carpentry, electrical, concrete & masonry, painting, earthwork & heavy equipment and plumbing & HVAC. Quarterly cost updates are available at no charge on the Internet. You'll also have the *National Estimator* program — a stand-alone estimating program for *Windows*™ that *Remodeling* magazine called a "computer wiz," and *Job Cost Wizard*, a program that lets you export your estimates to *QuickBooks Pro* for actual job costing. A 60-minute interactive video teaches you how to use this CD-ROM to estimate construction costs. And to top it off, to help you create professional-looking estimates, the disk includes over 40 construction estimating and bidding forms in a format that's perfect for nearly any *Windows*™ word processing or spreadsheet program.
CD Estimator is $108.50

Contractor's Guide to *QuickBooks* 2012

QuickBooks 2012 has many new features that simplify a building contractor's bookkeeping work. You'll wonder how you managed without them. To help you make the most of these new features, or to make getting set up with *QuickBooks* almost painless, this user-friendly manual walks you through *QuickBooks'* detailed setup procedure and explains step-by-step how to create a first-rate accounting system. You'll learn in days, rather than weeks, how to use *QuickBooks* to get your contracting business organized, with simple, fast accounting procedures. But setting up *QuickBooks* from scratch can be time-consuming. On the free download included with purchase of the book you'll find a *QuickBooks* file preconfigured for a construction company. Open it, enter your company's data, add info on your suppliers, subs and customers, and you're up and running. The setup's done for you. **288 pages, 8½ x 11, $58.50**

See checklist for other available editions.

National Painting Cost Estimator

A complete guide to estimating painting costs for just about any type of residential, commercial, or industrial painting, whether by brush, spray, or roller. Shows typical costs and bid prices for fast, medium, and slow work, including material costs per gallon; square feet covered per gallon; square feet covered per manhour; labor, material, overhead, and taxes per 100 sq. feet; and how much to add for profit. Includes a CD-ROM with an electronic version of the book with *National Estimator*, a stand-alone *Windows*™ estimating program, plus an interactive multimedia video that shows how to use the disk to compile construction cost estimates.
440 pages, 8½ x 11, $73.00. Revised annually

National Construction Estimator

Current building costs for residential, commercial, and industrial construction. Estimated prices for every common building material. Provides manhours, recommended crew, and gives the labor cost for installation. Includes a CD-ROM with an electronic version of the book with *National Estimator*, a stand-alone *Windows*™ estimating program, plus an interactive multimedia video that shows how to use the disk to compile construction cost estimates. **672 pages, 8½ x 11, $72.50. Revised annually**

Roofing Construction & Estimating

Installation, repair and estimating for nearly every type of roof covering available today in residential and commercial structures: asphalt shingles, roll roofing, wood shingles and shakes, clay tile, slate, metal, built-up, and elastomeric. Covers sheathing and underlayment techniques, as well as secrets for installing leakproof valleys. Many estimating tips help you minimize waste, as well as insure a profit on every job. Troubleshooting techniques help you identify the true source of most leaks. Over 300 large, clear illustrations help you find the answer to just about all your roofing questions. **432 pages, 8½ x 11, $38.00**

Concrete Construction

Just when you think you know all there is about concrete, many new innovations create faster, more efficient ways to do the work. This comprehensive concrete manual has both the tried-and-tested methods and materials, and more recent innovations. It covers everything you need to know about concrete, along with Styrofoam forming systems, fiber reinforcing adjuncts, and some architectural innovations, like architectural foam elements, that can help you offer more in the jobs you bid on. Every chapter provides detailed, step-by-step instructions for each task, with hundreds of photographs and drawings that show exactly how the work is done. To keep your jobs organized, there are checklists for each stage of the concrete work, from planning, to finishing and protecting your pours. Whether you're doing residential or commercial work, this manual has the instructions, illustrations, charts, estimating data, rules of thumb and examples every contractor can apply on their concrete jobs. **288 pages, 8½ x 11, $28.75**

Profits in Buying & Renovating Homes

Step-by-step instructions for selecting, repairing, improving, and selling highly profitable "fixer-uppers." Shows which price ranges offer the highest profit-to-investment ratios, which neighborhoods offer the best return, practical directions for repairs, and tips on dealing with buyers, sellers, and real estate agents. Shows you how to determine your profit before you buy, what "bargains" to avoid, and how to make simple, profitable, inexpensive upgrades. **304 pages, 8½ x 11, $24.75**

Renovating & Restyling Older Homes

Any builder can turn a run-down old house into a showcase of perfection — if the customer has unlimited funds to spend. Unfortunately, most customers are on a tight budget. They usually want more improvements than they can afford — and they expect you to deliver. This book shows how to add economical improvements that can increase the property value by two, five or even ten times the cost of the remodel. Sound impossible? Here you'll find the secrets of a builder who has been putting these techniques to work on Victorian and Craftsman-style houses for twenty years. You'll see what to repair, what to replace and what to leave, so you can remodel or restyle older homes for the least amount of money and the greatest increase in value. **416 pages, 8½ x 11, $33.50**

Estimating Home Building Costs Revised

Accurate estimates are the foundation of a successful construction business. Leave an item out of your original estimate and it can take the profit out of your entire job. This practical guide to estimating home construction costs has been updated with Excel estimating forms and worksheets with active cells that ensure accurate and complete estimates for your residential projects. Load the enclosed CD-ROM into your computer and create your own estimate as you follow along with the step-by-step techniques in this book. Clear, simple instructions show how to estimate labor and material costs for each stage of construction, from site clearing to figuring your markup and profit. Every chapter includes a sample cost estimate worksheet that lists all the materials to be estimated. Even shows how to figure your markup and profit to arrive at a price.
336 pages, 8½ x 11, $38.00

Craftsman's Construction Installation Encyclopedia

Step-by-step installation instructions for just about any residential construction, remodeling or repair task, arranged alphabetically, from *Acoustic tile* to *Wood flooring*. Includes hundreds of illustrations that show how to build, install, or remodel each part of the job, as well as manhour tables for each work item so you can estimate and bid with confidence. Also includes a CD-ROM with all the material in the book, handy look-up features, and the ability to capture and print out for your crew the instructions and diagrams for any job. **792 pages, 8½ x 11, $65.00**

National Electrical Estimator

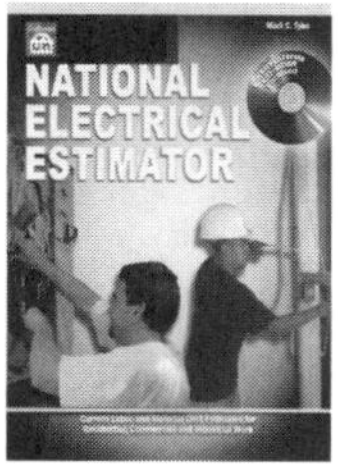

This year's prices for installation of all common electrical work: conduit, wire, boxes, fixtures, switches, outlets, loadcenters, panelboards, raceway, duct, signal systems, and more. Provides material costs, manhours per unit, and total installed cost. Explains what you should know to estimate each part of an electrical system. Includes a CD-ROM with an electronic version of the book with *National Estimator*, a stand-alone *Windows*™ estimating program, plus an interactive multi-media video that shows how to use the disk to compile construction cost estimates.
552 pages, 8½ x 11, $72.75. Revised annually

Roof Framing

Shows how to frame any type of roof in common use today, even if you've never framed a roof before. Includes using a pocket calculator to figure any common, hip, valley, or jack rafter length in seconds. Over 400 illustrations cover every measurement and every cut on each type of roof: gable, hip, Dutch, Tudor, gambrel, shed, gazebo, and more. **480 pages, 5½ x 8½, $26.50**

Wood-Frame House Construction

Step-by-step construction details, from the layout of the outer walls, excavation and formwork, to finish carpentry and painting. Packed with clear illustrations and explanations updated for modern construction methods. Everything you need to know about framing, roofing, siding, interior finishings, floor covering and stairs — your complete book of wood-frame homebuilding. **320 pages, 8½ x 11, $25.50. Revised edition**

National Renovation & Insurance Repair Estimator

Current prices in dollars and cents for hard-to-find items needed on most insurance, repair, remodeling, and renovation jobs. All price items include labor, material, and equipment breakouts, plus special charts that tell you exactly how these costs are calculated. Includes a CD-ROM with an electronic version of the book with *National Estimator*, a stand-alone *Windows*™ estimating program, plus an interactive multimedia video that shows how to use the disk to compile construction cost estimates.
488 pages, 8½ x 11, $74.50. Revised annually

Steel-Frame House Construction

Framing with steel has obvious advantages over wood, yet building with steel requires new skills that can present challenges to the wood builder. This book explains the secrets of steel framing techniques for building homes, whether pre-engineered or built stick by stick. It shows you the techniques, the tools, the materials, and how you can make it happen. Includes hundreds of photos and illustrations, plus a FREE download with steel framing details, a database of steel materials and manhours, with an estimating program. **320 pages, 8½ x 11, $39.75**

National Repair & Remodeling Estimator

The complete pricing guide for dwelling reconstruction costs. Reliable, specific data you can apply on every repair and remodeling job. Up-to-date material costs and labor figures based on thousands of jobs across the country. Provides recommended crew sizes; average production rates; exact material, equipment, and labor costs; a total unit cost and a total price including overhead and profit. Separate listings for high- and low-volume builders, so prices shown are specific for any size business. Estimating tips specific to repair and remodeling work to make your bids complete, realistic, and profitable. Includes a CD-ROM with an electronic version of the book with *National Estimator*, a stand-alone *Windows*™ estimating program, plus an interactive multi-media video that shows how to use the disk to compile construction cost estimates. **496 pages, 8½ x 11, $73.50. Revised annually**

Building Code Compliance for Contractors & Inspectors

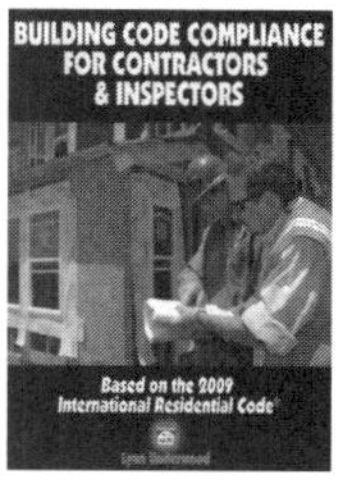

An answer book for both contractors and building inspectors, this manual explains what it takes to pass inspections under the 2009 *International Residential Code*. It includes a code checklist for every trade, covering some of the most common reasons why inspectors reject residential work — footings, foundations, slabs, framing, sheathing, plumbing, electrical, HVAC, energy conservation and final inspection. The requirement for each item on the checklist is explained, and the code section cited so you can look it up or show it to the inspector. Knowing in advance what the inspector wants to see gives you an (almost unfair) advantage. To pass inspection, do your own pre-inspection before the inspector arrives. If your work requires getting permits and passing inspections, put this manual to work on your next job. If you're considering a career in code enforcement, this can be your guidebook.
8½ x 11, 232 pages, $32.50

Builder's Guide to Accounting Revised

Step-by-step, easy-to-follow guidelines for setting up and maintaining records for your building business. This practical guide to all accounting methods shows how to meet state and federal accounting requirements, explains the new depreciation rules, and describes how the Tax Reform Act can affect the way you keep records. Full of charts, diagrams, simple directions and examples to help you keep track of where your money is going. Recommended reading for many state contractor's exams. Each chapter ends with a set of test questions, and a CD-ROM included FREE has all the questions in interactive self-test software. Use the Study Mode to make studying for the exam much easier, and Exam Mode to practice your skills.
360 pages, 8½ x 11, $35.50

Painter's Handbook

Loaded with "how-to" information you'll use every day to get professional results on any job: the best way to prepare a surface for painting or repainting; selecting and using the right materials and tools (including airless spray); tips for repainting kitchens, bathrooms, cabinets, eaves and porches; how to match and blend colors; why coatings fail and what to do about it. Lists 30 profitable specialties in the painting business.
320 pages, 8½ x 11, $33.00

Estimating & Bidding for Builders & Remodelers

This 5th edition has all the information you need for estimating and bidding new construction and home improvement projects. It shows how to select jobs that will be profitable, do a labor and materials take-off from the plans, calculate overhead and figure your markup, and schedule the work. Includes a CD with an easy-to-use construction estimating program and a database of 50,000 current labor and material cost estimates for new construction and home improvement work, with area modifiers for every zip code. Price updates on the Web are free and automatic.
272 pages, 8½ x 11, $89.50

Home Inspection Handbook

Every area you need to check in a home inspection — especially in older homes. Twenty complete inspection checklists: building site, foundation and basement, structural, bathrooms, chimneys and flues, ceilings, interior & exterior finishes, electrical, plumbing, HVAC, insects, vermin and decay, and more. Also includes information on starting and running your own home inspection business.
324 pages, 5½ x 8½, $24.95

Insurance Restoration Contracting: Startup to Success

Insurance restoration — the repair of buildings damaged by water, fire, smoke, storms, vandalism and other disasters — is an exciting field of construction that provides lucrative work that's immune to economic downturns. And, with insurance companies funding the repairs, your payment is virtually guaranteed. But this type of work requires special knowledge and equipment, and that's what you'll learn about in this book. It covers fire repairs and smoke damage, water losses and specialized drying methods, mold remediation, content restoration, even damage to mobile and manufactured homes. You'll also find information on equipment needs, training classes, estimating books and software, and how restoration leads to lucrative remodeling jobs. It covers all you need to know to start and succeed as the restoration contractor that both homeowners and insurance companies call on first for the best jobs.
640 pages, 8½ x 11, $69.00